Genus IV. *Gluconobacter*
Family II. *Azotobacteraceae*
Genus I. *Azotobacter*
Genus II. *Azomonas*
Genus III. *Beijerinckia*
Genus IV. *Derxia*
Family III. *Rhizobiaceae*
Genus I. *Rhizobium*
Genus II. *Agrobacterium*
Family IV. *Methylomonadaceae*
Genus I. *Methylomonas*
Genus II. *Methylococcus*
Family V. *Halobacteriaceae*
Genus I. *Halobacterium*
Genus II. *Halococcus*
Genera of Uncertain Affiliation
Genus *Alcaligenes*
Genus *Acetobacter*
Genus *Brucella*
Genus *Bordetella*
Genus *Francisella*
Genus *Thermus*

PART 8.
GRAM-NEGATIVE FACULTATIVELY ANAEROBIC RODS

Straight and curved rods; some are nonmotile; others are motile by polar or peritrichate flagella; all members are nonsporeformers; some have special growth requirements

Family I. *Enterobacteriaceae*
Genus I. *Escherichia*
Genus II. *Edwardsiella*
Genus III. *Citrobacter*
Genus IV. *Salmonella*
Genus V. *Shigella*
Genus VI. *Klebsiella*
Genus VII. *Enterobacter*
Genus VIII. *Hafnia*
Genus IX. *Serratia*
Genus X. *Proteus*
Genus XI. *Yersinia*
Genus XII. *Erwinia*
Family II. *Vibrionaceae*
Genus I. *Vibrio*
Genus II. *Aeromonas*
Genus III. *Plesiomonas*
Genus IV. *Photobacterium*
Genus V. *Lucibacterium*
Genera of Uncertain Affiliation
Genus *Zymomonas*
Genus *Chromobacterium*
Genus *Flavobacterium*
Genus *Haemophilus*
   (*H. vaginalis*)
Genus *Pasteurella*
Genus *Actinobacillus*
Genus *Cardiobacterium*
Genus *Streptobacillus*
Genus *Calymmatobacterium*
Parasites of *Paramecium*

PART 9.
GRAM-NEGATIVE ANAEROBIC BACTERIA

Strict (obligate) anaerobic, non-spore forming organisms; some members are motile; pleomorphism (variation in shape) occurs

Family I. *Bacteroidaceae*
Genus I. *Bacteroides*
Genus II. *Fusobacterium*
Genus III. *Leptotrichia*
Genera of Uncertain Affiliation
Genus *Desulfovibrio*
Genus *Butyrivibrio*
Genus *Succinivibrio*
Genus *Succinimonas*
Genus *Lachnospira*
Genus *Selenomonas*

PART 10.
GRAM-NEGATIVE COCCI AND COCCOBACILLI

Cocci, characteristically occuring in pairs; adjacent sides of the cells may be flattened; organisms are not flagellated

Family I. *Neisseriaceae*
Genus I. *Neisseria*
Genus II. *Branhamella*
Genus III. *Moraxella*
Genus IV. *Acinetobacter*
Genera of Uncertain Affiliation
Genus *Paracoccus*
Genus *Lampropedia*

PART 11.
GRAM-NEGATIVE ANAEROBIC COCCI

Cocci of variable size and characteristically in pairs; they are not flagellated

Family I. *Veillonellaceae*
Genus I. *Veillonella*
Genus II. *Acidaminococcus*
Genus III. *Megasphaera*

PART 12.
GRAM-NEGATIVE, CHEMOLITHOTROPHIC BACTERIA

Pleomorphic rods, these organisms use inorganic materials for energy

a. Organisms oxidizing ammonia or nitrite
Family I. *Nitrobacteraceae*
Genus I. *Nitrobacter*
Genus II. *Nitrospina*
Genus III. *Nitrococcus*
Genus IV. *Nitrosomonas*
Genus V. *Nitrosospira*
Genus VI. *Nitrosococcus*
Genus VII. *Nitrosolobus*
b. Organisms metabolizing sulfur
Genus *Thiobacillus*
Genus *Sulfolobus*
Genus *Thiobacterium*
Genus *Macromonas*
Genus *Thiovulum*
Genus *Thiospira*
c. Organisms depositing iron or manganese oxides
Family I. *Siderocapsaceae*
Genus I. *Siderocapsa*
Genus II. *Naumanniella*
Genus III. *Ochrobium*
Genus IV. *Siderococcus*

PART 13.
METHANE-PRODUCING BACTERIA

Rods or cocci; some members are Gram-positive, others are Gram-negative; all are anaerobic and produce methane

Family I. *Methanobacteriaceae*
Genus I. *Methanobacterium*
Genus II. *Methanosarcina*
Genus III. *Methanococcus*

PART 14.
GRAM-POSITIVE COCCI

Various arrangements of cocci that are aerobic, facultative, or anaerobic

a. Aerobic and/or facultatively anaerobic
Family I. *Micrococcaceae*
Genus I. *Micrococcus*
Genus II. *Staphylococcus*
Genus III. *Planococcus*
Family II. *Streptococcaceae*
Genus I. *Streptococcus*
Genus II. *Leuconostoc*
Genus III. *Pediococcus*
Genus IV. *Aerococcus*
Genus V. *Gemella*

INTRODUCTORY
MICROBIOLOGY

INTRODUCTORY
MICROBIOLOGY

FREDERICK C. ROSS

Delta College

CHARLES E. MERRILL PUBLISHING COMPANY

A Bell & Howell Company

COLUMBUS □ TORONTO □ LONDON □ SYDNEY

Published by Charles E. Merrill Publishing Co.
A Bell & Howell Company
Columbus, Ohio 43216

This book was set in Souvenir.
Production Editor: Rex Davidson
Cover Photo by John Hanson and
    Robert Pfister, Department of
    Microbiology, Ohio State University
Cover Design by Tony Faiola
Text Designer: Ann Mirels

Library of Congress Catalog Card Number: 82–081120
International Standard Book Number: 0–675–20003–2
Printed in the United States of America
    2 3 4 5 6 7 8 9 10—87 86 85 84 83

*I*ntroductory Microbiology is a survey of the comparative biology of microbes, including the bacteria, viruses, fungi, algae, and protozoa. This text is directed toward those students interested in careers in diverse fields of allied health, such as nursing, home economics, dental hygiene, and surgical technology, and the biological sciences, such as biology and the animal sciences. The approach is a general one directed toward those students entering these areas with little or no previous study in biology or chemistry.

In the writing of this text, a number of important considerations have been given attention. First, the student should find *Introductory Microbiology* accurate in microbiological content. Both fundamental concepts and their applications are presented in order to provide a foundation in microbiology that will enable the student to venture into more advanced and applied fields. The presentation of applications is also designed to aid the student in integrating the concepts into a basic framework of biological principles. Second, this text was written for the student. Presenting microbiological principles in a way that is interesting has been the goal throughout. The manuscript has been class tested and many students' suggestions have been incorporated. Too often, textbooks are not read because the writing style and the massive amount of factual material are burdensome to the student. Third, it is hoped that this text will serve to stimulate a continuing interest in the field of microbiology.

The parts and chapters of *Introductory Microbiology* are arranged into five groups. Part 1, Organizing the Microbial World, introduces the student to the fundamental characteristics and types of microbes as we know them today. Basic cell chemistry, a comparison of cellular types, and the tools used by microbiologists to study these small creatures are presented. Part 2, How Microbes Function, examines in detail microbial metabolism, growth and reproduction, and genetics. The collection, cultivation, identification, and control of microbes is also dealt with in this part. Host-Parasite Relationships, the subject of Part 3, presents information on the various interactions that can occur between microbes and the

Preface

multicellular organisms with which they may live. Special attention has been given to the principles of immunology, since this area of microbiology has become of such great importance. The principles of disease transmission are also presented in Chapter 15, Epidemiology. Part 4, Medical Microbiology, is a review of the major disease-causing microbes. These seven chapters are organized taxonomically because it was felt by the author that most instructors are microbiologists first and "physicians" only by necessity. Chapters include information from the areas of bacteriology, mycology, protozoology, and virology. *Bergey's Manual of Determinative Bacteriology*, 8th edition, is the key reference for this taxonomic approach. Part 5, Environmental Microbiology, presents four chapters describing the areas of food, agriculture, water and wastewater, and industrial microbiology. They are intended to serve as an overview of the fascinating diversity of microbial life as well as displaying the microbes' interactions with humans and our environment.

Many have knowingly or unknowingly helped in the writing of this text. My family was patient and provided the support so necessary to bring this project to its completion. Thousands of students have given feedback over the years concerning the material and its relevancy. They were an invaluable pool for review and criticism. I also wish to thank all the people at the Charles E. Merrill Publishing Company for their concern, effort, and encouragement, particularly Rex Davidson, Pam Cooper, Meg Malde, and Bob Lakemacher. Marilyn Plowdrey and Florence Kehrer assisted with the typing of the manuscript. I gratefully acknowledge the invaluable assistance of many reviewers throughout the development and preparation of the manuscript including: Louis Baron; Robert Rychert, Boise State University; Joseph Price, Trenton State College; Marjorie Sharp, University of Texas; and Bruno Kolodziej, Ohio State University.

TO THE STUDENT

Learning is a continuing process of combining the familiar with the unfamiliar. The role of the instructor is to set the stage for learning and encourage students to gather meaning out of new terms, to perform new skills in different situations, and to solve problems in new ways. The content and organization of this textbook is designed to help you learn the concepts and facts that relate to a particular subject, microbiology. The success you have will depend on how well you learn and understand the material. In most cases, your understanding will be evaluated by examinations of one form or another. Keep in mind that in most science courses, instructors insist on exact answers to questions. The answers you need are presented in the text, or are covered in the lectures or laboratory. The concepts and principles that science instructors want you to learn are described in a language that may be very foreign to many students. Learn new vocabulary words as you would any foreign language. Learn to speak the language of your instructor.

This text has a variety of aids to the student that are valuable and can further your knowledge of microbiology. At the beginning of each section is an overview that describes in a general way the content of the chapters within the unit. This overview can provide you with a preview of the chapters and will serve to orient you to a fairly substantial amount of material. The section overview differs from the Chapter Introductions. Chapter introductions direct the student more specifically to the chapter contents—where the chapter is going and how it fits in with the others previously covered. Both the section overviews and chapter introductions can be used to review for examinations.

Another study aid that can be beneficial are the Learning Objectives provided at the beginning of each chapter. The objectives will tell you in a general way what material you should be able to master. Keep these topics in mind as you read and study to help focus on major items.

There are a number of illustrations throughout the textbook. The illustrations should do more than just attract your attention. Each has been chosen carefully to help you understand a point or to help you tie a point to something you already know.

Also dispersed throughout the chapter are short text questions. These have been placed in the body of the text to serve as a quick but pointed review of material presented in the preceding paragraphs.

Each chapter ends with a Summary. As you finish your study of a chapter, read the summary sentence-by-sentence. Make sure that there is no new information in the summary. If something is new to you, return to that area in the chapter.

Following the chapter summary, there is a thought-provoking question called "With A Little Thought." This question allows you to use your newfound knowledge and place your previous experiences on the situation. Most often, there is no one correct answer. It is a device that can stimulate you to think something through and to raise points for discussion. One of the most important and valuable aspects of a microbiology course is not the tidbits of factual information that you learn, but the new way in which you see yourself and your environment. It is hoped that these thought questions will give you practice in using microbiological information and allow you to apply microbiological concepts to real-life situations.

After each chapter there are a series of Study Questions. These can be used in a number of ways; you might use them to help channel your attention as you study a chapter, or as a review to tell you when you are well prepared for a test over the chapter material. Each of these questions is directly answered in the chapter narrative or in the illustrations.

In addition, each chapter has a list of Key Terms along with some phonetic pronunciations. The words in this list are not only used frequently in the chapter but are likely to be found in other chapters. The Pronunciation Guide for Organisms should help you to speak the language of microbiologists and better communicate the knowledge you acquire in this science. As new organisms are introduced, their pronunciations will be given at the end of that chapter.

F.C.R.

INTRODUCTION   **1**

Chapter 5   Organelles and Their Importance   107

Chapter 6   Laboratory Equipment and Procedures   127

PART 1

Organizing the Microbial World   *22*

Chapter 1   Fundamental Characteristics of Microbes   24

Chapter 2   The World of Microbiology   35

Chapter 3   Basic Cell Chemistry   61

Chapter 4   Cell Anatomy and Physiology   84

PART 2

How Microbes Function— Physiology   *154*

Chapter 7   Metabolic Pathways   157

Chapter 8   Growth and Reproduction   181

Chapter 9   Collection, Cultivation, and Identification of Microbes   200

Contents in Brief

Chapter 10   The Control of Microbes   227
Chapter 11   Microbial Genetics   249

PART 3

Host-Parasite Relationships   *272*

Chapter 12   Symbiotic Relationships and Normal
              Flora   274
Chapter 13   Immunology I: Host Defense
              Mechanisms and Immunity   296
Chapter 14   Immunology II: Hypersensitivity,
              Transplants, and Tumors   328
Chapter 15   Epidemiology   356

PART 4

Medical Microbiology   *384*

Chapter 16   Bacterial Pathogens and Diseases (I)   387

Chapter 17   Bacterial Pathogens and Diseases (II)   408
Chapter 18   Fungal and Protozoan Diseases   430
Chapter 19   Viruses   450
Chapter 20   Viruses and the Disease Process   473
Chapter 21   Viral Diseases I: The DNA Viruses   482
Chapter 22   Viral Diseases II: The RNA Viruses   489

PART 5

Environmental Microbiology   *508*

Chapter 23   Food and Dairy Microbiology   510
Chapter 24   Soil and Agricultural Microbiology   533
Chapter 25   Water and Wastewater Microbiology   551
Chapter 26   Industrial Microbiology   572
Appendix A   Prefixes and Suffixes   597
Appendix B   Metric Conversions   600
Appendix C   Four-Place Logarithms   602
Glossary/Index   604

INTRODUCTION   1

   Science and the Study of Microbes   1
   Microbes Influencing Our Lives   2
   Of Microbiologists and Microbes   6
   A Founding Father   8
   Methods and Media   10
   Jenner and the Milkmaid   11

PART 1

*Organizing the Microbial
World*   **22**

**CHAPTER 1   FUNDAMENTAL
CHARACTERISTICS
OF MICROBES**   **24**

   Essential Life Characteristics   24
   Growth and Reproduction   26
   Adjusting to a Changing World   28
   Maintaining a Balance   29
   Enzymes in Real Situations   31

**CHAPTER 2   THE WORLD OF
MICROBIOLOGY**   **35**

   Grouping for a Better Understanding   36
   The Binomial System   39
   The Major Groups of Cellular
     Procaryotic Microbes   39
   The Photosynthetic Procaryotes   46
   Protista—the Eucaryotes   47
   Noncellular Microbes   55

CHAPTER 3   BASIC CELL CHEMISTRY   **61**

   Differences Among Elements   62
   Combining Atoms   63

Contents

Of Lemons and Lye   64
Chemical Reactions   65
Getting It Going   67
Carbon-Containing Molecules   69
Shape Is Important   70
Nucleic Acids   74
Coding for Proteins   75
Life and Death of Cells   79

**CHAPTER 4   CELL ANATOMY AND
PHYSIOLOGY   84**

Cell Components   84
Organelles   88
Cell Membranes   89
Passive Movement: No Metabolic Energy Required   93
Proteins of the Unit Membrane   99
Carbohydrates and Endocytosis   103

**CHAPTER 5   ORGANELLES AND THEIR
IMPORTANCE   107**

The Difference is Obvious   108
The Cell's Work Surface   108
Further Surface Specialization   110
Separate but Equal   111
The Greening of a Cell   112
External Organelles of Locomotion   114
Pili (Fimbriae)   115
Enduring Changes   116
The Ribosomes: Translator of the Cell   117
The Control Center   118
Outer Coverings   119
Capsules   122

**CHAPTER 6   LABORATORY EQUIPMENT AND
PROCEDURES   127**

Pure Cultures   128
Pure Culture Technique   129
Culture Preservation   131
Aseptic Technique   132
Sterilization: The Absence of Life   136
Chemical Sterilization   138
Staining Techniques   139
Microscopy   144
Visible Light Microscopy   145
Ultraviolet and Electron Microscopes   147

PART 2

How Microbes Function— Physiology   *154*

CHAPTER 7   METABOLIC PATHWAYS   157

Life Needs   158
Growth Factors   158
The Environment   159
Classification: A Physiological Method   161
Energy Production   164
Aerobic Cellular Respiration   166
Fat and Protein Metabolism   168
Biosynthesis and Metabolism   172
Controlling Metabolism   173

**CHAPTER 8   GROWTH AND
REPRODUCTION   181**

Increasing Mass and Populations   182
The Population Growth Curve   183
The Growth Curve: Home, Industry, and Hospital   184
Counting Methods   186

**CHAPTER 9   COLLECTION, CULTIVATION,
AND IDENTIFICATION
OF MICROBES   200**

Laboratory Work   201
Symptoms   204
Collecting and Handling Specimens   206
Media Selection and Preparation   209
Balancing the Environment   215
Oxygen Requirement for Culturing   219
Culture and Susceptibility   221

CHAPTER 10   THE CONTROL OF MICROBES   227

Different Approaches to Control   228
Factors that Influence Success   229
Surface-Active Agents   232
The Methods and Materials of Control   233
The Ideal Antibiotic   241
The Development of Drug Resistance   244
Overcoming Resistant Pathogens   245

CHAPTER 11   MICROBIAL GENETICS   249

The Fundamentals of Genetics   250
Mutagenesis and Carcinogenesis   252
The Effects of Selection   253
Transformation   254
Gene Transfer, Plasmids, and Drug Resistance   259
Relatedness and Genetic Engineering   260

PART 3

Host-Parasite Relationships   *272*

**CHAPTER 12   SYMBIOTIC RELATIONSHIPS
                    AND NORMAL FLORA   274**

We Are Not Alone   275
Shifting Relationships   276
The Skin   278
The Oral Cavity and Upper Respiratory Tract   281
The Gastrointestinal Tract   284
The Genitourinary Tract   288
Virulence: Toxins and Invasiveness   290
The Hemolysins and Other Substances   291

**CHAPTER 13   IMMUNOLOGY I: HOST
                    DEFENSE MECHANISMS AND
                    IMMUNITY   296**

Nonspecific Host Defense Mechanisms   297
The Reticulo-Endothelial System and
    Inflammation   299
Immunity—A Specific Host Defense   301
Basics of the Immune System   302
Immunoglobulins and Antigens   303
Lymphocytes and the Immune System   306
Antibody-Antigen Reactions   307
Other Serological Tests   314
Cell-Mediated Immunity (CMI)   318
The Anamnestic Response and Immunization   320

**CHAPTER 14   IMMUNOLOGY II:
                    HYPERSENSITIVITY,
                    TRANSPLANTS, AND TUMORS   328**

Hypersensitivity   329
Type I Hypersensitivity   329

Sensitivity Testing and Treatment   332
Type II Hypersensitivity   336
Type III Hypersensitivity   338
Type IV Hypersensitivity   340
Transplantation Immunology   342
The ABO and Rh Blood Systems   343
Transfusion Reactions   344
Histocompatibility Antigens   347
Cancer and Immunology   348
Immunotherapy and Cancer Control   352

CHAPTER 15   EPIDEMIOLOGY   356

Epidemiology: Scope and Terminology   357
Reservoirs and Epidemics   357
Carriers and Vectors   358
An Organizational Approach   360
Classification of Diseases   361
How Diseases Spread   362
Transmission Routes   369
Herd Immunity and Disease Cycles   370
The Control of Communicable Diseases   373
Nosocomial Infection   377

PART 4

Medical Microbiology   *384*

**CHAPTER 16   BACTERIAL PATHOGENS AND
                    DISEASES (I)   387**

Pyogenic Cocci   388
Gram-Positive, Spore-Forming Bacilli   396
Enteric Gram-Negative Bacilli   398

**CHAPTER 17   BACTERIAL PATHOGENS AND
                    DISEASES (II)   408**

Small Gram-Negative Bacilli   408
Corynebacteria   413
Acid-Fast Bacteria   415
Spirochetes   420
Gram-Negative Anaerobes   423
Mycoplasmas   424
Rickettsias   425
Chlamydia   426

CHAPTER 18 **FUNGAL AND PROTOZOAN DISEASES** 430

Medical Mycology 431
Subcutaneous Mycoses 435
Systemic Mycoses 437
Actinomycetes: Filamentous Bacteria 441
Protozoa of Medical Importance 442

CHAPTER 19 **VIRUSES** 450

General Characteristics of Viruses 451
Cultivation and Counting 454
Classification of Viruses 455
Bacteriophage Life Cycles 456
Animal Virus Life Cycles 458
Transformation 464
Transmission of Transforming Viruses 466
Cancer: Virus-Human Interaction? 467
Unconventional Viruses 469

CHAPTER 20 **VIRUSES AND THE DISEASE PROCESS** 473

Cytopathic Effects and Symptoms 473
Interferon 474
Immunity and Serology 476
Immunization and Chemotherapy 477

CHAPTER 21 **VIRAL DISEASES I: THE DNA VIRUSES** 482

Parvoviruses 482
Papovaviruses 483
Poxviruses 483
Herpesviruses 483
Adenoviruses 487

CHAPTER 22 **VIRAL DISEASES II: THE RNA VIRUSES** 489

Picornaviruses 489
Rhabdoviruses 494
Hepatitis Viruses 496
Paramyxoviruses 499
Arboviruses 502
Orthomyxoviruses 503

PART 5

Environmental Microbiology **508**

CHAPTER 23 **FOOD AND DAIRY MICROBIOLOGY** 510

Microbes and Their Growth in Food 511
Microbial Problems: Spoilage, Poisoning, and Infection 512
Natural Preservation 518
Drying 518
Refrigeration and Freezing 519
Irradiation 520
Canning 520
Chemical Preservatives 522
Food Microbiology: Fermented Foods 523
The Staff of Life 529

CHAPTER 24 **SOIL AND AGRICULTURAL MICROBIOLOGY** 533

Soils, Soil Formation, and Microbes 534
Decomposers and Biodegradable Materials 536
Biogeochemical Recycling 538
Nitrogen Cycle 540
Crops and Microbes 544
Animals and Microbes 546

CHAPTER 25 **WATER AND WASTEWATER MICROBIOLOGY** 551

Types of Water 552
Classification of Water 554
Water Quality Standards 557
Identifying and Counting Contaminants 559
Water Purification 561
Home Sewage Disposal 562
Public Wastewater Treatment 563

CHAPTER 26   INDUSTRIAL MICROBIOLOGY   572

Cultivation of Microbes   573
Cell Cultures   575
Alcoholic Fermentation   576
Alcoholic Beverages   577
Microbial Enzymes in Industry   587
Vitamins and Amino Acids   588
Antibiotics and Steroids   589

INTRODUCTORY
MICROBIOLOGY

MICROBES are part of our lives in more ways than most understand. They have shaped our present environment and their activities will greatly influence our future. Microbes should not be considered apart from humans, but should be considered a part of our world. Microbiology is one of the most applied of all the biological sciences. Knowledge gained in this field has led to the development of many concepts, which in turn resulted in action. The more we learn, the better we can improve the quality of human life as well as that of all living forms. Our goal should be to understand the environment and better stabilize it to the benefit of all.

SCIENCE AND THE STUDY OF MICROBES

Most people know that biology deals with the study of plants and animals, and probably recognize it as a science. Textbooks define **biology** as the science that deals with the study of life. This appears to be a simple, straightforward definition until the words **science** and **life** are considered.

Science is the study or collection of knowledge of natural events and materials in an orderly fashion for the purpose of learning the basic laws that govern these events. The information (fact) is collected and the scientist uses certain rules in order to place these facts into a sensible framework. Rules, laws, and principles are developed as scientists begin to see patterns or relationships among a number of

Introduction

isolated facts. Some rules are very old while others are being constructed today. The method of collecting information and the way in which it is organized is really what makes something a science. The laws and rules are continually tested by the addition of new bits of information. If the new information can fit into the constructed framework, it reinforces the framework.

Biology is a broad science that draws on chemistry and physics for its foundation and applies these basic physical laws to living things. Because there are millions of kinds of living organisms, there is a very large number of special study areas in biology. Practical biology such as medicine, agriculture, plant breeding, and dentistry is balanced by more theoretical biology—evolutionary biology, molecular genetics, and "recreational" biology like insect-collecting and bird-watching. Biology is a science that deals with living organisms and how they interact with the environment around them.

Since the prefix "micro-"means *small,* microbiology must be the biology of small living things. In fact, **microbiology** is the branch of biology that deals mainly with the study of **microscopic** organisms called microbes, which are composed of only one cell. Organisms not usually included in this study are those that are composed of many cells. Trees, grass, humans, and animals have many cells arranged into tissues, organs, and organ systems. These other life forms are studied in the related sciences of botany and zoology. Because of the small size of microbes, microbiologists must study their cells using special equipment and procedures. Microscopes, test tubes, and chemicals are very important tools to a microbiologist. Much of the work is done to understand better the chemical reactions that occur in living cells. This strong connection between chemistry and biology has led to the development of a separate science called **molecular biology.** Molecular biology focuses on the kinds of molecules found in living cells, their behavior, and how they work together to make a cell alive. Information gained from each of these three sciences (microbiology, chemistry, and molecular biology) is frequently shared. Since the cell is the basic unit of all life as we know it, the study of microbes provides much information in the fields of genetics (inheritance), physiology (function), and biochemistry (chemistry of life). As a result, there has been a great deal of knowledge gained, and we have come to understand better the influence of microbes on humans and the environment.

There are many places in our environment where microbes and the results of their activities can be seen. Both our natural and manmade worlds contain microbes of many kinds. Each type of microbe has special qualities which enable it to survive in such unique places as the soil, oceans, rivers and streams, ice, water pipes, concrete, hot springs, the human intestine, roots of plants, and even oil wells. Because many types of microbes have become specialized to live in unique environments, microbiologists have specialized the study of certain groups of microbes. Microbiology is subdivided into a number of diverse fields. **Bacteriology** is the study of the bacteria, **virology** the study of viruses, **protozoology** the study of protozoans, **phycology** the study of algae, and **mycology** is the study of fungi such as yeast and molds (Figure I.1). These fields center around the kinds of organisms under study. Other fields of microbiology center around where the microbes grow and have the most significant effects. These include such areas as medical microbiology, food microbiology, agriculture, waste treatment, and industrial microbiology.

---

1 What does the prefix "micro-" mean?
2 What type of equipment do microbiologists use?
3 What is the basic unit of life?

MICROBES INFLUENCING OUR LIVES

The science of microbiology can be divided into theoretical and applied fields. Farmers are applied scientists. They use microbiological principles to get the best yields from their farms. Doctors are applied microbiologists, too, because their primary interest is to keep people healthy through the use of scientific knowledge. The theoretical scientist does not have in mind any specific use for the new knowledge gained. The purpose is to obtain new information to see how it fits the "old laws," and write "new laws" if necessary.

While investigating the cause of wine spoilage in the vineyards of France, Louis Pasteur became interested in the theoretical problem of whether life could be generated from nonliving material. Much of his theoretical work led to very practical applications. His theory that there were very small organisms causing diseases and decay led to the development of vaccinations and the preservation of foods by pasteurization.

At this point, however, it is important to note that the study of microbiology is much like the study of a foreign language. The vocabulary includes a great many new words. Once you learn the meaning of these words and how to use them properly, the science of microbiology will be much easier. Many terms refer to the organism being studied; others refer to the activities of a particular organism and using them improperly could lead to a misunderstanding. For example, the words John Smith tell you who a person is, while the word microbiologist tells you what a person does.

A brief look at each of the applied fields of microbiology may give some understanding of how microbes influence our lives.

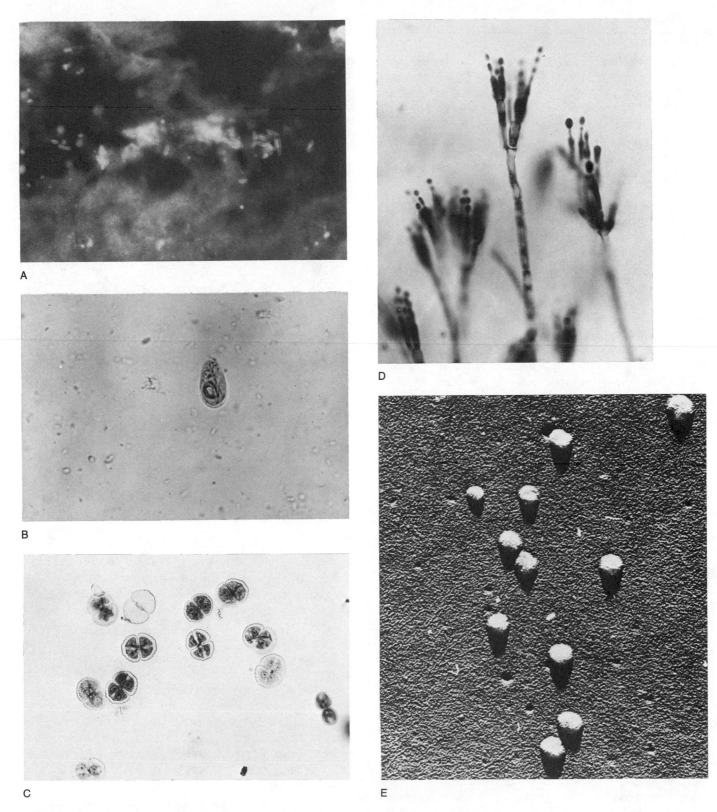

FIGURE I.1

Representative microbes. (A) *Escherichia coli* is a common bacterium found in the intestinal tract of humans and many domesticated animals. (B) *Giardia* is a protozoan, or single-celled animal. (C) *Cosmarium* is a green alga. (D) Downy mildew is a filamentous fungus that is a parasite on plants. (E) Influenza viruses. (Photos A, B, and E from the Center for Disease Control, Atlanta. Photos C and D courtesy of Carolina Biological Supply Company.)

3

Medical microbiology is probably a very familiar field since it deals with microbes which cause diseases in humans, animals, and many plants. When asked of their knowledge of microbiology, people will mention something about a sickness or disease that they have had themselves. The "flu" (influenza), strep throat, tetanus, and malaria are all examples of diseases that are caused by the activities of microbes. In medical microbiology, the microbes that cause illness are called **pathogens.** The term **disease** means a process or event that results in illness or harm to a living organism. The words **infectious disease** refer to the fact that some microbes are able to go through the process of entering another organism, growing, and causing harm. Some of these microbes may cause only minor damage, while infection by others may result in quick death.

Some microbes can spread through a population and cause disease. Over one hundred years of work was needed to show that a particular illness was caused by the actions of a particular microbe that was able to be transmitted from one individual to another. The bacteria that cause the disease cholera *(Vibrio cholerae)* were not identified with the symptoms they produced until 1883. Today our understanding that microbes can spread through a population and cause a particular disease is known as the **germ theory of disease.** Using this theory as a pattern for research has led to a greater understanding of the disease process. The germ theory of disease has become a unifying concept in that it defines the role of microbes in the disease process. Before the germ theory, many diseases were suspected to have causes ranging from those caused by "evil spirits" to unknown "miasmas" emanating from the soil or earth. Today we know a particular microbe may cause a number of different illnesses. An infection by one microbe may show different symptoms or signs depending on where it becomes located in the body. When a bacterium is located in the lung it may cause pneumonia, while in the joints it may cause arthritis, or meningitis in the spinal cord or brain. Medical microbiology also deals with disease prevention, the body's resistance to disease-causing organisms, and ways in which sick persons may be helped to recover (Figure I.2).

Foods must be transported great distances and stored for long periods for the world to have a safe, nutritious food supply. To accomplish this, the role of microbes in food poisoning, spoilage, and preservation must be investigated. Contributions to the field of **food and dairy microbiology** have helped resolve many food problems. When microbes enter a food they may either cause it to spoil, make it dangerous to eat, or change it to another form that is still acceptable as a food. Scientists have had to find out how the microbe enters the food, what action it has, and how to control the microbe. The whole idea of preserving foods is based on preventing microbial growth (Figure I.3). **Pasteurization** is one of the best-known methods used to pre-

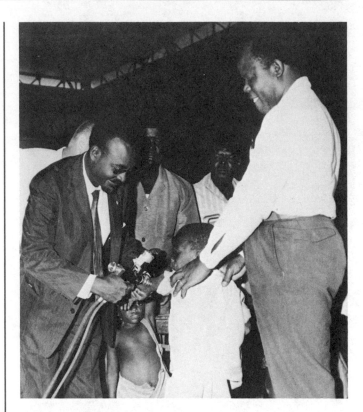

FIGURE I.2

The practical results of medical microbiology can be seen as this West African child is being vaccinated against the terrible effects of smallpox. Research and control have reduced smallpox to the point of extinction. (Courtesy of the Center for Disease Control, Atlanta.)

vent spoilage and the transfer to disease. This heating procedure was first used in the wine industry in an attempt to keep wine from being changed into vinegar. This worked so well that other food industries now use the process. Cheeses, beer, and milk are usually pasteurized to reduce the number of harmful microbes in the food.

In some cases, microbes are intentionally mixed with foods to change the food to another form. When certain bacteria are added to milk and encouraged to grow under controlled conditions, the milk is converted to cheese. The food value of cheese is very high, and it is much easier to store than milk. The many different kinds of cheeses found in the world are also the result of the actions of different microbes. The flavor, smell, and texture of a cheese is determined by the kinds of microbes that are grown in the milk.

1 Who developed the pasteurization process?
2 The words "infectious disease" refer to what types of illness?
3 Name one beneficial activity of microbes in food.

A   B

FIGURE I.3
(A) Food preservation through canning is a relatively recent development. (B) Brining of cucumbers. Foreground, the surface of a tank of fermenting cucumbers being purged with nitrogen to remove dissolved carbon dioxide. Background, a tank being filled. (Photo A courtesy of the Campbell Soup Co., Camden, N.J. Photo B compliments of H. P. Fleming, *Pickle Pak Science* 6:8–22, 1979.)

Water in our environment commonly contains microbes. The kinds of microbes and their method of entering the water is of great concern since many waterborne microbes can cause human disease. The field of **water and wastewater microbiology** explores all of these areas. The ultimate source of water is from rain or precipitation. The water moves over the surface of the ground as rivers and streams or through the earth as groundwater. Microbes and chemicals may enter the water from the air as the water passes through it or from untreated or poorly treated sewage being dumped into a body of water. Drinking water and industrial water are pumped from lakes, rivers, and wells. This water must be cleared of harmful bacteria, viruses, and other microbes to control diseases and prevent fouling of industrial equipment. Cholera and typhoid fever are diseases that are able to be transferred in water. Purification plants throughout the country treat water before it is used. Wastewater treatment plants treat water before it is again released into the environment. In some cases, water becomes clouded with large amounts of chemicals, mud, and plant material. Unless this water is fast moving and well aerated, microbes will use these materials for food and produce poisons or foul odors.

Soil contains great numbers of many microbe types. Bacteria, algae, and fungi are commonly found in rich, fertile soils. The more fertile the soil, the more likely it is to contain a thriving population of microbes. An effort to maintain good farm production has led microbiologists to develop the field of **soil and agricultural microbiology.** Microbes are responsible for the decomposition and decay of dead plants and animals. If it were not for microbial decomposition, the earth would be covered with all types of decaying organisms. The molecules in these would stay locked-up and unavailable for reuse by other, younger organisms. The recycling of materials through decay activities in the soil is vital to all life. Sewage treatment plants are, in part, giant microbial cultures for speeding decay of waste products. The waste material put into compost piles is broken down as a result of microbial activity and becomes rich, soil-fertilizing material for plants. The increased amount of solid waste throughout the world along with the increase in human population has made it essential to explore new ways of controlling decay. Agricultural microbiology deals with those microbes associated with animals. For example, cattle and other animals with rumens, or "second stomachs," are laiden with microbes. These are of great value to the health of the animal. The food they eat, silage, is grain that has been preserved by the action of microbes while the grain has been stored in a silo.

Due to the small size of microbes, simple test tubes and covered dishes can provide enough space to grow them for tests and explore their activities. However, many microorganisms are capable of producing products of value to humans, if they can be produced in very large quantities. Microbiologists, engineers, and business have come together for this purpose and developed the field of **industrial microbiology.** This area involves the large-scale growth of particular microbes. Yeast, bacteria, and molds may be grown in 5, 10, or even 50,000 gallon containers. For example, yeast cells are grown for use in ironized yeast tablets for human use and in even larger quantities as a supplement to farm animal feed. In other instances, the by-product of the growth and activities of a microbe is the most useful. These include enzymes, amino acids, antibiotics, alcohol, and organic acids. Handling large quantities of growing microbes is not easy. Expensive equipment and well-trained personnel are essential to the smooth operation of an industrial process such as antibiotic or beer production (Figure I.4). Industrial microbiolgists must closely monitor the microbes' chemical activities and be prepared to change or stop the activities of the organism to prevent product loss. Careful attention must also be paid to ensure that only one kind of microbe is being grown. Successful control of these factors has resulted in the development of multimillion dollar microbial industries.

---

1 What action can microbes have on silo-stored grain?
2 List two products industrially produced by microbes.

OF MICROBIOLOGISTS AND MICROBES

A vast amount of chemical and microbiological information has been gained from studies in the various microbiology fields. These efforts have resulted in solutions to many problems, and have also revealed new and more challenging areas of concern. A better understanding of the scope of microbiology may be gained by taking a brief look at the history of microbiology and some of the questions explored by famous microbiologists.

For centuries, one of the most intriguing questions related to the origin of life. Even though the answer still continues to be an area of speculation, research and experimental attempts to explain the origin of life have inadvertently been prime movers in expanding our knowledge of microbiology. In simpler times, the origin of life from nonliving things was never doubted. The Greeks, Romans, Chinese, and many other ancients believed that maggots, lice, frogs, and even mice could spontaneously arise from mud. They thought they saw these events happening every day. It was thought that mice could be produced from a sweaty shirt if it was kept in a dark, cool room with several grains of wheat. Many prominent scientists believed in this concept. Only through the efforts of scientific investigators like Redi, Spallanzani, and Pasteur was this classical concept of **spontaneous generation (abiogenesis)** discarded (Figure I.5).

The argument between the supporters of spontaneous generation and those of **biogenesis** has lasted over 300 years. People believing in biogenesis thought that all living things came from pre-existing life. During this period, the cleverness and imagination of those involved were used to the fullest in the attempt to disprove the others' position. Even though microbes could not be seen, a number of people suspected that such small living things existed. In the first century B.C., Varro suggested that diseases were due to invisible organisms. Later in 1546, Fracastorius wrote three

FIGURE I.4
This large copper brew kettle contains the "wort" to be fermented by yeast in the production of beer. Yeast has been controlled for beer production since ancient Egyptian times. (Courtesy of the G. Heileman Brewing Company, Inc.)

FIGURE I.5
This illustration portrays a primitive story of life from nonlife. A destroyed army is recreated by throwing stones on the ground and giving a special prayer. As the stones strike the earth, they become living soldiers.

books, *De Contagione.* They contained the first scientific statements as to how infections were transmitted. Fracastorius was medical adviser to the Council of Trent (Italy), which had to be transferred to Bologna because of an outbreak of typhus fever. He wrote, "Contagion is a precisely similar putrefaction which passes from one thing to another: its germs (seminaria) have great activity, they are made up of a strong and viscious combination, and they have not only a material but also a spiritual antipathy to the animal organism." Fracastorius believed even at this early date that consumption (tuberculosis) was able to be transferred from person to person. He commented, "it is extraordinary to see in families up to the fifth and sixth generation all the members die of phthisis (tuberculosis) at the same age."

But it was Francesco Redi (1668) who set up the first controlled experiment to disprove those who supported spontaneous generation. This is the best type of experimental setup from which to draw conclusions, since there is only one unknown factor in question. Redi used two sets of dishes and varied only one part of the experiment (Figure I.6). These experiments ended the concept of spontaneous generation for only a short period of time. In 1676, the idea of spontaneous generation was accidentally revived, due to the tinkerings of a Dutch clothier, Anton van Leeuwenhoek. His discovery of "animalcules," little animals, while using lenses reopened the debate. Leeuwenhoek's successful method of grinding lenses allowed him to construct over 200 microscopes (all disposable), capable of magnifying a specimen to 300 times its normal size. The images were so clear that Leeuwenhoek made drawings of what we now know to be bacteria. These were submitted to the influential Royal Society of London and were so impressive that many others

FIGURE I.6
The two sets of jars used in Redi's experiment are identical in every way with one exception—the gauze covering. The set with gauze is called the control group, and the set without gauze is called the experimental group. Any differences between the control and the experimental groups are the result of a single variable. In this manner, Redi concluded that the presence of maggots in meat was due to flies laying their eggs on the meat, and not to spontaneous generation. (From Eldon D. Enger et al, *Concepts in Biology,* 3rd ed. © 1982, 1979, 1976 by Wm. C. Brown Company Publishers, Dubuque, Iowa. Reprinted by permission.)

began exploring the world of microbes (Figure I.7) and attempting to resolve the questions of spontaneous generation.

Where did these new life forms come from? Another scientist, Lazaro Spallanzani (1729–99) benefited from Redi's use of controlled experiments and successfully put to rest the spontaneous generation theory once more. In response to a challenge by Joseph Needham, an English priest and naturalist, Spallanzani devised an experiment that not only settled the argument, but demonstrated that heating a container and using an airtight seal would prevent spoilage of foods. No "animalcules" were generated in his experiment. This experiment on the origin of life ultimately led to the basic preservative methods used in the commercial canning industry (Figure I.8).

Spallanzani's evidence supported the theory of biogenesis until 1775 when Lavosier and Priestly discovered oxygen. This discovery brought on another wave of interest in spontaneous generation. It was suggested that since oxygen was excluded from Spallanzani's experiment, the spontaneous generation of life was prevented. With this new chal-

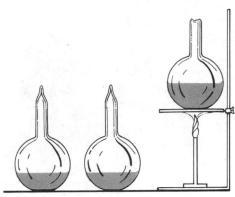

FIGURE I.8

Spallanzani carried the experimental method of Redi one step further. By sealing the flasks after they had been boiled, he demonstrated that spontaneous generation could not occur unless the broth was exposed to the germs in the air. (From Eldon D. Enger et al, *Concepts in Biology,* 3rd ed. © 1982, 1979, 1976 by Wm. C. Brown Company Publishers, Dubuque, Iowa. Reprinted by permission.)

lenge to the theory of biogenesis, Louis Pasteur (1869) began to experiment with the origin of bacteria and their need for oxygen. He successfully defended the biogenic theory. Even though Pasteur's original work was designed to investigate biogenesis, he gathered much other information related to microbes and their activities.

A FOUNDING FATHER

Louis Pasteur (1822–95) was one of the outstanding scientists of his day. His achievements served as stepping-stones for many others. Pasteur confirmed that certain microbes were directly responsible for the formation of such different kinds of molecules as acetic acid and lactic acid. The wine industry of France relied on him to solve the problem of wine changing into vinegar as the alcohol disappeared and acetic acid was produced. His work demonstrated how microbes were able to form not only acetic acid but also lactic acid, butyric acid, and alcohol. As a result of these investigations, he developed a process of heating wine to kill harmful microbes and prevent them from ruining the wine (1864). Today, we call the process **pasteurization** and have adapted it for use in controlling the quality of many other food products. Pasteur became interested in the growth or culturing of certain microbes and developed a process of transferring selective organisms from one batch to another to maintain good quality wine. In addition, he also demonstrated the presence of bacteria in the air. Pasteur found that airborne organisms could be kept out of sterile materials by plugging the tops with sterile cotton and still have air circulate in the

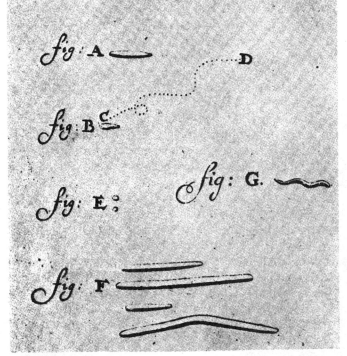

FIGURE I.7

This drawing was made by Leeuwenhoek of "animicules" he saw through his microscope. (From W. Bulloch, *The History of Bacteriology.* Oxford University Press, 1938.)

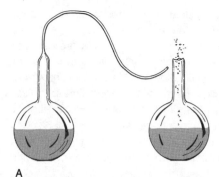

A

FIGURE I.9

(A) The gooseneck flasks used in Pasteur's experiment allowed for the flow of air into the flasks, but not airborne organisms. Pasteur demonstrated that germ-free air with its oxygen does not support spontaneous generation of life. (B) The autoclave is designed to destroy all life forms and was developed on the basis of Pasteur's experimentation. (Part A from Eldon D. Enger et al, *Concepts in Biology,* 3rd ed. © 1982, 1979, 1976 by Wm. C. Brown Company Publishers, Dubuque, Iowa. Reprinted by permission. Photo B courtesy of American Sterilizer Company.)

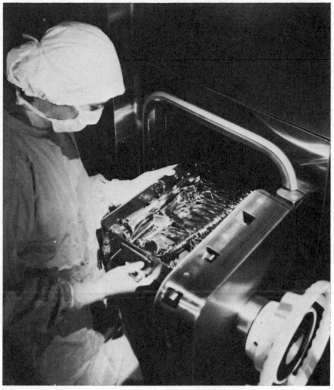

B

bottle. This technique is still used today in microbiological laboratories as a part of regular sterilizing techniques (Figure I.9). Later, his work expanded to involve diseases of animals and humans.

Pasteur was familiar with the work of another famous microbiologist, Robert Koch (1843–1910). Koch had discovered the bacterium that caused anthrax in cattle and demonstrated the progression of the disease. The excitement generated by Koch's work encouraged Pasteur to investigate anthrax. As a result, Pasteur demonstrated that he could prevent anthrax in cattle by injecting healthy animals with live anthrax bacteria that had been specially treated to reduce their disease-causing abilities. His success with anthrax led him to investigate hydrophobia, or rabies. He had worked for several years to prevent this disease in animals by using much the same techniques that had worked with anthrax. In 1885, his efforts were put to the test when a frightened mother sought out Pasteur for help. Her son, Joseph Meister, had been bitten by a rabid animal and Pasteur began treating the boy for rabies. The boy survived and Pasteur was again praised for his brilliant work.

Louis Pasteur is known the world over as one of the great men of science. He is credited with starting microbiologists down a path of research in the area of preventive medicine. His concern for the prevention of diseases by inoculation eventually led to the founding of the science of immunology. During this same period in history, Robert Koch began a comparable career investigating microbes. His efforts led to achievements as exciting and important as those

of Pasteur. Koch began research into the isolation and identification of individual microbes, especially bacteria. Because of this work, he is regarded by many as the first true bacteriologist. His work eventually led to the founding of the sciences of medical microbiology, bacteriology, and virology. Koch was educated as a physician and chose to practice in a small German town. While serving as a country doctor, he became intensely interested in the cause of the disease anthrax. The disease begins with symptoms similar to a cold and causes itching skin. Blisters or vesicles form which later turn black and swell. The bacteria may move into the blood and cause fever, shock, and eventually death. The research he did was praised as brilliant even though he used the most basic of techniques and equipment. Koch worked long hours and successfully isolated the bacterium that caused anthrax. He used house mice to show how the disease moved through animals and speculated on the formation of a resistant form of the bacterium, an **endospore.** Later he did special work on the endospores and demonstrated how they were able to survive in dead animals and in soil where they could serve as a source of infection for other animals. He studied the anthrax bacterium in frogs and horses, and saw how the disease organisms became concentrated in the lymph sacs and spleen. This information led to a better understanding of how the lymph system functions to control disease-causing microbes in the body.

When Koch presented this information to the Royal Society of London, the praise and support he received was so great that he shifted his work from medicine to research.

METHODS AND MEDIA

While involved in his research, Robert Koch developed a great variety of tools and techniques to examine bacteria (Figure I.10). Most of these we still use today and are the basis of modern laboratory procedures. It was Koch who first adapted staining methods to better see bacteria under the microscope. Basic dyes like methylene blue will chemically bond to bacteria and make them more visible. The hairlike projections, **flagella,** that make some bacteria move were first seen by Koch using a special staining process. He used a special oil on his slides to improve lighting at higher magnification and added another lens developed by Abbé to the microscope to better channel the light through the lenses. We use both oil and the Abbé condenser lens on microscopes in microbiological work today. The technique of smearing bacteria on a clean glass slide followed by a slight heating to kill and stick the cells to the glass was developed by Koch. In 1881, he presented the Royal Society with another breakthrough so grand that Pasteur met with him after the presentation with the greeting, "C'est un grand progre's, Monsieur!" ("This is great progress, sir!").

Until this point, bacteria had been grown in the lab in broth or on the surface of potato slices sterilized by hot air (also invented by Koch!). Koch found that by adding the protein gelatin to beef broth it could be solidified and the bacteria more clearly seen and separated. Because this medium could contain a greater variety of nutrients than potato medium, it was possible to grow a greater variety of bacteria. The major problem with his method was that a number of bacteria could digest gelatin. When grown on the gelatin media, these bacteria would form little puddles which made it difficult to identify individual characteristics. This problem was solved when a coworker's wife told Koch about a solidifying agent she used in cooking called **agar-agar.** This chemical comes from seaweed, algae, and was the perfect answer to the problem because most microbes cannot digest agar-agar. Today this material is used worldwide to solidify bacteriological media. It is so common that many use only the term "agar" to refer to the solid nutrient agar-agar growth medium.

> **1**   Louis Pasteur did research on what bacteria?
> **2**   What function does a flagellum serve?
> **3**   Agar-agar is used for what purpose?

However, nobody is perfect, not even Robert Koch! In 1882, Pasteur developed a theory and a method for vaccinating animals against anthrax. He presented his ideas at a scientific meeting which included Koch. Koch vigorously opposed Pasteur's idea in person and through correspondence. The argument between them continued even while Pasteur began field testing his vaccine and method. Koch's letters of opposition soon faded as Pasteur's successes continued to rise. However, this minor setback did not discour-

FIGURE I.10
Robert Koch's own laboratory.
(From W. Bulloch, *The History of Bacteriology.* Oxford University Press, 1938.)

age Koch. He continued his work and discovered the bacterium that causes cholera (*Vibrio cholerae*) and isolated the bacterium that causes tuberculosis (*Mycobacterium tuberculosis*). The following appeared in the May 1882 edition of *Scientific American* and recounts Robert Koch's efforts, which confirmed that the bacterium *Mycobacterium tuberculosis* is the cause of the disease once known as the "wasting disease."

> "Professor Tyndall has communicated to the London *Times* an account of results obtained by Dr. Koch of Berlin in the investigation of the etiology of tubercular disease, as set forth by him in an address delivered on March 24 before the Physiological Society of Berlin. In pursuing these investigations Dr. Koch subjected the diseased organs of a great number of men and animals to microscopic examination, and he found in all cases the tubercles were infested with a minute, rod-shaped parasite, which, by means of a special dye, he differentiated from the surrounding tissue. Transferring directly, by inoculation, the tuberculous matter from diseased animals to healthy ones, he in every instance reproduced the disease. Dr. Koch has examined the matter expectorated from the lungs affected with phthisis and found in it swarms of bacilli, whereas in matter expectorated from the lungs of persons not thus afflicted he has never found the organism. Guinea pigs infected with expectorated matter that had been kept dry for two, four or eight weeks were smitten with tubercular disease quite as virulent as that produced by fresh expectoration."

Koch also worked with the bacterium that causes boils (*Staphylococcus aureus*), and with the cause of cattle plague (a virus infection). His single most outstanding contribution to all science has become known as **Koch's Postulates.** The Postulates state: (1) a particular organism can always be found in association with a particular disease, but not in a healthy individual; (2) the organism can be grown in the laboratory by itself; (3) this pure growing culture will produce the same disease when placed back into a new, susceptible animal; and (4) it is possible to recover the organisms from this sick animal and grow them in pure culture. By using these postulates as guidelines, many microbiologists have proven that a particular organism is in fact the cause of a particular disease.

JENNER AND THE MILKMAID

Among other researchers involved in the study of disease prevention was the well-known British physician Edward Jenner (1749–1823). Jenner first developed the technique of vaccination in 1795, well before the time of Pasteur or Koch. This was the result of a twenty-six-year study of two diseases, cowpox and smallpox. Cowpox was known as **vaccinae,** and from this word Pasteur developed the pres-

ent terms "vaccination" and "vaccine." Jenner observed that milkmaids rarely became sick with smallpox, but they did develop pocklike sores after milking cows infected with cowpox. This led him to perform an experiment in which he transferred pus material from the cowpock to human skin. Since the two disease organisms are so closely related, the person vaccinated with cowpox developed an immunity to the smallpox virus as well. The reaction to cowpox was minor compared to the more serious smallpox. Public reaction was mixed; some people thought that the process of vaccination was a work of the devil. However, many European rulers supported Jenner by encouraging their subjects to be vaccinated. Napoleon and the Empress of Russia were very influential, and in the United States, Thomas Jefferson had a number of his family vaccinated. Today, almost two hundred years after Jenner developed a vaccination against smallpox, the disease has been eliminated. The World Health Organization announced that smallpox is the first disease to become extinct through human efforts!

The names and personalities of the people described are but a few in the history of microbiology. Many others have made equally outstanding contributions. Table I.1 lists chronologically a brief history of microbiology from the first century B.C. to the present. During the early 1900s, advances were made and formed the foundation for the field of molecular biology. Microbes were very important in this research. Our increased understanding of the microbe has ushered in a period of rapid advancement in biology. One of the first major contributions came in 1952 when Alfred Hershey and Martha Chase demonstrated by using bacteria and viruses that DNA (deoxyribonucleic acid) is the controlling molecule of cells. Their work with the viruses that infect bacterial cells, **bacteriophage,** was so significant that the **phage** became a standard laboratory research organism. Just one year following this work, James Watson and Francis Crick used this information, and that of others, to propose the now famous double helix molecular structure for DNA. Ten years later, Watson, Crick, and a coworker, Wilkins, were awarded a Nobel Prize for the work. In 1958, George Beadle and Edward Tatum won a Prize for their discovery that genes act by regulating definite chemical reactions in the cell, the "one gene-one enzyme" concept. The chemical reactions of the cell are controlled by the action of enzymes and it is the DNA that chemically codes the structure of these special protein molecules.

At first glance, some research by microbiologists may seem irrelevant or unrelated to everyday life. But it is a rare occasion when such research ideas do not make their way into our lives in some practical, beneficial form. The work of Watson, Crick, Beadle, and Tatum has been applied in hospitals and doctors' offices. Their basic research into DNA has provided the information necessary to develop medicines that control disease-causing organisms and others that regulate basic metabolic processes in our bodies. The ease with which such theoretical research information becomes a part of science and moves into our lives makes microbiology one of the most applied of all the biological sciences. A quick look at Table I.2 illustrates this point further. Advances in microbiology were responsible for eliminating tuberculosis as the number one cause of death in humans. Between 1937 and 1967, pneumonia and influenza dropped from the second to the fifth position as a cause of death. Future researchers will be engaged constantly to find cures for diseases and improving life in many other ways.

TABLE I.1

The history of microbiology—an overview.

Date	Event	Historical Perspective
1st century B.C.	Varro suggests diseases due to invisible organisms.	
		Caesar conquers Gaul.
1546	Fracastor's *De Contagione* makes first scientific statement of how infections are transmitted.	
		Henry VIII succeeds Edward VI.
1555	First use of word "physiology" in modern sense (Jean Fernel).	
1590	First compound microscopes.	
		Shakespeare writes *Henry VI.*
1658	Kircher sees "innumerable worms" under microscope.	
1660	Royal Society founded.	
		Rembrandt: "The Syndics of the Cloth Hall."
1665	*Philosophical Transactions* of the Royal Society first published. Robert Hooke's *Micrographia.* First drawing of cell (Hooke).	
		Delaware becomes a separate colony.
1676	Van Leeuwenhoek discovers "little animalcules"; perfects lenses to magnify 300 times.	
1688	Redi publishes book on spontaneous generation of maggots from putrid flesh.	
1720	Bradley's germ theory.	London's Royal Academy of Music names George Frederick Handel as its director and presents Handel's oratorio *Ester.*
1740	Buffon's "organic molecules" as infective agents floating in the air.	
1765	First drawing of cell division (Trembley).	
1765	Spallanzani shows that Buffon's "organic molecules" are distinct organisms.	
1770	Hill introduces new methods of staining and preserving specimens for microscopic study.	A spinning jenny that automates part of the textile industry is patented by English weaver-mechanic James Hargreaves. The Black Death strikes Russia and the Balkans in epidemic form.
1796	Jenner inoculates James Phipps with cowpox.	
		Beethoven writes the "First Symphony."
1802	First use of word "biology" (Treveranus).	
1807	First achromatic microscope.	
1821	First international congress of biology (organized by Oken).	

1835	Bassi's theory of "living contagion" in silkworm disease.	
1838	Liebig establishes biochemistry.	The word "protein" is coined by Dutch chemist Gerard Johann Mulder, 36, who adapts a Greek word meaning "of the first importance."
		Gas ovens are installed at London's Reform Club. Coal or wood is the common cooking fuel in most of the world, but Arab nomads use camel chips, American Indians use buffalo chips, and Eskimos use blubber oil.
1838–39	Schwann and Schleiden found modern cell theory: plants and animals composed of basically identical units.	
1844	Bassi asserts smallpox, bubonic plague, syphilis, spotted fever due to living parasites.	Potato crops fail throughout Europe, Britain, and Ireland as the fungus disease caused by *Phytophthora infestans* rots potatoes in the ground and also those in storage. Irish potatoes are even less resistant than potatoes elsewhere—up to half the crop is lost.
		Portland is founded in Oregon Territory near the junction of the Columbia and Willamette Rivers. The town is named after the 213-year-old city in Maine as two New Englanders let a flip of a coin decide in favor of Portland rather than Boston.
1845	Siebold recognizes protozoa as single-celled organisms.	
1848	Semmelweiss demonstrates childbed fever is a form of septicemia and becomes a pioneer in aseptic technique.	
1850	Davaine asserts that anthrax is due to "bacterides" which he sees in the blood of dead sheep.	
		Dickens writes *David Copperfield*.
1854	Davaine sees "monads" in stools of cholera patients.	
1857	Pasteur demonstrates that lactic acid fermentation is due to a living organism.	
1858	Virchow's Doctrine *"Omnis cellula e cellula"* declares that "all cells come from cells."	Iowa State College is founded at Ames.
		Oregon State University if founded at Corvallis.
		Anatomy of the Human Body, Descriptive and Surgical by London physician Henry Gray, is published.
1860	First selective biological staining.	
		U.S. Civil War begins.
1864	Pasteur invents pasteurization (for wine).	

1867	Lister publishes work on antiseptic surgery.	
1869	Miescher discovers nucleic acid.	Washington D.C.'s Pennsylvania Avenue is paved with wooden blocks for a mile between 1st street and the Treasury Department building at 15th Street.
1869		"Ecology" is coined to mean environmental balance by German zoology professor Ernst Heinrich Haeckel, 35, who is the first German advocate of Charles Darwin's organic evolution theory.
1876	Koch gives three-day demonstration of his work on anthrax, in which he had discovered the sequence of development.	
1877	Koch describes techniques of fixing, staining, and photographing bacteria. Also discovers *Bacillus anthracis* as cause of anthrax.	
1879	Albert Neissen discovers *Neisseria gonorrhea* as cause of gonorrhea.	
1880	Laveran sees malarial parasite but is disbelieved.	
1880	Typhoid bacillus and leprosy agent discovered.	
1881	Koch works out method of culturing bacteria on gelatin. Pasteur, spurred by Koch's work, turns to study anthrax; publicly inoculates sheep at Melun with his "attenuated culture."	Dostoevsky writes *Brothers Karamazov.*
1882	Koch discovers tubercle bacillus, and enuciates "Koch's Postulates."	
1882	Mechnikov launches phagocytic theory: "cellular theory of immunity."	
1883	First apochromatic microscopes.	
1885	Pasteur inoculates Joseph Meister for rabies. Theodor Escherich discovers *E. coli.*	The first ready-to-use surgical dressings are introduced by Johnson and Johnson. Phagocytosis is discovered by Russian zoologist bacteriologist Ilya Ilich Mechnikov, 40.
1887	Buist, Edinburgh infirmary superintendent, sees pox virus and believes it is a form of bacteria.	
1888	Richet confers immunity on rabbits accidentally with serum from an infected dog.	
1897	Buchner discovers that cell-free yeast converts sugar to carbon dioxide and alcohol.	
1897	Buchner demonstrates that cell-free yeast extract will catalyze glucose breakdown. Van Ermengem discovers *Clostridium botulinum.*	

1898	Beijerinck discovers and names tobacco mosaic virus; viral cause of foot-and-mouth disease demonstrated.	
1898	Benda discovers and names the mitochondria, previously seen by Altmann.	
		First wireless communication between Europe and America.
1902	Richet discovers anaphylaxis.	
1902	Landsteiner investigates agglutination when blood from different human donors is mixed.	
1903	Sir Almroth Wright and others discover "opsonins" (i.e., antibodies) in blood of immunized animals.	
1905	Harden and Young show inorganic phosphate responsible for fermentative ability of yeast juice.	Helen Keller is graduated magna cum laude from Radcliffe and begins to write about blindness.
1906	Schaudinn and Hoffman discover *Treponema pallidum* as cause of syphilis.	
1906	Von Wassermann develops test for syphilis.	
1912	Ehrlich demonstrates first chemotherapeutic agent for a bacterial disease (syphilis).	
1915	D'Herelle and Twort independently show existence of bacteriophage—viruses which destroy bacteria.	World War I begins.
1923	Landsteiner shows M and N factors in blood.	
1925	Keilin discovers cytochrome.	Al Capone takes over as boss of Chicago bootlegging from racketeer Johnny Torrio, who retires after sustaining gunshot wounds.
1926	Ultracentrifuge (Svedberg).	
1928	Elford demonstrates size of viruses (from 10 to 3000 micrometers).	
1929	Lohmann identifies adenosine triphosphate (ATP) as necessary for the phosphorylation of sugar. Fleming describes penicillin.	The Great Depression begins.
1932	Sir Hans Krebs describes and names the citric acid cycle.	
1933	First electron microscope (Ruska).	Popular songs "Basin Street Blues" by Spencer Williams whose work was published in small orchestra parts 4 years ago; "Only a Paper Moon" by Harold Arlen, lyrics by E. Y. Harburg, Billy Rose; "Lazybones" by Hoagy Carmichael, lyrics by vocalist Johnny Mercer, 23; "Love is the Sweetest Thing" by Ray Noble; "Stormy Weather-Keeps Rainin' All the Time" by Harold Arlen, lyrics by Ten Koehler.

1935	Stanley crystallizes virus.
1941	Beadle and Tatum establish "one gene-one enzyme" theory.
1944	Avery discovers "blueprint" function of DNA.
1945	Role of mitochondria revealed.
1948	Electrophoretic methods (Tiselius).
1952	Partition chromatography (Synge and Martin).
1952	Hershey and Chase prove that DNA injected by bacteriophage is what disorganizes bacterium. Phage becomes the fruit fly of the molecular biologists.
1952	
1952	Sexual recombination discovered in bacteria.
1952	Waksman discovers streptomycin, the first antibiotic effective against tuberculosis.
1953	Lipmann discovers coenzyme A and its importance for intermediary metabolism.
1953	Phase-contrast microscope.
1953	Medawar shows tolerance to grafts can be conferred by inoculating newborn animal or embryo with antibodies from future donor.
1953	Lederberg and Zinder discover transduction.
1953	Crick and Watson propose a structure for deoxyribonucleic acid—a double spiral.
1954	Enders, Weller, and Robbins discover poliomyelitis viruses in cultures of various types of tissue.
1957	Virus structure determined. Interferon discovered.

World War II begins.

Aerosol spray insecticides begin a revolution in packaging. The commercial "bug bombs" employ a Freon-12 propellant gas developed by two U.S. Department of Agriculture researchers in 1942. They have been used during the war to protect troops from malaria-carrying mosquitoes.

Korean conflict begins.

The American Bandstand debuts in January on ABC network stations. Host Dick Clark, 22, will continue to emcee the show for more than 30 years.

Films: Fred Zinneman's *High Noon* with Gary Cooper, Grace Kelly; John Huston's *The Red Badge of Courage* with Audie Murphy, Bill Mauldin.

Sputnik, the first earth satellite, launched by Russia.

1958	Lederberg makes discoveries concerning genetic recombination and the organization of the genetic material of bacteria.	Cocoa Puffs breakfast food, introduced by General Mills, is 43 percent sugar. Transatlantic jet service is inaugurated by Pan American World Airways and British Overseas Airways (BOAC).
1959	Kornberg and Ochoa awarded Nobel Prize for discovery of enzymes that produce artificial DNA and RNA.	
1961	Nirenberg, using artificial DNA, synthesizes a protein molecule.	
1962	Role of thymus in immunity established.	
1962	Crick, Watson, and Wilkins make discoveries concerning the molecular structure of nucleic acid and its significance for information transfer in living material.	
1964	Bloch and Lynen work out the mechanism and regulation of the cholesterol and fatty acid metabolism.	President Kennedy killed. U.S. Marines land in Vietnam.
1966	Rous discovers tumor-inducing viruses.	
1966	Huggins identifies hormonal treatment of prostatic cancer.	
1968	Holley, Khorana, and Nirenberg win a Nobel Prize for interpreting the genetic code and its function in protein synthesis.	U.S. natural gas consumption begins to exceed new gas discoveries and reserves for interstate pipelines begin falling. U.S. first class postal rates climb to 6 cents; up from 5 cents per ounce in 1963.
1969	Delbruck, Hershey, and Luria win a Nobel Prize for the replication mechanism and genetic structure of viruses.	
1971	T. O. Diemer identifies viroids.	First man on the moon.
1972	Edelman and Porter: Nobel Prize for work concerning the chemical structure of antibodies.	U.S. *Apollo 16* astronauts Charles M. Duke, Thomas K. Mattingly, and John W. Young blast off April 16 from Cape Kennedy. A human skull found in northern Kenya by Richard Leakey and Glynn Isaac allegedly dates the first humans to 2.5 million B.C.
1974	Claude, DeDuve, and Palade receive a Nobel Prize for discoveries about the structural and functional organization of the cell.	
1975	Baltimore, Dulbecco, and Temin awarded a Nobel Prize for researching the interaction between tumor viruses and the genetic material of the cell.	

1976	Gajdusek and Blumberg did research leading to Nobel Prize for a test to show hepatitis viruses in donated blood and to an experimental vaccine against the disease.	U.S.A. Bicentennial. Carter becomes President.
1978	Arber, Smith, and Nathans given Nobel Prize for discovery of restriction enzymes and their application to the problems of molecular genetics. Austrial wins award for first vaccine against pneumococcal pneumonia.	
1979	Henle identified first virus regularly associated with human cancer.	
1980	First U.S. patent issued for the process of producing biologically functional molecular chimeras; inventors: Stanley N. Cohen and Herbert W. Boyer.	
1980	Nobel Prize for Chemistry; Paul Berg, Walter Gilbert, and Frederick Sanger for development of a rapid way to determine the chemical makeup of DNA.	
1982	Epstein and Barr receive award for showing relationship between EBV and Burkitt's lymphoma.	World's Fair in Knoxville, Tennessee.

TABLE I.2
Leading causes of death in the United States.

	1870s	1900s	1930s	1960s	1970s
1	Tuberculosis	Tuberculosis	Heart disease	Heart disease	Heart disease
2	Pneumonia	Pneumonia	Pneumonia and influenza	Cancer	Cancer
3	Accidents	Diarrhea and enteritis	Cancer	Stroke	Stroke
4	Diarrhea and enteritis	Heart disease	Stroke	Accidents	Accidents
5	Scarlet fever	Nephritis	Accidents	Influenza and pneumonia	Influenza and pneumonia
6	Infant cholera	Diseases of infancy	Nephritis	Diseases of infancy	Cirrhosis of the liver
7	Circulatory disease	Stroke	Diabetes mellitus	Arteriosclerosis	Diabetes mellitus
8	Encephalitis	Accidents	Suicide	Diabetes mellitus	Arteriosclerosis
9	Convulsions	Cancer	Diseases of infancy	Cirrhosis of the liver	Diseases of infancy
10	Measles	Bronchitis	Appendicitis	Suicide	Emphysema

SUMMARY

The science of microbiology involves the study of living, microscopic organisms and their interactions with their environment. The field is closely related to many others, including chemistry and molecular biology. New data from one area usually finds a place in the others. This information has led to advancements in our general knowledge of life. The field of microbiology can be readily subdivided. One such division concerns the specific nature of the microbes. As a result, the fields of bacteriology, mycology, protozoology, phycology, and virology have evolved. A second way to organize the study of microbes is to group them according to their influence on their environment. This division has created such areas as medical microbiology, food and dairy microbiology, water and wastewater microbiology, and industrial microbiology. The evolution of each of these fields has resulted from many years of research. Probably the most-asked question in the development of microbiology relates to the origin of life as we know it. The theory of spontaneous generation was supported by many great scientists and proposed that living things could arise from nonliving material. The theory of biogenesis proposed that living things could only arise from living parents. The work that has been done to prove each of these theories has contributed greatly to our knowledge of microbes. Van Leeuwenhoek developed fine microscopes, Pasteur identified certain disease-causing organisms and began the process of pasteurization. Koch became a leader in the development of basic laboratory techniques in the area of bacteriology. More recently, Beadle, Tatum, Watson, and Crick have been involved in some of the most exacting and relevant work ever in the areas of molecular biology and microbology. Work from the past has allowed present researchers to explore and apply information in many areas of our daily lives.

STUDY QUESTIONS

1 List three ways in which microbes influence our lives.

2 How does microbiology relate to science in general?

3 List three fields of microbiology and describe the activity of each.

4 State the germ theory of disease.

5 What is pasteurization?

6 Describe two activities of microbes studied in the area of soil and agricultural microbiology.

7 List four microbial products manufactured industrially.

8 What is the basic difference between the theories of spontaneous generation and biogenesis?

9 Describe two achievements of Louis Pasteur.

10 Describe two achievements of Robert Koch.

SUGGESTED READINGS

BULLOCH, W. *The History of Bacteriology.* Oxford University Press, London, 1938.

DE KRUIF, P. *Microbe Hunters.* Harcourt Brace Jovanovich, New York, 1966.

DOBELL, C. *Anton Van Leeuwenhoek and His "Little Animals."* Dover Publications, New York, 1960.

GABRIEL, M. L., and S. FOGEL, editors. *Great Experiments in Biology.* Prentice-Hall, Englewood Cliffs, N.J., 1955.

LECHEVALIER, H., and M. SOLOTOROVSKY. *Three Centuries of Microbiology.* Dover Publications, New York, 1974.

PENN, M., and M. DWORKIN. "Robert Koch and Two Visions of Microbiology." *Bacteriological Reviews,* 40(2):276–83, 1976.

PORTER, J. R. "Anton van Leeuwenhoek: Tercentenary of His Discovery of Bacteria." *Bacteriological Reviews,* 40(2):260–69, 1976.

POSTGATE, J. *Microbes and Man.* Penguin Books, Baltimore, 1975.

TAYLOR, G. R. *The Science of Life.* McGraw-Hill, New York, 1963.

KEY TERMS

biology

microbiology

bacteriology

virology (vi-rol'o-je)

protozoology (pro"to-zo-ol'o-je)

phycology (fi-kol'o-je)

mycology

pathogen

disease

germ theory of disease

Koch's Postulates

bacteriophage (bak-te're-o-faj")

Pronunciation Guide for Organisms

Vibrio (vib're-o)
  cholerae (kol'er-ee)

Mycobacterium (mi"ko-bak-te're-um)
  tuberculosis (tu-ber"ku-lo'sis)

Staphylococcus (staf"i-lo-kok'us)
  aureus (aw-re'us)

LIVING matter is structurally different from matter that has never been alive. This is true at the most fundamental level since the chemistry of the living is different from the chemistry of the nonliving. The basic structural unit, called the cell, is found only in things that are alive or were alive at one time. Cells in all kinds of living organisms have basic similarities, but may differ in the way they function. The next few chapters review some of the basic chemical and physical characteristics of living beings. Understanding these characteristics may help to explain what life is. Various classification systems will be discussed, which organize new facts into logical patterns. Taxonomy is the orderly classification of living things according to their natural relationships. There are many ways to sort living things by function, chemistry, or shape. Some systems are based on very general characteristics, while others are very specific. A knowledge of taxonomic systems will help to understand the cell structure, evolutionary relationships, and nutritional needs of microbes.

One of the most functional classification systems divides cellular life into procaryotic and eucaryotic forms (see Chapter 2). Much of the information that has been gained from the study of these two basic cell types was originally developed in the area of theoretical science but is now used in applied science areas. Attention will be given to these applied areas including the need for specialized tools and procedures in microbiology. Microscopes, dyes, and other related equipment are required to see microbes and gather information. In many sciences, standard laboratory procedure is to dissect the organism under study to better understand its structure and function. Dissecting microbes is also of great value, but scalpels and scissors cannot be used. The dissection must be done with special chemicals and microscopes. Chemical reactions are performed to highlight very small components of microbial cells and make these cell parts more visible through the microscope. Many of the basic methods used today were developed by Robert Koch over one hundred years ago. However, modern advances in science and technology have provided a number of new methods that have greatly added to our knowledge of microbes.

◄ *Common soil microbes (Streptomyces sp.) as seen through a phase-contrast microscope. Magnification: ×1600. (Courtesy of the Upjohn Company.)*

Fundamental Characteristics of Microbes

ESSENTIAL LIFE CHARACTERISTICS
GROWTH AND REPRODUCTION
ADJUSTING TO A CHANGING WORLD
MAINTAINING A BALANCE
ENZYMES IN REAL SITUATIONS

Learning Objectives

- [ ] To list several characteristics which classify matter into the categories living and nonliving.
- [ ] Understand the structure-function concept as a basic relationship found in all living organisms.
- [ ] Identify the eight characteristics of life.
- [ ] Be familiar with examples, both microbial and nonmicrobial, of the eight characteristics of life.
- [ ] Understand what a biochemical pathway is and how it relates to each of the characteristics of life.
- [ ] Be familiar with examples of the three levels of responsiveness.
- [ ] Relate the Second Law of Thermodynamics to the life characteristic of waste production.
- [ ] Correlate the characteristics of life with common activities seen in familiar living beings.

Most students begin the study of microbes with very little previous knowledge of these organisms. For this reason, it may be more difficult to organize thoughts and apply them to the information in the field of microbiology than in other fields. Nevertheless, many common observations are as useful in understanding microbes as they are in understanding the larger organisms that are seen every day. The characteristics of life demonstrated by people, animals, or trees are basically the same as those found in microbes. However, microbes must carry out all the characteristics common to life in just one single cell. The study of life at the cellular level reveals two important items. First, the study shows that all life forms are basically the same. There is a unity of life in the nature of structure and function of the cell. Second, we gain a more clear understanding of what is minimally necessary for something to be "alive."

ESSENTIAL LIFE CHARACTERISTICS

Living organisms show several unique features associated with their structural and functional organization, which the

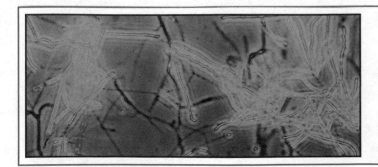

1

nonliving do not exhibit. The ability to manipulate both energy and matter from the environment is the prime criterion for life. There are eight ways in which this occurs: (1) ingestion, (2) assimilation, (3) growth, (4) reproduction, (5) waste elimination, (6) responsiveness, (7) coordination, and (8) regulation.

All living things expend energy to take in nutrients (food) from their environment. This process is called **ingestion**. Large, multicellular organisms may ingest by eating or swallowing other organisms. Microbes are for the most part unable to feed like this, but have their own system of ingestion. Microbes ingest by absorbing their food from the environment through the cell boundaries. They may feed on both dead and living materials. Organisms that ingest in this way are called saprotrophs, or saprobes (Figure 1.1). Microbes release chemicals (digestive enzymes) into the surrounding environment that break down food into smaller components. These digested molecules are then taken into the cell by different methods. These methods will be explored more specifically later.

The cell is constantly expending energy to manufacture these digestive enzymes and take in the products of diges-

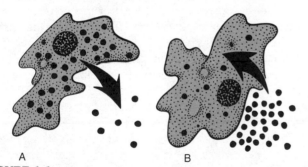

FIGURE 1.1

Saprotrophs and ingestion. (A) Microbes release digestive enzymes into their surroundings through cell boundaries. (B) These enzymes break down nutrients into smaller particles which can then be taken into the cell.

tion. Once inside the cell, these digested molecules may enter a series of chemical reactions known as **assimilation**. These reactions are controlled by the enzymes of the cell and result in the use of the incoming nutrients to help maintain the cell's structure and function. Assimilation consists of the major chemical reactions that occur in living things. Because these reactions are dependent on one another and occur in specific series, they are commonly referred to as **biochemical pathways**. A knowledge of these pathways will help to understand how energy and matter are utilized within a microbe. This will help distinguish one microbe from another.

The chemical bonds that hold the atoms of nutrient molecules together contain energy. If a bond is broken, the energy is released and may be converted into other forms. The conversion of energy from one form to another is quite common. Living cells require chemical energy to accomplish the many complex reactions occurring every second. Food has large amounts of chemical-bond energy. The energy can be released when the bonds in the molecules are broken. Microbes cannot directly use this type of released energy, but must convert it to a usable form such as ATP (adenosine triphosphate) (Figure 1.2). As the chemical bonds in a molecule of food are broken, a portion of the released energy is incorporated into special high-energy bonds of ATP. When ATP bonds are broken, a form of energy usable to the cell is released.

Most bacteria are capable of **aerobic cellular respiration**. In this metabolic pathway, food and oxygen combine to yield energy and release carbon dioxide and water. In chemical shorthand, this may be written:

sugar + oxygen $\longrightarrow$ carbon dioxide
+ water + energy (ATP + heat)

Some bacteria that are of significance to the health professions are *anaerobic*. The process they carry out is called **anaerobic cellular respiration** because atmospheric ox-

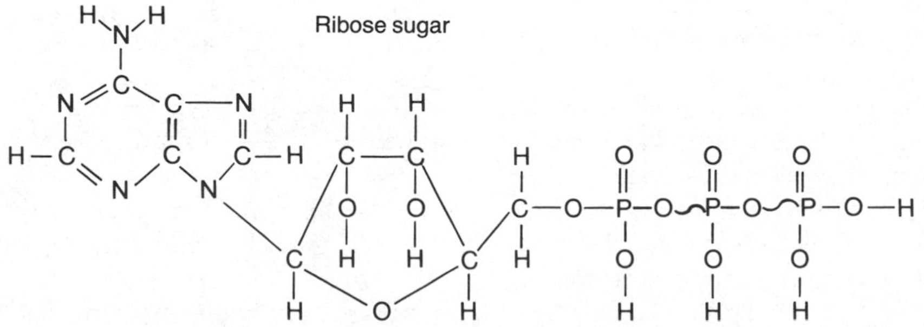

FIGURE 1.2

Adenosine triphosphate (ATP) is an important biological molecule that has a simple sugar as part of its structure. (From Eldon D. Enger et al, *Concepts in Biology*, 3rd ed. © 1982, 1979, 1976 by Wm. C. Brown Company Publishers, Dubuque, Iowa. Reprinted by permission.)

ygen is *not* used in the reactions but is replaced by some other molecule. The chemical shorthand for these types of reactions is written:

$$\text{sugar} \longrightarrow \text{organic molecule} + \text{energy (ATP} + \text{heat)}$$

One bacterial genus that functions anaerobically is *Clostridium.* Clostridia are unable to grow in the presence of atmospheric oxygen. Special equipment and techniques must be used to rid the environment of oxygen if they are to grow successfully. One member of *Clostridium, C. botulinum,* produces a very powerful nerve poison that, if ingested by humans, causes the disease botulism and may cause death very rapidly. Another bacterium, *C. perfringens,* is able to grow anaerobically in wounds, thus causing a great deal of gas and tissue destruction known as gas gangrene. If unchecked, this infection may be fatal (Figure 1.3). The specific biochemical pathways present in various microbes provide each organism with all of its essential life needs. The exact nature of these pathways will differ for different groups of organisms.

---

1 What is ATP?
2 Name two diseases caused by bacteria of the genus *Clostridium.*
3 What does the prefix "an-" mean?

GROWTH AND REPRODUCTION

The third major characteristic involving the manipulation of energy and matter is **growth**. Biological growth may be defined as an increase in size as a result of assimilation. Cells add to their structure, repair parts, and store nutrients for later use. When these synthetic reactions occur faster than destructive reactions, the cell grows. However, cells cannot continue to grow indefinitely. As they get larger, they become inefficient. The result could be chemical chaos with the cell wasting energy by producing excess DNA and proteins. Cells respond to this problem by dividing.

Living things show the ability to produce more individuals like themselves. In many ways **reproduction** is one of the most important characteristics of life because it is the only way that new living organisms give rise to progeny. Microbes devote a large portion of their activities to reproduction. Prior to dividing, cells must reproduce all component molecules. When a microbe divides into two new cells, each must have a complete set of genetic information. Since the original cell had only one set, there must be a doubling of this DNA information if the two new cells are to have a complete set. The process of DNA duplication is a major

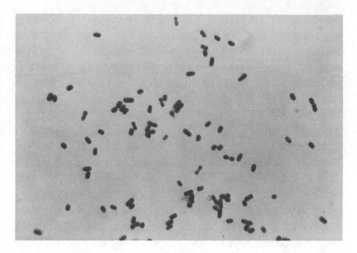

FIGURE 1.3
This photomicrograph shows *Clostridium perfringens,* the bacterium responsible for the disease known as gas gangrene, taken from a twenty-four hour growth in chopped meat. (Courtesy of the Center for Disease Control, Atlanta.)

function of DNA. The accuracy of duplication is essential to guarantee the continued existence of the correct gene sequence in all future generations. This involves the splitting of the cell and the distribution of the same genetic information to the new daughter cells (Figure 1.4). In this way each new cell has the necessary information to control its activities. The mother cell does not die, but ceases to exist, since it divides its contents between the smaller daughter cells.

This means that the young daughter cells can use "fresh" building materials to cope with environmental changes and reproduce themselves. In this way, cell types continue to exist unless they are killed. A cell does not really die when it reproduces itself; it merely starts over again. This is called the **life cycle** of a cell (Figure 1.5). There are a number of ways that cells can divide and redistribute genetic information to the next generation. These processes will be dealt with later.

---

1 What is DNA?
2 Name a method by which cells reproduce.

Not all nutrient molecules entering a living cell as food are valuable to it. There may be portions of molecules that are valuable because they contain energy or building materials essential to the organism. But the rest of the molecule may not be valuable, or it may actually be harmful to the cell. In that situation organisms eliminate excess material or energy in some way. The process of **eliminating waste products** is the fifth major characteristic of living things.

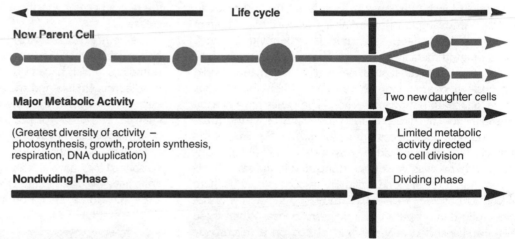

FIGURE 1.4
These are the generalized events in a cell during the process of DNA duplication and reproduction. Notice that the final cells have the same kind of DNA found in the parent cell. Each is identical to the other and the original.

DNA

Parent cell

Daughter cell

Daughter cell

FIGURE 1.5
Although there is much variation, reproduction is apt to be a very brief part of the life cycle of a cell. (From Eldon D. Enger et al, *Concepts in Biology,* 3rd ed. © 1982, 1979, 1976 by Wm. C. Brown Company Publishers, Dubuque, Iowa. Reprinted by permission.)

Life cycle

New Parent Cell

Two new daughter cells

Major Metabolic Activity

(Greatest diversity of activity – photosynthesis, growth, protein synthesis, respiration, DNA duplication)

Limited metabolic activity directed to cell division

Nondividing Phase

Dividing phase

A basic law of energy states that some usable energy is lost whenever energy is converted from one form to another. Living systems obey this **Second Law of Thermodynamics** and are therefore constantly losing some of their useful energy. If living organisms do not receive a constant supply of energy, usually in the form of chemical-bond energy, they will die.

Excess energy lost from a cell may take the form of heat or light. Some bacteria and fungi actually glow as they lose energy (Figure 1.6). Many of the excess materials released by microbes may be beneficial to humans. The antibiotics penicillin and streptomycin are good examples. Waste products from the chemical activities of microbes build up in the environment of the cells that produce them. If these cannot be removed by the actions of other organisms or the environment, the waste may eventually kill the microbes. Yeast cells release alcohol as a by-product of their activities. If the alcohol concentrates around the cells and is not washed away, the yeast pollute their own environment and are killed.

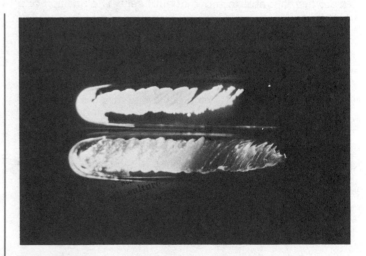

FIGURE 1.6
The bacteria, *Vibrio fischeri,* growing on this surface were photographed in their own light. As the bacteria metabolize, they release enough light as waste energy (luminescence) to expose the film. (Courtesy of Carolina Biological Supply Company.)

ADJUSTING TO A CHANGING WORLD

The sixth category of function associated with the manipulation of energy and matter is the ability of living things to **respond** to external and internal changes in their environment. All living things react. This is easily seen when animals respond to changes in light intensity, sound, touch, and other kinds of changes. This is not seen as easily in plants or microbes, but they also respond to changes. The range of response in living beings is very great. Three levels of response can be identified: (1) **irritability,** (2) **adaptation,** and (3) **evolution.**

Irritability responses are changes that involve individual organisms. Only the individual, not a group, responds to a stimulus, and gene action is not directly involved in the response. These actions are usually very quick, and the mechanism for the response is already set up and ready to operate. An example of an irritability response in a human is the knee-jerk reflex. The stimulus of tapping the knee results in an immediate response. The muscles, bones, connective and nervous tissues have all been manufactured previously and are ready to respond. When the stimulus occurs, genes do not go into action. The genes have already done their job by directing the production of all the structures and chemicals necessary to function at the time of the stimulus. No other individual responds to the stimulus unless they themselves have been stimulated in the same way.

In the microbial world, irritability responses may be more difficult to observe. Cells which move by the action of flagella easily demonstrate this type of response. When these cells are suspended in water and placed on a microscope slide, they are able to move about freely. The addition of a harsh chemical such as an acid will stimulate the cells to move away from the acid toward the other side of the slide. Individual cells respond to the stimulus very quickly and no gene action is required to carry out the response. The reaction gives those cells capable of responding a better chance of surviving in their changing environment.

A second way in which living things react to stimluli is through a process called **adaptability.** This response is usually slower than an irritability response. Adaptability resembles irritability in that it involves the individual but differs since there is genetic activity. Individual microbial cells have the ability to adapt to changes in their surroundings. Bacteria contain genes that do not operate in one environment but work at very high levels in another environment. The cells are able to regulate gene action. By making this kind of adaptation, they can conserve energy and regulate their chemical activities to better survive in a changing environment. The bacterium *Escherichia coli* is found in the human intestine and is commonly used in microbiology laboratories (Figure 1.7). A great deal of biochemical information has been gained from exploring the chemical activities of this particular bacterium. *Escherichia coli* contains genes that code for enzymes which control the digestion, ingestion, and assimilation of lactose sugar (*see* Chapter 3). If a culture of *E. coli* is grown in a solution of nutrients containing no lactose sugar and only glucose sugar, it is very difficult to find the enzymes that are coded for by these lactose genes. The bacteria do not expend energy to manufacture these proteins but direct their activities to utilize the available glucose molecules as a food source. However, when the glucose has been completely used and lactose sugar is added to the broth culture, the amount of enzymes coded by the lactose genes increases about 1000 times! The stimulus for this new gene activity is the presence of lactose sugar in the environment. The genes remain active as long as the lactose sugar is available as a food source. When the supply is exhausted, the genes become inactive and the amount of enzymes decreases.

Another characteristic of living organisms is their ability to change so that they are still able to cope with their surroundings, which have also undergone change. This change is called **evolution** and may take millions of years. Individuals do not evolve because evolution is the result of slow changes in a group or population of organisms over many generations. Each individual in a population carries with it a unique set of genetic information, which is unique for two reasons. First, natural changes called spontaneous mutations have occurred in the DNA of the parent cell. Mutations are known to be caused by many factors. Chemicals, cosmic

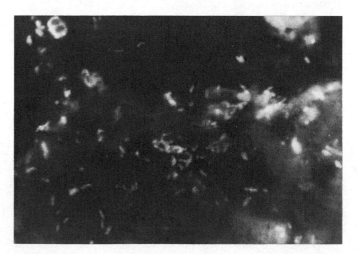

FIGURE 1.7
Escherichia coli are small, rod-shaped bacteria that normally inhabit the intestinal tract of humans and many domesticated animals. While most types of *E. coli* are not harmful, a few do cause disease. (Courtesy of the Center for Disease Control, Atlanta.)

radiation, and manmade radiation alter the genes of cells and thus change the proteins coded from these genes. If a cell is exposed to such an agent, it may undergo a change that can be transferred to the next generation. The second reason for each individual's genetic uniqueness has to do with the process of sexual reproduction, or gene transfer. Variety may be generated as a result of a reshuffling of genes as the genetic material is sorted out and sent into another cell at the time of reproduction. The exact type of genes going into the offspring (next generation) will vary according to the type of sexual transfer carried out by a particular organism. When sexual gene transfer occurs, genes donated by each of the parent cells are recombined to form a new cell. This new individual will have new genes and new gene combinations not found in either of the parents.

In a population, there are many different kinds of individuals, some of which have advantages over others. These will survive better and reproduce greater numbers of offspring than those without the favorable characteristics. This is the theory of natural selection proposed by Charles Darwin and Alfred Wallace. This theory was proposed in 1859 by Darwin in his book *On the Origin of Species by Means of Natural Selection, or the Preservation of Favored Races in the Struggle for Life*. This title needs clarification on two points. First, **natural selection** means the same as the phrase **differential reproduction,** and both mean that some organisms have more offspring than others. Therefore, a gene that better suits an individual to its environment is one that allows the individual to have more offspring than other organisms. This gene will become an increasingly important part of future populations. Second, the phrase **struggle for life** is not the same as the term **conflict.** Natural selection is a creative process and results in a better population. It is not necessary for fighting to occur during the natural selection process.

When a population of microbes has experienced gene change and natural selection for many generations, it has gone through the process of evolution. In the microbial world, there are many examples of evolution. Pathogenic bacteria change to better suit their environment, the human body. Although bacteria have benefited from this change, health care professionals have found this response to be very troublesome. There has been a gene change and evolution in the bacterium *Neisseria gonorrhoeae*, which causes the venereal disease gonorrhea (Figure 1.8). In the past, a relatively small amount of the antibiotic penicillin-G was able to control a gonorrheal infection. Penicillin is measured in International Units (IU) that are a measure of the effectiveness or activity of the drug. In the early 1940s, an injection of 0.3 million units of penicillin-G was effective in controlling gonorrhea. Today, about 4.8 million units is required to control a comparable infection. More recently another form

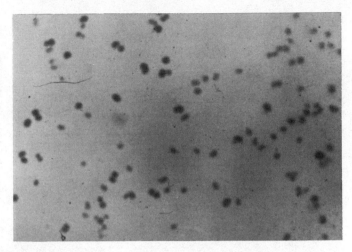

FIGURE 1.8
Neisseria gonorrhoeae are spherically-shaped bacteria that normally group in pairs. They cause the sexually transmitted disease gonorrhea and have evolved in a direction making them resistant to the antibiotic penicillin. (Courtesy of the Center for Disease Control, Atlanta.)

of gonorrhea has emerged that actually grows better in the presence of penicillin than it does in an environment lacking this drug. The gene change that has occurred allows the bacteria to produce a new enzyme called penicillinase. Microbiologists refer to the enzyme as P-ase*.

This is an example of long-term genetic change in a population of microbes. The organism is not responding to an environmental change by turning an existing gene "on" or "off" as is found in adaptation. A new, different kind of gene has come into the population due to mutation, and through natural selection the microbes with this gene have become more common than those without the gene.

---

1 Where can you commonly find the bacterium *Escherichia coli?*
2 What causes mutations?
3 What kind of a microbe is *Neisseria gonorrhoeae?*

MAINTAINING A BALANCE

The totals of all the different complex chemical reactions within a cell are called **metabolic processes** and are used to classify something as alive. While any single reaction may be duplicated in a test tube, only living organisms are ca-

---

*The suffix "-ase" means *enzyme.* An **enzyme** is a special kind of protein produced by living cells and is able to speed up the rate at which chemical reactions occur under otherwise normal conditions.

pable of performing them in complex patterns that are self-sustaining. It is also important to realize that these reactions must take place in an organized series that benefits the cell and allows all the other characteristics of life (growth, assimilation, etc.) to be accomplished. When biochemical pathways are established and linked by the specific activity of particular enzymes, the cell is demonstrating **coordination,** the seventh fundamental characteristic of life.

The structure of a particular enzyme molecule is determined by a cell's genes, which have their own three-dimen-sional structure. A given enzyme has a unique shape that allows it to combine with another molecule. The molecule with which an enzyme reacts is called the **substrate.** A new temporary molecule, an **enzyme-substrate complex,** is formed by this physical joining of the original enzyme with the substrate molecule. While in this complex, the substrate molecule will be changed into a new molecule known as the **end product.** At the end of the reaction, the enzyme will leave unchanged (Figure 1.9). A given enzyme only works on one type of substrate or on substrates having

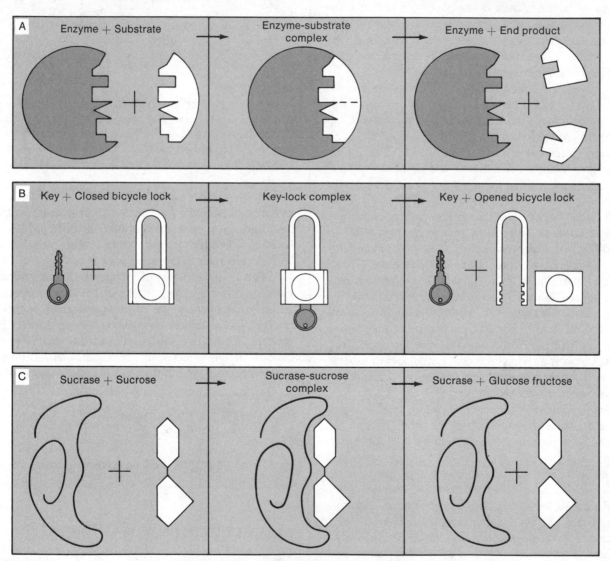

FIGURE 1.9
(A) An enzyme and its substrate physically join together to form the enzyme-substrate complex. The enzyme emerges from this complex unaltered; the substrate changes into the end product. (B) If the "key" doesn't fit the "lock," no change results in the lock. (C) When together, sucrase and sucrose are a specific enzyme-substrate complex. (From Eldon D. Enger et al, *Concepts in Biology,* 3rd ed. © 1982, 1979, 1976 by Wm. C. Brown Company Publishers, Dubuque, Iowa. Reprinted by permission.)

a somewhat common shape. This is known as **enzymatic specificity.** The enzyme sucrase (saccharase) is specific for the substrate sucrose because the molecules will physically fit together. Enzymes have specific portions of their structure that interact with the substrate. These areas are known as **active sites.** The active site may only be a small portion of a large enzyme molecule, but this is where the chemical reaction really occurs.

In linked biochemical reactions, each enzyme is specific for the end product of the previous reaction.

$$A \xrightarrow{a} B \xrightarrow{b} C \xrightarrow{c} D \xrightarrow{d} E$$

The capital letters in the series represent the substrates while the lowercase letters represent those enzymes capable of combining with each of the substrates. This kind of coordination allows chemical reactions to be performed in an orderly manner from A to E and guarantees to a high degree that the cell will manufacture the molecules it needs for life.

The linkage of biochemical reactions is of great importance to a cell, but equally important is the regulation of the rate of these reactions. Regulation is the eighth essential characteristic of living things. The production of excesses of any molecule in a cell may result in the cell's death. Waste products could accumulate and interfere with metabolic pathways. The **regulation** of the rate of chemical activities in cells is also carried out by the enzymes. The number of substrate molecules that one molecule of enzyme can react with in a given period of time is the **turnover number**. Under ideal conditions the turnover number is usually large, typically up to 50,000 per minute. This means that one molecule of enzyme could react with 50,000 different molecules of a substrate in one minute. Without the enzyme, less than 5 substrate molecules might be altered. Many environmental factors can change this turnover rate, either in the test tube or a living cell. Figure 1.10 illustrates a generalized graph of the influence of temperature on the turnover number. An increase in the amount of heat energy results in an increased turnover number until the maximum turnover number is reached. After this temperature has been reached, the turnover number will decrease. In the cell, the **optimum turnover number** (best for the cell) is not usually the maximum turnover number. Since all enzymes are proteins, a high temperature will change the structure of the enzyme by breaking some of the chemical bonds that hold the enzyme in its particular shape. The new shape will not allow the enzyme to form an enzyme-substrate complex; they will no longer fit together. As a result of the higher temperature, the turnover rate will decrease. When excess heat is applied to a protein in this way, the protein is **denatured**—the protein structure changes. An example of the denaturing of proteins is the change in consistency of eggs

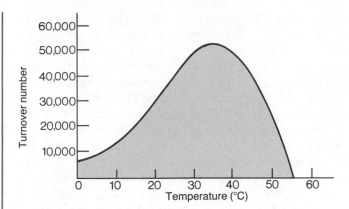

FIGURE 1.10
As the temperature increases, the rate of enzymatic reaction increases until the heat changes the shape of the enzyme. When the enzyme does not fit the substrate, the enzymatic reactions decrease and eventually cease. (From Eldon D. Enger et al, *Concepts in Biology*, 3rd ed. © 1982, 1979, 1976 by Wm. C. Brown Company Publishers, Dubuque, Iowa. Reprinted by permission.)

when they are cooked. The egg white, which is protein, changes its molecular structure. The opposite environmental change—lowering the temperature—also has a direct effect on an enzyme's activity and turnover number. For example, foods are stored in a refrigerator because the low temperature reduces the turnover number of bacterial enzymes and thus inhibits the spoiling of food by these contaminating bacteria.

Another environmental factor that will alter the turnover number is the hydrogen ion concentration, or **pH**. Each enzyme has its own optimum pH at which it attains the highest turnover number. Many function best at a pH close to neutral (neither acid nor alkaline), which is represented by the number 7 on the pH scale.* However, do not be misled: many enzymes perform at a pH quite different from 7. Temperature and pH are not the only factors that influence the rate of enzyme activity. The concentration of enzymes, the concentration of substrate, and many other factors also influence the enzymatic reaction rate.

ENZYMES IN REAL SITUATIONS

Any microbial cell has thousands of different kinds of enzymes. Each is sensitive to changing environmental conditions and controls specific chemical reactions. These en-

---

*A solution of pH 0 to 7 is acid, pH of 7 is neutral, and pH over 7 to 14 is alkaline.

zymes cooperate to satisfy the cell's needs for immediate energy, growth, reproduction, and storage of materials for future use. All of these needs may be met by using the same nutrients. The result is enzymatic competition for specific substrates. The success of any one of these enzyme systems during competition will depend on the number of enzymes available and a suitable environment for their optimum operation. For example, if a microbe requires immediate energy to move away from danger, it can shift the flow of nutrients to generate this energy. This can be done by increasing the number of enzymes involved in the reaction or by changing the environment so that the enzymes operate at a faster rate. Figure 1.11 illustrates such a situation. Acetyl is a molecule that can be a substrate for three different reactions. When the need for energy is greater than the need for protein and fat production in the microbe, the enzyme citrate synthetase does its job more efficiently than either fatty acid synthetase or malate synthetase. This means that more acetyl molecules are used in the synthesis of citric acid, and the energy needs of the cell are met. On the other hand, if the cell is not expending a great deal of energy, the acetyl is utilized in the synthesis of either protein or fat. The number of enzymes produced is regulated by genes. Some molecules can help speed up the activity rate of reactions occurring in the cell. For example, coenzymes and vitamins (niacin, riboflavin, etc.) become part of the complete enzyme molecule structure. Should these coenzymes or vitamins not be available or not chemically combine to complete the enzyme structure, the enzyme will not function at all. Minerals are also important in the operation of enzymes. Magnesium, calcium, and cobalt are minerals that play a part in the efficient operation of many enzymes.

Enzymatically controlled reactions may be slowed. If worn-out enzymes are not replaced, the chemical reactions that they speed up will be slowed because of a decrease in enzyme availability. Slowing of reactions may also be brought about by specific inhibitor molecules.

Some **inhibitors** are so specific that they only work on certain types of enzymes. These will occupy the active site of enzymes, so that there can be no normal enzyme-substrate complex formed (Figure 1.12).

Some inhibitors have a shape that closely resembles the normal substrate of the enzyme. The enzyme cannot distinguish the inhibitor from the normal substrate and combines with the inhibitor and not the substrate. These enzyme-inhibitor complexes are sometimes permanent. Such an inhibitor completely removes the enzyme as a functioning part of the cell, and the end product is no longer formed. Even if the inhibitor does not form a permanent enzyme-inhibitor complex, the formation of such a temporary complex results in a lower turnover number for that enzyme.

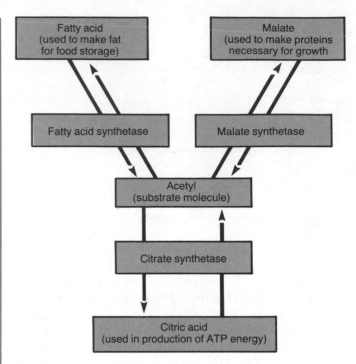

FIGURE 1.11
Acetyl serves as a substrate for a number of competing enzymes. The requirements of the cell help determine which of the enzymes react with the substrate. (From Eldon D. Enger et al, *Concepts in Biology,* 3rd ed. © 1982, 1979, 1976 by Wm. C. Brown Company Publishers, Dubuque Iowa. Reprinted by permission.)

In disease control, the principles of *enzyme inhibition* are put to use. The sulfa drugs are used to control different kinds of bacteria such as *Streptococcus pyogenes*—the cause of strep throat and scarlet fever. This drug, resembling one of the bacterial cell's necessary substrates, is picked up by the organism instead of the needed substrate, and thus prevents some of the enzymes from producing essential cell components. As a result, the normal metabolism of the bacterial cell cannot be maintained, and the cell dies. All of the chemical reactions involved in the manipulation of matter and energy require the use of enzymes. These molecules are very important to the smooth operation of any cell. A disruption in enzyme formation or function will inhibit characteristic life functions (ingestion, assimilation, growth, waste production, reproduction, responsiveness, coordination, and regulation), and the cell will die.

1 What is meant by pH?
2 What is an active site?
3 Give an example of an enzyme inhibitor.

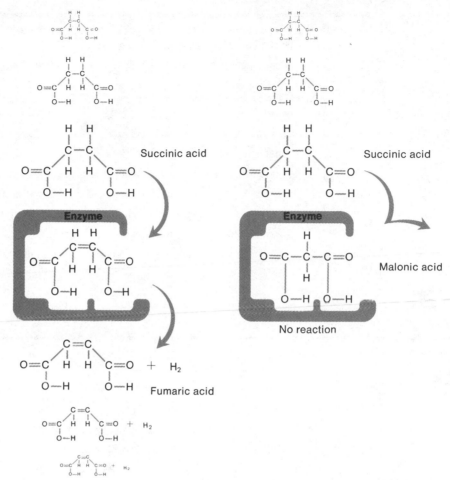

FIGURE 1.12

Malonic acid is similar in shape to succinic acid. Because malonic acid is capable of reacting with the enzyme, the enzyme is inhibited from reacting with its usual substrate, succinic acid. Thus, fewer enzymes are available and the amount of fumaric acid molecules produced decreases. (From Eldon D. Enger et al, *Concepts in Biology,* 3rd ed. © 1982, 1979, 1976 by Wm. C. Brown Company Publishers, Dubuque, Iowa. Reprinted by permission.)

SUMMARY

A limited knowledge of microbiology need not be a stumbling block to understanding microbes. Common experiences and information about familiar organisms can be used to understand the characteristics of microbial life. All living things share a basic structure that support these functions necessary for life. Truly living organisms demonstrate the eight methods of matter and energy manipulation. Ingestion is the method used to take in nutrients. Assimilation processes in the cell are biochemical pathways that utilize nutrients to maintain the life of the cell. Growth occurs when the cell manufactures more new materials than it destroys for energy production. Some of the matter and energy used during assimilation is not useful to the cell and is eliminated in the process of waste production. Reproduction in a cell occurs when the cell divides into daughter cells. All living things maintain themselves in a changing environment by responding to stimuli. Individuals may respond by an irrita-bility reaction that does not involve gene action, or by adaptation, which occurs when the genes are either turned "on" or "off." Populations of cells may evolve in response to a stimulus when new combinations of genes occur in the population and the group experiences natural selection. All cell life activities are coordinated and regulated by enzymes. The information gained from the study of the characteristics of life has found such practical applications as the development of antibotics, and the production and control of microbial growth.

WITH A LITTLE THOUGHT

Recent advancements in electronic technology have raised the question of building a robot that would be "alive." What characteristics would have to be built into such a creature? If progress in technology did advance to this point, do you think it possible to create "life" in the form of a single cell?

STUDY QUESTIONS

1   Which feature of a virus might indicate that it is a living thing? Which feature does a sugar crystal have that might indicate that it is alive?

2   What characteristic allows biologists to separate living things from dead and nonliving things?

3   What processes occur in a living organism that involve the manipulation of matter and energy?

4   Give an example of a saprotroph and explain how it ingests.

5   What is meant by a biochemical pathway?

6   Give an example of a biochemical pathway that operates during assimilation.

7   What is the difference between growth in crystals and growth in a living cell?

8   How does the Second Law of Thermodynamics relate to waste production?

9   Describe what is meant by the life cycle of a cell.

10   From your own experience, give an example of an irritability response and an adaptive response.

SUGGESTED READINGS

BAKER, J. J. W., and G. E. ALLEN. *Matter, Energy, and Life,* 2nd ed. Addison-Wesley, Reading Mass., 1981.

LURIA, S. E. *Life: The Unfinished Experiment.* Charles Scribner's Sons, New York, 1973.

MARQUAND, J. *Life: Its Nature, Origins, and Distribution.* W. W. Norton, New York, 1971.

MAYR, E. *Animal Species and Evolution.* Belknap Press, Cambridge, Mass., 1963.

MERRELL, D. *Evolution and Genetics.* Holt, Rinehart and Winston, New York, 1962.

SIMPSON, G. G. *The Meaning of Evolution.* Yale University Press, New Haven, Conn., 1967.

WALLACE, R. A., J. L. KING, and G. P. SANDERS *Biology, The Science of Life.* Goodyear Publishing Co. Santa Monica, Calif., 1981.

KEY TERMS

ingestion	waste elimination	evolution
assimilation	laws of thermodynamics	natural selection
aerobic and anaerobic cellular respiration	responsiveness	enzyme
growth	irritability	coordination
reproduction	adaptability	regulation

Pronunciation Guide for Organisms

Clostridium (klos-trid'-ee-um)
   botulinum (bot-u-lin'um)
   perfringens (per-frin'jens)
Escherichia (esh-er-ik'-ee-a)
   coli (kol'i)

Neisseria (nye-seer'ee-ah)
   gonorrhoeae (gon"o-re'ee)

Streptococcus (strep-to-kok'us)
   pyogenes (pi'o-jen-ez)

The World of Microbiology

GROUPING FOR A BETTER UNDERSTANDING

THE BINOMIAL SYSTEM

THE MAJOR GROUPS OF CELLULAR
PROCARYOTIC MICROBES

THE PHOTOSYNTHETIC PROCARYOTES

PROTISTA—THE EUCARYOTES
Algae
Fungi
Protozoa
NONCELLULAR MICROBES

Learning Objectives

☐ Define the terms "heterotroph" and "autotroph" and know how these types of organisms are related.

☐ Describe procaryotic and eucaryotic cell types and relate their evolutionary relationship to the kingdom system of classification.

☐ Be familiar with the bases of the four kingdom classification system and know examples of each kingdom.

☐ Know the major groups of cellular procaryotic microbes as described by *Bergey's Manual of Determinative Bacteriology* and give examples of each.

☐ Know the basic morphological types of procaryotic microbes.

☐ Be able to describe the fundamental characteristics of fungi and give an example of each of the four major classes.

☐ Be familiar with some of the positive and negative effects of fungi.

☐ Know the characteristics of the protozoa, the basis of their classification, and list examples of these eucaryotic microbes.

☐ Understand the structural and functional features unique to the viruses.

If one of each kind of organism on earth were to be assembled in one place, there would be such diversity that it would seem impossible for all of these individuals to have evolved from one common organism. But keep in mind that according to the theory of organic evolution, the organisms seen today were not always present on earth. Present-day organisms are the end product of approximately three billion years of evolution. Evolution is a series of changes that occur in a population over a number of generations. These changes are the result of natural selection. For natural selection to occur, three factors are needed: (1) a reproducing population, (2) genetically inheritable differences among individuals, and (3) some environmental factor that favors the survival and reproduction of certain individuals because they

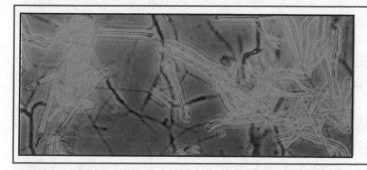

2

possess favorable genes. Although millions of life forms exist on earth today, this was not always so. At one time there was that first type of living microbe from which millions of organisms have evolved. Some of these are present today, but most are extinct. It is possible to trace the evolutionary changes that are thought to have occurred from the first living cell through the three-billion-year period. The results of this evolutionary process can be seen in the various kinds of plants, animals, and microbes living today.

GROUPING FOR A BETTER UNDERSTANDING

The first microbes are thought to have been organisms that could not manufacture their own food. These organisms required complex organic molecules to fuel their metabolism. They probably utilized the organic nutrients in their ocean home. For millions of years prior to the appearance of the first form of life, organic nutrients had been produced by chance contact of atoms and molecules. Microbes which require complex molecules as food and have animal-like characteristics are called **heterotrophs.** For these first microbes, as in all present forms, there were genetic variations within the population. Some individuals were better able to compete for food in their environment. As the number of individuals increased, the existing supply of food decreased. The organisms were living in a changing world—one with a decreasing food supply. The increasing competition for food would favor those microbes that required little pre-formed

organic food. As a result of various mutations, some forms of early life may have synthesized enzymes that allowed them to produce their own complex organic molecules from simple inorganic molecules. These organisms were the world's first **autotrophs.** Thus, two types of life forms occurred on the earth, the original heterotrophs and the new autotrophs. Separating organisms into these two categories requires a knowledge of biochemical pathways and how nutrients are used in the cell. The evolution of autotrophs was extremely important since they became an efficient biological assembly line for the constant production of large amounts of organic molecules. Because heterotrophs eat these organic molecules, they ultimately became dependent on the autotrophs.

Based on various lines of evidence, these first autotrophs and heterotrophs were probably small, uncomplicated single cells. They may have been something like the modern bacteria. These present-day organisms lack a distinct nucleus and many other complex cell structures, and are called **procaryotes.** As competition continued and different genes were introduced into the population by mutations, some of the mutations may have resulted in a more complex cell structure. This could have allowed these cell types to expand into new niches and avoid competition. Such cell types may have resembled the present-day, single-celled protozoans, algae and fungi. These cells contain a complex array of cell structures not found in the procaryotic bacteria and are called **eucaryotes.** The prefix "eu-" means *true,* and "caryote" means *nut,* or *nucleus.* Eucaryotes are "true nuts," or cells with a nucleus. This classification scheme is based on the most fundamental differences between cell

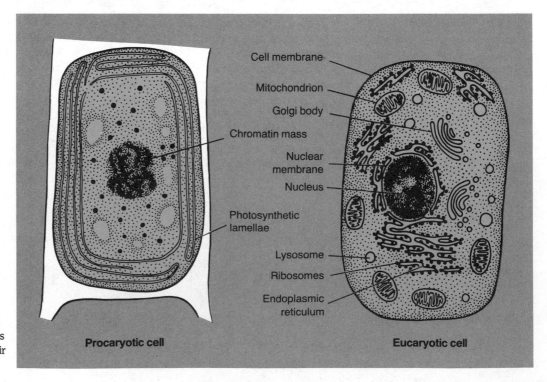

FIGURE 2.1

Note the absence of a nucleus and other membrane structures in the procaryotic cell, and their presence in the eucaryotic cell.

Cell membrane

Mitochondrion

Golgi body

Chromatin mass

Nuclear membrane

Nucleus

Photosynthetic lamellae

Lysosome

Ribosomes

Endoplasmic reticulum

Procaryotic cell

Eucaryotic cell

types found in all living forms (Figure 2.1). Chapters 3, 4, and 5 will deal specifically with the similarities and differences between the procaryotic and eucaryotic cells.

Living things are usually not categorized as either procaryote or eucaryote by nonscientists but are grouped as either plant or animal. There is a basic difficulty with this casual way of classifying living things. Some organisms just will not fit neatly into these two categories. Therefore, biologists have chosen to categorize all organisms into major groups called **kingdoms.** These kingdoms are the largest, highest level of categories and contain the greatest variety of organisms. The criteria used to place an organism into a kingdom include general characteristics such as relative size, chemical composition, complexity, and theoretical relationships. Because these criteria and the decision for their placement are artificial, a number of difficulties have developed. Many new features continue to be revealed by in-depth research. This new information leads taxonomists to change their minds and shift an organism from one kingdom to another or establish new kingdoms. As a result, a number of different classification systems have been developed and are used in the biological community (Table 2.1). This text will use the four kingdom system. The first kingdom contains the single-celled bacteria and cyanobacteria, and is called the **Procaryotae.** In the kingdom **Protista** there are three major types or organisms: the protozoans, eucaryotic algae, and fungi. These protists may have been the ancestors of the more complicated, higher plant and animal kingdoms, **Plantae** and **Animalia** (Figure 2.2).

Two and one-half billion years of evolution has resulted in great variety. Some of this variety can be seen in present-day organisms and in the fossils of corals, seaweeds, protozoa, starfish, bacteria, worms, oysters, and reptiles.

To illustrate the relationship among these taxonomic names easily, they are listed in Figure 2.3 with examples from each of the four kingdoms.

Notice the several blanks in Figure 2.3. Consider the reason for filling in some blanks and not others in the group called subdivision. Within this division of the animals, the subdivision space is filled in with the name Vertebrata. It just happens that animals included in the division Chordata can be divided into several groups, one of which has vertebrae, or backbones. This is an important difference from other organisms within the Chordata. Its placement here is a statement that this difference was a major development in the evolution of chordates. There is no comparable difference in the division Bacteria of the kingdom Procaryotae. A subdivision name is unnecessary if there is no special need for that classification.

The last names, **genus** and **species,** are both used to make up the scientific name of an organism. The name *Streptococcus pyogenes* is the scientific name of the bacterium commonly known to cause strep throat and scarlet fever. The scientific name has the advantage of being recognized all over the world, regardless of the native language.

1 What is the basic difference between a procaryotic and eucaryotic cell?
2 List two criteria used in the four kingdom classification system.
3 Using the four kingdom system, list two members of each kingdom.

TABLE 2.1
Classification systems.

Two Kingdom System	Three Kingdom System	Three Kingdom System	Four Kingdom System	Five Kingdom System
Plantae	**Monera**	**Protista**	**Procaryotae**	**Procaryotae**
Bacteria	Bacteria	Bacteria	Bacteria	Bacteria
Cyanobacteria	Cyanobacteria	Cyanobacteria	Cyanobacteria	Cyanobacteria
Chrysophytes	**Plantae**	Protozoa	**Protista**	**Protista**
Green algae	Chrysophytes	Slime molds	Protozoa	Protozoa
Brown algae	Green algae	**Plantae**	Chrysophytes	Chrysophytes
Red algae	Brown algae	Chrysophytes	Green algae	Slime molds
Slime molds	Red algae	Green algae	Brown algae	**Plantae**
True fungi	Slime molds	Brown algae	Red algae	Green algae
Bryophytes	True fungi	Red algae	Slime molds	Brown algae
Tracheophytes	Bryophytes	True fungi	True fungi	Red algae
Animalia	Tracheophytes	Bryophytes	**Plantae**	Bryophytes
Protozoa	**Animalia**	Tracheophytes	Bryophytes	Tracheophytes
Multicellular	Protozoa	**Animalia**	Tracheophytes	**Fungi**
animals	Multicellular	Multicellular	**Animalia**	True fungi
	animals	animals	Multicellular	**Animalia**
			animals	Multicellular
				animals

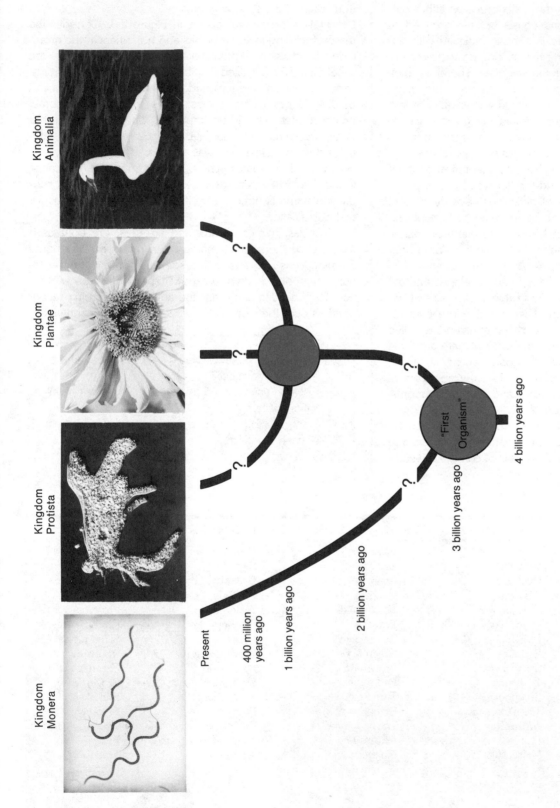

FIGURE 2.2

The evolution of the four major kingdoms of life can be visualized in this figure. The question marks indicate there are still many unknowns. (Photos courtesy of Carolina Biological Supply Company.)

	Bacteria commonly associated with "Strep" throat	A pond water organism with simple structure	Sugar maple tree	Coyote or brush wolf
Kingdom	Procaryotae	Protista	Plant	Animal
Division	Bacteria	Protozoa	Pterophyta	Chordata
Subdivision		Sarcomastigophora		Vertebrata
Class	Schizomycetes	Sarcodina	Angiospermae	Mammalia
Subclass			Dicotyledoneae	
Order	Eubacteriales	Amoebiformes	Sapindales	Carnivora
Family	Lactobacillaceae	Amoebidae	Aceraceae	Canidae
Genus	*Streptococcus*	*Amoeba*	*Acer*	*Canis*
Species	*pyogenes*	*proteus*	*saccharum*	*latrans*

FIGURE 2.3

Representatives of the four kingdoms of life. Procaryotae is represented by the bacterium *Streptococcus;* protista by one-celled mobile *Amoeba;* plants by the sugar maple leaf; and animals by the coyote. (From Eldon D. Enger et al, *Concepts in Biology,* 2nd ed. © 1979, 1976 by Wm. C. Brown Company Publishers, Dubuque, Iowa. Reprinted by permission.)

THE BINOMIAL SYSTEM

The system of naming organisms is the **binomial system of nomenclature** since the scientific name of an organism is composed of both a genus and species. The names of the organisms and the groups to which they belong have been derived from Latin and Greek. Remembering them may be difficult at first, but with practice, they will become more familiar. To aid your memory, the root word for each term will be given as it is first used in the text. Knowing the root words can also be of help in understanding something about the organism being discussed. For instance, the term "pyogenes" means pus forming; streptococcus means a chain (strep-) of spheres (coccus); procaryotae means a cell without a nucleus, or "nut." Linking all this information together tells that this microbe is a bacterial cell that lacks a nucleus, forms pus, and grows in a chain of spheres.

Not all the individual bacterial cells of a species are exactly the same. Individual microbes may differ just as people exhibit differences. For instance, *Escherichia coli (E. coli)* is a bacterium that is normally found in the intestine of all people. Some types of *E. coli,* however, are able to cause a disease in infants known as infantile diarrhea. These harmful *E. coli* can be identified from the more normal kind by very specific chemical tests that highlight the chemical differences between their cell walls. These different microbes are then classified as subspecies or more commonly as races, types, or strains. Knowing the subspecies or type of *E. coli* that is causing infantile diarrhea allows medical personnel to identify the source of the infection and more quickly bring the disease under control.

Now that "our" classification scheme has been organized (Figure 2.4), we see that some organisms do not quite fit into this system. By noting that it is "our" system, we point out the major difficulty of any classification system. "Ours" means that it is artificial, or manmade. It doesn't completely show how the various organisms are related. Even worse is that many systems of classification, including the one used here, actually hide some relationships. This point will become clearer as the various groups of microorganisms are presented.

THE MAJOR GROUPS OF CELLULAR PROCARYOTIC MICROBES

A great deal of information has been found to strongly support a theory of evolution. The first life forms most likely resembled present-day bacterial cells (Figure 2.5). Since all of the members of the kingdom Procaryotae have certain general characteristics in common, it is assumed that they have a similar background and are related. All of these microbes are able to exist as independent, single cells, and none contains genetic material in a "nut," or cell nucleus.

The more advanced kinds of reproduction, meiosis and mitosis, do not occur in these microbes. They reproduce by a more basic method called binary fission.

The procaryotes ingest nutrients by absorbing food molecules, not by the engulfment of large particles or whole organisms. The most well-known division of the procaryotes is the Bacteria. There is an enormous variety of bacteria in the world. *Bergey's Manual of Determinative Bacteriology* (8th ed.) lists over 1700 species of bacteria and describes the subtle differences between each. This manual is a useful

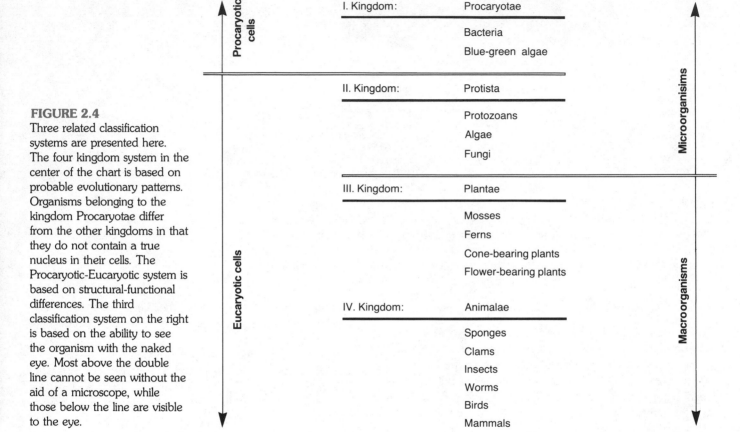

FIGURE 2.4
Three related classification systems are presented here. The four kingdom system in the center of the chart is based on probable evolutionary patterns. Organisms belonging to the kingdom Procaryotae differ from the other kingdoms in that they do not contain a true nucleus in their cells. The Procaryotic-Eucaryotic system is based on structural-functional differences. The third classification system on the right is based on the ability to see the organism with the naked eye. Most above the double line cannot be seen without the aid of a microscope, while those below the line are visible to the eye.

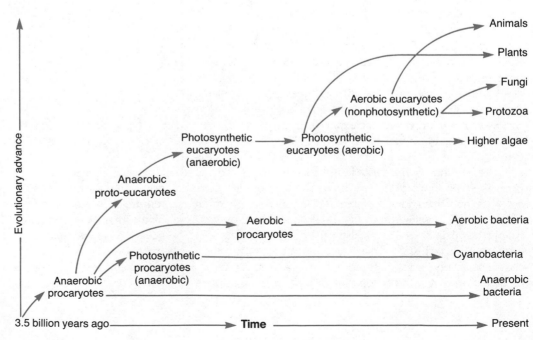

FIGURE 2.5
For the past three and one-half billion years, life forms have changed in response to a changing environment. Those groups on the right side of the chart are known to exist today. This pattern is based on our understanding of biochemical pathways, cell structure and function, and research from such areas as astrophysics. "Given so much time, the impossible becomes possible, the possible probable, and the probable virtually certain." (George Wald, Harvard University.)

reference for bacteriologists interested in the specific nature of these microbes. In this manual, the division Bacteria has been divided into 19 separate groups known as Parts (Table 2.2). Because bacteriologists have reorganized the kingdom Procaryotae into new Parts, it has become necessary to ad-

just the classification system. Table 2.3 shows the new classification of the bacterial species *Streptococcus pyogenes* according to *Bergey's Manual*. Compare this system to that presented in Figure 2.3. The characteristics which distinguish one Part from another include such features as reac-

TABLE 2.2
Division: Bacteria.

Part*	Category	General Description	Number of Genera	Representative Genera**
1	Phototrophic bacteria	Gram-negative, spherical or rod-shaped bacteria; multiplication is by binary fission and/or budding; they are photosynthetic without producing oxygen; pigments are purple, purple-violet, red, orange-brown, brown, or green	18	*Chromatium, Rhodomicrobium*
2	Gliding bacteria	Gram-negative rods typically embedded in a tough slime coat; they are capable of a slow gliding movement; reproduction is by binary fission; gliding bacteria sometimes form colorful fruiting bodies	27	*Beggiatoa, Cytophaga, Leucothrix*
3	Sheathed bacteria	Gram-negative rods that occur in chains within a thin sheath; they sometimes have a holdfast cell for attachment to surfaces	7	*Crenothrix, Leptothrix, Sphaerotilus*
4	Budding and/or appendaged bacteria	Bacteria with rod-, oval-, egg-, or bean-shaped filamentous growth; multiplication is by budding or binary fission; these bacteria sometimes have a holdfast cell	17	*Caulobacter, Hyphomicrobium*
5	Spirochetes	Slender, flexible, coiled cells; they may occur in chains and exhibit transverse fission	5	*Borrelia,** Cristispira, Leptospira,**,† Treponema**,†*
6	Spiral and curved bacteria	Rigid, helically curved rods with less than one complete turn to many turns	6	*Bdellovibrio,† Campylocabacter,† Microcyclus, Spirillum†*
7	Gram-negative aerobic rods and cocci	Rods that are usually motile with polar flagella; bluntly rod-shaped to oval cells, some of which are motile by polar or peritrichous flagella and some of which are cyst formers; and rods and cocci that require high concentrations of sodium chloride for growth	20	*Acetobacter, Alcaligenes,† Agrobacterium, Azotobacter,** Bordetella,**,† Brucella,**,† Francisella,†,* Halobacterium, Pseudomonas**,†*
8	Gram-negative facultatively anaerobic rods	Straight and curved rods; some are nonmotile; others are motile by polar or peritrichous flagella; all members are non-sporeformers; some have special growth requirements	26	*Citrobacter,† Edwardsiella, Enterobacter,† Escherichia,**,† Haemophilus,**,† Klebsiella,**,† Pasteurella,**,† Salmonella,**,† Serratia,† Shigella,**,† Streptobacillus,† Vibrio,**,† Yersinia**,†*
9	Gram-negative anaerobic rods	Strict (obligate) anaerobic, non-spore-forming organisms; some members are motile; pleomorphism (variation in shape) occurs	9	*Bacteroides,**,† Desulfovibrio, Fusobacterium,† Leptotrichia*
10	Gram-negative cocci and coccal bacilli	Cocci, characteristically occuring in pairs; adjacent sides of the cells may be flattened; organisms are not flagellated	6	*Acinetobacter,† Branhamella,† Moraxella,**,† Neisseria**,†*
11	Gram-negative anaerobic cocci	Cocci of variable size and characteristically in pairs; they are not flagellated	3	*Veillonella†*
12	Gram-negative chemolithotropic bacteria	Pleomorphic rods; these organisms use inorganic materials for energy	17	*Nitrobacter,** Nitrococcus,** Thiobacillus**'*
13	Methane-producing bacteria	Rods or cocci; some members are Gram-positive, others are Gram-negative; all are anaerobic and produce methane	3	*Methanobacterium,** Methanococcus, Methanosarcina*

14	Gram-positive cocci	Various arrangements of cocci that are aerobic, facultative, or anaerobic	12	*Aerococcus*,† *Micrococcus*,**·† *Peptococcus, Sarcina, Staphylococcus*,**·† *Streptococcus***·†
15	Endospore-forming rods and cocci	Members are aerobic, facultatively anaerobic, or anaerobic; most members are Gram-positive	6	*Bacillus*,** *Clostridium*,**·† *Sporosarcina*
16	Gram-positive asporogenous (non-spore-forming) rod-shaped bacteria	Members may be aerobic, facultatively anaerobic, or anaerobic	4	*Erysipelothrix*,† *Lactobacillus*,** *Listeria***·†
17	Actinomycetes and related organisms	Rods or pleomorphic rods, with filamentous and branching filaments; included are aerobic, facultatively anaerobic, and anaerobic rods; these organisms are usually Gram-positive, and some are acid-alcohol-fast (acid-fast)	39	*Actinomyces*,† *Arachnia, Arthrobacter, Bifidobacterium*,**·† *Corynebacterium*,**·† *Mycobacterium*,**·† *Nocardia*,† *Propionibacterium, Streptomyces***
18	Rickettsia	The majority of cells are Gram-negative coccoid or pleomorphic rods; most are obligate intracellular parasites transmitted by arthropods	18	*Chlamydia*,**·† *Cowdria, Coxiella***·† *Ehrlichia, Neorickettsia, Rickettsia*,**·† *Rickettsiella, Rochalimaea*,† *Symbiotes*
19	Mycoplasma	Highly pleomorphic, Gram-negative organisms that contain no cell wall; they reproduce by fission, by production of many small bodies, or by budding; members may be aerobic, facultatively (adaptable) anaerobic, or anaerobic	4	*Acholeplasma, Mycoplasma*,**·† *Spiroplasma, Thermoplasma*

*Based on the divisions and descriptions in R. E. Buchanan and N. E. Gibbons, editors. *Bergey's Manual of Determinative Bacteriology,* 8th ed Baltimore, Williams & Wilkins, 1974.

**Genera discussed in various chapters of this text.

†Medically important species are contained in this genus.

SOURCE: Reprinted with permission of Macmillan Publishing Co., Inc. from *Microbiology and Human Disease,* 3rd ed. George A. Wistreich and Max D. Lechtman; Copyright © 1980 by Glencoe Publishing Co., Inc.

TABLE 2.3

Classification of *Streptococcus pyogenes* according to *Bergey's Manual of Determinative Bacteriology,* 8th ed.

Kingdom	Procaryotae
Division	Bacteria
Part number	14. Gram-positive cocci
Family	Streptococcaceae
Genus	*Streptococcus*
Species	*pyogenes*

> 1 Which grouping contains a greater variety of organisms: class or order?
> 2 What is meant by the term "binomial"?
> 3 Of what value is *Bergey's Manual?*

tion to stains, genetic ability to manufacture a cell wall, and cell shape (Box 1).

Bacteria may come to lack cell walls in two ways. Some cells may only temporarily lose their walls. The wall may be digested away by some external agent or the bacterium may not make a wall. Bacteria of this type are called **L-forms** (after Joseph Lister). In the second category, the other forms lack a cell wall because they are genetically unable to produce this structure. This second group of bacteria are the Mycoplasma (Figure 2.6). These microbes are probably the smallest cells able to grow independently. Some microbiol-

ogists refer to these as the PPLOs (pleuropneumonia-like organisms) because they resemble the bacteria that cause pleuropneumonia in cattle. This type of infection in humans usually lasts about twenty-one days and is recognizably different from symptoms caused by other microbes.* More accurate and complete identification of bacterial types results largely from biochemical tests to determine details of their metabolism. How a particular type of bacterium gains usable energy, deals with waste products, produces structural components, and copes with environmental changes all give clues to the underlying metabolic activity on the organism. Bacterial DNA serves as the genetic material which codes

*The term "pneumonia" refers to a set of symptoms, not the cause of the disease. Many different microbes may cause pneumonia including viruses, bacteria, and mycoplasmas.

BACTERIA can be identified by their shape (morphology). They can be placed into one of four morphological groups: **coccus** is spherical, **bacillus** is rod-shaped, **spirillum** is twisted or bent, and **pleomorphic** has many, or varied, shapes. The spiral bacteria can be further divided into three

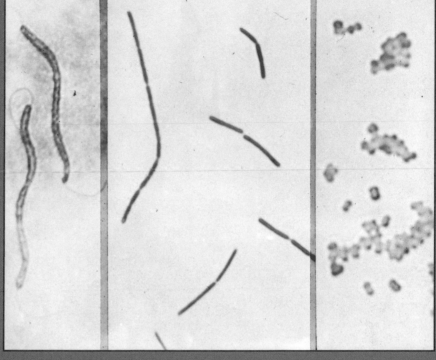

Courtesy of Carolina Biological Supply Company.

BOX 1
MORPHOLOGY
OF
MICROBES

subgroups on the basis of their morphology. One subgroup is called the **vibrios.** These bacteria have only a slight twist, as seen in the bacterium *Vibrio cholerae,* which causes the disease cholera. The second category of spiral bacteria is known as the **spirilla,** which are corkscrewlike in shape, and very rigid. Members of the genus *Campylocabacter* have this morphology and are responsible for intestinal infections that resemble cholera. The third group of spiral bacteria, the **spirochetes,** are also corkscrew in shape, but are flexible and able to wiggle or twist. This is made possible by an axial filament that spirals from one end of the cell to the other. The venereal disease syphilis is caused by *Treponema pallidum,* which looks like a wiggling corkscrew under the microscope.

Bacteria sometimes adhere to one another in characteristic forms. Those that form chains are named using the prefix ''strep-.'' For example, *Streptococcus pyogenes,* a bacterium that can cause scarlet fever, forms chains of spheres. Cells which can stick together in a random cluster are named using the prefix ''staph-.'' The cells of *Staphylococcus aureus* are spheres that adhere to one another and resemble a bunch of grapes. Streptobacillus is another possible formation. A group's shape is determined by the way the bacilli remain connected to one another after they have gone through division. The bacterium *Bacillus subtilis* forms long chains after it divides. When it reproduces rapidly, cell separations are difficult to see.

Another pattern used to classify the shape of bacteria is designated by the prefix "diplo-." Bacteria with this formation are usually linked together two at a time. A bacterium that typically takes this form is *Neisseria gonorrhoeae*, the cause of gonorrhea.

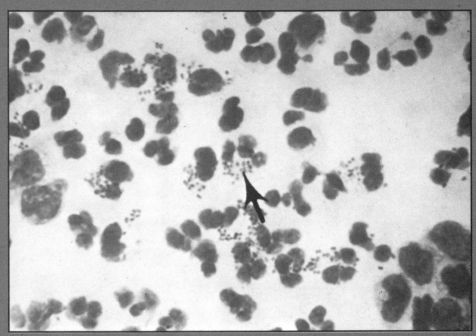

Courtesy of the Center for Disease Control, Atlanta.

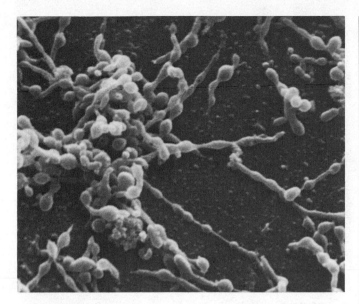

FIGURE 2.6
The *Mycoplasma* are slightly smaller than the other bacteria. The "bubbly" appearance of the cell results because these microbes have no cell wall, only a very elastic cell membrane. (Courtesy of Dr. M. G. Gabridge.)

for the production of the enzymes necessary to run the cell. The identification of these enzymes or the detection of their activity can be carried out in the laboratory. This is an accurate way to distinguish one type of bacterium from another.

An organism that does not produce certain enzymes may become dependent on other life forms for these essential molecules. When such a dependency becomes extreme, the organism may enter into a parasitic relationship. Although some kinds of bacteria are parasites, these same bacteria are not obligated to be intracellular parasites as are the viruses. *Neisseria gonorrhoeae* (found in Part 10) is a bacterium which is able to live inside or outside a host cell. Therefore, it is not an obligate parasite, but is called a **facultative** intracellular parasite. It has the facility to perform its life functions in a test tube containing special nutrient medium or in a human cell. For the most part the bacteria are susceptible to antibiotics. In some cases, the antibiotic susceptibility of a group of bacteria is greatly reduced by the heavy use of antibiotics and the changing genetic makeup of the cells (refer to chapters 1, 11, and 12).

Most rickettsias (Part 18) are obligate intracellular parasites (Figure 2.7). Because viruses are also obligate intracellular parasites, many microbiologists in the past classified the Rickettsia as being "somewhere" between the viruses and bacteria. Today, however, they are classified as true procaryotic bacteria and may be found in the cells of certain insects such as fleas, body lice, and ticks. These host animals spread the microbes to humans through biting. Only

one of the rickettsias is known to be spread by a method other than biting insects. *Coxiella burnetii* can be passed from insects to humans through cow's milk after the cow has been infected. Proper pasteurization of milk destroys these pathogens. Rickettsias may be bacilli, cocci, or pleomorphic, and they also form chains. The cells have a wall similar to other bacteria and reproduce by binary fission. There are five groups of rickettsias that cause diseases. Typhus fever, spotted fevers, scrub fever, Q fever, and trench fever are all rickettsial diseases. Even though present-day antibiotics are able to control these microbes, epidemic typhus (*Rickettsia prowazekii*) was an important human disease in the past. The number of deaths caused by this microbe parallels those caused by "black death" (bubonic plague). Body lice harbor *R. prowazekii* and may transmit them to people having poor personal hygiene. The disease symptoms include headache, chills, fever, and general body pains. Blood poisoning and darkened patches on the skin may develop in the course of the disease, which lasts about two weeks. Death occurs in as many as 40 percent of those who exhibit the symptoms. Vaccinations are available to people throughout the world to fight this disease. This is one of the microbes that is carefully monitored by the World Health Organization (WHO) in hopes of eliminating it as a world problem. Although epidemic typhus has practically been eliminated in the United States, Britain, and many Scandinavian countries, poor living and nonsanitary conditions in Africa, Asia, and Mexico make those areas prime candidates for major outbreaks of the disease.

The chlamydia (Figure 2.8) are a type of rickettsia which are so simple that for years many believed them to be large viruses. However, close examination showed them to be true cells. They are spherical cells that function as obligate intracellular parasites and get their energy from the host cell.

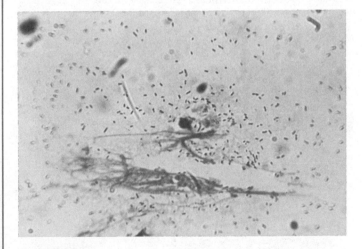

FIGURE 2.7
The *Rickettsia* are responsible for the disease known as epidemic typhus. (Courtesy of the Center for Disease Control, Atlanta.)

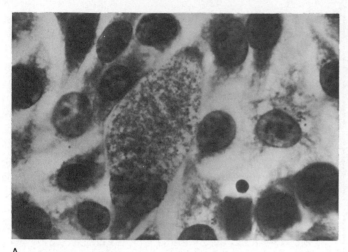
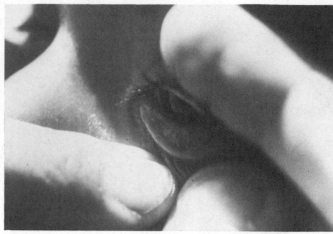

A

B

FIGURE 2.8
Chlamydia trachomatis is responsible for the eye infection known as TRIC, or conjunctivitis.
(Courtesy of the Center for Disease Control, Atlanta.)

Diseases that may result from an infection by the chlamydia are ornithosis (psittacosis), trachoma and inclusion conjunctivitis (TRIC), and lymphogranuloma venereum, a venereal disease. Current research indicates that a chlamydia is responsible for legionnaire's disease. Only thorough research may confirm this suspicion. Ornithosis, or parrot fever, is caused by *Chlamydia psittaci* and may infect humans who inhale the microbes that have been blown into the air from dried bird droppings. Pigeons, ducks, canaries, and chickens may be sources of this chlamydia infection. People working in pet shops or on poultry farms have a higher risk than most people of contracting this disease. The infection causes headache, fever, cough, constipation, and loss of appetite, although death from this disease is rare. All organisms in the division Bacteria are closely interrelated, and it has taken years of intense study to separate them into distinct Parts.

---

1 List the four basic morphological types of procaryotes.
2 What does the term "facultative" mean?
3 How do members of the genus *Rickettsia* differ from those in *Chlamydia*?

THE PHOTOSYNTHETIC PROCARYOTES

Within the division Bacteria, Part 1 contains true bacteria capable of carrying out the biochemical pathways of pho-

tosynthesis. This Part includes such bacteria as the *Rhodospirilla* and the *Chlorobia*. However, it is not the only group of photosynthetic procaryotae. Organisms in the division Cyanobacteria used to be called blue-green "algae," but more recent classification systems have changed this name to indicate the close relationship of these microbes to the Bacteria (Figure 2.9). The term "algae" is only a term of convenience, not a scientific term for the simpler types of organisms with chlorophyll. Cyanobacteria occur singly, as various clusters of cells, or as a series of cells organized into filaments. Like all algae, they have chorophyll, but the chlorophyll makeup of the Cyanobacteria is not the same as

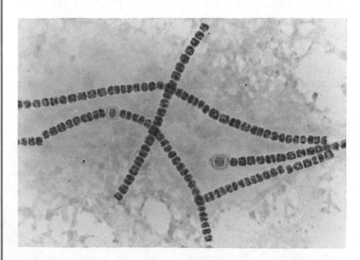

FIGURE 2.9
Nostoc is a blue-green alga. The blue-greens, or cyanobacteria, are now classified as bacteria because they are true procaryotes. (Courtesy of Carolina Biological Supply Company.)

that of all other algae. In addition to the green chlorphyll, most of them have a unique bluish pigment in their cells known as phycocyanin. Unfortunately, the blue-green color is not a reliable characteristic for easy sight identification. There are other algae that imitate this color, and there are some Cyanobacteria that show such other colors as red, yellow, purple, brown, and black.

When the Cyanobacteria photosynthesize, they are able to release oxygen (O_2) into the atmosphere. Some forms of photosynthetic bacteria do not do this, but release other by-products, such as sulfur. The photosynthetic pigment is found in the cells in special layers of membranes called **thyla-coids,** which are structures unique to the Cyanobacteria. Another unique structure is the **heterocyst cell.** The biochemical activity taking place in this structure is thought to be the process of nitrogen fixation. Nitrogen from the air (N_2) is converted to a usable form (NH_3) in these specialized cells. Very few types of organisms are able to carry out this vital conversion. The Cyanobacteria also have a special way of moving called "slime motion" or "gliding." The cells are thought to have the ability to move their sticky capsule material over the cell surface. This motion causes the entire cell to move slowly. The Cyanobacterium *Oscillatoria* was given its name because of the oscillating, back-and-forth motion it shows under the microscope.

Members of the Cyanobacteria have received increased attention as the interest in environmental quality has risen. The presence of some indicate organic pollution, while others are pollutants themselves. Many cause foul odors or tastes in drinking water and may produce poisons that kill animals. Occasionally a sudden increase in the population (a bloom) of a species of Cyanobacteria will occur and cause trouble for water-living plants and animals. This bloom can shut off the light to submerged plants, thus causing their

death and decay. The decay may, in turn, remove enough oxygen from the water so as to suffocate fish and other animals (Figure 2.10). It should be noted that the Cyanobacteria are found in a variety of habitats, not just in freshwater. A small quantity are found in salt water; they are common in soil, on damp rocks, on tree bark; and they even grow in the hot springs of Yellowstone National Park.

Members of the photosynthetic bacteria include *Rhodospirilla* and the *Chlorobia*. Both are able to photosynthesize, but unlike the Cyanobacteria, they do not need oxygen in the environment and they do not produce O_2 as an end product of photosynthesis. Most of these types of microbes are found in water. The cells may be cocci, bacilli, vibrios, or spirilla. The special chlorophyll pigments used in the photosynthesis reactions are located on stacks of membranes in the cells. These stacks are called **photosynthetic membranes,** or **lamellae.**

PROTISTA—THE EUCARYOTES

The kingdom Protista is surely the most artificial category in the classification scheme described earlier. Even from the brief description given, it is easy to see that there are great differences among the organisms of this kingdom. According to evolutionary theory, great differences in structure should mean great differences in relationships. However, the Protists have only two things in common: (1) they are all eucaryotic cells, and (2) they lack complex structures with differentiated cells. The common name for the green Protista is "algae," and the nongreen are called fungi and protozoa. The algae with their chlorophyll can carry on photosynthesis, while the fungi and protozoa must obtain their food from organic materials such as living or dead organisms.

FIGURE 2.10
Algal blooms occur when the warm summer sun shines on a pond rich in nutrients and causes the rapid multiplication of algal cells. (Courtesy of A. H. Gibson.)

1 In what kinds of microbes are heterocysts found?
2 In what types of microbes do you find photosynthetic lamellae?
3 List the basic characteristics of eucaryotic algae that distinguish them from procaryotic algae.

ALGAE

The algae are separated into divisions on the basis of the different kinds of photosynthetic pigments, compounds used for nutrient storage, and details of their reproductive cycles. The six important divisions in the eucaryotic algae are the Chlorophyta (green algae), Euglenophyta (euglenoids), Chrysophyta (diatoms), Pyrrophyta (dinoflagellates), Phaeophyta (brown algae), and Rhodophyta (red algae). The common names indicate that the differences among these groups are based primarily on the color of the cells. The algae are "pond scum" or seaweed and are generally considered to be a great nuisance by many people. They can cause problems by fouling beaches and drinking water, as well as by producing poisonous substances, but it is also true that they have tremendous importance for all water-living animals. The algae occupy a key ecological position, in that they are the first link in the food chain within all bodies of water. They are the producer organisms and as such deserve attention, since we depend on them in many hidden ways.

The green algae (Chlorophyta) grow mostly in fresh water, but there are some in damp soil, salt water, and even in snowfields high in the mountains. In addition to being single cells, some occur as long filaments of single cells, as spherical clusters, or as lettucelike sheets (Figure 2.11). Most live up to their common name, green algae, since they are grass-green. The cells contain two kinds of chlorophyll, *a* and *b*, and they store their food as starch.

The euglenoids are an interesting group of single-celled algae that have the plant characteristic of chlorophyll *a* and *b* in addition to some animal characteristics. They pull themselves through the water by lashing flagella back and forth. The flagellum is a long, flexible, whiplike filament. Euglenoids ingest food particles in addition to carrying on photosynthesis. They are equipped with a light sensitive spot that enables them to move toward or away from light, and they lack a cell wall. Most of this small group live in fresh water, but a few live in damp soil, and some in the food tubes of animals (Figure 2.12). Food in these algae is stored in the form of a starchlike carbohydrate molecule called **paramylum.**

The diatoms (Chrysophyta) are the most important producers in salt water, yet they are commonly found in fresh water. The statement that diatoms are producers means that the survival of sea animals depends on these organisms, which are so small that a microscope is needed to see them. They occur primarily as individual cells, but some remain attached to each other after division and form a cluster called a **colony.** The diatoms are unique in that they deposit silicon dioxide in their cell walls. When the organisms die, everything breaks down and disappears except the silicon dioxide. These cell walls are arranged in pairs that resemble a dish with a cover (Figure 2.13). The glassy walls

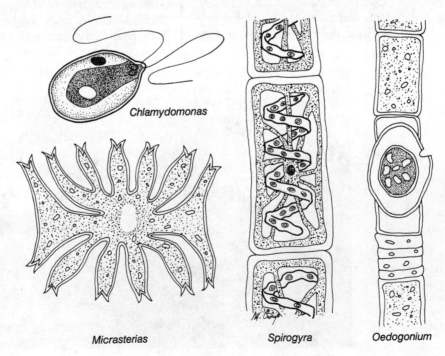

FIGURE 2.11
These are examples of the variety found in the green algae. (From Eldon D. Enger et al, *Concepts in Biology*, 3rd ed. © 1982, 1979, 1976 by Wm. C. Brown Company Publishers, Dubuque, Iowa. Reprinted by permission.)

Chlamydomonas

Micrasterias

Spirogyra

Oedogonium

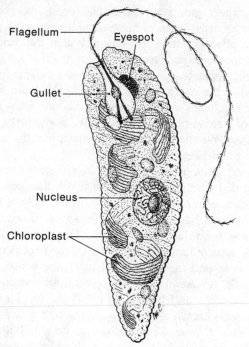

Flagellum

Eyespot

Gullet

Nucleus

Chloroplast

FIGURE 2.12

Euglena is an organism having both plant and animal characteristics. (From Eldon D. Enger et al, *Concepts in Biology*, 3rd ed. © 1982, 1979, 1976 by Wm. C. Brown Company Publishers, Dubuque, Iowa. Reprinted by permission.)

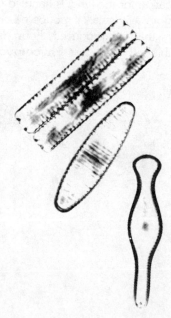

FIGURE 2.13

A few of the seemingly endless variety of diatoms. (From Eldon D. Enger et al, *Concepts in Biology*, 3rd ed. © 1982, 1979, 1976 by Wm. C. Brown Company Publishers, Dubuque, Iowa. Reprinted by permission.)

of diatoms are decorated with grooves and holes that are so small they are used to check the quality of microscopes. Some markings are too small to be seen except under an electron microscope. The glassy remains of these algae have been deposited on ancient sea floors in such quantity that it is commercially profitable to remove them with power shovels. Since diatoms are so small, the mass has only the texture of talcum powder and is therefore used as a very fine abrasive in metal polishes. The cell walls of diatoms are able to trap small particles of impurities in liquid and thus can be used as filters. The brewing industry has made use of diatoms to clarify beer. They can also serve as an excellent insulator and are not destroyed by extremely high temperatures.

The dinoflagellates (Pyrrophyta) are next in importance to the diatoms as a producer in the oceans. Blooms of these organisms are so striking they are called **red tides,** since the cells (*Gonyaulax*) are present in such quantities that they color the water. If it were only for the pinkish color added to ocean waves, they would be given little notice,—they might even be an attraction. Unfortunately, this particular species of dinoflagellate produces a poison deadly to fish, many smaller organisms, and humans. These blooms kill fish, which litter the beach and drive the tourists away. Oysters or clams growing in such waters concentrate the poison in their bodies as they filter food out of the water. This concentration can be enough to cause illness to persons eating these seafoods. There is an old saying that oysters and clams are in season only during the months containing the letter R—"Oysters R in season." The statement was true only when refrigeration was unreliable, and to some extent because the red tide occurred more frequently during the warmer months (none of the summer months contain the letter R). Fortunately, red tides are confined to relatively small areas. Either the blooms have been better reported in recent years or there has been an increase in these "red tides."

The brown algae (Phaeophyta) are mostly saltwater organisms. The seaweed, called kelp, that covers the rocks along the ocean coasts is mostly of this group. Some grow to quite a large size—up to one hundred meters in length. Kelp is harvested mechanically and may be refined to a commercially valuable product depending on the particular species involved. Labels of many foods list the algae product they contain, either algin or alginates. These substances from brown algae are found in ice cream, cake frosting, pudding, cream-centered candies, and many cosmetics. Kelp is also an excellent source of iodine since these algae are able to concentrate this element from seawater. Many people utilize kelp in a dry, crumbled form in their cooking as a source of this necessary element.

The red algae (Rhodophyta) are also saltwater organisms, and they grow on rocks along the seacoast. Red algae have attracted scientific attention not only because they are unlike other algae but also because they differ from the higher plants; thus, they have been referred to as an evolutionary dead end. The red algae are also used as an additive to thicken and flavor food. The genus *Porphyra* is even "farmed" on wooden racks along the coast of Japan. A red algae product named carrageenan is used as a stabilizer to keep solids suspended in liquids, such as the particles of chocolate in chocolate milk. A high-protein cattle food has been produced from the red algae grown along Scandinavian coasts. Agar, another product of red algae, has been used worldwide in hospital and bacteriological laboratories, and is commonly sold as a granular powder that is added to broth. When the mixture cools, it becomes semisolid, like gelatin, and is used principally as a medium on which to grow bacteria and other microbes. Although some people eat agar, this is not generally the case, because agar acts as a laxative.

---

1 What type of chlorophyll is found in all types of eucaryotic algae?
2 Of what ecological significance is *Gonyaulax*?
3 How is the algal product carrageenan used?

FUNGI

Fungi are members of the Protista that lack chlorophyll. Many have cell walls composed of a very resistant material called **chitin,** rather than the more common plant cell material **cellulose.** The more familiar examples of this non-

photosynthetic group are the molds, yeasts, bracket fungi, and mushrooms. None of these organisms has the plantlike features of roots, stems, or leaves. Within the fungi, there are two morphological types, the molds and the yeasts. These are easily separated since the yeasts are essentially unicellular, while the molds are multicellular. The basic structural unit in the molds is the **hypha,** a slender filament composed of a series of cells (Figure 2.14). Some fungi have so many filaments crowded into a firm mass that the individual filaments are difficult to recognize.

In other fungi, the filaments form a vast network extending down into the material on which they are growing and appear as a cottonlike growth. An example of this type of growth can be seen in bread mold or sometimes as a fuzzy growth on dead insect bodies. This mass of filaments, in either type of growth habit, is called the **mycelium.**

The second major group of fungi, the yeasts, exist as single cells. An example of this type is *Saccharomyces cerevisiae,* brewer's and baker's yeast. These fungi reproduce by a process known as budding. As the bud grows from the parent cell, it forms what appears to be a bubble or outpouching. In a newly dividing cell, this bud is easy to recognize since it is smaller than the parent cell. Yeast and molds are not official taxonomic groups, and many fungi may shift from one of these forms to the other. Under certain environmental conditions, yeast may change and grow into a cottonlike mass. The fungi that are capable of making this change are called dimorphic (two forms). The more acceptable classification for the fungi is handled by subdividing them into taxonomically smaller groups based on the kind of sexual spores that are produced. Four classes of fungi have been established. The class **Phycomycetes** includes

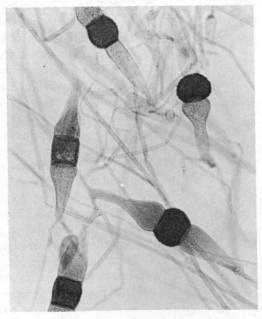

FIGURE 2.14
(A) Rhizopus. Many forms of fungi are filamentous. The basic structure is called a hypha, which is a series of cells. (B) Aspergillus. When the hyphae mat together, the group is called the fungal mycelium. (Courtesy of Carolina Biological Supply Company.)

A

B

water molds and other forms that are commonly found in very moist environments (Figure 2.15A). These fungal filaments have no separations (cross-walls) between their cells, so the cell contents may move from one cell to another. Biologists call these nonseptate hyphae. The sexual reproduction of the phycomycetes results in the formation of a structure known as a **zygospore.** *Mucor* and *Rhizopus,* molds found in household dust, are members of this group. A second class of fungi is the **Ascomycetes** (Figure 2.15B). These have septate hyphae and reproduce sexually by spores formed in characteristic oval sacs called asci. The spores are known as **ascospores.** If a fungus produces just one ascus in a parent cell, it is called a yeast. But, not all members of this group grow in such an uncomplicated way. *Penicillium,* which is the source of the antibiotic penicillin, produces a very complex series of asci. The third class of fungi are the

Basidiomycetes (Figure 2.15C). This group also contains septate hyphae, but the spores are not the same as those produced by the other two groups. Mushrooms, bracket fungi on trees, and puffballs produce their spores in a club-shaped structure known as a basidium. The spores are called **basidiospores.** These can be seen as a cloud of dustlike particles leaving a mushroom when it has been dried and shaken. The last group of fungi is known as the imperfect fungi since the sexual stage of reproduction has not yet been discovered (Figure 2.15D). When this process is identified, the members of this group will be shifted into one of the other three classes. The technical name for this class is the **Deuteromycetes.**

The Molds as Harmful Fungi   Fungi are commercially important to us as destroyers as well as producers.

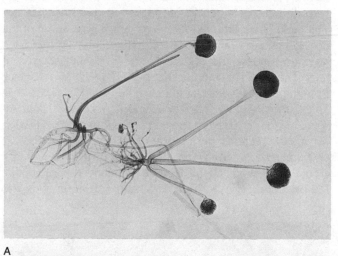

A

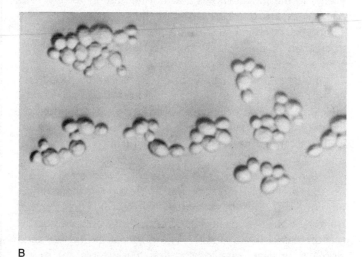

B

C

D

FIGURE 2.15
The representative fungi shown here are (A) *Rhizopus,* a bread mold; (B) *Saccharomyces,* yeast; (C) A puffball; and (D) *Aspergillus,* a common mold. (Courtesy of Carolina Biological Supply Company.)

They cause losses of millions of dollars annually by destroying food and manufactured materials, and expensive chemical dusts and sprays are needed to control them. Because they produce millions of spores that can be easily and widely spread, their control is difficult. Many kinds of fungi have the ability to grow in very strange environments. Storing food in the refrigerator helps to keep it fresh and prevents spoilage, but many fungi are able to grow even in this cold environment. That fuzzy growth on the green beans on the back shelf of a refrigerator is from spores that contaminated the food. The fungi release millions of spores into the atmosphere where they are rapidly distributed by the wind currents. These spores may linger in the air or be inhaled by animals and humans. Many people show the symptoms of what is commonly referred to as "hay fever" when they are in contact with too many of these spores. This is an allergy to the fungal spores and not necessarily to hay. *Mucor* spores are a major cause of these upper respiratory symptoms. Allergic reactions are the leading chronic disease in children in the United States; thirty-one million children have some form of allergy related to fungal spores.

Fungi are able to grow on the surface of the body and cause diseases such as athlete's foot and ringworm. Fungal infections can also occur inside the body. Liver, lung, and kidney infections cause severe damage and may be deadly if not quickly brought under control by proper medical therapy.

A number of fungi produce deadly poisons called **mycotoxins.** There is no easy way to distinguish those which are poisonous from those that are safe to eat. Fungi that are nonpoisonous are called mushrooms; those that are poisonous are called "toad stools." The origin of the name "toad stool" is unclear; one idea is that toad stools are mushrooms on which toads sit, while another states that the word is derived from the German "todstuhl"—seat of death. The most deadly of these, *Amanita verna,* is known as the "destroying angel" (Figure 2.16) and can be found in woodlands during the summer. Mushroom hunters must learn to recognize this deadly, pure white species. This mushroom is believed to be so dangerous that food accidentally contaminated by its spores can cause illness and possibly death. Another mushroom, *Psilocybe mexicana,* has been used for centuries in religious ceremonies by certain Mexican tribes because of the hallucinogenic chemical that it produces. These mushrooms have been grown in culture, and the drug psilocybin has been isolated. In the past, it was used experimentally to study schizophrenia. Potatoes can be infected by a fungus known as *Phytophthora infestans.* When this fungus grows, it forms a scaly material on the surface of the potato that makes it inedible. In the past, this fungus was responsible for the loss of millions of dollars worth of potatoes and affected the lives of many people. The Irish potato famine of 1845 was the result of potato crop losses due to this fungus.

---

1. The basic threadlike unit of a fungus is given what name?
2. On what basis are the different classes of fungi separated?
3. What is a mycotoxin? Name a fungus that produces one.

Some Advantageous Molds Although mold on potatoes may spoil them, many cheeses, such as blue cheese and gorgonzola owe their distinct flavors to molds. Molds are also beneficial to such industries as brewing, baking, and drug-making. Antibiotics like penicillin and other mold byproducts are a great blessing.

In 1928, Dr. Alexander Fleming was working at St. Mary's Hospital in London. As he sorted through some old petri dishes on his bench, he noticed an unusual situation. The mold *Penicillium notatum* was growing on some of the petri dishes. Apparently, the mold had found its way through an open window and onto a culture of *Staphylococcus aureus.* The bacterial colonies that were growing at a distance from the fungus were typical but there was no growth close to the mold (Figure 2.17). Fleming isolated the agent responsible for this destruction and named it penicillin. Through his research efforts and those of several colleagues, the chemical was identified and used for about ten years in microbiological work in the laboratory. Many suspected that the penicillin might be used as a drug, but the fungus was

FIGURE 2.16

Amanita. This fungus produces a poisonous chemical known as mycotoxin. (Courtesy of J. Robert Waaland, University of Washington, Biological Photo Service.)

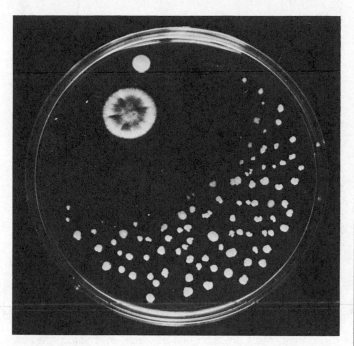

FIGURE 2.17
A beneficial effect of fungi is illustrated here as *Penicillium* mold inhibits the growth of the bacterium *Staphylococcus,* as seen by the zone of clearing around the fungal colony. (Courtesy of Ronald Hare and reprinted with permission from *Chemistry* (now *SciQuest*), vol. 51, no. 7, 1978, by the American Chemical Society.)

not able to produce enough of the chemical to make it worthwhile. When World War II began and England was being fire bombed, there developed a great urgency for a drug that would control bacterial infections in burn wounds. Two scientists from England were sent to the United States to begin research into the mass production of penicillin. Their research in isolating new forms of *Penicillium* and purifying the drug were so successful that cultures of the mold now produce over one hundred times more of the drug than the original mold discovered by Fleming. In addition, the price of the drug dropped considerably, from a 1944 price of $10,000 per pound to a current price of less than 65¢! The species of *Penicillium* used to produce all the antibiotics today is *P. chrysogenum*. This was first isolated in Peoria, Illinois, from a mixture of molds found growing on a cantaloupe. The species name, *chrysogenum,* means golden and refers to the fact that the mold produces golden-yellow droplets of antibiotic on the surface of the mycelium. The spores of this mold were isolated and irradiated with high dosages of ultraviolet light which caused mutations to occur in the genes. When some of these mutant spores were germinated, the new mycelia were found to produce much greater amounts of the antibiotic.

There are over one hundred species of *Penicillium* and each characteristically produces spores in a brushlike bor-

der; the word "penicillus" means *little brush*. Other members of this group do more than just produce antibiotics. Many people are familiar with the blue, cottony growth that sometimes occurs on citrus fruits. The *P. italicum* growing on the fruit appears to be blue because of the pigment produced in the spores. The blue cheeses, such as Danish, American, and the original Roquefort cheese, all have this color. Each has been aged with *Penicillium roquefortii* to produce the color, texture, and flavor (Figure 2.18). Differences in the cheese are determined by the kind of milk used and the conditions under which the aging occurs. Roquefort cheese is made from sheep's milk and aged in Roquefort, France, in special caves. American blue cheese is made from cow's milk and aged in many places around the United States. The blue color has become a very important feature of these cheeses. The same research laboratory that first isolated *P. chrysogenum* also found a mutant species of *P. roquefortii* that would produce spores having no blue color. The cheese

FIGURE 2.18
In the manufacture of cheese, bacteria placed in the milk aid both in curd formation and flavor production during the ripening process. This illustration shows the piercing of a block of "blue-veined" cheese known as blue cheese. Holes enable air to enter the cheese so that the fungus *Penicillium* can grow actively to produce the proper flavor, aroma, and texture. (Courtesy of Pfizer, Inc.)

made from this mold is "white" blue cheese. The flavor is exactly the same as "blue" blue cheese, but commercially it is worthless because people want the blue color.

A last example of a very beneficial fungus is *Aspergillus niger.* This mold produces very dark brown or black spores, which are in chains and may commonly infect and spoil refrigerated foods. But more importantly, this organism is the basis for the soft drink industry. The citric acid that gives a soft drink its sharp taste was originally produced by squeezing juice from lemons and purifying the acid. Today, however, *Aspergillus niger* is grown on nutrient media with table sugar (sucrose) to produce great quantities of citric acid.

Fungi and algae come together in a peculiar group called **lichens.** Alga is believed to furnish food for the fungus, and the fungus supplies moisture and protection to the alga. Together, this association is important as a pioneer in ecological succession. Lichens are capable of growing on bare rock, where they begin the soil-making process. The so-called "reindeer moss" is actually a lichen that provides caribou with a source of food in northern Canada and Alaska.

PROTOZOA

The third major group in the kingdom Protista is the division **Protozoa.** The name derives from the Greek words meaning "first animal." It is still classified by many as an animal and is studied in zoology classes, just as algae and fungi are studied in botany classes. These eucaryotes were first observed by van Leeuwenhoek in 1675 and have been studied and observed very intensely ever since. Presently, more than 20,000 species have been identified. They require a moist environment for survival and are found commonly in lakes, rivers, stagnant pools, and in moist soil. The cells usually exist as single units, but some species are able to live in

small, characteristic colonies. Some of the cells may be green, yellow, or gold in color, but they do not have chlorophyll. The protozoans have a tough outer layer that aids in maintaining their shape. Some cells are oval, spherical, or elongated, while others lack any definite shape. These cells appear as "blobs" and are known as the amoebas. The protozoans feed on both living and nonliving material in their environment: some actually surround their food and engulf it, while others function more like the bacteria and absorb food. Sexual and asexual reproduction in the group is complex and varied. A very resistant structure called a cyst serves the same purpose as the endospores of some bacteria, although it is not formed in the same manner. Most protozoans can move under their own power at some stage in their life cycle, and the particular method of locomotion is used as a basis for classifying subdivisions.

One class of protozoans is called the **Sarcodina** (Figure 2.19). The genus *Amoeba* and its relatives flow without having any permanent body form. This type of flowing movement is so characteristic that it is called amoeboid movement. Most amoeba live in water, and some are parasitic. One such parasite, *Entamoeba histolytica,* causes amoebic dysentery in humans, and is responsible for death or serious illness in disaster-stricken areas where drinking water has become contaminated. The symptoms include diarrhea, fever, and chills. Many people in underdeveloped countries have the protozoans in their intestine, yet do not show any symptoms of the infection. These people may serve as a source of infection for others, especially in areas where living conditions are poor and sanitary disposal of human feces is inadequate.

Another class of protozoans, the **Mastigophora,** moves by means of flagella. These protozoans have one or more flagella and a relatively simple structure as illustrated by

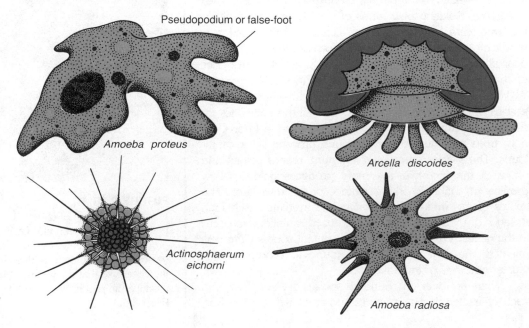

FIGURE 2.19
Sarcodina are one-celled organisms that move by leglike extensions of the protoplasm called pseudopods. (From Eldon D. Enger et al, *Concepts in Biology,* 3rd ed. © 1982, 1979, 1976 by Wm. C. Brown Company Publishers, Dubuque, Iowa. Reprinted by permission.)

Pseudopodium or false-foot

Amoeba proteus

Arcella discoides

Actinosphaerum eichorni

Amoeba radiosa

Hexamita (Figure 2.20). The organism *Trypanosoma gambiense,* which causes the deadly African sleeping sickness, is a flagellated protozoan, as is *Trichomonas vaginalis,* the cause of the venereal disease trichomoniasis. *T. vaginalis* is frequently called vaginitis ("itis" means *infection*) and the disease is seen as a yellow foamy discharge with a foul odor that is easy to identify under the microscope because of the relatively large size of the protozoans. Similar symptoms may occur in men, but to a lesser degree.

The single largest class of Protozoa, the **Ciliata** (Figure 2.21), moves by means of **cilia,** which are short, flexible filaments that frequently cover the cell completely. Cilia beat in an organized, rhythmical manner and propel the cells so rapidly that they are difficult to keep in focus under a microscope. Ciliates are commonly seen in cultures of pond water and hay. In a freshwater ecosystem, the ciliates feed on algae, bacteria, yeasts, and other protozoa. The only ciliated pathogen of humans is *Balantidium coli.* This causes a disease of the colon resulting in diarrhea (dysentery) and sometimes nausea and vomiting. Stool samples from a patient may contain blood, mucus, and cysts of this protozoan. This is a rare disease in humans but can be transmitted in epidemic numbers in areas with poor sanitation facilities. Antibiotics can be used to treat infections of balantidial dysentery.

The organism *Plasmodium,* which causes malaria in humans, is representative of a fourth class of Protozoa (Figure 2.22). All members of this class, known as **Sporozoa,** are parasites. One of their major methods of reproducing is by means of resistant cells called spores. Many important disease organisms are in this group and are often transmitted from person to person by mosquitos, ticks, and fleas.

The sporozoan *Toxoplasma gondii* is the cause of the disease toxoplasmosis. An infection of this protozoan may show no symptoms at all or it could be fatal. The most dangerous infection in humans occurs when the microbe is transmitted to pregnant women. This infection may pass across the placenta and cause great harm to the unborn child. If an infection of toxoplasma does not cause death of the fetus or a stillbirth, it may result in "water on the brain" (hydrocephaly) or damage to the central nervous system. This parasite can be transmitted to pregnant women from infected cats through special cysts in the cat's feces deposited in cat litter. To help reduce the chances of this transmission, cat litter should be changed frequently and the litter box should be kept outside.

---

1 What type of microbe is responsible for the production of penicillin?
2 Name a useful product obtained from the fungus *Aspergillus.*
3 List the four classes of protozoa and cite an example of each.

NONCELLULAR MICROBES

Viruses are peculiar—their structure is unlike all the other microbes mentioned so far. These parasites are not consid-

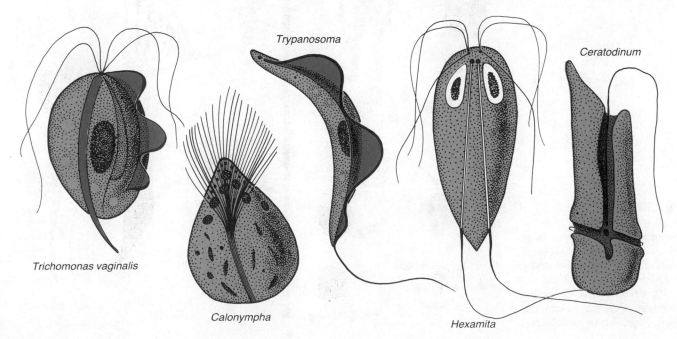

FIGURE 2.20
Mastigophora are protozoans that move by the back-and-forth motion of long, hairlike appendages called flagella.

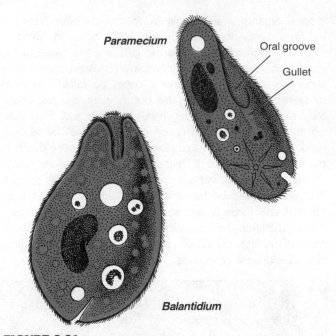

FIGURE 2.21
These two protozoans are members of the group Ciliata, which move through the environment by means of their numerous cilia.

ered to be organisms at all since they are not cellular (Figure 2.23). Viruses consist of a nucleic acid enclosed by a covering of protein. Other components, typically found in procaryotic or eucaryotic cells, are absent. One could say that they have no life until their nucleic acid enters another cell. Viruses are called **obligate intracellular parasites** since they are unable to carry out any of the typical life functions until they are inside a host cell. The virus sets up a new relationship with this **host** cell in which the virus (parasite) benefits and the host is harmed. Such a relationship is known a parasitism.

The virus parasite enters the host cell which contains all the materials needed by the virus. When the virus is still outside the host cell, it is called a **virion,** or virus particle. This word is comparable to the word **cell** in that both represent a single unit of that kind of microbe. Upon entering the host cell, the virion loses its protein coat. The nucleic acid portions may remain free in the cell or this viral nucleic acid may link with the host's genetic material. Some virions contain as few as three genes while other, larger virions contain as many as 500 genes. Viral genes take command of the host's metabolic pathways and direct it to carry out the work of making new copies of the original virus. The virus

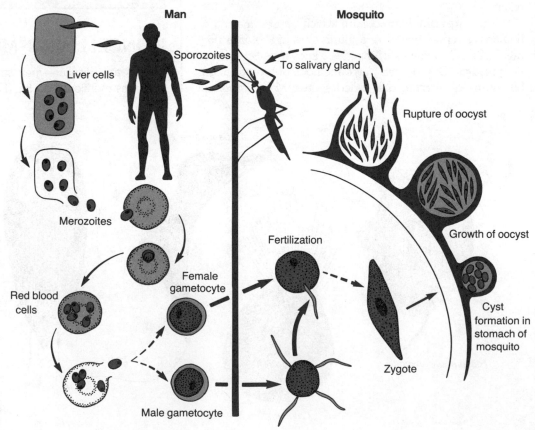

FIGURE 2.22
A generalized life cycle of the protozoan parasite *Plasmodium,* which causes malaria. (From Eldon D. Enger et al, *Concepts in Biology,* 3rd ed. © 1982, 1979, 1976 by Wm. C. Brown Company Publishers, Dubuque, Iowa. Reprinted by permission.)

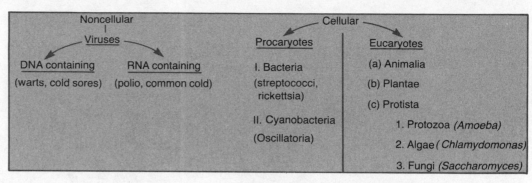

FIGURE 2.23

All viruses lack typical cell structure; therefore, they are classified in their own separate group. Viruses can be subdivided into those that contain DNA as their genetic material and those that contain RNA as genetic material.

makes use of the available enzymes and ATP to start producing copies of itself. When enough new viral nucleic acid and protein coat are produced, complete virus particles are assembled and released from the host (Figure 2.24). The number of virions released ranges from tens to thousands. The virus that causes polio releases about 10,000 new virions after it has invaded its human host cell.

Figure 2.25 shows one type of virion, but the size and shape vary greatly. These two characteristics are of help when classifying viruses. Some are rod-shaped, others are round, while still others are in the shape of a coil or helix. Viruses are some of the smallest infecting agents known to mankind. None can be seen with a standard laboratory microscope; they require an electron microscope to make them visible. The electron microscope has been of great value for this purpose, but it is not very practical in most college class-

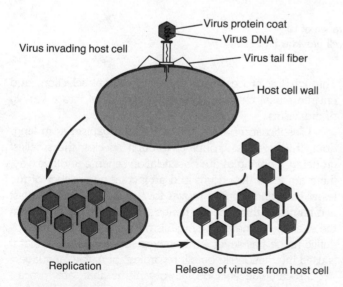

FIGURE 2.24

During viral invasion of a cell, the viral nucleic acid takes control of the living cell that it invades. The result of this infection is the bursting of the host cell. While the illustration shows the bursting of the host cell, not all virus infected cells are destroyed during the release process.

rooms. A great deal of work is necessary to isolate virions from the environment and prepare them for observation with an electron microscope. For this reason, most viruses are more quickly identified by their activities in host cells.

When viruses infect host cells, they alter the normal metabolism of the cell. If the alteration is extensive and the cell is harmed, the virus causes disease symptoms. A few of the diseases caused by viruses in animals and humans include cold sores, warts, polio, measles, and the "flu." In plants, viruses cause a mosaic pattern on tobacco leaves, color patterns in tulip petals, and celery leaf rot. Even microbes can be infected with microbes. Bacteria are infected by viruses called bacteriophage. As mentioned earlier in the Introduction, viruses have been of great value in research related to cell biochemistry and molecular biology. The fundamental concept that deoxyribonucleic acid (DNA) serves as the genetic material was derived from work performed with certain viruses that infect bacteria. The names of bacteriophages are very different from the scientific names of plants and animals. T_2, ϕ X-174 (phi "x" - 174), and phage λ (lambda) are viruses that infect bacterial cells. This naming system is different from the binomial system of nomenclature mentioned earlier because scientists are not sure of the evolutionary relationship of the viruses to cellular forms of life. However, scientists are attempting to group viruses in a uniform way by using fundamental chemical characteristics.

Some believe that the viruses were at one time parts of primitive procaryotic or eucaryotic cells. These parts displayed no functional characteristics until they again became located inside another whole cell. A change in their function allowed them to reproduce and then move on to another cell. Thus, the viruses may have been groups of genes that gained the ability to multiply independently of other genes in procaryotic and eucaryotic cells. Their independent nature allowed them to move from one host to another producing more of their own kind. Others believe that the viruses have evolved from some of the first bacterial cells. As time passed, these cells became specialized and no longer manufactured their own surrounding cell structures; they relied on the cells of others to perform the necessary life func-

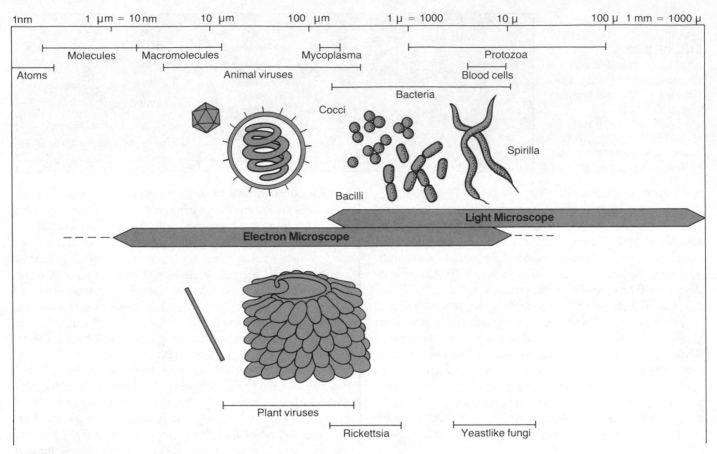

FIGURE 2.25
The viruses are among the smallest microbes. They border on the size range of large macromolecules such as nucleic acids and proteins. The human blood cell provides a basis of comparison to show how small these virions are.

tions. Since no one was around at the time, it may never be known which theory is likely to be true. More detailed information on viral structure, function, and disease will be presented in chapters 20, 21, and 22.

---

1 What is the basic structure of a virus?
2 What macromolecules are found in viruses?
3 Why is it so difficult to classify viruses?

SUMMARY

Interactions among living cells have occurred ever since the appearance of the first heterotrophs and autotrophs. The first organisms to evolve were single-celled organisms of the kingdoms Procaryotae and Protista. From these simple beginnings, more complex, many-celled organisms progressed, creating the kingdoms Plantae and Animalia.

Through the processes of mutation, natural selection, and environmental change, these groups evolved into a variety of organisms.

Classification names categorize an organism from kingdom, the broadest grouping, through species, the smallest grouping. Looking at the classificiation scheme another way, there are a few kingdoms and an increasing number of the lesser categories as one moves toward the species. The last two names of this scheme, genus and species, together form the scientific name of the organism being classified. The scientific name has several advantages, one of which is that it is used internationally by all, regardless of native language.

Procaryotae include all organisms that are procaryotic—the Bacteria and Cyanobacteria. Some members of the Procaryotae are important as disease agents, for carrying on fermentation, as decay organisms, and for their lowering of water quality.

The Protista include those eucaryotes lacking specialized cells. Eucaryotes are organisms whose cells possess a nucleus with a nuclear membrane as well as other structures. The Protists include the fungi, the protozoa, and all the eucaryotic algae. In general, algae and fungi are more plant-like and the protozoans are more animal-like. Algae produce their own food by photosynthesis with their chlorophyll, while fungi and protozoa depend on organic materials produced by some other living thing. Algae are restricted in their living situation to a lighted area because they photosynthesize, but otherwise are wide ranging from water to moist-land habitats. Fungi are not restricted by light or its absence and grow on almost any organic material, natural or artificial. Protozoa are largely aquatic, and some are parasites. The importance of protists to humans ranges from their role in causing diseases to industrial processes, food production, and recycling of organic materials.

Viruses are protein-nucleic acid particles that function only as intracellular parasites and display no cellular form.

WITH A LITTLE THOUGHT

A patient has just arrived at the hospital, and a diagnosis of his illness is as follows:

Begins as a small superficial, solid elevation of the skin (papule) and spreads peripherally, leaving scaly patches of temporary baldness. Infected hairs become brittle and break off easily. Occasionally boggy, raised, and pus-filled lesions called kerions developed. Examination of the scalp under ultraviolet light for yellow-green fluorscence is helpful in diagnosing this disease.

What characteristics of this infectious agent would you look for in the lab to specifically identify the cause of this ringworm? How would you classify this organism in the systems described in this chapter?

STUDY QUESTIONS

1 Cite two differences between an autotroph and a heterotroph.

2 Give two examples of organisms for each of the four kingdoms.

3 How can you identify the scientific name of an organism?

4 Name two factors which may contribute to the formation of a new species.

5 Which of the microbes are obligate intracellular parasites?

6 What is the difference between a virion and a cell?

7 Name a disease that may be caused by microbes belonging to the following groups: Rickettsia, Mycoplasma, and Protozoa.

8 What characteristics make the Cyanobacteria more closely related to the Procaryotae than the eucaryotic algae?

9 What is the difference between hyphae and mycelia?

10 Name two positive results of fungal growth and activity.

SUGGESTED READINGS

CHRISTENSEN, C. M. *The Molds and Man,* 3rd ed. McGraw-Hill, New York, 1965.

EDMONDS, P. *Microbiology, An Environmental Perspective.* Macmillan Inc., New York, 1978.

GOODHART, C. R. *An Introduction to Virology.* W. B. Saunders Company, Philadelphia, 1969.

PHAFF, H. J., M. W. MILLER, and E. M. MARK,. *The Life of Yeasts.* Harvard University Press, Cambridge, Mass., 1978.

WILMAR, B. I. *A Classification of the Major Groups of Human and Other Animal Viruses* 4th ed. Burgess Publishing Company, Minneapolis, Minn., 1969.

KEY TERMS

heterotroph

autotroph

procaryote (pro-kar'e-ot)

eucaryote (u-kar'e-ot)

binomial system of
nomenclature

Bergey's Manual

coccus (kok'us)

bacillus (bah-sil'lus)

spirillum (spi-ril'um)

pleomorph (ple'o-morf)

spirochete (spi"ro-ket')

algae

fungi

protozoa

virus

virion

Pronounciation Guide for Organisms

Campylocabacter (kam"pi-lo-ka-bak'ter)

Treponema (trep"o-ne'mah)
  pallidum (pal'li-dum)

Bacillus (bah-sil'lus)
  subtilis (suh'til-is)

Coxiella (kok"se-el'lah)
  burnetii (bur-net'e-e)

Rickettsia (ri-ket'se-ah)
  prowazekii (prow-ah-zek'ee-ee)

Chlamydia (klah-mid'e-ah)
  psittaci (sit-ah'see)

Rhodospirilla (ro"do-spi-ril'ah)

Chlorobia (klo-ro'be-a)

Gonyaulax (gon"e-aw'laks)

Porphyra (por'fi-rah)

Psilocybe (si"lo-si'bee)

Phytophthora (fi"to-fi'tor-ah)

Penicillium (pen"i-sil'e-um)
  chrysogenum (kry"sog-ah'num)
  roqueforti (rok-fort'i)

Trichomonas (trick-oh-moan'us)

Balantidium (bal"an-tid'e-um)

Plasmodium (plaz-mo'de-um)

Saccharomyces (sak"ah-ro-mi'seas)
  cerevisiae (ser"e-viz'e-eye)

Phycomycetes (fi"ko-mi-se'tez)

Mucor (mu'kor)

Rhizopus (rise'oh-pus)

Ascomycetes (as"ko-mi-se'tez)

Basidiomycetes (bah-sid"e-o-mi-se'tez)

Deuteromycetes (du"ter-o-mi-se'tez)

Amanita (am"ah-ni'tah)

Aspergillus (as"per-jil'us)

Sarcodina (sar"ko-di'nah)

Entamoeba (en"tah-me'bah)
  histolytica (his-to-li'tik-ah)

Hexamita (hek-sam'i-tah)

Trypanosoma (tri"pan-o-so'mah)

Toxoplasma (toks"o-plaz'mah)

Basic Cell Chemistry

DIFFERENCES AMONG ELEMENTS
COMBINING ATOMS
OF LEMONS AND LYE
CHEMICAL REACTIONS
GETTING IT GOING
CARBON-CONTAINING MOLECULES
SHAPE IS IMPORTANT
NUCLEIC ACID
CODING FOR PROTEINS
LIFE AND DEATH OF CELLS

Learning Objectives

☐ Describe the structure of atoms and molecules in relation to the energy of chemical bonds.

☐ Be able to list some of the characteristics of acidic, basic, and neutral substances in addition to carbohydrates, proteins, lipids, and nucleic acids.

☐ Understand the Laws of Thermodynamics and be able to apply them to chemical reactions, namely, oxidation-reduction.

☐ List the basic subunits that make up proteins, nucleic acids, carbohydrates, and lipids.

☐ Describe the process of protein synthesis.

☐ Be familiar with several types of chemical reactions.

☐ Understand the importance and the events involved in the DNA duplication process.

All the matter you can think of—the sidewalk, water, a cell—is made up of one or more types of substances called **elements.** Many of these elements are already familiar to you—oxygen, iron, aluminum, silver, carbon, and gold. Elements can be subdivided into even smaller pieces, the smallest of which is an **atom.** The atom consists of a central region called the **atomic nucleus,** which is surrounded by smaller parts that fly about it at predictable distances (Figure 3.1). The nucleus is composed of two types of particles, **neutrons** and **protons.** The particles flying about the nucleus are called **electrons.** Neutrons have no electrical charge, protons have a positive electrical charge, and electrons have a negative charge. An atom has an equal number of protons and electrons, so the number of positive charges equals the number of negative charges; thus, atoms have no overall electrical charge (Table 3.1).

The text, tables, and illustrations of this chapter originally appeared in *Concepts in Biology,* 3rd ed. Eldon D. Enger et al. Copyright © 1982, 1979, 1976 by Wm. C. Brown Company Publishers, Dubuque, Iowa. Reprinted by permission.

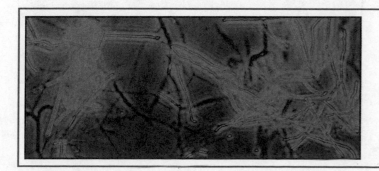

3

DIFFERENCES AMONG ELEMENTS

The atoms of each kind of element have a specific number of protons. For example, oxygen always has eight protons, and no other element has that number. Carbon always has six protons. The **atomic number** of an element is based on the total number of protons in an atom of that element; therefore, each element has a unique atomic number. Since oxygen has eight protons, its atomic number is eight. Every oxygen atom will have eight protons, and only oxygen atoms will have eight protons. Looking at it from another point of view, since nitrogen has the atomic number seven, you can conclude that nitrogen has seven protons, but also that it has seven electrons because the number of electrons and protons is the same (Table 3.2).

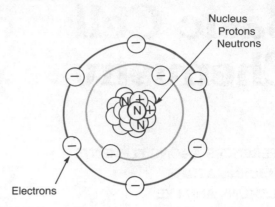

FIGURE 3.1

An oxygen atom is composed of eight protons, eight neutrons, and eight electrons.

TABLE 3.1
Comparison of atomic particles.

	Protons	Electrons	Neutrons
Location	Nucleus	Outside nucleus	Nucleus
Charge	Positive (+)	Negative (−)	None (neutral)
Number Present	Identical to the atomic number	Equal to number of protons	Mass number minus atomic number
Mass	1	1/1836 proton mass	1

Although all atoms of the same element have the same number of protons, they do not always have the same number of neutrons. In the case of oxygen, over 99 percent of the atoms will have eight neutrons, but there will be others with more or less than this number. These atoms of an element that differ only in the number of neutrons they contain are called **isotopes.** The **mass number** of an atom is customarily used to compare different isotopes of the same element. The mass number is found by adding the number of neutrons and protons in the particular atom you are considering. Therefore, you can speak of an isotope of oxygen with a mass number of sixteen, or another isotope of oxygen with a mass number of seventeen (Figure 3.2).

Since isotopes differ in their structure, some isotopes will show characteristics that are different from the most common form of the element (Table 3.3). For example, there are two isotopes of carbon, ^{12}C is the most common isotope and has a mass number of 12. ^{14}C differs in mass number, and it is also radioactive. **Radioactive** isotopes are unstable and break down into smaller and more stable atoms. The radioactive isotope of carbon releases an amount of energy large enough to be recognized as a change on photographic film or a Geiger counter. It is possible to follow the movement of carbon through the metabolic pathways of microbes by using the radioactive form of carbon as a nutrient source. As the carbon-containing molecules are

TABLE 3.2
Table of protons and electrons.

Elements	No. of Protons = (+) Atomic No.	No. of Electrons = (−)
Carbon (C)	6	6
Nitrogen (N)	7	7
Oxygen (O)	8	8
Sodium (Na)	11	11
Phosphorus (P)	15	15
Chlorine (Cl)	17	17
Potassium (K)	19	19
Calcium (Ca)	20	20

TABLE 3.3
Comparison of isotopes of oxygen and iodine.

	^{16}O	^{17}O	^{127}I	^{131}I
Protons	8	8	53	53
Electrons	8	8	53	53
Neutrons	8	9	74	78
Mass Number	16	17	127	131

chemically changed to meet the needs of the cell, the location of the radioactivity will change from one type of organic molecule to another within the cell (Figure 3.3).

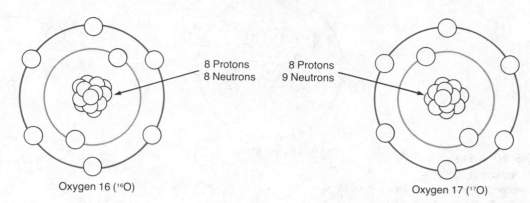

Nuclei

8 Protons
8 Neutrons

8 Protons
9 Neutrons

Oxygen 16 (^{16}O)

Oxygen 17 (^{17}O)

FIGURE 3.2
Two isotopes of oxygen: note that they differ in the number of neutrons.

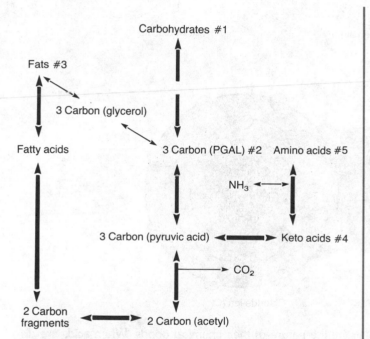

Carbohydrates #1

Fats #3

3 Carbon (glycerol)

Fatty acids

3 Carbon (PGAL) #2   Amino acids #5

NH$_3$

3 Carbon (pyruvic acid) ◀━━▶ Keto acids #4

CO$_2$

2 Carbon fragments ◀━━▶ 2 Carbon (acetyl)

FIGURE 3.3
The radioactive isotope of carbon is ^{14}C. This numbered pathway shows how the carbon moves from its introduced form, carbohydrate (#1), to a three-carbon molecule, **PGAL** (#2), on to fats (#3), and finally to amino acids (#5) produced within the cell. The sequence can be determined by killing the cell quickly after the ^{14}C is given and analyzing the various types of molecules for radioactivity.

COMBINING ATOMS

Many atoms combine with each other to form **molecules.** For example, oxygen atoms combine with iron atoms to form molecules of iron oxide, or rust. Molecules are units of matter made up of two or more atoms. When molecules are composed of atoms from different kinds of elements, these combinations are called **compounds.** Salt, water, and carbon dioxide are all compounds.

The forces that combine atoms into molecules are called **chemical bonds**, and these bonds vary in strength. The two types of chemical bonds that hold atoms together to make molecules are **covalent bonds** and **ionic bonds.** The covalent bond is formed when two atoms share a pair of electrons. Figure 3.4A shows the separate diagrams of two hydrogen atoms and one oxygen atom. Figure 3.4B shows the oxygen atom sharing two electrons with each of the two hydrogen atoms. Each chemical bond should be thought of as belonging to each of the atoms involved. Covalent bonds are analogous to people shaking hands—the people represent the atoms, and the handshake is the bond. Generally, this sharing of a pair of electrons is represented by a single straight line between the atoms $\left(\begin{smallmatrix}H & & H \\ & O & \end{smallmatrix}\right)$

An ionic bond is formed when electrons are actually transferred from one atom to another. An atom with an extra electron is a negatively charged particle since it no longer has an equal number of positive and negative charges. Similarly, the atom losing an electron has lost a negative charge and, therefore, has one more positive charge than negative charge. An atom with either a negative or a positive charge is called an **ion.** The ion that has gained an electron is a negatively charged ion. The other, which has lost an electron, is a positively charged ion. Since opposite charges attract each other, an ionic bond is formed between these oppositely charged ions. Figure 3.5 shows this relationship between a sodium ion and a chloride ion in common table salt. Of course, it is possible for atoms to gain or lose *more* than one electron, in which case, the ions total charge reflects the number of electrons gained or lost.

Still another kind of attraction is one called a **hydrogen bond,** although these bonds are typically very weak. These bonds may be formed between hydrogen and molecules that contain an atom that attracts electrons better (is electronegative) than one of the atoms that is already attached to it. The lopsided attraction tends to keep the negatively charged electrons moving around one end of the molecule a great portion of the time and creates a molecule

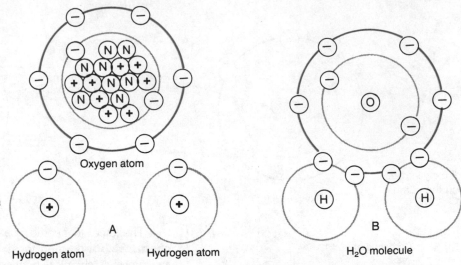

FIGURE 3.4
(A) A water molecule is composed of one oxygen atom and two hydrogen atoms. (B) One oxygen atom is covalently bonded to two hydrogen atoms.

Oxygen atom

Hydrogen atom   A   Hydrogen atom

H_2O molecule

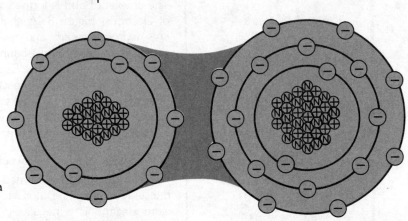

FIGURE 3.5
Table salt, sodium chloride, is a molecule bonded by the attraction between a positive ion and a negative ion.

Sodium ion (Na^+)   Chloride ion (Cl^-)

that has weak magnetic poles. Molecules that take this form are called **polar.** Two of the most common elements that are important in hydrogen bonding are oxygen and nitrogen (Figure 3.6). The weak magnetic force *between molecules* determines important characteristics of the many materials in which they are formed. Water, proteins, nucleic acids, and other biologically important molecules have internal hydrogen bonds.

> **1** An atom that has gained an electron becomes known as what type of particle?
> **2** How does a molecule become polar?
> **3** How does an element become radioactive?

OF LEMONS AND LYE

Acids and **bases** constitute another class of biologically important compounds, and their characteristics are determined by the nature of their chemical bonds. When acids are dissolved in water, hydrogen ions (H^+) are set free to float around. A hydrogen ion is positive because the atom has lost its electron and now has only the positively charged proton. Compounds that behave this way are called acids. Two common examples of acids are sulfuric acid (H_2SO_4), which is used in your car's battery, and acetic acid (CH_3COOH), which is present in vinegar.

Some other compounds when dissolved in water release hydroxyl ions (OH^-). Each of these ions is negatively charged because it consists of an oxygen atom, a hydrogen atom, and an additional electron. Some cans of drain cleaner contain the strong base caustic soda (NaOH). If you have read the directions on the can, you know that such a base can be as dangerous as a strong acid. Some acids and bases are stronger than others, and the strength of an acid or a base is represented by a number called its **pH.** A pH of seven indicates that the solution is neutral and that an equal number of hydrogen and hydroxyl ions are floating around. As the pH numbers get smaller, the strength of the acid increases, and more hydrogen ions (H^+) are released. Sim-

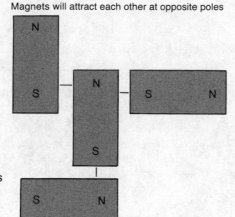

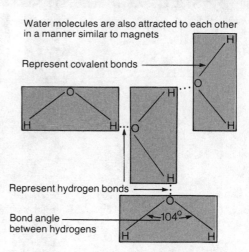

Magnets will attract each other at opposite poles

Water molecules are also attracted to each other in a manner similar to magnets

Represent covalent bonds

Represent hydrogen bonds

Bond angle between hydrogens

104°

FIGURE 3.6
Hydrogen bonds cause molecules to orient themselves with respect to each other. These bonds are represented by three dots (. . .).

ilarly, as the pH numbers increase, the strength of the base increases. Notice the pH values of the various substances in Figure 3.7. This illustration shows that a strong acid, such as sulfuric acid, has a high concentration of hydrogen ions and a low concentration of hydroxyl ions. This balance between hydroxyl and hydrogen ions can be produced either by adding hydrogen ions or trapping hydroxyl ions. For this reason an acid can be a molecule that not only releases hydrogen ions, but also captures hydroxyl ions. The opposite situation is true for bases. A base can be a molecule that either releases hydroxyl ions or captures hydrogen ions and removes them from the solution.

CHEMICAL REACTIONS

Strong acids and bases have the ability to enter into situations in which parts of molecules are exchanged. This rearrangement of parts is called a **chemical reaction.** Perhaps when you were younger you tried stirring together a bunch of chemicals with the hope that something spectacular would happen. Only certain types of reactions are possible, and it is fortunate for those of you who have tried this that most of your "experiments" did not react. A commonly seen

chemical reaction is the burning of natural gas to produce heat and light (Figure 3.8). This burning involves a breaking of the chemical bonds in the gases methane and oxygen, followed by the formation of new bonds in the carbon dioxide and water molecules. Since the energy found in the chemical bonds of the carbon dioxide and water molecules is less than the energy that was present in the methane and oxygen molecules, the excess energy is released in the form of heat and light. Such energy-yielding reactions are called **exergonic,** or **exothermic.** An exothermic reaction in "bioluminescent" bacteria and fungi occurs when molecules are rearranged by a series of chemical reactions and light is released; however, very little heat is produced in these cases. The glowing spots on many of the deep-sea fish are from such exothermic reactions by the bacteria that live on the skin of these fish. Other reactions do not release excess energy but require an input of energy to enable the reactions to occur. Such energy-requiring reactions are called **endergonic,** or **endothermic.** An example of this type of reaction is photosynthesis, in which sunlight energy combines carbon dioxide with water to form sugar.

In each of these examples, certain starting molecules are changed during the chemical reaction to form new arrangements of atoms. The molecules that enter the reaction

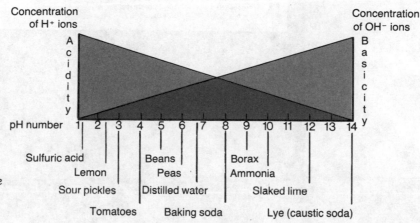

Concentration of H⁺ ions

Concentration of OH⁻ ions

A c i d i t y

B a s i c i t y

pH number   1 2 3 4 5 6 7 8 9 10 11 12 13 14

FIGURE 3.7
The concentration of hydrogen ions is greatest at low pH, while the concentration of hydroxyl ions is greatest at high pH.

Sulfuric acid
Lemon
Sour pickles
Tomatoes
Beans
Peas
Distilled water
Baking soda
Borax
Ammonia
Slaked lime
Lye (caustic soda)

TRADITIONALLY elements are represented in a shorthand form by letters rather than writing out the complete name of the molecule (i.e., water = H_2O), which means that a molecule of water consists of two atoms of hydrogen and one atom of oxygen. These chemical symbols can be found on any Periodic Table of Elements. The Periodic Table is useful to us in that we can determine the number and position of the various parts of atoms by using information it contains. Also the arrangement of the table is such that atoms number 3, 11, and 19, etc. are underneath each other in column one. The atoms in this column all act in a similar way since they all have one electron in their outermost layer. In the next column, Be, Mg, Ca, etc. all act alike because these metals all have two electrons in their outermost electron layer. Similarly atoms number 9, 17, and 35, etc. all have seven electrons in their outer layer. Knowing how fluorine, chlorine, and bromine act, you can probably predict how iodine will act under similar conditions. At the far right in the last column, helium, neon, etc. all act alike. They all have eight electrons in their outer electron layer. Atoms with eight electrons in their outer electron layer seldom form bonds with other atoms.

BOX 2
PERIODIC TABLE

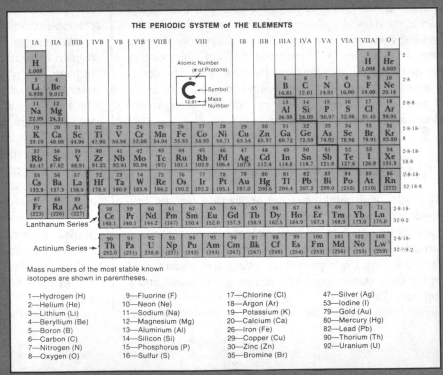

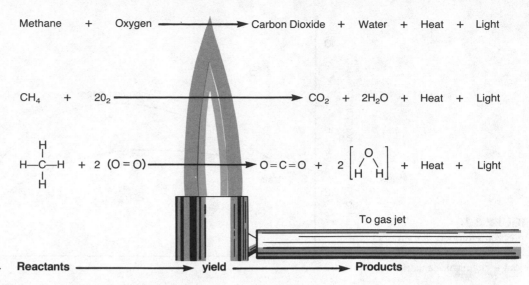

FIGURE 3.8
Oxidation of methane is one example of a chemical reaction.

are called the **reactants,** and the newly formed molecules are called the **products** of the reaction. As the reaction takes place, the amount of reactant decreases and the amount of product increases. As long as the new products are in close contact with one another, they, too, may react to again form the original molecules. This type of back-and-forth chemical reaction results in a balance of the reactants and products. When this balance is reached, the reaction is said to be in **chemical equilibrium** (Figure 3.9). However, if any one or all of the products are removed or lost during the reaction, it will continue until all of the reactants have been changed into product molecules. In photosynthesis, the reaction results in the formation of oxygen molecules that leave the cell, thereby preventing the conversion of sugar and oxygen back to carbon dioxide and water. This results in the accumulation of sugar molecules, which are used in growth, reproduction, and other cell activities.

GETTING IT GOING

Most chemical reactions require an initial input of energy, called **activation energy,** to get them started. Some reactions occur naturally at room temperature, since there is sufficient activation energy present. Hydrogen peroxide normally breaks down into water and oxygen ($2H_2O_2 \longrightarrow 2H_2O + O_2$). You have probably experienced this reaction when some hydrogen peroxide was placed on a cut to prevent infection. The bubbles that came off were oxygen bubbles. The rate of this reaction can be increased when the chemical manganese dioxide (MnO_2) is mixed with the hydrogen peroxide. The manganese dioxide is a **catalyst,** an inorganic substance that speeds up this chemical reaction and is recovered unchanged.

If you were to put some saliva into a test tube of hydrogen peroxide, there would also be a rapid increase in

Energy + $6CO_2$ + $6H_2O$ ⇌ $C_6H_{12}O_6$ + $6O_2$

Reactants **Products**

Energy + 6 carbon dioxide + 6 water ⇌ Sugar + 6 oxygen

Products **Reactants**

FIGURE 3.9
Most reactions are reversible; that is, they can proceed in either direction. The direction depends on the energy that is available and the relative amounts of materials involved. (From Eldon D. Enger et al, *Concepts in Biology,* 2nd ed. © 1979, 1976 by Wm. C. Brown Company Publishers, Dubuque, Iowa. Reprinted by permission.)

the breakdown of hydrogen peroxide. This occurs because the saliva contains an enzyme (see Chapter 1) that interacts with hydrogen peroxide much like manganese dioxide does. Enzymes are types of catalysts made of protein that speed up the rate of chemical reactions. The genes of the cell control the production of enzymes.

For a substrate to be changed into the end product, a certain amount of activation energy must be supplied to start the reaction. For instance, the rate of a chemical reaction may be increased by applying more heat. But, in a living cell, the addition of large amounts of heat would raise the temperature to a point that would result in the death of the cell. Enzymes allow the rate of a reaction to increase without the addition of heat energy. Thus, an enzyme can bring about the desired increase in reactions without having a destructive increase in the temperature of the cell.

In Figure 3.10, the original substrate molecule contains four units of energy and the end products contain one unit of energy. In order for this reaction to start, energy must be added to make the substrate more reactive. Without an enzyme, three units of energy would have to be added to the

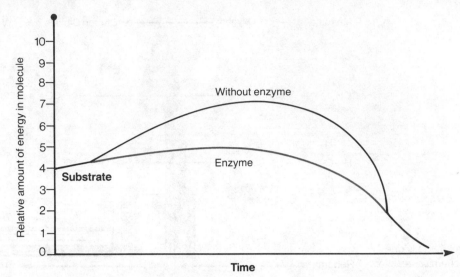

FIGURE 3.10

An enzyme reduces the amount of energy needed to bring about a chemical reaction.

substrate to get this particular reaction started; however, with an enzyme only one unit of energy is needed. The peak of the top line in Figure 3.10 represents the energy needed to start the reaction without an enzyme. The peak of the lower line represents the energy needed to start the reaction in the presence of an enzyme. Thus, an enzyme speeds up a reaction by lowering the amount of activation energy required to start the reaction. Enzymatic reactions may aid in breaking down a substrate into smaller pieces, or the reaction may be one in which small molecules are bonded to one another forming a more complex molecule.

---

1 Why is bioluminescence an example of an exergonic reaction?

2 What does activation energy have to do with catalysts?

3 How do catalysts affect the rate of chemical reactions?

Let us consider some common types of chemical reactions that happen in living things. Such a process is **dehydration synthesis,** which results in the joining of two small molecules to make one larger molecule with the removal of a molecule of water. The water molecule from this dehydration-synthesis reaction was not originally present as a molecule of water, but was manufactured from a hydro-

gen from one of the smaller molecules and a hydroxyl group from the other molecule, as shown in Figure 3.11. In this reaction, the H^+ from one molecule and OH^- from the other molecule join to form the molecule of water. This is the sort of reaction that happens when a bacterial cell produces organic compounds for storage. The opposite kind of reaction is called **hydrolysis.** In this reaction, a large molecule is split into two smaller molecules by the addition of a water molecule. This kind of reaction takes place when the stored organic material is utilized by the cell for energy or building materials (Figure 3.12).

Still another kind of reaction that is important to cells involves the transfer of electrons. This reaction may result in the combination of elements or the breakdown of large molecules. Another term for electron-transfer reaction is **oxidation-reduction** reaction. The most important characteristic of *redox* (*reduction + oxidation*) reactions is that energy-containing electrons are being transferred from one molecule to another. These reactions enable cells to produce useful chemical-bond energy in the form of ATP (cellular respiration) and to synthesize the energy-containing bonds of carbohydrates (photosynthesis). Oxidation means the loss of electrons, and reduction means the gain of electrons. (Do not associate oxidation with oxygen, many different elements may enter into redox reactions.) Molecules that lose electrons (or serve as electron donors) in a redox reaction usually release this chemical-bond energy and are broken down into more simple molecules. Molecules that

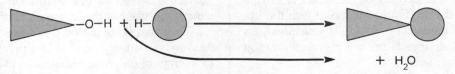

FIGURE 3.11

A dehydration-synthesis reaction occurs when two molecules are bonded together. H^+ is removed from one molecule and OH^- is removed from the other. The H^+ and OH^- ions bond together to form water.

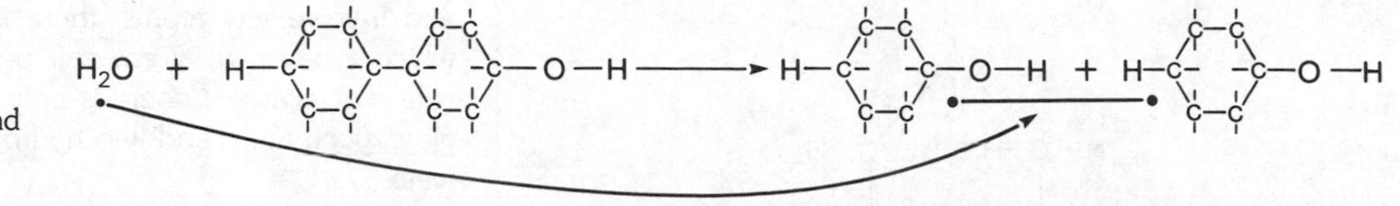

FIGURE 3.12
Hydrolysis is the opposite of dehydration synthesis. H^+ and OH^- from water split a molecule into two pieces.

gain electrons (or serve as electron acceptors) in a redox reaction usually gain electron energy and are enlarged, forming a more complex molecule (Figure 3.13). Since electrons cannot exist apart from an atomic nucleus for a long period, both oxidation and reduction occur in a redox reaction; whenever an electron is donated, it is quickly gained by another molecule. A simple way to help identify a redox reaction is to use the mnemonic memory phrase "L.E.O. the lion says G.E.R." L.E.O. stands for "*loss of electrons is oxidation*" and G.E.R. stands for "*gain of electrons is reduction.*" An additional helpful fact is that many important redox reactions in microbiology (e.g., cellular respiration and photosynthesis) involve the transfer of hydrogen electrons; therefore, in many situations, it is easier to follow the movement of hydrogen (H) from one molecule to another.

CARBON-CONTAINING MOLECULES

The chemistry of living things really centers around the chemistry of the carbon atom and a few other atoms that are able to combine with carbon. In order to understand some aspects of the structure and function of microbes, it is important to be exposed to some basic organic chemistry.

Microbes are composed of rather complex molecules that contain carbon atoms as basic building blocks. These are called **organic molecules.** In contrast to these complex carbon-containing molecules, most of the ones discussed earlier were **inorganic molecules.**

The original meaning of the terms "inorganic" and "organic" related to the fact that organic materials were thought either to be alive or to be produced only by living things. Inorganic materials were neither alive nor could they be produced by the aid of living things. Therefore, a very strong

---

1 Give an example of an oxidation-reduction reaction.
2 Why are oxidation-reduction reactions important to cells?
3 What makes carbon a unique element?

---

link exists between organic chemistry and the chemistry of living things, called biological chemistry. Modern chemistry has altered this original meaning considerably, since it is now possible to manufacture unique organic molecules that cannot be produced by living things.

All organic molecules, whether they are natural or manmade, have certain common characteristics. The carbon atom, which is the central atom in all organic molecules, has some unusual properties and is unique in that it can combine with other carbon atoms to form long chains:

$$-C-C-C-C-C-C-C-C-$$

In many cases, the ends of these chains may join together to form ring structures (Figure 3.14). Only a few other atoms have this ability. What really is unusual is that these bonding sites are all located at equal distances from one another. If you were to take a rubber ball and stick four nails into it so that they were equally distributed around the ball, you would have a good idea of the geometry involved.

Carbon atoms usually bond covalently. Compare carbon, which can form four covalent bonds, with nitrogen, which can form three covalent bonds; oxygen, which can form two covalent bonds, and hydrogen, which can form only one covalent bond. Since carbon has four places where it can bond, the carbon atom can have up to four other atoms combined with it. This is the case with the methane molecule, which has four hydrogen atoms attached to a single carbon atom, and is a colorless, odorless gas usually found in natural gas (Figure 3.15).

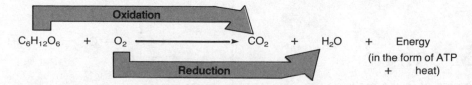

FIGURE 3.13
Oxidation-reduction is illustrated here by aerobic cellular respiration. Notice that the more complex glucose molecule ($C_6H_{12}O_6$) is being broken down as it donates electrons (H) to oxygen (O_2) in the oxidation process. The oxygen is gaining the electrons (H) and is becoming reduced to a larger hydrogen-containing molecule, water (H_2O). The energy stored in the sugar is released and used to produce the molecule ATP or is released as waste heat.

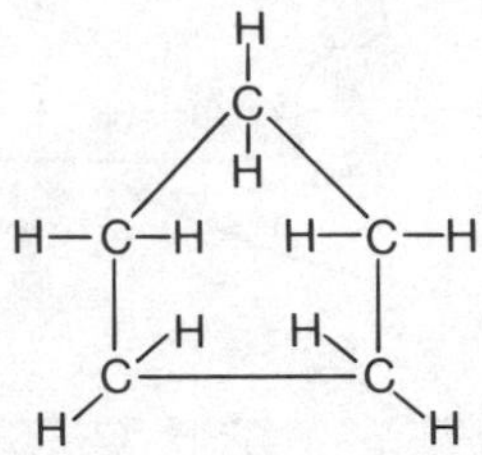

FIGURE 3.14
The ring structure shown here is formed by joining the two ends of a chain of carbon atoms.

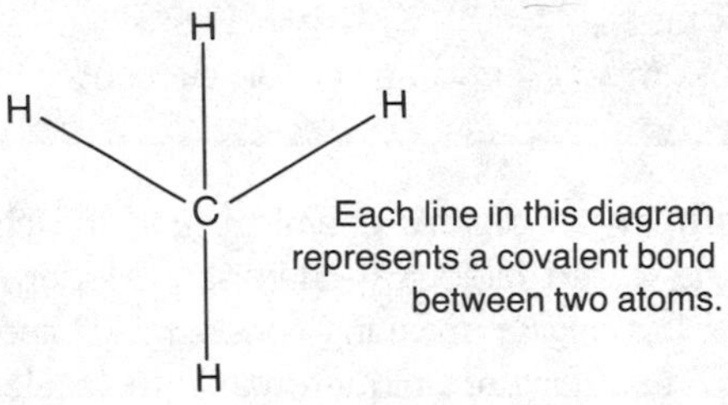

FIGURE 3.15
A methane molecule is composed of one carbon atom bonded with four hydrogen atoms. These bonds are formed at the bonding sites of the carbon.

Some atoms may be bonded to a single carbon atom more than once to result in a slightly different arrangement of bonds around the carbon atom's center. Generally, this kind of covalent bond is referred to as a double bond and is denoted in Figure 3.16 by the two lines ($=$) between the carbon and oxygen atoms $\left(\begin{array}{c} | \\ -C=O \end{array}\right)$. The two carbon atoms in Figure 3.16 each share four bonds, and each of the four hydrogen atoms shares a single bond. The oxygen atom is double bonded with the carbon atom since two pairs of electrons are being shared between the two atoms. Only one pair of electrons is being shared between the hydrogen and carbon atoms. Although most atoms can be involved in the structure of organic molecules, a few are most commonly found—hydrogen (H) and oxygen (O) are almost always present, while nitrogen (N), sulfur (S), and phosphorus (P) are also very important in specific types of organic molecules. Most organic molecules are quite large and are called macromolecules. The size of organic molecules is directly related to the fact that carbon atoms are able to form long chains or rings.

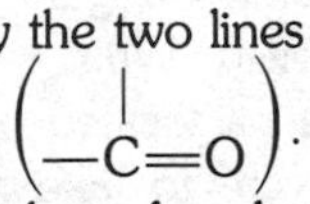

SHAPE IS IMPORTANT

The types of atoms present in the molecule are important in determining the properties of the molecule. The three-dimensional arrangement of the atoms within the molecule is also important. Since most inorganic molecules are small

and involve few atoms, there is usually only one way in which a group of atoms can be arranged to form a molecule. For example, there is only one arrangement for a single oxygen atom and two hydrogen atoms in a molecule of water.

In a molecule of sulfuric acid, there is only one arrangement for the one sulfur atom, the two hydrogen atoms, and the four oxygen atoms.

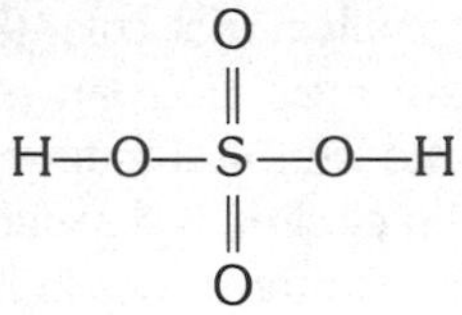

However,

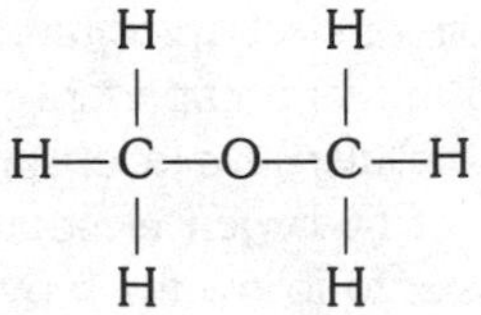

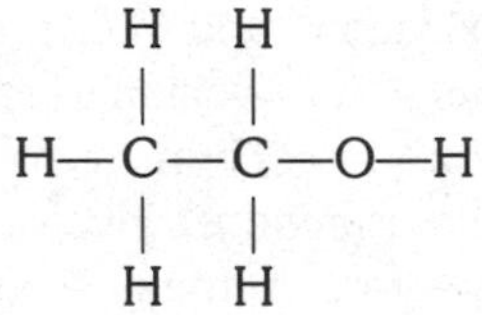

both contain two carbon atoms, six hydrogen atoms, and one oxygen atom, but they are quite different in the arrangement of atoms and in the chemical properties of the molecules.

While 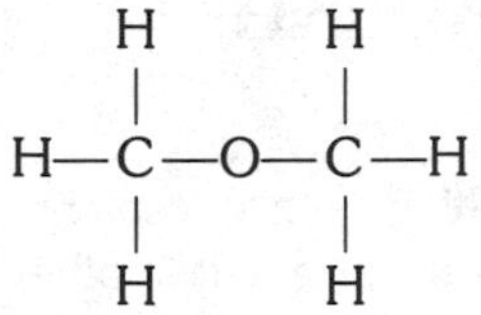 is an ether,

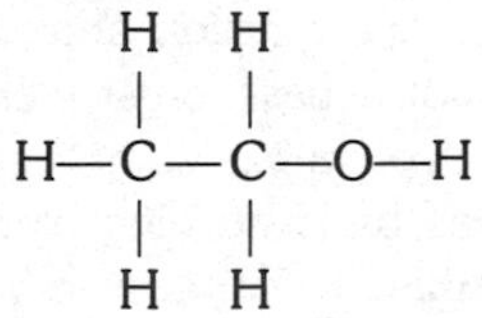

is an alcohol

—the kind of alcohol found in beer, wine, and liquor. Since the ether and the alcohol molecules both have the same number and kinds of atoms, they are said to have the same chemical formula, which in this case is written C_2H_6O. This formula simply indicates the number of each kind of atom within the molecule. However, it is obvious that their **structural formula** is not the same (Box 3). As the number of

FIGURE 3.16
These diagrams show two different ways of illustrating double bonds. Each shows an oxygen atom sharing two bonds (pairs of electrons) with a carbon atom.

FREQUENTLY structural formulas are simplified by removing the carbons and just connecting lines at the points where the carbons would have been.

For example:

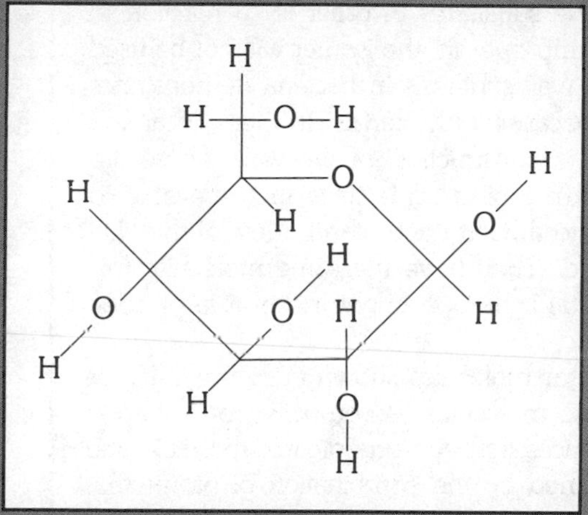

can also be used to indicate a sugar like glucose.

**BOX 3
STRUCTURAL
FORMULA**

carbon atoms increases, the number of different structural arrangements increases. For example, there are at least five different structural formulas for all the molecules having the same chemical formula, $C_6H_{12}O_6$ (Figure 3.17).

Molecules that have the same **molecular formula** (tells how many of each kind of atom are in a molecule) but differ in their structural formula are called **isomers.** Isomers can be recognized by microbes since their enzymes must geometrically "fit" the substrates in order for a reaction to occur. If there is an improper fit, the isomer will not be used in the reaction. Cell-wall synthesis in bacteria demonstrates this specialization, because only certain isomers of amino acids are used in the construction of the wall. These are identified as "L" form amino acids since they are able to bend polarized light to the left (levorotary). Most of the "D" isomers are not used. They have the same molecular formula but will bend the light to the right (dextrorotary). (See Box 4.)

Each of the sugar molecules shown in Figure 3.18, as well as other organic molecules, is composed of a carbon skeleton. The differences between one organic molecule and another are determined by the combination of atoms that are chemically bonded to the carbon skeleton. Specific combinations of atoms are called **functional groups** since their presence on the carbon skeleton determines the chemical activities of the molecule. By learning to recognize functional groups, it is possible to identify an organic molecule and know something about its activities. There are only a few functional groups of importance to us at this point:

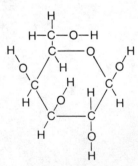

None of these exists as an individual unit; each is a part of an organic molecule.

Since there is such a large number of different organic molecules, it is helpful to organize them into groups on the

---

1 What makes one organic molecule different from another?
2 How does a molecular formula differ from a structural formula?

FIGURE 3.17
Some structural arrangements of six-carbon sugars. Each of these molecules has the same number of carbon, hydrogen, and oxygen atoms; but they are arranged differently.

Glucose $C_6H_{12}O_6$

Mannose $C_6H_{12}O_6$

Galactose $C_6H_{12}O_6$

Sorbose $C_6H_{12}O_6$

Fructose $C_6H_{12}O_6$

BOX 4
COMMON AMINO ACIDS

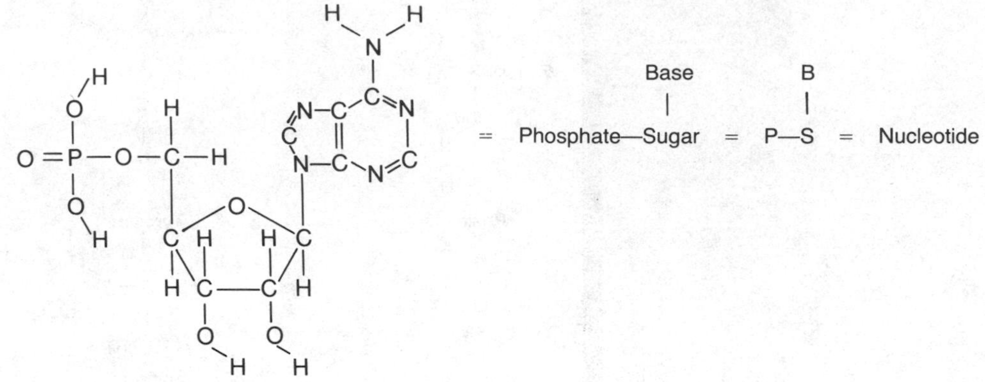

FIGURE 3.18
All nucleotides are constructed in this basic way. The nucleotide is the basic structural unit of all nucleic acid molecules.

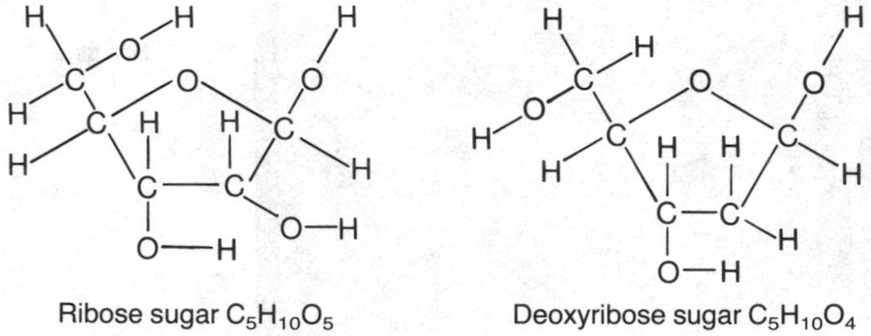

basis of the similarity of structure and/or the chemical properties of the molecules. Microbes are made up of a large variety of molecules, therefore, it is important to be familiar with a few types of organic molecules that are especially common or have particular significance to microbiology. The proteins, carbohydrates, and lipids all contain carbon, hydrogen, and oxygen, but differ in the other elements they contain and the specific arrangement.

NUCLEIC ACIDS

Like the other groups of organic macromolecules, the nucleic acids are made up of specific chemical elements, namely, carbon, oxygen, hydrogen, nitrogen, and phosphorous. These elements are arranged into basic types of building blocks called **nucleotides,** each of which is composed of a sugar molecule (S) containing five carbon atoms, a **phosphate** group (P)[*], and a kind of molecule called a **nitrogenous base** (B) (Figure 3.18).

There are eight common types of nucleotides available in a cell for building larger nucleic acids. Nucleotides differ in the kind of sugar and nitrogenous base they contain, and due to these differences, it is possible to classify the nucleic acids into two main groups: ribonucleic acid (RNA) and deoxyribonucleic acid (DNA). The name of each gives information about the structure of the molecules. For example, the prefix "ribo-" in RNA indicates that the sugar part of this nucleic acid is ribose. Similarly, DNA contains a ribose sugar that has been *deoxygenated* (lost an oxygen atom) and is called deoxyribose (Figure 3.19). The nucleotide units contain nitrogenous bases that are either large or small. The larger ones are called **purines** and are adenine (A) and guanine (G), which differ in the kinds of atoms attached to their double-ring structure (Figure 3.20). The smaller molecules are called **pyrimidines** and are cytosine (C), thymine

[*]The phosphate group will be represented in the illustrations by a single capital (P).

FIGURE 3.19
Notice the difference between the structures of the two nucleic acid sugars, ribose and deoxyribose.

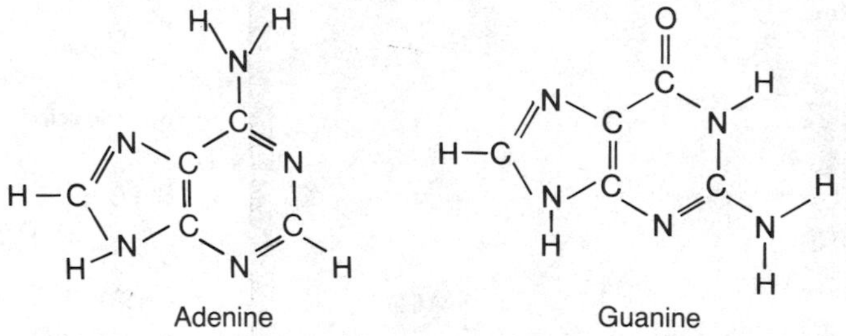

Ribose sugar $C_5H_{10}O_5$ Deoxyribose sugar $C_5H_{10}O_4$

FIGURE 3.20
These are the two large nitrogenous base molecules, adenine and guanine. Note their structural differences. This group is known as the purines.

Adenine Guanine

(T), and uracil (U). Each differs from the other by the atoms attached to its basic single-ring structure (Figure 3.21). These differences in size are important, as you will see later. Table 3.4 shows the basic differences between the makeup of RNA and DNA.

Figure 3.22 illustrates an example of a nucleotide commonly found in cells. These nucleotide building blocks can be bonded to one another in a specific manner by a dehydration synthesis reaction involving a specific enzyme. The result of this nucleic acid synthesis is a long-chain macromolecule, which resembles a comb. The protruding "teeth"

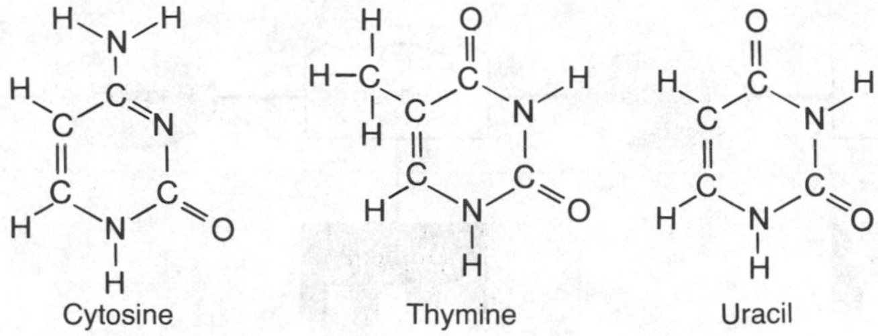

FIGURE 3.21
The three small nitrogenous bases (cytosine, thymine, and uracil) used in nucleic acid molecules. This group is known as the pyrimidines.

TABLE 3.4
Comparison of DNA and RNA. The composition of the RNA molecule differs from that of DNA in both the categories of base type and sugar type.

Nucleic Acid Type	RNA	DNA
Base type	AGCU	AGCT
Acid type	Phosphoric acid	Phosphoric acid
Sugar type	Ribose	Deoxyribose

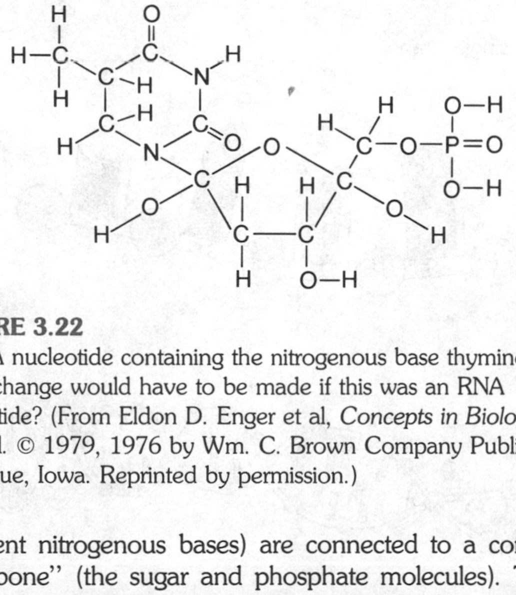

FIGURE 3.22
A DNA nucleotide containing the nitrogenous base thymine. What change would have to be made if this was an RNA nucleotide? (From Eldon D. Enger et al, *Concepts in Biology*, 2nd ed. © 1979, 1976 by Wm. C. Brown Company Publishers, Dubuque, Iowa. Reprinted by permission.)

(different nitrogenous bases) are connected to a common "backbone" (the sugar and phosphate molecules). This is the basic structure of both RNA and DNA (Figure 3.23). Notice in Figure 3.24 that it is possible to make sense out of the sequence of nitrogenous bases. If read from left to right in groups of three, three "words" are seen: CAT, ACT, and TAG. It is possible to "write" a message in the form of a stable DNA molecule by combining different nucleotide units in particular sequences. The four DNA nucleotides can be used as an alphabet and are restricted to three-letter words. Realize, also, that in order to make sense out of such a code, it is necessary to read in a consistent direction: Reading the sequence in reverse doesn't always make sense. DNA language is the language of the cell. The coded DNA serves as a central library; DNA contains all the information the cell needs to sustain, grow, and reproduce itself. To accomplish all these processes, the library (DNA) must contain all the specifications necessary to manufacture the tools needed to perform every chemical job of the cell. If such information is missing or inaccurate, the cell will not work properly and may even die. Enzymes, the tools that control the chemical reactions of the cell, sometimes are constructed from DNA specifications. For this reason, DNA is called the "blueprint of life." The necessary specification for making a particular enzyme is called a **structural gene.** In most cases, structural genes are chemically linked to form a large threadlike molecule. Each strand may have tens or hundreds of genes linked end to end.

1 What is the difference between −OH and OH⁻?
2 List the purines.
3 List the nucleotides.

CODING FOR PROTEINS

The blueprint for any essential tool must be protected from damage or the job for which it is to be used might never be completed. Such protection could be accomplished in a number of ways. For example, a protective cover could be placed over the valuable blueprint and then rolled into a tube. Another method to protect the blueprint would be to use a copy instead of the original.

Living systems have adopted all of these protective measures to one degree or another, which, as a result, have better ensured their continued existence. The protective "cover" over the genetic material takes the form of a second long strand of DNA. This strand's structure is designed to provide maximum protection by forming a smooth, parallel molecular cover that can be removed easily when the blueprint needs to be read. The cover DNA is kept in place by the formation of weak hydrogen bonds between certain **complementary bases** of the gene and the cover. Three such bonds are formed between guanine and cytosine, and two such bonds are formed between adenine and thymine (Figure 3.25). This double-stranded DNA molecule is stabilized because its sides are parallel, just as a ladder with parallel sides is more stable than one having bulges or constrictions located at random from one end to the other. This parallel structure is maintained in the DNA molecule, since the large bases (G and A) always pair with the small bases

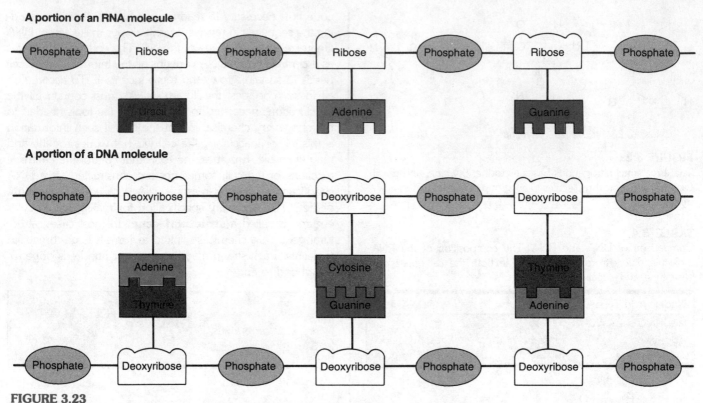

A portion of an RNA molecule

Phosphate — Ribose — Phosphate — Ribose — Phosphate — Ribose — Phosphate

Uracil   Adenine   Guanine

A portion of a DNA molecule

Phosphate — Deoxyribose — Phosphate — Deoxyribose — Phosphate — Deoxyribose — Phosphate

Adenine / Thymine   Cytosine / Guanine   Thymine / Adenine

Phosphate — Deoxyribose — Phosphate — Deoxyribose — Phosphate — Deoxyribose — Phosphate

FIGURE 3.23
DNA and RNA differ in structure in that DNA is a double strand and RNA is a single strand.

FIGURE 3.24
A single strand of DNA resembles a comb. The molecule is much longer than pictured here and is composed of a sequence of linked nucleotides.

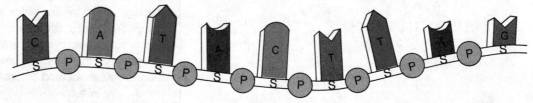

FIGURE 3.25
DNA is found in the three-dimensional form of a double helix. One strand is a chemical code while the other may be thought of as a cover strand of DNA.

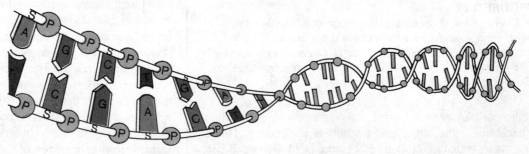

(C and T). A is complementary to T, and C is complementary to G. Referring again to our blueprint idea, the DNA becomes coiled or twisted when not in use. This provides protection for the molecule since the sugars and phosphates of the backbone cover the bases. This means that the most important part of the molecule, the chemical code sequence of bases, is tucked inside this **double helix** away from the potentially damaging effects of its environment. Another protective device involves the production of a copy of the blueprint. This is the first step in the process of using the blueprint information for **protein synthesis.**

For the cell to be efficient, it must produce all the chemicals required to sustain, grow, and reproduce itself. To do this the cell requires enzymes to operate on substrates. After much use, an enzyme becomes worn-out or damaged and must be replaced by another enzyme just like it. The DNA that codes for that enzyme is separated from the cover DNA. The DNA is not used directly by the cell to manufacture enzymes, but a copy of the gene is made to accomplish this. Many copies will be made if the cell has a great need for a specific enzyme. If the enzyme is rarely used, few copies will be made. This copying process is the first event in protein synthesis and is called **transcription.** It involves the four types of RNA nucleotides: A, G, C, and U. (Remember, there is no thymine in RNA, but the base uracil is found in its place.)

In order to guarantee accuracy, each nucleotide in the structural gene is copied or paired with an individual RNA nucleotide (Figure 3.26). These attach to one another by hydrogen bonding between the single RNA nucleotides and the nucleotides of the DNA strand. (Remember, the cover DNA has been removed and is now at a distance from the gene.) None of the RNA nucleotides base-pairs with the cover DNA nucleotides. When the copying process is complete, the RNA nucleotides are attached to one another, and they form a single strand of RNA. This strand is removed with the aid of an enzyme, and the blueprint is recovered and

coiled into the double helix. This newly formed molecule is RNA, not DNA. It is a copy of the message of the structural gene and is called **messenger RNA,** or **mRNA** (Figure 3.27).

---

1 What complementary base pairs enable a DNA double helix to be a smooth, parallel molecule?
2 What nucleotide differences exist between DNA and mRNA?

The mRNA molecule is a coded message written in nucleic acid language. This code must be read and the information used to assemble amino acids into protein. This second event in the process of protein synthesis is called **translation.** In order to translate mRNA language into protein language, it is necessary to have a dictionary. The translation must account for the fact that there are only four let-

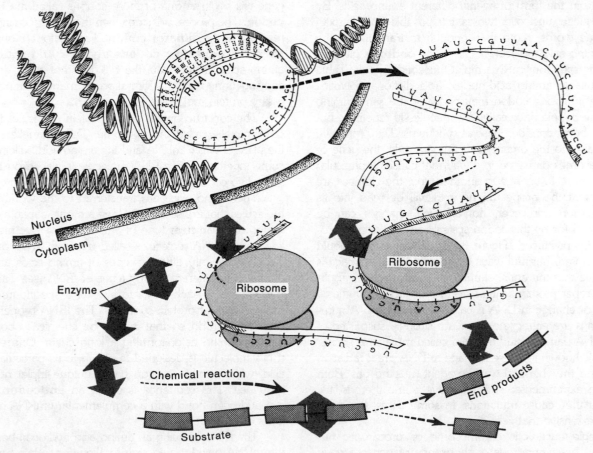

FIGURE 3.26
The assembly line of the cell is very complex and requires DNA, RNA, ribosomes, amino acids, and a number of specific enzymes to function efficiently. The cell is not just a sac of water in which occurs a random assortment of chemical reactions. (From Eldon D. Enger et al, *Concepts in Biology,* 2nd ed. © 1979, 1976 by Wm. C. Brown Company Publishers, Dubuque, Iowa. Reprinted by permission.)

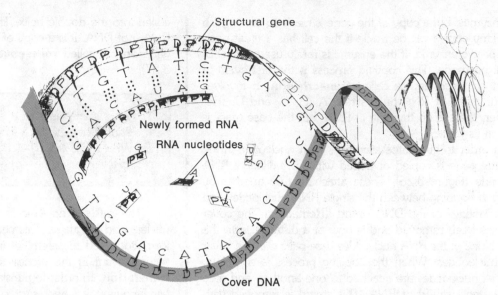

FIGURE 3.27
During the first stage of protein synthesis (transcription), individual RNA nucleotides base-pair on the chemical code of DNA to form an exact copy of the DNA nucleotide sequence.

ters in the nucleic acid alphabet, while the protein language has twenty in the form of twenty different amino acids. By using three-letter nucleotide words, it is possible to write sixty-four unique words, more than enough to translate for the twenty amino acid molecules. The name **codon** is given to each of the sixty-four mRNA triplet nucleotide words. Table 3.5 is a usable amino acid-nucleic acid dictionary. Notice that more than one codon may code for the same amino acid. Some people have called this needlessly repetitive, but you might better consider these as synonyms. They can have survival value to the organism. If, for example, the gene or mRNA becomes damaged in a way that causes a particular nucleotide base to change to another type, the chances are still good that the proper amino acid will be read into its proper position. However, not all such changes can be compensated for by this codon system, and an altered protein may be produced (Figure 3.28). Changes that would show up in very harmful ways can occur. Some damage is so extensive that the entire strand of DNA is broken, resulting in improper protein synthesis or a total lack of synthesis. This kind of change in DNA is called a **mutation.** A number of things are either known or are strongly suspected of causing DNA damage, and they are called mutagenic agents. Two agents known to cause damage to DNA are x-radiation (x-rays) and the chemical warfare agent mustard gas. Both have been experimented with extensively, and there is little doubt that they cause mutations. In some cases, the damage is so extensive that cells die.

A single eucaryotic chromosome or procaryotic nucleoid may have hundreds or thousands of genes aligned end-to-end. Therefore, it is important to know where the blueprint for one protein molecule ends and the next begins. The codon table shows that such "punctuation marks" do exist in the nucleic acid language. These are also co-

dons. The process of transcribing a mRNA from a particular gene will begin at one of these codons called the **initiator codon.** The process will stop when the newly forming mRNA reaches the **terminator codon.** Terminator codons are often called **nonsense codons** since they do not code for amino acids. In this way, the cell does not waste energy by copying unnecessary information or manufacturing unnecessary protein molecules (Figure 3.29).

The construction site of the protein molecules is on the cell organelle called the ribosome. This organelle serves as the place at which mRNA and the amino acid building blocks come together. The mRNA molecule is placed on the ribosome two codons (six nucleotides) at a time (Figure 3.30).

The amino acids are transferred to the workbench by molecules that are so specific they are only capable of transferring one particular type of amino acid. These are cloverleaf shaped RNA molecules called **transfer RNA,** or **tRNA.** There are twenty different types of amino acids and more than twenty different coding types of tRNA (see Table 3.5). It is the job of each tRNA to transfer a specific free amino acid to a site of protein synthesis. The tRNA properly aligns each amino acid so that it may be chemically bonded to another amino acid, forming a long chain. One end of a tRNA molecule is designed to attach to its particular amino acid, while another section has a unique triplet nucleotide sequence. This sequence is called an **anticodon** since it can hydrogen-bond with a complementary mRNA codon as it sits on the ribosome.

The tRNA carrying an amino acid hydrogen-bonds with the mRNA only long enough to allow for certain reactions:

1 The ribosome-tRNA-mRNA complex is formed. Both RNA molecules combine first with the smaller of the ribosome units, then the larger unit is added.

TABLE 3.5
Amino acid-nucleic acid dictionary. A dictionary can come in handy for learning any new language. This one can be used to translate nucleic acid language into protein language.

Amino Acid	mRNA Codons	tRNA Anticodons	Amino Acid	mRNA Codons	tRNA Anticodons
Phenylalanine	UUU	AAA	Tyrosine	UAU	AUA
	UUC	AAG		UAC	AUG
Leucine	UUA	AAU	Histidine	CAU	GUA
	UUG	AAC		CAC	GUG
	CUU	GAA	Glutamine	CAA	GUU
	CUC	GAG		CAG	GUC
	CUA	GAU	Asparagine	AAU	UUA
	CUG	GAC		AAC	UUG
Isoleucine	AUU	UAA	Lysine	AAA	UUU
	AUC	UAG		AAG	UUC
	AUA	UAU	Aspartic acid	GAU	CUA
Methionine	AUG	UAC		GAC	CUG
Valine	GUU	CAA	Glutamic acid	GAA	CUU
	GUC	CAG		GAG	CUC
	GUA	CAU	Cysteine	UGU	ACA
	GUG	CAC		UGC	ACG
Serine	UCU	AGA	Tryptophan	UGG	ACC
	UCC	AGG	Arginine	CGU	GCA
	UCA	AGU		CGC	GCG
	UCG	AGC		CGA	GCU
	AGU	UCA		CGG	GCC
	AGC	UCG		AGA	UCU
Proline	CUU	GGA		AGG	UCC
	CCC	GGG	Glycine	GUU	CCA
	CCA	GGU		GGC	CCG
	CCG	GGC		GGA	CCU
Threonine	ACU	UGA		GGG	CCC
	ACC	UGG	Terminator	UAA	
	ACA	UGU		UAG	
	ACG	UGC		UGA	
Alanine	GCU	CGA	Initiator	AUG	
	GCC	CGG			
	GCA	CGU			
	GCG	CGC			

2 The molecules containing tRNA-amino acid move in to combine with the mRNA on the ribosome. Each is sequentially bonded to one another to form a protein. As each is bonded in order, the unit is translocated (moved) along the mRNA to allow the next tRNA-amino acid to fit into position.

3 Once the final amino acid is bonded into position, the molecules are released from the ribosome. The termination codons signal this action.

4 The newly synthesized chain of amino acids folds into its typical three-dimensional structure.

In this way, the mRNA moves through the ribosome and its specific codon sequence allows for the chemical bonding of a specific sequence of amino acids. Remember that the sequence was originally determined by the DNA.

LIFE AND DEATH OF CELLS

When a cell divides into two new cells, each must have a complete set of genetic information. Since the original cell had only one set, there must be a doubling of this DNA information if the two new cells are to have a complete set.

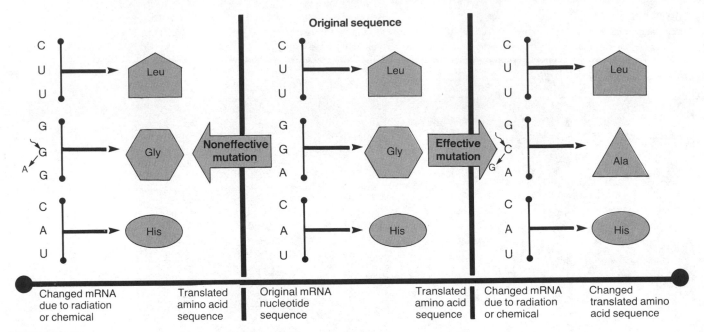

FIGURE 3.28

A nucleotide substitution will result in a change in the genetic information only if the changed codon results in a different amino acid being substituted into a protein chain. (From Eldon D. Enger et al, *Concepts in Biology,* 2nd ed. © 1979, 1976 by Wm. C. Brown Company Publishers, Dubuque, Iowa. Reprinted by permission.)

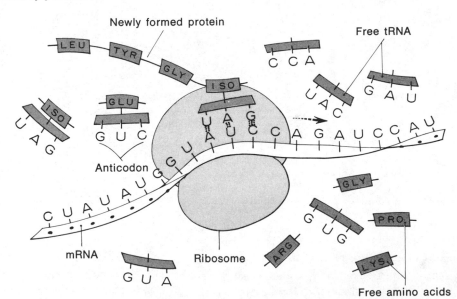

FIGURE 3.29

The second stage of protein synthesis (translation) occurs at the ribosomes. Each needed amino acid is brought into its proper position by a particular tRNA molecule.

The process of **DNA duplication** is the second major function of DNA. The accuracy of duplication is essential to guarantee the continued existence of the correct nucleotide sequence in all future generations.

The DNA duplication process begins much the same as protein synthesis. An enzyme breaks the hydrogen bonds between the bases of the genes and the DNA covers. It "unzips" the halves of the DNA double helix (Figure 3.31). As this enzyme proceeds down the length of the DNA, new individual DNA nucleotides are moved into position. The complementary bases pair with both exposed DNA strands by forming new hydrogen bonds. Once properly aligned, a covalent bond occurs between the newly positioned nucleotides to form the strong backbone of sugar and phosphate molecules (Figure 3.32). A new cover DNA is formed on the old DNA genes, and new genes are formed on the old cover DNA. In this way, the original DNA serves as **templates,** or patterns, for the formation of the new DNA. As

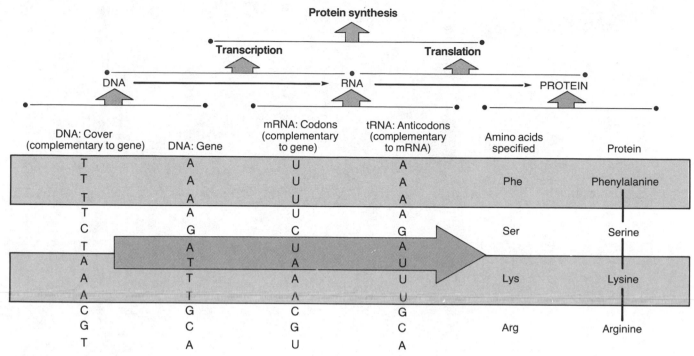

FIGURE 3.30
Steps involved in protein synthesis.

the new DNA is completed, it twists into its protective double helix shape. The completion of the process yields two double helices identical in their nucleotide sequences. Now the cell contains twice the amount of genetic information and is ready to divide.

SUMMARY

All matter is composed of atoms, which contain an atomic nucleus having neutrons and protons. This nucleus is also surrounded by moving electrons. There are over one hundred different kinds of atoms, called elements, which differ from one another by the number of protons and electrons they contain. Each is given an atomic number based on the number of protons in the nucleus and a mass number determined by the total number of protons and neutrons. Atoms of an element that have the same atomic number but different mass numbers are called isotopes. Some isotopes fall apart, releasing energy and smaller particles, and are called radioactive elements. Atoms may be combined into larger units called molecules. There are two kinds of chemical bonds, covalent and ionic, which allow atoms to form molecules. A third bond, the hydrogen bond, is a weaker bond which holds molecules together and also may help large molecules maintain a specific shape.

An ion is an atom that is electrically unbalanced. Those compounds that release hydrogen ions when dissolved in

water are called acids, while those that release hydroxyl ions are called bases. A measure of the hydrogen ions in relation to the hydroxyl ions present in a solution is known as the pH of the solution. Molecules that interact and exchange parts are said to undergo chemical reactions. The changing of chemical bonds in a reaction may release energy or require the input of additional energy. Three important biological reactions are dehydration synthesis, hydrolysis, and oxidation reduction.

The chemistry of living things involves a variety of large, complex molecules and is based on the chemistry of the carbon atom, and the fact that carbon atoms can be connected together to form long chains or rings. This results in a vast array of molecules. The structure of each molecule is important in determining how the particular molecule functions. Molecules with the same molecular formula but different structural formula are called isomers. Some of the most common types of organic molecules found in living things are carbohydrates, lipids, proteins, and nucleic acids.

The successful operation of a living cell depends on its ability to control chemical reactions. This is done directly by the enzymes, but the production of protein molecules is under the control of the nucleic acids, the primary control molecules of the cell. It is the structure of the nucleic acids, DNA and RNA, that determines the structure of the proteins, and it is the structure of the proteins that determines their function in the cell's life cycle. The process of protein synthesis is a decoding of the DNA into specific protein molecules

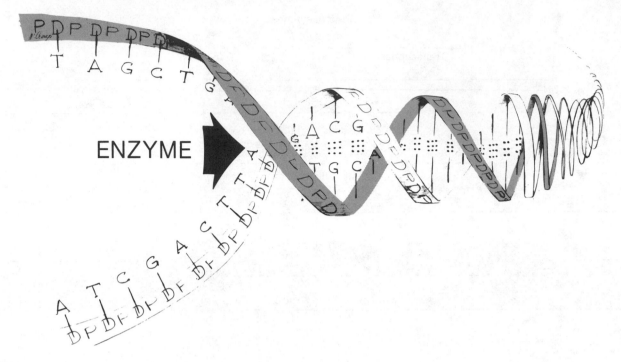

FIGURE 3.31
Separation of the double helix of DNA is shown here as the action of an enzyme opens the DNA molecule, separating the two strands. This is the beginning of DNA duplication.

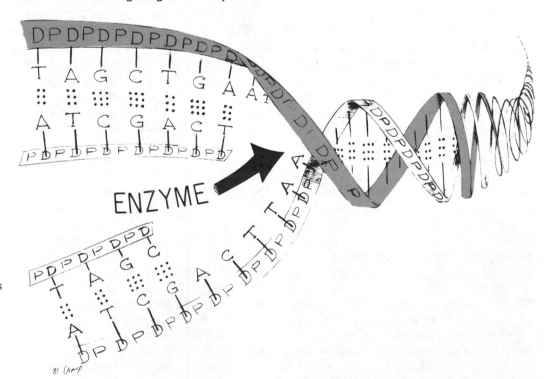

FIGURE 3.32
DNA duplication is illustrated as incoming DNA nucleotides base-pair with the separated strands of DNA. In this fashion, a new DNA cover strand is formed and a new DNA chemical code is formed.

and involves the use of the intermediate molecules mRNA and tRNA at the ribosome. Errors in any of the codons of these molecules may produce observable changes in the cell's functioning and lead to the death of the cell.

The process of DNA duplication results in the exact doubling of the genetic material. This occurs in a way that to a high degree guarantees identical strands of DNA being passed on to the next generation of cells.

WITH A LITTLE THOUGHT

Hydrogen peroxide (H_2O_2) is commonly found in the medicine cabinet and is used as an antiseptic. Its antiseptic properties result from the breakdown of this molecule into water and oxygen (O_2), which can destroy germs.

$$2H_2O_2 \longrightarrow 2H_2O + O_2$$

This reaction can take place right in the bottle over a period of time and results in the loss of germ-killing properties. Describe what is happening to the molecules of hydrogen peroxide and to the water and oxygen. Why will the bottle finally contain nothing but water and oxygen? In your description, include diffusion, possible changes in chemical bonds, and kinetic energy.

STUDY QUESTIONS

1 How many electrons, protons, and neutrons would be in an atom of potassium (K) having a mass number of 39?

2 Diagram an atom showing the relative positions of electrons, protons, and neutrons.

3 Define the term "chemical reaction" and give an example.

4 What is the difference between atom and element; molecule and compound?

5 How do acids and bases differ?

6 Give an example of each of the following classes of organic molecules: carbohydrate, protein, lipid, and nucleic acid.

7 What does it mean if a solution has a pH number of 3; 9?

8 What two characteristics of the carbon atom make it unique?

9 What is the difference between inorganic and organic molecules?

10 How does DNA duplication differ from the manufacture of an RNA molecule? How do these two molecular types function in a cell?

SUGGESTED READINGS

HOLUM, J. R. *Elements of General and Biological Chemistry,* 4th ed. John Wiley & Sons, New York, 1979.

KORNBERG, R. D., and A. KLUG. "The Nucleosome." *Scientific American,* Feb., 1981.

LEHNINGER, A. L. *Biochemistry,* 2nd ed. Worth Publishing, New York, 1975.

MASTERTON, W. L., and E. J. SLOWINSKI. *Chemical Principles,* 4th ed. Holt, Rinehart & Winston, Philadelphia, 1977.

STRYER, L. *Biochemistry.* Freeman Press, San Francisco, Calif., 1981.

WATSON, J. D. *Molecular Biology of the Gene,* 3rd ed. W. A. Benjamin, Inc., Menlo Park, Calif., 1976.

KEY TERMS

atom	chemical reaction	oxidation-reduction
compounds	endergonic (en″der-gon′ik)	nucleic acid
chemical bond	exergonic (eks″er-gon′ik)	transcription
molecule	activation energy	translation
acids	catalyst	DNA duplication
bases		

Cell Anatomy and Physiology

CELL COMPONENTS

ORGANELLES

CELL MEMBRANES

PASSIVE MOVEMENT: NO METABOLIC ENERGY REQUIRED

PROTEINS OF THE UNIT MEMBRANE

CARBOHYDRATES AND ENDOCYTOSIS

Learning Objectives

☐ Recognize that the cell is a structural and functional unit and that all cells have similarities.

☐ List cellular inclusions and describe their structure and function.

☐ List cellular organelles and describe their structure and function.

☐ Be able to describe the structure of cell membranes and explain the fluid mosaic model.

☐ Be familiar with the roles of proteins, phospholipids, and carbohydrates in cell membranes.

☐ Understand the various methods by which materials move through membranes and the significance of each method.

The cell is the simplest structure capable of existing as an individual living unit, and within this unit there are certain chemical reactions required for maintaining life. These reactions do not occur at random; instead, they are associated with the specific parts of *all* the many different kinds of cells. The next few chapters will deal with certain cellular structures found within most types of cells and discuss their functions. By developing an understanding of cell structure and function, a better understanding will result as to how microbiologists are able to control microbes in our environment.

CELL COMPONENTS

Cells are classified into two types, procaryotic and eucaryotic, based on the kinds of specific structural components they contain (Box 5). Because they contain a nucleus, eucaryotic cells are easily distinguished from the procaryotic cells. The nucleus of a eucaryotic cell is made more visible by selectively dying or staining these cells. The stain used (usually basic fuchsin) chemically combines with the genetic material, deoxyribonucleic acid, that is concentrated in this portion of the cell. When viewed through the light microscope, the nucleus appears as a colored body distinct from the unstained cytoplasm (Figure 4.1). All the specific com-

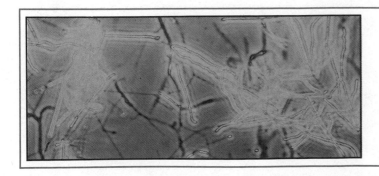

4

BOX 5

CELL
CONCEPT

THE concept of a cell did not arise spontaneously, but has been added to and modified over many years. This process of modification continues today.

Several events associated with particular individuals represent milestones in the construction of the cell concept. Anton van Leeuwenhoek is generally given credit for clearly seeing the first microbes. When he discovered that he could see things moving in pond water, his curiosity stimulated him to look at a variety of things through his crude microscopes, among them blood, semen, feces, pepper, and tarter. Van Leeuwenhoek was the one who first saw individual cells and recognized them as living units. He did not, however, call them cells; instead he called them "animalcules"—little animals.

The honor of first using the term *cell* goes to Robert Hooke (1626–1703) of England, who was also interested in how things looked when magnified. He chose to study thin slices of cork from a tree. What he saw was a mass of cubicles that fit neatly together, reminding him of the barren rooms in a monastery. Hence, he called them by the same name—cells. Ironically, we have come to use the word "cell" with respect to a living unit, when in fact what Hooke saw was only the walls that had surrounded the living portions of cells.

Hooke's use of the term "cell" was really only the beginning. Soon after the term caught on, it was determined that the vitally important portion of the cell was inside. This was called **protoplasm,** which literally means "first juice" or "juice of the first importance." "Protoplasm," then, allows us to distinguish between the nonliving cell wall and the living portion inside. As with any new field of study, this early attempt at terminology was not a complete success. Very soon, microscopists were able to distinguish two different regions of protoplasm. One type of protoplasm was more viscous and darker than the other, and appeared as a central body within a more fluid juice. Therefore, two new terms were introduced: **cytoplasm** (literally "cell juice"), and **nucleus** (the kernel, or central body). Both the cytoplasm and the nucleus are parts of the protoplasm.

As laboratory techniques continued to improve, the chemical nature of cytoplasm became more clearly understood. The molecules found in a "typical" living bacterial cell are described in Table 4.1. Better light microscopes and ultimately the electron microscope revealed that these molecules are organized into discrete cellular structures that were not previously seen.

TABLE 4.1

Many molecules are found in a living cell. Each displays a characteristic structure and function. The molecules listed in this table are typical of those found in the bacterium *E. coli.*

Molecular Type	Structure	Function	Estimated Number in Cell	Number of Different Kinds	Percent of All Molecules in Cell
Proteins	Composed of amino acids (C,H,O,S,N)	Enzymes and structural parts of the cell wall	1,000,000	2000–3000	16
Carbohydrates	Composed of simple sugars (C,H,O in 1:2:1 ratio)	Energy storage and a support material in cell wall	200,000,000	200	3
Nucleic acids: DNA	Composed of nucleotides in chains (C, H, O, N, P)	Genetic material	4	1	1
: RNA		Protein synthesis	461,000	1000+	6
Lipids	Composed of glycerol and fatty acids (C, H, O, possibly phosphorous)	Energy storage and structure in cell membranes	25,000,000	50	2
Other organic molecules	Breakdown products of foods		15,000,000	200+	1
Inorganic molecules	Na^+ K^+ Mg^{+2} Ca^{+2} Fe^+ Cl^- SO_4^{-2}	Various biochemical pathways	250,000,000	20+	1
Water	H_2O	All pathways	40,000,000,000	1	70

SOURCE: Adapted by permission from J. D. Watson, *Molecular Biology of the Gene*, 3rd ed. Menlo Park, California. The Benjamin/Cummings Publishing Co., table 3–3, p. 69.

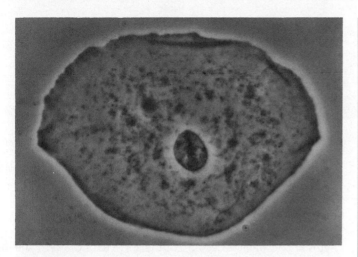

FIGURE 4.1
This human cheek cell has been stained to highlight the nucleus. (Courtesy of Carolina Biological Supply Company.)

ponents of either cell type are divided into two main categories. **Inclusions** are cellular components that are not directly involved with the life functions of the cell; they usually function as storage materials within the cell. The **organelles** ("-elle" means *little*) are components directly involved with life functions, and each has a particular structure related to its function. By knowing the specific nature of the inclusions and organelles found in various cell types, growth, reproduction, or other cell functions can be more effectively controlled. For example, the disease "strep throat" is caused by the activities of bacterial cells in a host organism composed of eucaryotic cells. By knowing the special structural and functional features that make these procaryotic parasites different from the eucaryotic host, it is possible to use chemicals **(antibiotics)** that will selectively interfere with the normal function of the procaryotes but not with the eucaryotic cells of the host.

Inclusions are found in both procaryotic and eucaryotic cells, but usually only one kind of storage material is formed by a particular kind of microbe. These storage materials have been identified through the use of special stains and biochemical tests. The presence of certain inclusion types may help identify a particular cell type. Some of these inclusions may be familiar. **Starch** is typically found in plant cells as a form of stored carbohydrate, while animals form the inclusion **glycogen** as their stored carbohydrate. When energy-containing nutrients are not available to cells containing these inclusion types, the cells metabolize these molecules to gain the needed energy or provide themselves with building materials. Bacteria that form the inclusion **polyhydroxybutyric acid (PHB)** function in this way. PHB inclusions in bacterial cells will disappear if the bacteria are starved by reducing available carbon-containing molecules. If the necessary nutrients are added to the growth medium, the inclu-

sions are reformed. Bacteria that are able to function in this way can store material for future use; this adaptation helps them survive adverse environmental conditions. Another procaryotic inclusion, **volutin,** (Figure 4.2) is composed of phosphate molecules chemically combined in long chains. These inclusions are sometimes referred to as **metachromatic granules,** since they stain a different color from what is typically expected. After applying a blue stain, these inclusions appear red. Members of the genus *Corynebacterium* typically contain volutin. The bacterium *Corynebacterium diphtheriae,* which causes the disease diphtheria, belongs to this group. Phosphate molecules are necessary in the cell's metabolism and are used in the formation of such important molecules as DNA and ATP. It has been speculated that phosphate molecules may serve as a form of "primitive enzyme" in some chemical reactions; in fact, these may have been the first kind of molecules that accelerated chemical reactions in primitive cells.

Some inclusions result from metabolic activities other than energy storage. The microbes, *Beggiatoa* and *Thiothrix* use H_2S (hydrogen sulfide) as a source of energy. When the energy has been extracted from the molecule, sulfur is deposited as an inclusion. Protein "crystal" inclusions are produced by some members of the genus *Bacillus* (e.g., *B.*

---

1 Why are the terms "protoplasm" and "cytoplasm" of little value today?

2 Why is it of great value to know the specific differences between procaryotic and eucaryotic cell types?

3 Name an inclusion composed of carbohydrate that is found in plant cells but not animal cells.

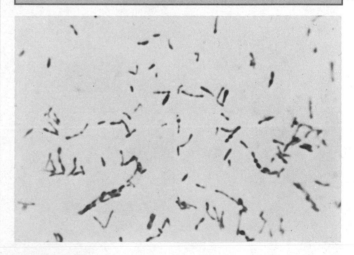

FIGURE 4.2
These bacteria, *Corynebacterium diphtheriae,* have been stained to highlight the volutin inclusions. (Courtesy of the Center for Disease Control, Atlanta.)

thuringiensis) (Figure 4.3). These unique bacteria produce a diamond-shaped inclusion that has been found to be deadly to a number of different kinds of insect pests; for example, the infamous gypsy moth has been successfully kept under control in some areas of Spain by allowing the caterpillars of this moth to eat these bacteria. When the bacteria are digested, the crystal inclusions are released into the caterpillars and cause death.

Other insect pests that may be controlled by these bacteria include tomato hornworms and fruitworms, cabbage

worms and loopers, grape leaf rollers, corn borers, cutworms, and tent caterpillars. These bacteria are grown for sale commercially and sold in a powdered spore form at nurseries and garden supply stores under the trade names of Dipel, Biotrol, or Thuricide.

ORGANELLES

The organelles of a cell are a complex array of components that must be maintained if a cell is to continue its life functions.

These cell components are constantly involved in biochemical pathways necessary for the survival of the cell. The most familiar organelle is the nucleus, but there are many others. Most organelles are not found as single units within a cell, but occur as numerous, clearly identifiable structures. Figure 4.4 illustrates a "typical eucaryotic cell" and some of the organelles found in the cytoplasm.

The organelles of a cell are divided into two main groups based on their structure. One group, called the **membranous organelles,** is composed of a material known as unit membrane, a sheetlike material that can be fashioned into many shapes. Just as it is possible to fashion sheets of paper maché into many shapes, cells fashion sheets of unit membrane into many differently shaped organelles. Examples of organelles with the unit membrane structure include the plasma (cell) membrane, endoplasmic reticulum, plas-

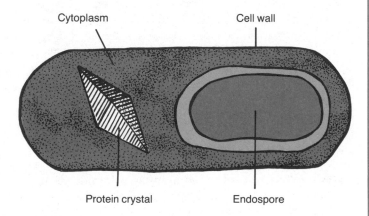

Cytoplasm   Cell wall

Protein crystal   Endospore

FIGURE 4.3

The diagram illustrates the bacterium *Bacillus thuringiensis* and the protein "crystal" inclusion. If these bacteria are fed to cattle or chickens, the crystals are excreted in the manure. When fly larvae, maggots, ingest the crystals, they are killed.

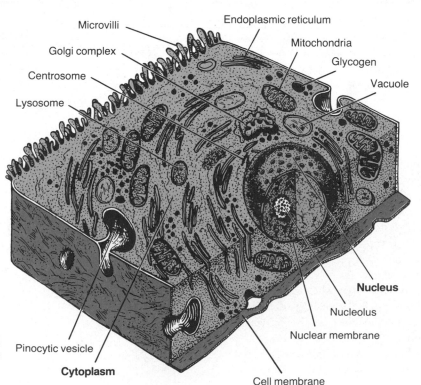

Microvilli

Golgi complex

Centrosome

Lysosome

Endoplasmic reticulum

Mitochondria

Glycogen

Vacuole

Nucleus

Nucleolus

Nuclear membrane

Pinocytic vesicle

Cytoplasm

Cell membrane

FIGURE 4.4

This illustrates many of the inclusions and organelles found in eucaryotic cells. Some of the structures are also found in procaryotic cells.

tids (mitochondria and chloroplasts), the Golgi apparatus, lysosomes, vacuoles, vesicles, and mesosomes.

The second group of organelles is composed of a variety of other materials not fashioned from unit membrane. There is no common structural component for the **nonmembranous organelles.** Proteins, nucleic acids, lipids, and carbohydrates may all be used singly or in combination as structural material for these cell components. Nonmembranous organelles include such structures as chromosomes, ribosomes, cilia, flagella, and pili. Since all organelles play a vital role in the life of a cell, it is important to examine the structure and function of each in order to more clearly understand the nature of microbial life.

Before exploring the structure of each organelle in detail, it is necessary to understand the structure and function of the unit membrane. At one time, it was thought that membrane was nothing more than an elastic outer covering of a cell, but detailed biochemical research has revealed this view to be inaccurate.

---

1 Of what value are inclusions to a cell?
2 Of what value are organelles to a cell?
3 What is the major difference between membranous and nonmembranous organelles?

CELL MEMBRANES

All membranous organelles have a basic structure similar to the **plasma (cell) membrane,** and they have the ability to regulate the passage of material through their boundaries. The plasma membrane is more than a simple barrier that separates the cytoplasm from the surrounding environment.

It is a highly active, functional organelle which regulates the passage of many different molecules such as water, carbohydrates, lipids, proteins, and also many inorganic ions, such as Mg^{+2}, Na^+, K^+, and Ca^{+2}. By controlling the movement of these materials, the membrane's actions prevent the cell from swelling or shrinking, concentrate necessary nutrients inside the cell, and permit the secretion of large molecules such as digestive enzymes. To understand how the plasma membrane is able to perform these vital functions, we need to examine its structure.

In 1935, a model of membrane structure was proposed by J. F. Danielli and H. Davson to explain membrane function. The membrane was pictured as a sheet of material composed of a center layer of lipid coated on either side by a sheet of protein (Figure 4.5). The lipid layer was composed of special phospholipid (Box 6) molecules in two layers organized so that the water-repelling, or **hydrophobic,** ends of these molecules faced each other in the center of the membrane. The other ends of the phospholipid molecules were water-attracting, or **hydrophilic,** faced away from the center, and were coated with a layer of protein. The most important part of the membrane was assumed to be the lipid layer, which was responsible for the structural stability and biological activities thought to occur at cell surfaces. As time passed, it became clear that the protein on the inside of the lipid differed from the protein on the outside, and these protein molecules played as important a role as the lipids. The structure and function of the membrane could not be explained by a model that represented the protein as flat layers or sheets. Investigators used a variety of techniques to explore the plasma membranes of a great many kinds of cells.

In 1972, S. J. Singer and G. L. Nicolson proposed a new working model called the **fluid mosaic model** that better explains the known functions of the membrane. This model is in agreement with photographs of the cell mem-

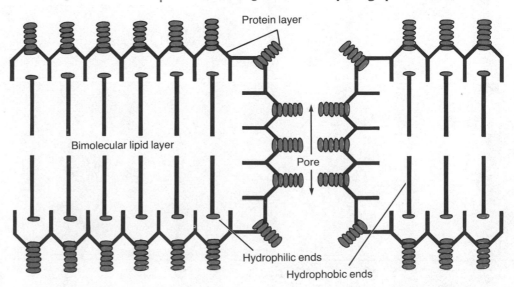

FIGURE 4.5
The Danielli-Davson membrane model consisted of two layers of lipids sandwiched between two layers of protein. Pores in the membrane were thought to be lined with protein. Carbohydrate was believed to coat the outside surface like a layer of paint.

THE group of macromolecules known as lipids has a variety of structures and functions. Lipids are large, organic molecules that are insoluble in water, but soluble in fat solvents such as alcohol. Among the many different kinds of lipids are fats, phospholipids, and steroids. Fats are made up of two smaller molecules called glycerol and fatty acids. **Glycerol** is an alcohol with the following structure:

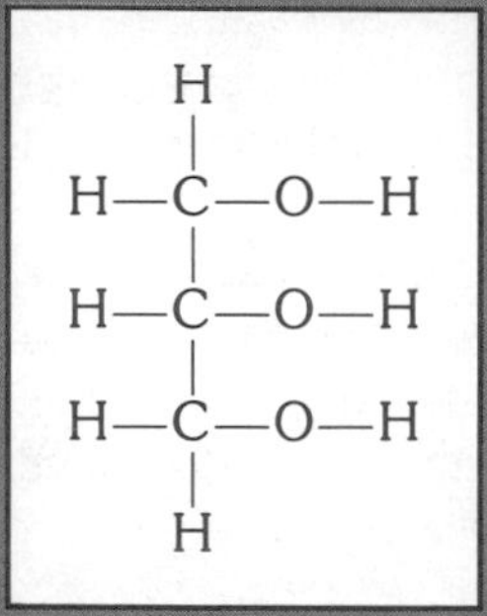

Fatty acids consist of carbon chains of varying lengths having a group at one end called a carboxylic acid group.

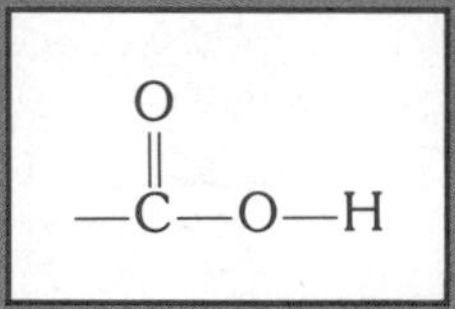

An example of a fatty acid is shown in the following structure:

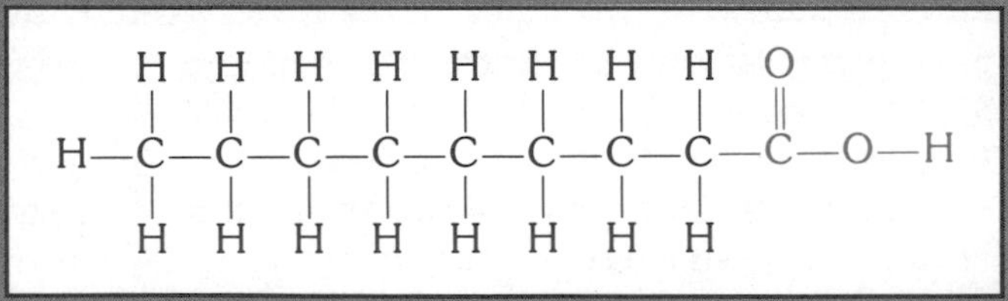

Fats such as those present in animals do not usually have double bonds between any of the carbons in the fatty acid portion of the molecule. These kinds of fats are called **saturated fats** and are usually solid at room temperature. Lard, whale blubber, and suet are saturated animal fats. Most fats of plants, however, do have double bonds between carbons in the fatty acid portion of the molecule. These kinds of fats are **unsaturated fats** and are usually liquid at room temperature. Olive oil, linseed oil, corn oil, and cottonseed oil are all plant fats.

Fats are important molecules for storing energy: there is twice as much energy in a gram of fat as in a gram of sugar. This is important to a microbe, since fats can be stored in a relatively small space and still yield a high amount of energy. However, if the cells of an organism need energy, they tap the carbohydrate source before the stored fat. This is very evident when a person tries to lose weight. A successful diet is one in which the carbohydrate and fat intake is greatly reduced. This places a strong demand for energy on

BOX 6

THE LIPIDS

the stored fat molecules. Since fat has twice as much energy per gram as carbohydrate, the process of losing weight is often slow.

Phospholipids are a particular class of water-insoluble molecules that resemble fats but contain phosphate groups (PO_4) at some place in their structure.

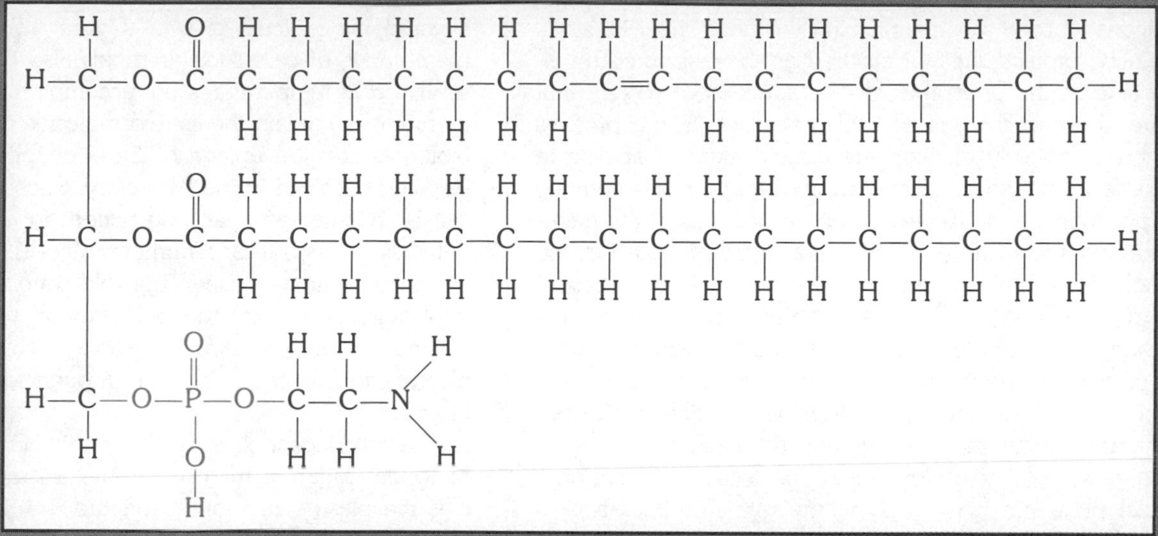

Phospholipids are involved in the structure of all cells. Nerve and brain cells especially contain many different kinds of phospholipids, all of which are essential to the operation of an animal's nervous system.

Steroids are a group of complex molecules that usually consist of ringlike arrangements of atoms. They are often hormones that aid in regulating body processes. One steroid molecule that is probably most familiar is cholesterol. This steroid has been implicated in many cases of atherosclerosis, or clogging of the arteries. On the other hand, cholesterol is necessary for the manufacture of vitamin D. The cholesterol molecules are found in the skin and react with ultraviolet light to produce vitamin D, which is used to assist in the proper development of bones and teeth:

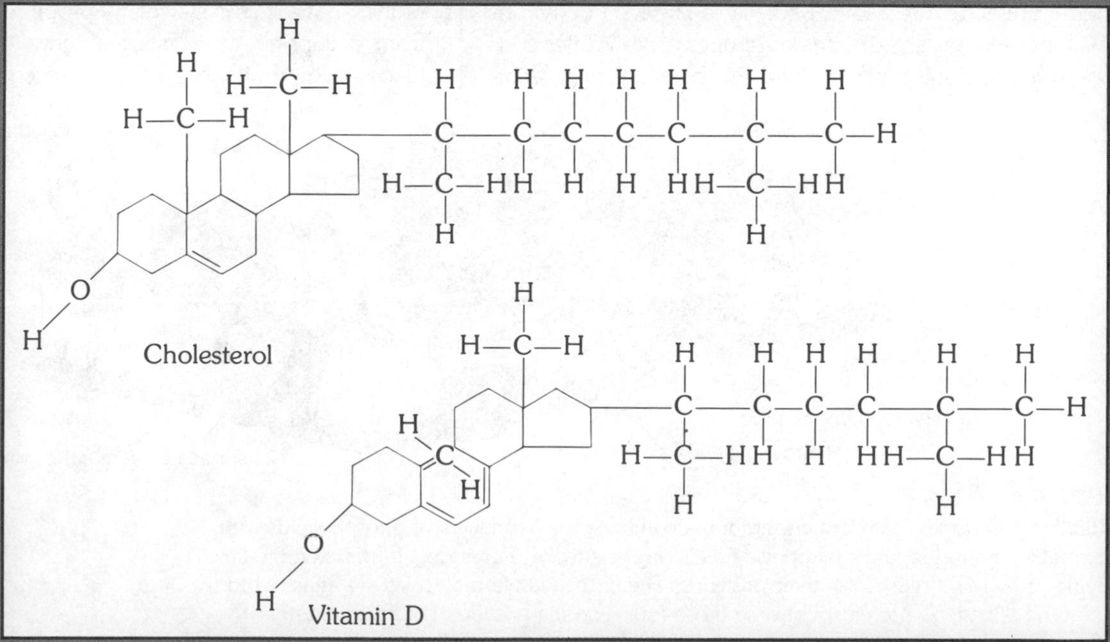

Cholesterol

Vitamin D

brane taken with an electron microscope. Their model presents proteins "floating" in a "sea" of lipid (Figure 4.6).

The description of the membrane is that of a dynamic system. The components are in constant motion, shifting positions while always maintaining the basic sheetlike structure. The proteins are not pictured as sheets layered on either side of the phospholipid, but as globs of protein that move about in the sea of lipid molecules. Some proteins extend completely through the double layer of lipid from one side to the other. Others are smaller and nestled among the phospholipid molecules, some on one side of the membrane and some on the other. A third type of molecule, the carbohydrate (Box 7), has also been found to be a part of the membrane structure. Carbohydrates extend out from the moving surface of the membrane and are attached to either the proteins or the lipids. When carbohydrates and proteins combine, they are **glycoproteins;** when lipids and carbohydrates combine, they are **glycolipids.** Each of these three main types of molecules (lipids, proteins, and carbohydrates) plays important roles in the dynamic operation of the cell. A great deal can be learned about membrane function in living cells by examining how these nonliving molecules interact in experimental situations.

Research has shown that in procaryotic cells, the fatty acid part of the phospholipid layer is composed of short-chain, unsaturated molecules. Most eucaryotic membranes have long-chain, saturated fatty acids. The difference in chain length and the amount of hydrogen in the chain is related to the physical state of the molecules (see Box 6). Bacteria contain shorter, unsaturated fatty acids that have a lower "crystallization" point. This means they do not freeze solid and break until the temperature is very low. This difference in membrane structure allows bacteria to survive in environments that have great extremes of temperature. Winter conditions will not significantly reduce the bacterial population

in soil as a result of membrane freezing and breaking. Their ability to survive can be demonstrated by sprinkling a sample of frozen soil on nutrient agar and incubating at 37°C (98.6°F). In only twenty-four hours, the nutrient agar will be covered by bacterial growth (Figure 4.7). The eucaryotic membranes of cells such as molds also have the ability to survive cold temperatures but accomplish this because their membranes contain the steroid molecule cholesterol. Microbiologists take advantage of these adaptive features when studying microbes in the laboratory. Such microbial cultures can be maintained over long periods in a refrigerator without losing a significant number of cells. Few will die of membrane damage due to the cold conditions. These structural features are not the only survival adaptations of the plasma membrane. Another survival characteristic of the phospholipid layer can be seen through some simple experiments.

When phospholipid is placed in water and shaken, the molecules organize themselves into a structure that resembles the plasma membrane (Figure 4.8). This is the most energetically stable or "comfortable" form. When disrupted again, the molecules will quickly reorganize into the same arrangement. Because the hydrophobic, or water-repelling, ends of these lipid molecules are facing one another, they form a double-layered sheet that is two molecules thick. In all membranous organelles, this arrangement of lipid restricts the movement of water molecules from one side of the membrane to the other and is responsible for the barrier properties of the membrane. If the double layer is punctured with a small pin, the phospholipid molecules move back into position and seal the hole, preventing the loss of the essential nutrients inside the cell. All biological membranes have this self-sealing property which protects cells from collapsing or bursting (Figure 4.9). However, some places in the phospholipid layer reorganize into an arrange-

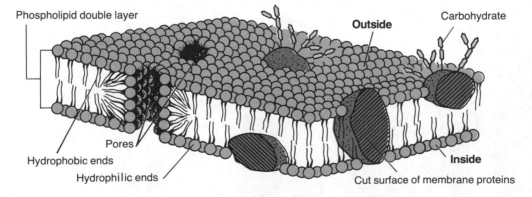

FIGURE 4.6

The fluid mosaic model portrays a membrane composed of two layers of phospholipids with hydrophobic tails inside and hydrophilic heads on the surface. Pores may form when the lipid molecules move and reorganize their positions. The carbohydrate molecules are attached to some of the "floating" proteins, but only those found on the outside of the membrane.

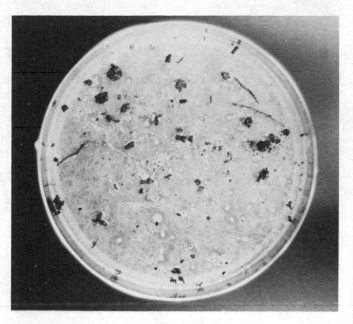

FIGURE 4.7

The common soil microbes growing on the surface of this petri plate were obtained by sprinkling a small amount of frozen soil on the nutrient agar surface and incubating the plate for only twenty-four hours.

ment that forms a temporary hole or pore in the membrane, rather than forming a smooth surface. Because of the dynamic nature of the membrane, these pores may form and close very quickly. Their diameters may be no wider than a water molecule, and this small size and temporary nature may allow only small water molecules or ions, such as Ca^{+2}, to move through the membrane.

> **1** List two important functions of proteins and lipids found as components of cell membranes.
> **2** What are the elemental differences between carbohydrates and lipids?

PASSIVE MOVEMENT: NO METABOLIC ENERGY REQUIRED

The plasma membrane acts as a barrier to separate the inside from the ouside of the cell; therefore, materials must

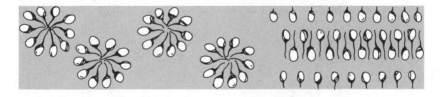

FIGURE 4.8

If lipids and water are shaken, the lipid molecules organize themselves into specific patterns. In this illustration, the hydrophobic tails of the molecules are oriented toward one another, and the hydrophilic heads extend out from the center of these "synthetic" membranes.

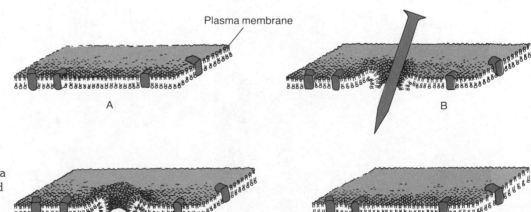

FIGURE 4.9

This sequence illustrates the self-sealing nature of a plasma membrane. The phospholipid layer is reorganized into a smooth sheet after being punctured with a sharp object.

CARBOHYDRATES contain three kinds of atoms: carbon, hydrogen, and oxygen. In general, the number of carbon and oxygen atoms are equal, while the number of hydrogen atoms is about double that number. There is a large number of different molecules that fit this basic pattern, which can be represented by the ratio $C_nH_{2n}O_n$, where n is a positive integer (e.g., $C_6H_{12}O_6$).

Some common carbohydrates include many kinds of sugars, corn syrup, cornstarch, and wood fiber. These different carbohydrate molecules are used in many ways by cells. For example, **simple sugars** such as glucose, fructose, and galactose provide the chemical energy necessary to keep organisms alive. The following are some structural arrangements of six-carbon sugars. Each of these molecules has the same number of carbon, hydrogen, and oxygen atoms, but each is arranged differently.

BOX 7

THE CARBOHYDRATES

Glucose
$C_6H_{12}O_6$

Mannose
$C_6H_{12}O_6$

Galactose
$C_6H_{12}O_6$

Sorbose
$C_6H_{12}O_6$

Fructose
$C_6H_{12}O_6$

Simple sugars may also combine by dehydration synthesis to form larger molecules. Sucrose (common table sugar) consists of a glucose and a fructose molecule hooked together.

Glucose + Fructose → Sucrose + Water

Glucose + Glucose + Glucose + Glucose → Starch molecule which is about 1,000 glucose units long + Water

Glucose + Glucose + Glucose + Glucose → Cellulose molecule which is about 1,500 glucose units long + Water

Simple sugars can also be used by the cell as a component of other more complex molecules such as adenosine triphosphate (ATP). The sugar present in ATP is a five-carbon sugar known as ribose. This simple sugar molecule is located in the center of the ATP molecule and has attached to it many other types of atoms including oxygen, hydrogen, nitrogen, and phosphorus.

Ribose sugar

cross this barrier to enter or leave the cell. **Passive diffusion,** which results from the random movement of molecules, is an important process that allows materials to cross the membrane. Molecules are not always evenly distributed. Many times there may be a high concentration of a particular type of molecule in one place and a lower concentration of that molecule in another. When this occurs, the random motion of the molecules will eventually result in the net movement of the molecules from the area of highest concentration to the area of lowest concentration. This difference in concentration is referred to as the diffusion gradient. When the molecules are equally distributed, no such gradient exists. Diffusion can only take place as long as there are no barriers to free movement. In the case of a cell, the membrane permits some molecules to pass through while others are held back. In the case of passive diffusion, however, the cell membrane does not distinguish between outside and inside. The restriction of molecules passing into the cell is just as effective for those leaving the cell. The cell membrane allows lipid soluble molecules to pass while lipid insoluble molecules are prevented from traveling through. Membranes that function in this way are called **differentially permeable.**

The process of diffusion is an important means of exchanging materials between cells and their surroundings. Because the cell has no control over this process, the molecules move at random. For example, cells that require oxygen constantly use up this gas in their cells. The cells, then, will have a low concentration of oxygen compared to the oxygen level outside the cell. This diffusion gradient goes from the outside of the cell to the inside. Thus, these cells will constantly gain oxygen as long as they use it.

Water is one of the molecules that easily diffuses through the pores in cell membranes. The net movement of water molecules through a differentially permeable membrane is a special case of diffusion known as **osmosis.** A proper amount of water is required for a cell to function efficiently; yet, too much water in a cell may dilute the cell contents and interfere with the chemical reactions necessary to keep the cell alive. Too little water in the cell may result in the buildup of a high concentration of poisonous waste products. As with the diffusion of other molecules, the cell has no control over the diffusion of water molecules. This means

that most cells can only remain in balance in an environment that does not cause the cell to lose or gain too much water. If a cell does not contain the same concentration of water as its surroundings, it will either gain or lose water. Many organisms have a concentration of water that is equal to their surroundings. This is particularly true of microbes living in the ocean. If a microbe is going to survive in an environment that has a concentration of water different from its cells, it must expend energy to maintain this difference. Microbes that live in fresh water have a lower concentration of water than their surroundings. Since these cells tend to gain water by osmosis, they must expend energy to eliminate this excess if they are to keep from swelling and bursting.

Many microbes will not burst because they are surrounded by a rigid **cell wall.** If the water concentration is higher on the inside of the cell than on the outside, the cell will collapse because more water molecules are leaving than are entering (Figure 4.10). This action is called **plasmolysis** ("plasma" means *fluid;* "lysis" means *effecting decomposition*) and has been used to investigate membrane function. By changing the concentration gradient of water that surrounds the cell, the cell can be seen under the microscope to undergo plasmolysis. This process can be reversed in some microbes; the cell contents can be made to shrink and swell. This does not harm the cell if the plasma membrane is not attached to the cell wall. In some pathogenic bacteria (e.g., *Pseudomonas aeruginosa*), the cell membrane is attached to the wall, and concentration changes in the surrounding environment may result in extensive damage to the membrane as plasmolysis occurs. We can take advantage of this information and use the process to kill these kinds of bacteria by changing the osmotic pressure in their environment. For example, beef may be preserved in the form of beef jerky by cooking salt into the meat and concentrating it by drying. Bacteria that fall on the surface are inhibited from growing and spoiling the meat because of the high osmotic pressure.

The water loss that occurs with plasmolysis can inhibit the metabolism of many microbes, but the cell wall and membrane characteristics of some bacteria allow them to grow in environments that contain very high concentrations of salts. Members of the genus *Halobacterium* may grow in

FIGURE 4.10
This sequence illustrates what happens to the contents of a cell placed into a solution containing less water than is found in a cell, a reaction known as plasmolysis.

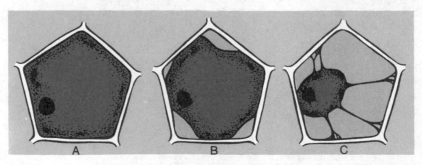

solutions that are ten times more concentrated (15% NaCl) than bacteria such as *E. coli* (1.5% NaCl). This ability allows these salt-requiring bacteria, **halophiles,** to grow in highly salted foods.

The structural stability of the membrane is not always able to be maintained. Its dynamic nature can be altered to the point that it is unable to seal and prevent the cell contents from spilling out. A number of bacteria produce toxins (poisons) which severely damage the membranes of other cells. Members of the genera *Streptococcus* and *Staphylococcus* release such toxins, which are responsible for many of the disease characteristics associated with infections. Pus formed in these infections is composed of leukocytes and necrotic (dead) tissue that have been killed by these toxins (Figure 4.11). The bacterium *Clostridium perfringens* is the cause of gas gangrene, and one symptom of this infection is the destruction of red blood cells—a process known as hemolysis ("hemo" means *blood;* "lysis" means *destroy*). This particular symptom is the result of the action of a phospholipid-destroying enzyme known as **phospholipase** released by *C. perfringens.* As another example, special human white blood cells (T-lymphocytes) release enzymes that destroy cell membranes. These enzymes are very specific in their action and, therefore, do not cause self-destruction of the lymphocytes. The lymphocytes are able to destroy bacterial cell membranes and are an important part of the body's defense mechanisms.

If toxins damage a membrane and the self-sealing mechanisms are not able to correct the problem, metabolic pathways for repair are brought into play. Eucaryotic cells use the vitamins ascorbic acid (vitamin C) and tocopherol (vitamin E) to aid in the repair operation. If these vitamins are not available, the repair operation will be hampered, and the cell may die. These special reactions require use of many other kinds of membrane components, especially proteins (Box 8).

PROTEINS OF THE UNIT MEMBRANE

The word **protein** means molecule of "first importance," and it usually plays two main roles in living cells. Proteins may serve as either structural materials in cells or as enzymes. However, this separation of roles is not always clear-cut: proteins in the unit membrane may interchange roles. Their presence in the phospholipid layer is thought to provide support and structural stability to the membrane, and they may be involved in the transportation of lipid insoluble molecules from one side of the membrane to the other. Detailed investigation has pointed out that the proteins found on one side of the membrane are different from those found on the other. This characteristic of "sidedness" is very important to a living cell. Molecules that are vital to the functioning of the cell must be selectively taken across the membrane into the cell (ingested), while waste materials that may be harmful must be moved out of the cell (waste elimination). The direction of the movement and the recognition of which molecules are nutrients and which are wastes is in part controlled by the proteins associated with the unit membrane. Proteins have three characteristics that are associated with this transport role. First, the structure of the

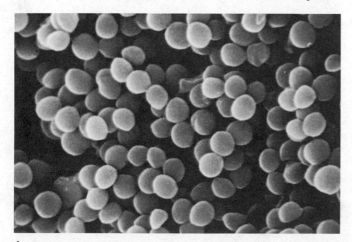

A

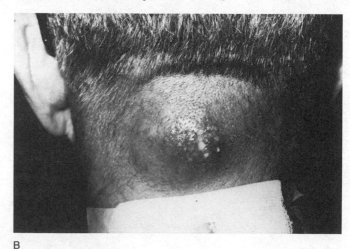

B

FIGURE 4.11

(A) Staphylococci tend to form clusters of cells. (B) The infection shown is a carbuncle, which is a staphylococcal infection of many hair follicles. This type of infection often occurs on the nape of the neck. (Part A electron micrograph by F. Siegel. Courtesy of Burroughs Wellcome Co., Research Triangle Park, N.C. Photo B from *Bacterial Infections of the Skin, I.* Abbott Laboratories.)

PROTEINS are molecules made up of subunits known as amino acids. Amino acids contain nitrogen in an amino group at one end of the molecule and an acid group at the other end of the molecule.

There are about twenty kinds of amino acids that are identical except for the side chain, which differs from one amino acid to another. These amino acids can be combined to form chains. Just as the twenty-six letters of the alphabet can be combined to make millions of words, these amino acids can be arranged in millions of different ways. When the chain is short (about 100 amino acids or less), it is called a polypeptide chain. When the chain is longer than 100 amino acids, it is usually called a protein. Proteins are constructed by attaching amino acids end-to-end.

The structure of a protein is intimately related to the function of the protein. In most cases, a change in a protein's structure will alter its function. We will consider two aspects of the structure of proteins: the sequence of amino acids within the protein and the overall three-dimensional shape of the molecule. Any change in the arrangement of amino acids within a protein can have far-reaching effects on the protein's function. For example, normal hemoglobin found in red blood cells consists of two kinds of polypeptide chains called the **alpha** and **beta** chains. The beta chain has about 140

BOX 8
PROTEINS

amino acids along its length. If a particular amino acid is changed from one form to another, the hemoglobin molecule does not function properly, and thus results in a disease condition known as sickle-cell anemia.

In addition to the sequence of amino acids in the protein, the long chains of amino acids can coil and be folded into rather complex three-dimensional shapes. This three-dimensional structure is partly determined by the sequence of amino acids: some of the amino acids at different places in the chain may chemically bond with one another. This results in a three-dimensional folding of the protein. For example, some amino acids have sulfur in them, and two sulfur atoms will form covalent bonds to hold the chain of amino acids in a particular shape. But other forces, such as hydrogen bonds, are also important in maintaining a particular shape for a protein molecule.

Thousands of kinds of proteins can be placed into two broad categories. Some proteins are important for holding cells and organisms together and are usually referred to as **structural proteins.** The other kinds of proteins are either hormones or enzymes, and the enzymes carry out the activities of cells.

molecules is highly specific. They function like enzymes and will combine in the transport process only with substrate molecules that will "fit" together with them. If a molecule that is able to be transported across the membrane is replaced with another of similar (but not exact) structure, the replacement will not be moved into the cell. Second, the rate of transport is greatly increased when these proteins are present in the cell membrane. If the proteins are selectively destroyed, molecules that would normally be transported inside the cell remain on the outside. These modified cells eventually die because they are unable to obtain essential nutrients. Third, the proteins remain unchanged after they have performed their role as transport molecules. They are able to be reused to carry molecules across the membrane.

The two transport processes associated with membrane proteins are **facilitated diffusion** and **active transport.** Recall that passive diffusion is the movement of molecules from an area of high concentration to an area of lower concentration. If the membrane is not a barrier to the molecule (the moving molecule is lipid-soluble), passive diffusion may occur and small molecules move into or out of the cell depending on the diffusion gradient. Facilitated diffusion also involves the movement of molecules from an area of high concentration to an area of lower concentration. However, this process is facilitated, or helped, by proteins of the unit membrane. These proteins are called **permeases** (enzymes that increase permeability). The kind of molecules usually moved in facilitated diffusion are lipid-insoluble, such as sugar. The way in which facilitated diffusion occurs is not exactly known, but one idea suggests that as the substrate fits into the transport protein, the shape of the permease molecule changes (Figure 4.12). The change automatically results in a rotation of the protein-substrate unit in such a way that

the substrate becomes moved to the other side of the membrane. When on the other side, the two separate and the transport is complete. There is no metabolic energy (e.g., ATP) used in this process. The transport protein helps the molecule across until the concentration on both sides of the membrane are equal, and the process stops.

The active transport process is similar to facilitated diffusion in that a transport protein combines with the substrate and moves the molecule through the membrane. However, this process differs in two major ways. First, *metabolic energy is required* for the operation. ATP energy must be expended in order for the action to occur. The energy is used to change the permease so that it can combine with the substrate molecule. The molecule is transported through the membrane and released on the opposite side, and the membrane protein is able to be reused another time (Figure 4.13). Second, active transport is responsible for the movement of molecules from one side of the membrane to the other regardless of the concentration gradient. This transport allows cells to accumulate essential ions or molecules that would normally tend to leave the cell. Sodium (Na$^+$) and potassium (K$^+$) ions are actively transported in this manner. Some bacteria utilize this system so effectively that they can concentrate potassium ions as much as 1000 times higher than the concentration outside the cell! Because this activity resembles the action of pumping water uphill, the process of active transport is referred to as a molecular, or ion, pump. The energy for cellular active transport comes from the metabolism of sugars, such as glucose or lactose, for the production of ATP. In *E. coli,* the sugar lactose is actively transported into the cell to be metabolized as a source of energy. This bacterium is thought to have about sixty types of membrane proteins specialized in "pumping" different ions, sug-

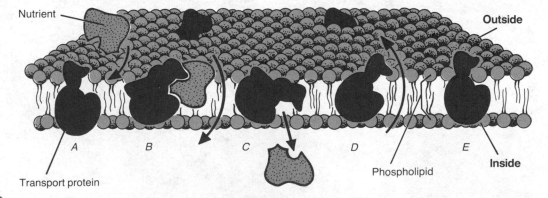

FIGURE 4.12

Facilitated diffusion consists of three parts: (1) nutrients move from an area of high concentration to an area of lower concentration; (2) the nutrients pass through a membrane; and (3) membrane proteins (permeases) assist in the passage. *(A)* represents the protein before the nutrient has "fit" onto the surface. After the nutrient and protein have combined, the protein's structure is altered *(B).* This shape change causes the protein-nutrient combination to rotate, moving the nutrient inside. After the nutrient is released *(C),* the protein resumes its original shape and rotates back into position *(D* and *E).*

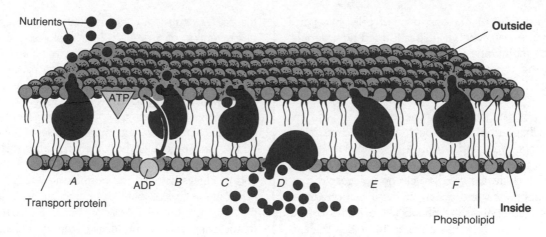

FIGURE 4.13

Active transport is the movement of nutrients across a membrane with the help of ATP energy and membrane proteins called permeases. The expenditure of energy is necessary to move the molecule against a concentration gradient, i.e., from a low to a higher concentration. The original protein (A) is altered by ATP (B) so that it may combine with the nutrient (C). The transport protein, a permease, then moves the nutrient into the cell (D). Following this action, the protein moves back into position (E and F).

ars, and amino acids. The active transport of these molecules is vital to a cell, and disruption of this activity will result in the cell's death. Toxins known as **colicins** are able to interfere with the energy utilization portion of active transport in bacteria. They destroy the pumping of such important molecules as lactose and certain amino acids. The continuous operation of all components of the plasma membrane is vital to the cell: all three molecular types (phospholipids, proteins, and carbohydrates) are in some way involved with regulating the movement of molecules into and out of the cell.

---

1 What elements are unique to the proteins in comparison to lipids? In comparison to carbohydrates?

2 What are the two primary differences between active transport and facilitated diffusion?

CARBOHYDRATES AND ENDOCYTOSIS

The phospholipids of the plasma membrane regulate diffusion and osmosis. Membrane proteins control active transport and facilitated diffusion. The carbohydrate components of the membrane, glycoproteins and glycolipids, are linked to the movement of materials by the process of **endocytosis.** The term "endocytosis" is used to describe the enclosing of materials outside of the cell in a membrane pouch called a vacuole or vesicle. Two forms of endocytosis known

to occur are called phagocytosis and pinocytosis. **Phagocytosis** refers to the endocytosis of relatively large particles of food. For example, bacteria or other microbes may be phagocytosed by protozoans. The material that is taken into the cell is enclosed in a vacuole that may be seen through a microscope on higher magnification. **Pinocytosis** refers to the endocytosis of water or large molecules in solution. The fluid is taken into the cell in very small vacuoles called pinocytotic vesicles, which require an electron microscope to be seen. Both phagocytosis and pinocytosis require metabolic energy, the presence of Ca^{+2} ions, and membrane carbohydrates. A great deal of information about these processes has been gained from the study of the familiar protozoan *Amoeba* and some human phagocytes known as **leukocytes** ("leuko" means *white;* white blood cells). Phagocytosis by these cells is usually called engulfment. Portions of the membrane **(pseudopods)** extend out to surround the organism being ingested, and the membrane fuses together to enclose the food inside a vacuole. The process of phagocytosis is one of the body's most important defense mechanisms. There are many types of phagocytes capable of engulfing foreign materials or microbes, including the **neutrophils, macrophages,** and **monocytes.** The activity of the phagocytes helps prevent some microbes from causing harm to the body. As one of these cells engulfs a microbe, the phagocyte releases a highly reactive form of oxygen, singlet oxygen, into the newly forming vacuole, and this oxygen chemically combines with the pathogen to destroy its activity. Enzymes are then released into the vacuole that will break down the microbe into simpler components. The vacuole membrane is dismantled, leaving the nutrients free in the cytoplasm of the phagocyte.

Phagocytosis and pinocytosis in other cells take place in much the same way. The vacuole is formed by a depression in the cell membrane which encloses the nutrient and seals itself at the top (Figure 4.14). The sealing action is similar to the self-sealing action mentioned earlier (see Figure 4.9).

Cells do not phagocytose surrounding materials constantly or in a haphazard way. The molecules to be taken in must be recognizable and bound to specific sites on the membrane before the action will occur. These **recognition binding sites** include the glycoprotein and glycolipid molecules. The importance and action of these carbohydrates can be seen by comparing the phagocytic activities of two different cells. One cell type has a cover that hides the carbohydrate molecules, while the other has had the cover removed. The cell with the exposed carbohydrates is able to phagocytose protein more rapidly than the cell that is covered. Macrophages and neutrophils can recognize materials that can be phagocytosed. The surface of many materials and cells contain molecules that "fit" the binding site of the macrophage membranes, so that if the two substances make contact, phagocytosis may occur. Some pathogenic bacteria have their surface recognition molecules covered, allowing them to avoid being taken in. The bacterium *Streptococcus pneumoniae* has a capsule that covers some of its important surface recognition molecules and makes it extremely resistant to phagocytosis by macrophages. Some malignant cancer cells have very little covering material; thus, their surface recognition molecules are easily identified. This may be why some cancers are destroyed by the body before they become well established.

Many chemicals are able to interfere with the normal, smooth operation of cell membranes and cause death of the cell. A **drug** is a chemical that is used in the treatment or prevention of a disease. Microbiologists have discovered or created a variety of drugs that are used to inhibit cell membrane functions of pathogenic microbes. Examples of these drugs include polymyxin, novobiocin, and nystatin. The polymyxins are able to destroy the ability of some bacterial membranes to regulate osmotic pressure. Bacteria like *Pseudomonas aeruginosa,* a common cause of infections in skin burns, become coated with the drug after it is given to the patient. The polymyxin concentrates in the phospholipid layer and produces holes in the membrane which allow essential cell contents to leak out. This particular drug does not work this way on fungal infections, since these eucaryotic cell membranes contain protective steroid molecules that are absent in bacteria like *Pseudomonas.*

SUMMARY

The concept of the cell has developed only after many years of study. At one stage only two regions, the cytoplasm and the nucleus, were identified. At present, numerous cell organelles are recognized, some of which are nonmembranous and others are membranous. The basic structure of the unit membrane, now known as the fluid mosaic model, is based on the idea that the membrane is a sheet of two layers of phospholipid. Protein molecules are found on both sides of the membrane; some float on one side or the other; other protein molecules penetrate through the double layer of lipid. These proteins provide structural support to the lipid and are involved with the transportation of molecules through the surface. Facilitated diffusion utilizes these proteins in moving materials from one side of the membrane to the

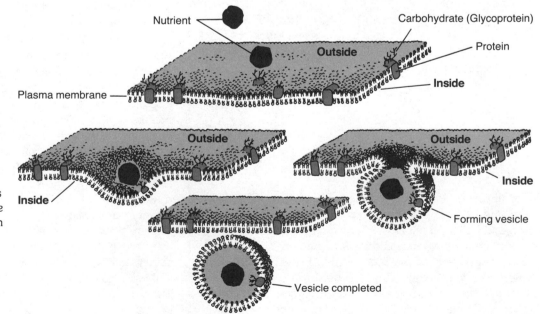

FIGURE 4.14
There are three types of endocytosis: engulfment, phagocytosis, and pinocytosis. In phagocytosis, the cell "eats" nutrients such as carbohydrates and proteins; in pinocytosis, the cell "drinks" water or ions from the surrounding environment. As the newly forming vacuole (large pouch) or vesicle (small pouch) depresses into the cell, the phospholipid layer fuses to seal the membrane.

other. The molecules are moved from an area of high concentration to an area of lower concentration. No metabolic energy is required for this transport. Energy is required for active transport. This transporting method utilizes membrane proteins to move materials through the membrane from an area of low to an area of higher concentration or from an area of high to an area of lower concentration. The membrane also influences the movement of molecules by diffusion and osmosis. All transport processes aid in maintaining the stability of the cell and the biochemical reactions that occur within the cell. The carbohydrates found on the outside of membranes are involved with the recognition and binding of material to be taken into the cell by the process of endocytosis. Engulfment, phagocytosis, and pinocytosis are different forms of endocytosis. The maintenance of the structure and function of the membrane is vital to the cell's operation. Some disruptions of the membrane may be so harmful that the natural self-sealing operation may not enable repair before the cell contents leak out and the cell dies. A number of drugs can cause such harm and may be used to control microbes.

WITH A LITTLE THOUGHT

Phagocytosis by white blood cells is one method of controlling infection in the human body. A number of infectious microbes avoid being destroyed by this process because they have a protective cover. This prevents these white cells from "recognizing" the invaders since they cannot make effective contact with their cell membranes; that is they do not "see" the microbes. What methods might be used to improve this recognition process in the case of an active infection, or to prevent an infection?

STUDY QUESTIONS

1   Distinguish among cytoplasm, nucleus, and protoplasm. How have these concepts changed over the years?

2   What is the difference between an inclusion and an organelle? Give an example of each.

3   Describe the structural arrangement of the following molecules in the Danielli-Davson membrane model: protein, phospholipid, and carbohydrate. How does this differ from the fluid mosaic model?

4   What does it mean to say something is hydrophobic? hydrophilic? How do these traits affect membrane formation?

5   Describe a hypothetical situation in which passive diffusion would occur and benefit the cell.

6   What characteristic of the membrane could be related to differential permeability?

7   How do the processes of passive and facilitated diffusion differ?

8   In what ways do the processes of facilitated diffusion and active transport resemble one another?

9   Describe endocytosis. What role might the surface carbohydrates play in this process?

10   In what way might a drug be used to control a microbe? Base your answer on membrane structure and function.

SUGGESTED READINGS

BRANDON, D. "Freeze-etching Studies of Membrane Structure." *Philosophical Transactions of the Royal Society,* B261, 133, 1971.

LODISH, H. F., and J. E. ROTHMAN. "The Assembly of Cell Membranes." *Scientific American,* Feb., 1979.

LUCY, J. A. "The Plasma Membrane." *Oxford Biology Reader,* Oxford University Press, 1975.

LURIA, S. E. "Colicins and the Energetics of Cell Membranes." *Scientific American,* Dec., 1975.

NOVIKOFF, A. B., and E. HOLTZMAN. *Cells and Organelles,* 2nd ed. Holt, Rinehart & Winston, New York, 1976.

SINGER, S. J., and G. L. NICOLSON. "The Fluid Mosaic Model of the Structure of Cell Membrane." *Science,* 175:720–31, 1972.

KEY TERMS

inclusion	plasma (cell) membrane	osmosis
organelle	hydrophobic	plasmolysis
PHB (polyhydroxybutyric acid)	hydrophilic	halophile
volutin (metachromatic granule)	fluid mosaic model	facilitated diffusion
membranous organelle	glycoprotein	active transport
nonmembranous organelle	glycolipid	endocytosis
	passive diffusion	phagocytosis
	differentially permeable	pinocytosis

Pronunciation Guide for Organisms

Corynebacterium (ko-ri"ne-bak-te're-um)
   diphtheriae (dif-the're-eye)

Beggiatoa (bej"je-ah-to'ah)

Thiothrix (thi'o-thriks)

Bacillus
   thuringiensis (thu-rin-jen'sis)

Pseudomonas (su"do-mo'nas)
   aeruginosa (a"er-u-jen-o'-sah)

Halobacterium (hal"o-bak-te're-um)

Streptococcus
   pneumoniae (nu-mo'ne-eye)

Organelles and Their Importance

THE DIFFERENCE IS OBVIOUS
THE CELL'S WORK SURFACE
FURTHER SURFACE SPECIALIZATION
SEPARATE BUT EQUAL
THE GREENING OF A CELL
EXTERNAL ORGANELLES OF LOCOMOTION
PILI (FIMBRIAE)
ENDURING CHANGES
THE RIBOSOMES:
TRANSLATOR OF THE CELL
THE CONTROL CENTER
OUTER COVERINGS
CAPSULES

Learning Objectives

☐ Understand the basic concept of the eucaryotic nucleus in contrast to the procaryotic nucleoid.

☐ Be familiar with the structure and function of the endoplasmic reticulum and know how the procaryotic cell types accomplish these functions using alternative surfaces.

☐ Understand the structure and function of the eucaryotic Golgi apparatus and lysosomes, and how these affect phagocytosed bacteria and other material.

☐ Be able to describe mitochondria, chloroplasts, and the fundamental metabolic pathways associated with these plastids.

☐ Explain the basic structural differences between eucaryotic and procaryotic organelles of locomotion.

☐ Be familar with the function of two types of procaryotic pili.

☐ Be able to define the terms "endospore" and "vegetative cell," and know the significance of each.

☐ Contrast eucaryotic and procaryotic ribosomes and state their functions.

☐ Understand the relationship among chromosomes, chromatids, chromatin, nucleoprotein, and DNA.

☐ List the major events of binary fission.

☐ Be familiar with the basic materials found as cell wall material.

☐ Be able to describe the structure of procaryotic cell walls.

☐ Be familiar with the nature of procaryotic capsules and their significance.

Cells contain a variety of organelles, each specialized to perform a particular function necessary to maintain life processes. These organelles are not simply passive structures for chemical reactions but are essential to the life of the cell. Each organelle is dynamic in its operation, changing shape and size as it works. They move throughout the cell, and some even produce duplicates of themselves. There is a

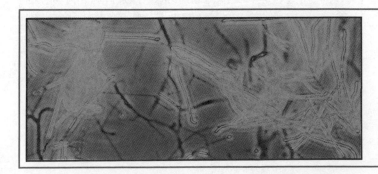

5

constant turnover of organelles within a cell as they continuously undergo change during their particular function. The organelle may be destroyed during the life of the cell; however, these lost organelles are rapidly replaced, and the total number of a particular organelle remains relatively constant.

Among the variety of organelles present in a cell, the **membranous organelles** are all constructed of phospholipid and protein. This is the sheetlike membrane that is the building material for the plasma (cell) membrane. The plasma membrane is a true organelle with specific abilities, and many of its characteristics are found in other membranous organelles. The unique shape and composition of each membranous organelle enables it to specialize in its function. However, organelles may be interconverted during the normal operation of the cell due to the common membrane structure.

In addition to the specialized surfaces of the membranous organelles, there is another essential group of cell structures called the **nonmembranous organelles.** The organelles in this group are *not* composed of the basic unit membrane structure; each has a different structure that sets it apart from both membranous organelles and other nonmembranous organelles.

The functions carried out by all cells are basically the same. Procaryotic and eucaryotic cells, however, differ greatly in the kinds of organelles they contain. This chapter will concentrate on the structure and function of organelles, and how that information has been put to use by microbiologists and the medical profession. The structure and function of eucaryotic organelles will be described first since they display a greater degree of specialization. Following that description, a comparison will be made with a "typical" procaryotic cell. Be sure that you separate the two cell types as you read and note the important differences.

THE DIFFERENCE IS OBVIOUS

The most significant difference between procaryotic and eucaryotic cells, the **nucleus,** was discovered in 1883 by Robert Brown. At that time, the nucleus was described simply as a central "body." Brown's staining method enabled him to see a colored body inside the cells, but he was unaware that the stain actually colored the genetic material inside the nucleus and not a surrounding membrane. Until advances were made in microscopic techniques, the true nature of the nucleus was not clear. The nucleus is a *place* in the cell, not a solid mass. Just as a room is a place created by the presence of walls, floor, and ceiling, the nucleus is a place in a cell created by the presence of the **nuclear membrane.** If the membrane were not formed around the

genetic material, the nucleus would not exist. Procaryotic cells have no membrane enveloping their nuclear material; therefore, they lack a true nucleus. The place in a procaryotic cell containing the DNA is called the **nucleoid** ("-oid" means *like,* or *similar to*), or nuclear body. The DNA is not confined by a surrounding membrane (Figure 5.1). In eucaryotic cells, the nuclear membrane is formed from flattened sacs of membrane. This is seen in the electron microscope as a double layer of unit membrane. The sacs come together to form a relatively smooth cover which envelops the genetic material. Eucaryotic cells have pores in the nuclear membrane, and it is believed that these pores regulate the movement of macromolecules such as RNA (involved in the synthesis of protein) into and out of the nucleus. One form of RNA, called ribosomal RNA, is gathered together in the nucleus to form the **nucleolus** in eucaryotic nuclei.

THE CELL'S WORK SURFACE

Many times electron micrographs show the nuclear membrane to be connected to another important membranous organelle, the **endoplasmic reticulum (ER).** The endoplasmic reticulum serves as a channel system for the movement of molecules throughout the cell and also as a surface on which reactions may occur. The membranes of the endoplasmic reticulum are arranged in stacked layers to form what appears to be channels or isles through the cytoplasm of eucaryotic cells. This complex series of membranes is unique to eucaryotic cells and is not found in any of the procaryotes. Even though the endoplasmic reticulum appears in electron micrographs as thin lines (Figure 5.2), each line is actually composed of a sheet of membrane. The enormous surface area provided by this structure is related to its function as a **surface catalyst,** since many metabolic reactions occur on its surface. After the reactions are completed, the ER is available to be reused.

Close examination of many electron micrographs has revealed that the ER is an interconnected series of membranes forming large sacs that have inside and outside surfaces. The outside surface of one form of ER appears "rough" because of the presence of **ribosomes,** one kind of nonmembranous organelle. These little, protein-nucleic acid (ribosomal RNA) units are anchored to the ER. The second form of ER does not have anchored ribosomes and appears as "smooth" little tubules. As will be described later, ribosomes are the sites of protein synthesis. Cells with extensive amounts of **rough ER** produce large amounts of protein, while the **smooth ER** is a place where other chemical reactions such as lipid synthesis and detoxification processes occur. The detoxification reactions convert harmful sub-

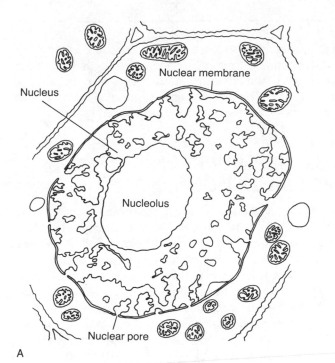

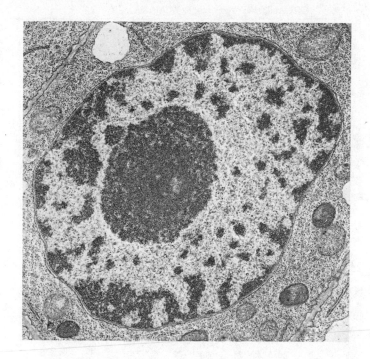

Nucleus

Nuclear membrane

Nucleolus

Nuclear pore

A

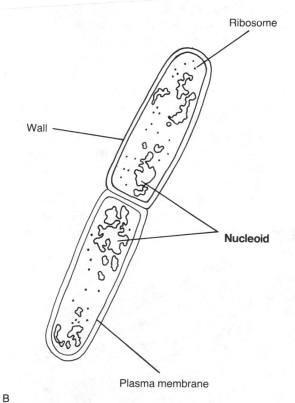

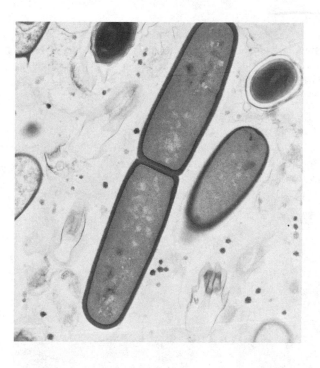

Ribosome

Wall

Nucleoid

Plasma membrane

B

FIGURE 5.1

(A) A eucaryotic cell and the surrounding nuclear membrane. The dense mass within the nucleus is the nucleolus, a concentrated area of RNA; (B) A bacillus baterium showing the nucleoid, where the genetic material is located. There is no surrounding nuclear membrane in procaryotic cells. (Photo A courtesy of Biophoto Associates/Myron C. Ledbetter/Brookhaven National Laboratory. Photo B courtesy of Carolina Biological Supply Company.)

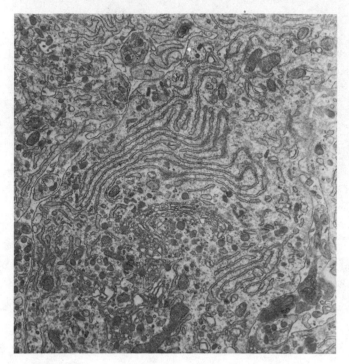

FIGURE 5.2
Eucaryotic cells contain an endoplasmic reticulum, a membranous organelle composed of sheets of unit membrane. The increased surface area created by this organelle increases the metabolic efficiency of the cell. (Courtesy of Biophoto Associates.)

stances into harmless chemicals normally found within the cell. Procaryotic cells have no endoplasmic reticula, but the same kinds of metabolic reactions still occur in these cells. Similarly organized sequences of reactions are carried out in order to keep procaryotic cell types alive. Protein synthesis,

lipid synthesis, and detoxification reactions in bacteria occur both inside the plasma membrane or free in the cytoplasm. The plasma membrane of procaryotes is relatively smooth with few folds. The total surface area provided by the plasma membrane is far less than the area provided by ER.

FURTHER SURFACE SPECIALIZATION

Another organelle is the **Golgi apparatus,** which resembles the ER in many ways. Its basic structure is composed of unit membrane, and the Golgi's function is closely related to the activities that occur on the endoplasmic reticulum.

Cells produce both useful and useless chemical products that must be released from the cell in a manner that does not interfere with other metabolic processes. These products are often assembled in and released through the operation of the Golgi apparatus of eucaryotic cells (Figure 5.3). Cells cannot afford to lose a great deal of water during the elimination of cell products, since all the chemical reactions in a cell are basically carried out in water. If too much water is lost, the cell becomes dehydrated and may die. The cell cannot simply open its pores and release materials to the environment. The structure and function of the Golgi is able to help overcome this problem.

The Golgi is composed of a stack of three to seven membranous sacs that resemble a stack of pancakes. The edges of this stack show the formation of vesicles which eventually "bud," or pinch off, from the main sac. Types of molecules secreted by Golgi include mucus, proteins, glycoproteins, hormones, and polysaccharides.

The specialized surfaces of the Golgi are not found in all cells: procaryotes function in a more basic manner. They,

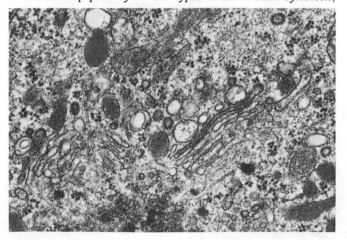

FIGURE 5.3
The Golgi apparatus is a membranous organelle in eucaryotic cells. (Courtesy of B. Nichols, D. Bainton, and M. Farquhar. *Journal of Cell Biology,* 50:498, 1971. By copyright permission of The Rockefeller University Press.)

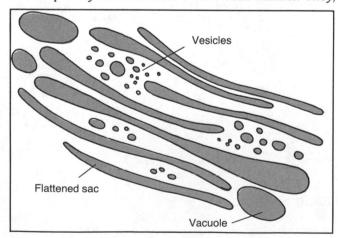

Vesicles

Flattened sac

Vacuole

too, eliminate products and activated enzymes, but must accomplish these processes without the help of this membranous organelle. Bacterial cells rely on the plasma membrane to regulate the movement of products to the outside. A number of enzymes that are potentially dangerous to the cell are synthesized only when necessary and in a position on the membrane that allows them to be quickly moved through to the outside, safeguarding the interior of the cell. The lack of a Golgi apparatus results in fewer types of vacuoles and vesicles in procaryotic cells, and places greater importance on maintaining the structure and efficient functioning of the plasma membrane.

> **1** What are the differences among a nucleus, nucleoid, and nuclear membrane?
> **2** What happens on the endoplasmic reticulum?
> **3** How does the Golgi differ from the ER?

Water, food, wastes, and minerals are usually stored in one large, centrally located **vacuole** in most algae. In protozoans, many smaller, individual vacuoles typically contain food or other materials. A vacuole is composed of a single surrounding membrane and its contents. Vacuoles may form by endocytosis from the cell membrane, Golgi apparatus, or ER. A specialized type of vacuole has been identified that is formed by the Golgi and is called **lysosome** ("bursting body"). The lysosome contains the proteolytic (protein-destroying) and lipolytic (lipid-destroying) enzymes. Because these enzymes are so highly concentrated in these small vacuoles, they stain very heavily and are easily recognized by the electron microscope (Figure 5.4).

Lysosomes help destroy pathogenic bacteria and fungi. When a person becomes infected by one of these microbes, neutrophils can phagocytose the pathogen and encase it in a vacuole. In a very selective manner, the neutrophil's lysosomes fuse with the vacuole containing the bacteria and allow the contents of both to mix (Figure 5.5). The powerful lysosomal enzymes react with the microbe and destroy it. One of the enzymes aiding in this destruction is called **lysozyme,** which is able to destroy the cell walls of certain bacteria and leave the phagocytosed cell exposed to the effects of other destructive enzymes.

SEPARATE BUT EQUAL

Plastids are a group of membranous organelles with the following basic characteristics:

1 All are double-membraned structures that seldom, if ever, interchange segments.

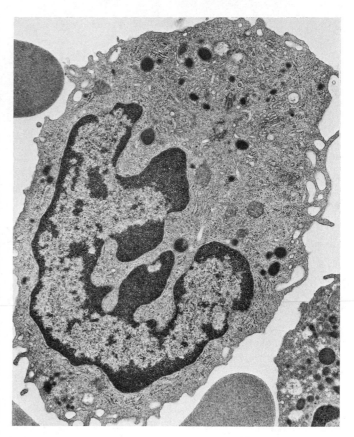

FIGURE 5.4
The dark circular areas in the cytoplasm of this guinea pig blood monocyte are single-membraned vesicles called lysosomes. Lysosomes are formed by the Golgi apparatus. Each of these vesicles contains a collection of powerful enzymes that is used in the digestion of many kinds of substances, including invading microbes. (Courtesy of B. Nichols, The Proctor Foundation for Research in Ophthalmology, University of California, San Francisco.)

2 Each contains special pigments such as **chlorophyll, carotene,** or **cytochrome.**

3 Each of these organelles is able to control its own reproduction within the host cell.

Plastids contain their own DNA, which differs from the nuclear DNA. The presence of genes within the plastid enables these organelles to synthesize some of the enzymes necessary for their functioning. The action of this extra-nuclear (outside the nucleus) DNA is not random but is coordinated with the genes in the nucleus. The two most familiar plastids are the green-chlorophyll-containing **chloroplasts** found in plant cells and the **mitochondria** found in all eucaryotic cells (Figure 5.6). Both organelles are closely associated with energy-producing reactions in the cell.

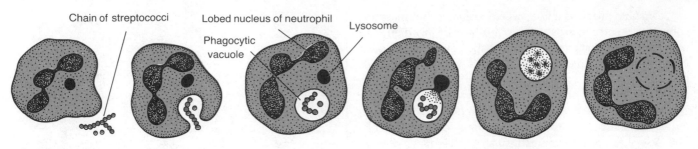

(1) Neutrophil moves toward bacteria   (2) Phagocytic vacuole forms around bacteria   (3) Lysosome moves toward completed vacuole   (4) Lysosome and vacuole fuse   (5) Bacteria are destroyed in vacuole   (6) Vacuolar membrane fragments releasing contents

FIGURE 5.5

The illustrations show the phagocytosis of a chain of streptococci by a white blood cell (1–3), and the fusion of the newly formed phagocytic vacuole with a lysosome. After the lysosomal enzymes inside the vacuole kill the bacteria, the vacuole is fragmented and the contents freed in the leukocyte's cytoplasm.

Eucaryotic cells contain from 1000 to 10,000 mitochondria, which are usually found where very large amounts of energy are being used by the cell. As the mitochondria carry out their reactions, they utilize oxygen and change shape, but most of these double-membraned organelles are seen as elongated ovals. The outer membrane is smooth and covers an inner, folded membrane, the folds of which are called **cristae.** They contain short, rodlike projections with small particles attached to their surfaces and pigment molecules known as **cytochromes.** The space between the cristae is filled with a jellylike material called the **matrix,** in which are found ribosomes used in protein synthesis.

Contained within the mitochondria are specific enzymes responsible for the chemical reactions that release usable energy (ATP). These enzymes are not arranged in a haphazard manner, but are in a sequence that speeds the breakdown of food molecules. The orderly breakdown of food by these enzymes furnishes the energy required for the proper functioning of the cell. The energy-releasing reactions carried out in the mitochondria are a part of the **aerobic cellular respiration** of glucose. It is in the mitochondria that oxygen is used in aerobic respiration: no other place in eucaryotic cells is used for this purpose. Eucaryotic cells that lose their mitochondria cannot use free oxygen and will die.

The lack of mitochondria is not always deadly to cells. These double-membraned organelles are not found in any of the procaryotes, but cellular respiration still occurs in many of them because important energy-yielding reactions are carried out by alternative methods. Some procaryotes use other membranes to generate their ATP. In some bacteria, the plasma membrane folds into the cytoplasm to form a special structure known as a **mesosome** (Figure 5.7). This folded extension increases the inner surface area of the plasma membrane. When studied biochemically, the mesosome shows a great amount of enzymatic activity associated with ATP formation.

1 Lysosomes contain which enzymes?
2 Name two kinds of plastids.
3 Mesosomes are found in what types of cells?

THE GREENING OF A CELL

Another energy-converting organelle is the plastid known as the **chloroplast,** which is a double-membraned sac found in the cells of algae and higher plants, and contains chlorophyll. In these organelles, light energy is converted to chemical-bond energy by the process called photosynthesis. This form of energy is found in food molecules. Electron micrographs of chloroplasts reveal that the entire organelle is enclosed by a membrane, and other membranes are folded and interwoven throughout. In some areas, concentrations of these membranes are stacked-up or folded back on themselves. Chlorophyll molecules are attached to these membranes and resemble daubs of chlorophyll placed on sheets of membrane. The membrane is then folded in such a way that the daubs of chlorophyll are stacked on top of one another. These areas of concentrated chlorophyll-containing membranes are called the **grana** of the chloroplast. The space between the grana, which has no chlorophyll and few membranes, is known as the **stroma.** The increased surface area provided by the grana increases the efficiency of its operation, just as the cristae increase the efficiency of mitochondria. However, chloroplasts are found in a greater variety of shapes and sizes than mitochondria. Some are oval, others are star-shaped, while still others are spiral.

No chloroplasts are found in bacteria, but extensive membrane systems have been identified. The cyanobacteria

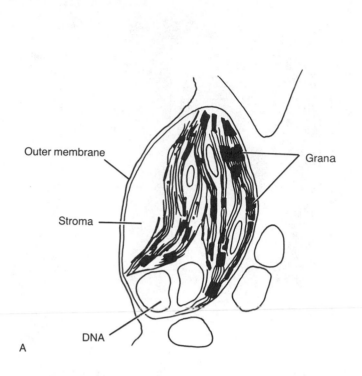

Outer membrane

Grana

Stroma

DNA

A

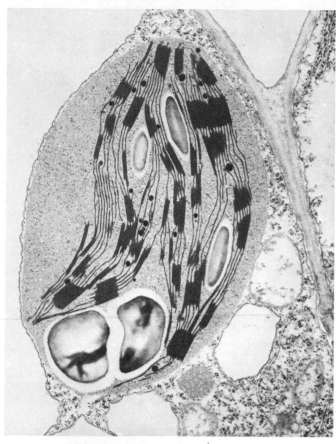

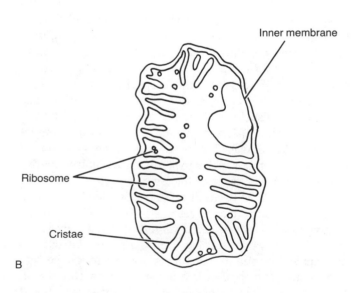

Inner membrane

Ribosome

Cristae

B

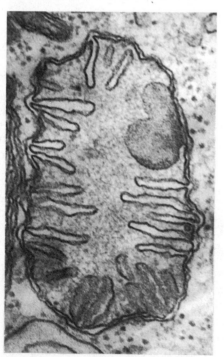

FIGURE 5.6

Mitochondria and chloroplasts are two membranous organelles which belong to a group known as the plastids. Both have a smooth outer membrane and an internal membrane that is specialized for certain biochemical pathways. The chloroplast (A) is the site of photosynthesis, while the mitochondria (B) is the site of aerobic cellular respiration. (Photo A courtesy of Biophoto Associates/Myron C. Ledbetter/Brookhaven National Laboratory. Photo B courtesy of Biophoto Associates.)

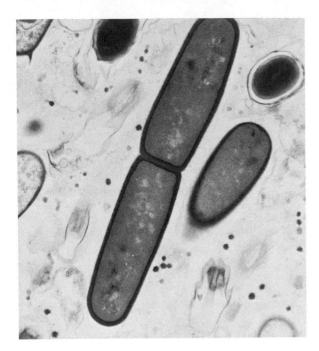

FIGURE 5.7
This photograph shows a well-developed mesosome in a bacillus bacterium. This particular cell is dividing. The mesosome is attached to the plasma membrane next to the dark constriction of the cell center. (Courtesy of Carolina Biological Supply Company.)

have an internal network of membranes on which the chlorophyll molecules are located. This system of flattened membranous sacs is called a **thylacoid** and is connected at only a few points to the cell membrane (Figure 5.8). Other photosynthetic bacteria contain a membranous network similar to the mesosome mentioned earlier. In these cells, the photosynthetic membrane is an extension of the plasma membrane and increases as photosynthetic activity increases.

EXTERNAL ORGANELLES OF LOCOMOTION

An arrangement of **microtubules** ("little tubes") found in some eucaryotic cells is associated with hairlike structures that extend out from the cell surface. These are the non-membranous organelles, **cilia** or **flagella.** Cilia are more numerous and shorter than flagella. Ciliated cells are known to line the oviduct of a female and move an egg from the ovary to the uterus. Each cell that lines the human trachea has over 250 cilia. As foreign materials, such as bacteria, become trapped in the thin mucus lining of the trachea, the

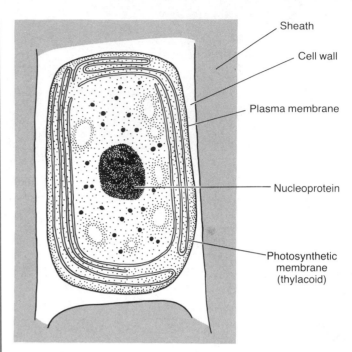

FIGURE 5.8
The cyanobacteria carry out photosynthesis using a special bacterial chlorophyll molecule. This photosynthetic pigment is located on the thylacoid, a special membrane system found throughout the cell. The thylacoid has no surrounding membrane, and it is attached at only a few places to the plasma membrane. (From E. Gantt and S. F. Conti. *Journal of Bacteriology,* 97:1486, 1969.)

ciliary action moves the potentially dangerous microbes away from the lungs. An entire class of protozoans, the Ciliata, propel themselves through their environment by the stroking motion of their cilia. The protozoan class known as the Mastigophora (Flagellata) is made up of cells that move as a result of the whipping motion of one or a few flagella. Both cilia and flagella of eucaryotic cells have a similar structure. Each unit appears as a tube with an internal framework of microtubules. The outer cover is an extension of the plasma membrane; internally the microtubules are arranged in a 9 + 2 pattern* (Figure 5.9).

The flagella of bacterial cells are considerably different than those found in eucaryotic cells. Bacterial flagella are composed of a protein called **flagellin.** These flagella are so small that special staining techniques must be used to accentuate their size to make them visible through the light microscope. The flagellin is arranged in a coiled helix attached to the cell by a complex structure known as a **basal body.** The basal body is found beneath the cell wall and may have two sets of rings depending on the type of bac-

*9 outer doublets + 2 center microtubules.

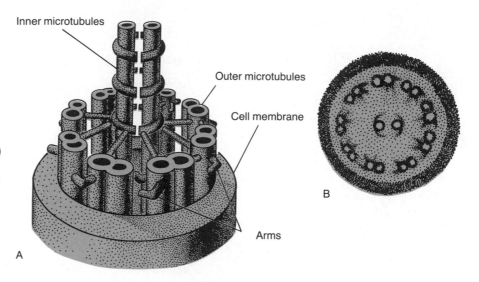

FIGURE 5.9
Cilia and flagella of eucaryotic cells both have the same basic 9 + 2 microtubular pattern. (A) The complex arrangement of the tubulin protein microtubules inside these hairlike appendages. (B) A cross-section of a cilium. (From B. Afzelius. "A Human Syndrome Caused by Immotile Cilia." *Science*, 193:317–19, 1976.)

teria; one set is positioned in the plasma membrane, while the other (if present) is in the outer portion of the cell wall. Bacteria swim by rotating their flagella as if they were stiff propellers. The motion of the flagella is the result of a selective exchange of ions in the basal body that causes the entire unit to rotate. This function is similar to the operation of an electric motor. The rate and pattern of motion of flagellated bacteria depend on how many flagella are on a cell and their arrangement. **Monotrichous** (one flagellum per cell) bacteria generally move faster than **peritrichous** (many flagella uniformly around the cell) or **lophotrichous** (tufts of flagella) bacteria (Figure 5.10). **Amphitrichous** bacteria (those with flagella on opposite ends) are intermediate in speed.

Flagellated bacteria have chemical-sensing devices on their cell surface (chemoreceptors) that control the action and motion of the moving cell. The cells may be attracted or repelled by chemicals in the environment. The movement of a microbe toward or away from a chemical is called **chemotaxis.** Bacteria respond in a very complex manner to changes in their environment. They have the ability to sense changes in surrounding chemicals over a period of time (a temporal change) and also have what some call a "memory device," which allows them to compare past conditions to present conditions, then change this direction of movement toward the more favorable area. The cells "run" in one direction, stop and "twiddle" randomly in a tight area,

then "run" again. This irritability response allows the microbe to locate the best environment and increases its chances of survival.

PILI (FIMBRIAE)

Some bacteria also have tubelike projections known as **pili** (*sing.,* "pilus" means *hair*). These are nonmembranous, hollow protein tubes that may be either short or long. The long pili are involved in sexual gene transfer, **conjugation,** and are known as the F, or sex, pili (Figure 5.11). During the conjugation the "male," or donor cell, pairs with the "female," or recipient cell. The donor cell makes contact by having the sex pilus join with the surface of the recipient cell. The sex pilus (only found on the donors) has a specific recognition site on its surface that reacts with a comparable site on the recipient. When contact is made, the sex pilus joins the two to form a tubular connection between the cells. Once securely joined, the DNA of the donor passes through the pilus to the recipient.

The second type of pili are shorter and more numerous. These have been associated with the ability of cells to stick or adhere to other cells or surfaces. The presence of adhesive pili is directly related to a bacteria's ability to initiate disease. For example, it has been demonstrated that *Neisseria gonorrhoeae* which produce no pili are less likely to establish an infection, since they are unlikely to adhere to the cells lining the urogenital tract.

ENDURING CHANGES

Some types of bacteria are able to exist in two different forms (Table 5.1). The **vegetative cell** is the growing form

---

1 What kind of cells contain microtubules?
2 How do bacterial flagella differ from eucaryotic flagella?
3 What does the term "chemotaxis" mean?

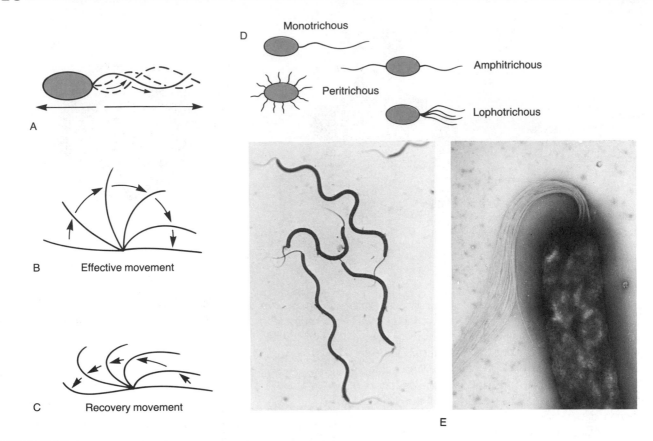

FIGURE 5.10

(A) The wave motion of the flagellum may move the microbe so that the "hair" either trails or precedes the cell. Ciliary action may resemble a swimming stroke. (B) The effective stroke is the stiff, power stroke that causes motion. (C) The recovery stroke is flexible and repositions the cilium for another power stroke. (D) The distribution and number of bacterial cells are classified as monotrichous, peritrichous, amphitrichous, or lophotrichous. (E) Because the flagella are so small, a special staining technique that causes the flagella to swell must be used to highlight these nonmembranous organelles. (Photos courtesy of Carolina Biological Supply Company.)

that exhibits all of life's characteristics and may be easily destroyed by physical or chemical changes in the environment. The other form is a nongrowing, heat-resistant structure that may be formed within the vegetative cell and is called an **endospore.** The chemical makeup of the endospore makes it very difficult to stain, and under the microscope it may appear as a shiny, unstained white spot inside the rod-shaped cell. Special staining techniques must be used to color the endospore and make it more visible under the light microscope.

The formation of this structure is an adaptive feature and takes place in response to changes in the environment (Figure 5.12). The endospores of the bacteria *Bacillus* and *Clostridium* are unique in that they are more heat resistant than spores formed by other types of microbes. Bacterial endospores are able to withstand the *effects* of strong chemicals, such as acids and disinfectants, and resist the destructive action of radiation.

When the environment of the vegetative cell begins to change in a direction that is harmful to the cell (e.g., higher temperatures or fewer nutrients), a sequence of observable changes occurs, and the endospore is formed. Layers of material accumulate around DNA and other cell components, and a unique molecule, **dipicolinic acid (DPA),** is deposited in the endospore. It is the DPA that accounts for the heat-resistant quality of the endospore. A calcium-DPA complex is deposited in the endospore wall and maintains the dehydrated, hydrophobic quality of these resistant structures. Once spore formation is complete, the surrounding vegetative cell is nothing but a nonliving shell. Endospores may last for years. Spores isolated from the dry dust inside the pyramids of Egypt were able to change back into their vegetative form after being dormant for thousands of years. The **spore germination** process may take place in a matter of minutes and occurs when the environment is again favorable for growth in the vegetative form.

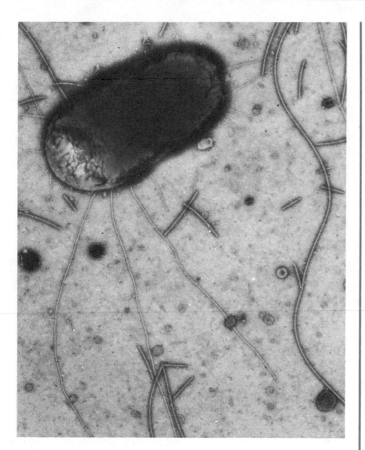

FIGURE 5.11

Some bacteria have tubular projections called sex pili, which are hairlike projections that form a conjugation tube between two cells. The DNA then travels though this tube during sexual reproduction. The donor cell has pili, but the recipient cell does not. (Courtesy of D. Bradley, Faculty of Medicine, Memorial University of Newfoundland.)

THE RIBOSOMES: TRANSLATOR OF THE CELL

The **ribosomes** of procaryotic and eucaryotic cells may be either free or attached to the surfaces of membranes (Figure 5.13). Unattached ribosomes synthesize proteins that serve as enzymes, while attached ribosomes manufacture structural proteins or those that are excreted from the cell. Ribosomes are complexes of protein and a nucleic acid. The nucleic acid molecule, unique to these nonmembranous organelles, is called **ribosomal ribonucleic acid (rRNA).** All ribosomes are *not* the same: procaryotic ribosomes are smaller than eucaryotic ribosomes and are composed of different types of protein and rRNA. However, the ribosomes found in eucaryotic mitochondria and chloroplasts are similar to procaryotic ribosomes.

TABLE 5.1
Characteristics of bacterial endospores and vegetative cells.

Characteristic	Endospore	Vegetative Cell
Chemical makeup:		
DPA	Present	Absent
CA^{+2}	High	Low
PHB	Absent	Present
H_2O	Low	High
Enzyme activity level	Low	High
O_2 use in metabolism	Low	High
Growth	No	Yes
Heat resistance	High	Low
Radiation resistance	High	Low
Disinfectant resistance	High	Low

Cells maintain a relatively constant number of ribosomes. Highly active, large eucaryotic cells may have more than 100,000 ribosomes, while bacterial cells have between 5000 and 10,000 ribosomes available for protein synthesis. These nonmembranous organelles are the sites where **messenger ribonucleic acid (mRNA)** and the amino acid

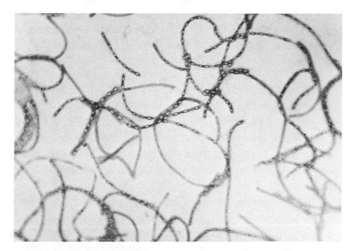

FIGURE 5.12

The endospores formed by bacteria are unique in that they are much more heat resistant than those formed by other microbes. As the vegetative cell experiences changes in the environment, the interior of the cell changes its metabolism to produce the endospore. Layers of material are selectively added to encase the DNA and other cell components in the newly forming endospore. Spore germination takes place when environmental conditions again become favorable, and the endospore breaks from its encasement to return to the vegetative form. (Courtesy of Carolina Biological Supply Company.)

A

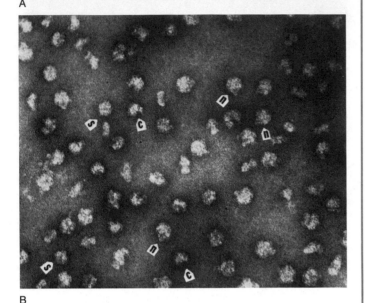

B

FIGURE 5.13
The electron micrographs illustrate a number of bacterial ribosomes separated into their components. (A) A sample of purified 30 S subunits. (B) Dissociated ribosomes: crescent *(c)*, nose *(n)*, and slit *(s)* views of 50 S subunits. (Courtesy of M. Lubin, Dartmouth Medical School.)

building blocks come together. The mRNA molecule attaches itself to the smaller portion of the ribosome to begin the process. The amino acids are transferred to the ribosome by molecules that are so specific they are only capable of transferring one particular type of amino acid. These are cloverleaf-shaped RNA molecules called **transfer RNA (tRNA).**

All proteins synthesized in a cell are coded for by the genetic material, DNA. These macromolecules have very specific structures and are essential to the life of the cell. As in the case of many other nonmembranous organelles, there is a difference between the genetic material of procaryotic and eucaryotic cells.

THE CONTROL CENTER

As stated earlier, one of the first structures to be identified in cells was the nucleus. The nucleus of eucaryotic cells was referred to as the cell center. It has been demonstrated that eucaryotic cells cannot live for an extended period of time

or reproduce without the nucleus. Although procaryotic cells lack a nucleus, they still contain the genetic material required for reproduction.

When nuclear structures were first identified, it was noted that the material in the nucleus became more stained by certain dyes than the rest of the cell. This nuclear material was called colored material, or **chromatin.** Chromatin is composed of long molecules of deoxyribonucleic acid (DNA) in association with proteins known as **histones.** These nucleoprotein molecules, unique to eucaryotes, are the information center for the cell. Unwound and loosely organized DNA in the nucleus is chromatin material. When the nucleoprotein coils into a shorter structure, it is a nonmembranous organelle called a **chromosome** (colored body). Note that chromatin and chromosomes are really different forms of the same thing (Figure 5.14). Both contain the blueprints for the construction and maintenance of the rest of the cell. Each eucaryotic chromosome is made of two parallel, threadlike parts lying side by side. These two parallel threads, called **chromatids,** were made earlier during interphase of the mitotic division process, when the DNA molecules were duplicated. Each of these chromatid pairs is held together at the **centromere**—the point at which the spindle fibers attach during cell division.

The terms **diploid** and **haploid** are used to indicate the number of chromosomes present in eucaryotic cells. Some cells have only one set of chromosomes, and therefore, only one set of genetic data. They are haploid, or have the haploid number of chromosomes. However, other cells have two sets of genetic data and are called diploid.

The two most familiar types of reproduction among eucaryotic cells are mitosis and meiosis. Since procaryotic cells have no true nucleus, no form of nuclear division occurs. The distribution of the duplicated genetic material must take place in an alternative manner. This division process is known as **binary fission** (Figure 5.15) and occurs in three main phases: (1) duplication of the genetic material, (2) separation of the DNA into newly forming daughter cells, and (3) cytokinesis (cell splitting).

The DNA duplication process in procaryotic cells is basically the same as that found in eucaryotic cells. However, since bacteria have a single loop of DNA as their genetic material, the duplication occurs in a unique pattern. The duplication process begins at several points on the loop and proceeds in one direction until the entire gene series is complete. This is different from the duplication of eucaryotic chromosomes, where duplication of DNA occurs simultaneously at several points but proceeds in both directions.

Bacterial DNA contains no attached histone protein and does not coil into tight bodies after duplication. No true chromosomes can be seen in these cells. Because of the difference in DNA and its arrangement as a loop in microbes, the terms chromosome, haploid, and diploid have

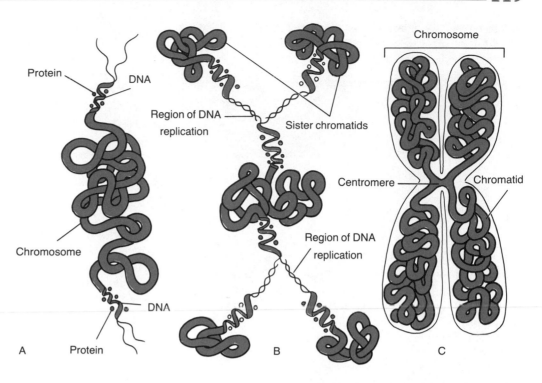

FIGURE 5.14
This illustrates the relationships among DNA, chromatids, and other chromosomal structures. (A) The coiling of an unduplicated strand of DNA and its attached protein, histone. (B) As duplication proceeds, the single strand splits and becomes two separated units known as chromatids. (C) The completion of the process results in the formation of a whole chromosome composed of two chromatids joined at the centromere. (From E. J. DuPraw, *Cell and Molecular Biology,* 1968. Courtesy of Academic Press.)

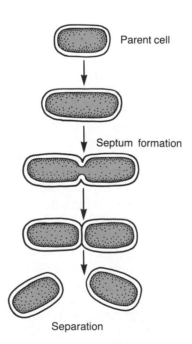

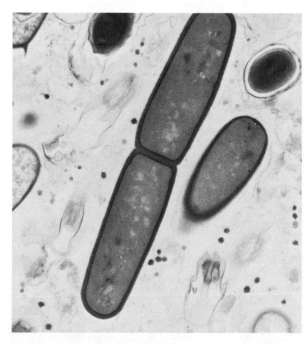

FIGURE 5.15
Binary fission, the division of procaryotic cells illustrated here by the bacterial cell, is more rapid but less organized than either mitosis or meiosis. (Photo courtesy of Carolina Biological Supply Company.)

no real application to procaryotic cells. When a procaryotic cell has completed the duplication process and grown to its maximum size, a cross-wall, or **septum,** begins to form, separating the "mother cell" into two daughter cells. Since no spindle apparatus is present in these cells, there is no assurance that the two loops of genetic material will be separated equally into the daughter cells. If binary fission results in one cell lacking a loop and the other having two loops,

both will probably die. The one cell will have no genes to control its metabolism, and the other will have so many genes that it, in all likelihood, will be hampered from conducting its normal metabolism efficiently. Bacteria are very efficient at segregating their DNA into daughter cells. However, cells that contain a mesosome have an advantage in that they have one of the duplicated DNA loops attached to this membranous organelle. When septum formation occurs, the mesosome and its attached loop become located in only one of the daughter cells. The other loop is free to be moved into the other daughter cell.

---

1. Where is the compound DPA found?
2. Name two bacterial types that form endospores.
3. Name the type of cell division found in bacteria.

OUTER COVERINGS

The **cell walls** of microbes, plants, and fungi appear to be rigid, solid layers of material, but are really loosely woven layers resembling a leaky basket through which many types of molecules easily pass (Figure 5.16). The arrangement of living material inside the confining cell wall lends strength and protection to the cell but hampers its flexibility and movement. However, in some microbes the cell wall may not be present. A bacterial cell, for example, may have the cell wall removed by the action of enzymes in the environment but will not die if the surroundings are osmotically balanced. If the proper nutrients are available, the bacterium may regenerate a new cell wall and appear as it did before the loss.

All cell walls are composed basically of layers of complex carbohydrate molecules. The exact type of material depends on the kind of cell which forms the wall. There are three kinds of material typically used for cell walls. All the higher plants and most algae have **cellulose** as their wall material. It surrounds each individual cell, and when found in large amounts it is known as wood. Most organisms are incapable of breaking down cellulose into its glucose components. However, a few bacteria are able to produce the

enzyme **cellulase** that will do this job. These bacteria are normally found in the intestinal tract of ruminant animals, such as cattle and buffalo, and are responsible for the digestion of the grasses the animals eat.

The second cell wall material is **chitin** and is found in the fungi. One may be more familiar with chitin as exoskeleton material in insects. The outer shell of a beetle or a shrimp is chitin. However, in these animals, the chitin surrounds masses of tissue and is not around individual cells as in mushrooms or molds. The fungi are not crunchy or brittle like a shrimp skeleton because the chitin surrounding each cell is very thin.

The procaryotes have a different cell wall material known as **peptidoglycan** (or mucopeptide). Most bacteria have this material as part of their wall to provide support and shape to the cell.

Peptidoglycan is very thin and difficult to see when viewed through the light microscope. It is composed of both carbohydrates (N-acetylglucosamine and N-acetylmuramic acid) and amino acids (e.g., L-alanine, D-alanine, D-glutamic acid, and either lysine or diaminopimelic acid—DAP). The arrangement of these molecules is shown in Figure 5.17 and consists of layers of the carbohydrate connected by short chains of amino acids (peptides). The shape of a bacterium may be determined by the lengths of peptidoglycan chains and how they are interlinked. The peptidoglycan has never been found in eucaryotic cells but is present in all procaryotic cells with walls. The molecules that make the peptidoglycan layer unique are N-acetylmuramic acid, diaminopimelic acid (DAP) and D-form amino acids. Not all bacteria have the same arrangement or combination of molecules in their cell walls (Figure 5.18). This fact has been put to use in the laboratory and has resulted in one of the most important identification procedures in bacteriology.

On the basis of the **Gram staining reaction,** bacteria are divided into two major groups: those that are not decolorized after staining with dye crystal violet, the Gram-positive bacteria, and those that are decolorized, the Gram-negative bacteria. The differences in cell wall structure and composition are easily identified by the Gram staining method. In most Gram-negative bacteria, the peptidoglycan layer is covered with an "outer membrane" known as the **lipopolysaccharide layer,** or **LPS.** This layer contains a complex arrangement of lipids, carbohydrates, and proteins, and is linked to the outer layer of the true cell wall (Figure 5.19). Between the LPS and the peptidoglycan layer is an area called the **periplasmic space.** The LPS restricts the movement of certain enzymes released from the cell and keeps them in the periplasmic space. These enzymes are thought to be involved in the active transport of molecules into the cell by binding nutrients before they are moved into the cell. In addition, the LPS contains molecules that are toxic to humans when the molecules are released from the bacterium after it dies and fragments. These toxins are known as **endotoxins** and will be discussed in a later chapter.

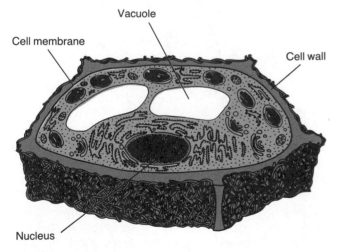

FIGURE 5.16
Plant cell walls consist of a layer of cellulose outside of the cell wall that supports and protects the protoplasm.

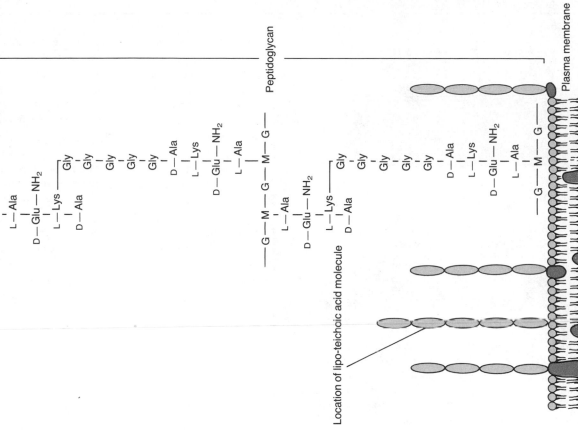

FIGURE 5.17

These two models show proposed arrangements of the components of the peptidoglycan layer in bacterial cell walls. (A) A portion of a Gram-negative bacterium wall such as *E. coli*. (B) The possible arrangement for a Gram-positive bacterium such as *Staphylococcus aureus*. (G = N-acetylglucosamine; M = N-acetylmuramic acid.)

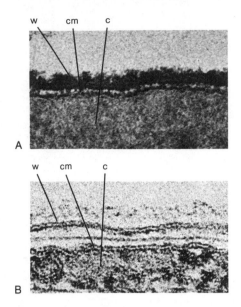

FIGURE 5.18
The electron micrographs show the two basic types of peptidoglycan cell walls found in procaryotic cells. (A) Gram-positive bacteria. (B) Gram-negative bacteria. (c = cytoplasm; cm = cell membrane; w = wall). The Gram-positive wall is a single thick layer, while the Gram-negative wall is thinner and more complex. (From R. Stanier, E. Adelberg, and J. Ingraham, *The Microbial World,* 4th ed, 1976. Reprinted by permission of Prentice-Hall, Inc., Englewood Cliffs, N.J.).

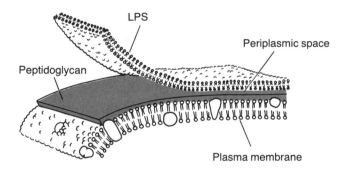

FIGURE 5.19
Relationships among the LPS layer, bacterial cell wall, and cell membrane.

Gram-positive bacteria lack an LPS layer, periplasmic space, and binding enzymes, but they do have a different feature not found in the Gram-negative bacteria. Attached to their cell walls are carbohydrate molecules known as **teichoic acids.** These molecules are able to regulate the action of enzymes called **autolysins.** The autolysins are responsible for loosening the bonds of the peptidoglycan layer so that new cell wall material can be added as growth occurs. If the autolysins were to operate unchecked, they would weaken the cell wall and make the bacterium susceptible to bursting as a result of osmotic pressure changes in the environment.

Drug manufacturers have taken advantage of this information to develop certain antibiotics. **Penicillin** and **cy-**

closerine are antibiotics that interfere with the metabolic pathways involved in the production of new cell wall materials. Destruction of actively growing bacteria occurs when the holes created by autolysin cannot be plugged because the antibiotic has interfered with the synthesis of needed cell wall building blocks. Bacterial cells that grow by adding new wall material in a band at the equator of the cell are much more susceptible to these antibiotics than bacteria that add new cell wall material randomly in patches over the entire cell surface. Gram-positive bacteria grow at the equator, and Gram-negative bacteria grow by the patch method.

Since the Gram staining procedure is easily performed, it only takes a matter of minutes to discover a great deal about a bacterium. Table 5.2 lists some of the differences known to exist between the Gram-negative and Gram-positive bacteria. Many other differences probably occur but have not as yet been identified.

CAPSULES

The outermost covering of a procaryotic cell is not always the cell wall. Many have an additional layer of material known as a **capsule.** This layer not only provides benefits to the microbe, but also provides information to researchers about the microbe that produces it.

Bacteria capsules are produced by the cell and may be composed of complex carbohydrates (e.g., dextran), organic acids (e.g., hyaluronic acid), or protein (including the amino acid glutamic acid). Capsules are slimy or sticky substances and require a special staining procedure to be seen under the light microscope (Figure 5.20). The bacterium *Leuconostoc mesenteroides* produces a thick dextran capsule. This bacterium is used to "culture" buttermilk and helps

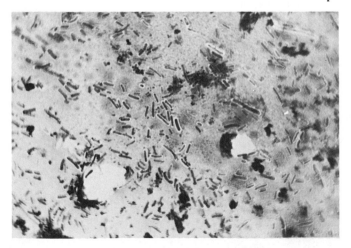

FIGURE 5.20
These *Klebsiella pneumoniae* bacterial cells have a capsule. The cells have not been specifically stained to highlight the capsule, so it appears as a clear halo around the cell. (Courtesy of Carolina Biological Supply Company.)

TABLE 5.2
Differences between Gram-positive and Gram-negative bacteria.

Feature	Gram (+)	Gram (−)
Example	*Streptococcus* sp.	*Pseudomonas aeruginosa*
Gram reaction	Retains crystal violet dye	Decolorized, able to accept counterstain (safranin)
Color	Violet	Red
Peptidoglycan	Single, thick layer	Multilayer, complex
Mechanical strength	Stronger, resistant to destruction	Weak, easily damaged
LPS	None	Present (determines antigenicity, toxigenicity, and susceptibility to virus infection)
Susceptibility to lysozyme	Easily removed by enzyme	Cell wall only mildly affected
Presence of mesosome	Many with mesosomes	Few with mesosomes
Nature of toxin	Primarily produces exotoxins	Primarily produces endotoxins
Antibody activity	Not killed by specific antibody, with or without complement	Complement is necessary for bactericidal action with antibody
Lipid	Small amount	Large amount
Lipoprotein	Small amount	Large amount
Lipopolysaccharide	Small amount	Large amount
Location of new wall material	Equatorial	Patches
Amino acids	Limited variety in cell wall	Full range in cell wall
Flagellar structure	2 rings in basal body	4 rings in basal body
Periplasmic space	Absent	Present
Binding enzymes	Absent	Present
Susceptibility to penicillin and cycloserine	More susceptible	Less susceptible
Resistance to drying	More resistant	Less resistant

give the buttermilk its thick, creamy body. *Streptococcus pneumoniae* produces a capsule composed of complex polysaccharides. Depending on the amount and exact nature of this capsule material, encapsulated *S. pneumoniae* may be able to resist phagocytosis by certain leukocytes. When these bacteria are grown on nutrient agar, two types of colonies can be identified. One form is shiny, has a round edge, and is sticky; the other is dull, has a wavy edge, and is not as sticky. The sticky, encapsuled streptococci are called "smooth," or "S," type colonies. The less sticky, nonencapsulated are called "rough," or "R," colonies. As few as ten encapsulated cells injected into the body cavity of a mouse will cause death; however, an injection of 100,000 nonencapsulated streptococci will not cause harm. The degree to which these bacteria cause the host harm is determined by the presence or absence of the capsule. Any material produced by a pathogenic organism that is presumed to be associated with its ability to cause disease is called a **viru-lence factor.** This factor is usually determined by comparing two forms of the same organism as in the case of the "S" and "R" forms of *Streptococcus pneumoniae.* The encapsulated form is more resistant to phagocytosis, more virulent, and able to grow in the host and cause damage. Conversely, the lack of a capsule leaves the bacteria less virulent and susceptible to phagocytosis.

SUMMARY

Living cells contain a variety of components. One group is composed of unit membrane and molded into a number of structures known as membranous organelles. The other group contains structures that do not have membrane structure but are equally essential to the life of the cell. Membranous organelles found in eucaryotic cells include nuclear membrane, endoplasmic reticulum, Golgi apparatus, lysosomes, vacuoles and vesicles, and plastids. The nonmembranous organelles include such structures as cilia, flagella, endospores, and ribosomes. Organelles found in eucaryotic cells may not be found in procaryotic cells. The biochemical pathways associated with a particular eucaryotic organelle are carried out in association with another type of organelle, or handled in another fashion, in procaryotes (Table 5.3).

The nuclear membrane identifies the location of the genetic material in the eucaryotic cells, is double-membraned, and has pores through which large molecules may pass. This membrane is not found in procaryotes. The endoplasmic reticulum of eucaryotic cells is found in two forms—smooth and rough. The inclusion of this membrane system in a cell greatly increases the surface on which vital chemical reactions may be carried out and increases their efficiency. The Golgi is a series of stacked membranous sacs responsible for the concentration and activation of molecular products that are synthesized within

the cell. Many types of containers, known as vacuoles and vesicles, are formed by the Golgi. One special vesicle, the lysosome, is a container of powerful enzymes used by the cell in digestive processes and to destroy phagocytosed microbes.

The plastids, mitochondria and chloroplasts, are also found in eucaryotic cells. Both are typically double-membraned structures and are the sites of aerobic cellular respiration and photosynthesis, respectively. Since procaryotes do not contain plastids, these reactions may be carried out on other specialized surfaces, such as mesosomes or thylacoids. Many of the membranous organelles are able to interchange as they function.

Cilia and flagella are nonmembranous organelles that contain microtubules. The arrangement of tubules is in a 9 + 2 pattern for eucaryotic cells. Procaryotic cells have flagella, but their hairlike appendages are constructed in a totally different manner from the eucaryotes by using a pro-

TABLE 5.3
Summary of structural-functional relationship.

Organelle	Location	Structure	Function
Plasma membrane	Procaryote Eucaryote	Typical membrane structure; phospholipid; protein.	Controls passage of some materials to and from the environment of the cell.
Nuclear membrane	Eucaryote	Typical membrane structure. Double layer of sacs.	Separates the nucleus from the cytoplasm.
Endoplasmic reticulum	Eucaryote	Folds of membrane forming sheets of canals. Smooth and rough.	Surface for chemical reactions and canal systems.
Golgi apparatus	Eucaryote	Membranous stacks; specialized folded region.	Associated with the production of secretions.
Vacuoles and vesicles	Eucaryote Procaryote	Membranous sacs.	Containers of materials.
Lysosome	Eucaryote	Membranous container.	Isolates very strong enzymes from the rest of the cell.
Mitochondria	Eucaryote	Large membrane folded inside of a smaller membrane.	Associated with the release of energy from food; site of cellular respiration.
Mesosome	Procaryote	Plasma membrane folded inside of the cell.	Associated with the release of energy from food; site of cellular respiration.
Chloroplast	Eucaryote	Double membranous container of chlorophyll.	Site of photosynthesis or food production in green plants.
Thylacoid	Procaryote	Complex series of single membranes.	Site of photosynthesis in cyanobacteria.
Cilia and flagella	Eucaryote Procaryote	Flagellin or tubulin protein.	Cell mobility.
Endospore	Procaryote	Contains DPA and other chemicals.	Adaptation to severe environmental changes.
Ribosome	Eucaryote Procaryote	Protein and rRNA.	Site of protein synthesis.
Genetic material	Eucaryote Procayote	DNA	Regulates metabolism of the entire cell.
Cell wall	Eucaryote Procaryote	Cellulose, chitin, or peptidoglycan.	Provides protection, support, and shape.
Capsule	Eucaryote Procaryote	Carbohydrate or protein.	Protects cell against phagocytosis, allows cells to stick together.
Pilus	Procaryote	Protein filament; short adhesive type and long sex pili.	Short adhesive pili serve for attachment; longer sex pili serve as conjugation tube.

tein called flagellin. The motion of the flagella allows a microbe to move toward or away from a chemical depending on whether or not the microbe is chemotactically attracted or repelled.

Many microbes are capable of existing in two life forms: vegetative cells and endospores. The endospore is a heat-resistant form of the cell and enables the microbe to withstand severe environmental changes. The endospore is formed within the vegetative cell.

Two types of protein projections may also be found on the surface of some bacterial cells. The adhesive pili serve to stick the cell to other cells and inanimate objects, while the sex pili serve in the sexual transfer of genetic material in a process known as conjugation.

Ribosomes are protein-nucleic acid organelles that serve as the site of protein synthesis. As mRNA travel through the ribosomes, tRNA carries specified amino acids to the ribosome for linking into a protein chain.

The genetic material of the cell may be considered to be a nonmembranous organelle because of its size and complexity. Eucaryotes have a complex of DNA and protein, known as nucleoprotein, as their genetic material. During certain phases of the eucaryote's life cycle, the nucleoprotein coils into typical chromosomal configurations. Procaryotes have DNA but no attached protein. No coiling takes place and chromosome formation does not occur. Distribution of duplicated genetic material takes place by mitosis or meiosis in eucaryotic cells and by binary fission in procaryotic cells.

The outer covering of some microbes is known as the cell wall and may be composed of cellulose, chitin, or peptidoglycan. The bacteria and other procaryotic cells are unique in that their cell walls are composed of the peptidoglycan material. Because of the difference in chemical composition of this layer, bacteria may be divided into the two major groups of Gram-positive and Gram-negative. This is accomplished by performing the Gram staining procedure. The Gram stain is based on the ability of the cell wall to retain the dye crystal violet after rinsing with alcohol. The outermost cover on many microbes is known as the capsule and may be composed of carbohydrate, protein, or organic acids. This layer enables the cell to resist phagocytosis, and its presence is related to the microbe's ability to cause disease.

WITH A LITTLE THOUGHT

Some research speculates that present-day mitochondria and chloroplasts were at one time free-living bacteria and cyanobacteria. What structural and functional similarities might be used as evidence to support this hypothesis? Propose an alternative hypothesis that might explain how bacteria and cyanobacteria may have evolved from mitochondria and chloroplasts.

STUDY QUESTIONS

1 What is the difference between a nucleus and a nucleoid?

2 What type of materials are found in lysosomes, and how do these organelles function?

3 How do the plastids differ from other membranous organelles?

4 Describe how the Golgi and ER interact as membranous organelles.

5 Describe two differences between eucaryotic and procaryotic flagella.

6 Differentiate among monotrichous, peritrichous, and lophotrichous.

7 Describe two major differences between the vegetative cell and the endospore.

8 What is the significance of the Gram stain?

9 How is the presence of a capsule related to the virulence of a microbe?

10 What is the difference between DAP and DPA?

SUGGESTED READINGS

BAUER, W. R., F. H. CRICK, and J. H. WHITE. "Supercoiled DNA." *Scientific American,* July, 1980.

BERG, G. C. "How Bacteria Swim." *Scientific American,* Offprint, July, 1980.

CAIRNS, J. "The Bacterial Chromosome." *Scientific American,* Offprint #1030.

DUSTIN, P. "Microtubules." *Scientific American,* Aug., 1980.

HOLTZMAN, E. *Lysosomes, A Survey.* Springer- Verlag, Vienna, 1976.

LLOYD, D. *The Mitochondria of Microorganisms.* Academic Press, New York, 1975.

LURIA, S. "Colicins and the Energetics of Cell Membranes." *Scientific American.*

MAZIA, D. "How Cells Divide." *Scientific American,* Offprint #93, 1961.

MILLER, O. L., Jr. "The Visualization of Genes in Action." *Scientific American,* Offprint #1267, 1973.

MOHRI, H. "The Function of Tubulin in Motile Systems." *Biochemica et Biophysica Acta,* 456:85, 1976.

REINERT, J., and H. URSPRUNG, ed. *Origin and Continuity of Cell Organelles.* Springer-Verlag, New York, 1971.

RUBIN, E., and C. S. LEIBER "Alcoholism, Alcohol and Drugs." *Science,* 172:1097–1102, 1971.

SATIR, B. "The Final Steps in Secretion." *Scientific American* Offprint #1328, Oct., 1975.

SOIFER, D., ed. "The Biology of Cytoplasmic Microtubules." *Academy of Science,* 253:1, 1975.

TZAGOLOFF, A. *Membrane Biogenesis: Mitochondria, Chloroplasts and Bacteria.* Plenum Press, New York, 1975.

KEY TERMS

nucleus	cilia	ribosome
nucleoid	flagella	binary fission
endoplasmic reticulum (ER)	pili	peptidoglycan
	vegetative	Gram staining
Golgi apparatus	endospore	capsule
mesosome		

Pronunciation Guide for Organisms

Leuconostoc (lu″ko-nos′tok)
mesenteroides (mes-en′ter-oid-ez)

Laboratory Equipment and Procedures

PURE CULTURES
PURE CULTURE TECHNIQUE
CULTURE PRESERVATION
ASEPTIC TECHNIQUE
STERILIZATION: THE ABSENCE OF LIFE
CHEMICAL STERILIZATION
STAINING TECHNIQUES
MICROSCOPY
VISIBLE LIGHT MICROSCOPY
ULTRAVIOLET AND ELECTRON MICROSCOPES

Learning Objectives

☐ Be able to define "pure culture" and "mixed culture" and understand the importance of each.

☐ Know the value and methods of obtaining pure cultures of microbes.

☐ Be able to list several methods of preserving microbial cultures.

☐ Be able to define, give examples of, and explain the value of aseptic technique.

☐ Understand the various methods of generating sterile materials, including steam sterilization, dry-heat sterilization, and membrane filtering.

☐ Be familiar with the methods and materials of chemical sterilization and disinfection.

☐ Know the basic microscopic slide preparation techniques, how they are performed, and the advantages and disadvantages of each.

☐ Write the sequence of steps in the Gram staining procedure and know the difference between Gram-positive and Gram-negative bacteria.

☐ Define "magnification" and "resolution" and understand their relationship to microscopy.

☐ Know the basic operating mechanics of the different kinds of light microscopes and the advantages and disadvantages of each.

☐ Know the basic operating mechanics of ultraviolet and electron microscopes and the advantages and disadvantages of each.

The microbial world is composed of a complex array of life forms, each maintaining the characteristics of life in its own unique way. To develop an understanding of the useful and harmful actions of microbes and apply this knowledge in beneficial ways, a clear and detailed understanding of their structure and function must be developed. The extremely small size of these organisms makes this task very difficult and demands special equipment and techniques. Since most microbes are too small to be seen with the naked eye, spe-

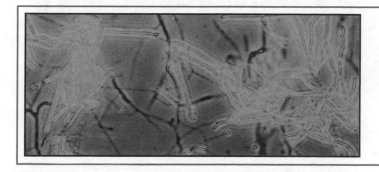

6

cial microscopes have been developed that produce enlarged, clear images. Viewing specimens through a light microscope provides information on size and motility. However, this type of microscope has limits when investigating live microbes. The high percentage of water in cells (70–90 percent) allows visible light rays to pass through the cell very easily and little contrast is developed between the cell and its surroundings. To overcome this problem and highlight special cell components, dyes or stains are used to color the cell before viewing. In addition, there are types of microscopes which may be used to obtain other perspectives of microbes. Ultraviolet light microscopes and electron microscopes have advantages not found in standard light microscopes.

The small size of microbes also limits a researcher's ability to gather information about a single microbe. Laboratory methods used to study large multicellular organisms such as animals and plants are not readily adapted to the microbiology lab. An individual microbe is not easily dissected and studied apart from the whole organism. For this reason, microbiologists must approach the study of microbes using different, more effective, techniques. Microbes are usually studied in populations and not as individuals. Large numbers of a particular microbial species are grown in pure culture and studied collectively. Our understanding of a species is based on what the "group" does or how it behaves in a particular environment. If microbes of a different species become mixed with the study group, the results of the study may be inaccurate. Therefore, microbiologists spend much time and energy maintaining pure cultures for study. Many special procedures have been developed to prevent both contamination of pure cultures and the lab worker by invisible bacteria. The tools and techniques used in microbiology laboratories to investigate the nature of microbes will be described in this chapter, and as in previous chapters, specific applications will be made wherever possible.

PURE CULTURES

A **pure culture** is sometimes called an **axenic culture** and refers to the growth of a single type of microbe in an environment free of any other kind of living thing (Figure 6.1). Many microbiologists prefer to use the term "axenic" in order to avoid the misunderstanding that *pure* cultures are genetically pure. A pure, or axenic, culture is not *genetically* pure. It is a population of microbes of the same species but may contain some individuals with mutations. Such genetically different individuals occur in populations of all kinds. However, this small degree of genetic variation does not usually interfere with the identification or understanding of a species because the mutants are such a small portion of the group being studied.

Pure cultures are used in the laboratory to study several characteristics that identify and classify microbes. By placing cultures in various environments, microbiologists can identify the nutrients required for growth and the physical conditions that best suit the microbe. These are called the **cultural characteristics** of the microbe. Microscopic examination of the microbes will reveal such traits as size, shape, cell arrangement, and motility. These are **morphological features.** Very precise measurements of the nutrients used and wastes produced can be carried out with the cultures, and identification of the **metabolic characteristics** can be made. Killing the microbes and fragmenting them is often done to separate their components, which in turn are analyzed to discover their **chemical composition,** unique differences among cell types **(antigenic features),** and **genetic characteristics.** When all this information is combined, a very complete description and understanding of the microbe is obtained. However, there are drawbacks to the pure culture method.

No organism lives alone. Populations of organisms are constantly influenced by other populations in their environ-

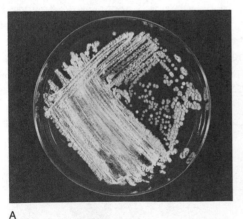

A

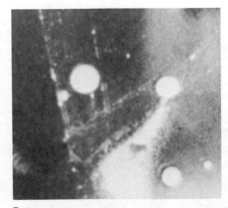

B

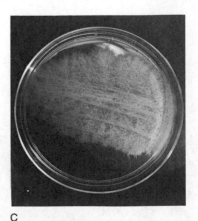

C

FIGURE 6.1

(A) *Streptomyces* sp. (B) *E. coli.* (C) *Bacillus cereus* var. *mycoides.* Each species has its own unique growth characteristics. (Photo A courtesy of the Upjohn Company.)

ment. Their "typical" behavior is, to a large extent, the result of various interactions with other species. For example, animals that are isolated from others for long periods demonstrate "abnormal behavior patterns"; some even become schizophrenic! The behavior of microbes is also influenced by the presence of other species in their environment. Limiting lab characterization and identification to pure culture methods provides very valuable information, but does not give a complete picture of the microbe as it "normally" functions in an environment of mixed species. To overcome this problem and expand our understanding, new cultural methods are being developed called **mixed culture techniques.** By growing a number of different organism types in the same controlled laboratory environment, it is possible to simulate a "real-life" situation and obtain information that will help describe the microbe. Special tools and techniques are required to develop and maintain pure cultures or cultures of known mixtures.

PURE CULTURE TECHNIQUE

Obtaining pure cultures for identification is one of the most important laboratory procedures in microbiology. When samples from soil, water, food, or body are placed on nutrient-containing **petri plates,** it quickly becomes clear that each sample contains a variety of microbes. This is particularly true of specimens collected from the skin or mucous membranes of patients with infections. The clinical specimens contain both harmless microbes normally found on the body (normal flora) and the microbes causing the disease. In order to accurately identify which organism is responsible for the illness, the many different types of microbes must be separated. After separation, the individual microbes must be placed on a medium that will allow them to reproduce into a clearly visible population, or **colony,** composed of only that one species. Samples from this colony can then be "picked" out and transferred to a separate culture for further growth and identification. The microbiologist will then be able to study such characteristics as morphology, metabolism, and antibiotic sensivitity. There are several techniques used to separate mixed cultures; however, the **streak plate** method and the **pour plate** method are the two most common.

1   What is a pure culture?
2   What is a pure colony?
3   Why is "purity" of such great importance in microbiology?

Skill is required to properly perform both methods, but the streak plate method is probably the more practical. A nutrient agar medium is prepared that is known to support the active growth of the organism isolated from the mixed specimen. This medium is sterilized to remove all foreign microbes and ensure that only microbes from the specimen will grow on the petri plate. Special wires called **inoculating needles** or **inoculating loops** are used to transfer the specimen to the petri plate. These must also be sterilized to prevent contamination. This is accomplished by placing the wire in an open Bunsen burner flame or other heating apparatus to destroy all microbes by **incineration** (Figure 6.2). Inoculating needles are used to transfer microbes growing on solid media since the needle will carry fewer cells than the loop. The loop is used to transfer microbes from liquid since it will hold more of the microbes dispersed in broth culture. The streaking patterns used to separate the microbes vary from laboratory to laboratory (Box 9), but the object of each pattern is the same. The microbes are placed on the agar surface in a small concentrated area and the inoculating loop is flamed. The loop is then used to spread a small sample, or **inoculum,** into another section of the

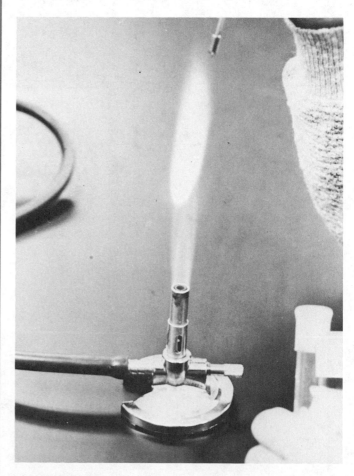

FIGURE 6.2
A quick way to destroy all contaminating microbes on an inoculating loop is to dip the wire into an open flame. This method is known as sterilization by incineration.

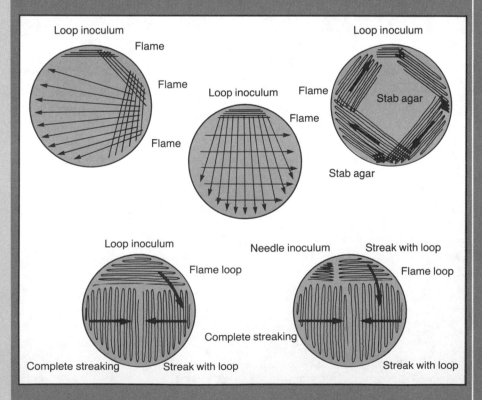

TWO methods are usually used in laboratories to separate mixed cultures into their components. Both the pour plate and streak plate techniques have their advantages; however, streak plates are much easier and quicker to perform, and will result in the formation of surface colonies. An inoculating loop is sterilized by incineration in a Bunsen burner flame and used to transfer a small portion of the culture, the inoculum, to the petri plate. With the loop held lightly in the hand, the inoculum is skated over the solid agar surface, which allows the individual microbes to be deposited at ever increasing distances from each other. By making use of the entire surface area of the petri plate, individual microbial cells will be separated far enough to allow each to reproduce and form a discrete colony. Many streaking patterns are used in laboratories. The particular pattern used may depend on the source of the inoculum (solid or liquid medium), the density of the original culture (heavy or light), or special information that might be needed from the streak plate (e.g., blood-destroying capabilities).

Before any streak pattern is performed, the agar plate should be poured with the necessary medium and allowed to dry. The formation of small water drops on the agar surface will result in "puddling" if the loop is passed through these droplets. Plates can be room dried by turning the plate upside-down. Oven drying requires a special apparatus that warms the plate to skim off the condensation droplets by evaporation. The illustrations show just a few of the patterns used for obtaining pure cultures by streaking.

BOX 9

STREAKING FOR ISOLATION AND IDENTIFICATION

plate, thus dispersing fewer microbes into a larger area. This basic technique is repeated until individual cells have been moved so far apart that each cell may grow into a separate colony. Since each colony has developed as a result of binary fission, all cells within it are of the same species. Each pure colony can then be "picked" with an inoculating needle and cells transfered to new, sterile media for further tests.

The pour plate method does not involve separation of microbes on the surface of agar plates, but separation within liquid, cooled media. Sterile agar in test tubes is kept at about 48°–50°C. At this temperature, the agar will be cool enough to prevent most bacteria from being killed, yet warm enough to remain in liquid form (agar begins to solidify at 42°C). The mixed inoculum is placed into the first of a series of agar-containing media tubes with an inoculating loop and shaken thoroughly. This disperses the cells throughout the medium. The loop is then flame sterilized and a sample is taken from the first tube for transfer into a second tube. This procedure is repeated a third, and possibly, a fourth time—a process called **serial dilution.** Each tube in the series contains fewer microbes than the previous tube. If the technique is performed properly, the last tube in the series will contain the fewest separated cells. The tubes' contents are then poured into sterile petri plates and allowed to solidify. Since the cells have been dispersed throughout the medium, colonies may grow on the surface of the agar, within the agar, and under the agar (Figure 6.3). The surface colonies may be picked for isolation of pure cultures, but those within the agar are more difficult to reach for this purpose. In addition to pure isolation, the pour plate method may be used to determine the number of viable (living) cells in the original specimen. The greatest accuracy in counting is obtained when plates have between 30 and 300 colonies. Plates with more than 300 colonies are likely to be inaccurate because of errors in technique or contamination. A more detailed explanation of viable cell counting will be given in Chapter 8.

---

1 Why should melted agar media be kept at between 48° and 50°C?

2 What are the advantages of streak plating?

3 Which tube in a serial dilution should contain the least number of cells?

The presence of a single colony on the surface of a petri plate does not guarantee that the colony is pure. Several procedures must be carried out to determine that the pure culture techniques have been successful. Microscopic analysis, and determination of physiological and colony characteristics must all be performed to confirm that the microbes have been successfully separated from the original mixture. The Gram staining technique plays a vital role as a confirmation test, as do other staining methods described later in this chapter. The physiological characteristics of a microbe are found by placing pure culture samples in different nutrients and biochemically determining their actions in these environments (see Chapter 9). Colony characteristics are determined by the appearance of a population of microbes as it grows on the solid medium surface. Special terminology may be used by microbiologists to describe these features. For example, the amount of growth may be described as slight, moderate, or abundant. Colonies that grow very well may appear opaque, transparent, or translucent. The form the colony takes on the surface of the agar may be said to be round, filamentous, rhizoid, or complex. Other terms used to describe growth characteristics and their appearances are shown in Figure 6.4. Even though many bacteria may demonstrate any one of these growth characteristics, no two bacterial species display the same combination of characteristics. This fact enables microbiologists to uniquely characterize a particular microbe with relative ease.

CULTURE PRESERVATION

Once a pure culture has been obtained from the mixture, the microbes may be maintained in the laboratory over long periods. Several techniques are used to preserve cultures, although **refrigeration** is the most common. By decreasing the environmental temperature to about 5°C, microbial metabolism will be slowed to the point that the rates of growth, reproduction, and death will be extended over very long periods. Living cells may be removed from refrigerator-stored cultures over a period of several months in some cases. For long-term storage, alternative methods must be used. Some resistant microbes may be preserved by **freezing.** The culture tubes are placed in acetone dry-ice baths (−70°C) or liquid nitrogen (−200°C). However, the cells must be protected from ice crystal formation. If crystallization occurs inside the cells as the temperature drops, the sharp edges of the ice will slice organelles and membranes, which will lead to the death of the cell. By adding such agents as glycerol or sucrose sugar to the culture, cells will be inhibited from forming harmful ice crystals, and the rate of survival will be greater.

Another preservative method that has found application in the food industry is **freeze-drying,** or **lyophilization.** The culture to be preserved is placed in a special lyophilization chamber and the temperature is dropped to about −70°C as a high-vacuum pump draws off the water from the cells (Figure 6.5). This enables the culture to remain frozen as the cells are dried. Since the amount of water in the cells has been reduced to a minimum, the chance of ice crystal formation decreases, ensuring a relatively high survival rate. Skimmed milk, sucrose, and blood serum can be

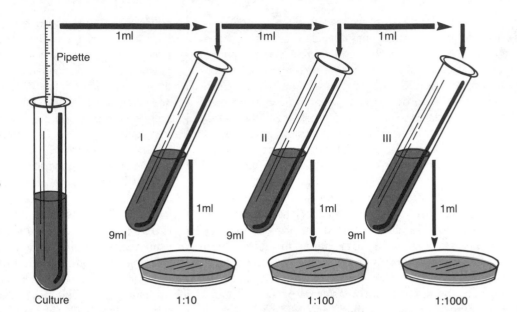

FIGURE 6.3
A series of test tubes is used to dilute a culture of microbes. The final tube in the serial should contain the fewest cells. When these tubes are poured into petri plates and incubated, each separated cell will grow into an easily identifiable colony.

used to protect the microbes being lyophilized. To rehydrate the cells for culturing, small amounts of culture medium is added to the preserved microbes and placed in a favorable growth environment. Lyophilization is not a typical technique performed in medical laboratories but is used by research companies, industrial firms, and other organizations whose business is to provide pure microbial cultures for study. Most pure **stock cultures** are maintained by refrigeration in hospital and teaching laboratories.

Not all pure cultures need be obtained from the environment by streaking or pour plating. Many biological supply companies keep stock cultures of microbes that can be purchased at relatively low prices. These cultures are composed of only one species, although they may contain a variety of subspecies (strains). For many types of experimental work, this genetic variation is acceptable. However, some research demands that this subtle variation be eliminated. To develop and maintain such "super-pure" cultures, a great deal of control must be exercised, which results in higher costs. One source of these pure cultures is the **American Type Culture Collection (ATCC)** (Figure 6.6). The ATCC was organized in 1925 as a unique national resource for the aquisition, preservation, and distribution of authentic cultures of living microorganisms. ATCC serves a broad spectrum of microbiology through its departments of Bacteriology, Cell Culture, Computer Science, Mycology, Protozoology, and Virology. Currently, there are more than 30,000 different strains of authentic cultures of animal cells and microbes available. This is the most diverse collection of strains maintained at one facility anywhere in the world.

ATCC meets the demand of the scientific community for authenticated strains. More than 40,000 cultures were distributed to scientists around the world in 1981, and one-quarter million were distributed during the last 10 years.

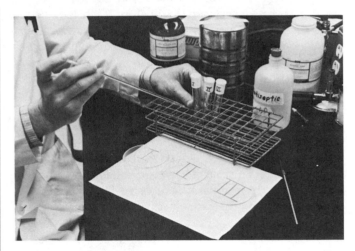

ATCC has a history of successful preservation of strains by freezing and freeze-drying. In order to accomplish this, special handling techniques have been developed.

ASEPTIC TECHNIQUE

Any technique or procedure that prevents both the mixing of cultures and the contamination of the worker or working area with pathogenic microbes or their toxins is called an **aseptic technique.** When aseptic technique is not followed and contamination does occur, a condition of **sepsis** develops. In medically related situations, this term usually refers to an infection by disease-causing organisms or tissue damage due to the toxins produced by microbes. Aseptic techniques must take into consideration both the work and the worker. Procedures that guarantee the complete absence of any living thing on material and equipment used

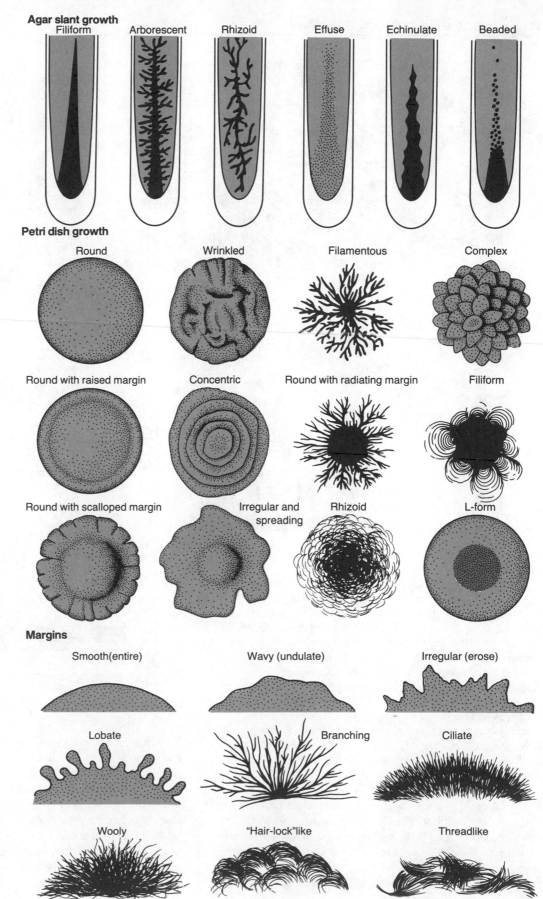

Agar slant growth
Filiform Arborescent Rhizoid Effuse Echinulate Beaded

Petri dish growth
Round Wrinkled Filamentous Complex

Round with raised margin Concentric Round with radiating margin Filiform

Round with scalloped margin Irregular and spreading Rhizoid L-form

Margins
Smooth(entire) Wavy (undulate) Irregular (erose)

Lobate Branching Ciliate

Wooly "Hair-lock"like Threadlike

FIGURE 6.4
The colony form of a growing population of microbes is characteristic of a particular microbe. These diagrams illustrate some of these surface features.

133

A

B

C

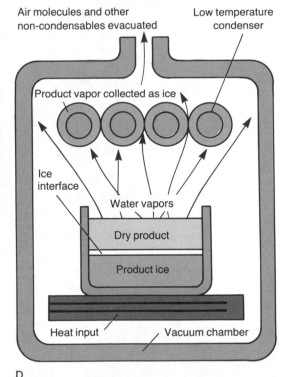

D

FIGURE 6.5

The method of drying biologicals by sublimation of ice in a vacuum has been known for over fifty years, when Shackell (1909) applied vacuum pumps to his experiments to accelerate the process, but it wasn't until shortly before the Second World War that primitive designs of laboratory freeze dryers were made commercially available. During this war, much attention was given to the development of equipment and techniques for the purpose of supplying enormous quantities of dried blood plasma and penicillin to the Armed Forces. By the end of the war, the technique had become accepted as one of the most perfect methods of preserving biological materials. Photo A shows a large-scale industrial unit. (B) A chamber for research and light production work. (C) Ice plug removal. (D) Diagram of how a unit operates. (Courtesy of VirTis Corporation.)

in the laboratory are known as **sterilization procedures** and will be described shortly. Procedures that ensure the absence of microbes from the worker are called **safety precautions.** All jobs have their unique hazards, and special precautions must be followed if the worker is to avoid being harmed on the job. Working with microbes is no different

except for the fact that the danger cannot always be seen. Therefore, it becomes essential to follow rules that will allow the person to control the location of the microbes in the working environment.

Table 6.1 lists several safety precautions that should be followed in a microbiology laboratory. However, each worker must develop an attitude about microbes and how they must be handled in the lab. Once this "second sense" is developed, the aseptic techniques become a habit and are performed automatically. Until that time, it is essential to pay very close attention to each step performed. Probably the

FIGURE 6.6
Cultures of many kinds of microbes can be stored frozen in liquid nitrogen. Laboratories can purchase microbes grown from such stock cultures. These subcultures are usually on agar slants. (Courtesy of American Type Culture Collection.)

most basic of all procedures is called the **transfer technique,** or **inoculation routine.** This procedure enables the safe movement of a microbe sample without contamination (Figure 6.7). Before a sample of broth culture is transferred, the test tube is examined to be sure the microbes are evenly distributed within the liquid medium. If they have settled to the bottom, the microbes are mixed by "rolling" the tube between the hands. Never shake a liquid culture up and down with your thumb on top! Most test tube caps are designed to allow air to pass in and out of the tube but restrict the movement of microbes (Figure 6.8). The fluid may be easily shaken out through this air space, or a mist of small airborne droplets known as an aerosol can move out through this gap. Once mixed, the top is removed (but not put down!) and the test tube opening is passed through a Bunsen burner flame to kill all contaminants around the rim. This action also begins a hot air current that moves microbes in the air away from the opening to prevent their accidentally falling into the tube. The inoculating loop is flamed and dipped into the culture. A sample is carefully removed, so as not to touch the inside of the culture tube. If this accidentally happens, the culture may "flick" off the loop onto the working area or the worker. After the transfer is complete, both the loop and test tube are reflamed before the cap is replaced and the loop set down.

The greatest hazard of transferring microbes is that they become highly concentrated in culture. After a specimen has

TABLE 6.1
Safety in the microbiology laboratory is essential. Good aseptic technique means keeping control of the microbes to prevent contamination of the work by foreign microbes, contamination of the working area and materials, and infection of the worker. Each of the following precautions is basic to all microbiological work.

1. Know the procedure beforehand. Being unfamiliar with the work will only cause delays that may result in inaccurate results; microbes do not live forever.
2. Organize your work area. Accidents are more likely to occur if notebooks, equipment, and cultures are cluttering the lab bench.
3. Laboratory coats or aprons will help protect personal clothing from possible contamination. These are easier to remove and sterilize than personal clothes.
4. All surfaces should be scrubbed with a suitable disinfectant before work is begun. Microbes from the air may have fallen on the work area and could serve as a source of contamination to your work.
5. Encircle your work space with all the needed equipment for easy access and eliminate all unnecessary materials such as books and personal belongings.
6. Be sure that all open flames are at a distance from any flammable materials such as alcohol or paper.
7. Do not eat in the lab. Foods that you find nutritious are excellent growth media for microbes.
8. Do not smoke in the lab. Keeping all things away from your face will greatly reduce the chances of ingesting microbes.
9. All microbial cultures should be maintained in an appropriate container, such as a test tube rack or petri dish container.
10. If a spill occurs, (1) notify the instructor or lab supervisor immediately; and (2) carry out the appropriate disinfection procedures for that particular problem.

The old proverb "a place for everything, and everything in its place" should be used as the most basic safety rule in any microbiological work.

been taken, it is inoculated into a growth medium that contains all the necessary conditions for growth and reproduction; a few relatively innocent microbes become millions of potentially dangerous pathogens. One, or even a few, of these microbes can be handled by the body's defense mechanisms if it were to become infected; however, if thousands were to be accidently transferred, this high number of microbes might cause severe disease. Many microbiologists have died as a result of laboratory accidents after they had become infected with the microbe they were studying.

Test tubes are usually used to culture microbes in the laboratory. The tube may contain microbes growing in liquid medium, **broth culture,** or they may be growing on medium that has been solidified on a slant in the test tube. These tubes are known as **agar slant cultures.** Petri plates or dishes are also used to grow microbes (Figure 6.9). These

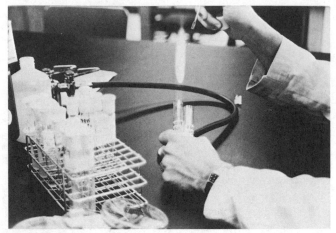

FIGURE 6.7

A wire inoculating loop is first flamed and then used to transfer a small sample of culture from one tube to the other. The opening to both tubes should also be flamed to prevent airborne microbes from falling into the opening.

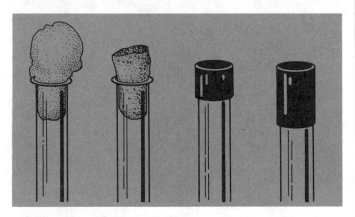

FIGURE 6.8

Many types of caps or closures are used in the microbiology lab. Each has advantages and disadvantages depending on the kind of work being performed and the nature of the microbe being grown.

flat glass or plastic dishes also contain a solid medium. There are advantages and disadvantages to each of these culture containers (Table 6.2). In addition to plates and tubes, pipettes, forceps, swabs, and syringes are used in microbiol-

---

1 What is the difference between freezing and freeze-drying?
2 Name one aseptic technique used in a microbiology laboratory.
3 What is the greatest hazard of working with microbial cultures?

---

ogy laboratories. Each of these tools must also be sterilized and handled in an aseptic manner.

STERILIZATION: THE ABSENCE OF LIFE

All lab equipment and tools may be placed in one of three categories, depending on the nature of their contamination. Equipment that is **clean** has been washed to remove grease and dirt, and to limit the amount of chemical contaminants that might interfere with the growth of microbes. Cleaning means the removal of soil, but never means that the equipment has been sterilized. **Dirty** equipment is soiled with contaminants that may be chemical or biological in nature. In a hospital situation, this equipment has been used often, and has in all likelihood come in contact with patients. Equipment that is **sterile** contains no living organisms to interfere with the laboratory analysis of a specimen or contaminate others should they come in contact with the equipment after it has been discarded. Because of the need to successfully grow pure cultures of pathogenic microbes for identification and analysis, it is essential to clean *and* sterilize all laboratory equipment before it is used. Several chemical and physical methods have been developed to achieve sterile conditions. Such physical methods include moist-heat sterilization (autoclaving), dry-heat sterilization, and filtration. Chemical sterilization is usually performed by using a gas or chemicals.

An **autoclave** operates on the same basic principle as a home pressure cooker. Steam under pressure penetrates the materials in the autoclave more throughly and rapidly than dry heat. In the operation of an autoclave, the chamber that contains the materials to be sterilized (culture me-

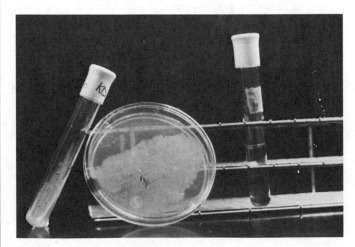

FIGURE 6.9

Containers used for growing microbes: agar slant (left), petri plate (center), and broth tube (right).

TABLE 6.2
Culture containers. Three basic containers are used for growing microbes in the lab: test tubes of liquid medium, test tubes of solid medium, and petri plates of solid medium. Each has advantages and disadvantages.

Container	Advantage	Disadvantage
Test tube, broth culture	Tube is easy to handle, microbes are distributed throughout fluid as individuals or typical clusters, contamination less likely.	No colony characteristics can be seen, liquid is easy to spill.
Test tube, slant culture	Tube is easy to handle, microbes are concentrated on a solid surface, some colony characteristics are easily seen, spills are eliminated.	Cells are clumped together, individuals harder to see, individual colonies usually are not formed.
Petri plate	Large surface area, cells may be isolated, colony characteristics may be viewed.	Large opening to container easily contaminated, medium dries very rapidly, more difficult to handle.

dia, glassware, cotton stoppers, and paper materials) is filled with steam as the air is flushed out (Figure 6.10). Once this exchange has occurred, additional steam is forced into the chamber and the pressure is increased. Sterilization is accomplished when the materials have been exposed to 121°C at 15 to 20 pounds per square inch (psi) pressure for about 15 to 20 minutes, or in a high-speed vacuum autoclave at 132°C for 3 minutes. The time, temperature, and pressure can be varied depending on the material to be sterilized and its thickness. However, these conditions are usually sufficient to destroy resistant endospores of bacteria and all vegetative cells. When the process has been completed, the steam is released from the chamber and the materials removed. Liquid materials such as broth and water must have the steam slowly released from the autoclave to prevent boiling over. The heat energy that has been forced into the fluid by the high pressure will cause the fluid to boil if the pressure is reduced too rapidly. This process is usually done

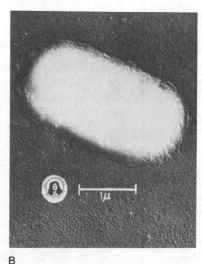
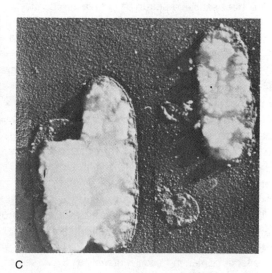

A   B   C

FIGURE 6.10
An autoclave operates much like a household pressure cooker. (A) The materials to be sterilized are placed inside and steam is forced into the chamber under pressure. The high heat and pressure result in the coagulation of cell proteins and the destruction of the cells. (B) The photomicrograph is an *E. coli* cell before heating. (C) Similar cells after being heated at 50°C for only ten minutes. The extreme temperatures and pressures created during the autoclaving process ensure the destruction of bacterial endospores as well as vegetative cells. (Photo A courtesy of Daria Smith. Parts B and C reproduced from C. G. Heden and R. W. G. Wyckoff. "The Electron Micrography of Heated Bacteria." *Journal of Bacteriology,* 58:153, 1949.)

automatically by the autoclave in the slow exhaust cycle. Caution should also be exercised when removing liquid materials from the autoclave since many types of media will boil when shaken, even after the slow exhaust cycle. Nonliquid materials such as empty flasks, test tubes, and bottles can be fast exhausted since there is no danger of boiling over. Containers plugged with cotton or foam caps should be run through a drying cycle following exhausting. Microbes may penetrate the wet caps after the material is taken from the autoclave and contaminate the interior of the tube or flask.

Many materials cannot withstand the high heat of the autoclave, but still must be sterilized. To sidestep this problem, John Tyndall introduced in 1877 an alternative technique now know as **Tyndallization.*** This process allows heat-resistant endospores to be destroyed at only 100°C by exposing them to a repeated cycle of heating and cooling. The materials to be sterilized are exposed to live steam for a period of about 30 minutes, then cooled to room temperature for a day. During the cooling process, many of the endospores that might be present germinate into their more heat-sensitive vegetative form. On the following day, the process is repeated, killing newly germinated cells and stimulating the germination of the remaining endospores. The steam is applied for a last time as a precaution and kills all remaining life forms. Even though this process takes three days it makes possible the sterilization of culture media containing heat-sensitive sulfur or certain amino acids. For materials that may be damaged by the moisture used in autoclaving, dry-heat sterilization may be a good alternative.

Dry-heat sterilization is a very useful procedure for materials that are able to withstand very high temperatures. This process requires that glassware, instruments, or hypodermic needles be heated to between 160°C and 170°C for about 90 minutes. No special equipment is needed to dry-heat sterilize, and a kitchen oven may be used if set above 356°F for at least one hour. Do not sterilize cotton or paper materials in this fashion, since this high temperature will cause them to char and possibly burn. The problems related to high temperatures and moisture may make it impossible to sterilize many special types of culture media commonly used in a medical laboratory, and an alternative to moist and dry heat must be used.

Media containing such materials as urea, serum, plasma, or other body fluids may be sterilized using the **membrane filter method** (Figure 6.11). A special funnel apparatus is used to filter out the microbes found as contaminants in these types of media. The filter must have pores so small that the microbes will be captured on its surface as the liquid is pulled through the filter by a vacuum machine. These filters may be made of cellulose, finely granulated glass, asbestos, or diatomaceous earth. The pore size of many com-

---

*This process is also known as fractional sterilization or spore shocking.

1 At what time, temperature, and pressure is steam sterilization usually performed?
2 Why can the process of Tyndallization be carried out at temperatures lower than those of autoclaving?
3 What types of materials cannot be dry-heat sterilized?

mercially available filters can be specified when the filter is purchased, but most viruses will readily pass through. This is rarely a problem in bacteriological work since both viruses and rickettsias are obligate intracellular parasites and require host cells for growth. Culture media should be sterilized and not contain cells of any type. The entire apparatus is autoclaved before the fluid is suctioned through and collected in a flask. It is then transferred, using aseptic technique, to another sterile container for use.

CHEMICAL STERILIZATION

Much of the equipment and many containers used in microbiology laboratories are made of disposable plastic or rubber. None of this material is able to withstand the heat sterilization processes nor may they be filter sterilized. To sterilize thermolabile plasticware such as petri plates, syringes, and pipettes, gases such as ethylene oxide, methyl bromide, or ozone can be circulated through the packaged materials. This is done in a chamber that resembles an autoclave. The most widely used **gas sterilization** method utilizes ethylene oxide, which is usually mixed with carbon dioxide to reduce its flammability and caustic action on skin and mucous membranes. The materials to be sterilized are wrapped and placed in an airtight chamber (Figure 6.12), and the gas is circulated for 3 to 12 hours, depending on the material. After the exposure, the gas is released and air is allowed to circulate in the chamber for 24 hours. This is a safeguard for those who will be handling the material. Any residual gases may cause skin blistering or eye irritations.

In addition to gases, other chemicals may be used to kill or inhibit the growth of microbes. These may be added to culture media or washed on the surfaces of heat-sensitive materials. The killing of microbes by using chemicals that are applied directly to an inanimate object is known as **disinfection,** and the chemical is called a **disinfectant.** Thermometers, plastic tubing, and metal instruments are commonly disinfected by chemicals such as ethyl alcohol, phenol (carbolic acid), sodium hypochlorite (bleach), and quaternary ammonium compounds (cationic detergents). (Refer to Chapter 10 for details.) Since many of these disinfectants are easily inactivated by organic materials, all dirt and grease should be removed from the equipment before

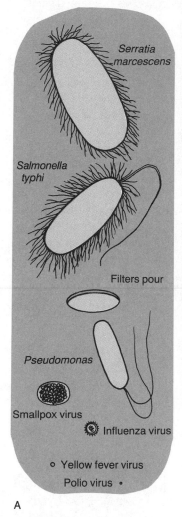

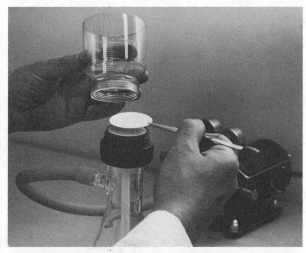

FIGURE 6.11
(A) Microbes of a certain size can be filtered by specifying the pore size of the filter. However, most viruses are so small that they will pass through even the smallest filter pores. (B) A special funnel base and filter holder are used during the filtration process. A vacuum pump draws the fluid through the filter into a sterile container. (Courtesy of Millipore Corporation.)

A

B

it is disinfected. The only chemical currently recognized by the Food and Drug Administration (FDA) as a chemical sterilizer is activated glutaraldehyde (cidex). Material to be sterilized must be soaked in this agent for at least ten minutes. However, not all disinfectants destroy the resistant endospores of microbes. Therefore, a truly sterile environment may not always be created by the use of disinfectants. Due to their harmful nature, chemical disinfectants are not used on animate objects. Those chemicals that are able to be used on the surface of living tissue to kill microbes are known as **antiseptics.**

To sterilize culture media, chemicals such as crystal violet and phenol may be added to a very concentrated form of the medium. The combination is allowed to stand at room temperature for about 24 hours while the chemical destroys any microbes in the solution. After this period, the medium is diluted to offset the effect of the chemical so that a pure culture of microbes can be grown. All of the aseptic techniques, sterilization, and safety procedures discussed to this point are designed to ensure the growth of microbes in a pure culture. Once this has been accomplished, the microbes may then be transferred for further analysis by either biochemical methods (see Chapter 9) or microscopic methods.

STAINING TECHNIQUES

The point of preparing slides of microbes and selectively staining them is for microscopic examination. A great deal of information can be gained by looking at enlarged or magnified microbes. Not only is the microbiologist able to determine the size and shape of a microbe, but many chemical characteristics may be ascertained from the staining reactions of different microbes. The two types of microscopic slide preparations generally made in the laboratory are wet mounts and dry mounts (smear preparations). A **standard wet mount** specimen enables the microscopist to examine large microbes suspended in a drop of water. The flat, glass microscope slide is cleaned thoroughly to re-

FIGURE 6.12

Gas sterilization units come in many sizes to accomodate a variety of materials. This unit is operated by a microcomputer, which enables hospital personnel to sterilize numerous articles used by patients with infectious diseases. (Courtesy of American Sterilizer Company.)

move any grease film, and a drop of water is placed on the slide. The organisms to be examined are transferred to the slide using aseptic technique, and a coverglass is placed on top. Living fungi, protozoans, and algae can be studied to determine whether or not they are motile, and what cell arrangements might be formed. Because of the small size of microbes such as the bacteria, it is very difficult to bring the cells into sharp focus using a standard wet mount. To resolve this problem, an alternative is used called the **hanging drop slide preparation** (Figure 6.13). To prepare this slide, small amounts of petroleum jelly are placed on the corners of a clean cover glass, and an inoculating loop is used to aseptically transfer a sample of culture to the center of the glass. A special microscope slide with a depression in its surface is then placed over the coverglass to center the specimen in the depression. The completed slide is then flipped right-side-up and placed on the microscope for examination. Slides prepared in this manner will not dry out as rapidly, and they allow the microscopist to easily magnify the microbes one thousand times their normal size.

Dry slides may also be prepared in two ways. The **negative smear preparation** may be used to examine

microbes that are difficult to dye with the usual laboratory stains. This preparation technique provides little information about the chemical nature of the cells but is very useful in determining the accurate size and shape of the cells. A small amount of the chemical nigrosin or India ink is placed on the end of a clean, flat glass slide. The microbes are mixed into this chemical with an inoculating loop, and a second glass slide is used to spread the suspended cells over the slide's surface. As the suspension covers the slide, it quickly dries to a dark film that may be examined without a coverglass. Nigrosin and India ink do not stick to the cell surfaces but only surround the microbes, which appear as shiny little rods, cocci, or spirals in a dark gray-purple background when viewed through the microscope.

The other dry slide technique is the **smear preparation,** which is a method of placing a sample of microbes on a clean glass slide in preparation for staining. Culture from liquid media is placed on the slide with an inoculating loop, while those from solid media (petri plates, or slants) are transferred by inoculating needle into a drop of sterile water. The specimen is smeared over the surface of the slide and allowed to air dry. In order to prevent the cells from being washed off during the staining procedure, the slide is quickly passed through a Bunsen burner flame to heat-fix the cells to the slide. Heat-fixed smear preparations may be kept for a long time, since the cells have been killed and fixed to the surface of the glass. However, since the cells are small and contain a high percentage of water, very little

1 What is the difference between an antiseptic and a disinfectant?
2 Why does gas sterilization take so long?
3 Why use an inoculating needle instead of a loop when transferring culture from solid media?

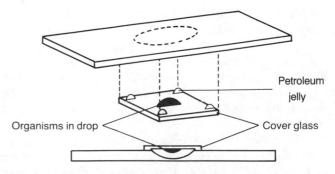

Petroleum jelly

Organisms in drop

Cover glass

FIGURE 6.13

The hanging drop slide has the advantage of allowing the microscopist to see cell arrangement and motility. The microbes are free to move about in the hanging drop and will not be disturbed by the movement of the microscope as the focus is adjusted.

information may be gained by examining the killed, unstained bacteria. To identify internal structures and learn something about the chemical nature of the microbes, stains may be added to the smear before microscopic examination.

Two staining methods commonly used in the lab are the **simple stain** and the **differential stain.** Only a single dye is used in the simple staining procedure, while the differential technique makes use of more than one stain to highlight the differences between cell parts or between cell types. The most commonly used dyes in microbiological work are known as **cationic dyes,** since the colored portion of their ions has the positive charge. The stain methylene blue is methylene chloride (CH_2Cl_2). The color-containing portion of the molecule is called the **chromatophore,** while the nonpigment-containing ion is called the **auxochrome.** When dye is washed over the surface of bacterial cells, the postively charged methylene ion (chromatophore) will chemically combine with the negative electrical charge that normally occurs on the cells' surface. More commonly, these dyes are called basic dyes. The acidic dyes (e.g., sodium eosinate) have a negatively charged chromatophore and will not attach to the surface. Simple staining is a very quick method. After the smear preparation has been heat-fixed, the dye is poured onto the surface and allowed to remain there for about 45 seconds. Then it is washed off with water, and the slide is gently blotted dry. Other simple stains commonly used for bacterial cells include crystal violet, safranin, and carbolfuchsin.

These same dyes may be used for differential staining methods when applied in combination. Each particular dye is chosen on the basis of its ability to selectively react with certain cell parts or certain cell types. One such staining method has already been described in the previous chapter. The **Gram stain** uses crystal violet, Gram's iodine, alcohol, and safranin to show the differences between two major bacterial cell types, Gram-positive and Gram-negative. Keep in mind that there is no single dye called a "Gram stain,"

but that this is a differential staining technique requiring the use of several chemicals (Box 10). Three other useful methods are the acid-fast (Ziehl-Neelsen), endospore (Schaeffer-Fulton), and capsular (Anthony) staining. All of these methods are performed on smear preparations in a specified sequence. The **acid-fast stain method** is similar to the Gram method in that it is based on the ability of cell types to be decolorized after staining and enables microbiologists to categorize microbes. Acid-fast bacteria (AFB) are not easily decolorized with acid-alcohol (HCl + 95% ethyl alcohol) after staining with hot carbolfuchsin dye since these bacteria have a thick, waxy lipid material that prevents the dye from being removed through the cell membrane. If these AFB cells are damaged after the dye is added, the acid-alcohol will remove the red dye and the cell will appear to be non-acid-fast. This staining method is used in the clinical laboratory for the identification of bacteria such as *Mycobacterium tuberculosis* (the cause of tuberculosis) and *Mycobacterium leprae* (the cause of leprosy) (Figure 6.14). A counterstain is used in the acid-fast technique to color those cells that are non-acid-fast and have been decolorized by the acid-alcohol. Since the acid-fast cells are carbolfuchsin red, the counterstain usually used is methylene blue. If a mixture of acid-fast *(Mycobacterium)* and non-acid-fast (e.g., *Streptococcus*) bacteria are stained on the same slide, mycobacteria will appear as red bacilli and the streptococci as blue cocci. This staining procedure demonstrates differences in cell types, whereas the endospore and capsular stains demonstrate differences in cell parts.

Endospores produced by the genera *Bacillus* and *Clostridium* are chemically resistant to most dyes. Therefore, a special staining method must be used to highlight their presence inside vegetative cells. The Schaeffer-Fulton method utilizes hot malachite green stain to color the resistant endospores and cold safranin (red) to color the surrounding vegetative cell. When this staining precedure has been completed, cells that have formed endospores will show pink cytoplasm and green endospores. In order to display the

FIGURE 6.14

Acid-fast bacteria are so named because they hold the dye carbolfuchsin *fast* (tightly) to their cells despite washing with acid-alcohol solution. Photo A is *Mycobacterium tuberculosis;* Photo B is *M. leprae.* (Courtesy of the Center for Disease Control, Atlanta.)

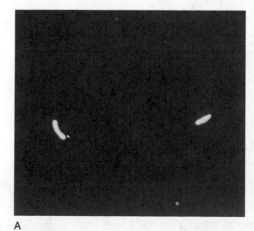

A

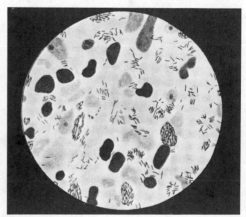

B

THE Gram staining procedure was first developed in 1884 by the Danish bacteriologist Christian Gram, who discovered that most bacteria could be divided into two main groups based on their staining reactions. This technique is called a **differential staining** since it allows the microbiologist to highlight the differences between cell types. Bacteria which are not easily decolorized with 95 percent ethyl alcohol after staining with crystal violet and iodine are said to be Gram-positive. Those that are decolorized are Gram-negative and will be very difficult to see through the microscope. For this reason another stain, called a **counterstain,** is added to make these cells more visible. A number of different stains may be used as a counterstain, but safranin is preferred since it provides the greatest contrast. In order to ensure a successful combining of the crystal violet with the cell wall, iodine is added to better complex the dye with the wall. Any chemical that is capable of intensifying or deepening a reaction is called a **mordant;** Gram's iodine serves this purpose in this differential staining method.

The sequence of chemicals used in the Gram procedure and their actions are outlined in the following flowchart:

BOX 10

**THE
GRAM
STAINING
PROCEDURE**

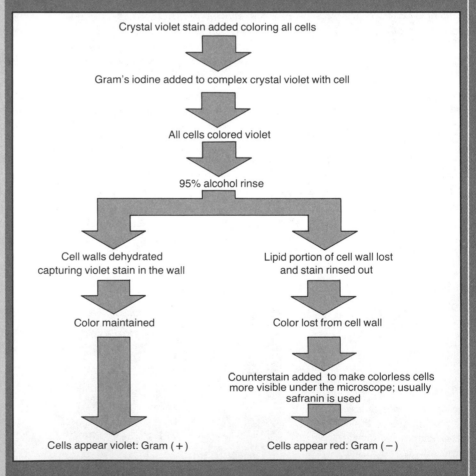

The reactions of some pathogenic bacteria to Gram staining are listed on the next page. Knowing this information is of great value in determining how to handle these microbes in cases of infection.

Gram-positive	Gram-negative
Cocci	Cocci
Streptococcus pyogenes	*Neisseria meningitidis*
Staphylococcus aureus	*Neisseria gonorrhoeae*
Streptococcus pneumoniae	
Bacilli	Bacilli
Corynebacterium diphtheriae	*Escherichia coli*
Mycobacterium tuberculosis	*Shigella dysenteriae*
Bacillus anthracis	*Salmonella typhi*
Clostridium tetani	*Klebsiella pneumoniae*
Clostridium botulinum	*Haemophilus influenzae*
Clostridium perfringens	*Bordetella pertussis*
Mycobacterium leprae	*Pseudomonas aeruginosa*
	Proteus vulgaris
	Brucella abortus

capsules of bacteria by the Anthony method, crystal violet is used to stain cells in a smear preparation containing skimmed milk. Copper sulfate is then washed over the slide to color the milky background, and the slide is dried in the air. The capsule can be seen as an unstained layer surrounding the violet-colored cell (Figure 6.15). Both simple and differential stains make it easier to identify and classify microbes seen through the microscope. The staining procedures just discussed are for use with the compound light microscope, but other types of microscopes may be useful in gathering information about microbes.

MICROSCOPY

A **microscope** is a tool or machine with the ability to increase the visual size of an object so that it is easier to see. All types of microscopes must perform two important func-

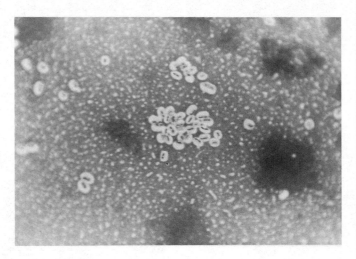

FIGURE 6.15
These bacterial cells have been specially stained to highlight the thick mucilaginous capsules that surround them. (Courtesy of Carolina Biological Supply Company.)

tions: they must **magnify** (enlarge) the specimen to a size that can be seen by the human eye, and they must provide a clear image that will enable the microscopist to distinguish the component parts of the specimen, a feature known as **resolution.** These may be accomplished by using visible (white) light, ultraviolet light, or electron beams. Various forms of energy share certain qualities. One is that they behave as if they were waves. The wavelike forms of energy such as radio waves, x-rays, and light make up the electromagnetic spectrum. A wave consists of a high point (crest), followed by a low point (trough), and again rises to a high point. The distance from one crest to the next is one wavelength. In the electromagnetic spectrum, wavelength is usually measured in nanometers (nm). A nanometer is one-billionth meter (10^{-9} m) long. Figure 6.16 shows various energy types, along with their wavelengths and the energy forms used for the light, ultraviolet, and electron microscopes.

In order to see the components of an object, light rays coming from the object must be separated. If two very small objects are moved closer and closer, there will be a point at which the two will be seen as one. The smallest distance between the two objects at which they may be seen as separate objects is the **resolving power,** or resolution, of the lens system. The human eye has a resolving power of about 0.1 mm, or 100 micrometers. This means that a person can tell by looking that there are two separate objects if they are separated by a distance of at least 0.1 millimeters. If they were any closer together, they would appear as a single object. The resolving power of a lens system depends on the ability of the light to pass between the objects being viewed. The shorter wavelengths of light are able to pass between a pair of close objects much easier than the longer wavelengths. This means that the components of a specimen will be seen more clearly as the wavelength of the light being used decreases. (The shorter the wavelength, the greater the resolving power.) Shorter wavelengths of blue light provide the most clear images when viewed with a light microscope. A typical light microscope is able to enlarge a

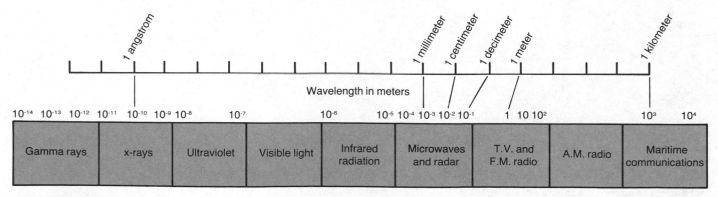

FIGURE 6.16
The various kinds of energy in the electomagnetic spectrum all are wavelike in nature. Notice that the difference between one kind and another is the wavelength.

specimen and provide a resolved image 1000 times its actual size. Resolution depends on the kind of materials used in the lens system and the way in which the lenses are made. A quality light microscope becomes more expensive as the magnification increases because of the greater difficulty in providing resolution, not magnification. Since the spectrum of visible light is relatively narrow, the limits of resolution are reached very quickly when using a light microscope. To see specimens clearly at higher magnifications, other microscopes have been developed that use even shorter wavelengths of energy from the electromagnetic spectrum. The ultraviolet light microscope and the electron microscope utilize energy that cannot be seen directly with the naked eye; therefore, special viewing screens and cameras have been added to these microscopes to make the images visible. Electron microscopes are able to enlarge a specimen and provide a resolved image 200,000 times or more depending on the kind of equipment used. Many types of micro-

scopes are used to investigate the structure of microbes. Each type varies in the kind of energy used and the way the energy is passed through the specimen. There are three basic types of microscopes that use visible light energy: (1) the bright-field, or compound, light microscope, (2) the dark-field microscope, and (3) the phase-contrast microscope. Ultraviolet energy may be used in either direct ultraviolet microscopy or indirect ultraviolet microscopy. Controlled electron beams are used as the energy source in the transmission electron microscope and the scanning electron microscope.

VISIBLE LIGHT MICROSCOPY

The **bright-field microscope** is used for most routine examinations (Figure 6.17). Specimens are prepared using the hanging drop or smear methods described earlier and may

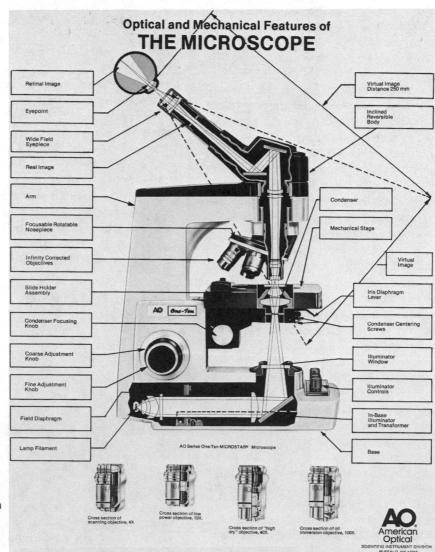

FIGURE 6.17
This illustration points out the major components of the bright-field compound light microscope and the path taken by the light as it passes through the lens system. (Courtesy of the American Optical Corporation.)

be stained before viewing. This scope contains sets of lenses, which in the proper combination may magnify up to 1000 times. For this reason the bright-field microscope is also known as a **compound microscope.** The top lens closest to the eye is the eyepiece, or **ocular lens,** while those closest to the slide are called the objective lenses. These lenses may be rotated into position to change the total magnification. A microscope which maintains a clear, focused image as each different lens is rotated into position is said to be **parfocal.** The total magnification provided by a particular combination of ocular and objective lenses is determined by multiplying the magnification potential of each of these lenses. If the ocular is capable of magnifying a specimen 10 times (10×) more than an objective lens that can magnify 45 times (45×), the total magnification will be 450× (ocular × objective = total magnification). For microbiological work, the platform, or stage, of the microscope on which the slide is placed, usually has a device to control the movement of the slide. This **mechanical stage** is operated by two control knobs located below the stage. Also below the stage is a light-regulating unit known as the **substage condenser** which serves two functions. The **iris diaphragm** found in the substage unit can be used to regulate the size of the opening through which the light passes. Opening the diaphragm increases the light and is valuable in viewing dark-stained specimens; closing the diaphragm decreases the light and provides greater contrast when viewing living specimens. However, closing the diaphragm also decreases the resolving power. The **Abbé condenser lens** in the substage unit may be moved away from or toward the slide to focus the light at different points on the specimen much like a hand-held magnifying glass can be adjusted to focus light into a point (Figure 6.18). Since the image of the specimen is only two-dimensional (length and width), and cannot be picked up and viewed from different directions, the variation in lighting patterns is very valuable in providing many different views of the same specimen. As the objective lenses are changed to increase magnification, the size of the lens opening decreases and the amount of light passing through the lens system decreases. Since light is necessary to see, conserving all the light possible by condensing the rays through the lenses is essential. The substage Abbé lens helps condense the light before it reaches the specimen, and a special oil serves this same purpose after the light leaves the specimen. **Immersion oil** helps channel the light rays as they leave the specimen and enter the objective lens (Figure 6.19). This oil is placed directly on the slide, and the oil immersion lens (usually 100×) is moved into the oil drop, functioning as an additional "liquid lens." Other lighting modifications may be added to the bright-field microscope to change the kind of image.

The **dark-field microscope** has a special condenser that makes the specimen appear bright against a dark background. The consenser only allows light rays from the side

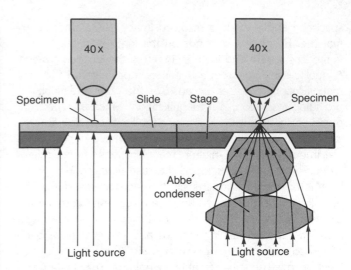

FIGURE 6.18
The drawing on the left shows how light rays travel to the specimen when there is no Abbé lens below the stage of the microscope. When the Abbé lens system is in place, the light rays are not lost but bent by the lenses into the field of view.

to pass into the specimen. When these rays pass through the specimen, they are reflected into the objective lens and make the microbe appear bright and glowing. Since no light rays pass directly through the slide, the background (field of view) appears dark. This type of microscope is very useful when viewing microbes that are difficult to stain. *Treponema pallidum,* the cause of syphilis, is one such bacterium and is easily identified in the fluid material (exudate) released from this type of infection (Figure 6.20). This examination is done with a wet-mount slide so that the motility of the spirochetes can also be observed.

The **phase-contrast microscope** also utilizes visible light in its operation. This microscope takes advantage of the fact that when light passes through an object, it is slowed

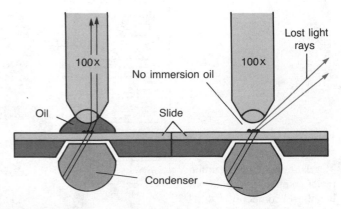

FIGURE 6.19
Light rays that would normally be bent out of the lens system when they pass from glass to air can be captured by immersion oil. This oil acts like an additional lens in the system and prevents the loss of necessary light rays.

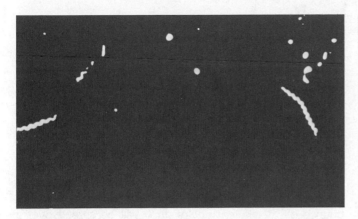

FIGURE 6.20
The spirochetes seen in this field are the bacteria *Treponema pallidum* and have been photographed through a dark-field microscope.

down. A dense material will slow the light more than a less dense material. Light that has been slowed is said to be out of phase in comparison to light that has come through the object unaffected. The phase-contrast microscope highlights these density differences by showing the more dense material as dark objects and the less dense material as light objects (Figure 6.21). This scope is especially useful in examining living cells since many stains are poisons, and examining a dead microbe does not give the same information as a live one. This technique is also valuable in observing bacteria in tissue sections in host cells. Bright-field, dark-field, and phase-contrast microscopy all utilize energy within the visible spectrum, and the total magnification and resolution

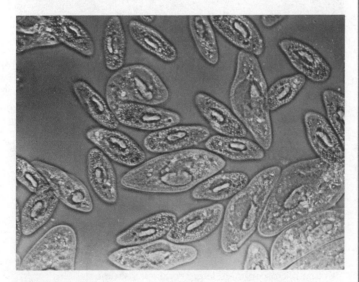

FIGURE 6.21
Protozoa photographed through a phase-contrast microscope appear shiny. The bright areas are less dense than the darker areas inside the cell. (Courtesy of the American Optical Corporation.)

is limited by the wavelength of light in that range. Greater magnification and resolution can only be achieved by using energy with shorter wavelengths.

> **1** What purpose is served by using immersion oil?
> **2** What does a phase-contrast light microscope enable you to see that a compound light microscope does not?

ULTRAVIOLET AND ELECTRON MICROSCOPES

The use of ultraviolet light enables microscopists to magnify and resolve specimens from two to three times more than with visible light. The two ultraviolet methods used in labs are direct and indirect, or **fluorescent,** microscopy. In the direct method, the ultraviolet light is focused through a system of quartz lenses onto a special photographic film plate. Special quartz lenses are necessary because this wavelength of energy will not pass through regular glass. The film must be used since the human eye is not able to "see" this light. Ultraviolet (UV) light is invisible and dangerous to the eye; viewing UV light could result in a damaged retina. Fluorescent microscopy uses ultraviolet light to highlight a specimen, not to magnify it. The UV light is projected onto a specimen that has been specifically stained with a chemical called a **fluorochrome.** This chemical will glow, or fluoresce, when UV light strikes it and emits rays of visible light. The microscopist may look *directly* through the microscope to see this specimen. The staining reaction can be very specific and allows for quick identification of particular disease-causing microbes. This technique may be used in the identification of streptococci, pathogenic *E. coli,* and other bacteria (Figure 6.22). Many details of internal cell structure are not visible with fluorescent microscopy, since the magnification is limited. In order to overcome this problem, energy rays of even shorter wavelength must be used.

Electrons move in a wavelike pattern and have a natural short wavelength of about 1 nm. Engineers have made use of the natural physical qualities of electrons in developing two types of electron microscopes, the transmission electron microscope and the scanning electron microscope. The scopes have become very important tools in the identification and classification of microbes and their component parts. Many organelles previously thought only to be present in cells have had their existence confirmed by use of these microscopes (e.g., the Golgi apparatus). The extremely high magnifying ability of these instruments (200,000×) has also made it possible to see microbes that were only suspected of existing (e.g., viruses). The electron microscope has become an essential laboratory tool, and in many ways resem-

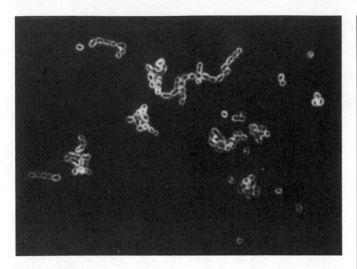

FIGURE 6.22

This photomicrograph shows chains of streptococci that have been stained with a fluorochrome dye. They appear to glow because the ultraviolet light striking the bacteria causes the dye to emit visible light. (Copyright © 1970 by the New York State Department of Health. Reproduced by permission.)

bles the bright-field microscope in its operation. The body of a **transmission electron microscope** resembles a long tube (Figure 6.23). The source of the electrons is at the top, where they are beamed down through the scope to a viewing screen or film plate. Since electrons are small, negatively charged particles, they will not pass through glass or quartz lenses and must be controlled by electromagnets. These magnets are positioned in the microscope so they are able to direct the moving electrons down through the scope to the specimen. Since even small molecules may cause interference and result in the electrons being scattered in all directions, the air molecules in the tube must be removed with a vacuum pump. In order to prepare cells or tissues for examination in the electron microscope, very thin sections must be cut and readied using special techniques and tools (i.e., a cell that is too thick will block all electrons and show no contrast). The specimen is placed on a "slide" made of copper wire (a grid) and inserted into a chamber that is in line with the electron beam. The more dense areas of the cell block the electrons and prevent them from reaching the viewing screen; the more electrons blocked out, the darker the specimen will appear. The density differences in the specimen can be highlighted by special staining techniques much like those used to prepare cells for the bright-field microscope. Chemicals, such as lead, gold, or osmium tetroxide, are added to the cells which will selectively combine with the organelles. In addition, procedures may be used to show cell features not easily seen with these regular staining techniques. The freeze-fracture method described in Box 11 is one such method. The specimen may be viewed on the screen at the base of the scope or a picture may be taken

for later study. Since the picture is produced by electrons beaming on a piece of film and not from visible light rays, it is not a "photograph" but is called an **electron micrograph.** The electron micrographs from the transmission electron microscope are two-dimensional only, while those from a scanning electron microscope appear to be three-dimensional. The electron beam produced in the **scanning electron microscope** is also directed and controlled by electromagnets. However, the electrons are not transmitted through the specimen as in the case of the transmission electron scope, but are bounced off its surface into an apparatus called a **collector.** Since the reflected electrons bounce off at different angles, they will be spread over a large surface area, rather than being collected at the same point. The collector then translates the pattern of scattered electrons into an image that appears on a viewing screen or film plate. Specimens may be magnified by this technique from $15\times$ to about $10,000\times$ and must be specially prepared by coating their surface with a heavy metal, such as gold or platinum. This hard surface will allow the electrons to bounce into the collector. Since just the surface of the specimen may be coated, the scanning electron microscope is used only to examine the outside structures of an object. Internal details may only be viewed if the cell is split and then coated with the metal.

FIGURE 6.23

Electron microscopes are large, complex instruments. This machine requires a great deal of laboratory space and a specially trained operator. The operator sits at the console and views the specimen through the small lens system. (Courtesy of Biophoto Associates.)

A very important tool used in the study of microbes is the transmission electron microscope. The design of this microscope makes it possible to magnify specimens up to 200,000 times their actual size. Specimens are prepared using special techniques, placed in the microscope, and viewed on a screen that resembles a television screen. Photographs of the specimen, **electron micrographs,** can be taken and studied at a later time. Two techniques of specimen preparation have been very important in the development of the fluid mosaic model of membrane structure. **Freeze-fracture** and **freeze-etching** are ways of freezing the membrane into a

BOX 11

FREEZE-FRACTURE AND FREEZE-ETCHING

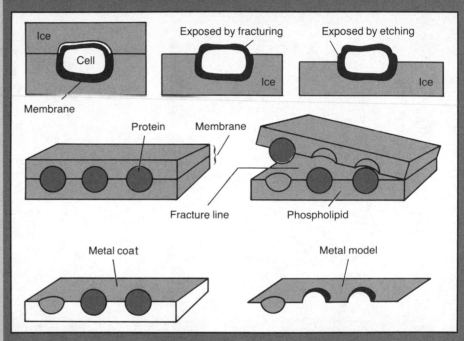

The freeze-etch micrograph on the left is *Penicillium claviforme* (×17,500). Notice the characteristic "rodlets" on the surface. Compliments of W. M. Hess, Brigham Young University, Electron Optics Laboratory. The electron micrograph on the right is a freeze-fractured red blood cell. By permission of Daniel Branton, Harvard University.

block and then breaking the frozen membrane into sections with a sharp knife. This separates the two surfaces of the phospholipid layer, but the position of all the component parts is maintained. The fractured surface is etched by removing a small amount of the block's surface in a vacuum container. This etched surface is coated with a layer of metal (usually platinum and carbon) to form a thin, metal model of the surface. This model is placed in the electron microscope for viewing. When the electrons from the microscope pass over the model, they produce an image on the screen that shows many little "bumps" that are presumed to be the protein molecules nestled in the smooth phospholipid layer.

The images produced by different microscopes all contribute to our understanding of microbes. Limiting the investigation of microbes by using only one type of microscope and one staining method would place the results in doubt. All laboratory procedures are designed to produce accurate, reliable information that enables microbiologists to identify, classify, and control microbes for the benefit of mankind.

SUMMARY

The extremely small size of microbes makes their study more difficult than large organisms. Special lab techniques and procedures must be followed to gather information to identify and classify an unknown microbe. Many of these techniques require microbiologists to work with populations of cells instead of individual microbes. Our understanding of a particular microbe is based on how a group of individuals is behaving as it grows in different cultural environments. To ensure that only one type of microbial population is being studied in the laboratory, pure culture techniques have been developed. The two most frequently used methods are the pour plate and streak plate procedures. Once microbes are separated from one another, they must be maintained in the laboratory using aseptic techniques. These handling procedures ensure that the microbes will not be contaminated with other species, nor will the work area or worker become contaminated. Special safety precautions must be taken in a microbiology lab and sterilization methods used on all equipment before and after use. Sterilization may be accomplished by using moist heat in the autoclave, or Tyndallization at lower temperatures. Dry heat may be a good alternative for materials that can withstand higher temperatures. Chemical sterilization with gases such as ethylene oxide or disinfection with liquids may also be used.

Special wet or dry slides (hanging drop or smear) are made in the lab to microscopically check the cultures for purity. These slide preparations may be stained to highlight cells or cell parts using either simple or differential staining methods. Basic or cationic dyes work best with bacteria since their chromatophores are negatively charged and may chemically combine with the cell. Simple stains include such chemicals as methylene blue, carbolfuchsin, and safranin. Differential staining techniques include the Gram, acid-fast, and endospore stains. Microscopic examination may be done by using a variety of equipment. There are three basic kinds of microscopes, and they differ in the wavelength of energy used to produce an enlarged, resolved image of the specimen. The light microscope (bright-field, phase-contrast, and dark-field) use visible light. The ultraviolet microscopes (direct and fluorescent) use the shorter ultraviolet light wavelengths, and the electron microscopes (transmission and scanning) use the shortest wavelength of electron beams.

WITH A LITTLE THOUGHT

Why would you *not* do any of the following?

1 Autoclave a sealed container of nutrient broth.
2 Have lunch in the lab.
3 Place your inoculating loop on the table immediately after transferring a culture.
4 Use sheets immediately after they have been gas sterilized with ethylene oxide.
5 Shake a broth culture tube with your thumb on top.
6 Look directly at an ultraviolet light source.
7 Neglect to use the immersion oil when magnifying $1000\times$.
8 Use a disinfectant when procedure calls for sterilization.
9 Look for a bottle of Gram's stain in the stock room.
10 Stain a smear preparation without first heat-fixing.

STUDY QUESTIONS

1 Define the following terms: pure culture, aseptic technique, sterilization, and streak plating.
2 Why is it necessary to work in the lab with populations of microbes instead of individual cells?
3 List and describe four characteristics of microbes that may be identified through routine lab procedures.
4 Describe how you would go about obtaining a pure culture of a particular bacterium from a teaspoon of ordinary soil.
5 What are the advantages and disadvantages of the pour plate and streak plate methods for obtaining pure cultures?
6 Why is it necessary to flame an inoculating loop before use; flame a tube before transfer; and autoclave nutrient broth before it is inoculated?

7 List and describe three methods of preserving cultures.

8 Define or explain the following: resolving power, electromagnetic spectrum, magnification, immersion oil, and Abbé condenser.

9 Compare and contrast the bright-field, dark-field, and phase-contrast microscopes.

10 Compare and contrast the light, ultraviolet light, and electron microscopes.

SUGGESTED READINGS

BARER, R. *Lecture Notes on the Use of the Microscope*. Blackwell Scientific Publications, Oxford, 1959.

BARTLETT, R. C., and G. O. CARRINGTON. "How to Avoid Hazards in Microbiology." *Medical Laboratory Observer*, 1:46, July, 1969.

BAUER, J. D., et al, *Clinical Laboratory Methods*, 9th ed. The C. V. Mosby Company, St. Louis, 1982.

CORRINGTON, J. D. *Adventures With the Microscope*. Bausch and Lomb Optical Co., Rochester, N.Y., 1934.

HAMMON, W. M. "Human Infection Acquired in the Laboratory." *Journal of the American Medical Association*, 203:647, 1968.

KLEIN, A. E. *The Electron Microscope*. McGraw-Hill, New York, 1965.

LENNETTE, E. H., E. H. SPAULDING, and J. P. TRUANT, editors, *Manual of Clinical Microbiology*, 3rd ed. American Society of Microbiology, Washington, D. C., 1980.

OHNSORGE, J., and R. HOLM. *Scanning Electron Microscopy*. Publishing Sciences Group, Inc., Acton, Mass., 1978.

PROVINE, H., and P. GARDNER. "The Gram-Stained Smear and Its Interpretation." *Hospital Practice*. 9:85, Oct., 1974.

SNYDER, B. "Pitfalls in the Gram Stain." *Lab. Med.*, 1:41, July, 1970.

STEIN, H. J. "Caution: Biology May Be Hazardous to Your Health." *BioScience*, 21:80, 1971.

KEY TERMS

pure culture

petri plate

colony

pour and streak plate

inoculum (i-nok′u-lum)

serial dilution

lyophilization (li-of′i-li-za′shun)

aseptic technique

slant culture

autoclave

sterilization

disinfectant

antiseptic

smear preparation

simple and differential stains

acid-fast staining

resolution

Pronunciation Guide for Organisms

Mycobacterium leprae (lep′ree)

PART 2
How Microbes Function— Physiology

I N many biological sciences, simply observing an organism can reveal a great deal about its anatomy, morphology, and physiology. Understanding a microbe, however, requires more than its examination under a microscope. In the field of microbiology, a basic knowledge of chemical and physical principles is essential, since microbes are best characterized by their biochemical pathways and interactions with the environment. The study of atoms, molecules, compounds, and their interactions is one of the foundations required to understand microbiology. The exact nature of interlinked biochemical pathways is unique to each species of microbe, so that an understanding of the different pathways helps in the identification, classification, and control of microbes.

All of the chemical reactions that take place in a microbe may result in an increase in the cell's mass. This process is known as cell growth and will only occur if certain conditions are available to the microbe. The cell must have adequate energy and building materials, and must be living in a suitable environment. Microbes grow very rapidly when these conditions are available, and as a result, only live a short time. Because of microbes' short life cycles in comparison to humans, their rapid rate of reproduction can have very significant effects on the environment. To control individual cell growth and reproduction, it is necessary to understand what happens in the life of a cell and how a population of cells increases. This information has far-reaching applications in the laboratory, industry, and health care fields.

Each type of microbe has its own metabolic pathways and growth characteristics. Pathogenic microbes are able to produce a variety of symptoms depending on how they interact with the patient. Therefore, it is essential to isolate and grow the suspected microbe in pure culture to perform those tests that will lead to the identification and control of the infectious agent. Special care is required from the time the specimen is taken until the completion of the testing. A specimen that is not handled properly will be of little value to the laboratory technician, doctor, or patient.

◀ *Scanning electron micrograph of* Mycoplasma pneumoniae *growing on nutrient-coated glass plate. Magnification: ×2550. (Courtesy of M. G. Gabridge.)*

To culture a microbe in the laboratory, an optimum balance of nutrients and a suitable growth environment must be provided. The media used for cultivation must contain all the materials necessary for the microbe to move quickly into the log phase of growth. The proper laboratory diagnosis of an infectious disease is vital to help the physician properly treat the infection. The diagnosis also provides the health care staff with information and guidance that may prevent others from becoming ill. Thus, the cultivation of a suspected pathogen is of utmost importance to successful treatment.

Microbes that benefit mankind have their growth and actions encouraged by providing them with an optimum environment containing nutrients, a source of energy, and needed growth factors. Microbes that are harmful to humans or the environment are also controlled by inhibiting their growth and actions, and destroying or removing them from the environment. Antimicrobial agents capable of controlling microbes include chemicals, radiation, and physical methods. The choice of a particular antimicrobial agent depends on the kind of material to be treated, the kind of microbe to be controlled, and the environmental conditions at the time of its use.

Genetics is the study of genes and how they produce characteristics which are inherited. The proposed double-helical structure for DNA has become the cornerstone for explaining gene function, gene replication, and the nature of mutations. Many geneticists use microbes as research organisms. The ease of culturing microbes and their short generation times enable geneticists to produce large populations quickly and trace the flow of inheritable characteristics through hundreds of generations. Although there are many advantages to using these organisms, microbes also demonstrate a number of genetic characteristics and inheritance patterns not previously found in eucaryotic cells. As these patterns emerged, it became clear that they would have far-reaching implications in medicine and in the field of eucaryotic genetics.

Metabolic Pathways

LIFE NEEDS

GROWTH FACTORS

THE ENVIRONMENT

CLASSIFICATION: A PHYSIOLOGICAL METHOD

ENERGY PRODUCTION

AEROBIC CELLULAR RESPIRATION

FAT AND PROTEIN METABOLISM

BIOSYNTHESIS AND METABOLISM

CONTROLLING METABOLISM

Learning Objectives

☐ State the first and second laws of thermodynamics and relate them to the three basic needs of all living things.

☐ Define and give examples of growth factors and vitamins.

☐ Classify microbes according to their optimum growth temperature and pH.

☐ Differentiate between photoautotrophic and chemoautotrophic metabolism.

☐ Understand the differences among facultative anaerobes, obligate aerobes, obligate anaerobes, and microaerophiles.

☐ Be familiar with the concepts of bonding, activation, and useful free energy, and know how they relate to energy production in microbes.

☐ Recognize the processes involved in carbohydrate metabolism including glycolysis, Krebs cycle, and the electron transporting system.

☐ Recognize the processes involved in the oxidation of fat and protein.

☐ Explain both aerobic and anaerobic oxidation.

☐ Recognize the intermediary metabolic pathways responsible for the interconversion of materials that occurs among carbohydrates, lipids, and proteins.

☐ Be able to explain the three methods of controlling metabolism, including alternative pathways, feedback inhibition, and the Operon Theory of Gene Regulation.

Microbes are independent cells capable of controlling the flow of matter and energy. They channel both matter and energy through specific, linked sequences of enzymatically controlled chemical reactions in order to perform the activities of life. The exact nature of these interlinking biochemical pathways is unique to each species of microbe. An understanding of the different biochemical pathways helps identify, classify, and control microbes. Metabolic pathways may be constructive and lead to the building of those cell materials used for activities such as growth and reproduction. This

7

creative metabolism is known as **anabolism.** Other pathways result in the destruction or breakdown of complex molecules. These processes are known as **catabolism** and result in the release of energy and component parts of complex molecules. If the anabolic reactions can be made to operate faster than the catabolic, the cell will grow and reproduce; if the catabolic reactions operate too fast, the cell will self-destruct and die. A healthy microbe has both of these processes under control and is able to adjust its biochemical pathways as the environment changes.

LIFE NEEDS

All microbes must meet three basic life needs in order to stay in balance with the environment. They must have (1) a source of usable energy, (2) a source of usable materials, and (3) a suitable environment in which to function. The loss of any of these will ultimately result in the death of the microbe.

Two basic forms of energy of use by microbes are visible light and chemical-bond energy. The immediate source of energy for cells is found in the chemical bonds of food molecules. Even those cells that can use the sun's energy must first convert sunlight into chemical-bond energy if they are to sustain life. The chemical bonds that hold the atoms of a molecule together contain energy. If a bond is broken, the energy is released and may be converted into other forms. The conversion of energy from one form to another is a common occurrence. In these reactions, matter (atoms) and energy are not created or destroyed but are converted from one form to another. This statement is known as the **First Law of Thermodynamics.**

All energy forms can be interconverted, and it is important to know that whenever such a conversion takes place, there is always a loss of some "usable" energy. For example, in an ordinary light bulb, electrical energy is converted to usable light energy; however, some heat energy is lost as unusable energy. The idea that whenever energy is converted from one form to another, some useful energy is lost is called the **Second Law of Thermodynamics.** Living systems obey the Second Law of Thermodynamics, and are, therefore, constantly losing some of their useful energy in the form of heat. If living things do not receive a constant supply of useful energy, they die.

Materials needed for a cell to function also take many forms. The two general types are simple inorganic molecules and the more complex organic molecules. Simple elements usually do not enter a cell's metabolic system in pure form but are in some chemical combination of use to the microbe. For example, nitrogen is essential to the efficient functioning of all cells, but only a very few microbes are

capable of assimilating atmospheric nitrogen (N_2) directly into their systems. Table 7.1 lists several biologically important elements, their main reservoir in the environment, and how they are used within most microbes. These elements are used in relatively large amounts. This is only a partial list since many others are used but only in very small amounts. Most of these **micronutrients,** or trace elements, are readily available to microbes since they are dissolved naturally in water and include such elements as cobalt (Co), zinc (Zn), and copper (Cu). Simple inorganic molecules not only serve as construction materials but also function in energy manipulation processes in microbes. Inorganic molecules in a reduced form (e.g., NH_3, H_2S, NO_2^-) may be oxidized by certain microbes to release their potential energy for the manufacture of ATP.

$$NH_4^+ \xrightarrow[\;H\;]{\;O\;} NO_2^- \xrightarrow{\;O\;} NO_3^- + energy$$

(ammonium ion)　　(nitrite ion)　　(nitrate ion)

$$H_2S \xrightarrow[\;H\;]{} S \xrightarrow{\;O\;} SO_4^{-2} + energy$$

(hydrogen sulfide)　　(sulfur)　　(sulfate ion)

> **1** What evidence is there that microbes obey the Second Law of Thermodynamics?
> **2** Where do microbes find necessary micronutrients?

GROWTH FACTORS

Organic molecules may be broken down and rearranged to meet the energy and building material requirements of a living cell. These metabolic pathways are made possible by the presence of specific enzymes produced within the cell. However, many chemical reactions would not proceed efficiently enough to maintain life were it not for growth factors. **Growth factors,** which may belong to any of the four basic organic groups, are essential in only small amounts and cannot be manufactured by the cell itself. For a cell to live, these molecules must be supplied as a part of its regular diet. The most familiar growth factors are **vitamins** and **amino acids.** The vitamin niacin is used by cells to manufacture the hydrogen-carrying molecule NAD^+. Another hydrogen carrier, FAD^+, is a component of the vitamin riboflavin. Without a proper amount of niacin or riboflavin,

TABLE 7.1
A few biologically important elements.

Element	Main Reservoir	Use
C (carbon)	CO_2 (air)	The basic element in all organic molecules; autotrophs convert CO_2 to reduced C (e.g., CH_4, $C_6H_{12}O_6$); heterotrophs oxidize it to CO_2.
H (hydrogen)	H_2O (water)	Found in organic molecules (e.g., carbohydrates); used as reducing agent for $N \rightarrow NH_3$, $S \rightarrow H_2S$, $C \rightarrow CH_4$, and $NAD^+ \rightarrow NADH_2$.
O (oxygen)	O_2 (air)	Found in most organic molecules; used as an electron acceptor in aerobic cellular respiration; an end product of some photosynthetic pathways.
P (phosphorus)	PO_4^{-2} (rocks and minerals)	Found in nucleic acids, proteins, coenzyme hydrogen carriers, and energy transporting molecules (e.g., ATP).
N (nitrogen)	N_2 (air)	Inorganic forms found as NH_3, NO_2^-, NO_3^-; organic forms found as amino acids and nucleotides; enters from atmosphere (N_2) by nitrogen fixation in certain microbes; essential for nucleic acid and protein synthesis; found in many electron carriers (e.g., FAD^+).
S (sulfur)	SO_4^{-2} (soil)	Found in some amino acids (cysteine, methionine); functions in reduced form (H_2S) as energy source or in oxidized form (S, SO_4^{-2}) as electron acceptor; functions as electron carriers in bacteria as iron-sulfur proteins.
Mg (magnesium)	Mg^{+2} (soil)	Necessary for metabolism of ATP; found in chlorophyll and used in surface recognition sites on bacterial cell walls.
Fe (iron)	Fe^{+2} or Fe^{+3} (soil)	Used in cytochromes and as electron acceptor in some bacteria.
Ca (calcium)	Ca^{+2} (soil)	Found in association with DPA in endospores; used in surface recognition sites, endocytosis, amoeboid movement, cell division, and microtubule production.

cells cannot synthesize the required NAD^+ or FAD^+, and aerobic respiration becomes inefficient. Table 7.2 lists various vitamins required by some cells to function properly and their use in a cell. Other growth factors that may be required by bacterial cells are the amino acids. Remember, there are twenty common amino acids, but not all may be synthesized by interconversion in biochemical pathways. A number of these may be provided to the cell if it is to remain healthy. Examples of amino acids that are growth factors for certain bacteria include glutamic acid, alanine, and asparagine.

In some cells, the nitrogenous bases (purines and pyrimidines) are unable to be synthesized and must be provided as growth factors. Without these essential organic molecules, the cell would not be able to construct DNA, RNA, and many molecules involved in the transfer of electrons during oxidation and reduction reactions. Another unique growth factor required by some pathogenic bacteria is the colored portion of the hemoglobin molecule, **heme.** This iron-containing organic molecule may be used by the bacteria to produce enzymes or to form iron-containing electron carriers. Both usable energy and materials must be supplied to microbes if they are to live; however, even these are not enough. A suitable environment must also exist if the cells are to successfully utilize available energy and materials.

THE ENVIRONMENT

Each organism type requires specific environmental conditions for growth, including such factors as heat, moisture, salt concentration, oxygen content, and pH. Microbes have these same requirements. Their microenvironments may change in a direction that either supports or inhibits growth. For example, the high temperature of the autoclave, high

TABLE 7.2
Uses of vitamins in microbes. These are only a few of the vitamins used in metabolism.

Vitamin	Use in Cell
Thiamine (B$_1$)	Coenzyme used in Krebs cycle
Riboflavin (B$_2$)	Part of coenzyme used in electron transfer system (FAD$^+$)
Niacin (nicotinic acid)	Part of coenzyme used in electron transfer system (NAD$^+$)
Pyridoxine (B$_6$)	Coenzyme used in synthesis of amino acids
Cyanocobalamin (B$_{12}$)	Used in protein synthesis and DNA production
V factor	Probably involved in electron transfer reactions
Folic acid	Part of biochemical pathway leading to production of purines and pyrimidines
Biotin	Probably involved in breakdown of certain amino acids and oxidation of pyruvic acid and lactic acid
Pantothenic acid	Part of biochemical pathway leading to production of aspartic acid, an amino acid, and coenzyme A

concentration of sugar in jellies and jams, and low pH of yogurt make the growth of microbes in these environments very difficult. But the optimum moisture content, near-neutral pH, and balanced salt concentration of unrefrigerated beef stew make the growth of spoilage organisms very likely. Microbes can be classified according to the range of environmental conditions under which they are able to grow. Those that prefer to grow in cold temperatures are called **psychrophiles,** or "cold lovers," those that favor the middle temperatures are **mesophiles,** and "heat lovers" are known as **thermophiles** (Figure 7.1). Thermophiles actively grow at higher temperatures, but **thermoduric** microbes only endure high temperature, waiting until more favorable conditions develop. Many spore-formers (*Bacillus* and *Clostridium*) are thermoduric but not thermophilic because they have the ability to produce endospores.

Microbes that grow well in acid environments are known as **acidophiles,** those that favor more neutral conditions may be called **neutrophiles,** and those that favor more alkaline environments are **alkalophiles.** A last group of

1 In what form do members of the genus *Bacillus* endure high temperatures?
2 In what type of environment might acidophiles be found?

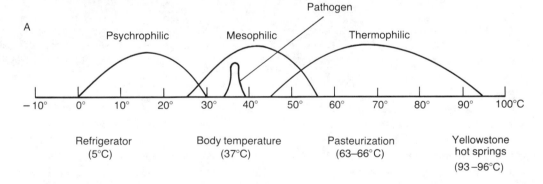

FIGURE 7.1
The two graphs illustrate the growth range of different types of bacteria in environments that vary in temperature (A), and in hydrogen ion concentration, pH (B). These graphs do not point out that most bacteria favor the mesophilic and neutrophilic environments. Pathogens such as *Treponema pallidum,* staphylococci, and streptococci have very narrow growth ranges, and are easily destroyed by relatively minor changes in environmental conditions.

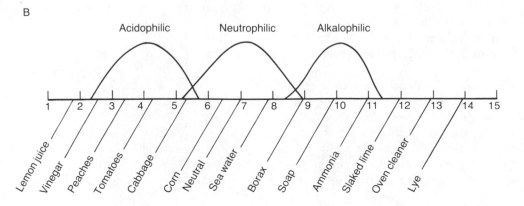

microbes that favors environments that are high in salt concentration are the **halophiles** ("halo" means *salt*).

CLASSIFICATION: A PHYSIOLOGICAL METHOD

Microbes have been grouped according to probable evolutionary relationships, morphology, staining characteristics, colony characteristics, and environmental preferences. Each grouping provides a different insight into the nature of microbes. Microbes can also be classified according to their basic metabolic processes (Table 7.3). The two major categories are the **autotrophs** and the **heterotrophs.** They differ from one another in their ability to utilize simple or complex materials and their ability to "see" light or chemical-bond energy. Autotrophs may be subdivided into two groups, the photoautotrophs and the chemoautotrophs. **Photoautotrophs** use visible light energy in the biochemical pathway of photosynthesis and simple inorganic materials such as CO_2, H_2O, and NH_3. Green plants, algae, and a few bacteria fall into this group. **Chemoautotrophs** also use simple materials in their metabolic pathways, but gain their energy from the oxidation of simple inorganic molecules such as NH_3. These reactions are referred to as **cellular respiration.** If respiration involves the use of molecular oxygen (O_2), it is called **aerobic,** and if a substitute for O_2 is used, it is called **anaerobic.** Heterotrophs may also carry out aerobic and anaerobic cellular respiration; however, their biochemical pathways depend on complex

organic molecules for chemical-bond energy and construction materials. Many microbes function both aerobically and anaerobically. These are **facultative anaerobes;** i.e., they have the facility to grow using atmospheric oxygen (O_2), or an alternative pathway such as fermentation. Some microbes do not have a choice. These **obligate aerobes** must have oxygen to function and stay alive. Others must have oxygen removed from their environment, otherwise the oxygen would rapidly kill them. These microbes are **obligate anaerobes.** Special care and equipment must be used to successfully grow these types of microbes. A last group of organisms is called the **microaerophiles** ("little-oxygen-lovers"). This group requires oxygen for respiration but prefers a lower concentration (2–10 percent) of oxygen than is found in the air (20 percent). Whether a cell functions aerobically or anaerobically, the membrane structure on which these biochemical pathways occur is essential to the production of sufficient ATP to sustain the cell.

All microbes can be represented on a scale. At one end are a few kinds of self-sustaining photoautotrophs that are able to use visible light energy and simple inorganic molecules (Box 12). The other end is represented by microbes that must have nearly all their material needs supplied to them in the form of complex organic molecules. These may have such diverse and complex requirements that they may only be able to exist as parasites on other living organisms. Between the two extremes is an increasing percentage of microbes that have lost the ability to synthesize certain complex molecules and must have them supplied from the environment. Most microbes are in the middle of this scale and are heterotrophs. Whether the energy for biochemical

TABLE 7.3
Classification according to metabolism.

	Biochemical Pathway	Energy	Materials
Autotroph			
Photoautotroph	Photosynthesis	Visible light	Simple inorganic molecules (e.g., CO_2, H_2O)
Chemoautotroph	Respiration	Oxidation of simple inorganic molecules	Simple inorganic molecules
Heterotroph			
	Respiration	Oxidation of complex organic molecules; use O_2; aerobic	Complex organic molecules (e.g., sugar)
	Fermentation	Oxidation of complex organic molecules; do not use O_2; anaerobic	Complex organic molecules

BACTERIA are grown in a solidified (agar-containing) nutrient medium. This medium is a mixture of carbohydrates, proteins, lipids, and growth factors. To demonstrate the differences in oxygen requirements, an oxygen gradient must be established for the bacteria being grown. This shows whether or not the bacteria were successful growing in an environment with a large amount of oxygen, or whether they were better able to grow in an environment that lacked the gas.

This procedure is done in a test tube. In addition to the agar and nutrients listed above, these special media also contain the chemicals **sodium thioglycolate** and **resazurin**.

BOX 12
OXYGEN REQUIREMENTS OF BACTERIA

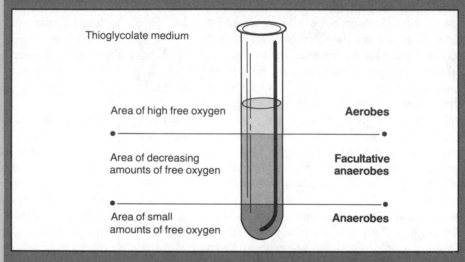

Thioglycolate is a molecule that is able to capture oxygen from the medium and bind it tightly. This makes the oxygen unavailable to the microbes that are inoculated into the tubes. The resazurin is a color indicator. When the oxygen is removed from the nutrient medium by the thioglycolate, the resazurin changes color, showing that the tube is ready to be inoculated.

Notice that the upper portion of the agar medium contains the greatest amount of free oxygen (O_2) because the gas may easily diffuse into the medium from the atmosphere. The larger center zone contains a decreasing

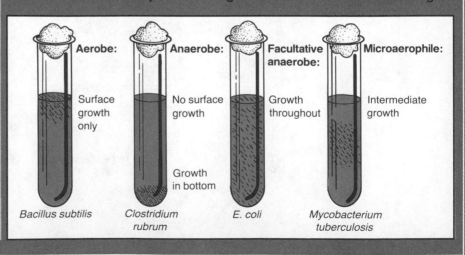

amount of oxygen because the gas has more difficulty reaching this portion of the medium. The bottommost portion lacks oxygen. Any bacteria that would grow well in the upper portion of this tube culture would be aerobic. Those growing in the center portion of the tube would be facultative anaerobes, and those growing only at the bottom of the tube would be anaerobes.

> **1**   What special type of environment is needed to grow obligate anaerobes?
>
> **2**   What purpose does the chemical thioglycolate serve in culture media?

reactions comes from light or the oxidation of molecules, most of it is converted to the usable form of chemical-bond energy, ATP (Figure 7.2).

The ATP manufactured in the cell is used for all the work required to sustain life. For example, ATP energy is used to synthesize and degrade fats, carbohydrates, proteins, and nucleic acid molecules. ATP energy is also used to move molecules into and out of the cell. As the ATP is

generated by a cell, other molecules may be produced. These vary from one cell type to another and may depend on what food molecules are available to the cell. These by-product molecules may be waste materials, or they may be useful to the cell in its growth and repair processes.

ENERGY PRODUCTION

Nearly all organisms rely on the same types of chemical reactions to release energy. In most aerobic heterotrophs, the three main pathways are glycolysis, the Krebs cycle, and the electron transmitting system (ETS). Figure 7.3 shows the generalized pathways of aerobic cellular respiration and may be read from left to right (reactants to products), and from

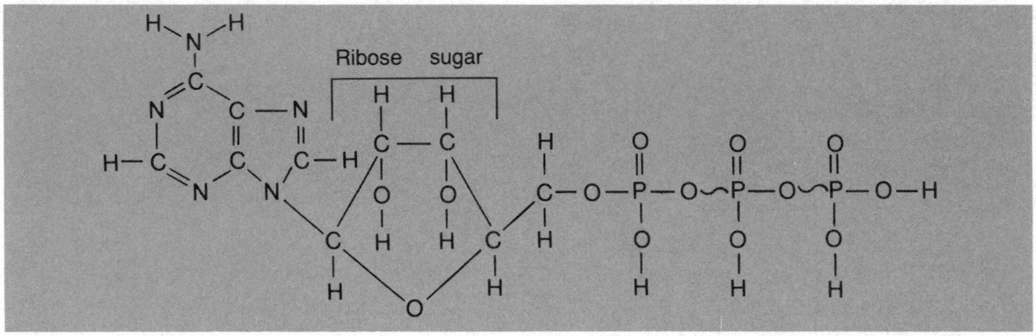

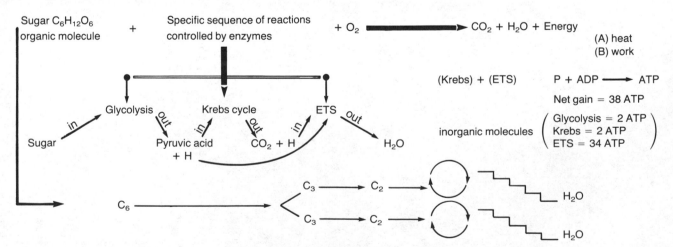

FIGURE 7.2

The adenosine triphosphate (ATP) molecule is able to transfer the last phosphate group to other molecules to increase the likelihood of their reaction. Notice that there are two "special" covalent bonds holding two phosphate groups to the rest of the molecule. These are called "high-energy phosphate" bonds.

FIGURE 7.3

This sequence of reactions in the aerobic oxidation of glucose is an overview of the energy-yielding reactions of a cell. The first line presents the respiratory process in its most basic form. The next two lines expand on the generalized statement and illustrate how sugar (glucose) moves through a complex series of reactions to produce usable energy (ATP). Note that both CO_2 and H are products of the Krebs cycle, but only the H enters the ETS.

top to bottom (simple to more detailed description). Chemical reactions can only take place in a direction that leads to lower free energy. **Free energy** is the amount of energy in a chemical system (e.g., carbohydrate + O_2) that is available to do work: in this case, the formation of ATP. When a carbohydrate molecule reacts with oxygen, a great amount of stored free energy is released. The resulting CO_2 and H_2O molecules contain less free energy than that found in the original carbohydrate and oxygen molecules. ATP may be synthesized in cells by a flow of electrons through a series of electron carrier molecules that drives hydrogen protons to a lower free energy state. This occurs down a gradient from a chemical system containing high free energy (e.g., glucose + O_2) to one of lower free energy (e.g., CO_2 + H_2O). The gradient of the electron transmitting system (ETS) is formed across closed, saclike membranes such as mitochondria, chloroplasts, and microbial plasma membranes (Figure 7.4). Usually only the last high-energy phosphate bond is used in chemical reactions where ATP serves

as an energy source. Therefore, only the third phosphate group needs to be added to ADP (adenosine *di*phosphate) to synthesize ATP (adenosine *tri*phosphate):

$$ADP + P \longrightarrow ATP + H_2O$$

A great deal of energy is required to form this bond and it becomes stored in the ATP molecule.

The amount of activation energy needed to break the last phosphate bond on ATP is very low compared to a normal covalent bond between a carbon and a phosphate group. Because only a small amount of energy is needed to break the high-energy bond, much usable free energy can be made available for use in other chemical reactions:

bonding energy − activation energy
= useful free energy

By donating this last phosphate on ATP to other molecules, energy can be made available to drive chemical reactions that could not otherwise occur due to their low initial free

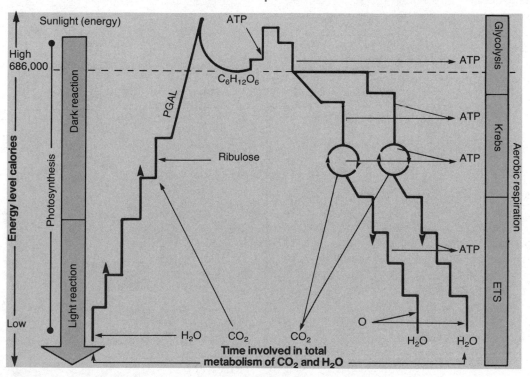

FIGURE 7.4

Living organisms manipulate both energy and matter. This graph illustrates how energy from the sun moves through photosynthesis into the chemical-bond energy of glucose and out again in the form of ATP. The time involved in these processes may vary from a millisecond to a million years and the flow of materials is always cyclical. Carbon dioxide and water molecules combine through photosynthesis to increase their free energy. Glucose ($C_6H_{12}O_6$) is shown in a "valley," since this molecule is very stable, and requires the action of enzymes and activation energy to lower it to the energy gradient of aerobic cellular respiration. As the oxidation-reduction reactions occur, free energy is released to form ATP molecules, and, ultimately, oxygen is combined with hydrogen to form water. Since these reactions follow the Second Law of Thermodynamics, useful energy is lost during each reaction and released as excess heat.

energy. An example of this last reaction takes place during the dark reaction of photosynthesis when sugar is formed from CO_2 and a five-carbon ribulose molecule. The addition of phosphate (from ATP) to the five-carbon molecule increases the free energy of the system enough to allow for the chemical combination of carbon from CO_2 to make a new six-carbon sugar molecule:

$$\underset{\text{compound}}{\text{5-carbon ribulose}} \xrightarrow[\displaystyle CO_2]{\overset{\displaystyle ATP \quad ADP + P}{\curvearrowright}} \underset{\text{compound}}{\text{6-carbon sugar}}$$

Energy is measured in units of kilocalories per mole. One **kilocalorie** is the amount of energy needed to raise the temperature of one kilogram of water one degree Celsius. Do not confuse a **calorie** with a kilocalorie! People have, over the years, misused these terms. When people talk about a 1000 calorie diet, they actually mean a 1000 *kilo*calorie diet. The kilo, or *k*, prefix (1 kilocalorie = 1000 calories) has been dropped in common usage. A 1000 kilocalorie diet (that *they* call a 1000 calorie diet) is actually limiting food intake to 1,000,000 calories. A mole is not the same as a molecule of a substance. A **mole** is the amount of a material measured in grams that is equal to its molecular weight. For example, a mole of glucose is 180 grams since the molecular weight of glucose ($C_6H_{12}O_6$) is 180.

$$180 = (6 \times {}^{12}C) + (12 \times {}^{1}H) + (6 \times {}^{16}O)$$

A mole of water equals 18 grams or

$$H_2O = (2 \times {}^{1}H) + (1 \times {}^{16}O) = 18.$$

When a mole of glucose is synthesized by photosynthesis, a sizable amount of material has been produced and contains about 686,000 calories (686 kilocalories) of free energy. The complete oxidation of this mole of glucose through aerobic cellular respiration can theoretically result in enough free energy to produce about 46 moles of ATP. However, not all of this energy is channeled into the synthesis of these molecules since all cells follow the Second Law of Thermodynamics. Only about 38 ATP moles are formed and the remaining energy is found in CO_2 and H_2O molecules or as waste heat. All biochemical pathways presented in this chapter show a single formula for each chemical substance. These formulas do not represent single molecules but only the general nature of the reactions. The numbers of ATP, glucose, or other chemicals described in the pathways are in moles.

---

1 In what way might the activation energy of a molecule be lowered?

2 How is free energy related to activation energy?

3 How many grams is a mole of glucose?

AEROBIC CELLULAR RESPIRATION

The first stage in carbohydrate metabolism involves the breakdown of a sugar molecule without the use of molecular oxygen. This process is known as the **glycolytic pathway** (or Embden-Meyerhof pathway). This pathway is the first series of biochemical reactions in the oxidation of glucose (Figure 7.5). Because the glucose molecule is very stable and difficult to hydrolyze, a series of chemical reactions is required to start the process. In two of the first four reactions, ATP is added to increase the free energy of the system. Two other reactions involve the restructuring of the molecules to lower the activation energy required to begin the energy extraction process. These reactions occur in the cytoplasm of the cell. Any molecule that gains phosphates

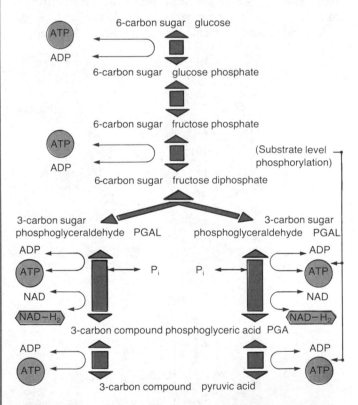

FIGURE 7.5
The glycolytic pathway is an enzymatically controlled series of chemical reactions that begins with a six-carbon glucose molecule and results in the formation of two pyruvic acid molecules. During this breakdown by oxidation-reduction reactions, free energy is released for the formation of four ATPs. However, since two ATP molecules were required to initiate this series, the net gain of useful energy is only two ATPs. The hydrogen atoms removed from the organic molecules are transferred to the hydrogen-carrying molecule NAD^+ which may either carry the potential energy of the electrons to the ETS or to other organic molecules in the cell. (P_i = inorganic phosphate from a pool of these ions in the cytoplasm.)

is said to be phosphorylated. Although this added energy is necessary to get the energy-releasing reactions started, it will be regained later in the pathway. These first reactions result in the splitting of a glucose molecule into two three-carbon molecules. These phosphoglyceraldehyde (PGAL) molecules acquire a second phosphate from a "phosphate pool" normally found in the cell, (P_i), and each now has two phosphates. A series of reactions follows in which energy is released by the breaking of chemical bonds, and each of these compounds loses its phosphate. These high-energy phosphates are combined with ADP already found in the cell to form two ATP molecules for each three-carbon molecule. These glycolitic reactions that result in the formation of ATP are called **substrate-level phosphorylations.** At this point, the two ATPs that were necessary to begin the reaction have been reclaimed (two used and two recovered), and the rest of the reaction sequence is in a direction that will result in the release of more usable energy. In addition, four hydrogen atoms are removed from the carbon skeletons and attached to two hydrogen-carrying molecules, nicotinamide adenine dinucleotide (NAD^+). This is an oxidation-reduction reaction that transfers a great amount of potential energy to the NAD^+ molecules, reducing them to $NADH_2$. The $NADH_2$ potential energy may be converted to the more usable ATP form later in the electron transmitting system. The result of these reactions is the formation of

phosphoglyceric acid (PGA), which is then converted to pyruvic acid. This reaction again releases enough free energy to form two more ATP molecules. The process of glycolysis starts with a six-carbon sugar molecule and undergoes reactions that lead to the formation of four ATPs, two $NADH_2$s, and the two remaining three-carbon molecules of pyruvic acid. Since two ATPs were used to start the process and a total of four ATPs were formed, each simple sugar molecule that undergoes glycolysis results in a net production of two ATPs. Glycolysis is an anaerobic process because no atmospheric oxygen (O_2) is used.

If atmospheric oxygen is available, and the cell is genetically capable of producing the necessary enzymes, the pyruvic acid may continue to be oxidized to release even more energy. This is an aerobic process since oxygen (O_2) will ultimately be used as the final acceptor of hydrogen. The next series of reactions which follows glycolysis is known as the **Krebs cycle** and is illustrated in Figure 7.6. The Krebs cycle is the second series of biochemical reactions and is the oxidation of acetyl coenzyme A derived from pyruvic acid. It begins with pyruvic acid and ends with the formation of carbon dioxide, two ATP molecules, and the release of hydrogen.

Pyruvic acid is an "all-purpose molecule" in the cell because many different molecules useful to a cell can be manufactured from pyruvic acid. If O_2 is available to a cell

FIGURE 7.6
The Krebs cycle is a series of reactions also known as the tricarboxylic acid cycle since this form of organic acid is found in the cycle. The cycle begins after pyruvic acid (from the glycolytic pathway) has been oxidized to CO_2, $NADH_2$, and acetyl. The two-carbon acetyl molecule is chemically combined with coenzyme A and then moved into the cycle. The series results in the complete oxidation of the organic molecule to CO_2 and H; the original sugar molecule does not exist in any organic form after the Krebs cycle has been completed. Even though only two ATPs are generated by substrate-level phosphorylation, several NAD^+ and FAD^+ molecules are reduced with hydrogen for later transfer to the ETS. (P_i = inorganic phosphate from a pool of these ions in the cytoplasm.)

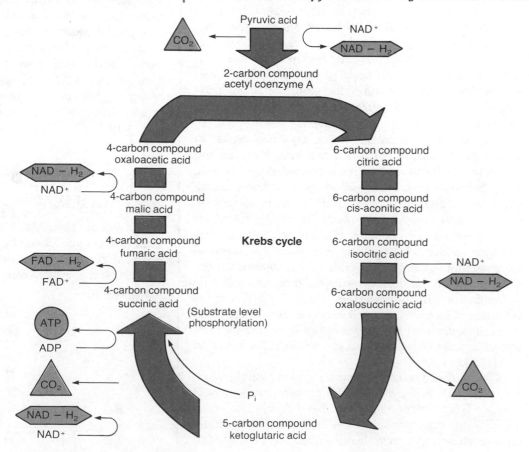

with the necessary enzyme system, each pyruvic acid molecule will lose carbon dioxide and hydrogen to reduce NAD^+ to $NADH_2$. With the loss of these carbon dioxide molecules, the three-carbon acid now becomes a two-carbon acetyl compound. This carbon dioxide is a waste product that the cell will release into the atmosphere. The acetyl compound temporarily combines with a large molecule called coenzyme A (CoA), forming an acetyl-CoA.

Many enzymes do not work by themselves but require the cooperation of some sort of a helper. This helper is a molecule that aids in the reaction by removing one of the end products, or it may aid by bringing in part of the substrate. Such helpers are called **coenzymes.** Vitamins are a principle part of many coenzymes. Without the coenzyme, enzymes do not function as well as they are able.

Acetyl-CoA reacts with a previously existing four-carbon molecule. In this reaction, the two-carbon acetyl is transferred to the four-carbon compound to form a new six-carbon molecule, citric acid. After making the transfer, the CoA may return to pick up a new acetyl compound. The function of CoA is to assist in the combining of the two-carbon compound with the four-carbon compound.

The new six-carbon compound is broken down in a series of reactions to a four-carbon compound by the release of two carbon dioxide molecules and a number of hydrogen atoms. The resulting four-carbon molecule is now used to combine with a newly entering two-carbon acetyl compound, and the chemical reactions are repeated. Each time an acetyl group moves through this Krebs cycle, an ATP is formed. Thus, each pyruvic acid molecule has lost a carbon dioxide to form an acetyl group and additional carbon dioxide in the cycle. All of the six-carbon atoms originally found in the simple glucose sugar are released into the atmosphere as the gas carbon dioxide.

At this point, the breakdown of sugar has resulted in the net production of four ATPs: two from glycolysis and two from the Krebs cycle. The hydrogen atoms released during the Krebs cycle are all picked up by hydrogen acceptors such as NAD^+ and FAD^+. FAD^+ is flavin adenine dinucleotide and operates very much like NAD^+ but is part of the growth factor riboflavin. $NADH_2$ and $FADH_2$ carry hydrogen to the third pathway known as the **electron transporting system (ETS).** This system converts the potential energy in the hydrogen-carrying molecules into ATP by **oxidative phosphorylation** (Figure 7.7). This is the last series of biochemical reactions in aerobic cellular respiration. It begins with the movement of hydrogen electrons through a series of cytochromes and results in the formation of ATP and water.

The greatest potential energy is found in $NADH_2$, which enters the ETS at the top. $FADH_2$ contains the equivalent of one ATP's worth of energy less than $NADH_2$, and enters the ETS one step below $NADH_2$. In the first oxidation-reduction reaction, the entire hydrogen atom is transferred to a lower free energy state and enough energy is made available to synthesize one ATP. From this point, only the electron from the hydrogen is transmitted through the series. Two additional ATP molecules are synthesized as the electrons flow through a series of iron-containing electron transmitting molecules known as **cytochromes.** This flow drives the freed hydrogen protons (H^+) to a lower free energy state. When the electrons combine with oxygen to form a negatively charged ion, two positively charged hydrogen protons chemically combine with it to form water.

---

1 During which stage of aerobic cellular respiration is the original glucose completely oxidized?

2 What purpose does a coenzyme serve?

3 Name two hydrogen carriers found in the ETS.

This system converts the potential energy in the hydrogen-carrying molecules into ATP. The reactions yield enough energy to form thirty-four additional ATPs. Add to this the two ATPs from glycolysis plus two ATPs from the Krebs cycle reactions, and the complete aerobic breakdown of a simple sugar yields a total of thirty-eight ATPs, if oxygen is available (Figure 7.8).

Oxygen is used in the ETS and is ultimately combined with hydrogen to form molecules of water. The hydrogen atoms found in the original molecule of simple sugar are now part of water. At this point, the simple sugar has undergone many reactions and been changed from sugar to carbon dioxide and water, and released enough energy to form thirty-eight ATPs. Simple sugars are not the only type of organic molecules that may be used as an energy source by microbes. Fats and proteins may also be oxidized to produce ATP.

FAT AND PROTEIN METABOLISM

Before fats can be oxidized to release energy, they must be broken down into glycerol and fatty acids. Refer to Figure 7.9 as fat oxidation is discussed.

Glycerol is a three-carbon compound with the following structure:

$$
\begin{array}{c}
H \\
| \\
H-C-O-H \\
| \\
H-C-O-H \\
| \\
H-C-O-H \\
| \\
H
\end{array}
$$

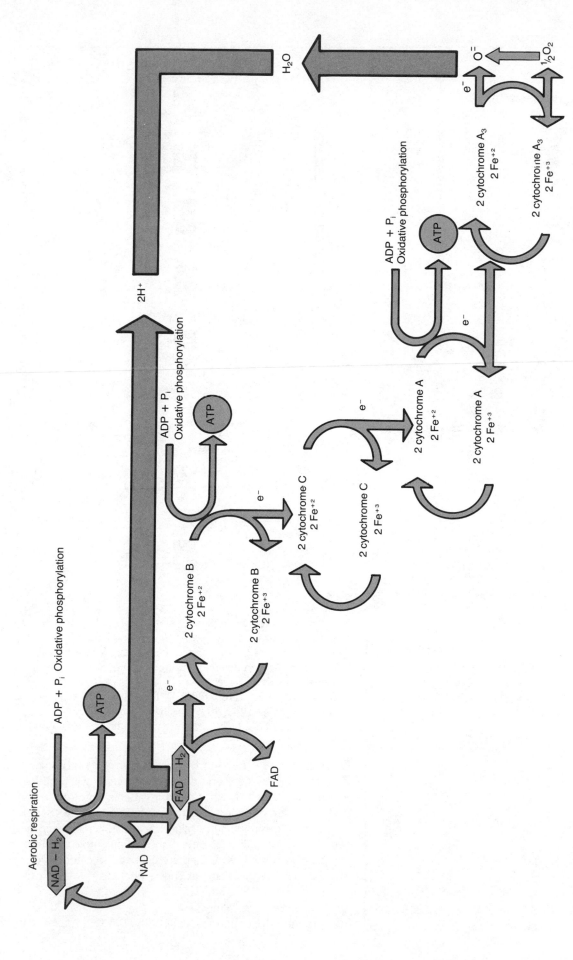

FIGURE 7.7

The electron transporting system (ETS) is a series of oxidation-reduction reactions also known as the cytochrome system. The movement of electrons down this biochemical "wire" establishes a kind of electrical current that drives the H^+ protons to atmospheric oxygen. As the electrons flow through the system, ATPs may be produced by oxidative phosphorylation. Electrons entering the system from $NADH_2$ result in the synthesis of three ATPs. If they enter from $FADH_2$, only two ATPs will be produced. When oxygen is not available, this system closes down completely as does the Krebs cycle, leaving only the glycolytic pathway as a ready source of ATP.

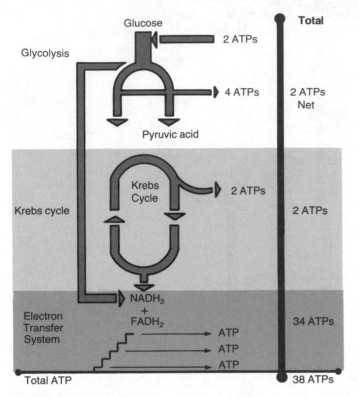

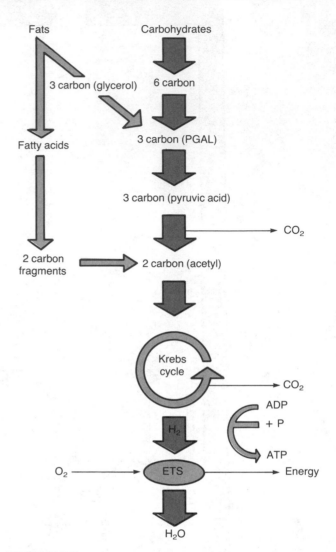

FIGURE 7.8

In the respiration of a six-carbon sugar, two ATPs are put into the reaction. Under anaerobic conditions, four ATPs are produced. The net profit to the cell is two ATPs. Under aerobic conditions, two ATPs are put into the reaction and forty ATPs are produced (four from glycolysis, two from the Krebs cycle, and thirty-four from the electron transfer system). The net profit to the cell is thirty-eight ATPs. (From Eldon D. Enger et al, *Concepts in Biology,* 3rd ed. © 1982, 1979, 1976 by Wm. C. Brown Company Publishers, Dubuque, Iowa. Reprinted by permission.)

Once the glycerol is converted to pyruvic acid, it proceeds through the Krebs cycle in the same way as did the pyruvic acid formed from simple sugars.

Fatty acids are long-chain carbon compounds that can break apart into two-carbon units:

$$\begin{array}{ccc} H & H \\ | & | \\ H-C-C-H \\ | & | \\ H & H \end{array}$$

Enzymes can split off these two-carbon units into acetyl molecules:

$$\begin{array}{ccc} O & H \\ \| & | \\ H-C-C-H \\ & | \\ & H \end{array}$$

FIGURE 7.9

Fats must be digested into glycerol and fatty acids before they can undergo oxidation. The glycerol is converted into PGAL and then proceeds through the respiratory pathways. The fatty acid molecules are broken down into two-carbon fragments; these are converted into acetyl, which then proceeds through the respiration pathways. (From Eldon D. Enger et al, *Concepts in Biology,* 3rd ed. © 1982, 1979, 1976 by Wm. C. Brown Company Publishers, Dubuque, Iowa. Reprinted by permission.)

Once the acetyl is formed, it is oxidized in the Krebs cycle. Each two-carbon fragment of a fatty acid chain will produce a number of ATPs. If all of these fragments from one molecule of fat are considered, a mole of fat yields more energy than a mole of simple sugar, since fat contains more hydrogen. When both the fat and carbohydrate molecules are completely broken down, they are changed to carbon dioxide and water. In addition to sugars and fats, proteins can be broken down to form ATP.

For proteins to release energy, they must first be broken into individual amino acids. The amino acid must then have the amino group removed. This amino group is converted into ammonia or another nitrogen-containing compound. The rest of the amino acid molecule (called a **keto acid**) will be changed and entered into the respiration pathway as pyruvic acid or as one of the types of molecules found in the Krebs cycle (Figure 7.10).

If no oxygen is present, the ETS system of many bacteria stops working. The Krebs cycle then stops, and acetyl is not formed from pyruvic acid. As a result, only glycolysis will occur and only two ATPs are formed. Many microbial cells can survive without atmospheric oxygen by making use of the glycolytic pathway and additional reactions that convert the pyruvic acid to molecules other than CO_2 and H_2O

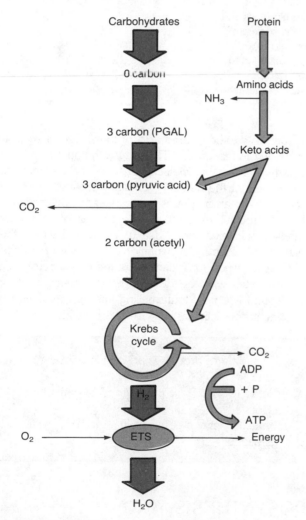

Carbohydrates

Protein

0 carbon

Amino acids

NH_3

3 carbon (PGAL)

Keto acids

3 carbon (pyruvic acid)

CO_2

2 carbon (acetyl)

Krebs cycle

CO_2

ADP

+ P

H_2

ATP

O_2 → ETS → Energy

H_2O

FIGURE 7.10

Proteins must be digested into amino acids before they can undergo oxidation. The amino acids are converted into various keto acids. The keto acids enter the respiration pathway as pyruvic acid or one of the keto acids of the Krebs cycle. (From Eldon D. Enger et al, *Concepts in Biology*, 3rd ed. © 1982, 1979, 1976 by Wm. C. Brown Company Publishers, Dubuque, Iowa. Reprinted by permission.)

(Figure 7.11). These processes are called **fermentation.** Fermentation is the anaerobic oxidation of an organic compound in which gaseous oxygen is not used as the hydrogen acceptor but replaced by another organic compound. Organisms that use fermentation to produce ATP can only receive two ATP molecules from each sugar molecule they break down. Therefore, fermentation is less efficient than aerobic respiration. All fermentation may involve the glycolytic pathway, yet the final steps in the processes differ because other organisms produce different enzymes.

Anaerobic cellular respiration of organic molecules is an oxidation process. Since no O_2 is used in this biochemical pathway, other organic molecules must serve as final electron acceptors. Many of the microbes that ferment to produce ATP energy have no ETS (cytochrome system). Lactobacilli, which are normally found in the mouth and are partially responsible for tooth decay, convert glucose to pyruvic acid by glycolysis. The hydrogen that is removed from the carbon-containing sugar is used to reduce NAD^+ to $NADH_2$ (Figure 7.12). Since there is no ETS through which the electrons may be passed, the hydrogen is transferred from $NADH_2$ to the pyruvic acid molecules that have been produced from glycolysis. This reaction converts the pyruvic acid to lactic acid. Microbes that produce acid as a product of fermentation are called **acidogenic** (acid-generating). The NAD^+ is regenerated in this reaction and is free to return to the glycolytic pathway to be reduced again to $NADH_2$. Cycling of NAD^+ and $NADH_2$ enables the reactions to continue, because the carrier molecules are driving the hydrogen protons to a lower free energy state.

Not all bacteria, even within the same genus, produce the same fermentation products. The hydrogen from $NADH_2$ may be deposited on pyruvic acid or other molecules produced from pyruvic acid to form a variety of end products. For example, there are two major groups within the genus *Lactobacillus*. The **homofermentative** ("homo" means *same*) lactobacilli produce only lactic acid as a fermentation product. The **heterofermentative** ("hetero" means *different*) lactobacilli produce lactic acid plus other, different organic molecules from the reduction of pyruvic acid. *Escherichia coli* is a bacterium that is also capable of producing many products from pyruvic acid. This acidogenic bacterium may yield succinic acid, lactic acid, formic acid, acetic acid, ethyl alcohol, or hydrogen and carbon dioxide gases.

Glucose is not the only organic molecule that may be fermented for the production of ATP. Many bacteria have the ability to oxidize amino acids and sugars. In order for this to occur, protein must first be broken down into component parts by enzymes released from the cell. *Bacillus* and many other bacteria hydrolyze protein in this process known as **putrefaction** (spoilage). The end products of this reaction result in odors that are very unpleasant. In the fermentation reaction, amino acids that contain a high free energy are oxidized with the help of NAD^+, and the hydrogen

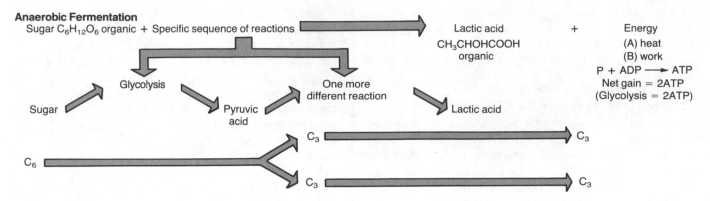

FIGURE 7.11

This is only one example of anaerobic fermentation. The chart is read from left to right and top to bottom. The specific sequence of reactions controlled by enzymes includes glycolysis and "one more different reaction." Fermentation reactions vary in the nature of this "different" reaction. In the case of yeast cells, after pyruvic acid has been produced from glycolysis, it changed to acetylaldehyde, which may serve as a hydrogen acceptor and be converted to alcohol and CO_2. The complete oxidation of glucose to CO_2 and H_2O never occurs in fermentation reactions. The product always includes a complex organic molecule other than glucose.

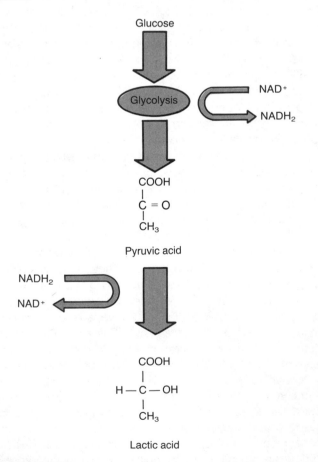

FIGURE 7.12

The formation of lactic acid in fermentation occurs when the $NADH_2$ molecule is used to reduce pyruvic acid. Following this reaction, the oxidized NAD^+ is able to return to the glycolytic pathway to pick up more hydrogen.

is carried to another amino acid of lower free energy. The result is the formation of ATP along with various amino acids and other organic molecules. Amino acid fermentation by members of the genus *Clostridium* produces such foul-smelling chemicals as putrescine, cadaverine, hydrogen sulfide, and methyl mercaptan. The names of these compounds should give some idea of when and where these clostridia are active in fermentation. *Clostridium perfringens* and *C. sporogenes*, for instance, are the two anaerobic bacteria associated with the disease gas gangrene. A gangrenous wound is a very foul-smelling infection resulting from the fermentation activities of these two bacteria.

1 What type of process is anaerobic cellular respiration?
2 Name two different products of bacterial fermentation.
3 What bacteria are normally associated with gas gangrene?

BIOSYNTHESIS AND METABOLISM

So far we have seen that the same basic metabolic pathways of glycolysis, the Krebs cycle, and the ETS are involved in the breakdown of many kinds of organic molecules to produce energy. Not all molecules entering these pathways are destroyed. Some are used in the reactions that build new cells or cell parts. To accomplish these **bio-**

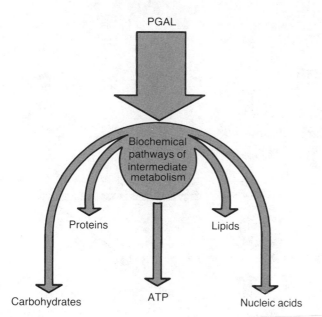

FIGURE 7.13

PGAL can follow various pathways and be converted into a number of different products. (From Eldon D. Enger et al, *Concepts in Biology*, 2nd ed. © 1979, 1976 by Wm. C. Brown Company Publishers, Dubuque, Iowa. Reprinted by permission.)

synthesis reactions, one kind of molecule may be converted to another. This interconversion of molecular types will occur if a necessary building block happens to be in short supply or must be stored for later use.

A cell's diet may include all the materials of the four major classes of organic compounds: carbohydrates, proteins, lipids, and nucleic acids. Depending on the building materials that are required at the time, the cell has the capability of converting one organic molecule to another by modifying that molecule at various places in the glycolytic pathway or the Krebs cycle. One molecule that has many uses in the cell is PGAL (phosphoglyceraldehyde). After PGAL is produced in the glycolytic pathway, the microbe may use these molecules as an energy source or convert this raw material into other organic molecules such as carbohydrates and fats. With the addition of ammonia as a raw material, the PGAL may be converted to protein (Figure 7.13). Because many of the chemical reactions are carried out in either direction, it is possible to change carbohydrates (e.g., simple sugars) to proteins (e.g., amino acids) or to lipids (e.g., fatty acids). This type of interconversion is going on constantly within the cell to maintain all the cell's structural components in good repair and increase the size of the cell in preparation for reproduction. This interconversion may also be referred to as **intermediary metabolism.**

If all the nutrients taken in by a microbe were utilized for nothing but energy production, the cell would release more energy than it could expend in moving, reproducing, or other activities. In addition, the microbe would begin to break down because no building materials would be available to combat the natural aging process and the destruction of molecules. A balance must be attained between the process of energy release and the formation of building materials throughout the glycolytic pathway, the Krebs cycle, and the ETS. Understanding what happens in a cell can help determine what type of nutritional balance must be attained to control the microbe's activities. A cell whose energy intake exceeds its daily energy requirements will only convert a portion of the food into the needed amount of energy. The excess food may be converted into fats or proteins depending on the cell's needs and the enzymes present at that particular time (Figure 7.14). The interconversion of one organic molecule to another is vital to the survival of any cell; however, these processes cannot go on in an uncontrolled manner. Several intracellular control mechanisms have been discovered that prevent the cell from either destroying too many essential molecules or producing a deadly excess of others.

CONTROLLING METABOLISM

The three primary regulating systems that operate inside cells to control both anabolic and catabolic pathways are alternate pathways, feedback inhibition, and gene regulation. Catabolic pathways for a molecule may be very different from anabolic pathways. Anabolic reactions do not interfere with, or neutralize, catabolic reactions in the manner which acids and bases neutralize one another. By using different enzymes and pathways for these processes, the cell is better able to regulate the production and destruction of essential component parts. The chances of a head-on collison between anabolic and catabolic reactions is reduced.

The second regulation system controls the synthesis rate of many molecule types. The time required to produce a molecule is very important to a cell. If molecules are manufactured too rapidly, they could accumulate, interfere with other reactions, and ultimately destroy the cell. On the other hand, if synthesis is too slow, essential molecules may not be available to continue the life of the microbe. Since enzymes control the rate of reactions, any interference with their normal, smooth operation would also reduce the rate of end-product formation. **Feedback inhibition** takes place when the end product reaches a high enough concentration. As the end product builds up, it combines with a special **"allosteric site"** on a critical enzyme in the pathway. A three-dimensional change occurs in the configuration of the enzyme, the rate of reaction is greatly reduced or halted, and the amount of product manufactured is greatly decreased. When the concentration of the end product decreases to the point that there are not enough molecules to interfere with the enzyme operation, the reaction rate once again increases. The overall result is a steady manufacture

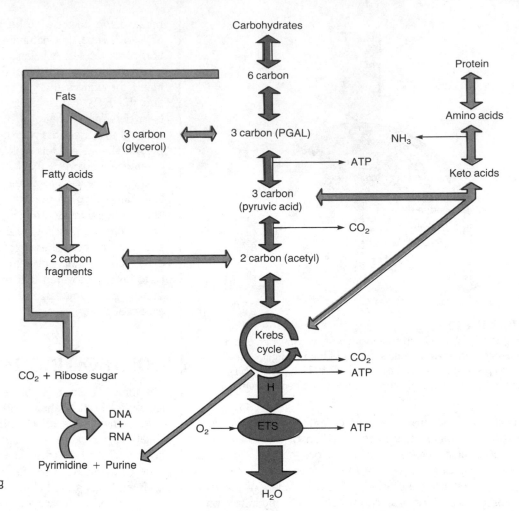

FIGURE 7.14
The cells do not necessarily utilize all the food as energy. It is possible to change one type of food into another, depending on the requirements of the cell.

of the end product (Figure 7.15). Feedback inhibition takes place in the biochemical pathways of the microbe, while the control mechanism known as gene regulation occurs on the DNA.

> 1 Why do cells not convert all incoming nutrients into energy?
> 2 Name two types of nutrients that are interconvertible through intermediary metabolism.
> 3 How does the three-dimensional shape of a molecule regulate its production and action?

Many genes in a microbe are continuously synthesizing enzymes for use in metabolic pathways. These are called **constitutive enzymes.** Most of the enzymes involved in the glycolytic pathway belong to this category. However, not all enzymes are produced on a continuous basis. Many have their synthesis regulated to ensure the most efficient functioning of the cell.

Gene regulation involves the "turning off" of a gene's function by **repression** and the "turning on" by **induction** (Box 13). These mechanisms regulate the enzyme's synthesis by DNA and are influenced by the presence or absence of the enzyme's substrate. For example, if a necessary amino acid is available to a microbe in the surrounding environment, the microbe conserves energy by avoiding the biochemical pathway which would produce that amino acid. The genes that control the synthesis of these enzymes are turned "off," or **repressed,** by the presence of the amino acid substrate. When the amino acid is not available to the cell, the gene is no longer repressed and synthesis of the enzymes begins so that the cell can manufacture the amino acid.

Induction is the opposite of repression in that genes are turned "on" in the presence of substrate. To synthesize enzymes if there is no substrate upon which to operate is a waste of energy. Therefore, inducible genes remain turned "off" as long as there is no substrate available to the microbe. When the substrate is added, its presence in the cell induces the genes to turn "on." The DNA synthesizes the

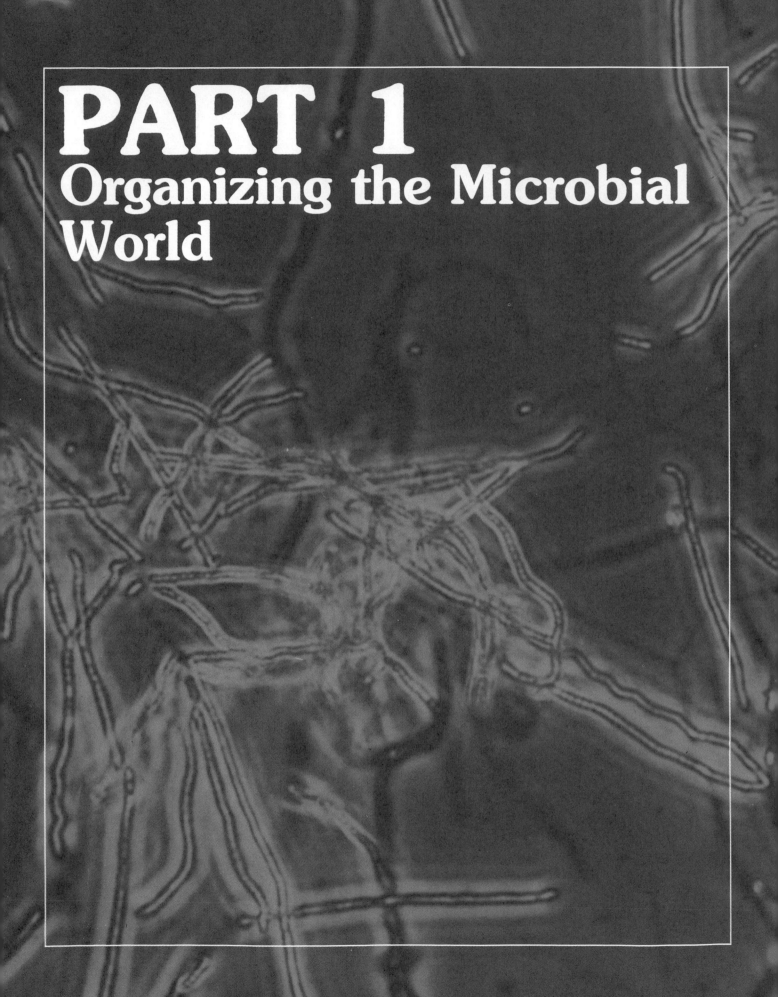

PART 1
Organizing the Microbial World

THE concept of gene regulation by induction was first explored by Francois Jacob and Jacques Monod in 1961 and is now known as the **Operon Theory of Gene Regulation.** The term **operon** refers to a cluster of enzyme-producing **structural genes** that are all under the control of another gene called the **operator gene.** The entire operon is affected by the functioning of a **regulator** gene. The regulator gene is not a part of the operon cluster and may be at a distance from the operon. One of the first operon systems to be studied was the **lac** (lactose sugar) operon system found in *E. coli.* This operon is composed of an operator gene and three structural genes. The operator gene controls the structural genes, which are responsible for the synthesis of three separate enzymes. The first structural gene codes for an enzyme that hydrolyzes lactose sugar outside the cell to galactose and glucose. The second codes for an enzyme that is involved with the transport of the simple sugar molecules into the cell, while the third gene codes for an enzyme that is responsible for the conversion of the galactose to a second molecule of glucose.

BOX 13
INDUCTION AND REPRESSION OF PROTEIN SYNTHESIS

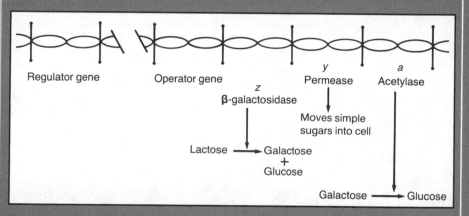

The **lac** operon works as an inducible system. Other operon systems are repressible. These regulation mechanisms can be explained best by illustration (pp. 176–77).

SOURCE: Modified from P. W. Davis and E. P. Solomon, *The World of Biology.* McGraw-Hill, 1979. Reprinted by permission.

Induction

1. Normal situation, operon "off." By transcription and translation, regulator gene produces a repressor substance that locks up the operon preventing synthesis of enzymes from structural genes *x, y, a.*

2. Inducer (from environment), which is usually a substrate of enzyme *x, y,* or *a,* inactivates repressor. Operator gene now is free to turn operon "on."

3. When enzymes have completed their task and the substrate (inducer) is consumed, the repressor is free to combine again with the operator gene, which then turns the operon "off" again.

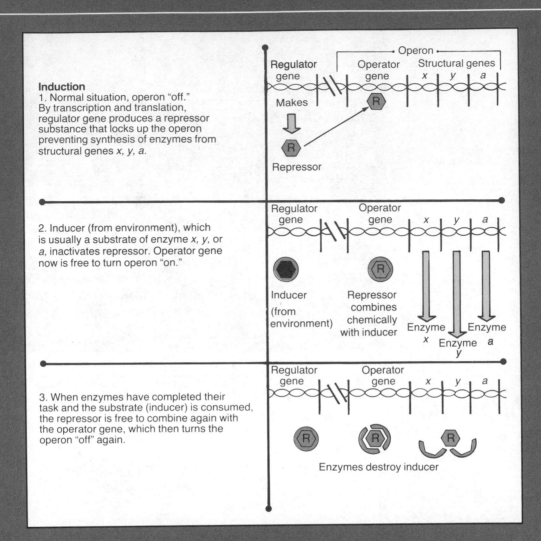

GENE REGULATION BY REPRESSION

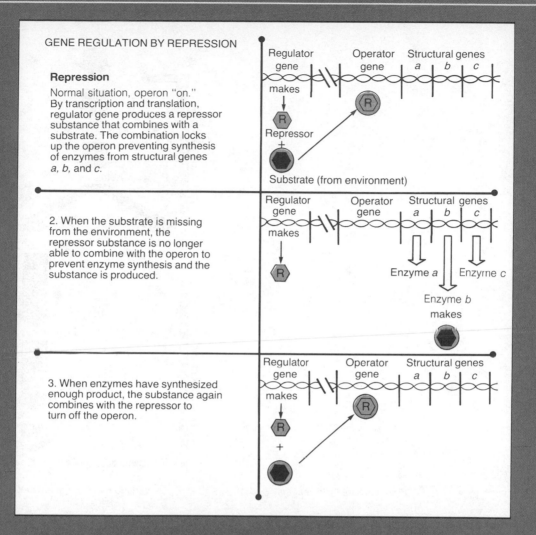

Repression

Normal situation, operon "on."
By transcription and translation,
regulator gene produces a repressor
substance that combines with a
substrate. The combination locks
up the operon preventing synthesis
of enzymes from structural genes
a, b, and *c.*

2. When the substrate is missing
from the environment, the
repressor substance is no longer
able to combine with the operon to
prevent enzyme synthesis and the
substance is produced.

3. When enzymes have synthesized
enough product, the substance again
combines with the repressor to
turn off the operon.

Regulator gene • Operator gene • Structural genes *a b c*

makes

R

Repressor

+

Substrate (from environment)

Regulator gene • Operator gene • Structural genes *a b c*

makes

R

Enzyme *a* • Enzyrne *c*

Enzyme *b*

makes

Regulator gene • Operator gene • Structural genes *a b c*

makes

R

+

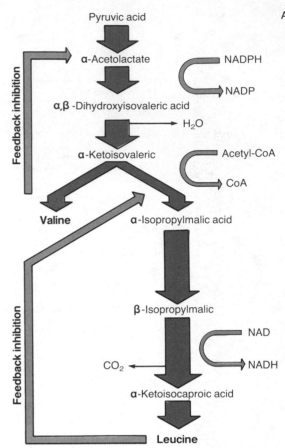

A

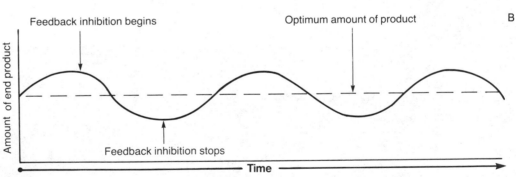

B

FIGURE 7.15
(A) This pathway leads to the synthesis of two important amino acids, valine and leucine. Both amino acids are able to inhibit their own production by interfering with a specific enzyme that leads to their synthesis (B) The result is a steady synthesis of amino acid.

mRNA molecules required to produce enzymes and the biochemical pathway is run. When the substrate has been metabolized and is no longer present, the genes turn "off." Both induction and repression regulate protein synthesis so that necessary enzymes are produced in the most energy efficient manner.

SUMMARY

The metabolism of all microbes can be divided into two main categories. Anabolic reactions are energy-requiring and result in the synthesis of molecules for use in the cell. Catabolic reactions are energy-releasing and result in the break-

down of complex molecules. All biochemical pathways are involved in providing the microbe with usable energy and building materials. Since all living things follow the Second Law of Thermodynamics, energy and matter must be available to the cell in a usable form throughout its life. The successful use of energy and matter depends on whether the microbe is in an environment that will allow it to perform assimiliation reactions. One group of molecules that is essential to the cell and must be supplied from the environment is known as growth factors. Vitamins and certain amino acids are two of the most common; however, nitrogenous bases and heme may also have to be supplied. Microbes may be classified according to the range of environmental conditions under which they are able to grow. Temperature classifications include psychrophiles, mesophiles, and thermophiles. pH classifications include acidophiles, neutrophiles, and alkalophiles. The halophiles are a group of bacteria that find high salt concentrations in the environment favorable for growth.

Microbes may also be classified as either autotrophs or heterotrophs depending on their energy and material needs. Autotrophs that use visible light energy and simple building materials are known as photoautotrophs. Those that use the chemical-bond energy of certain inorganic, reduced molecules are known as chemoautotrophs. The oxidation of molecules by microbes can occur with or without the use of molecular oxygen. Microbes that use oxygen (aerobes) produce useful energy in the form of ATP primarily by the metabolic pathways of glycolysis, the Krebs cycle, and the electron transmitting system. Those that do not use oxygen (anaerobes) may use glycolysis but do not have functioning Krebs or ETS pathways. In all cases, useful ATP energy is synthesized in cells by a flow of electrons through a series of electron carrier molecules that drives hydrogen protons to a lower state of free energy. Free energy is the amount of energy in a chemical system that is available to do work. Most ATP is generated in aerobic respiration as oxidative phosphorylation occurs in the ETS. A few additional ATPs are produced in glycolysis and the Krebs cycle by substrate-level phosphorylation. These systems function primarily by the oxidation of glucose; however, other types of organic molecules may also be used as energy sources.

Glycolysis and the Krebs cycle serve as a molecular interconversion system. Fats, proteins, and carbohydrates can all be interchanged depending on the needs of the cell. Regulation of all the biochemical pathways may be accomplished (1) by the fact that anabolic and catabolic systems do not follow the same pathways, (2) by negative feedback inhibition, and (3) by gene regulation mechanisms known as repression and induction.

WITH A LITTLE THOUGHT

A patient has just been brought in who was in a serious automobile accident . . . a week ago! His symptoms include fever, rapid heartbeat, severe pain, foul-smelling wound with watery discharge, creaking joints, and balloonlike skin around the wound. Since the patient's occupation is that of a septic tank cleaner, the doctor suspects gas gangrene. Knowing what you do about the difference in the cellular metabolism of the patient's cells and that of the *Clostridium perfringens*, what would be a logical recommendation for treatment to prevent the further destruction of the patient's tissues and his ultimate death?

STUDY QUESTIONS

1   What are the three basic needs of all living things? Describe each.

2   What are the differences among a nutrient, micronutrient, and growth factor?

3   Explain what the Second Law of Thermodynamics has to do with metabolism and the concept of free energy.

4   Distinguish among psychrophile, thermodure, acidophile, acidogene, and halophile.

5   Of what value is a cytochrome system to a microbe? How does it work?

6   What are the basic differences among aerobic cellular respiration, anaerobic cellular respiration, and fermentation?

7   What does the term "intermediary metabolism" refer to, and why is it of such great importance to the life of a microbe?

8   "Energy is measured in kilocalories per mole." What does this statement mean?

9   Explain feedback inhibition. How might it be possible to regulate more than one end product with this method?

10   What are the characteristics of a repressible gene system? How does this differ from an inducible gene system?

SUGGESTED READINGS

GOVINDJEE, R. "The Primary Events of Photosynthesis." *Scientific American,* Offprint #1310, Dec., 1974.

GREEN, D. E. "The Synthesis of Fat." *Scientific American,* Offprint #67, Feb., 1960.

HINKLE, P. C., and R. E. MC CARTY. "How Cells Make ATP." *Scientific American,* Offprint #1383, March, 1978.

JACOB, F., and J. MONOD. "Genetic Regulatory Mechanisms in the Synthesis of Proteins." *Journal of Molecular Biology,* 3:318–56, 1961.

MANIATIS, T., and M. PTASHNE. "A DNA Operator-Repressor System." *Scientific American,* Offprint #1333, Jan., 1976.

MORRISON, R., and R. BOYD. *Organic Chemistry,* 3rd ed. Allyn & Bacon, Inc., Boston, 1973.

WATSON, J. D. *Molecular Biology of the Gene,* 3rd ed. W. A. Benjamin Co., Menlo Park, Calif., 1976.

KEY TERMS

growth factor

psychrophile (syk′roh-fyl)

mesophile (mes′oh-fyl)

thermophile (ther′moh-fyl)

thermodure

acidophile

neutrophile

alkalophile

photoautotroph

chemoautotroph

facultative anaerobe

obligate aerobe

obligate anaerobe

microaerophile

free, activation, and bonding energy

glycolytic pathway

substrate-level phosphorylation

Krebs cycle

electron transporting system (ETS)

oxidative phosphorylation

fermentation

intermediary metabolism

feedback inhibition

induction and repression

Pronunciation Guide for Organisms

Lactobacillus (lak″to-bah-sil′us)

Growth and Reproduction

INCREASING MASS AND POPULATIONS

THE POPULATION GROWTH CURVE

The Lag Phase
The Log (Logarithmic) Phase
The Stationary Growth Phase
The Death Phase

THE GROWTH CURVE: HOME, INDUSTRY, AND HOSPITAL

COUNTING METHODS

Total Cell Count
Viable Cell Count
Biomass Determination
Biological Assay

Learning Objectives

☐ Define the terms "growth" and "growth rate" as they refer to individual cells.

☐ Recognize the concept of doubling time and how it correlates with the virulence of certain bacteria.

☐ Explain logarithmic growth as related to population growth.

☐ Draw a population growth curve, label its parts, and explain the events occurring in each part.

☐ Be familiar with how the population growth curve can be used to explain practical events associated with microbial activities in the home, industry, and hospital.

☐ Explain the four main methods used in determining a population of microbes.

☐ Be able to do a standard plate count.

☐ Explain the relationship among a standard plate count and optical density, biomass, and dry weight

☐ Explain the advantages and disadvantages of biological assay.

All chemical reactions in a microbe may result in an increase in the cell mass. This process is known as cell growth and will only occur when certain conditions are available to the microbe. The cell must have an adequate supply of energy and building materials, and must be living in a suitable environment. Microbes grow very rapidly when these conditions are available, and as a result, they only live a short time. Because they have short life cycles in comparison to humans, their rapid reproduction rate can have very significant effects on the environment. For example, soils that were unable to support plant growth one year may be fertile the next. Foods that were fresh and tasty only a few hours ago may now be spoiled and worthless. A healthy person infected with only a few pathogenic microbes may quickly become ill after these cells reproduce to become millions of damaging, harmful organisms. To control individual cell growth and reproduction, it is necessary to understand what happens in the life of a cell and how a population of cells increases in number. This information has far-reaching applications in the lab, industry, and health care fields. Special

8

methods have been developed to measure populations and to either discourage or encourage their growth in selected environments.

INCREASING MASS AND POPULATIONS

Growth has been defined as a characteristic of life in which new cell components are added through the processes of assimilation. When a cell comes into existence as a result of the reproduction of a parent cell, the new cell begins to carry out all the reactions characteristic of that species. The mass of the individual cell increases as new parts are synthesized through the biochemical pathways discussed earlier. Eucaryotic cell life cycles may be divided into several distinct parts based on their assimilation activities (Figure 8.1). Procaryotic microbes also have a life cycle in which their mass is increased; however, they lack the eucaryotic nuclear organization. The procaryotic cycle does not include a separate DNA duplication state (the S stage), but this activity begins very early in the life of a procaryote and extends throughout its life. The **growth rate** of a cell refers to how rapidly a cell increases in mass. Knowing a cell's growth rate can determine how long it is likely to live before producing another generation of cells. The time it takes to go through its life cycle is called its **generation time,** or **doubling time.** Generation times are usually shorter for procaryotes than eucaryotes, and shorter for smaller than for larger cells

since the growth rates are proportional to the energy metabolism of the cell. The faster a cell metabolizes nutrients, the shorter its generation time (Figure 8.2).

The generation time of some pathogenic bacteria can be correlated with their virulence and the treatment needed to control an infection caused by such organisms. Pathogens like *Salmonella typhi* have relatively high metabolic rates and short generation times. After these pathogens have been ingested, they synthesize toxins very rapidly and quickly increase in number. Therefore, symptoms of a *Salmonella* infection soon become apparent. Antibiotic therapy with drugs such as ampicillin or chloramphenicol should be given immediately to counteract the harm being done by these microbes. Because of their high metabolic rate, the antibiotic rapidly enters the metabolic pathways of *Salmonella* and kills them. In other situations, an extremely long generation time and lower metabolic rate result in the slow development of symptoms and the need for prolonged antibiotic therapy. *Mycobacterium tuberculosis* has a generation time of about 800 minutes (refer to Figure 8.2). Symptoms of this infection may not be seen until weeks after initial infection. Antibiotic therapy requires that a combination of drugs such as streptomycin, isoniazid (INH) and/or para-aminosalicylic acid (PAS) be given daily for a minimum of 12 months. This long therapy time is needed to effectively interfere with the slow, steady metabolism of these pathogens.

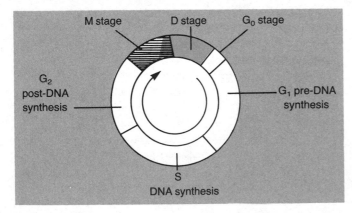

FIGURE 8.1

The life cycle of a eucaryotic cell may be divided into several typical stages. The M stage is the nuclear division state (M = mitosis) and overlaps into the D (cytokinesis, or cell division) stage. The G_0 stage (zero growth) may be a lengthy process in cells that do not immediately proceed toward another division. Muscle and liver cells have long G_0 stages. Microbial cells move directly toward another division beginning with the G_1 (first growth). This is followed by the DNA synthesis stage (S) and a second growth stage (G_2). G_1, S, and G_2 together are commonly known as the interphase of the life cycle.

Bacterium	Growth temperature, °C	Generation time, min.
Bacillus mycoides	37	28
B. thermophilus	55	18.3
Escherichia coli	37	12.5
Lactobacillus acidophilus	37	66–87
Mycobacterium tuberculosis	37	792–932
Rhizobium japonicum	25	344–461
Salmonella typhi	37	25
Staphylococcus aureus	37	27–30
Streptococcus lactis	37	26
Treponema pallidum	37	1980

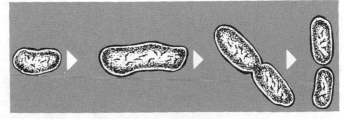

FIGURE 8.2

The time required for a cell to complete its life cycle is usually very short; however, some bacteria have extremely long generation times. By comparison, *Mycobacterium tuberculosis* lives 47 times longer than *E. coli*. (By comparison, this would be the difference between living 75 years or 3525 years.) (Data from W. B. Spector, ed., *Handbook of Biological Data,* table 75. W. B. Saunders Company, 1956.)

The end result of individual cell growth is usually reproduction. Microbiologists use the same terms to refer to **population growth** (increase in the number of cells by reproduction) and individual **cell growth** (increase in the mass of a cell). Populations have doubling times, generation times, and growth rates. The growth of an entire population is affected by its environment just as a single cell might be affected. Under optimum conditions, microbes increase in number by binary fission or other methods. Reproduction results in two new offspring from each parent cell. One cell divides into two cells, two into four, four into eight, etc.:

$$1, 2, 4, 8, 16, 32, 64, 128, 256, 512, \ldots$$

This increase in numbers by a constant doubling of cells in a population is known as exponential or **logarithmic growth.** This form of population growth can have devastating consequences if allowed to continue unchecked. A single cell in an optimum environment may increase to over 33,000,000 cells* in only 25 generations! If the generation time of this microbe is 15 minutes, this increase to 33,000,000 could occur in only 6.25 hours!

---

1 What is the difference between arithmetic growth and logarithmic growth?
2 Write the following numbers in their exponential form: 5670   2,300,000   87,100,000

THE POPULATION GROWTH CURVE

A population of microbes does not always increase in an exponential or logarithmic fashion. The environment greatly influences the rate and timing of cell reproduction. Knowing how a population of cells increases through time has many practical implications in the home, lab, and clinical setting. Microbes demonstrate a typical population growth pattern that can be represented by a **population growth curve** (Figure 8.3). After a small sample population of microbes has been inoculated into fresh culture medium, the population of live cells is counted at regular intervals. The curve may be determined by comparing the total number of living microbes present in the population over a period of time. The population growth curve has four stages or phases: (1) lag, (2) log, (3) stationary, and (4) death.

---

*Because microbiologists must deal with extremely large figures, the **exponential notation system** of writing large numbers is used when making population counts or estimates. Figures are written in the base 10; therefore, $10 = 10^1$, $100 = 10^2$, $1,000 = 10^3$, $10,000 = 10^4$, $100,000 = 10^5$, $1,000,000 = 10^6$, etc. The number 33,000,000 used in the example above can be written as 3.3×10^7 in exponential notation.

THE LAG PHASE

When an organism is inoculated into fresh culture medium, the population does not immediately double as might be expected. The total number of living cells remains fairly constant for a long period. During the lag phase of growth, cells are making major adaptations to new environmental conditions. Cells that lack certain enzymes, coenzymes, and other essential molecules must synthesize these molecules to survive. This requires time and energy not available for reproduction. If the cells are able to turn "on" or "off" genes that enable them to survive in this new environment, they will reproduce and the medium will become cloudy with growth. If the microbes lack suitable genes, there will be no growth and the culture will die. During the lag phase, there is no population increase since the reproductive rate equals the death rate $(R = D)$. The number of cells entering the population by reproduction equals the number of cells leaving the population by dying. If the environment is constantly changing, a population may continue in the lag phase for a long period. The end of the lag phase and beginning of the log phase occur when the microbes have made the necessary adaptations and begin to divide at a constant rate.

THE LOG (LOGARITHMIC) PHASE

Cells in the log phase are reproducing at the exponential rate and demonstrate most of the "typical" features of that species. Cellular respiration, protein synthesis, and many other metabolic pathways are being operated in their most effi-

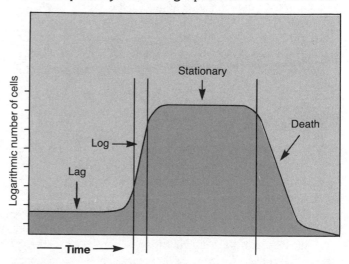

FIGURE 8.3
When a culture medium is inoculated with a sample of unicellular microbes, four separate stages of population growth can be graphed. The lag period is a time of adaptation to the new environment (reproduction rate = death rate). The log phase is a period of exponential growth rate (reproduction rate > death rate). The stationary phase occurs when reproduction rate equals death rate. During the death phase little or no reproduction occurs to replace the dying population.

cient manner. Reproduction is at the maximum rate during this phase and the death of cells is at its minimum $(R > D)$. The number of cells entering the population by reproduction is much greater than the number of cells leaving the population by dying. As the population continues to grow at its maximum rate, the microbes utilize the resources in their environment and produce waste products at an ever faster rate. Ultimately the loss of nutrients and the pollution of their environment begin to influence the population, resulting in a decrease in the growth rate. The growth curve shows a turn into the third major phase called the **stationary growth phase.**

THE STATIONARY GROWTH PHASE

The decreasing essential nutrients and increasing harmful waste products discourage the maximum, constant growth rate of the log phase. The material or combination of materials that first stops the microbe's growth is called the **limiting factor.** This may result from the loss of a growth factor such as a vitamin or the loss of an essential element in a usable form such as an amino acid. Growth may also slow due to a severe change in temperature or pH. In the stationary growth phase, the reproductive rate equals the death rate $(R = D)$.

When the cells do not reproduce or die, the total number of viable cells remains constant. The length of the stationary phase varies greatly. If the microbial population reaches a balance with the environment, the stationary phase may be very long. If the changes in the environment are too severe for the successful adaptation of the population, the number of cells dying will begin to exceed the number entering the population by reproduction. When this begins, the curve turns and enters the last portion of the population growth curve, the death phase.

THE DEATH PHASE

The death phase of the curve is basically the reverse of the log phase. Cells die and are destroyed by lysis at an exponential rate. Their death and lysis are primarily the result of an ever-increasing amount of acids and other harmful wastes in the environment. However, after most of the cells have died, the **death rate** may decrease rapidly at the last minute. This results in the survival of a few hardy cells that may live in the culture for months or years. Their survival may be based on the fact that the dead, disintegrated cells are

1 In which phases of the population growth curve does the reproductive rate equal the death rate?
2 What is a limiting factor?
3 How does the lag phase differ from the stationary phase?

no longer using nutrients, but their dead bodies are serving as a source of energy-containing compounds, building materials, and growth factors for the few remaining live cells.

THE GROWTH CURVE: HOME, INDUSTRY, AND HOSPITAL

The population growth curve may be used to explain many of the changes that occur in the environment as a result of microbial activities. In the home, milk may only be kept for a limited time before it spoils. Spoilage occurs when the bacteria remaining in the milk after pasteurization metabolize the protein to foul-smelling and foul-tasting chemicals. During pasteurization at the dairy plant, only the pathogens in milk are destroyed. This heating kills microbes that cause such diseases as staphylococcal food poisoning, brucellosis, tuberculosis, diphtheria, and Q fever. Many saprotrophic microbes survive the process and are normally found in the milk when bought at the store. According to state law, acceptable pasteurized Grade A milk may contain 20,000 microbes per milliliter. However, because of high quality control standards on the dairy farms and in the plant, many dairies are able to ship milk to the retail store with a cell count of only 300 per milliliter. After the initial opening of the milk container, the surviving saprobes and psychrophilic contaminants begin to go through the lag phase of growth. Each time the milk is removed from the refrigerator, it warms enough to allow the microbes to adapt to their environment. The longer it is left out, the more quickly the lag phase will proceed. Milk should be kept at 40°F (5°C) since the shelf life will be cut in half for every 5°F (3°C) increase in the temperature above 40°F. If the lag phase is completed while the milk is out for breakfast, it may be spoiled by lunch time. The lag phase may be lengthened by keeping the milk refrigerated, but once this phase has been completed, the exponential growth will very quickly destroy the product. Dairies realize this and label their products with "expiration dates" to help prevent this problem. For example, cheeses have an expiration date twelve days after packaging, while sour cream packages show a date one month after their production. The date indicates the most likely time when the lag phase will end and the log phase of growth will begin (Figure 8.4).

The growth curve is also of value in the microbiology laboratory. In order to accurately identify and classify a microbe, pure cultures must be grown in different environments. How the "average" microbe behaves will yield information about its genetic makeup. However, some cells will only display limited characteristics at certain stages of their growth. For example, spore-forming bacteria such as *Bacillus subtilis* will produce resistant endospores when the environment becomes depleted of certain nitrogenous compounds. When a culture of this bacterium is made for endospore staining, the technician must wait for the culture to

Product	Maximum number of microorganisms per milliliter or gram*
Grade A raw milk prior to commingling with other milk	100,000
Grade A raw milk commingled, prior to pasteurization	200,000–300,000
Grade A pasteurized milk and milk products (except cultured products)	20,000 (5–10 coliforms)
Grade A pasteurized cultured products	10 coliforms
Ice cream and other frozen desserts	50,000 (10 coliforms)
Liquid ice cream mix for machines	25,000 (5 coliforms)
Grade B raw milk	600,000
Grade B pasteurized milk	40,000

*The recommended standards vary slightly from one community to another.

FIGURE 8.4

These are the recommended temperatures for storing milk and ice cream. The U.S. Public Health Service has standards for grading milk as Grade A or Grade B based on bacterial cell counts. Direct microscopic counts are performed to make these determinations.

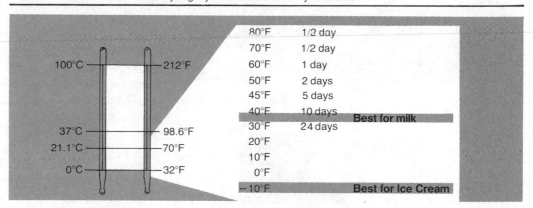

grow through the lag and log phases before examining the cells in the late stationary or early death phase (Figure 8.5).

In the hospital setting, laboratory technicians are responsible for the accurate identification of disease-causing microbes taken from an infected patient. Their precise identification will determine which type of therapy should be given to help a patient recover. Samples or specimens from the patient are taken to the lab for growth in pure culture. The culture must be incubated long enough for the microbes to move from the lag into the log phase of growth. Only during the log phase can an accurate identification be made. The time required to get the population into the log phase can be reduced if the specimen is taken from the patient when the microbe is actively reproducing and sent to the lab for culture as quickly as possible. In order to do this, the technician must inoculate the microbes into a suitable growth medium. By placing the organism in the appropriate nutrient medium, the cells may continue in the log phase of growth and be prevented from reentering the lag phase. This will shorten the culture time before identification can be made.

The practice of growing microbial cultures in the log phase also has applications in the industrial lab. The maintenance of cultures in the exponential (log) phase of growth is known as **steady-state** or **balanced growth** and allows

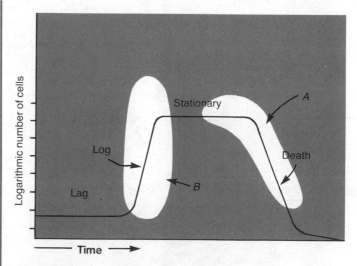

FIGURE 8.5

During certain stages of the population growth curve, microbes demonstrate different physiological, morphological, and chemical features. Endospore-forming bacteria, such as members of the genera *Bacillus* and *Clostridium,* form spores after certain nitrogenous compounds have become scarce (point A). Most microbes demonstrate the greatest variety of metabolic activity during the log phase (point B). During this growth phase, the identification of a particular microbe can best be made.

microbes to be cultured on a continuous basis. **Continuous cultures** are maintained in an apparatus called a **chemostat.** Many antibiotics, organic acids, and steroid compounds (hormones) are produced by microbes growing in continuous culture. The log phase of growth is maintained by continuously adding fresh, sterile medium to a culture vessel at a rate equal to the removal of culture and products (Figure 8.6). The exponential growth is regulated by maintaining the precise level of a limiting factor in the medium. This practice keeps the cells from either shifting into a lag or stationary phase. Continuous cultures have an advantage over **batch cultures** (culturing in a single container of medium) since the product can be constantly manufactured in large amounts. Ensuring that the culture is growing according to plan requires that cell counts be made on a regular basis. There are several methods regularly performed on cultures to identify a culture's location on the population growth curve.

> **1** How can bacterial growth in foods be slowed?
> **2** What must be done to achieve a steady-state growth?
> **3** What is a batch culture?

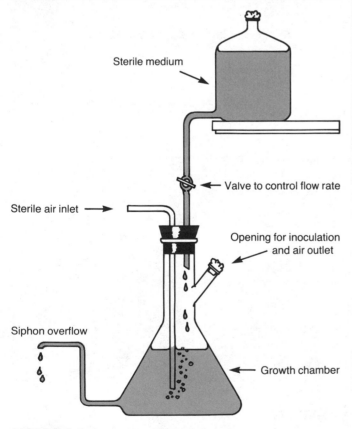

FIGURE 8.6

This diagram illustrates the basic idea of the continuous culture. The fungus *Penicillium notatum* may be cultured in balanced growth by the constant addition of fresh, sterile media, while the excess culture is drained into a receiving vessel for chemical isolation of the antibiotic penicillin. The air inlet maintains aerobic culture conditions and stirs the culture.

Labels: Sterile medium; Valve to control flow rate; Sterile air inlet; Opening for inoculation and air outlet; Siphon overflow; Growth chamber

COUNTING METHODS

Four main methods may be used to determine if a population is growing in the lag, log, stationary, or death phase. These methods include (1) total cell counting, (2) viable cell counting, (3) biomass determination, and (4) bio-assay.

TOTAL CELL COUNT

The **total cell counting** procedure may be used to determine the number of cells in milk, soil samples, vaccines, etc. In this method, both living and dead cells are counted. Because no time is spent distinguishing between the live and dead cells, population estimates can be made very quickly. However, there may be little value in knowing that a material contains two million *dead* cells per ml. If these cells are not active, a high count may not be significant. On the other hand, if the dead cells are *Clostridium botulinum,* a high count may indicate the presence of botulism toxins produced at an earlier time. Total cell counts may be made by using three basic techniques: dry-slide technique, Petroff-Hausser counting chamber, and electronic cell counter.

The **dry-slide technique** is performed by making a smear preparation of the material to be studied and staining the cells with a suitable dye. However, the slide must contain a known amount of material if a population estimate is to be accurate. In order to do this, the original culture is diluted and transferred to the slide with a special inoculating loop, which holds only a specified amount of medium. The population estimate is made by counting all the cells in several different microscopic fields of known area, averaging the counts, and determining the original cell count per milliliter of sample. For example, a culture may be diluted by pipetting a 1 ml sample into a sterile tube containing 9 ml of water. This results in a 1 in 10 (1:10) dilution, since 1 ml of culture has been thoroughly mixed in a total of 10 mls of fluid (9 ml water + 1 ml culture). From this well-mixed tube, an inoculating loop that holds only 0.01 ml of fluid is used to make a smear slide. By taking only 0.01 ml of culture from the tube, the sample is diluted 100 more times (0.01 = 1/100 of the sample, or 100 times fewer cells). The total dilution of the original culture sample being transferred to the slide is $1:10 \times 1:100 = 1:1000$. If the average cell count made through the microscope is 52, the cell count found in the original culture is

(count × dilution factor) = 52 × 1000

$$= 52,000 \text{ cells/ml.}$$

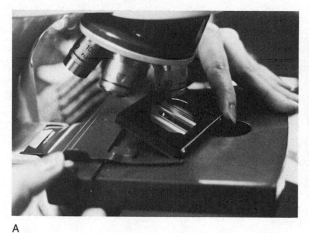

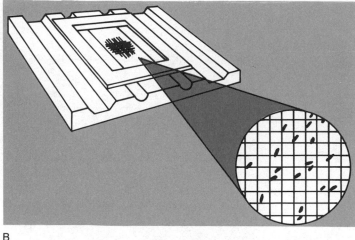

A

B

FIGURE 8.7

(A) The Petroff-Hausser counting chamber is a special glass microscope slide used to make counts of bacteria. (B) Its special design prevents cells from stacking so that counts are more accurate.

This procedure is used as a quick check of milk as it enters the dairy to determine the level of contamination before pasteurization. It is also used in pharmaceutical labs to determine the relative effectiveness of antibiotics against pathogenic bacteria such as *Mycobacterium tuberculosis*. This is done by counting the cells that remain after a culture has been exposed to a particular antibiotic.

The **Petroff-Hausser counting chamber** method (Figure 8.7) is very much like the dry-slide procedure; however, the sample of culture is not allowed to dry on the slide. A special slide and coverglass are used to contain a sample of culture of known volume. The slide has a grid or graph etched on its surface to make counting easier and more accurate. When the count has been made, the total is multiplied by the dilution factor of the sample in order to determine the number of cells per milliliter in the original culture. Blood cell counting is done on a similar slide known as a **hemocytometer,** which is larger than the Petroff-Hausser chamber. It is not able to be used for bacterial cell counting since the larger volume of sample allows cells to clump under the coverglass. This clumping makes the count less precise. To help reduce work and avoid the inaccuracies of these methods, another total cell count technique may be used.

Electronic cell counting machines are similar to the blood machines used in hospital labs. A sample of culture is injected into a reservoir and conducted through a narrow capillary tube (Figure 8.8). The tube is only large enough to allow one cell at a time to pass an electric-eye light beam. Every time an individual cell passes through the light beam, an electronic circuit is triggered to register a count. Blood cell counting machines are very accurate and efficient because the cells are large enough to flow through the capillary tube without difficulty. Counting microbial cells is more difficult because the cells are so much smaller. If the tube is

narrowed to allow only one cell at a time to pass, the culture does not flow well, and clogs. This method of counting is very valuable for solutions that cannot be sterilized by ordinary methods. Certain intravenous solutions (IVs) contain compounds that are very heat sensitive, making steam sterilization impossible. These solutions must be prepared aseptically from components that have been separately sterilized. Electronic cell counts may be used to check their sterility after preparation. This works well since a large number of samples may be counted very quickly. The drawback to this total cell counting method is that both living and nonliving cells are counted. In order to avoid this, a number of viable cell counting methods have been developed.

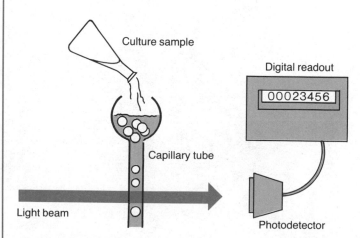

FIGURE 8.8

The diagram illustrates the basic operation of an electronic cell counter. As the cells pass through the capillary tube, they prevent the light beam from making contact with the photodetector. This acts as a signal to the counting apparatus to register the passage of a single cell.

VIABLE CELL COUNT

There are three **viable cell counting** procedures used by microbiologists to determine the number of living cells in a culture: (1) the standard plate count, (2) membrane filter, and (3) the most-probable number methods. The **standard plate count** is the most basic of all methods and is the foundation of many other indirect cell counting procedures. In this method, a known volume of culture is diluted through a series of flasks containing sterile media. This serial dilution disperses the individual living cells to such extremes that when they are placed in the optimum environment, each cell will grow into a separate colony. Therefore, each colony represents a single living cell that was present in the original sample. By counting the colonies (Figure 8.9) and multiplying this number by the dilution factor of the sample used in the series, the total number of living cells that were present in the original culture can be determined (Figure 8.10). If a petri plate has 46 colonies from a sample that was diluted 1:1,000,000, the original culture contained

$$46 \times 1{,}000{,}000, \text{ or } 46{,}000{,}000 \text{ bacteria/ml}$$

Many types of cultures or materials may be sampled using this method. Water may be tested for bacteriological contamination, milk populations counted to determine the

---

1 Name one drawback to the total cell counting method.
2 How are 1:100 dilutions obtained?
3 What is the chief advantage of a viable cell count?

---

likelihood of pathogenic contamination (Table 8.1), and vaccines checked for the presence of potentially dangerous microbes. A 1 ml sample is pipetted into a dilution blank bottle containing 99 ml of sterile broth or physiological saline (Figure 8.11). The bottle is shaken vigorously to break up clusters and disperse the cells evenly throughout the fluid. This blank now contains 100 ml of fluid. Since only 1 ml in the total 100 ml of fluid sample contained microbes from the original sample, the dilution factor for each 1 ml in this blank is 1:100. All 1 ml samples pipetted from this bottle will have 100 times fewer microbes than found in the original culture. If only 0.1 ml is removed, 10 times fewer cells will be sampled, that is

$$1:10 \times 1:100 = 1:1000 \ (10^3 \text{ dilution factor}).$$

In most cases, samples are not taken from this first dilution blank, since a cloudy bacterial culture usually contains between 10^5 and 10^6 cells per ml and seldom exceeds 10^9. (As the population approaches 10^9 per ml, 1 billion cells/ml, it enters the stationary growth phase, and cells are inhibited from reproducing by the accumulation of waste products in the environment.) Therefore, the dilution series is continued before it is sampled by pipetting into agar plates. A 1 ml sample is removed from the first dilution blank and transferred to a second 99 ml bottle. The total amount of fluid in this bottle is also 100 ml and should have a dilution factor of 1:100 (10^2); however, since the sample inoculated into this second bottle has already been diluted 100 times, the total dilution factor is

$$1:100 \times 1:100 = 1:10{,}000 \ (10^4 \text{ dilution factor}).$$

A

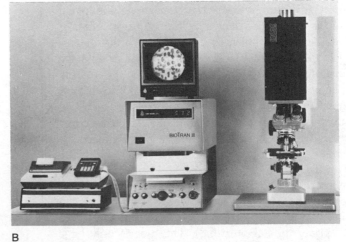

B

FIGURE 8.9

This equipment is used in many laboratories to determine the total number of cells in a culture sample. (A) A Quebec colony counter. (B) An electronic colony counter. The petri plate to be counted is placed on the machine and each distinct colony that is separated from its neighbors is counted as if it were a single cell. This operation is performed automatically and records the total plate count on the display panel. (Courtesy of the New Brunswick Scientific Company.)

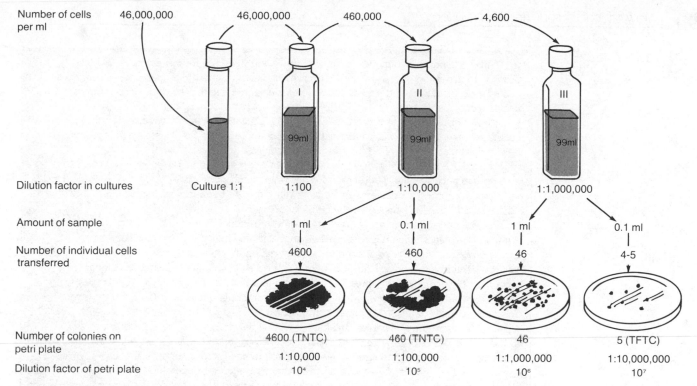

FIGURE 8.10

By transferring a sample of culture through a series of sterile dilution blanks, a concentrated culture may be dispersed into individual cells. This example shows how a culture originally containing 46,000,000 bacteria per ml is diluted and plated into petri dishes for incubation. Since the individual cells are spread out in each plate, they may grow into separated colonies, which are easily counted. Notice how the number of cells decreases as the dilution factors increase from left to right. (*TNTC = too numerous to count; TFTC = too few to count.)

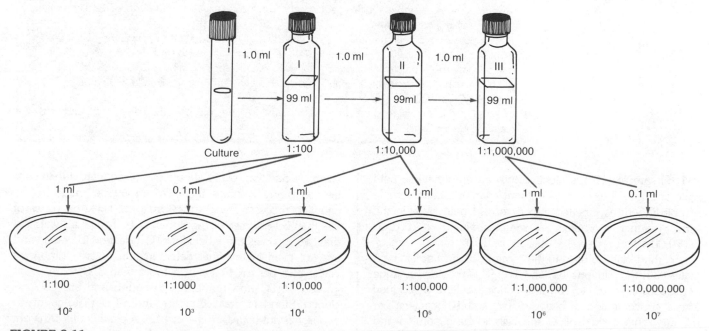

FIGURE 8.11

All dilution blanks may be sampled for counting; however, the first blank is usually eliminated since most cultures have between 10^5 and 10^6 bacteria per ml. An additional blank may have to be added if the culture is extremely cloudy with growth.

TABLE 8.1
P.I. test—what is it?

The dairy industry continues to develop and adopt new production practices and testing methods to assure consumers of the best milk and dairy products. High bacterial counts of milk are a concern to producers and processors. All precautions are taken to keep bacteria numbers at, or below, acceptable levels.

Many tests are used to determine if satisfactory sanitation practices have been followed. These include swab tests, rinse tests, water analysis, and chemical tests. The most important test has been the standard plate count (SPC) for enumerating bacterial numbers.

The standard plate count has been around for a long time and presently is considered by regulatory agencies as the best method for determining the numbers of bacteria in milk. The test is performed by mixing a small amount of milk with a growth medium and incubating the mixture at 32°C. (89.6°F) for 48 hours. Then the bacterial colonies are counted.

Federal and state regulations place a maximum bacterial limit of 100,000 per ml for Grade A raw milk for pasteurization. Counts higher than this clearly indicate a need for corrective action.

The present standard plate count procedure may not give a complete enumeration of the bacterial content of the milk. Thus, the number of bacterial colonies indicated by the test may be lower than the actual number present. There is little doubt that the highly efficient cooling procedure used on dairy farms today is one of the contributing factors to the counting error. Unsanitary conditions might be present on the farm and yet go undetected.

This is where the Preliminary Incubation, or P.I., test comes in. The P.I. test is not new. Dr. C.K. Johns developed the procedure in the late 1930s to provide a test that would indicate unsanitary conditions on the farm by revealing the presence of psychrotrophic bacteria, which are primarily made up of Gram-negative bacteria such as *Pseudomonas*, some coliforms, *Flavobacterium*, and *Alcaligenes*. These bacteria are capable of growing at temperatures below 15°C (59°F).

The P.I. test is conducted by incubating the raw milk sample at 13°C (55°F) for 18 hours and then a standard plate count (SPC) is run. Preincubation allows psychrotrophic bacteria to grow until numbers are high enough to be detected by the SPC. A significant increase in SPC after P.I. indicates a high level of psychrotrophic bacteria which probably came from unsanitary conditions on the farm. The P.I. test will point out that unsanitary conditions exist, and alert dairymen to take corrective measures.

Primary causes of high P.I. counts center around improper sanitation practices. Psychrotrophic bacteria come from unclean milking equipment, contaminated water, dirty udders, and improperly sanitized milking equipment.

Proposals have been made to the Food and Drug Administration to permit the use of the P.I. count in place of the SPC as a regulatory test. When a collaborative study is completed, the P.I. may become the alternative, or possibly, the required test for raw milk. A similar standard of 100,000 per ml is proposed, although some cooperatives and dealers already have a quality standard of 50,000 per ml.

And yes, it is possible to conduct the SPC and P.I. tests on the same milk sample and receive totally different results!

SOURCE: Klenzade, Division of Economic Laboratory, Inc. St. Paul, Minn.

A 1 ml sample (10^4) is pipetted into an empty sterile petri plate and mixed with liquid nutrient agar. A second sample of only 0.1 ml is pipetted into a second petri plate in the same manner to obtain a sample diluted 1:100,000 ($1:10,000 \times 1:10 = 1:100,000$, or 10^5).

A third blank is used in the same fashion to obtain petri plates with the dilutions 10^6 and 10^7. This provided a range of dilutions from 10^4 to 10^7. If the original culture was cloudy with a concentration of between 10^5 and 10^6 bacteria per ml, this range would allow the cells to be separated and grown into colonies that can be easily and accurately counted. If a more dense culture is used, an additional dilution blank may have to be added to the series. Errors in pipetting and transferring samples may occur at various points in the standard plate counting procedure. Therefore, it is very important to avoid counting plates that may be the result of such errors. In addition, plates should not be counted if they contain so many colonies that errors may be made in the counting process. These errors are most likely on plates of less than 30 colonies or more than 300. The plates with less than 30 colonies are labeled TFTC (too few to count) and are not used in the final determination of the number of viable cells per ml. Plates with more than 300 colonies are labeled TNTC (too numerous to count) and are also excluded. Standard practice is to round off all plate counts to two significant figures.* Therefore, if a plate count turns out to be 232,000,000, the number is changed to 230,000,000.

*The significant figures in a decimal number are the numbers reading from left to right, beginning with the first non-zero number and ending with the last digit written. For example, 12,678 has 5 significant figures; 12,600 has 3 significant figures; 12,000 has 2 significant figures. Numbers 5 or larger are rounded "up," while numbers from 1 to 4 are rounded "down." Thus, 145,000 becomes 150,000 and 144,000 becomes 140,000.

The standard plate count is very useful in isolating and counting viable cells from materials that contain only a very few cells. However, the method is time-consuming and requires a great deal of materials and preparation. Alternative methods have been developed to resolve these problems and are based on the accuracy of the standard plate count method. The **membrane filter** method also utilizes a dilution series. A sample from one or several of the dilution blanks is drawn through cellulose filters of known pore size to separate the cells from the medium. The filters are removed aseptically from the funnel apparatus and placed on the surface of soft nutrient agar. Once the nutrient pene-

trates the filter and makes contact with the cells, each separated cell will develop into a recognizable colony that can be identified and counted. **Sputum** is a thick, mucuslike material from the lungs. Samples of sputum from patients suspected of having tuberculosis may be membrane filtered to selectively detect the presence of *Mycobacterium tuberculosis* and determine the number of cells present in a known quantity of sputum (Figure 8.12).

A modification of the membrane filter technique has been developed, allowing doctors to quickly identify and count microbes from bodily fluids such as urine (Figure 8.13). A dried and compressed spongelike membrane filter is fixed

A

B

C

D

FIGURE 8.12

This sequence shows how membrane filtering can be used to identify and count *Mycobacterium tuberculosis*. (A) Typical colonies of *M. tuberculosis* from sputum after isolation and culturing on a Millipore filter. (B) A microfiber glass prefilter sterilizes air passing into the funnel during filtration. (C) Using forceps to avoid damaging the filter, a 47 mm filter is placed on the filter support. (D) The sputum-digestant mixture is aseptically poured from the collection tube into the funnel for filtering. (Courtesy of the Millipore Corporation.)

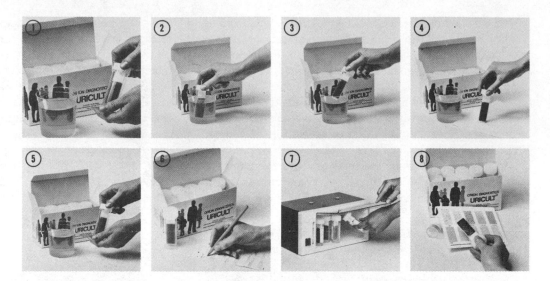

Detailed working procedure

(1) Unscrew the slide from the tube without touching the agar surfaces.

(2) Dip the slide into freshly voided urine so that the agar surfaces become totally immersed. If the quantity of urine is insufficient for this, the agar surfaces can also be wetted by pouring the urine over them.

(3) Allow the excess urine to drain from the slide...

(4) ...and remove the last drops with clean absorbant paper.

(5) Screw the slide back tightly into the tube.

(6) Fill in the accompanying label and attach it to the tube.

(7) Place the tube in an upright position for 16 to 24 hours in an incubator (approx. +37°C/99°F). Alternatively the tube may be sent to a laboratory for incubation. Incubation should not exceed 24 hours.

(8) Remove the slide from the tube and compare the colony density on the agar surfaces with the density model chart which indicates the respective bacterial count per ml of urine.

Uricult should preferably be incubated immediately after inoculation. If Uricult, however, is sent to a laboratory for incubation, satisfactory results may be obtained if incubation at 37°C/99°F is commenced within 72 hours from inoculation and provided that the slide has been protected from freezing. The Uricult dip-slide should be read after 16—24 hours' incubation.

FIGURE 8.13
A semiquantitative dip-slide culture test for bacterial counts in urine. (Courtesy of the Medical Technology Corporation.)

into the slot of a specially designed "culture stick." A patient suspected of having a urinary tract infection is asked to void into a specimen cup. The culture stick is inserted into the urine, causing the spongelike filter to swell as it takes up a 1 ml sample of the fluid. As the fluid is absorbed, microbes adhere to the filter and make contact with the culture medium in the stick. After the culture stick has been incubated, colonies that have grown on the surface can be counted and identified. This technique enables the doctor to gather information about the nature of the infection without having to send a specimen to a special lab for culturing and identification.

Viable cell counts can also be performed indirectly. The **most probable number (MPN)** method utilizes a series of dilutions. Three or four groups of three, four, or five test tubes may be used in the MPN test (Figure 8.14). The greater the number of dilution series and the more tubes used in each dilution series, the more sensitive and accurate the results will be. The purpose of this test is to statistically determine how many bacteria are in a culture by finding out which tubes in the dilution series show growth after incubation. If a culture contains only a few bacteria per ml, very

1 If a plate has a $1:10^6$ dilution of culture and a plate count of 45, how many cells per ml were in the original specimen?
2 What material is usually sampled with the MPN method?

few tubes in the series will show growth since the series will "dilute them out." If the culture contains a great many bacteria, more tubes will show growth. The data that has resulted from both the standard plate counts and MPN series from thousands of tests are entered into a computer. This data is then used to generate a table that, in turn, can be used to determine the number of bacteria in any culture using the MPN method (Table 8.2).

BIOMASS DETERMINATION

Most viable cell counting techniques are very accurate but time-consuming. In order to perform a more rapid population count, other indirect methods have been developed.

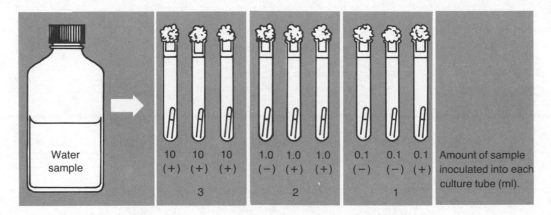

FIGURE 8.14

The three sets of lactose broth tubes are each inoculated with a different quantity of sample from the water to be tested. Each tube in the first set is inoculated with 10 ml of sample, the second set with 1 ml, and the third set with 0.1 ml each. This results in a triple dilution series. In this example, all three of the first set are positive (show growth and gas production), only two in the second set are positive, and only one of the third set is positive. This series of numbers, 3-2-1, is then checked on the most probable number (MPN) table to determine the total number of bacteria per 100 ml. in the original water sample.

One common method is to determine the **biomass** of a culture sample and correlate this value with the number of cells per ml previously measured by the standard plate count method. Biomass is the total weight of a particular kind of organism in a culture. It is used for comparison purposes when the number or size of organisms would lead to confusion. Biomass may be measured indirectly by using visible light to determine the **optical density (OD)** of a culture. Optical density is a measure of the amount of light that does *not* pass through a culture sample and is directly related to the number of cells in the sample. As the optical density of a culture increases, the number of cells in the culture increases. However, the amount of light that *does* pass through a culture may be easier to measure. This value is known as **percent transmittance (PT).** For example, if a fixed amount of light shows through three test tubes that contain different amounts of colored dye, the tube with the greatest amount of dye will have the lowest percent of light transmitted through it. The tube with the least amount of dye will have the greatest amount of light transmitted through it (Figure 8.15). By measuring the amount of light being projected onto the tube and the amount of light actually pass-

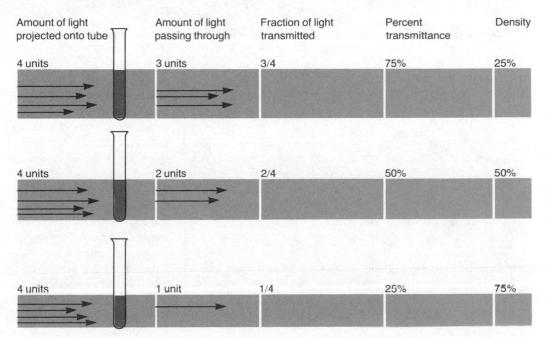

Amount of light projected onto tube	Amount of light passing through	Fraction of light transmitted	Percent transmittance	Density
4 units	3 units	3/4	75%	25%
4 units	2 units	2/4	50%	50%
4 units	1 unit	1/4	25%	75%

FIGURE 8.15

The three tubes have different concentrations of dye. By measuring the percentage of light (PT) that is able to pass through a tube, the relative density of the dye in the tube can be determined.

TABLE 8.2

MPN determination from multiple tube test. The highlighted series indicates that three of the first set of tubes contained enough bacteria to grow and produce gas, the second set contained two such tubes, and the third set contained only one. The *total* number of cells is recorded *per 100 ml sample*. The degree of confidence indicates that the count is accurate 95 percent of the time and will fall between the lower and upper limit number shown.

Number of Tubes Giving Positive Reaction			MPN Index per 100 ml	95 Percent Confidence Limits	
3 of 10 ml each	3 of 1 ml each	3 of 0.1 ml each		Lower	Upper
0	0	1	3	<0.5	9
0	1	0	3	<0.5	13
1	0	0	4	<0.5	20
1	0	1	7	1	21
1	1	0	7	1	23
1	1	1	11	3	36
1	2	0	11	3	36
2	0	0	9	1	36
2	0	1	14	3	37
2	1	0	15	3	44
2	1	1	20	7	89
2	2	0	21	4	47
2	2	1	28	10	150
3	0	0	23	4	120
3	0	1	39	7	130
3	0	2	64	15	380
3	1	0	43	7	210
3	1	1	75	14	230
3	1	2	120	30	380
3	2	0	93	15	380
3	2	1	150	30	440
3	2	2	210	35	470
3	3	0	240	36	1300
3	3	1	460	71	2400
3	3	2	1100	150	4800

SOURCE: *Standard Methods for the Examination of Water and Wastewater*, 12th ed. (New York: The American Public Health Association, Inc., 1965)

ing through, the percentage of light transmitted can be determined. The density of the dye can be calculated by subtracting the PT from 100%:

$$density = 100\% - PT$$

For example, if 30% of the light is actually transmitted through the tube (PT = 30%), a total of 70% is being blocked by the dye (density = 70%). However, microbiologists do not work with percentages when determining OD, but convert all numbers to their logarithmic form:

$$OD = \log 100^* - \log PT$$

(See Box 14 for an explanation of logarithms and Appendix C for a table of four-place logarithms).

---

*Note that since we are now using logarithms we are determining the logarithm of 100, not 100%.

The determination of the PT of a culture is performed in the lab by placing a sample in a machine known as a **photocolorimeter** (Figure 8.16). This machine projects a particular wavelength of light through the sample culture, measures the amount of light passing through, and displays the percent transmittance on a scale for easy reading. The optical density of the culture can then be determined. The total number of cells per ml is found by using a previously prepared graph correlating the OD with number of cells per ml (Figure 8.17). This method is very quick but does have some difficulties. The size of the cells in culture, the color of the media, and the cleanliness of the tube must all be carefully regulated if the results are to be accurate and reliable.

Two other methods are useful in determining the biomass of cultures. By measuring the **dry weight** of a culture and comparing it to a graph similar to the OD curve, an estimate can be made of the total number of cells in a cul-

A logarithm is a different way of writing a number. Just as numbers can be written in fractions (1/4) or decimals (.25), they may also be written in logarithms (-1.3979). All logarithms have two parts. The **characteristic** is written to the left of the decimal point and the **mantissa** is written to the right:

$$\underline{\text{(characteristic)}} \; . \; \underline{\text{(mantissa)}}$$

In order to change a number to its logarithmic form, two reference tables are needed. The first helps to determine the characteristic. For example,

Number Between		Characteristic
.1–0.9	=	-1
1–9.9	=	0
10–99.9	=	1
100–999.9	=	2
1000–9999.9	=	3

The second is a standard table of four-place logarithms used to find the mantissa of the number. Do not *memorize* this table, but *learn* how to use it!

In order to change the number 30.6 to its log form, first determine the characteristic. Since the number 30.6 falls between 10 and 99.9, its characteristic is 1. Now look up the number 30.6 in the log table (Appendix C). Read down the N column to find 30, then move to the right to column 6, since the number is 30.6. The number in the column (4857) is the mantissa of the number 30.6. Therefore, the logarithm of the number 30.6 = 1.4857. The log of 68.8 = 1.8376. The log of 100 = 2.0000.

The formula for determining the optical density of a culture is

$$OD = \log 100 - \log PT$$

Converting to logs

$$OD = 2.0000 - \log PT$$

If a culture of *E. coli* is placed in a photocolorimeter and found to have a percent transmittance of 68.4, the OD of this sample may be determined according to the formula

$$OD = 2.0000 - \log 68.4$$
$$OD = 2.0000 - 1.8351$$
$$OD = 0.1649$$

If the PT of a culture is 49.8, the OD is

$$OD = 2.0000 - \log 49.8$$
$$OD = 2.0000 - 1.6972$$
$$OD = 0.3028$$

Notice that the only figures that change in this equation are the mantissa and OD. The log of the PT is always subtracted from the log of 100. The PT of a sample culture will very seldom be less than 10 or greater than 100. (A culture with 100% transmittance is no culture at all since there are no cells in the fluid to block the light!)

BOX 14
LOGARITHMS AND OPTICAL DENSITY

FIGURE 8.16
A photocolorimeter is an instrument used to measure the percent transmittance of light through a sample of culture. The sample is placed in a special test tube called a cuvette and placed in the holding chamber. When the light passes through the fluid the meter registers the PT of the culture. (Courtesy of Bausch and Lomb.)

ture. To measure the dry weight, a culture sample is passed through a membrane filter and dried in an oven at 100°C. The total dry weight of the sample is determined by subtracting the original weight of the filter paper from the weight of the paper plus the added, dried cell material. The figure that is obtained is then correlated with the total number of cells per ml on a standard dry-weight graph (Figure 8.18A).

1 What is the difference between OD and PT?
2 What is the logarithm of the number 52.6?
3 How do you measure dry weight?

A more precise biomass may be determined by analyzing the chemical composition of the dried cell material. Usually the **nitrogen content** of the materials is measured, since nitrogen is typically found in the amino acids and nitrogenous bases of all cells. If the culture has been actively growing, new amino acids and nitrogenous bases would have been synthesized from components in the medium. The increase in nitrogen then would reflect the total number of cells in the culture medium (Figure 8.18B). Dry-weight and nitrogen-content methods of estimating the biomass of a culture allow microbiologists to determine the approximate number of cells per ml in a culture. These methods are not typically used in clinical labs, but are regularly used in research labs. A last method of determining where a population of microbes is on the population growth curve involves the measurement of chemical changes in the medium. This method is performed in many types of labs, including those in industrial microbiology.

BIOLOGICAL ASSAY

To determine directly the total number of microbes in a culture is very time-consuming and costly. An estimate of the population may be made more quickly and conveniently by

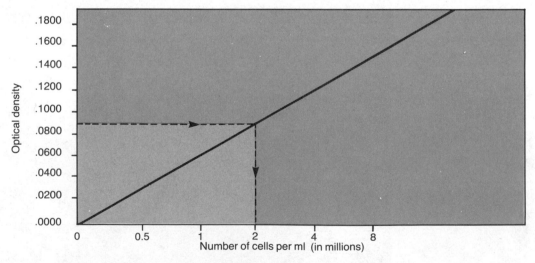

FIGURE 8.17
This graph shows the correlation between optical density and cells per ml of culture. As the optical density (OD) increases, the number of cells per ml increases. In the example plotted on the graph, a sample culture was found to have an OD of 0.0850. A line was drawn parallel to the horizontal axis until it intersected with the angled graph line. At that point, a second line was drawn straight down to the horizontal axis to indicate the number of cells per ml of culture.

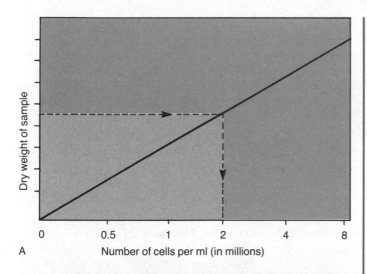

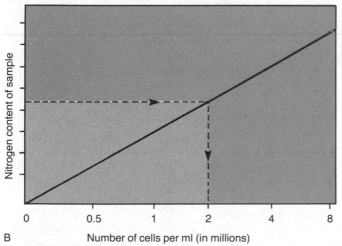

FIGURE 8.18
(A) The relationship between dry weight and number of cells. (B) The relationship between nitrogen content and number of cells. Notice that the population is in millions of cells per milliliter.

performing a **biological assay (bio-assay).** This measures the chemical changes in the medium and compares them to the population growth curve. The two chemical changes that are most easily assayed are the increase in waste products and the decrease in nutrients. After a medium has been inoculated, the microbes begin to produce wastes during the lag phase of growth. The rate of increase in waste products is steady but slow. As the population shifts into the log phase of growth, the rate of waste production also becomes logarithmic, accumulating at a maximum rate. When the cells are unable to cope with the extremely high amount of pollution, the population growth curve moves into the stationary phase. However, the amount of waste produced does not level off, but the rate of increase slows, since many cells are still alive and metabolizing the limited amount of nutrients from the medium. By knowing the met-

abolic rate of the microbes and assaying the amount of waste product being released, it is possible to correlate the two curves and determine the population (Figure 8.19). When milk is fermented into curds and whey, the bacterial population increases according to the population growth curve. The rate of acid production by the bacteria influences how fast the curd will form and what its quality will be. By measuring the amount of acid released over a period of time, the population growth rate can be determined, as well as the likelihood of successful cheese-making. By knowing where the culture is on the growth curve and how fast the growth is progressing, adjustments in growth conditions can be made to ensure a good batch of cheese.

Population growth can also be estimated by performing an assay on the medium to determine the amount of a particular nutrient that had been added before culturing. As the population proceeds through the phases of growth, there is a corresponding loss of the nutrient until it has been completely eliminated from the medium (Figure 8.20). By adjusting the amount of this nutrient in the medium, it may also be possible to extend or delay a particular phase of growth as desired. This practice is the standard method of estimating the size of a yeast culture used in the brewing process. After inoculating the medium with the yeast *Saccharomyces*, the rate of population growth and beer formation is monitored by measuring the amount of sugar remaining in the fermentation vessel. When the alcohol concentration increases at too fast a rate, it can be slowed by diluting the culture. If the fermentation rate needs to be increased, additional sugar is added, encouraging the yeast to continue in the log phase of growth. This kind of regula-

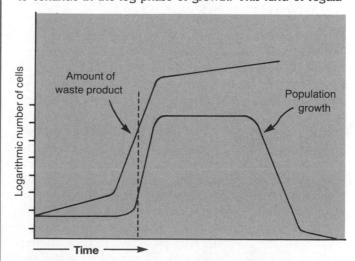

FIGURE 8.19
The two curves show how the release of waste products correlates with the growth of the population. By measuring the amount of waste in the medium, it is possible to determine where the population is on the growth curve. (Note the vertical dotted line).

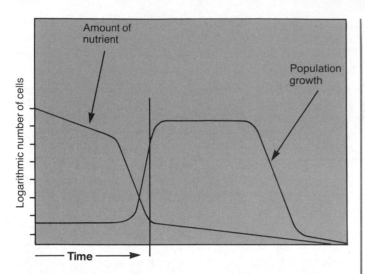

FIGURE 8.20
The two curves show how the amount of a particular nutrient in a growing population decreases through time. By measuring the amount of nutrient left in the medium, it is possible to determine where the population is on the growth curve. This is done by simply drawing a vertical line through both curves.

tion enables the brewmaster to produce a quality product that meets the taste standards set by the company. When the optimum fermentation has been achieved, the fermentation vessel is cooled to settle out the cells and stop the process.

SUMMARY

The metabolic activities within a single cell may lead to an increase in the mass of the microbe if environmental conditions are suitable. Since microbes have relatively short life cycles, they are able to reproduce very rapidly and have significant effects on their environment. The time for a single microbial cell to complete its life cycle is known as its generation time and may be as short as 10 minutes or over 2000 minutes. When a population of microbes grows, the group demonstrates a characteristic growth pattern. A graph of this pattern is the population growth curve. There are four main sections to the population growth curve. The lag phase represents a period when the population is adapting to the environment and the reproductive rate equals the death rate. The lag phase may be extended if the environment is constantly changing or if essential environmental factors, limiting factors, are in short supply. The second phase is the log phase of growth. During this period the population reproduces at its maximum rate and the number of cells increases at an exponential growth rate. During this phase

the species demonstrates its most typical genetic characteristics. The third phase, the stationary growth phase, begins when nutrients become limited and waste products build in the medium. During this period, the cells may either stop reproducing or the reproductive rate may equal the death rate. In the stationary growth phase, there is no increase in the living cell population. As wastes accumulate and food becomes short in supply, the population enters the final phase known as the death phase. During this period, there are more cells leaving the population by death than entering by reproduction.

Knowing where a culture is located on the population growth curve can provide valuable information that may lead to the identification, classification, and control of the microbes. Four methods may be used to determine the number of cells in a population. Total cell counts such as the dry-slide technique, Petroff-Hausser counting chamber, and electronic cell counters, enable the microbiologists to quickly determine the total number of both dead and living cells. Viable cell counting by the standard plate count, membrane filter, or most probable number methods only reveal the number of living cells in a culture. Indirectly, determination of the biomass of a culture may show whether the culture is in the lag, log, stationary, or death phase. Biomass determination may be done by optical density, dry weight, or nitrogen content of a culture. Bio-assay is performed by analyzing the chemical changes that the living cells have made in the medium. These changes may be correlated with the population growth curve to demonstrate in which phase of growth they are located.

Because the lab technician must work with such large numbers of microscopic cells, special equipment, mathematical procedures, and techniques must be used to ensure accurate and reliable results. These results may find application in the home, lab, and industry.

WITH A LITTLE THOUGHT

A new antibiotic has been discovered and is highly effective against a deadly bacterial infection. However, this antibiotic is produced by another microbe that is itself pathogenic. The only way this drug can be produced is by growing this pathogen in the lab under very controlled surroundings. You have been employed as the microbiologist to take charge of the entire operation. You must ensure that large quantities of this antibiotic are produced on a regular basis and that the product is safe for use by the general public. Knowing what you do about the growth and reproduction of microbes, design an effective system for the antibiotic's production.

STUDY QUESTIONS

1 What is the difference between the generation time of an individual cell and that of an entire population of cells?

2 What is exponential growth?

3 Write the following numbers in exponential notation: 456,000,000; 89,000; 325.

4 When a specimen is sent to the lab for identification, it is inoculated into fresh media and cultured. Why?

5 Describe how the stationary growth phase might be lengthened.

6 How does the presence or absence of growth factors affect the growth of a population of microbes?

7 What problems may prevent the growth of a culture in a condition of balanced growth?

8 Explain the differences between viable cell counting and total cell counting.

9 Explain how to go about performing a standard plate count on a pasteurized milk sample.

10 How are membrane filters used in the lab to estimate the population of microbes in a culture?

SUGGESTED READINGS

American Public Health Association. *Standard Methods for the Examination of Water and Wastewater,* 14th ed. Washington, D.C., 1976.

DAWSON, P.S., ed. *Microbial Growth (Benchmark Papers in Microbiology,* vol. 8). Academic Press, New York, 1975.

DONACHIE, W.D., N.C. JONES, and R. TEATHER. "The Bacterial Cell Cycle." Symposium for the Society of General Microbiology. 23:9, 1973.

LAMANNA, C., M. F. MALLETTE, and L. N. ZIMMERMAN. *Basic Bacteriology. Its Biological and Chemical Background,* 4th ed. Williams & Wilkins, Baltimore, 1973.

MANDELSTAN, J., and K. MC QUILLEN, editors. *Biochemistry of Bacterial Growth,* 2nd ed. Halstead Press, New York, 1973.

MEADOW, P. M., and S. J. PIRT, editors. *Microbial Growth.* Cambridge Book Company, New York, 1969.

MEYNELL, G. G., and E. MEYNELL. *Theory and Practice in Experimental Bacteriology,* 2nd ed. Cambridge University Press, New York, 1970.

NORRIS, J. R., and D. W. RIBBONS, editors. *Methods in Microbiology,* vol. 1. Academic Press, New York, 1969.

KEY TERMS

generation times	stationary growth phase	viable cell count
logarithmic growth	death phase	biomass
population growth curve	continuous and batch cultures	bio-assay
lag phase	total cell count	standard plate count
log phase		

Pronunciation Guide for Organisms

Salmonella (sal″mo-nel′ah)
typhi (ti′fi)

Collection, Cultivation, and Identification of Microbes

LABORATORY WORK

SYMPTOMS
Incubation Period
Prodromal Period
Acute Period
Convalescence

COLLECTING AND HANDLING SPECIMENS

MEDIA SELECTION AND PREPARATION

BALANCING THE ENVIRONMENT

OXYGEN REQUIREMENTS FOR CULTURING

CULTURE AND SUSCEPTIBILITY

Learning Objectives

☐ List the seven basic procedures that may be carried out in a microbiology laboratory.

☐ Understand the importance and value of proper specimen collection.

☐ Know the four clinical stages of infection and what typically occurs during each.

☐ Be familiar with the most common types of clinical specimens used in diagnosis, know how each is collected, and learn the types of pathogens that might be identified in each.

☐ Know the basic types of culture media used for microbial growth, the advantages of each, and how they aid in the identification of microbes.

☐ Be familiar with the various differential tests used to identify an unknown bacterium.

☐ Understand the importance of providing a balanced environment for microbial growth with respect to pH, osmotic pressure, and oxygen.

☐ Know how to perform a culture and susceptibility test and understand the value of this important clinical procedure.

Microbes can be presented on a nutritional spectrum ranging from the most self-sufficient autotrophs to the most dependent heterotrophs. Each type has its own unique metabolic pathways and growth characteristics. Pathogenic microbes are able to cause a variety of symptoms depending on how they interact with the patient. Therefore, isolation and growth of the suspected microbe in pure culture is essential to perform tests that will identify and control the infectious agent. The specimen must be handled extremely carefully throughout the testing phase or else the test results will be inaccurate.

An optimum balance of nutrients and a suitable growth environment are needed to culture a microbe in the laboratory. The culture media should contain all the materials necessary for the microbe to move quickly into the log phase of growth. Water, energy, carbon, nitrogen, minerals, and growth factors must be available in a usable form. Many growth media can be prepared for the microbiologist to iso-

9

late, grow, and identify a particular species of microbe in the laboratory. Some media have ingredients that encourage the growth of a suspected pathogen, while others include materials that discourage the growth of unwanted contaminants. Special media have also been designed to identify a suspected pathogen in preparation for biochemical testing.

The proper laboratory diagnosis of an infectious disease helps the patient successfully cope with an infection and provides the health care staff with information that may prevent others from becoming ill. Because the cultivation of a suspected pathogen is vital to successful treatment, it is very important that all those involved in the process be aware of what happens in the lab.

LABORATORY WORK

Some microbes are capable of causing infectious disease. The symptoms that result from the host-parasite interaction may vary depending on the virulence of the microbe, the susceptibility of the host, and the location of the infection in the host. For example, *Streptococcus pyogenes* is a microbe that can cause many symptoms. An infection of this bacterium may result in such diseases as impetigo, septic (strep) sore throat, endocarditis, rheumatic fever, septicemia, erysipelas, puerperal fever, urinary tract infection, endometritis, meningitis, respiratory distress syndrome, and glomerulonephritis. Of course, any one set of symptoms may be the result of an infection by more than one type of microbe. For example, meningitis (inflammation of the meningeal membranes that cover the central nervous system) may be caused by *Neisseria meningitidis* (meningococcal meningi-

tis), *Acanthamoeba* (amoebic meningitis), various viruses (viral meningitis), *Cryptococcus* (fungal meningitis), and a number of bacteria including *Streptococcus, Straphylococcus, Mycobacterium, E. coli, Pseudomonas,* and *Treponema* (all causing bacterial meningitis). The fact is that any one microbe may cause a variety of symptoms, and any one set of symptoms may be caused by a variety of infectious microbes. Therefore, it is vitally important to separate the microbes from other normal inhabitants, isolate the cause of the disease in pure culture, identify the microbe, and specifically determine what type of antimicrobial therapy should be used to help bring the infection under control. The initial step in this process is to take a specimen of the microbe to the lab for cultivation (Figure 9.1). Culturing is done to (1) isolate and identify a particular microbe, (2) perform population counts, (3) carry out serological tests, (4) determine the susceptibility of a microbe to antimicrobial agents, (5) produce biological products such as vaccines and antitoxins, (6) check sterility of essential materials, and (7) maintain stock cultures for use as control groups for teaching (Box 15).

Specimens are typically taken for microbial isolation and identification from many materials. Cheese manufacturers regularly extract core samples from ripening cheeses to determine the population and purity of microbes inoculated into the curd that are responsible for developing cheese flavor, texture, and aroma. These cores are taken to the laboratory where the microbes are freed from the curd mechanically. Once separated and diluted, the specimens may be examined under the microscope or grown on special media. In the field of industrial microbiology it is essential that specific strains of microbes be maintained in pure culture. Should mutations occur in the population's productivity, product quality and great financial losses might occur.

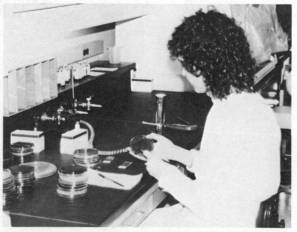

A   B

FIGURE 9.1
(A) A medical microbiology laboratory. (B) Transferring cultures for slide preparation and examination under the microscope are routine procedures. (Courtesy of Saginaw Medical Center, L. DeGuise.)

MICROBIOLOGY laboratories may be found in hospitals, dairy companies, sewage treatment plants, universities, pharmaceutical companies, government agencies, and chemical plants. One of the major responsibilities in all labs is the physical containment of microbes. This is especially important in labs that work with highly virulent microbes. To ensure that the best techniques, equipment, and procedures will be used to contain and control these microbes, the U.S. Department of Health, Education, and Welfare has provided guidelines for the classification of labs. These requirements present the basic safety criteria for each level of physical containment. Other microbiological practices and lab techniques which promote safety are encouraged by the Department.

BOX 15
LABORATORY CLASSIFICATION

Laboratory Classification: Level P_1

1 No biologically hazardous microbes used.
2 Doors closed during use.
3 Bench-top examination.
4 Lab not separated from traffic pattern.
5 Require decontamination or leakproof container for wastes.
6 Pipetting by mouth permitted.
7 No eating, drinking, or smoking.
8 Hand wash facilities must be available.
9 Gowns or lab coats discretionary.
10 Outside autoclaves acceptable.

Laboratory Classification: Level P_2

1 Minimal biological hazard.
2 Transfer cabinet for aerosols required.
3 Decontamination required before disposal.
4 No children permitted inside laboratory.
5 No mouth pipetting.
6 Lab coats required.

Laboratory Classification: Level P_3

1 Moderate biological hazard.
2 Not open to public.
3 Air-lock entry way and locker rooms necessary.
4 Autoclave inside lab.
5 Room should be under a negative air pressure.
6 All work is done in transfer cabinets or glove box.
7 Solid front on glove box required.

Laboratory Classification: Level P$_4$

1 Hazardous biological microbes.

2 Building designed so that microbes cannot hide.

3 Autoclave in wall, entered from the lab through glove box, and emptied from the other side.

4 Biological hazard signs posted.

5 Only authorized personnel permitted in lab.

6 All personnel instructed in special techniques and procedures.

7 Transfer cabinet or glove box connected to autoclave.

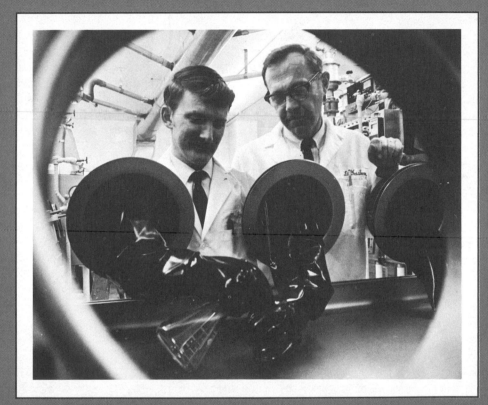

Courtesy of the Center for Disease Control, Atlanta.

For this reason, specimens are regularly taken from the large-scale culture vessels for laboratory checks. In the area of water and wastewater microbiology, water samples must be closely monitored to protect the public against microbial contamination. Specimens are taken from faucets and discharge outlets, and sent to the laboratory to determine the presence and number of pathogenic microbes. Should harmfully large numbers of microbes be identified, corrective action must be taken immediately.

None of these procedures may be successfully performed unless the entire process begins properly. The lab technician, doctor, and patient rely on the collection of specimens that have been selected correctly, taken at the right time, collected properly, and sent for culturing as quickly as possible. By following proper aseptic techniques during specimen collection, the chances of cross-infection among patients in the same hospital and among different hospitals is greatly reduced.

The kinds of specimens taken for culturing will depend on the nature of the infectious disease; therefore, the "course of infection" has to be thoroughly understood. This includes how the microbe has entered the patient, how it moves (if it does) through the patient's body, how the patient's body responds to the presence and actions of the microbes, and most importantly, how quickly these processes occur. This understanding will help in deciding what kind of material should be collected for laboratory analysis. Many different body tissues and fluids may serve as specimens. Blood, urine, sputum, pus, and feces may contain the microbes that are responsible for the symptoms of a disease (Figure 9.2). Spinal fluid, bile, pleural fluids, stomach fluids, and bone marrow may also be sampled. These specimens should be taken at the appropriate time to obtain the greatest number of microbes. This should be done according to the patient's symptoms; that is, specimens should be collected when the patient is displaying a "rise" in symptoms. If the specimen is collected after the **febrile period** (the period of infection in which a fever occurs), the patient's body will have begun to cope with the microbe and reduced the number of pathogens to a level so low that they may not be identified in the lab. Therefore, it is important to know the clinical stages of infection of a disease.

SYMPTOMS

Infections can be divided into two main types, acute and chronic. An **acute disease** lasts for a relatively short time and is severe. The symptoms usually last as long as the pathogen or its products are in the patient's body. A **chronic disease** lasts over a long period of time. Chronic diseases persist since the microbe is better able to defend itself against host defense mechanisms or because the host defenses are not able to respond normally. In either situation, the microbes are not killed but persist for a long time. The **clinical stages of infection** of an acute infectious disease occur in four typical phases: (1) the incubation period, (2) the prodromal period, (3) the acute period, and (4) the convalescent period.

INCUBATION

This period begins when the microbe makes contact with the host and attempts to establish a parasitic relationship. During this period, which may last from two days to several months, the infecting microbes must increase to a certain population level before symptoms become apparent. Many virulent bacteria have very short incubation periods (e.g., *Streptococcus pneumoniae,* one to three days), while viral and fungal diseases have more prolonged incubation periods (e.g., *Coccidioides immitis,* fungal coccidioidomycosis, which can last up to four weeks in primary infections) (Table 9.1).

PRODROMAL PERIOD

This is the period during which the patient first begins to feel that "something is wrong." Symptoms begin to be displayed during this two- to four-day period and may take the form of **malaise** (a vague feeling of discomfort), headache, nausea, or upset stomach. Not all diseases have a long prodomal period; many display themselves "full blown" almost immediately after incubation. The microbes are beginning

FIGURE 9.2
Inoculated petri plates and tubes are placed in a walk-in incubator for culturing. Typically these cultures will be examined in 24 to 48 hours.

TABLE 9.1
Incubation periods.

Disease	Cause	Usual Incubation Period
Adenovirus infections	Adenovirus	5 to 7 days
Amebiasis	*Entamoeba histolytica*	2 weeks (varies)
Brucellosis	*Brucella abortus*	5 to 30 days (varies)
Chickenpox	Herpes-zoster	14 to 16 days
Coxsackie virus infections	Coxsackie virus	2 to 14 days
Dengue fever	Arbovirus	3 to 15 days
Diphtheria	*Corynebacterium diphtheriae*	2 to 6 days
Encephalitis	Staphylococci, viral	3 to 15 days (varies as to type)
Gas gangrene	*Clostridium perfringens*	1 to 5 days
Gonorrhea	*Neisseria gonorrhoeae*	3 to 5 days
Hepatitis type A, infectious	Enterovirus (HAV)	15 to 50 days
Hepatitis type B, serum	Enterovirus (HBV)	2 to 6 months
Herpesvirus infections	Herpes virus	4 days
Histoplasmosis	*Histoplasma capsulatum*	5 to 18 days
Impetigo contagiosa	Streptococci	2 to 5 days
Infectious mononucleosis	Epstein Barr virus	4 to 14 days
Influenza	Influenza virus	1 to 3 days
Leprosy	*Mycobacterium leprae*	3 to 5 years (?)
Lymphopathia venereum	*Chlamydia trachomatis*	1 to 4 weeks
Measles	Exanthem virus	9 to 14 days
Meningitis, acute bacterial	*Neisseria meningitidis*	1 to 7 days
Mumps	Respiratory virus	14 to 28 days
Mycoplasmal pneumonia (primary atypical)	*Mycoplasma pneumoniae*	7 to 14 days
Pertussis (whooping cough)	*Bordetella pertussis*	5 to 21 days
Plague	*Yersinia pestis*	3 to 8 days
Poliomyelitis	Enterovirus	7 to 14 days
Psittacosis	*Chylamydia psittaci*	4 to 15 days
Q fever	*Coxiella burnetii*	14 to 26 days
Rabies*	Rhabdovirus	2 to 6 weeks (to 1 year)
Rocky Mountain spotted fever	*Rickettsia rickettsii*	2 to 12 days
Rubella	Exantham virus	14 to 21 days
Salmonellosis	*Salmonella* spp.	
Food poisoning		6 to 48 hours
Paratyphoid fever		1 to 10 days
Typhoid fever		7 to 21 days
Shigellosis (bacillary dysentery)	*Shigella* spp.	1 to 7 days
Smallpox	Poxvirus	12 days
Streptococcal infections	Streptococci	2 to 5 days
Syphilis, primary lesion	*Treponema pallidum*	14 to 30 days
Tetanus	*Clostridium tetani*	3 days to 3 weeks
Tuberculosis, primary lesion	*Mycobacterium tuberculosis*	2 to 10 weeks

*Rabies incubation in dogs is 21 to 60 days.

to increase in number and cause enough harm to result in the production of disease symptoms.

ACUTE PERIOD

During this time the disease is at its height and specimen collections can be made. The patient shows all the "typical" symptoms of the infection which may include fever, chills, rash, nausea, vomiting, diarrhea, and others more charac-

teristic of a particular infection. During this period the host begins to respond to the presence of the pathogen by bringing into play all its various defense mechanisms. In many cases, the acute period marks the time when the disease may be transferred directly or indirectly to other susceptible persons in the population. This is known as the **communicable period.** In some diseases, the communicable period begins in the incubation period and extends through the acute phase (e.g., diphtheria). However, in most situations, the communicable period corresponds with the acute phase.

CONVALESCENCE

This period occurs if the patient has not experienced enough damage to cause death. Convalescence may also be called the recovery period since it is during this time that the patient's body repairs damage and completes the destruction of the invading pathogen.

Successful specimen collection must also take two other factors into consideration. All specimens should be taken for analysis before any antibiotic therapy is given. If this does not happen, the antibiotic will move through the patient's body and contaminate the specimen. When the lab receives the specimen, the antibiotic may already have interfered with the normal growth and activities of the pathogen. Timing is also important in determining when to take blood samples for serological testing. **Serology** is the study of antibody-antigen reactions in a test tube *(in vitro)*. In these studies, blood **serum** (the fluid portion of the blood after clotting) is analyzed to detect the presence of antibodies. An **antibody** is a protein molecule manufactured by the body in response to the presence of a foreign molecule known as an antigen. An **antigen,** or **immunogen,** is usually a large protein molecule, such as a bacterial toxin or enzyme, a part of a cell capsule, or a whole virus particle. The antigen is specifically able to stimulate certain cells in the body to produce antibody. They then combine in an antibody-antigen reaction. When this reaction occurs, the dangerous effects of the foreign antigen are eliminated. Antibodies are produced when an infectious microbe makes effective contact with antibody-forming cells. This production may begin before symptoms of the infection are seen and may last for years after the convalescent period. Serological tests to determine the amount and types of antibodies found in the blood are performed to detect infections that do not display obvious symptoms or only display mild symptoms. These tests also help determine whether a patient has developed a sufficient level of protection (immunity) against a particular disease. A steady increase in the amount of antibody (the **antibody titer**) in serum samples taken over a given period may indicate that the patient is currently experiencing an infection. Therefore, blood samples must be taken at the appropriate time or at specified intervals to allow the lab technician to perform the proper serological tests.

---

1   List the four clinical stages of infection.
2   What is the usual incubation period for measles, rabies, and tetanus?
3   What is meant by an "antibody titer"?

COLLECTING AND HANDLING SPECIMENS

Specimens may be collected in a number of ways, as shown in Table 9.2. Each method is designed to ensure that the proper material will be sent to the lab using good aseptic technique. One of the most common tools used to collect specimens is the sterile **swab** (Figure 9.3). Swabs are used to gather samples from the eye, ear, nose, throat, open wounds, and fecal material. Special handling of the pre-sterilized package is necessary to prevent contaminating the swab. Each manufacturer has its own unique package and provides instructions for its proper use. Many swabs have an ampule (small container) of sterile physiological saline solution built into the package. After the specimen is taken and returned to the container, the ampule is squeezed and broken to release the fluid onto the surface of the cotton swab containing microbes. This is essential since the microbes have been removed from an optimum growth environment— the patient. If the microbes are not protected from drastic changes in moisture content and temperature, they will die before they reach the lab. For this same reason, specimens should be sent to the lab immediately for culturing. If this cannot be done, most specimens should be refrigerated to limit cell death. Exceptions to this rule are *Neisseria gonorrhoeae* specimens, which are maintained at room temperature in the lab.

Specimens may also be collected by **needle aspiration.** The most familiar sample taken using this method is blood. After proper aseptic technique has been followed to

FIGURE 9.3
Swabs are common tools used for obtaining specimens. These swabs are sterilized and packaged in two different types of containers. Each type must be opened according to the directions to prevent contamination before sampling, and they must be handled properly afterwards to assure survival of the desired specimens.

TABLE 9.2
Typical examination specimens.

Nature of Specimen	Method of Collection	Quantity to be Collected	Suspected Pathogens	Possible Normal Flora Contaminants
Blood	Venipuncture	10 ml/culture	Any microbe	None
Throat	Swab	Cover swab reasonably	Any microbe in excess of 100,000/ml	Lactobacilli, diphtheroids, yeast
Urine	Swab	Cover swab reasonably	Streptococci, staphylococci	Corynebacteria, streptococci, staphylococci
Feces	Clean container rectal swab	Small	*Salmonella* spp., *Shigella* spp., staphylococci *Campylobacter* spp.	All normal flora, too numerous to list
Sputum	Sterile container	3 ml	*Mycobacterium* spp., streptococci	Streptococci

prevent infection of the patient and contamination of the hypodermic needle, a sample of venous blood is drawn into a container that has been treated with an anticoagulant such as heparin or potassium oxalate (Figure 9.4). The presence of the anticoagulant prevents the pathogenic microbes from being entrapped in a fibrin clot, which would make a quick isolation impossible. Needle aspiration may also be used to obtain spinal fluid. A modification of the needle aspiration technique involves the use of a long, sterile tube attached to a syringe. This method is known as **intubation** and is also used to collect specimens from the stomach, gall bladder, chest, or peritoneal cavities. Tubes used to obtain specimens vary in size, shape, and material. The two most common are the Levin tube and catheter. The **Levin tube** is a

soft rubber tube 4–6 mm (12–18 French*) in diameter and comes prepackaged in a sterile wrap. The Levin tube may be swallowed by the patient or passed through the nostril into the stomach. Specimens may then be withdrawn in small amounts (10–15 ml) every 10 to 15 minutes.

There are three types of catheters used to collect urine specimens. The **hard catheter** is made of a more firm rubber or plastic material and is used when the urethra has strictures (narrowing of the tube). Special precautions must

---

*The unit of measure denoting the sizes of catheters and other tubular instruments is the **French.** Each French unit is roughly equal to 0.33 mm (i.e., 18 French = 6 mm).

FIGURE 9.4
Venipuncture is used to obtain blood samples for analysis in the microbiology laboratory. The presence of a certain species of microbe in the blood is characteristic of many infectious diseases.

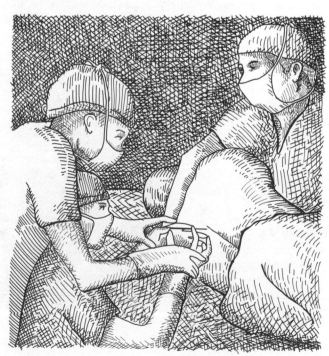

be taken when using this catheter to avoid tissue damage. The **French catheter** is made of soft material and is the type most frequently used to obtain single samples. If urine specimens or drainage are required for an extended period, a **Foley catheter** (indwelling catheter) is preferred. This is a tube within a tube. After the catheter has been introduced into the bladder, the tip of the outer tube may be blown up with air or water like a small balloon, to prevent it from being removed (Figure 9.5). Catheterization is one of the hospital procedures most likely to lead to infection. About 40 percent of all hospital-acquired infections (400,000 patients in the United States annually) occur as a result of catheterization. Urinary tract infections associated with catheters may be caused by such bacteria as *E. coli, Klebsiella, Enterobacter, Citrobacter,* and *Proteus.* These microbes are normally present in the colon or lower portion of the urethra. The bacteria may be moved into the urinary tract and bladder either during or after the catheter is inserted. Foley catheters are more likely to lead to infections in elderly female patients because of the patients' inability to properly cleanse themselves after defecation. The close anatomical position of the anus to the urethal meatus (opening) allows for an easy transfer of colon bacteria. To avoid these difficulties, noncatheterization specimens of urine may be taken from both males and females by the **clean-catch method.** After the patient has cleansed the tissue surrounding the meatus, a small test tube, cup, or bottle is used to catch the urine sample. In the **clean-catch midstream method,** the first portion of the voided urine should not be collected since it will be contaminated with bacteria normally occurring in the lower portion of the urethra. This first voided urine is discarded and only the second, or midstream, portion is saved for lab analysis, since it will likely contain mi-

crobes found in the bladder. The midstream specimen is quickly sent to the laboratory where the concentration of organisms is used to determine the presence or absence of a bladder infection (Figure 9.6).

Sputum is another material that may be taken from a patient without the use of special equipment. A distinction must be made between saliva and sputum. **Saliva** is the clear, alkaline secretion from the salivary glands that serves to moisten the mouth. **Sputum** is a thick, heavy material coughed up from the lungs. For instance, when collecting sputum samples for the detection of tuberculosis bacilli, the patient is instructed to rinse the mouth with sterile water and cough up all the material possible. Sputum collected in a special sterile cup will be thick and settle to the bottom (Figure 9.7).

Specimens collected for laboratory analysis should be properly labeled. In some hospitals, a plastic "charge card" is prepared for each patient and indicates the patient's name, unit number, hospital or case number, room and floor number, and other information. If such a system is not employed, the collection vessel label or a separate form is used to provide the lab with the necessary information (Figure 9.8). After proper identification, the sample is sent to the lab, where the technician begins a detailed investigation of the specimen. First, a microscopic check determines the possible nature of the microbe, and culturing follows for more complete identification. In order to perform all the necessary tests, the specimen is cultured in the appropriate media selected and prepared by the technician.

The importance of proper specimen collection cannot be overemphasized. Techniques described here are fundamentally the same as those used in other areas of microbiology and not limited to the medical microbiology. It is equally

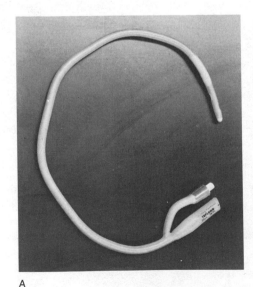

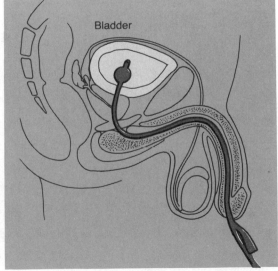

FIGURE 9.5
(A) A Foley catheter is used to obtain urine specimens. (B) When in place, a small "balloon" is inflated so that the catheter will remain in the bladder.

A

B

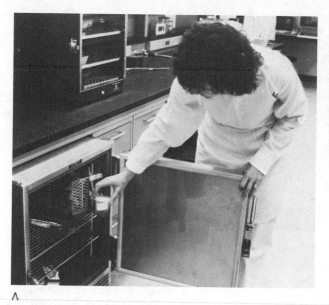

FIGURE 9.6
(A) Urine specimens may be refrigerated before examination. This inhibits bacterial growth and ensures the survival of pathogens. (B) Samples may be taken from the collection vessel with an inoculating loop for slide preparation and staining. (Photo A courtesy of Daria Smith. Photo B courtesy of Saginaw Medical Center, L. DeGuise.)

A

B

essential in the areas of water, food and dairy, and industrial microbiology to use proper equipment and aseptic technique to obtain specimens for analysis and identification. In many situations, special collection vessels have been designed to obtain and transport specimens. For example, sampling for microbes on the deep-ocean bottom is carried out using weighted cylinder samplers, which open to let in only certain materials after being triggered with a release cable attached to the research ship. These containers are then raised to the surface without being contaminated by surface microbes and taken to the laboratory for analysis.

MEDIA SELECTION AND PREPARATION

Culture media must contain all the essential nutrients required by the organism for it to grow and reproduce. A

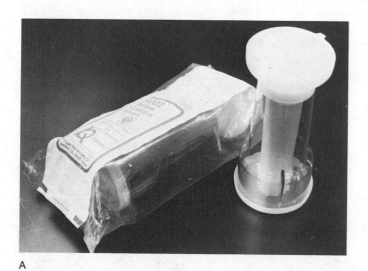

A

B

FIGURE 9.7
Handling of *Mycobacterium tuberculosis* in a sputum specimen examination. (A) Specially designed sputum collection cup will allow the patient to deposit a sputum specimen. (B) Once in the laboratory, the cup can be opened from the bottom to reduce the chances of technician contamination from pathogens that might be left on the top. (Courtesy of Saginaw Medical Center, L. DeGuise.)

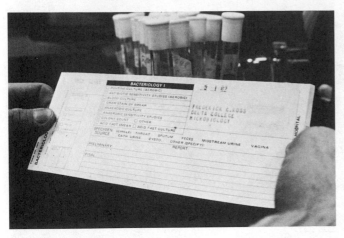

FIGURE 9.8
Cards such as this are regularly used in hospitals to more efficiently direct the medical technologist. (Courtesy of Saginaw Medical Center, L. DeGuise.)

suitable source of energy, building materials, and growth factors must be supplied in adequate amounts. Most pathogenic heterotrophs may have both their energy needs and building material requirements met by adding a carbohydrate to the medium. This may be a simple sugar such as glucose or a more complex form such as lactose or mannitol. Carbohydrates serve as sources of hydrogen, carbon, and chemical-bond energy. Nitrogen may be provided in an inorganic form, such as nitrate ions (NO_3^-) or ammonium ions (NH_4^+), or it may be organic, such as amino acids or nitrogenous bases. Minerals may be added to the medium in the form of inorganic salts (e.g., $NaCl$, $MgSO_4$). Micronutrients usually do not have to be added to the medium during preparation since many of these occur in sufficient amounts as contaminants in the water. For **fastidious** (difficult to satisfy) microbes, specific growth factors, such as vitamins and certain amino acids, may have to be added.

All these essential ingredients may be included in culture media by using either pure chemicals or extracting them from meat, milk, yeast, or other sources. Media that are prepared from chemical compounds that are highly purified and specific are called **defined,** or **synthetic, media** (Figure 9.9). Media that are prepared from ingredients that have not been precisely defined are called **complex,** or **nonsynthetic, media.** The most widely used complex medium is nutrient broth (Figure 9.10A). This medium is prepared by boiling beef to extract nutrients and adding an amino acid-nitrogen source known as peptone. **Peptone** is usually produced by the hydrolysis of beef protein, but other protein sources may also be used. **Casein** peptone and milk peptone are also used in complex media as sources of amino acids and nitrogen. Since these materials may contain a variety of unknown ingredients, nutrient broth is a complex, rather than synthetic, medium. If the terms "pep-

tone" and "casein" are found as ingredients in a prepared medium, they indicate that the medium is complex and not synthetic. Other ingredients derived from hydrolysis of macromolecules also show that the medium is complex. **Proteose** is another form of hydrolyzed protein, and **yeast extract** is made from boiled and dehydrated yeast cultures. Veal **infusion medium** is made from fresh, lean veal that has been ground and the nutrients extracted by steeping in hot water. The extracted nutrients are cooled overnight in a refrigerator and then gradually raised to the boiling point before filtering.

All liquid media, whether complex or synthetic, may be converted to solid media by adding either **gelatin** (a protein) or **agar-agar,** a complex polysaccharide extracted from red marine algae (Figure 9.10B). The use of agar has an advantage in that most bacteria are unable to hydrolyze this molecule into more simple components. Since gelatin is a liquid at room temperature, the use of agar allows the medium to remain in a solid form while the microbes are growing on its surface. However, a number of bacteria are able to hydrolyze gelatin. When this occurs, the solid medium becomes liquid and the advantages of culturing microbes on a solid surface are lost. Each solid and broth medium used in the lab is specially designed for certain purposes. There are four categories of media used in the lab: (1) enrichment, (2) selective, (3) differential, and (4) propagation.

Enrichment media are prepared with ingredients that will enhance the growth of certain microbes suspected of being in a specimen. Since most specimens arrive at the lab containing a variety of microbes from the infected patient, enrichment media encourage the growth of the suspected pathogen so that it will become the most predominant type of microbe in the culture. Two types of enrichment media

FIGURE 9.9
Microbiology laboratories maintain a stock of different types of culture media. In this laboratory, the dehydrated media are easily rehydrated for use in culturing microbes. (Courtesy of Daria Smith.)

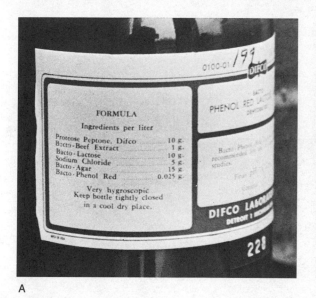
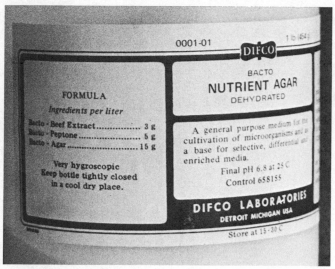

A   B

FIGURE 9.10

Each type of dehydrated medium has on its label a list of preparation instructions and ingredients. (Courtesy of Daria Smith.)

are blood agar and chocolate agar. Blood agar enriches the growth of such bacteria as *Streptococcus* and *Neisseria*. It is prepared by adding defibrinated human or sheep blood to melted medium.

> 1 What is the difference between sputum and saliva?
> 2 What does the term "fastidious" mean?
> 3 Why is agar-agar widely used as an agent of solidification for bacteriological media instead of gelatin?

Chocolate agar is blood agar that has been heated until the medium becomes brown, or chocolate colored. This medium is helpful in culturing bacteria such as *Neisseria* (gonorrhea), *Haemophilus* (influenza), *Yersinia* (plague), and *Bordetella* (whooping cough). Selenite F and Gram-negative (GN) broths are enrichment media used to separate *Salmonella* and *Shigella* from stool samples. The *Salmonella* grow better in these media than the *Shigella*.

Selective media are prepared with ingredients that inhibit the growth of unwanted microbes which might be in the specimen. The inhibitor may be an antibiotic, salt, or other chemical. Mixed cultures of microbes originally grown in enrichment media may be inoculated into selective media to isolate the desired microbe. The addition of the dye crystal violet to medium will inhibit the growth of staphylococci and other Gram-positive bacteria. However, mannitol salt agar is commonly used in the lab to identify staphylococci in mixed cultures. This bacterium is able to grow at the 7.5

percent salt (NaCl) concentration of this medium (Figure 9.11). All other bacteria normally found in association with staphylococcal infections are inhibited by this high salt concentration. Gram-negative organisms, such as *E. coli*, may be isolated by growing a mixed culture on EMB (eosin-methylene blue) medium since this selective medium contains chemicals that inhibit the growth of Gram-positive microbes.

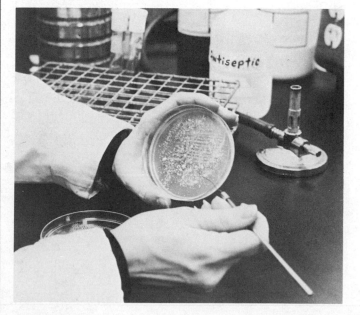

FIGURE 9.11

The streak plate shows how a highly concentrated sample of culture has been streaked over the surface of the agar plate to separate one cell from another. (Courtesy of Daria Smith.)

BOX 16

IDENTIFICATION
BY
FERMENTATION

BACTERIA are capable of fermenting a great many molecules including carbohydrates, amino acids, and organic acids. The fermentation of particular carbohydrates is a characteristic feature of certain species or genera of bacteria and can be used as an identification tool. This is a good characteristic upon which to base an identification since it demonstrates some of the metabolic capabilities of the microbe. These are in turn a reflection of the microbe's genetic makeup.

Identification of the fermentation abilities of a particular bacterium is carried out in the laboratory. The microbe is placed in a series of carbohydrate-containing test tubes, within which are **Durham tubes**. The nutrient medium of each tube contains beef extract, peptone (a nitrogen and protein source), vitamins, minerals, and a specific carbohydrate. One tube might contain glucose, another lactose, while still another, sucrose. In addition, a pH indicator such as phenol red is added. This indicator will change color if acid (e.g., lactic acid) is released as a result of the fermentation of a sugar. This setup consists of a tube within a tube. The smaller, inverted Durham tube inside the larger test tube collects any gases (e.g., CO_2, CH_4, H_2) that might be produced during the fermentation of the sugar. If gas is produced, it will rise into the Durham tube and be captured there, forming a clearly visible bubble. Some bacteria produce both acid and gas as fermentation products, while others produce only acid or only gas.

No two bacteria in this chart will ferment the same sugars and release the same combination of products.

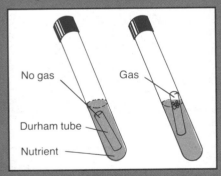

Bacteria	Phenol Red Glucose Broth		Carbohydrates Phenol Red Lactose Broth		Phenol Red Sucrose Broth	
	Acid	Gas	Acid	Gas	Acid	Gas
Bacillus subtilis	+	−	−	−	+	−
Escherichia coli	+	+	+	+	+	+
Proteus vulgaris	+	+	−	−	+	+
Staphylococcus aureus	+	−	+	−	+	−

The identification of fermentable sugars and the products formed can aid in identifying an unknown bacterium. For example, if an unknown bacterium was grown in all three of these sugar-containing tubes, and found to produce acid and gas in all the tubes, *E. coli* would be suspected as the unknown bacterium.

Differential media are designed to differentiate among microbes. Different bacterial species may produce dissimilar colony colors when grown on differential agar. While in differential broth cultures, the media change color. Differential media are used to confirm the identity of a microbe that has already been isolated by culturing in enrichment and selective media. One of the most important differential tests performed in the hospital lab is the **IMViC test**. This is not one test but a series designed to help identify members of the bacterial family Enterobacteriaceae. Members of this family and the diseases they cause include:

Bacterium	Name of Disease
E. coli	infantile diarrhea
Enterobacter	urinary tract infections
Klebsiella	pneumonia
Salmonella	typhoid fever and gastroenteritis
Shigella	bacillary dysentery

One group of Enterobacteriaceae is known as the **coliforms** because they are found in the colons of humans and many animals, and biochemically they resemble *Escherichia coli*. Members of the coliform group include *Klebsiella, Enterobacter, Serratia, Hafnia,* and *E. coli*. These organisms resemble one another not only in their Gram reaction (all are Gram-negative rods), but also in the symptoms they cause when found in a parasitic relationship. The two that are most similar are *E. coli* and *Enterobacter*. In order to identify the exact cause of a coliform infection, the IMViC test is run on specimens taken from the patient. The letters stand for:

I = Indole; demonstrates the production of indole (an organic molecule) by the bacteria from the amino acid tryptophan.

M = Methyl red; a pH indicator to determine whether the microbe has produced acid.

Vi = Voges-Proskauer; indicates the production of an organic molecule called acetylmethylcarbinol (acetoin).

C = Citrate; indicates whether or not the organism can utilize sodium citrate as a sole source of carbon.

The difference among these members of the coliform group can be shown in how each responds to the IMViC series:

	Test			
Organism	Indole	Methyl Red	Voges-Proskauer	Citrate
Escherichia coli	+	+	−	−
Enterobacter aerogenes	−	−	+	+

Another important series of differential tests performed in the laboratory demonstrates the fermentation abilities of bacteria.

1 What makes mannitol salt agar a selective medium? differential medium?
2 What is the significance of the IMViC test?
3 Of what value is the small inverted tube in the Durham tube setup?

Bacteria are capable of fermenting a great variety of molecules including carbohydrates, amino acids, and organic acids. The fermentation of particular carbohydrates is a unique feature of certain species of bacteria and can be used as an identification tool, since it demonstrates some of the metabolic capabilities of the microbe (Box 16). Other tests used to specifically identify unknown bacteria are described in Table 9.3.

Most microbes require about twenty-four hours culture time in order to produce observable, reliable results. Many

TABLE 9.3
Differential tests to identify an unknown bacterium.

Test	Description
Catalase	This test identifies whether a bacterium produces catalase to protect itself from the harmful effects of H_2O_2 by changing the peroxide to water and oxygen.
Oxidase	Indicates the presence of iron-containing enzymes that are able to reduce oxygen. Aids in the identification of *Neisseria*.
Nitrate reduction	Determines whether a bacterium can use nitrate as an electron acceptor: $NO_3^- \rightarrow NO_2^- \rightarrow N_2$.
Starch hydrolysis	Identifies the presence of the enzyme amylase, which is able to hydrolyze amylose, a component of starch.
Casein hydrolysis	Detects the presence of caseinase, an enzyme that causes milk protein (casein) to be hydrolyzed to a clear, transparent liquid. Also known as proteolysis or peptonization.
Fat hydrolysis	Detects the enzyme lipase, which breaks down fats into fatty acids and glycerol.
Urea hydrolysis	Indicates whether a bacterium is able to produce urease, an enzyme that breaks down urea into NH_3 and CO_2. Important in the identification of coliforms.

Litmus milk	A test to determine five characteristics: **1** Acid reaction turns pink to show fermentation of lactose. **2** Alkaline reaction turns blue to show no proteolysis. **3** White color shows litmus has acted as electron acceptor. **4** Coagulation (curd formation) shows acid produced from lactose or rennin production. **5** Peptonization shows curd digestion by casein hydrolysis.
Hydrogen sulfide production	Demonstrates the formation of H₂S from the amino acid cysteine by the presence of enzyme cysteine desulfurase; important in *Salmonella* identification.
Coagulase	Detects the production of coagulase enzyme, which causes blood clotting in plasma; an important test for pathogenic *Staphylococcus aureus*.
Gelatin liquefaction	Identifies whether a bacterium is able to produce an enzyme that hydrolyzes the protein gelatin.

hospital personnel become anxious when results are not made available to them only a few hours after the specimen has been sent for analysis. It is important to realize that if a mistake is made in the laboratory or if the microbes do not respond to culturing as they should, an additional twenty-four hours are required for the mistake to be rectified or the problem resolved. This difficulty emphasizes the importance of proper specimen collection and quick transmission to the lab. New culture methods and equipment are available that help speed up the identification process but in most cases, thirteen to eighteen hours are still required to obtain results (Box 17).

Propagation media are used to propagate, or keep microbes growing, for a long time. Samples grown on these media may be taken for analysis and the technician can be sure that there will be an abundant supply of cells. The most common propagation media are nutrient broth and agar. Table 9.4 lists a number of organisms and the propagation media on which they may best be grown.

Many of the media listed in the table have already been described as either enrichment, selective, or differential me-

dia. However, it is not unusual to use a particular type of nutrient for more than one purpose. For instance, EMB agar is both selective and differential at the same time. This medium inhibits the growth of Gram-positive bacteria and shows the difference between lactose-fermenting and nonlactose-fermenting Gram-negative bacteria by their appearance on

TABLE 9.4
Propagation media are used in the lab to increase the supply of microbes and maintain a particular type for use at a later time. Pure cultures kept in the lab for use at a later time are known as **stock cultures.** The bacteria listed may be maintained in the lab as stock cultures by growing them on the propagation media shown.

Microorganism	Propagation Media
Coliform group	Nutrient agar
Salmonella	Nutrient broth
Shigella	Cooked meat medium
Pneumonococci	Brain-heart infusion medium
Meningococci	Tryptose phosphate broth
Streptococci	Dextrose starch agar
(enterococci)	Tryptose agar
	Dextrose broth
	Dextrose agar
	AC medium
Staphylococci	Brain-heart infusion medium
	Tryptose phosphate broth
	Dextrose starch agar
	Tryptose agar
Neisseria	Dextrose starch agar
	GC medium base or proteose #3 agar enriched with bacto-supplement B
Corynebacteria	Loeffler's blood serum
Brucella	Tryptose agar
Listeria	Tryptose agar
	Cystine-heart agar with bacto-hemoglobin
Haemophilus	Proteose #3 agar with bacto-hemoglobin and bacto-supplement A or bacto-supplement B
	Brain-heart infusion with bacto-supplement B
	Bordet Gengou agar base with fresh blood
	Dextrose agar with fresh blood

Clostridia	AC medium
	Veal infusion medium
	Egg meat medium
	Cooked meat medium
Mycobacterium	Middlebrook medium, 7H10
tuberculosis	Dubos medium
	Long's asparagin medium
	Petragnani medium
	Lowenstein-Jensen medium
	Bovine TB medium
	Dorset medium
	Peizer TB medium
Fungi	Sabouraud maltose agar
	Littman oxgall agar
	Brain-heart infusion agar
Entamoeba	Entamoeba medium with
histolytica	horse serum saline 1:6
	and rice powder

the agar surface. The lactose-fermenting bacteria, such as *E. coli*, grow well on this medium and their colonies have a green metallic sheen produced by a change in the eosin-methylene blue dyes in the medium. All of the various types of media used in the lab may be prepared by the technician (Figure 9.12), but most are purchased in a dehydrated form from microbiological supply companies. In most instances, these media are complete and require only mixing with water and sterilization (Table 9.5).

BALANCING THE ENVIRONMENT

When dehydrated or fresh media are prepared, important factors must be considered, including pH, osmotic pressure, and oxygen requirements. A culture of microbes modifies its environment by producing waste products as it grows on agar or in broth. The accumulation of these products greatly

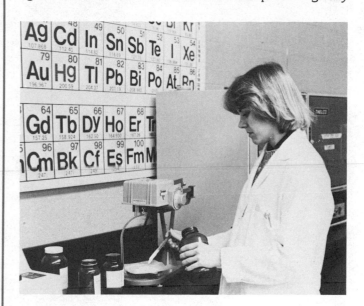

FIGURE 9.12
Dehydrated culture medium is weighed and mixed with water. The solution is then poured into test tubes for autoclaving. (Courtesy of Daria Smith.)

TABLE 9.5
The beginnings of dehydrated culture media. The article printed is an abstract of a paper first presented in 1909. This paper marked the beginning of a large industry centered on the production of microbiological culture media. It also marked the beginning of a period in history that saw the expansion of many sciences related to the study of cells, including bacteriology, histology, cytology, protozoology, and others.

DESICCATED CULTURE MEDIA

W. D. FROST

University of Wisconsin

Abstract of Paper at Boston (1909) Meeting of the Society of American Bacteriologists. *Science,* 31:555: (Apr. 8) 1910.

In order to overcome the generally recognized faults of bacterial culture media, such as variation in composition of small batches, time consumed in preparation, rapidity with which it deteriorates, its unavailability in small institutions or private practice, the preparation of culture media in large batches in establishments especially equipped for it and then desiccated is suggested.

The author's work on this problem, covering nearly a decade of time, is considered and samples are submitted.

There is apparently no reason why the different culture media cannot be put on the market in the form which requires merely the addition of water and sterilization to make it ready for use. Not only the ordinary, but probably most of the special media, can be prepared in this way and could be put up where desired, in the form of tablets, these to be of such size that they could be put directly in test tubes and when the proper amount of water is added they would be ready for sterilization and use.

CLINICAL microbiology labs handle hundreds of specimens daily. Each specimen may contain thousands of individual microbes of a number of different species. Obtaining accurate information about the nature of the pathogenic microbe depends on the very diligent efforts of a well-trained medical technologist. This person must set up and perform many special identification procedures that take into account the peculiarities of each species and culture test. The technologist must quickly manage and accurately interpret growth characteristics and biochemical tests. Preparing culture media, sterilizing glassware, autoclaving old cultures and new media take a great deal of valuable laboratory time that could otherwise be spent in handling and researching microbes. The time required for the identification of a bacterium has been greatly reduced by the use of any one of a number of commercially available "quick identification" procedures. All the methods are based on standardized tests and culturing procedures, which have been modified to speed the identification process.

API SYSTEM

Analytab Products Incorporated has developed an identification technique known as the API enteric system. This procedure enables the technician to perform 22 biochemical tests on a single bacterial colony in only 18 to 24 hours. The colony is transferred to distilled water and inoculated into the series of small plastic capsules on the API card. Each capsule contains special growth media used in the identification process. The card is placed in a special plastic tray, water is added to provide humidity, and the tray is incubated for 18 hours at 37°C. Changes in the colors of the capsules indicate positive or negative test results used to identify a Gram-negative enteric bacterium.

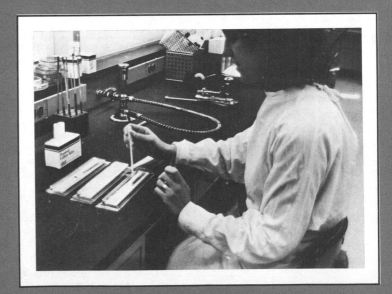

BOX 17
QUICK
METHODS
OF
IDENTIFICATION

ENTEROTUBE® II

This is a prepared, sterile multimedia tube used for the rapid differential identification of the Gram-negative bacteria in the family Enterobacteriaceae. The clear plastic culture tube contains twelve compartments of differential media and an enclosed inoculating needle. Fifteen standard biochemical tests may be run using

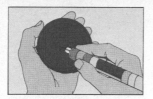

Remove both caps. Do not flame wire.

Pick: a well isolated colony directly with the tip of the ENTEROTUBE II inoculating wire. A visible inoculum should be seen at the tip and the side of the wire. Avoid touching agar with wire. Utilize one or more ENTEROTUBE II to pick additional colonies as experience dictates.

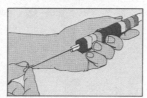

Inoculate: ENTEROTUBE II by first twisting wire, then withdrawing wire through all twelve compartments using a turning motion.

Reinsert: wire (without sterilizing) into ENTEROTUBE II, using a turning motion, through all 12 compartments. Withdraw wire until the tip is in H_2S/indole compartment. *Break wire* at notch by bending, discard handle and replace caps on tube loosely. The portion of the wire remaining in the tube maintains anaerobic conditions necessary for true fermentation of glucose, production of gas and decarboxylation of lysine and ornithine. The part of the wire in the H_2S/indole compartment will not interfere with these tests.

Strip off: blue tape after inoculation—but before incubation—to provide aerobic conditions in adonitol, lactose, arabinose, sorbitol, Voges-Proskauer, dulcitol phenylalanine, urea and citrate compartments. *Slide clear band* over glucose compartment to contain any small amount of sterile wax that may escape due to excessive gas produced by some bacteria. Incubate at 35 to 37 C for 18 to 24 hours with ENTEROTUBE II lying on its flat surface. Separate each ENTEROTUBE II slightly to allow for sufficient air circulation.

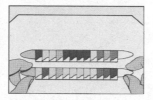

Interpret: and record all reactions with exception of indole and Voges-Proskauer. (For complete instructions on how to read results of ENTEROTUBE II see Results section.) All other tests must be read before the indole and Voges-Proskauer tests are performed as these may alter the remainder of the ENTEROTUBE II reactions.

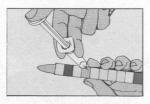

To perform indole test: place ENTEROTUBE II in rack with glucose compartment pointing downward or place horizontally and add 1 or 2 drops of Kovacs' reagent through plastic film of H_2S/indole compartment using either a needle and syringe or by melting a small hole using a warm inoculating needle or disposable pipette and dropping reagent into compartment. Allow reagent to contact the surface of the medium or the inner surface of the plastic film. A positive test is indicated by development of a red color in the added reagent on surface of media or mylar film within 10 seconds.

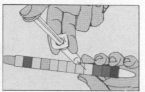

To perform Voges-Proskauer test: as for the indole test, place the ENTEROTUBE II in rack with glucose compartment pointing downward or place horizontally and add two drops of 20% potassium hydroxide solution containing 0.3% creatine and three drops of 5% alpha-naphthol in absolute ethyl alcohol. A positive test is indicated by the development of a red color within 20 minutes.

After reading dispose of ENTEROTUBE II by autoclaving.

Courtesy of Roche Diagnostics.

this system: (1) glucose fermentation, (2) gas production, (3) lysine decarboxylase production, (4) H_2S production, (5) indole formation, (6) lactose fermentation, (7) acetylmethylcarbinol formation, (8) citrate ultilization, (9) urea digestion, (10) dulcitol, (11) phenylalanine deaminase formation, (12) sorbitol fermentation, (13) arabinose fermentation, (14) adonitol fermentation, and (15) ornithine decarboxylase production.

AUTOMATED MICROBIOLOGICAL SYSTEMS

Several corporations market automated bacterial identification systems, each with its own set of unique features. Most are designed to handle over 200 specimens at any one time. Specimens are placed in special compartments of the machine for incubation at 35°C. The train of compartments is moved past a solid-state optics system that operates much like the photocolorimeter described earlier (see Chapter 8, Growth and Reproduction). As the culture passes through the light, the degree of light transmission and scattering are recorded in an on-board microcomputer. The recorded changes in light transmission and scattering over the incubation period (about 13 hours) are used to determine whether the microbe has been able to grow in a particular medium. The system may also be adapted for determining qualitative and quantitative (MIC) antimicrobial susceptibility results in about five hours. This is one day sooner than the conventional tube dilution method. The results are printed out (identification and test data), when enough bacterial growth triggers the mechanism.

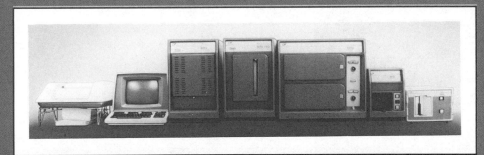

Courtesy of Vitek Systems, Inc.

1 What is catalase?
2 Of what value is the identification of coagulase in a bacterial culture?
3 What is the API system?

modifies the medium in which the microbe is growing and limits the growth of the population. For example, fermenting bacteria growing on glucose media produce large amounts of organic acid molecules. As the concentration of the hydrogen ions increases, the bacteria are inhibited and population growth will slow and eventually stop. In other cases, a culture of microbes may metabolize media ingredients forming alkaline, or basic, ions. The decomposition of proteins and amino acids in the medium makes the environment more alkaline and inhibits population growth. In order to overcome this problem, special chemicals are added to culture media to prevent drastic changes in the hydrogen ion concentration as the population grows. These chemicals are called **buffers.** The most widely used buffers in microbiology are the phosphates (K_2HPO_4 and KH_2PO_4). These control the pH of the media by capturing excess hydrogen ions as they are released during microbial growth or by releasing hydrogen ions to neutralize the effects of an increase in pH. By adjusting the amount of these phosphates in the media, a pH ranging from approximately 6.0 to 7.6 can be maintained. Many other inorganic agents have been tried as buffers; however, the phosphates are the most successful since they are nontoxic and may be metabolized as a source of phosphorus. They also buffer in the neutral range, a pH range that should be maintained when cultivating most pathogenic microbes.

The successful growth of a microbe also depends on how fast water enters and leaves the cell. If the medium contains a lower concentration of materials than is found inside the cell, the surrounding solution is said to be **hypotonic** ("hypo" means *below,* "tonic" means *tone*) (Figure 9.13). Cells placed in a hypotonic solution will tend to gain water from the environment by osmosis and swell. In most situations, this will not harm the microbes because the rigid cell wall will prevent the cell from lysing. Microbes that lack a wall are more likely to be lysed in a hypotonic environment and must be protected if they are to be grown in a culture medium. The mycoplasma (PPLOs) are microbes that are genetically unable to produce a cell wall and must have the tone, or **osmotic pressure,** of their environment regulated if they are to be cultured. This is accomplished by adding about 5 percent salt (NaCl) to mycoplasma medium. A solution that contains the same amount of dissolved materials that is found inside the cells is called an **isotonic solution** ("iso" means *equal*). Cells placed in an isotonic solution do not swell or shrink. However, shrinkage, or

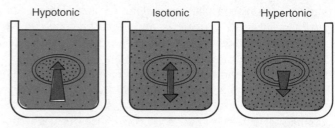

FIGURE 9.13
Osmotic pressure is the amount of pressure that must be placed on a solution to prevent the movement of water into a cell across a selectively permeable membrane. By measuring this pressure, it is possible to determine the relative concentration of dissolved materials in the surrounding medium in comparison to the concentration inside the cell. In a hypotonic solution, the water tends to move into the cell faster than it moves out; in hypertonic solutions the water tends to move out of the cell faster than it moves in; and in isotonic solutions the water moves into the cell at the same rate that it moves out.

plasmolysis, will occur if the "tone" of the surrounding medium is greater than that found in the cells. Solutions that contain a greater amount of dissolved materials than is found in the cells are called **hypertonic** ("hyper" means *above*). Hypertonic media can be made isotonic by diluting with water. By balancing the osmotic environment, the cells are able to metabolize the available nutrients efficiently. Another important factor necessary to attain a balanced culture is the oxygen requirements of microbes.

1 Why are buffers necessary additions to culture media?
2 What is the difference between a hypotonic and hypertonic solution?

OXYGEN REQUIREMENTS FOR CULTURING

Some of the microbes that cause infections are normally found on the human body. Specimens sent to the lab for identification will in all likelihood contain *Staphylococcus aureus, Streptococcus pyogenes,* coliforms (e.g., *E. coli* and *Enterobacter*), *Pseudomonas* spp., and *Proteus* spp.

Most of these microbes are aerobic and may be grown on petri plates, agar slants, and in broth culture. However, a number of other microbes which might be found on the surface of the body or in the intestinal, respiratory, and genitourinary tracts may function anaerobically. Microbes in this category include members of the genera *Clostridium, Bac-*

teroides, *Yersinia* (facultative), *Salmonella* (facultative), *Shigella* (facultative), *Peptococcus*, and *Peptostreptococcus*.

When resistance to infection is lowered or microbes become displaced, these microbes are able to cause severe and possibly fatal infections. In order to isolate and identify anaerobic pathogens, special lab techniques and procedures must be followed. As with any culture, successful identification begins with specimen collection. Anaerobes are suspected in infections that are foul smelling, such as gas gangrene. When a specimen is taken, it is sent to the lab in a special "gassed out" collection tube (Figure 9.14). This protects the suspected anaerobe from the deadly effects of atmospheric oxygen. To prepare a "gassed out" tube, the atmospheric oxygen normally found in a tube is replaced with CO_2 and stoppered before autoclaving. Swab and body fluid specimens are quickly placed in these special tubes to prevent the loss of CO_2 and the entrance of O_2. (Since CO_2 is a heavy gas, it will stay in the tube unless it is "poured out" by tipping the tube.) If these special tubes are not available, the specimen should *not* be refrigerated but inoculated into the proper culture media as soon as possible. Special media such as semisolid sodium thioglycolate may be used to create the necessary reducing conditions (thioglycolate removes oxygen from the medium), or the cultures may be placed in special containers that are kept oxygen-free. Anaerobic containers used in the lab are the Brewer's plate, spray plate, Brewer jar, and GasPak anaerobic jar (Figure

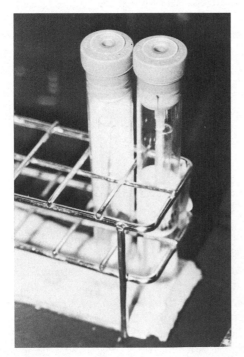

FIGURE 9.14
These special "gassed out" collection tubes have had their oxygen removed so that anaerobic bacteria can successfully be transported to the laboratory for anaerobic culturing.

9.15). The simplest of these is the **Brewer's plate.** This petri plate is filled with sodium thioglycolate agar and inoculated with the anaerobic specimen. The lid is designed with a deep lip to make contact with the surface of the agar when it is placed on the plate. This prevents the outside oxygen-containing air from making contact with the center portion of the culture. A **spray plate** looks like a petri dish with an exceptionally deep base divided into two compartments. Potassium hydroxide (KOH) is placed in one side and the chemical pyrogallol is placed in the other. The inoculated agar-containing lid is placed on top and sealed. By shaking gently, the KOH and pyrogallol are mixed in the base to produce a reaction that will remove oxygen from the atmostphere in the jar. The anaerobic bacteria will then be able to grow on the inverted agar surface. Plates incubated upside-down like this have an advantage in that the moisture that condenses in the plate will form on the upper surface. Since the plate is inverted, the culture will be "watered" automatically.

The **Brewer jar** is more complex and dangerous than either the Brewer's plate or spray plate. After the cultures have been placed inside and the lid sealed, a mixture of H_2, CO_2, and N_2 is flushed through the Brewer anaeobic jar. This exchange of gases is repeated two or three times before the gas tube is shut off. The oxygen remaining in the jar is then removed by electrically heating a catalyst in the lid. The catalyst is allowed to operate for about ten minutes before the entire jar is unplugged and placed in an incubator. A more practical and convenient method for culturing anaerobes is the **GasPak anaerobic jar** (Figure 9.16). This is basically the same as a Brewer jar; however, the anaerobic atmosphere is not generated in the same manner. A disposable envelope, which generates hydrogen and carbon dioxide, is torn open and placed inside. The hydrogen gas produced in the jar reacts with the oxygen on the surface of a catalyst to form water. As the oxygen disappears from the atmosphere inside the jar, a mist or fog will appear on the inner wall of the jar and the lid will warm. When the oxygen has been eliminated, a disposable anaerobic indicator inside the jar will change color. Anaerobic conditions should be generated in about 25 minutes, and the jar may then be placed in an incubator.

Some pathogenic bacteria are not anaerobic, but do require a reduced amount of oxygen in their growth environment. In order to successfully culture microbes such as mycobacteria and *Neisseria* in the lab, a modified environment must be established. These bacteria do not require anaerobic conditions but grow best when the atmosphere surrounding them has been enriched with about 5 percent carbon dioxide. This environment can be produced in the lab by using a candle jar, a biobag environmental chamber, or a carbon dioxide incubator (Figure 9.17). Any wide-mouth jar can be used as a candle jar. The cultures are placed inside with a lighted candle and the lid is sealed. As the

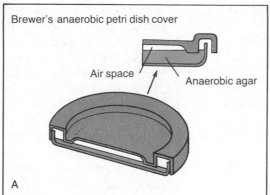

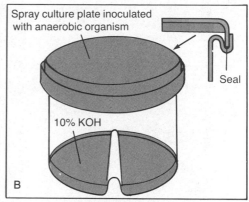

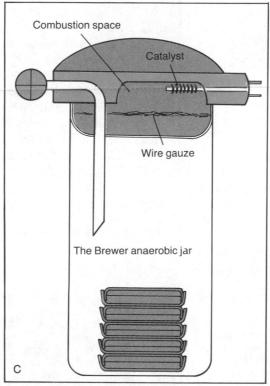

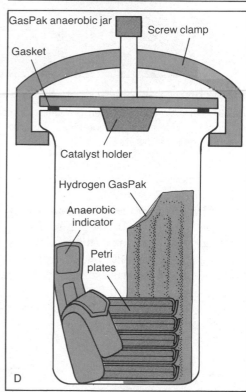

FIGURE 9.15
Several kinds of petri dishes and containers may be used to cultivate anaerobic microbes isolated from specimens. (A) Brewer's dish. (B) Spray dish. (C) Brewer jar. (D) GasPak anaerobic jar.

candle burns, the oxygen content is reduced from about 20 percent to 16 percent, and the CO_2 content is increased from about 0.4 percent to 4.0 percent, generating a favorable growth environment. The biobag environmental chamber method is similar to the candle jar; however, the decrease in oxygen and the increase in carbon dioxide is the result of the rapid aerobic metabolism of another plate of bacteria cultured along with the mycobacteria. Usually only one culture plate is needed for about a dozen mycobacteria plates. The aerobe utilizes the O_2 from inside the bag for its own respiration and releases CO_2 into the bag.

Carbon dioxide incubators are usually small units that have been specially adapted to maintain a constant temperature and CO_2 level. Cultures can be placed directly inside, and no special techniques are needed. After the suspected pathogen has been isolated, cultured, and identified by means

of these special techniques and equipment, the technician may then be asked to perform the antibiotic susceptibility test, one of the most important tests leading to the control of the infection.

CULTURE AND SUSCEPTIBILITY

Many microbiologists believe that the culture and susceptibility is the most important clinical test performed, because the results will provide the physician with significant information about the pathogen and how to control it. A **culture and susceptibility test** is the growth of a known pathogen in pure culture and its exposure to numerous antibiotics of different concentrations to determine which drug

FIGURE 9.16
GasPak anaerobic jars are anaerobic culture containers commonly used in medical laboratories. (Courtesy of Saginaw Medical Center, L. DeGuise.)

will kill or inhibit growth of the microbe. The smallest amount of the drug that prevents the microbe's growth is called the **minimal inhibitory concentration (MIC).** The drug with the lowest MIC is usually the one selected by the physician for use in treatment. The culture and susceptibility test, and determination of the MIC have become greatly important due to the increasing number of antimicrobial agents avail-

able to physicians, and because of an increase in the number and types of pathogens that have become resistant (unaffected) to these drugs. Two methods commonly carried out to determine the susceptibility of a microbe to an antimicrobial agent are the tube dilution method and the agar diffusion (disk susceptibility) method.

The **tube dilution method** tests the susceptibility of a bacterium to antibiotics and is performed by inoculating the bacterium into a serial dilution of the antibiotic (Figure 9.18). Each tube contains an appropriate culture medium and a specific concentration of the antibiotic. If the bacterium is inhibited by the drug, there will be no growth in the tube and the medium will remain clear. If the microbe is resistant to the effects of the drug, it will grow and produce a cloudy tube. Several antibiotics are compared using this method in order to determine which inhibits the bacterium in the lowest concentration. If more than one drug successfully controls growth, all the inhibiting drugs' names are sent to the physician for consideration. This testing procedure may be used to determine the susceptibility of microbes from blood cultures, the susceptibility of patients who have undergone a relapse during therapy, and the susceptibility of patients who have not responded to the use of an antimicrobial drug. This method is time-consuming and becomes fairly expensive when testing a number of antibiotics. Therefore, most labs are prepared to run a tube dilution series, but prefer the agar diffusion (disk susceptiblity) method.

The **disk susceptibility** (sensitivity) method is probably the most widely used and quickest susceptibility test used in hospital labs. In this method, the medium (usually

FIGURE 9.17
Some pathogens prefer to grow in environments that have lower oxygen levels and higher carbon dioxide levels than found in the atmosphere. The three methods pictured here are all able to maintain this special growth environment. (A) A candle jar. (B) Biobag environmental chamber. (C) Carbon dioxide incubator. (Photo courtesy of Saginaw Medical Center, L. DeGuise.)

FIGURE 9.18

In a culture and susceptibility test, a dispenser is used to correctly place antibiotic-containing paper disks onto the surface of a bacterial "lawn" made from bacteria obtained from a patient. The zones of clearing that result after incubation are measured to determine the relative effectiveness of the drugs. (Courtesy of Saginaw Medical Center, L. DeGuise.)

Mueller-Hinton) is inoculated by swabbing the bacteria over the entire surface so that they will grow into a "lawn." Small disks of paper impregnated with various antibiotics of different concentrations are placed on the surface of the plate. The plate is incubated overnight and checked the following day for growth inhibition. During the incubation period, the antibiotic molecules diffuse from the disks over the surface of the bacteria-coated agar. If the concentration of an antibiotic is strong enough to inhibit the bacteria's growth, there will be a zone of inhibition around the disk impregnated with that drug. If the bacteria are resistant to the drug, there will be little or no zone of clearing and the bacteria grow close to the disk. The diameter of the inhibition zone designates the relative effectiveness of the drug or the bacteria's susceptibility to the antibiotic. The zone is measured from the edge of the antibiotic disk to the edge of the cleared area that shows no visible growth.* The size is then compared to a table that has been developed from disk susceptibility tests run on known *E. coli* and *Staphylococcus aureus* cultures (Table 9.6). This is known as the Kirby-Bauer technique. The table lists the antibiotic agents, concentrations used in the disks, and the inhibition zone diameters measured from the standard *E. coli* and *S. aureus*. The inhibition zones are divided into three ranges: resistant, intermediate, and susceptible. Disk susceptibles that show clearing only to the resistant level have not been inhibited by the

---

*Also note that the size of the zone of clearing is related to the diffusion rate of the antibiotics. Since antibiotics have different molecular weights, they do not diffuse as the same rates. Therefore, a simple comparison of the zones of clearing will not be adequate unless the molecular weights of the antibiotics are taken into consideration.

1 Name two species of bacteria that can be cultured in an anaerobic environment.
2 What is meant by the minimal inhibitor concentration of an antibiotic?
3 Why is it necessary to use *E. coli* and *S. aureus* as bases of comparison when doing a culture and susceptibility test?

drug. Drugs that have cleared to the intermediate range may be used to control the pathogen, but may be required in high dosages. If this is the case, another drug may better be used. Drugs that have resulted in large zones of clearing on the agar surface and fall in the susceptible range are those that may be expected to inhibit the pathogen in the patient. They may be recommended for use in controlling an infection caused by the bacterium being tested. The disk susceptibility method is quick and accurate, but there are several sources of error in the test. These possible errors make it essential that a qualified technician perform and interpret the results:

1 Inoculum density not carefully standardized: larger zones of inhibition may result with a light inoculum, smaller zones with a more dense inoculum.

2 Using a mixed culture: the standardized test is based on the use of pure cultures, a mixed culture can yield completely erroneous susceptibility results.

3 Excessive moisture on the surface of the medium may yield very confluent growth and reduce the diameter of the inhibition zone even though the density of inoculum is correct.

4 A very dry surface may result in poor growth, yielding a larger inhibition zone.

5 Mishandling of susceptibility disks by permitting moisture uptake and prolonged exposure to room temperature results in deterioration of most antibiotics.*

All of the information gained from culturing microbes is sent to the physician. The physician then interprets this information and decides on an appropriate course of treatment in light of the patient's condition (both mental and physical), the available health care facilities, chemotherapeutic drugs, and the time needed for treatment. In order to ensure successful treatment of the patient and containment of the infection, the lab must perform another essential service known as **sterility testing.**

Hospitals use a great many sterile materials in the care and treatment of patients. Bandages, dressings, hypodermic

---

*Bacto-Sensitivity Disks, Difco Laboratories.

TABLE 9.6

This table illustrates the relative effectiveness of antibiotics using the disk susceptibility method. (Permission has been granted to use portions of the text of the Second Informational Supplement (M2-A2S2) to the "Performance Standard for Antimicrobial Disk Susceptibility Tests, Second Edition," by the National Committee for Clinical Laboratory Standards. NCCLS is not responsible for errors or inaccuracies. Copies of the complete standard may be obtained from NCCLS, 771 E. Lancaster Ave., Villanova, Pa., 19085.)

Antimicrobial Agent	Disc Content	Resistant	Zone Diameter, nearest whole mm intermediate[b]	Susceptible	Approximate MIC Correlates[a] Resistant	Susceptible
Amikacin[c]	30µg	≤14	15-16	≥17	≥32µg/mL	≤16µg/mL
Ampicillin when testing gram-negative enteric organisms and enterococci[d]	10µg	≤11	12-13	≥14	≥32µg/mL	≤8µg/mL
when testing staphylococci and penicillin G-susceptible microorganisms[d,e]	10µg	≤20	21-28	≥29	β-lactamase[e]	≤0.25µg/mL
when testing Haemophilus species[d,f]	10µg	≤19	—	≥20	≥4µg/mL	≤2µg/mL
Bacitracin	10 units	≤8	9-12	≥13	—	—
Carbenicillin when testing the Enterobacteriaceae[e]	100µg	≤17	18-22	≥23	≥32µg/mL	≤16µg/mL
when testing Pseudomonas aeruginosa	100µg	≤13	14-16	≥17	≥512µg/mL	≤128µg/mL
Cefamandole[g]	30µg	≤14	15-17	≥18	≥32µg/mL	≤8µg/mL
Cefoperazone[g]		Not established as of date of printing				
Cefotaxime[g]	30µg	≤14	15-22	≥23	≥64µg/mL	≤8µg/mL
Cefoxitin[g]	30µg	≤14	15-17	≥18	≥32µg/mL	≤8µg/mL
Cephalothin[g,h]	30µg	≤14	15-17	≥18	≥32µg/mL	≤8µg/mL
Chloramphenicol	30µg	≤12	13-17	≥18	≥25µg/mL	≤12.5µg/mL
Clindamycin[i]	2µg	≤14	15-16	≥17	≥2µg/mL	≤1µg/mL
Colistin[k]	10µg	≤8	9-10	≥11	≥4µg/mL	k
Doxycycline	30µg	≤12	13-15	≥16	≥16µg/mL	≤4µg/mL
Erythromycin	15µg	≤13	14-17	≥18	≥8µg/mL	≤2µg/mL
Gentamicin[c]	10µg	≤12	13-14	≥15	≥8µg/mL	≤4µg/mL
Kanamycin	30µg	≤13	14-17	≥18	≥25µg/mL	≤6µg/mL
Methicillin when testing staphylococci[l]	5µg	≤9	10-13	≥14	≥16µg/mL	≤4µg/mL
Mezlocillin	75µg	≤12	13-15	≥16	≥256µg/mL	≤64µg/mL
Minocycline	30µg	≤14	15-18	≥19	≥16µg/mL	≤4µg/mL
Moxalactam[g]	30µg	≤14	15-22[b]	≥23	≥64µg/mL	≤8µg/mL
Nafcillin when testing staphylococci[l]	1µg	≤10	11-12	≥13	≥8µg/mL	≤2µg/mL
Nalidixic Acid[m]	30µg	≤13	14-18	≥19	≥32µg/mL	≤12µg/mL
Neomycin	30µg	≤12	13-16	≥17	—	—
Netilmicin	30µg	≤13	14-16	≥17	≥32µg/mL	≤8µg/mL
Nitrofurantoin[m]	300µg	≤14	15-16	≥17	≥100µg/mL	≤25µg/mL
Oxacillin when testing staphylococci[l]	1µg	≤10	11-12	≥13	≥8µg/mL	≤2µg/mL
when testing pneumococci for penicillin susceptibility[n]	1µg	≤12	13-19	≥20	—	≤0.06µg/mL
Penicillin G when testing staphylococci[e]	10 units	≤20	21-28	≥29	β-lactamase[e]	≤0.1µg/mL
when testing other microorganisms[o]	10 units	≤11	12-21	≥22	≥32µg/mL	≤2µg/mL
when testing N. gonorrhoeae[j]	10 units	≤19	—	≥20	β-lactamase	≤0.1µg/mL
Piperacillin	100µg	≤14	15-17	≥18	≥256µg/mL	≤64µg/mL
Polymyxin B[k]	300 units	≤8	9-11	≥12	≥50 units/mL	k
Streptomycin	10µg	≤11	12-14	≥15	—	—
Sulfonamides[m,p]	250 or 300µg	≤12	13-16	≥17	≥350µg/mL	≤100µg/mL
Tetracycline[q]	30µg	≤14	15-18	≥19	≥12µg/mL	≤4µg/mL
Ticarcillin when testing P. aeruginosa	75µg	≤11	12-14	≥15	≥128µg/mL	≤64µg/mL
Trimethoprim[m,p]	5µg	≤10	11-15	≥16	≥16µg/mL	≤4µg/mL
Trimethoprim-sulfamethoxazole[m,p]	1.25µg, 23.75µg	≤10	11-15	≥16	≥8/152µg/mL	≤2/38µg/mL
Tobramycin[c]	10µg	≤12	13-14	≥15	≥8µg/mL	≤4µg/mL
Vancomycin	30µg	≤9	10-11	≥12	—	≤5µg/mL

a. These correlates are not meant for use as breakpoints for susceptibility categorization with dilution MIC tests as described in NCCLS M7-P.

b. The category "intermediate" should be reported. Infections with bacteria of "intermediate" susceptibility may be considered moderately susceptible and may respond clinically or bacteriologically to antimicrobial agents having a wide safe dosage range.

c. The zone sizes obtained with aminoglycosides, particularly when testing P. aeruginosa, are very medium dependent because of variations in divalent cation content. These interpretive standards are to be used only with Mueller-Hinton medium that has yielded zone diameters within the correct range shown in Table 3 when performance tests were done with P. aeruginosa ATCC 27853. Organisms in the intermediate category may be either susceptible or resistant when tested by dilution methods and should therefore more properly be classified as "indeterminant" in their susceptibility.

d. Class disc for ampicillin, amoxicillin, bacampicillin, cyclacillin, and hetacillin.

e. Resistant strains of S. aureus produce β-lactamase and the testing of the 10 unit penicillin disc is preferred. Penicillin G should be used to test the susceptibility of all penicillinase-sensitive penicillins, such as ampicillin, amoxicillin, bacampicillin, heta-cillin, carbenicillin, mezlocillin, piperacillin, and ticarcillin. Results may also be applied to phenoxymethyl penicillin or phenethicillin. The intermediate category contains penicillinase producing isolates and those strains should be considered resistant to therapy.

f. For testing Haemophilus use Mueller-Hinton agar supplemented with 1% hemoglobin (or 5% horse blood, chocolate) and 1% IsoVitaleX (BBL), Supplement VX (Difco) or an equivalent synthetic supplement. Adjust pH to 7.2. Prepare the inoculum by suspending growth from a 24-hour chocolate agar plate in Mueller-Hinton broth to the density of a turbidity standard. The vast majority of ampicillin-resistant strains of Haemophilus produce detectable β-lactamase.

g. **Cefamandole, cefoxitin, cefotaxime and moxalactam are recently released beta-lactams having a wider spectrum of activity against gram-negative bacilli than do other previously approved cephalosporins. Therefore, the cephalothin disc cannot be used as the class disc for these drugs.**

h. The cephalothin disc is used for testing susceptibility to cephalothin, cefaclor, cefadroxil, cefazolin, cephalexin, cephaloridine, cephapirin, and cephradine. Cefamandole, cefoxitin and cefotaxime or moxalactam must be tested separately. S. aureus exhibiting resistance to methicillin, nafcil-

lin or oxacillin discs should be reported as resistant to cephalosporin-type antimicrobics, regardless of zone diameter, because in most cases infections caused by these organisms are clinically resistant to cephalosporins. Methicillin-resistant S. epidermidis infections also may not respond to cephalosporins.

i. The clindamycin disc is used for testing susceptibility to both clindamycin and lincomycin.

k. Colistin and polymyxin B diffuse poorly in agar, and the diffusion method is thus less accurate. Resistance is always significant, but when treatment of infections caused by a susceptible strain is being considered, results of a diffusion test should be confirmed with a dilution method. MIC correlates cannot be calculated reliably from regression analysis.

l. Of the antistaphylococcal β-lactamase resistant penicillins, either oxacillin, nafcillin, or methicillin may be tested, and results can be applied to the other two of these drugs and to cloxacillin and dicloxacillin. Oxacillin is preferred due to more resistance to degradation in storage and its application to pneumococcal testing. Cloxacillin discs should not be used because they may not detect methicillin-resistant S. aureus. When an intermediate result is obtained with S. aureus, the strains should be further investigated to determine if they are heteroresistant.

m. Susceptibility data for nalidixic acid, nitrofurantoin sulfonamides and trimethoprim apply only to organisms isolated from urinary-tract infections.

n. The interpretation of penicillin susceptibility using the 1 µg oxacillin disc is as follows: ≥20 mm = susceptible, ≤12 mm = resistant or relatively resistant, and those strains with zones between 13 and 19 mm (rare) should be repeated preferably by another test method. Correlative MIC values listed in the table are for penicillin.

o. Intermediate category includes enterococci, and certain gram-negative bacilli that may cause systemic infections treatable with high parenteral dosages of penicillin but not of orally administered phenoxymethyl penicillin or phenethicillin.

p. The sulfisoxazole discs can be used for any of the commercially available sulfonamides. Blood-containing media, except media containing lysed horse blood, are not satisfactory for testing sulfonamides. The Mueller-Hinton agar should be as thymidine-free as possible for sulfonamide and/or trimethoprim testing. (See footnote e, Table 3.)

q. Tetracycline is the class disc for all tetracyclines, and the results can be applied to chlortetracycline, demeclocycline, doxycycline, methacycline, minocycline, and oxytetracycline. However, certain organisms may be more susceptible to doxycycline and minocycline than to tetracycline. (See footnote h, Table 1.)

NCCLS Vol. 2 No. 2

 From M2-A2 S Second Informational Supplement Performance Standards for Antimicrobic Disc Susceptibility Tests
Published March 1982

63

needles and syringes, glassware, scissors, catheters, and surgical instruments must all be checked for contamination by bacteria, fungi, and yeasts before they are used. Usually a sample from a batch of sterilized instruments is taken to the lab for testing. If the sample is found to be free of all living things, the entire batch is considered sterile and safe for use. The item to be sampled is opened using proper aseptic technique and sampled either with a sterile moist swab or with a RODAC plate. (RODAC is an acronym for Replicate Organism Detecting and Counting.) Swabs are typically used to sample hard surfaces such as utensils, glassware, or instruments while the RODAC plates are used to sample linens, gauze, and flat surfaces. The RODAC plate is a small petri dish filled with so much culture media that it can be pressed directly onto the surface of the material being tested. The plate is incubated for forty-eight hours after which the colonies may be counted and identified. Another technique is used to check the efficient operation of the autoclave and other hospital sterilization equipment. Special paper strips containing endospores of the bacteria *Bacillus stearothermophilus* and *Bacillus subtilis* are placed in the autoclave with the material to be sterilized. After the cycle has been completed and the material removed, the bacterial test strip is placed in a tube of culture medium and incubated. If the tube becomes cloudy with growth, the sterilization process was not successful.

SUMMARY

The microbiology lab performs many services, all related to the culturing of microbes. These cultures are grown in order to isolate and identify a microbe, perform population counts, do serological tests, determine the susceptibility of a microbe to an antimicrobial agent, produce products such as vaccines and antitoxins, check for the sterility of materials, and maintain stock cultures for controls and teaching. The success of the lab is dependent on the specimen taken from the infected patient. Specimen collection must take into account the type of material needed, the method of collection, aseptic technique, and the time the specimen is taken and how quickly it is sent to the lab for examination. Knowing the clinical stages of infection is of great help in determining when a specimen should be taken. The four stages include the incubation period, the prodromal period, the acute phase, and the convalescent phase. Several methods may be used to take specimens. Swabs may be used for specimens from the eye, ear, nose, throat, or fecal material. Needle aspiration may be used for blood and spinal fluid, and intubation may be used to obtain specimens of bile or peritoneal fluids. Catheters are tubes that are used for collecting urine specimens. Indwelling, or Foley, catheters must be watched carefully since their use may result in a urinary tract infection.

After the specimen has been taken and labeled properly, it is sent to the lab for identification. The medium used to culture the microbe may be either synthetic or complex depending on the nature of the ingredients used in its preparation. Four types of media are used in the lab. Enrichment media are used to encourage the growth of a desired microbe in a mixed culture. Selective media are used to inhibit the growth of undesirable microbes, and differential media are used to show the differences among microbes. There are many types of differential media, and they are used in combination to identify the microbe.

The IMViC test is an important series of tests performed in the lab to differentiate among members of the bacterial family Enterobacteriaceae. Propagation media are used to maintain stock cultures in the lab for controls and teaching. Careful attention must be paid in the preparation of all media to ensure that the proper pH, osmotic balance, and oxygen requirements are met.

After the microbe has been isolated, identified, and cultured, samples are used to perform a culture and susceptibility test. This test demonstrates which antibiotic is able to inhibit the pathogen's growth. The culture and susceptibility test may be done using either the tube dilution method or the agar diffusion method. The results are provided to the physician to determine which drug might be most effective in controlling the infection in the patient. Sterility tests are also performed by the lab to be sure that materials used throughout the hospital have been successfully sterilized and their use will not result in the infection of patients by contaminating microbes.

WITH A LITTLE THOUGHT

A patient is admitted to the hospital with an acute disease involving the large intestine. It is characterized by diarrhea, fever, cramps, and often vomiting. The stool displays blood, mucus, and pus. The attending physician orders that a fecal specimen be taken with a swab and sent to the laboratory for culturing and identification. The cause of the disease is suspected of being *Salmonella*, *Entamoeba*, or *Shigella*. Trace the path this specimen will take from the time it is taken until the physician prescribes treatment.

STUDY QUESTIONS

1. What types of specimens might be taken from a patient with an infectious disease? How might they be taken?

2. Why is it so important to send the specimen to the lab as soon as possible? If this is not possible, why would the specimen be refrigerated?

3. Why is it necessary to inoculate a specimen into enrichment medium when it arrives in the lab?

4. Why is it important to know the clinical stages of infection? How does this information relate to lab work?

5. What information might be gained from a serological analysis on blood serum?

6. What is the reason for a high correlation between urinary tract infections and catheterization in women?

7 Of what value is a culture and susceptibility test?

8 What are the differences among hypotonic, isotonic, and hypertonic?

9 What methods might be used to culture anaerobic pathogens?

10 What is the value of sterility testing?

SUGGESTED READINGS

BAILEY, W. R., and E. G. SCOTT. *Diagnostic Microbiology,* 4th ed. C. V. Mosby Co., St. Louis, Mo., 1974.

BARRY, A. et al. *Methods for Quality Control in the Clinical Lab.* University of California-Davis, Sacramento Medical Center, Sacramento, Calif., Aug., 1973.

COLLINS, C. H., and P. M. LYNE. *Microbiological Methods.* Butterworth & Co., London, 1976.

LINDAN R. "The Prevention of Ascending, Catheter-Induced Infections of the Urinary Tract." *Journal of Chronic Diseases* 22:321–32, 1969.

THORNSBERRY, C., and T. M. HAWKINS. *Agar Disc Diffusion Susceptibility Testing Procedure.* U.S. Department of Health, Education, and Welfare. Center for Disease Control, Atlanta. March, 1977.

U.S. Department of Health, Education, and Welfare; Public Health Service; National Center for Disease Control. E wing WH, *Isolation and Identification of* Salmonella and Shigella, July, 1974.

U.S. Department of Health, Education, and Welfare; Public Health Service; National Center for Disease Control. *National Nosocomial Infections Study Quarterly Report, Third Quarter,* 5–8. Summer 1979, issued March, 1982.

KEY TERMS

febrile period (feb'ril)	communicable	selective
chronic	serology	differential
incubation	antibody-antigen	IMViC test
prodromal	titer	culture and susceptibility test
acute	synthetic media	
convalescent	complex media	MIC
malaise (mal'az)	enrichment	

Pronunciation Guide for Organisms

Neisseria meningitidis (men"in-ji-ti-dez)

Coccidioides (kok-sid"e-oi'dez) *immitis* (i'mid-yis)

Klebsiella (kleb"se-el'lah)

Enterobacter (en"ter-o-bak-ter)

Citrobacter (sit"ro-bak-ter)

Proteus (pro'te-us)

Haemophilus (he-mof'i-lus)

Yersinia (yer-sin'e-ah)

Bordetella (bor"de-tel'lah)

Shigella (she-gel'ah)

Hafnia (haf'ne-ah)

Serratia (se-ra'she-ah)

Bacteroides (bak"te-roi'dez)

The Control of Microbes

DIFFERENT APPROACHES TO CONTROL
FACTORS THAT INFLUENCE SUCCESS
SURFACE-ACTIVE AGENTS
THE METHODS AND MATERIALS OF CONTROL
Heat
Drying
Radiation
Sound
Gas
Alcohol
Heavy Metals
pH
Dyes
Aldehydes
Oxidizing Agents
Chemotherapeutic Agents
THE IDEAL ANTIBIOTIC
THE DEVELOPMENT OF DRUG RESISTANCE
OVERCOMING RESISTANT PATHOGENS

Learning Objectives

- ☐ List the five modes of action which inhibit or kill microbes and give examples of each.
- ☐ Understand the death rate of a microbial population and how it occurs under ideal, controlled conditions.
- ☐ Be familiar with the environmental factors which influence the actions of an antimicrobial agent.
- ☐ Know definitions for the terms "surface tension," "surface-active agent," "soap," and "detergent."
- ☐ Name the twelve categories of control methods, give examples of each and their modes of action.
- ☐ Understand the concept of selective toxicity as it relates to chemotherapeutic agents.
- ☐ List the five main areas of microbial cell metabolism inhibited by various antibiotics.
- ☐ Describe the operation of a growth factor analogue.
- ☐ Know examples of the range of activity, affected structures, and possible side effects of some commonly used antibiotics.
- ☐ Be familiar with the concept of drug resistance and how this trait can become predominant in a microbial population.
- ☐ Know methods by which drug-resistant bacteria may be brought under control.

We have become acutely aware of the diversity of microbes in the environment and their impact on our health and well-being. Common observation led prehistoric humans to the practice of drying fish and meat to prevent spoilage. Aristotle (384–322 B.C.) recognized the importance of boiling drinking water and burying excrement to prevent the spread of illness. Today, research has made possible detailed information on the anatomy and physiology of microbes, and their interactions with other organisms. As knowledge has expanded, new ideas have been developed that have led to action. We use the knowledge and ideas gained from research to better control the microbes that influence us. Beneficial microbes have their growth and actions encouraged by providing them with an optimum environment containing nutrients, a source of energy, and needed growth

10

factors. For example, *Streptococcus cremoris* and *Leuconostoc citrovorum* are cultured in raw cream under special conditions to produce cultured sour cream, while some members of the genus *Pediococcus* are used in controlled meat fermentation to produce smoked sausage, summer sausage, pepperoni, and hot bar sausage.

Microbes that are harmful to us or the environment are also controlled. This is accomplished by inhibiting their growth and actions, destroying microbes, or removing them from the environment. There are many antimicrobial agents capable of controlling microbes, including chemicals (disinfectants and sterilizing ethylene oxide gas), light (ultraviolet), and physical agents (heat, detergents, and drying). The choice of a particular antimicrobial depends on the kind of material to be treated (e.g., living or nonliving), the kind of microbe to be controlled (e.g., bacteria or virus), and the environmental conditions at the time of its use (e.g., pH, microbe concentration, temperature). For example, the group of chemical control agents known as chemotherapeutic agents are specially designed to inhibit or destroy microbes on or inside living tissues. However, a chemotherapeutic agent would be a poor choice in attempting to control *Mycobacterium tuberculosis* if it were found as a contaminant on the surface of a stainless steel counter top. A variety of control agents is available, but there is no such thing as an ''ideal'' or ''all purpose'' agent. Therefore, this chapter will explore some of the more important microbial control methods and their uses.

DIFFERENT APPROACHES TO CONTROL

Control agents act in many ways to inhibit or kill unwanted microbes. The modes of action include coagulation, oxidation, cell wall damage, cell membrane damage, and interference with metabolic pathways. In order to maintain life processes, chemical reactions must be carried out in an orderly, rapid fashion. Any major change in the physical arrangement of the cell contents or its components will interfere with the smooth operation of metabolic pathways and may result in cell death. **Coagulation** destroys cells when a control agent comes in contact with the internal contents of the microbe and causes it to separate into a liquid portion and an insoluble mass. This process is usually irreversible and is most effective when cell proteins and nucleic acids are denatured in the process. The moist heat in an autoclave results in the coagulation of cell components.

Oxidation is the breakdown of large molecules into smaller molecules by the loss of electrons. Whole cells or individual enzymes may be destroyed by oxidation with such agents as hydrogen peroxide (H_2O_2), chlorine, dry-heat sterilization, or incineration (Figure 10.1). Oxidation reac-

tions are usually not specific but affect many susceptible cells or their components. A number of other control agents are more selective in their action. Damage to the cell walls, especially Gram-positive bacteria, may be caused by specific enzymes such as lysozyme and certain antibiotics such as the penicillins. Without the mechanical protection and support of a cell wall, the cell will lyse since the concentration of molecules inside a Gram-positive bacterium may be from 5 to 20 percent higher than that found in the surrounding environment. Gram-negative bacteria are less likely to be destroyed by lysis because the osmotic pressure is not as great. Death occurs indirectly because the cell wall has not provided the necessary support for the underlying cell membrane. Other agents operate directly on the structure or function of the cell membrane. Some directly interfere with the enzymatically controlled operations of the membrane (e.g., amphotericin B, an antifungal), while other agents destroy the structure of the membrane (e.g., some detergents) and result in lysis.

Interference with metabolic pathways is accomplished with agents known as **antimetabolites.** These work most effectively by disrupting enzymes of the respiratory pathway or enzymes of synthetic pathways that produce essential cell components. For example, sulfa drugs are antimetabolites that interfere with the normal production of essential folic

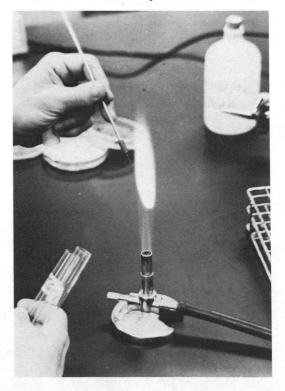

FIGURE 10.1

In the laboratory, the elimination of all life forms on an inoculating loop is guaranteed by holding the wire loop in an open bunsen burner flame until the wire becomes red-hot.

acid molecules in some bacteria. This interference leads to the death of the cell. In addition, an entire class of antibiotics works by injuring or destroying the nucleic aids of microbial cells. This class includes nalidixic acid and rifampicin. These drugs are able to interfere with the normal operation of procaryotic nucleic acids but have no effect on eucaryotes.

The successful use of any microbicide or microbistat (Box 18) depends on a firm understanding of the problem and the possible solutions available. The choice and use of a particular control agent or method must take into consideration such factors as the nature of the microbe to be controlled, temperature of the environment, time of exposure required for control, various environmental factors that may neutralize the desired effects, and the death rate of the microbes.

---

1 What are the major modes of action of microbial destruction and inhibition?

2 On what type of cellular molecule are antimetabolites most effective?

3 What is the difference between a microbicidal and a microbistatic agent?

FACTORS THAT INFLUENCE SUCCESS

The population growth curve helps explain the events that take place as a population increases in size over a period of time (Figure 10.2). It is also useful in understanding how a population dies, either naturally or when exposed to control agents. The death phase occurs as a result of a decrease in essential nutrients and a build-up of harmful wastes. Microbicides have the same negative effect on a population as naturally produced waste products. The main difference between natural death and microbicidal death is the rate at which it occurs. Natural death is usually much slower, while microbicides are deliberately designed to operate rapidly. A population of microbes does not die all at once but has an exponential, or logarithmic, death rate. Under ideal controlled conditions, the population is seen to decrease in numbers of live cells by a constant percentage per unit of time. For example, if 60 percent of a population of 1,000,000 cells is killed two minutes after being exposed to a disinfectant, 600,000 cells will die and 400,000 live microbes will be left. If the death rate is exponential, there will be 240,000 (60% × 400,000) cells killed in the next two minutes and only 160,000 will remain alive. In the next two minutes, 96,000 (60% × 160,000) will die and 64,000 will survive. If the example continues to an end, there will be a point reached when mathematically there is six-tenths of a live cell

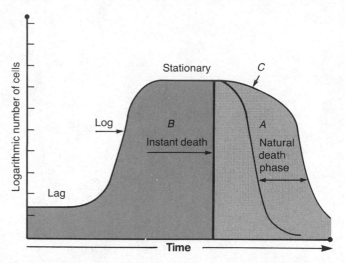

FIGURE 10.2
The decrease in cell numbers in a pure culture of microbes is illustrated in curve A, the natural death phase. All cells are in the same physiological state. The use of a microbicide *never* results in "instant" and complete death as represented by curve B. However, most microbial populations are mixed and are in various physiological states. They are more likely to follow death curve C.

left! This means that it is never certain that all cells have been destroyed. This is even more true when examining a population of microbes that is made up of individuals that are of different ages, physiological conditions, and forms (e.g., vegetative and spore). The death curve in this situation is not a straight line but may curve more gradually, which makes determining the point at which all cells have been killed much more difficult. Since it is very important to be sure that a population of microbes has, in fact, been destroyed, the concentration of microbicide may be increased and the time of exposure lengthened to produce a more rapid and complete kill. If the concentration of a disinfectant is low, the time required to destroy the microbe may be too long to assure the safety of those handling the material.

The effectiveness of a chemical control agent may be increased by raising the temperature of the agent. The increased heat energy results in more rapid destruction by disinfectants and better mechanical removal with soaps and detergents. The higher temperatures allow smaller amounts or lower concentrations of the agent to be used. The time required to remove or kill a population of microbes depends on their original concentration. Materials or surfaces that are heavily contaminated will require more exposure time to ensure proper disinfection. The disinfection of concentrated fecal material contaminated with pathogens, such as *Salmonella*, requires the use of a very effective disinfectant, such as phenol (carbolic acid), that should be kept in contact with the microbes for a long period.

BOX 18
THE
TERMINOLOGY
OF CONTROL

THE topic of controlling microbes requires an understanding of the structural and functional differences among microbes, some basic chemistry and physics, and a working knowledge of a great many technical terms. Over the years, many words have been misused and, as a result, there have been misconceptions about microbial control and the improper use of different agents. Some of these terms have already been introduced and discussed. However, others must be presented and defined now to prevent difficulties in the future.

Antimicrobial Agent Anything that kills or interferes with the multiplication, growth, or activity of microorganisms. The term usually refers to chemical agents such as disinfectants and antibiotics but may also be used to describe such physical agents of control as ultraviolet (UV) light or intense heat.

Antibiotic A chemical produced by a microbe that is able to kill or inhibit the growth or activity of another microbe. These chemicals should be nontoxic to the host, work at very low concentrations, and be nonantigenic.

Antiseptic An agent that opposes sepsis ("sepsis" means the presence of pathogenic microbes or their toxins in tissues). These are usually chemicals that inhibit the growth and activity of microbes and may be used on skin or other tissues. An antiseptic may be the dilute form of a disinfectant. (However, not all disinfectants are the concentrated form of antiseptics.)

Asepsis Any method that ensures freedom from infection by the prevention of contact with a microbe. **Surgical asepsis** is a procedure designed to exclude *all* microbes from the environment, while **medical asepsis** only prevents contact with microbes that cause communicable diseases.

Microbicide An agent that is designed to kill microbes. This is a general term used to describe many different agents such as **bactericides, sporicides, fungicides,** and **virucides.** The suffix "-cide" means *killer* or *"killing."* Microbicide and germicide may be used as synonyms.

Microbistat Any agent that is able to inhibit or halt the action of a microbe but does not necessarily kill the microbe. This is a general term used to describe other more specific agents that are able to induce **bacteriostasis, fungistasis,** or **virustasis.** The suffix "-stasis" means to *inhibit,* or *arrest.*

Disinfectant Any agent, usually a chemical, that is capable of killing microbes, when applied directly to an inanimate (nonliving) object. **Concurrent disinfection** refers to the use of the agent on inanimate articles as soon as possible after the microbes have left the patient's body. **Terminal disinfection** refers to the use of disinfectants on inanimate objects after a patient has either died, been discharged from the hospital, stopped being a source of infection, or been removed from isolation.

Sterilization   Any process that destroys *all* living organisms including the resistant forms of microbes and endospores. Sterile is an unqualified term that means the absence of any living thing; nothing is "almost sterile, partly sterile, or semisterile."

Sanitizer   An agent that is able to reduce the number of contaminating microbes to a safe level. The "safe level" is determined by public health agencies. Sanitation may be accomplished by cleaning with detergents, disinfectants, or antiseptics. This process is usually done on a daily basis to control microbes on eating utensils and food preparation equipment in restaurants, kitchens, and packaging plants.

Chemotherapeutic Agent   A chemical that is used in the treatment of a disease. These agents should harm the infectious agent but not the host. Antibiotics are chemotherapeutic agents.

Fumigant   A gaseous agent used in the killing of microbes or other life forms. The fumigation process is usually carried out in a closed container to confine the deadly fumes.

Personal Hygiene   Personal measures taken to promote health and limit the spread of disease-causing microbes. These measures usually include washing with soap and water, keeping the hands and unclean materials away from the face and mouth, and using a handkerchief or tissue to limit the spread of microbes from the nose and mouth.

All these terms are commonly used in allied health professions including nursing, respiratory therapy, dental hygiene, and surgical technology. Become familiar with these terms so that you will understand and use them properly.

The kind of organism to be controlled should be considered when selecting an antimicrobial. Many microbes vary in their susceptibility to control agents. Endospores of *Bacillus* and *Clostridium* are much more resistant than vegetative cells. A virus that causes hepatitis and the spores of pathogenic fungi are also resistant to the effects of many antimicrobial agents.

The environment surrounding the microbes greatly influences the actions of an antimicrobial. Many chemical control agents are readily inactivated by the presence of excess organic materials. The inhibiting organic molecules may (1) chemically combine with the agent and produce a new molecule that has no effect on the microbe, (2) coat the microbe and protect it from coming in contact with the disinfectant, or (3) physically combine with the agent and prevent it from acting by forming a precipitate, which settles out of solution. In order to prevent this inactivation, all excess organic material should be removed (thoroughly cleaned) before the disinfectant is applied. The inactivation of the antimicrobials may also result if the environment surrounding the microbes is too acid or alkaline. Therefore, it is important to know the pH of the surrounding medium. Some methods of control work very well in neutral solutions (pH 7) while others (moist heat) are more effective in acid environments (pH 5). An adverse pH interferes with the ability of the antimicrobial agent to make effective contact with the microbes. Effective contact between microbe and antimicrobial is essential, since many agents are carried to the microbe in a water solution. The more dense or viscous a material is, the more difficult it will be to penetrate to the microbe. This problem can be reduced by lowering the surface tension between the microbe and the agent.

---

1 Why is the death phase of the growth curve not a straight line down to zero living cells?
2 Why does an increase in environmental temperature in most cases speed microbial death?
3 What species of bacteria are most likely to survive high environmental temperatures?

SURFACE-ACTIVE AGENTS

The fact that water molecules are polar (have an uneven distribution of electrons) gives water some unusual properties. One of the most important is that the molecules tend to stick to one another with the positive pole of one molecule attracting the negative pole of another water molecule. The bond that holds them together is a hydrogen bond (refer to Chapter 3). This tends to make water molecules at the surface of a glass of water unbalanced in their attraction. The water molecules are more strongly attracted to the water beneath the surface than to the air above. The downward orientation of the water molecules results in the formation of a tight, strong layer at the surface capable of withstanding considerable force. This condition is known as **surface tension.** Many substances have high surface tensions that cause them to pull together and stay apart from their surroundings. The high surface tension of water prevents oil from spreading through water and causes oil droplets to form. The strong attractive forces among the water molecules exclude the nonpolar oil, causing the oil to form droplets. If the surface tension of a liquid is low, it will spread easily, or be "runny." These liquids will be **miscible,** or mix well, with other fluids. Alcohol is a liquid with a very low surface tension.

Liquids with high surface tensions make poor antimicrobial agents for three important reasons. First, high surface tension liquids may not be easily mixed with microbicidal chemicals. Many disinfectants operate because they are able to ionize in water. If this is not possible, they will be ineffective. Second, the high surface tension may prevent the antimicrobial from penetrating the outer covering of the microbe. If the antimicrobial cannot get to the microbe, it will not be effective. Third, liquids with high surface tensions do not mix well and may not be able to penetrate a cluster of cells to kill those inside, or get between microbes and the surface to which they may be attached. This last situation may prevent the microbes from being mechanically removed or "washed away."

Substances that are able to reduce surface tension by concentrating at the boundaries or interfaces between two materials are called **surface-active agents.** They make the material more miscible in solution and are more commonly known as **soaps** and **detergents.** These chemicals have the ability to "fit" between the lipid-containing covering of a bacterial cell and the surrounding medium, allowing them to mix. Soaps are called anionic detergents since they have a negative charge and are usually sodium or potassium salts of large fatty acids. One example of a soap is sodium lauryl sulfate:

$$Na^{+\ -}O—S—O—(CH_2)_{11}—CH_3$$
$$\underset{O\ \ O}{\swarrow\ \searrow}$$

Soap has a slippery feel and only limited antimicrobial properties because its primary purpose is to mechanically remove microbes from surfaces. Only a few microbes are easily killed by soaps. These include some species of *Streptococcus, Neisseria, Treponema,* and influenza viruses. Soaps break up oils and greases on skin into small droplets that are more easily washed away. The effectiveness of a soap may be improved by the addition of other true antimicrobials such as disinfectants. The deodorant soaps contain such chemicals and have significantly better antimicrobial properties compared to regular soap. However, it is important to realize that the excessive use of soaps may result in the removal of not only unpleasant odors and odor-producing bacteria, but they may also remove many natural antimicrobial agents produced by the body and found in oily secretions of the skin. Surface-active agents should be thoroughly removed in order to prevent soap buildup that might result in the inactivation of natural antimicrobials and the disinfectant found in some soaps. Before inserting an intravenous needle, or drawing blood from a vein, the skin should be washed with soap to remove surface microbes that might serve as a source of infection. The skin is then washed with alcohol to remove any remaining soap that might interfere with the action of the antiseptic that is swabbed over the vein before the needle is inserted.

Detergents are cationic surface-active agents since the organic portion of their molecule is positively charged. These are effective in both the mechanical removal of microbes and the destruction of cell membranes. Spores and viruses are not killed by detergents. Cationic detergents are commonly found in household products such as dishwashing powders, laundry products, and shampoos.

A popular detergent cream, **pHisoderm,** cleans more effectively than most soaps since it contains lactic acid (a microbial product), wool fat, and sulfonated petrolatum. These products will work better than soaps in "hard" (alkaline) water because they will not react with the "hard" ions to neutralize the effect of the detergent. One of the most effective groups of cationic detergents is known as the **quaternary ammonium compounds** (referred to as "quats") and includes products such as Zephiran, Ceepryn, Phemoral, and Diaparene. Roccal solution is commonly used in microbiology labs to remove contaminating microbes from counter tops and benches. These products work very well against Gram-positive and Gram-negative bacteria, fungi, and protozoans. Many of these detergents are used in hospitals as antiseptics or disinfectants to control microbial contamination on equipment such as thermometers and endoscopes. A 1:1000 dilution of Zephiran will kill most vegetative bacteria (although not mycobacteria) in only thirty minutes.

---

1 Give an example of a liquid with a very low surface tension.
2 How do detergents act as surface-active agents?
3 How does the alkalinity of water affect the operation of a detergent?

THE METHODS AND MATERIALS OF CONTROL

There are many methods and agents of chemical and physical control. Some are more effective when used in combinations than when used by themselves. Most control agents and their applications have become standardized for use in the allied health professions. Each has been tested for effectiveness and adaptability to specific situations requiring microbial control. The twelve categories of control methods are: (1) heat, (2) drying, (3) radiation, (4) sound, (5) gas, (6) alcohol, (7) heavy metals, (8) pH, (9) dyes, (10) aldehydes, (11) oxidizing agents, and (12) chemotherapeutic agents.

HEAT

Controlling the temperature of the environment is the simplest way to inhibit or destroy microbes. Several methods are regularly used (refer to Chapter 6). **Cooling** microbes to 5°C reduces their enzyme activity and acts as a bacteriostatic agent. Deep freezing at −20°C may be bactericidal, if the microbes are allowed to remain at this low temperature long enough to pass through the death phase. **Tyndallization** results in the sterilization of liquids as a result of repeated heating (100°C) and cooling. This repetitive process shocks contaminating spores into the vegetative state and then kills them at a relatively low temperature. Tyndallization allows very heat-sensitive materials to be sterilized without destroying their essential components. **Boiling** in water at 100°C for ten minutes will kill vegetative forms but have little effect on resistant endospores. The high temperature coagulates proteins and causes death. **Pasteurization** may also be performed at relatively low temperatures. The time required to kill all the target microbes at a specified temperature, the **thermal death time (TDT),** varies for each microbe (Figure 10.3). Pasteurization is carried out using the TDT of the most heat-resistant pathogen found in a particular food product. This TDT will produce the least amount of heat damage to the product. For example, milk, beer, and apple juice are all pasteurized at different TDTs to prop-

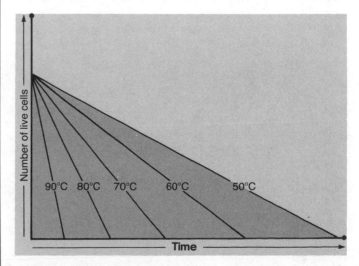

FIGURE 10.3
The time needed to kill all cells of a particular species at a specified temperature is illustrated in the thermal death time graph. Notice that the time required to kill all the cells (to reach the base line) increases as the temperature decreases.

erly preserve each. **Dry-heat sterilization** may be used on materials that cannot withstand the moisture of the autoclave but can resist high temperatures. Dry-heat sterilization is usually performed at 160°C for about 90 minutes and results in the oxidation of the microbes. **Incineration** of contaminated materials will also destroy by oxidation, but special care must be taken to be sure that pathogens do not accidentally escape the flame before they are killed. An overloaded incinerator may give off pieces of unburned contaminated material into the air. The **moist heat** of the autoclave coagulates the proteins and nucleic acids of the cells at 121°C and 15–20 psi, when cells are exposed for about 20 minutes. Moist, pressurized heat has the advantage of being able to drive the heat energy into flasks or other containers.

DRYING

Since the chemical reactions of microbes are basically carried out in water, drying cells by removing water from the environment will greatly inhibit their action. Many foods are dried to prevent spoilage by microbial action. The drying process is bacteriostatic for most microbes (although *Staphylococcus* and *Mycobacterium* may survive for months in a dry, dusty environment) or may be bactericidal for a few others (e.g., *Treponema pallidum*). Moisture may act as a transporter of microbes and therefore should be prevented from accumulating on sterile, paper-wrapped bandages, gauze, or other materials. If the wrapping becomes wet, it should be considered contaminated.

RADIATION

The electromagnetic spectrum contains three ranges of antimicrobial radiation (Figure 10.4). The extremely short wavelengths of **x-** and **gamma radiation** are referred to as ionizing radiation and they produce highly reactive units,

called free radicals (H and OH), when this form of radiation strikes water. The free radicals quickly combine with proteins and DNA to cause death. Since endospores have only a small amount of water, they are much more resistant to the killing effects of x-rays than vegetative cells. Gram-negative bacteria are more sensitive than Gram-positive bacteria. X-radiation is used to control microbial contamination in foods, drugs, and on the surfaces of plasticware and heat-sensitive surgical instruments. **Ultraviolet radiation** in the range of 265 to 280 nanometers is microbicidal. This wavelength does not penetrate glass or plastic but is absorbed by the proteins and nucleic acids of microbes. These molecules are easily destroyed when they absorb this energy. UV light is used to control microbes on the surfaces of transfer cabinets, hair brushes, barracks, and operating room tables. UV light must be carefully controlled since it may cause blindness and skin cancer. Wavelengths within the visible spectrum, called **sunlight,** have little or no direct sterilizing or disinfecting properties. Visible light, however, does increase the rate at which microbes are oxidized through a process known as **photooxidation.**

1 What is the difference between pasteurization and sterilization?
2 How is the TDT of a microbe determined?
3 What are the practical limits of UV light as an agent of microbial control?

SOUND

Ultrasonic sound (100,000+ frequency per second) may be used to destroy microbes. Although sound at this frequency cannot be heard, it is able to coagulate cell proteins and disintegrate cell components. This method of control may be used in research to cut microbes to pieces and iso-

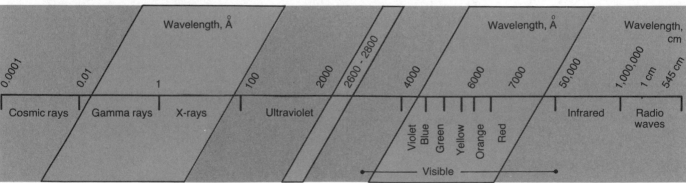

FIGURE 10.4
Three portions of the electromagnetic spectrum are antimicrobial. Gamma rays and x-rays destroy microbes through the formation of free radicals; ultraviolet (UV) light kills by destroying cell proteins and nucleic acids; and visible light increases the rate of photooxidation.

late their internal components. It may also be used in the hospital in the form of an ultrasonic "bath" for cleaning organic materials from instruments, making them easier to disinfect or sterilize.

GAS

Fumigation with ethylene oxide (ETO), beta propiolactone (BPL), or formaldehyde sterilizes both dry surfaces and some liquids. Ethylene oxide is able to penetrate fabrics and circulate over the surfaces of heat-sensitive plastics, rubber, and paper. This gas easily denatures cell proteins, destroying both vegetative and spore forms. Materials to be sterilized must be exposed for at least four hours in a special chamber since this gas is explosive, toxic if inhaled, and irritating to the skin. Therefore it is mixed with nitrogen or carbon dioxide gas to reduce these harmful characteristics. Beta propiolactone may be used to sterilize liquids in a similar manner; however, special precautions should be taken since this gas is not only irritating to the skin and eyes but is also carcinogenic. Formaldehyde gas is able to kill *Mycobacterium tuberculosis* and will readily destroy the endospores of *Clostridium botulinum*. Special transfer cabinets have been developed to subject expelled air to formaldehyde as it is released to the outside in order to control any airborne microbes that may have escaped previous disinfection.

ALCOHOL

The general formula for alcohol is $R—CH_2OH$. One alcohol differs from another in the nature of the R group. Alcohols are microbicidal; however, the larger alcohol molecules are usually more effective than the smaller ones. Those most frequently used are **ethyl alcohol** and **isopropyl alcohol.** Methyl alcohol is the smallest alcohol molecule and is not a good disinfectant when used by itself; in addition, it is very toxic. The maximum microbicidal activity of alcohol is achieved when it is at a concentration between 50 and 80 percent. Below this range, the alcohol has little effect and above this range it is unable to penetrate the cell to coagulate proteins. Since many antiseptics and disinfectants contain alcohol, they should be kept covered to prevent the quick evaporation of the alcohol below the 50 percent level. A solution of 70 percent is usually the standard concentration for antisepsis. Alcohol is not effective against spores and only slightly effective against most viruses. Ethyl alcohol is able to destroy *M. tuberculosis* if kept in contact for 30 minutes. However, because 70 percent alcohol solution evaporates so quickly, it is of little value as an antiseptic or disinfectant when trying to control microbes on large surfaces. All the alcohols have a drying effect on the skin and extensive use may result in damaged tissues that may actually support the growth of bacteria.

HEAVY METALS

Two of the most common heavy metals used as antimicrobials are **silver** and **mercury.** Both are capable of chemically combining with proteins to form poorly dissociable salts that inhibit or destroy cell activity. These metals may be bacteriostatic, bactericidal, and fungicidal but do not have spore-killing properties. Mercury may be used in inorganic compounds (e.g., mercuric chloride, mercuric oxide, ammoniated mercury) or in organic forms (e.g., mercurochrome, merthiolate, mercresin). The inorganic forms are very corrosive, toxic to animals, and are easily neutralized by the presence of excess organic materials. Mercuric chloride (bichloride of mercury) is used in ointment form as an antiseptic. The organic forms are more useful since they are less toxic or irritating and are commonly used as household antiseptics. The silver compounds include silver nitrate, silver picrate, and silver lactate, and are sold under many trade names (e.g., Argarol, Protargol, Silvol, Neo-Silvol). Silver nitrate is used in a 1 percent solution to prevent gonorrheal infection of the eyes of newborns (ophthalmia neonatorum) and in stick form to cauterize wounds.

Another common product used as an antiseptic and an astringent (causes contraction) is calamine lotion. This is a combination of zinc oxide and ferric (iron) oxide.

pH

All microbes have a pH range over which they may successfully grow. Most pathogens have a more narrow pH growth range than saprotrophs. By altering the hydrogen or hydroxyl ion concentration so that the pH falls outside the optimum range, proteins are denatured and microbes may be killed or inhibited. **Calcium hydroxide,** $Ca(OH)_2$, also called slaked lime, is a liquid derived from chalk ($CaCl_2$) and is used as a disinfectant to destroy pathogens in fecal material. A more dilute form of $Ca(OH)_2$ is known as **milk of lime** and is also used as a disinfectant. Many other alkalies are useful in controlling microbes and operate like $Ca(OH)_2$. Those that are most widely used include potassium hydroxide (KOH), sodium hydroxide (NaOH), and ammonium hydroxide (NH_4OH). The last of the group is found in household **ammonia.** It is the least microbicidal of all those listed since it does not dissociate into ions as readily as the others.

Sulfur dioxide (SO_2) is a gas used as a fumigant. When it dissolves in water, it forms sulfurous acid and is damaging to metals and fabrics. Fumigation with SO_2 must be carried out in a moist environment in order to generate the acid. The wine industry uses this gas to control undesirable yeasts and bacterial contaminants before fermentation with *Saccharomyces*. Many organic acids are commonly used in foods to control spoilage. **Acetic acid** (vinegar), **propionic acid,** and **benzoic acids** may be added to breads, fruit juices, and other foods to act as fungistatic and bacteriostatic agents. Homemade bread will mold faster than "store-bought" bread since it does not contain benzoic acid. Benzoic and salicylic acids are also found in Whitfield's ointment as a fungistat. The fungicidal action of Desenex (used to treat athlete's foot) is caused by the fatty acid-based undecylenic acid in the cream, and many eye washes contain boric acid as an antiseptic.

---

1 What two gases are typically used as sterilizing agents?

2 Why is alcohol used primarily at a 70 percent concentration?

3 Name a product that acts to inhibit microbes by unfavorably altering the pH of their environment.

DYES

Many of the dyes used for staining are antimicrobial. **Crystal violet** is added to media as a bacteriostatic agent since it is able to inhibit Gram-positive bacteria. A chemical relative called **gentian violet** is also antifungal and has been used in the treatment of fungal infections of the mouth; for example, thrush, caused by *Candida albicans*.

ALDEHYDES

These organic molecules all contain the $-\overset{\displaystyle O}{\underset{\displaystyle \|}{C}}-H$ group. The two most widely used antimicrobial aldehydes are **formalin** (liquid formaldehyde) and **glutaraldehyde.** Formalin is available in 37 percent concentrated solution but may be used in a more dilute 5 percent soluton provided the exposure time is lengthened. As a disinfectant, it is able to kill vegetative cells, *Mycobacterium tuberculosis,* spores, fungi, and resistant viruses such as hepatitis virus, by denaturing proteins. The destruction of spores requires about 18 hours contact with 37 percent formalin while the hepatitis virus requires about 12 hours. A quick wash with this chemical will not result in the sterilization of an object. Bard-Parker microbicide is one of the formalin-containing disinfectants used in hospitals. In addition to formalin, it contains isopropanol, methanol, and hexachlorophene. Formalin is very irritating to the skin and eyes, and some individuals display an allergic reaction such as a skin rash or asthma. Glutaraldehyde is a 2 percent alkaline solution that is bactericidal, sporicidal, and virucidal. It is marketed under the trade names Cidex and Sonicide. This agent is corrosive, irritating to the skin, and destructive to rubber and plastic. However, it is an excellent control agent for instruments that contain lenses and equipment used for inhalation therapy.

OXIDIZING AGENTS

Several agents control microbes by either releasing oxygen themselves, stimulating the release of oxygen from other molecules, or directly oxidizing cells and cell components. This group includes iodine, iodophores, chlorine and its derivitives, hydrogen peroxide, sodium perborate, and potassium permanganate. Because of its low solubility in water, iodine is marked as **tincture of iodine** solution (iodine-potassium, iodide-alcohol). It is effective at 2 percent against tuberculosis, amoeba, fungi, and other vegetative cells but not against spores. Tincture of iodine is readily inactivated by excess organic material that might be present on the skin. Therefore, cuts and wounds should be thoroughly washed to remove such materials before the tincture is applied as an antiseptic. Precaution should be taken when using tincture to be sure that the alcohol has not evaporated and concentrated the iodine. This concentrated solution may cause skin damage and stimulate allergic reactions. Diabetics may be warned not to use tincture of iodine on fingers or legs because of its damaging effects to tissue with poor circulation and slow healing properties. The benefits of iodine may be obtained and the dangers reduced by using iodophores. These are solutions of iodine in a surface-active agent. **Iodophores** are not irritating to the skin, do not sting when applied to an open wound, and do not form permanent stains on skin or fabrics. Depending on the composition and concentration of the solution, iodophores may be marketed as either antiseptics or disinfectants. Common solutions include Betadine, Iosin, Iclide, Hi-Sine, Wescodyne, and Prepodyne. Other members of the halogen family (fluorine, chlorine, bromine, and iodine) are also effective antimicrobials.

The most widely used halogen is **chlorine** and its derivitives. Chlorine gas (Cl_2) is used regularly to kill microbes in water purification and sewage treatment plants. Drinking water contains about 1 ppm chlorine (one part chlorine for one million parts of water; parts per million). This concentration will kill most bacterial and viral contaminants, however, *Entamoeba histolytica,* a protozoan that causes amoebic dysentery, is unaffected by this level of chlorination. To destroy this microbe, the water must be shocked with 9 ppm Cl_2 and then degassed to 1 ppm to make it potable, or useful for drinking. Sodium hypochlorite (NaOCl) is the form of chlorine found in household bleach and other laundry products. It is also an effective oxidizing agent in controlling microbes on floors, walls, and other surfaces. Placing 6–8 drops of chlorine bleach in a gallon of clear stream water, shaking, and waiting one hour will make the water potable. Tincture of iodine, 18–20 drops, has the same disinfecting effect, however, it will unfavorably alter the taste of the water. A more concentrated form of bleach is used to control microbes in swimming pools. Sodium hypochlorite is bactericidal and active against most viruses including hepatitis virus. Unfortunately, it is inactivated by organic compounds.

Hint: Household bleach will do a better job when added to the wash water to whiten the laundry *before* adding the organic detergent. Mixing both with the clothes will reduce the effectiveness of the bleach and detergent. *Caution:* Do not mix household ammonia with bleach or other cleaning products that contain NaOCl. There will be a chemical reaction that releases deadly chlorine gas ino the air. If this reaction takes place in a confined area such as a bathtub, it could be fatal!

Hydrogen Peroxide H_2O_2 is a liquid at room temperature and is available in 3 percent and 6 percent solutions. Sunlight readily decomposes this unstable molecule into water and free oxygen. For this reason, H_2O_2 is sold in brown bottles and should be kept in a cool, dark place. If H_2O_2 has been stored for a long time and does not foam or bubble when it is placed on a cut, it has probably decomposed and will not be an effective oxidizing agent. This agent is used as an antiseptic and has been recommended for use as a mouth wash by some dentists to control bacterial or fungal infection. However, note that *only* the 3 percent solution should be used and *only* according to the directions given by the dentist. The antiseptic is an indiscriminant oxidizing agent and does not recognize the difference between the gums and the infecting microbes. If used at too high a concentration or too frequently on mucous membranes, it can cause more harm than good. Peroxides better suited for this purpose are zinc peroxide and urea-hydrogen peroxide, which release their oxygen more slowly and safely.

1   What is the difference between tincture of iodine and an iodophore?
2   What factors will inhibit the effective killing action of chlorine?
3   What antimicrobial gas is evolved when hydrogen peroxide decomposes?

Phenol   Phenol, or carbolic acid, is a powerful disinfectant when used in a 5 percent solution; however, 1 to 2 percent solutions are usually standard. This agent is virucidal, bactericidal, and fungicidal. Phenol was originally used by Joseph Lister (1827–1912) to control microbial contaminants during surgery but has since been replaced by many other more effective and less expensive agents (Figure 10.5). Phenol is used to disinfect materials such as feces, sputum, and cultures. This agent operates generally on cells to destroy proteins, cell membrane, and other cell components. At lower concentrations, phenol may be more selective in its action and damage nucleic acids. Some bacteria actually grow in a 0.1 percent phenol solution. This agent has the ability to concentrate around microbes when it evaporates and provides a residual effect that is very beneficial over a long period. Pure phenol is very irritating and corrosive, and should not come into contact with the skin. However, phenolic derivities may readily be used as antiseptics, and remain active in soaps and detergents. **Pine oil** is found in many household detergent products (e.g., Pine-Sol) and is

A   B

FIGURE 10.5
(A) Joseph Lister developed this machine to spray carbolic acid over the patient in order to reduce the probability of surgical infection. (B) This is a re-creation of a nineteenth century operating room in Aberdeen, England. Notice that these surgeons worked dressed in suits, without the aid of masks and rubber gloves. (Courtesy of CIBA-GEIGY, Ltd. From *CIBA Zeitschrift,* no. 119, Basel, Switzerland, 1949.)

primarily effective against many Gram-negative bacteria such as *Salmonella, Shigella,* and *Enterobacter.* The **cresols** are more effective than pine oil or phenol and used in 5 to 10 percent solutions. They have bactericidal activity and are carried in surface-active agents. Commercially they are sold under such names as Lysol, Staphene, O-Syl, and Tricresol. Another class of phenolic compounds is **hexylresorcinol.** These are usually in detergent solutions to treat urinary tract infections. The **bis-phenols** are composed of two chemically combined phenol molecules in soaps or detergents. Their bactericidal activity remains very high but they are not sporicidal. Bis-phenols are sold under the trade names of pHisoHex and Hexagerm, and their most active antimicrobial ingredient is **hexachlorophene.** After its introduction, hexachlorophene was found to be extremely effective in controlling antibiotic-resistant staphylococcal infections in hospitals. Many infections in newborns were able to be controlled with this antiseptic that were otherwise impossible to control with some of the most common antibiotics. However, in 1972 the Food and Drug Administration (FDA) banned all nonprescription use of hexachlorophene and has since made it a prescription drug for use only as a surgical scrub and handwash product for health care personnel. This action was taken as a result of research indicating that hexachlorophene is able to induce brain lesions (damage) in rats, rhesus monkeys, and human newborns who receive topical application of the agent. The myelin sheaths surrounding the nerve fibers were badly damaged by the chemical. Recently, the FDA has also recommended that surgeons, nurses, and other health personnel who are, or could, become pregnant avoid hexachlorophene scrubs or powder as an antimicrobial. Research has indicated that it may be unsafe, since hexachlorophene use may result in birth defects. The loss of this control agent from more extensive use in hospitals has resulted in increased difficulty to control staphylococcal infections.

Phenol is also used as a standard for comparing the effectiveness of different disinfectants. An index showing the relative effectiveness of different agents in comparison to phenol is known as the **phenol coefficient,** or **PC.** Either *Staphylococcus aureus* or *Salmonella typhi* may be used as the test organism. Separate serial dilutions of phenol and test disinfectant are inoculated with a measured amount of the test organism. Samples from each tube are removed at regular intervals, inoculated into culture media, and incubated to determine whether any of the microbes have survived exposure to the agents. If the culture tube becomes cloudy with growth, the disinfectant did not kill the organisms. If it remains clear, the agent has been effective. The phenol coefficient is calculated by determining the dilution of the test disinfectant that kills all microbes in ten minutes but not in five minutes, and dividing this number by the dilution of phenol that kills all the same microbes in ten minutes but not in five minutes.

$$PC = \frac{\text{dilution factor of disinfectant}}{\text{dilution factor of phenol}}$$

For example, if a certain disinfectant kills all microbes at the proper time at a concentration of 1:150, and phenol is able to kill at a concentration of 1:75, the phenol coefficient is

$$PC = \frac{150}{75} = 2$$

As the PC increases above one, the effectiveness of the disinfectant increases. A phenol coefficient of one indicates a disinfectant that is equally effective in comparison to phenol, while a coefficient of less than one indicates that the disinfectant is not as effective as phenol. This index number can be a quick and important source of information when selecting a disinfectant for use to control microbes.

The final major category of control agents is unlike disinfectants in that these chemicals may be used to control microbes inside of living tissue. This group is known as the chemotherapeutic agents.

CHEMOTHERAPEUTIC AGENTS

Chemotherapeutic agents are not indiscriminant killers of both eucaryotic and procaryotic cells or inhibitors as are antiseptics and disinfectants. These chemicals are more selective in their actions. The chemical agents of control are specifically chosen for use because they are only able to kill or inhibit certain types of pathogenic microbes. The idea of **selective toxicity** was first explored by Paul Ehrlich (1854–1915) as he attempted to discover a control agent for venereal disease. In 1909 on his 606th experiment, Ehrlich found an organic arsenic compound (Salversan 606) that would act as what he described as a "magic bullet" in destroying *Treponema pallidum,* the cause of syphilis. It was Ehrlich who coined the term **chemotherapeutic agent.** These drugs may be produced in the laboratory by chemists or they may be harvested from pure cultures of microbes. Chemotherapeutic agents that are isolated from cultures of microbes are given the name **antibiotic.** Table 10.1 lists some of the microbes responsible for the production of several antibiotics. Chemotherapeutic agents work by inhibiting five main areas of microbial cell metabolism. They may disrupt (1) cell wall synthesis, (2) cell membrane functions, (3) protein synthesis, (4) nucleic acid functions, and (5) select metabolic pathways (Table 10.2).

The exact way a chemotherapeutic agent inhibits these vital processes varies; however, many function as **analogues.** An analogue is a molecule that resembles another molecule but is not exactly the same. The cell selects the analogue instead of the proper molecule because it "looks" the same. However, the cell is not able to use it correctly because the analogue does not have the same exact chemical or physical features. Use of an analogue instead of the

TABLE 10.1
The microbial source of some common antibiotics.

Group	Antibiotic	Microbe
Bacteria		
	Bacitracin	*Bacillus licheniformis*
	Colistin	*Bacillus colistinus*
	Polymyxin	*Bacillus polymyxa*
Streptomyces (filamentous bacteria)		
	Amphotericin B (Fungizone)	*Streptomyces* spp.
	Carbomycin (Magnamycin)	*Streptomyces halstedii*
	Chloramphenicol (Chloromycetin)	*Streptomyces venezuelae*
	Chlortetracycline (Aureomycin)	*Streptomyces aureofaciens*
	Erythromycin (Illotycin)	*Streptomyces erythreus*
	Kanamycin (Kantrex)	*Streptomyces kanamyceticus*
	Lincomycin (Lincocin)	*Streptomyces lincolnensis*
	Neomycin	*Streptomyces fradiae*
	Nystatin (Mycostatin)	*Streptomyces noursei*
	Oleandomycin	*Streptomyces antibioticus*
	Oxytetracycline (Terramycin)	*Streptomyces rimosus*
	Paromomycin (Humatin)	*Streptomyces rimosus*
	Ranimycin (antibiotic)	*Streptomyces lincolnensis*
	Rifampicin (Rimactane)	*Streptomyces mediterranei*
	Streptomycin	*Streptomyces griseus*
	Tetracycline	Certain *Streptomyces* species
	Vancomycin	*Streptomyces orientalis*
	Viomycin	*Streptomyces puniceus*
Fungi		
	Cephalothin (Keflin)	*Cephalosporium*
	Fumagillin	*Aspergillus fumigatus*
	Griseofulvin (Fulvicin)	*Penicillium griseofulvum dierckx, Penicillium janczewski,* and *Penicillium patulum*
	Nifungin (antifungal)	*Aspergillus giganteus*
	Penicillin	*Penicillium notatum* and *chyrsogenum*

TABLE 10.2
Antibiotics and their action sites.

		Inhibitory Effect on			
Cell Wall Synthesis	Cell Membrane Function	Protein Synthesis		Nucleic Acid Synthesis	Cell Metabolism
Bacitracin	Amphotericin B	Chloramphenicol	Chlortetracycline	Griseofulvin	Isoniazid
Cephalosporins	Benzalkonium chloride	Dihydrostreptomycin	Erythromycin	Idoxuridine	Para-aminosalicylic acid
Cycloserine	Colistin	Gentamicin	Lincomycin	Rifampicin	Sulfonamides
Penicillins	Nystatin	Kanamycin	Methacycline		
Ristocetin	Polymyxins	Neomycin	Oleandomycin		
Vancomycin		Paromomycin	Oxytetracycline		
Mefoxin		Puromycin	Tetracycline		
		Streptomycin	Nitrofurans		
		Trobicin			

proper molecule may result in poorly formed cell walls, membranes, or proteins. These improperly constructed components weaken the cell and make it more likely to fall apart or unable to carry out essential reactions. A **growth factor analogue** is a molecule that resembles an essential growth factor, such as an amino acid or vitamin, but it is not able to be used by the cell. These molecules inhibit cell metabolism by the process known as **competitive inhibition.** The sulfa drugs are good examples of this process. Many bacteria require para-aminobenzoic acid (PABA) in their diet as a growth factor (Figure 10.6). PABA is used by them in the pathway that leads to the synthesis of folic acid, which in turn is used in the synthesis of nucleic acids. If PABA is not available to these cells, nucleic acid synthesis will not take place efficiently enough to ensure the life of the cells, and the bacteria will die. Sulfonamides function as growth factor analogues of PABA. These manmade drugs compete with PABA for the active sites on the enzymes that are used in the formation of folic acid. As a result, the cells do not produce folic acid but rather a sulfa-containing non-functional molecule that cannot be used in the next path-way. Some bacteria and all animal and human cells are unable to make folic acid. They require folic acid as a growth factor from their environment; therefore, they are not inhib-ited by the sulfa drugs. The effects of sulfa drugs can be overcome by very high concentrations of PABA in the en-vironment. A form of reverse competitive inhibition will oc-cur, and the PABA-requiring microbes will be able to get all they need to survive despite the presence of the drug. PABA is an ingredient of many of the new "sun screen" sun lo-tions and may act as a topical source of PABA. If a sulfa-containing cream were to be applied to a skin infection, the high concentration of PABA in the suntan lotion could counteract its controlling effect. This kind of drug-environ-ment interaction is a very important consideration when ad-ministering or taking drugs. Antibiotics taken orally may have their effectiveness reduced or inhibited by many factors (Ta-ble 10.3). For example, penicillin G and erythromycin are easily inactivated by acid liquids such as cranberry juice, gin-ger ale, and lemon juice. People should avoid drinking more than eight ounces of acid beverages after taking these anti-biotics. Milk and dairy products interfere with the absorption of some tetracycline antibiotics since these foods are able to combine with or absorb the drugs and prevent them from entering the circulatory system. The effective use of many antibiotics is limited by these kinds of interactions and by the nature of the side effects that they may demonstrate if administered improperly. For example, bacitracin cream is used as a topical antibiotic for the control of Gram-positive

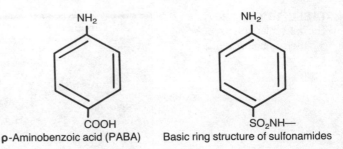

p-Aminobenzoic acid (PABA)   Basic ring structure of sulfonamides

FIGURE 10.6
PABA is a growth factor for many bacteria and must be supplied if they are to synthesize nucleic acids. The basic ring structure of the sulfonamide antimicrobials is very similar to PABA. Bacteria requiring PABA cannot distinguish the two molecules. This results in the death of the microbes, since the sulfa compound interferes with the synthesis of nucleic acids.

---

1 If a product has a phenol coefficient of 0.5, is this agent better or worse than phenol?

2 What is meant by the term "selective toxicity"?

3 How might an antibiotic work as a growth factor analogue?

---

bacterial infections. The cream should not be used internally since it is toxic to the kidney.

Many antibiotics have been introduced since penicillin was first made available in the 1940s (refer to Chapter 2). The quality of each drug has been tested for its effectiveness, selective toxicity, and side effects by the discoverer, manu-facturer, and the FDA. A number of antibiotics that have survived this extensive evaluation process are described in Table 10.4 and are in use today. Countless others have been abandoned because they were not able to meet rigid standards, or they have been returned to research labora-tories for modification. The **semisynthetic antibiotics** are drugs that have been isolated from culture and chemically altered to better control pathogenic microbes. In many in-

TABLE 10.3
Drug interactions of the gastrointestinal tract.

Drug	Interfering Agents	Interactions
Tetracycline Lincomycin	Antacids: Maalox, Silain-Gel, Mylanto Antidiarrheals: Kaopectate Donnagel	Combine to inhibit absorption
Penicillins (Ampicillin) Erythromycin Tetracycline	Presence of foods (liquid or solids)	Delay absorption or inactivate drug
Most oral antibiotics	Castor oil, citrate, magnesia, dulcolax	Decrease rate and extent of absorption

stances, a single antibiotic has evolved into a family of anti-biotics. A person taking "penicillin" could be taking any one of over five hundreds kinds of this drug (Figure 10.7). All antibiotics are selected or modified for use with certain goals in mind.

> **1** What factor might inhibit the action of an antibiotic after it has been given to a patient?
>
> **2** Name two antibiotics that operate on bacterial cell walls as their target structure.

THE IDEAL ANTIBIOTIC

When research is conducted on a new antibiotic, investigators set their standards as high as possible. They look for the optimum, or best, antibiotic that will be most effective for the longest period. The ideal antibiotic should (1) be selectively toxic to pathogenic microbes, (2) not stimulate the formation of antibodies, (3) not upset the normal microbial flora of the body, and (4) prevent the evolution of antibiotic resistant forms. These are ideal characteristics that researchers strive for with the understanding that they may never find or produce all of them in a single antibiotic. It is difficult to obtain a chemical agent that will interfere with only those biochemical pathways that are unique to the pathogen and also essential to its survival. Most antibiotics demonstrate some form of host harm if used at high levels or for prolonged periods. The best antibiotics are those that result in the fewest and mildest side effects. The antigenic properties of antibiotics must also be studied. These drugs should not be able to stimulate the immune system of the host, for if this should occur, the patient would respond to the presence of the drug by producing antibodies against the antibiotic. The antibiotic would be acting as an antigen. The drug would probably be *effective* the first time it was used to treat a patient but would be useless the second time. The patient's immune system would destroy the anti-

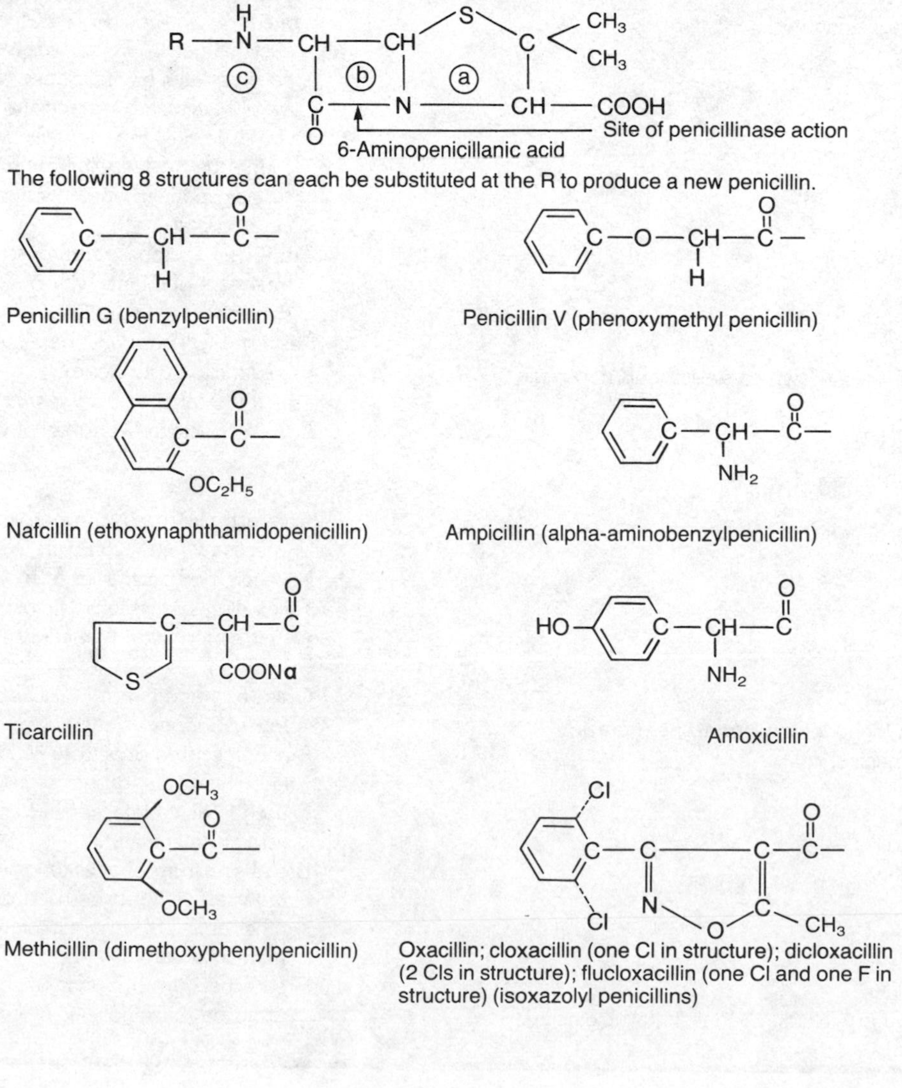

FIGURE 10.7

The basic structure of the penicillin family is illustrated here. The molecule is active in controlling many types of bacteria and has had its structure chemically modified (the R group) in order to produce different penicillins. Each variation has slightly different chemical characteristics that make it better suited for use in infections that are difficult to control.

TABLE 10.4
The antibiotics.

Drug	Range of Activity
Penicillin A family of antibiotics including penicillin F, G, X, K, O, and V; methicillin, oxacillin, ampicillin, nafcillin, and others.	Primarily against Gram (+) bacteria, e.g., α-β hemolytic streptococci and *S. pneumoniae;* with limited activity against Gram (−) bacteria, e.g., *Neisseria gonorrhoeae, E. coli, Salmonella.*
Streptomycin A member of the aminoglycoside family; also includes neomycin, gentamycin, and kanamycin.	Gram (+) and Gram (−) bacteria including *Yersinia, Brucella, Neisseria, Haemophilus influenzae, Klebsiella, Streptococcus faecalis;* urinary tract infections; *M. tuberculosis* in combination with other anti-T.B. drugs.
Tetracycline A family including chlortetracycline and oxytetracycline.	Procaryotic cells only; broad spectrum including rickettsia (Q fever, Rocky Mountain spotted fever, typhus fever), *E. coli, Enterobacter, Shigella;* used against *Strep. faecalis* and *S. pyogenes* only if found to be sensitive. Not effective against *M. tuberculosis, Vibrio, Treponema;* bacteriostatic.
Cephalosporin A group of chemically related molecules that resemble penicillin; includes cephalothin-Keflin and Cefazolin-Kefzol, or Ancefazolin.	Gram (+) and Gram (−) bacteria; especially against penicillinase-producing staphylococci; *E. coli, Salmonella, Shigella, Strep. pneumoniae.*
Erythromycin A member of the macrolides; includes spiramycin, oleandomycin, carbomycin.	Primarily against Gram (+) bacteria, but some Gram (−) bacteria also affected; especially against microbes in person found allergic to penicillin. *Strep. pyogenes, Staph. aureus, Treponema pallidum,* legionnaire's disease.
Lincomycin, Clindamycin	Same as erythromycin, but especially effective against the bacterial genera *Bacteroides* and *Clostridium;* bactericidal or bacteriostatic; few microbes resistant to this drug; not effective against *Neisseria, Haemophilus,* and *Strep. faecalis.*
Vancomycin	To be used *only* in potentially life-threatening infections that cannot be treated with less toxic drugs (penicillins and cephalosporins); primarily against staphylococci and the enterococci.
Bacitracin	Gram (+) microbes; intramuscular injections limited to treatment of infants with pneumonia caused by staphylococci shown to be susceptible.
Polymyxins B and E, also known as solistin and neosporin.	Restricted to Gram (−) bacilli, e.g., *Pseudomonas;* used in conjunction with neomyocin, bacitracin, and hydrocortisone as a topical antibiotic cream.
Isoniazid (INH)	Acid-fast bacteria *(M. tuberculosis).*
Para-aminosalicylic acid (PAS)	Acid-fast microbes *(M. tuberculosis);* bacteriostatic. Only effective when given with INH or streptomycin.
Griseofulvin	No effect on bacteria; used to treat fungi-caused skin, hair, and nail diseases; inhibits *Microsporum, Trichophyton, Epidermophyton* that cause ringworm, athlete's foot, barber's itch.
Nitrofurantoin Nitrofuraldehyde semicarbazone, Furacin	Gram (+) and Gram (−) *E. coli, Staph. aureus, Proteus, Pseudomonas, Klebsiella.*
Chloramphenicol	Broad spectrum of activity including Gram (+), (−), rickettsia and chlamydia; due to severe side effects, should only be used in acute infections of active typhoid fever or severe cases of meningitis.
Rifampicin	Broad spectrum of activity including the pox viruses and *M. tuberculosis;* must be used in conjunction with at least one other anti-T.B. drug; also for asymptomatic meningococcal meningitis carriers.
Sulfonamides Sulfa drugs, including gantrasin, azulfidine	Broad spectrum of activity including bacteria, chlamydia, some protozoa, bacteriostatic *E. coli, Proteus, Staphylococcus, Klebsiella.*

Affected Structure	Possible Side Effects
Interferes with peptidoglycan formation in cell wall.	Allergic reactions: skin rashes and anaphylaxis (shock), diarrhea.
Inhibitors of protein synthesis.	Ototoxicity (nausea, vomiting, and vertigo), fever and urticaria (skin rash), kidney damage.
Inhibitors of protein synthesis.	Nausea, vomiting, diarrhea, skin rashes, and mucous membrane lesions. Not to be used during tooth development (last half of pregnancy, infancy, and childhood to age 8). May cause permanent discoloration of the teeth (yellow, brown, gray); skin photosensitivity if exposed to direct sunlight.
Interferes with cell wall synthesis.	Anaphylaxis (shock), pain at site of injection, diarrhea, nausea, and vomiting.
Inhibitors of protein synthesis.	Fever, allergic reactions, abdominal cramping, skin rashes.
Inhibitors of protein synthesis.	Diarrhea and severe, possibly fatal enterocolitis.
Interferes with peptidoglycan formation in cell walls.	Thrombophlebitis (inflamation of a vein-associated clot), deafness, and kidney damage.
Inhibits cell wall synthesis.	Kidney damage, skin rash, nausea, and vomiting.
Disrupt osmotic stability of cell membrane.	Drowsiness, pain at site of injection, and kidney damage.
Competitive inhibition of essential enzymes.	Vitamin B_6 deficiency; fatal hepatitis may occur with therapy in older patients and alcoholics; degeneration of peripheral nerves; anorexia (loss of appetite).
Competitive inhibition; similar to PABA.	Gastrointestinal upset, goiter.
Inhibits the mitotic spindle microtubules.	Headache, drowsiness, skin rashes, mental confusion.
Inhibitors of protein synthesis.	Gastrointestinal upset, anemia, rash, anorexia, allergic reactions.
Inhibitor of protein synthesis.	Gastrointestinal upset; aplastic anemia is so severe that the use of this drug is restricted.
Inhibitor of DNA and RNA synthesis.	Color change in urine, rashes, liver and kidney malfunctions, dizziness, fever.
Competitive inhibition of essential enzymes (PABA analogue).	Allergic reactions, fever, skin rashes, gastrointestinal upset, anemia, liver and kidney damage.

biotic molecules with antibodies. To prevent this, all antibiotics must be thoroughly screened to detect this host response and modify the antigenic properties.

Another major host problem associated with the use of antibiotics is the loss of normal microbial flora from the intestinal tract. This may occur because many antibiotics are not selective enough or because the physician may choose a broad spectrum drug to bring an infection under control quickly. A **broad spectrum antibiotic** is one that is able to control a wide range of microbes including Gram-positive and Gram-negative bacteria (Table 10.5). When these nonspecific drugs are given orally, they may "accidentally" kill *E. coli* and many other beneficial bacteria that normally live in the intestine. These bacteria are essential to host health for a number of reasons. First, a large active population of *E. coli* is able to out-compete pathogens that enter the gastrointestinal tract for food. The large numbers of normal flora demand so much more from the environment that the few pathogens entering the intestine are unlikely to survive and cause harm. Only when the normal flora have been killed or their metabolism has been inhibited, do pathogens have the opportunity to take hold and establish an infection. Second, many of the normal flora release their own chemicals into the lumen (space) of the intestine, discouraging the growth of pathogens. Colicins and megacins are antibiotics that inhibit many pathogens. In addition, *E. coli* and *Lactobacillus acidophilus* normally release such organic acids as lactic and acetic acid as products of fermentation. The constant release of these acids has an inhibiting effect on acid-intolerant pathogens such as *Salmonella*. The presence of these acids at low, constant levels on the inner surface of the intestinal tract also optimizes the absorption of nutrients into the circulatory system. The loss of this acid layer due to antibiotic therapy reduces the health of the host because of a decrease in the movement of essential nutrients through the intestinal villi*.

The third advantage of maintaining normal intestinal flora relates to the fact that many are able to produce vitamins required by humans. These include vitamin B_{12} (cyanocobalamin) used in red blood cell formation and vitamin K needed for the synthesis of prothrombin necessary for blood clotting. In agriculture, many ranchers and farmers regularly supplement their cattle feed with a selective antibiotic that inhibits the bacteria of the intestine from destroying vitamin B_{12}. This practice leaves the intestinal tract of the cattle populated with vitamin B_{12} producers and increases the overall health of the animals. Healthy animals mean faster weight gain and less loss due to disease. A final positive effect of normal flora relates to the antigenic properties of the many different microbes found in the intestine. Each time there is a slight break in the lining of the intestine,

*A villus is a fingerlike projection from the surface of the intestinal lining.

TABLE 10.5
The spectrum of effectiveness of select antibiotics.

Broad Spectrum Antibiotics	Narrow Spectrum Antibiotics
Chloramphenicol (Chloromycetin)	Penicillin
Chlortetracycline (Aureomycin)	Streptomycin
Demethylchlortetracycline (Declomycin)	Dihydrostreptomycin
Oxytetracycline (Terramycin)	Erythromycin
Tetracycline	Lincomycin
Kanamycin (Kantrex)	Polymyxin B
Ampicillin (Polycillin)	Colistin
Cephalothin (Keflin)	Ristocetin (Spontin)
Rifampicin (Rimactane)	Oleandomycin
Gentamycin (Garamycin)	Vancomycin
	Nystatin (Mycostatin)
	Spectinomycin

some of these bacteria move into the tissue and may stimulate the immune system to produce antibodies. Many of the antibodies are able to cross-react with pathogenic bacteria because they also contain an identical or similar antigen. This constant boost to the immune system helps maintain high, active levels of antibody protection for the host.

THE DEVELOPMENT OF DRUG RESISTANCE

The last quality of an ideal antibiotic is that the use of the drug should not foster the development of **drug resistance.** A bacterium is resistant to the effects of a drug if it is not killed or inhibited by its presence (Figure 10.8). All microbes will experience a change in the gene frequency of their population over generations of time as a result of environmental change. Antibiotics act as natural selection agents (see Chapter 1) favoring those cells that have the genetic ability to withstand the effects of the drug. Those that survive become the grandparents of a new population of drug-resistant bacteria. Since one essential life characteristic is evolution, the prevention of drug resistance (drug fastness) is an impossibility. The development of resistance can only be slowed, not stopped. Microbes may become resistant to antibiotics in four ways. Either they (1) stop producing the drug-sensitive structure, (2) modify the sensitive structure, (3) become impermeable to the drug, or (4) are able to release enzymes that inactivate the antibiotic.

Some forms of *Staphylococcus aureus* have evolved the ability to survive antibiotics that inhibit cell wall synthesis (for example, penicillins) by not producing the cell wall. In the presence of these antibiotics, they change to the L-form as they temporarily lose part or all of their cell wall. Some bacteria are resistant to the effects of antibiotics that inhibit protein synthesis. As a result of a genetic change, they no longer produce the same type of proteins used in the manufacture of ribosomes. Since these new proteins are not affected by the antibiotic, protein synthesis continues without interruption. Other bacteria have mutated in ways that have resulted in the production of cell membranes that block drugs from entering. If the drug cannot get in, it cannot work. Many important pathogens have genes responsible for the production of enzymes that degrade antibiotics. The penicillinase gene (beta-lactamase) codes for the production of an enzyme that inactivates penicillin. The incidence of drug resistance to certain antibiotics is increasing at an alarming rate throughout the world. Resistance has been studied and confirmed in microbes such as *Shigella dysenteriae, Haemophilus influenzae, Salmonella typhi, S. panama, S. typhimurium, Neisseria gonorrhoeae, Staphylococcus aureus, E. coli, Klebsiella pneumoniae, Pseudomonas aeruginosa,* and *Bacteroides fragilis.* The loss of an antibiotic due to the development of a resistant pathogen could result in a devastating increase in disease and death. Some antibiotics found to be ineffective against drug-resistant bacteria include penicillin, erythromycin, PAS, INH, streptomycin, gentamicin, ampicillin, tetracycline, kanamycin, chloramphenicol, and clindamycin. A good example of how serious the problem has become is demonstrated by the development of penicillin resistance in *Neisseria gonorrhoeae.* In 1947, 0.2 unit of this drug was able to control a typical gonorrheal infection; in 1959, 0.6 unit was required. By 1966, the controlling dosage had increased to 2.4 units, and more recently 4.8 units are required. In addition, another drug, probenecid, is given with the antibiotic to reduce excretion of penicillin and increase its effectiveness two to four times the 4.8 level. The constant increase in dosage cannot continue. A point will be reached at which the concentration of the drug in the patient will result in severe side effects that will make its use impossible. When this occurs, the drug will be useless. Several steps must be taken to overcome this problem.

> **1** What is a broad spectrum antibiotic? What might be a good definition for the phrase "narrow spectrum antibiotic"?
>
> **2** How might a bacterium become resistant to the effects of an antibiotic?

OVERCOMING RESISTANT PATHOGENS

Everyone in the medical professions must become aware of the problem of resistant pathogens. Without understanding the long-term effects of increased drug resistance, few will take the actions necessary to ensure the effective use of the antibiotics currently available. The public should be re-educated to understand that the major responsibility of the medical profession is not to provide instant cures but to *help* the body cope with infections and heal itself. The practices of preventive medicine and early low-level treatment of infection places the patient's body in a more favorable position than delayed care and emergency high-level treatment. The body has many excellent host defense mechanisms to cope with infectious microbes. The immune system is one of these mechanisms; however, in many cases it must be stimulated into action by the presence of antigens (the microbes!). If antibiotics are given too quickly or are overused prophylactically (treatment given to prevent a possible future infection), the immune system may not receive this necessary stimulation, and the widest possible range of antibody activity may not be developed. Individuals who take antibiotics to prevent future infection (for example, rheumatic fever victims taking penicillin to prevent streptococcal infections that could trigger further heart damage) are more likely to be in this position.

Limiting the use of high levels of broad spectrum antibiotics is also important. The more favorable approach would be to use minimal amounts of an antibiotic that has been shown to control a specific infectious agent as a result of a culture and susceptibility test. Limitation is also important for those dispensing antibiotics on a nonprescription basis. Farm stores, pet stores, and druggists in many foreign countries

FIGURE 10.8
The colonies of bacteria seen in the zone of clearing around the antibiotic disk are mutated forms that are able to survive the effects of this drug. If they were to be cultured and the test repeated, there would be no zone of clearing visible around this same antibiotic of the same concentration.

currently sell large amounts of antibiotics without prescriptions (Figure 10.9). In 1977, the FDA ordered a cutback in the routine use of antibiotics in animal feed. These drugs (penicillin, chlortetracycline, and oxytetracycline) had been routinely added to animal feeds for the previous twenty-five years to speed weight gain and prevent disease. The FDA feared that continuing this practice would result in bacteria becoming resistant to these antibiotics and that this resistance could be transferred to bacteria in humans. The FDA allowed tetracycline to be used in animal feed for five specific diseases because there are no effective substitutes.

The problem of drug resistance may also be controlled by retiring an antibiotic from use for a period of time. This approach is based on the hope that by *not* exposing a resistant microbe to the drug, natural selection will no longer push evolution in the direction of increasing resistance. It is also hoped that genetic changes in the resistant population will take place during the retirement period that will result in the loss of this genetically controlled trait. Other options include modifying currently used drugs in ways that will "fool" the resistant microbes (i.e., make more semisynthetics), and isolating or synthesizing new antimicrobial agents that have never been used before. These last choices are being explored most extensively by pharmaceutical companies, university researchers, and government agencies (Figure 10.10).

The main source of drug-resistant microbes is from new genes that allow the bacterium to survive in the presence of the drug. Through natural selection as a result of exposure to the antibiotic, these genetically different microbes evolve and become a greater portion of the population. Though spontaneous mutations are very important in the develop-

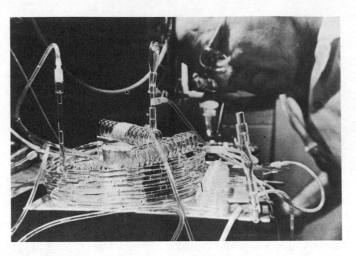

FIGURE 10.10
In the early 1960s this automatic analyzer was used by researchers at Eli Lilly to make chemical analyses of a series of sample fermentations of cephalosporin C submitted by microbiologists. The data obtained was helpful in selecting better strains of the antibiotic-producing fungus and in evaluating various nutrient media. (Courtesy of Eli Lilly and Company.)

ment of resistance, microbiologists also recognize that the rapid increase of resistance in microbial populations is due to the fact that these genes are able to move through a population in a number of unique ways. The study of microbial genetics has shed a great deal of light on gene transfer mechanisms.

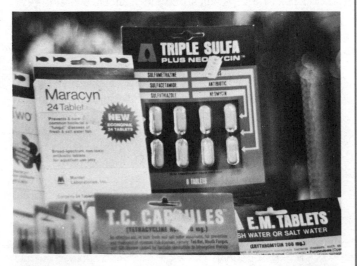

FIGURE 10.9
This photograph was taken in a pet store and shows the variety of commercially available antibiotics. Though the sale of antibiotics in this form is somewhat controlled in certain states, no prescription is needed for their purchase.

SUMMARY

The control of microbes has long been a concern of microbiologists. The major categories of control include chemical agents, physical agents, and light. These agents are able to kill or inhibit microbes by coagulation, oxidation, cell wall damage, cell membrane damage, or interference with metabolic pathways. Many special terms are used to discuss the physical and chemical agents of control. Several of these have been misused, leading to misconception about microbial control and the improper use of the agents. Therefore, it is essential to have a good working knowledge of terms such as disinfection, antiseptic, antibiotic, and chemotherapeutic agent.

Microbial populations that are exposed to microbicides die at an exponential rate depending on whether they are in pure culture and in the same metabolic condition. Because of the way microbes die, one can never be sure if they are totally dead. To cope with this problem, it is necessary to become familiar with the variety and specific applications of many disinfectants and antiseptics, if successful microbial control is to be carried out in hospital, laboratory, or home environments. The effectiveness of any agent may be increased by raising the temperature, removing excess organic material, increasing the concentration of the agent, and using surface-active agents to reduce surface tension. Soaps and detergents are two groups of effective surface-active agents.

The twelve categories of control agents presently recognized are heat, drying, radiation, sound, gas, alcohol, heavy metals, pH, dyes, aldehydes, oxidizing agents, and chemotherapeutic agents. The chemotherapeutic agents are only able to kill certain infectious microbes by selective toxicity. This group includes the manmade drugs (synthetics) and the microbe-made drugs (antibiotics). The use of antibiotics must take into consideration many factors including drug-environment interactions, destruction of normal flora, possible side effects, and the development of drug-resistant strains of the pathogen. Drug resistance in bacteria is reaching high levels throughout the world and is of major concern to microbiologists and the medical profession. To slow this evolutionary process, several steps have been taken. Expanded use of the culture and susceptibility test will pinpoint the nature of the microbe and help in selecting the most specific antibiotic for use in treatment of the infection. Limiting the use of these drugs to essential and low-level dosages will slow the development of resistance. Retiring drugs already known to be ineffective against certain microbes may help to genetically restructure the population of pathogens, and the use of new drugs may relieve natural selection pressures.

STUDY QUESTIONS

1 In what ways are microbes destroyed or inhibited by chemical agents?

2 Define antimetabolite, antibiotic, antiseptic, and antibody.

3 What is the difference between an antiseptic and a disinfectant? Give an an example of each.

4 Compare the log and death phases of the population growth curve. Why is it necessary to understand this concept to control microbes?

5 What is a surface-active agent and how does it work? Give two examples.

6 What is meant by the thermal death time of a microbe?

7 What wavelengths of light are antimicrobial? How does each destroy the cell?

8 Name four household products that are antimicrobial and describe their effects on microbes.

9 What is a growth factor analogue? How does it destroy microbes?

10 What is meant by drug resistance? How might this process be slowed?

SUGGESTED READINGS

ABRAHAM, E.P. "The Beta-Lactum Antibiotics." *Scientific American,* June, 1981.

"Battling Drug-Resistant Microbes." *Medical World News,* Feb. 1979.

BENESON, A. S., ed. *Control of Communicable Diseases in Man,* 13th ed. American Public Health Association, Washington, D.C., 1981.

BLACK, C. D., N. G. POPOVICH, and M. C. BLACK. "Drug Interactions in the GI Tract." *American Journal of Nursing,* Sept., 1977.

DIXON, R. E., R. A. KASLOW, D. C. MACKEL, C. C. FULKERSON, and G. F. MALLISON. "Aqueous Quaternary Ammonium Antiseptics and Disinfectants, Use and Misuse." *Journal of the American Medical Association,* vol. 236, no. 21, Nov., 1976.

"Evaluation of Useful Antimicrobial Chemicals." *American Operating Room Nursing Journal,* 23:1084, May, 1976.

FISHER, E. J. "Surveillance and Management of Hospital-Acquired Infections." *Heart Lung,* Sept.–Oct., 1976.

"Hexachlorophene—Interim Caution Regarding Use in Pregnancy." *FDA Bulletin,* Aug.–Sept., 1978.

Highlights in Microbiology. American Society for Microbiology, Committee on Public Services & Adult Education, 1976–77.

MARSHALL, E. "Scientists Quit Antibiotics Panel at CAST—Academics and Animal Feeds Do Not Mix." *Science,* vol. 203, Feb., 1979.

MIZER, H. E. "The Staphylococcus Problem Versus the Hexachlorophene Dilemma in Hospital Nurseries." *J. Obstet. Gynecol. Neo. Nurs.,* March-April, 1973.

SPAULDING, E. H. "Alcohol as a Surgical Disinfectant." *AORN Journal,* Sept.–Oct., 1964.

KEY TERMS

antimetabolite	thermal death time	selective toxicity
surface-active agent	photooxidation	antibiotics
quaternary ammonium compounds	iodophore	growth-factor analogue
	phenol coefficient	drug resistance

Pronunciation Guide for Organisms

Pediococcus (pe"de-o-kok'us)
Candida (kan'di-dah)
albicans (al'bi-kans)

Microbial Genetics

THE FUNDAMENTALS OF GENETICS

MUTAGENESIS AND CARCINOGENESIS

THE EFFECTS OF SELECTION

TRANSFORMATION

GENE TRANSFER, PLASMIDS, AND DRUG RESISTANCE

RELATEDNESS AND GENETIC ENGINEERING

Learning Objectives

- ☐ Be familiar with the fundamentals of genetics, including definitions of the terms "gene," "allele," "genotype," and "phenotype."

- ☐ Know the basic kinds of mutations and the factors that cause them.

- ☐ Name the two methods by which mutations may be repaired.

- ☐ Know the differences between mutagenic and carcinogenic agents.

- ☐ Understand the value of the Ames test and how to perform it.

- ☐ Be familiar with the concept of natural selection and how it applies to a bacterial population.

- ☐ Understand the mechanisms of genetic recombination in microbes and the advantages of the processes to a population.

- ☐ Be able to define the terms "exogenote" (extrachromosomal) and "endogenote" DNA.

- ☐ Explain the process of transformation.

- ☐ Define the term "plasmid" and relate it to F factors, R factors, and genetic recombination.

- ☐ Understand generalized and specialized transduction.

- ☐ Be able to explain the differences between a lytic and lysogenic life cycle.

- ☐ Know the theoretical relationships among plasmid, prophage, and nucleoid DNA.

- ☐ List advantages and disadvantages of genetic engineering.

Genetics is a relatively new field of biology and most associate this science with Johann Gregor Mendel (1822–84), who was the first person to formulate any laws about how characteristics are passed from one generation to the next. This kind of study is often called Mendelian genetics. His work was not generally accepted until 1900, when three men working independently rediscovered some of the ideas that Mendel had formulated thirty years earlier. However, genetics

11

was revolutionized in 1953 when James Watson and Francis Crick proposed a chemical structure for DNA. Their discovery made it possible to understand more clearly the chemical basis of heredity in both eucaryotic and procaryotic organisms. Since then the proposed double helix structure for DNA has become the cornerstone for explaining gene function, gene replication, and the nature of mutations. At this time many geneticists began to use microbes as research organisms. Laboratory mice and fruit flies were abandoned by many geneticists in favor of microbes such as *E. coli, Neurospora* (a fungus), and T$_4$ phage (a bacterial virus). The ease of culturing and short generation times of microbes enabled geneticists to quickly produce large populations and trace inheritable characteristics through hundreds of generations. Although there are a number of advantages to using these organisms, it was soon discovered that microbes demonstrate many genetic characteristics and inheritance patterns not previously found in eucaryotic cells. As these patterns were explored further, it became clear that transformation, conjugation, and transduction would have far-reaching implications in medicine and in the field of eucaryotic genetics.

THE FUNDAMENTALS OF GENETICS

In both eucaryotic and procaryotic cells, the molecule that serves as the ultimate agent of chemical control is deoxyribonucleic acid (DNA), while the inheritable material of viruses may be either DNA or ribonucleic acid (RNA). Knowing the Watson-Crick structure for DNA makes possible the definition of a gene both chemically and functionally. A **gene** is a portion of a DNA molecule composed of a specific series of nitrogenous bases that chemically codes for the production of a specific protein or RNA molecule, or serves as an operator in controlling the transcription of RNA within an operon unit (refer to Box 13). The chemical code for the placement of an amino acid is a specific triplet nucleotide sequence. Since protein molecules contain an average of 300 amino acids, the average gene is made up of about 900 nucleotide pairs. The DNA of *E. coli* is one of the most thoroughly investigated nucleoids and contains about 5×10^6 base pairs. This amounts to approximately 5000 genes, many of which have been identified in their proper sequence (Figure 11.1).

An organism's DNA constitutes a catalog of genes known as the **genotype** of the organism. The expression of these genes will result in a certain collection of characteristics referred to as the **phenotype.** While the phenotype of an organism consists of its observable characteristics, the genotype is not visible, since it is the DNA chemical code (formula) of an organism. There is not always a total expression of the gen-

otype. Particular genes may not express themselves for a variety of reasons. In some cases, the physical environment will determine if certain genes will have a chance to express themselves. For example, if lactose is not supplied to a bacterial population that can metabolize this sugar, the phenotype will not be seen because the presence of lactose is required to induce the formation of the enzymes needed for its breakdown.

DNA is very stable, thus it is an excellent molecule to serve as the transmitter of chemical codes through generations. The stability of DNA and its resistance to change ensures the continuation of a species even though alterations regularly occur in gene structure. Any permanent change in the nitrogenous base sequence of DNA is called a **mutation.** A single gene may mutate to many different forms. Those forms of a gene that affect the same characteristic but produce different expressions of that characteristic are called **alleles.** For example, in humans there are alleles for eye colors such as blue and brown. In bacteria, there are alleles for enzyme production. In some bacteria the gene that controls the operation of the **lac** operon (see Box 13) functions on an inducible basis ("off" and "on"), while others have a different allele of this gene which enables the same operon to function on a constitutive (always "on") basis.

Various mutations may be produced in DNA. A single nitrogenous base may be lost and replaced by a different base. This is known as a **point mutation** and may cause a single amino acid change in a protein. The sugar-phosphate backbone of the DNA may break, resulting in a change or loss of a sizable portion of the molecule. This kind of event in DNA changes more than just a nucleotide base. If the damaged section is lost and not repaired, the mutation is called a **deletion mutation.** If the damage is repaired by the insertion of the same piece of DNA in reverse order, it is called an **inversion mutation.** When a totally new base sequence is synthesized to fill the gap, it is called an **insertion mutation.**

Mutations may have any one of a number of effects on the cell in which they occur. In some cases, the change is not harmful. After a single base change (point mutation), the protein synthesized by the gene may still be functional and no phenotypic change may be seen. This may occur because of the nature of the nucleotide code system since, in some instances, the same amino acid can be coded for by more than one triplet codon sequence. Therefore, even if a point mutation takes place, the new triplet condon formed can still call for the positioning of the same amino acid in the protein being synthesized. The protein may also remain active if another very similar amino acid is coded for placement and serves the same function in the completed molecule. Other more extensive mutations may result in the severe reduction of enzyme activity. This reduction may be due to a distortion of the active site on the enzyme surface. Poor-

FIGURE 11.1

(A) Using special techniques, the nuclear material of *E. coli* can be extracted from a cell. This macromolecule is a double-stranded helix of DNA that contains approximately 5000 genes arranged in a linear pattern. (B) Other techniques have been used to identify and map the location of a number of these genes. The numbers on the inner circle represent the relative distances between the genes. (Photo A courtesy of the Center for Disease Control, Atlanta. Part B from A. L. Taylor and C. D. Trotter. "Linkage Map of *Escherichia coli* Strain K-12." *Bacteriological Review,* 36:504, 1972.)

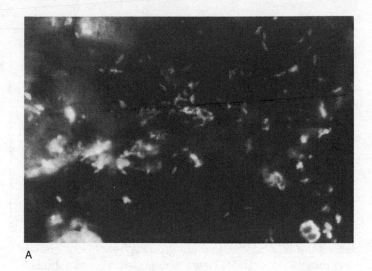

A

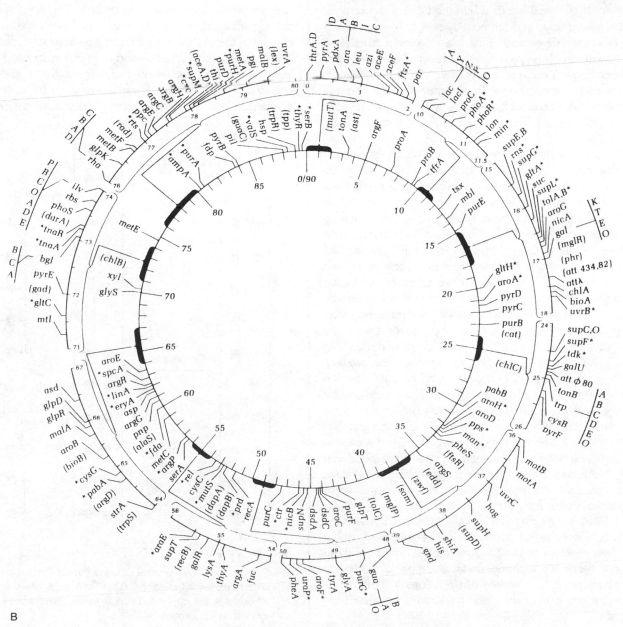

B

ly constructed enzymes may also be more susceptible to environmental changes. Minor fluctuations in temperature, ion concentration, or oxidizing agents may alter the enzyme's three-dimensional shape and cause inactivation. In some cases, the mutation may result in the complete loss of gene activity and death of the cell if an essential gene is inactivated.

Mutations may occur either spontaneously or be caused by agents such as radiation and chemicals. Anything that causes a permanent change in the DNA of a cell is called a **mutagenic agent.** Mutagenic agents include x-rays, mustard gas, and nitrous acid. Naturally-occurring spontaneous mutations are at a relative low rate in microbes. About one in one million (1×10^{-6}) bases may undergo a natural reorganization of chemical bonds within the DNA. For example, thymine may quickly and temporarily shift its internal bonds into a different arrangement. If this rearrangement takes place when thymine is base-paired with adenine (A-T), an error in base pairing will occur at the moment of DNA duplication. Because of this spontaneous bond rearrangement, normal base pairing (A-T) cannot take place, and the guanine-containing nucleotide is hydrogen bonded into the sequence by mistake (G-T).

Point mutations such as this may also be caused by mutagenic agents. Ultraviolet radiation induces mutations by causing the formation of bonds between thymine nitrogenous bases located next to one another on the same strand of the DNA double helix. These linked bases are known as thymine **dimers** (Figure 11.2). This change in bonding results in a **frameshift** mutation since the dimer will be skipped during replication and the transcription portion of protein synthesis. Reading-frameshift mutations might be compared to "skipping" the two letters *p* and *e* in the word "independence." It would then read "indendence" and make no sense. This type of mutation and others may also be caused by chemicals such as nitrous acid, acridine dyes, and alkylating agents. They induce changes that result in mispairing, reading-frameshifts, and deletions. However, many bacteria have the ability to repair some of these changes. The repair process known as **photoreactivation** requires the presence of visible light immediately after dimer formation. The visible light energy is able to break the dimer and reestablish the original base sequence before a permanent change is fashioned into the DNA. Another, more complex, repair system, called **dark repair,** may also correct ultraviolet damage. In this series of reactions, visible light is not required. The damaged section of DNA is "clipped out" of the strand enzymatically, the "hole" is enlarged to ensure a better "patch job," and a new segment of DNA is synthesized. Neither of these repair mechanisms is foolproof and many types of mutagenic agents may produce a variety of alterations in DNA structure and function that are easily recognized as phenotypic changes.

1 What is meant by a gene that operates on a constitutive basis?
2 Give an example of a mutagenic agent.
3 What is a dimer?

MUTAGENESIS AND CARCINOGENESIS

Mutations that show themselves as phenotypic changes are easily observed in microbial populations. The rapid growth rate and short generation time of microbes makes possible the identification of the few cells that have experienced a change in their DNA as a result of exposure to a mutagenic agent. By exposing a population of microbes to a test chemical and counting the number of mutants that appear after culturing, it is possible to determine whether the chemical is a mutagen and at what concentration permanent changes

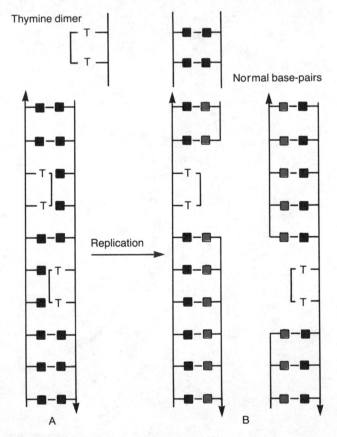

FIGURE 11.2
The DNA strand shows several thymine dimers formed by exposure to ultraviolet radiation. After DNA duplication, several "gaps" occur in the duplicated strands, since normal base pairs are not able to be formed.

are produced in the DNA. Because many mutagenic agents are also **carcinogenic** (cancer causing) to the eucaryotic cells of humans and animals, the carcinogenicity and mutagenicity of a chemical can be tested in much the same way. Such a carcinogenicity test has been developed by Bruce Ames and is known as the **Ames test.** The evidence that many cancers are caused by exposure to manmade and natural chemicals in our environment is mounting. Between 60 and 90 percent of cancer cases are believed to be environmentally induced by carcinogenic agents such as radiation and chemicals. Cosmetics, cleaning agents, food additives, industrial materials, adhesives, and many other synthetic materials all contain new chemicals that are constantly being introduced into the environment at levels never before experienced. Each year approximately 70,000 different chemicals become commercially available and ultimately may be ingested, inhaled, or absorbed into the body. The Ames test is a short-term test (2 days) of the potential carcinogenicity of many of these chemical compounds. Ames and his co-workers have produced a mutation in *Salmonella* that prevents its growth under normal conditions. However, if another mutation is caused by a test chemical, the microbe will be able to grow on agar plates. If the chemical does not cause this mutation, there will be no growth.

The test is run by exposing the bacteria to a number of dilutions of the suspect chemical, the microbes are mixed with a liver extract to simulate animal conditions, and the mixture is plated out on agar plates. The colonies growing on the plate are counted and compared to a control of a known carcinogen. A positive Ames test (appearance of colonies) indicates that the chemical can cause mutations and may be carcinogenic; however, a negative test may *not* mean that the chemical is safe to use because some carcinogenic agents do not cause mutations, and no carcinogenic test is perfect. Bacterial and human DNA are not the same and may not be affected by a chemical in the same way. To substantiate the results of the Ames test, long-term animal tests, which may take a year or longer, should be run on the suspect chemical. A number of chemical companies, including Dow and Dupont, have adopted the Ames test as part of their testing procedures. They have taken the position that new chemicals should be "guilty" of being carcinogens until proven "innocent." Tests on chemicals already in use are also being run; however, there is a resistance to ban them from use without running both the short-term Ames test and long-term animal test. For example, Tris, the flame retardant used in children's pajamas and other clothing, was identified by Ames and his associates in 1975 as a carcinogenic chemical that could be absorbed into the body. However, until animal tests confirmed the short-term Ames test, the chemical continued to be used. These animal tests required almost two years of research and evaluation. The same experience has also occurred with the chemicals used

in hair dyes. Pretesting new chemicals before they are mass produced and marketed may significantly reduce the incidence of cancer.

> **1** What natural methods are able to repair some types of mutations?
> **2** Of what value is the Ames test?

THE EFFECTS OF SELECTION

When a mutation occurs in a population of microbes, the total number of cells containing the change may be relatively small (for example, on the order of 1×10^{-10}). However, natural and unnatural (artificial) selection may quickly increase the percentage of the population that contains the new gene. Natural selection occurs as a result of differential reproduction. If a microbe containing the mutation is better able to survive and reproduce than those lacking this change, the mutant will increase naturally in frequency through time. Selection might also affect the frequency of mutants in a mixed culture of penicillin-resistant and susceptible microbes. By adding penicillin to the culture, the drug will act as a selective agent and rapidly destroy the susceptible cells. In only a few hours, the entire population might be composed of penicillin-resistant microbes.

The combined action of natural and artificial selection on resistant cells has resulted in an increase in drug-resistant pathogens throughout the world. The selection of resistant mutants, however, does not totally explain the emergence of resistant organisms. Other genetic mechanisms involving the acquisition of new genes play an important role in this phenomenon. Selective agents do not act directly on the genotype of microbes. Selection is based on how the agent interacts with the phenotype of the cell. If the genes are inactive for any reason, the microbe will be adversely affected by the selective agent. Thus, a mutation that is beneficial to the microbe, but not phenotypically expressed, can be lost from the population. Since the phenotype of a cell is the result of the expressions of its functioning gene combinations, natural selection operates most efficiently on new combinations of genes that are generated in a cell population. Any process that results in the integration of new combinations of genes together in a single cell is called **genetic recombination.** When fertilization (conjugation) occurs in eucaryotic organisms, genes donated by each of the parents are recombined into a single cell—the zygote. Since unique genes (due to mutations) and gene packages (due to crossing-over and independent assortment) are brought together in a single cell (genetic recombination), this new individual will have gene combinations (a genotype) not found in either parent.

In procaryotes there is no true fusion of cells. However, a portion of a DNA molecule from a donor may be transferred to a recipient cell to form a partial zygote called a **merozygote.** The incoming segment of DNA, the **exogenote,** or **extrachromosomal,** DNA, may join into the recipient DNA, the **endogenote,** by breakage and reunion (Figure 11.3). This process results in a recombinant strand of DNA since the original genes of the recipient have been replaced by the new exogenote genes. As a result, the genetic makeup of the cell has new genes and new gene combinations that may enable it to survive better in a changed environment. We now know that the exogenote DNA does not always join with the endogenote by genetic recombination, but may remain free in the cytoplasm of the recipient. This separate loop of DNA may function in the host without replicating. As a result, only one cell in the population will contain the exogenote at any particular time. As the population increases, the genetic uniqueness of this single microbe may be overshadowed by the other cells. However, self-replicating exogenotes may remain in the population indefinitely. These separate loops of genetic material replicate on their own and are distributed to the newly forming cells during binary fission in much the same manner as the host chromosomal genes. Cells which contain functional self-replicating exogenotes demonstrate a great amount of genetic variation not seen in cells that lack these extra genes. The possession of these genes may be a selective advantage because of increased genetic variety. It is also an advantage to be able to transfer exogenote and endogenote genes to other cells to maintain this variety. The three methods of transferring exogenote and endogenote DNA that have been identified in bacteria are transformation, conjugation, and transduction.

1 Why are new combinations of genes important to evolution?
2 What is the difference between extrachromosomal DNA and endogenote DNA?
3 Of what value is extrachromosomal DNA to a cell?

TRANSFORMATION

The **transformation** process occurs when the recipient cell takes in a segment of "naked" DNA from the environment. This extrachromosomal DNA may have been released from a donor cell while it was alive or after it died. The first evidence of this process was seen in 1928 when Fred Griffith noted that nonencapsulated *Streptococcus pneumoniae* ("R" strain) could be changed to the capsule-producing form ("S" strain). This was done by culturing live "R" strain cells in media containing dead capsule-producing streptococci (Chapter 5). Today a number of other genera have been identified that may also undergo transformation. *E. coli, Haemophilus, Bacillus,* and *Pseudomonas* are all able to be transformed. However, cells do not simply "take in" extrachromosomal DNA. Those microbes with the genetic ability to take up naked DNA and be transformed are called **competent cells.** Competent cells may operate naturally or they may be artificially stimulated to take in the DNA by altering the environment in which the culture is being grown. For example, *E. coli* can only become competent by being cultured in media with a high concentration of calcium ions.

Transformation occurs in three stages (Figure 11.4). A large segment of DNA is first bound to a special receptor site on the surface of the competent cell. The segment is then cut into smaller, more manageable pieces by a DNAase

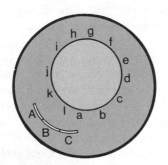
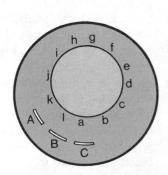
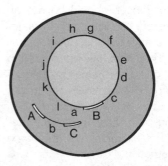
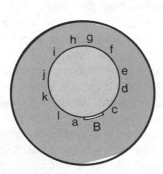

FIGURE 11.3
The process of genetic recombination involves the replacement of old genes with new genes from some outside source. In this sequence, the original bacterial DNA loop (called the endogenote) is first segmented by special enzymes at locations that are homologous to the new gene sequence (exogenote). A shift takes place which removes one of the original genes and replaces it with the new gene. The loop is repaired, and the extra genes found in the cytoplasm are metabolized by the cell in other biochemical pathways.

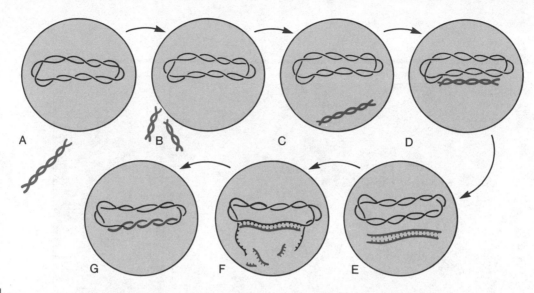

FIGURE 11.4

Transformation in bacteria occurs in a specific series of events. The exogenote DNA (A) attaches to a special receptor site on the cell wall. A DNAase enzyme breaks the DNA into smaller pieces (B) before the section of DNA is actively taken into the cell (C). Once inside, the new DNA pairs with homologous genes on the endogenote DNA (D). The transforming DNA is separated into single strands (E), and one is discarded. The host DNA opens (F) to allow the new nucleotide sequence to "fit into" the DNA loop (G).

enzyme released by the recipient. Finally, the attached segment of DNA is actively moved into the cell where it is prepared for recombination with the endogenote. The transformation process plays an important role in forming new gene combinations and creating genetic variety in microbes; however, conjugation probably plays a more important role.

CONJUGATION

As in the case of transformation, the ability to undergo conjugation is genetically determined. **Conjugation** in procaryotes is the transfer of genes from one cell to another by direct contact. The genes that control the process are located on an extrachromosomal piece of DNA called a plasmid. A **plasmid** is a small, circular piece of extrachromosomal DNA that is self-replicating and contains a limited number of genes (about 40) (Figure 11.5). Plasmids that control such characteristics as fertility are called F factors, and those that contain

genes for transferable drug resistance are known as R factors (Table 11.1). To date no Gram-positive cells have been shown to conjugate; however, many F factor-containing Gram-negative bacteria may undergo plasmid-controlled conjugation. Microbes which contain the F factor are desig-

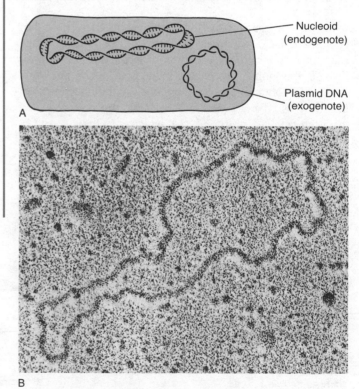

FIGURE 11.5

(A) Plasmids are "extra" loops of DNA that are not essential to the life of the cell. However, their presence may aid the cell in surviving the effects of an antibiotic. Plasmids that contain genes for drug resistance are called R factors, while those that determine fertility characteristics are called F factors. (B) The electron micrograph shows a plasmid known as pSC101 that was extracted from an *E. coli* cell. Some of the genes on this plasmid enable *E. coli* to resist the antibiotic tetracycline. (Electron micrograph courtesy of S. N. Cohen, Stanford University School of Medicine.)

TABLE 11.1
Types of plasmids.

Type	Organisms
Conjugative plasmids:	F factor, *Escherichia coli*, pfdm, K, *Pseudomonas*, P, *Vibrio cholerae*
R plasmids:	
Wide variety of antibiotics	Enteric bacteria, *Staphylococcus*
Resistance to mercury, cadmium, nickel, cobalt, zinc, arsenic	
Bacteriocine and antibiotic production:	Enteric bacteria, *Clostridium*, *Streptomyces*
Physiological functions:	
Lactose, sucrose, urea utilization, nitrogen fixation	Enteric bacteria
Degradation of octane, camphor, naphthalene, salicylate	*Pseudomonas*
Pigment production	*Erwinia*, *Staphylococcus*
Virulence plasmids:	
Enterotoxin, K antigen, endotoxin	*Escherichia coli*
Tumorigenic plasmid	*Agrobacterium tumefaciens*
Adherence to teeth (dextran)	*Streptococcus mutans*
Coagulase, hemolysin, fibrinolysin, enterotoxin	*Straphylococcus aureus*
Genera in which some evidence for plasmids has been obtained: *Achromobacter, Aerobacter, Aeromonas, Agrobacterium, Alkalescens, Arthrobacter, Bacteroides, Bacillus, Bartonella, Chromobacterium, Citrobacter, Clostridium, Enterobacter, Erwinia, Escherichia, Hafnia, Klebsiella, Micrococcus, Neisseria, Paracolobactrum, Proteus, Providencia, Pseudomonas, Rhizobium, Salmonella, Serratia, Shigella, Staphylococcus, Streptococcus, Streptomyces, Vibrio, Yersinia*	

SOURCE: T. D. Brock, *Biology of Microorganisms*, 3rd ed. Reprinted by permission of Prentice-Hall, Inc., Englewood Cliffs, N.J., © 1979.

nated F^+, or males, since they serve as donors. Those that lack the F factor are designated as F^-, or females, since they serve as recipients. When the F factor is donated to an F^- cell, the female becomes a male, or F^+. The first stage in plasmid-controlled conjugation involves the attachment of the two cells. The sex pilus, a filamentous structure extending from the cell wall of the F^+ cell, attaches to the cell wall of an F^- bacterium. A conjugation tube is formed between the cells and as this tube is formed, the plasmid is replicated inside the donor cell. This process takes place in the same way that the host nucleoid is duplicated. One of the F factor loops remains attached to the inner surface of the F^+ cell while the other plasmid is free to move through the conjugation tube into the recipient cell (Figure 11.6). After the transfer has been completed and the cells separate, both the recipient and the donor contain plasmids. The percentage of F factor-containing bacteria in a population increases if the microbes are crowded into close contact. Plasmid-controlled conjugation occurs more easily and successfully within the Gram-negative enteric species found as normal flora in

FIGURE 11.6
The F factor-containing donor cell makes contact with the F^- cell and forms a conjugation tube through which the newly replicating F factor may be transferred. When the F^- recipient takes in the F factor, it changes into an F^+ cell.

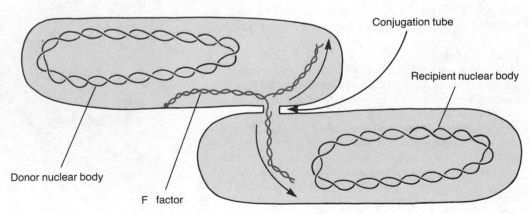

Conjugation tube

Recipient nuclear body

Donor nuclear body

F factor

the intestinal tract. These bacteria show a great amount of genetic variety. Bacterial populations that lack this close contact usually have a lower rate of conjugation, fewer F factors, and less genetic variety.

In some cases, the F factor plasmid may become integrated into the endogenote DNA (Figure 11.7). These cells have been carefully studied, and the frequency with which they genetically recombine with F⁻ cells is 1000 times greater in comparison to F⁺ cells. Therefore, these cells are

called **Hfr** bacteria, or **high-frequency recombinants.** During conjugation between an Hfr and an F⁻ cell, all of the genetic material in the Hfr undergoes DNA duplication. One of the loops is then broken within the Hfr gene sequence and begins to move through the conjugation tube into the F⁻ recipient (Figure 11.8). The entire length of DNA rarely moves into the recipient, since the two cells are being held together by such a fragile connection. Therefore, it is highly unlikely that the recipient will receive the entire

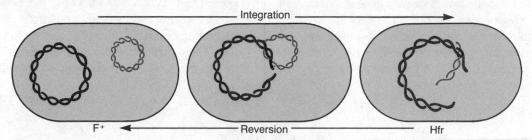

FIGURE 11.7

The F⁺ plasmid found in donor cells may attach and become integrated with the endogenote DNA. When this occurs, it is called an Hfr cell. The integrated plasmid may also revert to the F⁺ form if the DNA separates from the host loop.

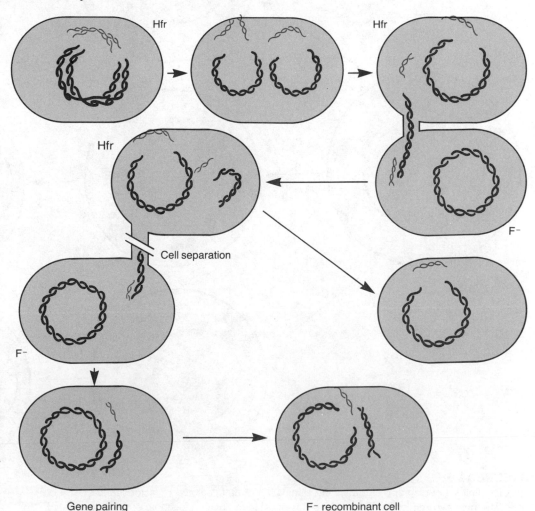

FIGURE 11.8

When conjugation occurs between an Hfr and an F⁻ cell, the result may be genetic recombination; however, there is little chance that the F⁻ cell will become an Hfr or an F⁺ cell. This is unlikely because the Hfr gene is not moved into the recipient as an intact unit. Notice in the diagrams that after conjugation, the Hfr cell remains Hfr while the F⁻ cell is genetically recombined.

1 What is the difference between conjugation and transformation?
2 What is a plasmid?
3 What is an F factor?
4 What is an R factor?

sion, a complete line of offspring will be produced that have these new characteristics. Any population of microbes produced by asexual reproduction from a single parent cell is called a **clone**.

Genetic variation may also be produced in conjugating populations if the Hfr cell converts to another form containing a plasmid known as F′ ("F prime"). An F′ plasmid is formed when the Hfr segment detaches from the host DNA loop and mistakenly carries with it some of the adjacent host genes (Figure 11.9). When this happens, the newly formed F⁻ plasmid contains endogenote genes from the host cell that may be transferred to other F⁻ cells during regular conjugation. These genes may then be recombined with the genes of the recipient cell. Plasmid-controlled conjugation generates

Hfr sequence and be converted to an Hfr or F⁻ cell. More importantly, the F⁻ cell will receive new genes from the Hfr cell that can be genetically recombined with the endogenote. When this recombination occurs, the recipient will have new genes and gene combinations not previously found in the population. After reproduction by binary fis-

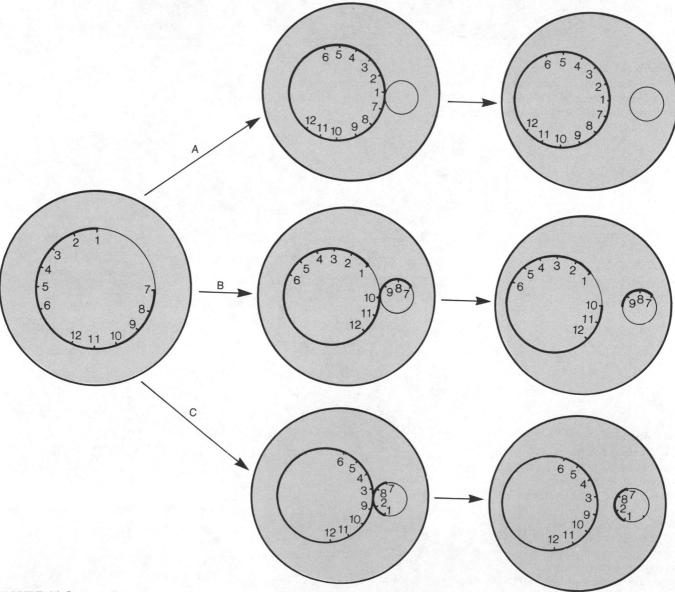

FIGURE 11.9
(A) The first set of diagrams illustrates the elimination of an integrated F′ factor strand in an Hfr cell. The next two series (B, C) show how it is possible to have endogenote genes incorporated into the F factor. F factors that contain genes from the host DNA are known as F′ plasmids.

genetic variation in a number of ways. The most important outcome of this form of gene transfer is the formation of new gene combinations. If these combinations provide the cell with a selective advantage, such as resistance to drugs, the recombinant microbe may produce a clone that could have a great medical significance.

GENE TRANSFER, PLASMIDS, AND DRUG RESISTANCE

The ultimate source of all alleles is the process of mutation. These changes may occur in either the endogenote or exogenote of a cell. Genes transmitted from one cell to another may increase in frequency in a microbial population. Genes for drug resistance that are found on the endogenote provide the cell with **chromosomal resistance.** Those found on plasmids provide **extrachromosomal resistance.** For example, chromosomal resistance to strepto-

mycin occurs with a frequency of about 10^{-10} and primarily centers on the cell's ability to modify ribosomal structure. Drugs such as streptomycin are unable to attach to these modified ribosomes and interfere with protein synthesis (Figure 11.10). Other chromosomal mutations exist which enable bacteria to resist the effects of kanamycin, erythromycin, and tetracycline. Extrachromosomal resistance is usually associated with the production of enzymes that are able to destroy such drugs as the penicillins, cephalosporins, and chloramphenicol. Also found on a plasmid are the genes that code for the production of the enzyme responsible for blocking the entrance of erythromycin. Gram-negative pathogens such as *E. coli* and *Pseudomonas aeruginosa* have between 60 and 90 percent of their resistance genes located on plasmids. Since many of these R factors (like F factors) are self-replicating and self-transmitting, they are able to be spread very quickly by conjugation (and transformation) through a population of bacteria. The likelihood of this transmission is greatly increased when Gram-negative

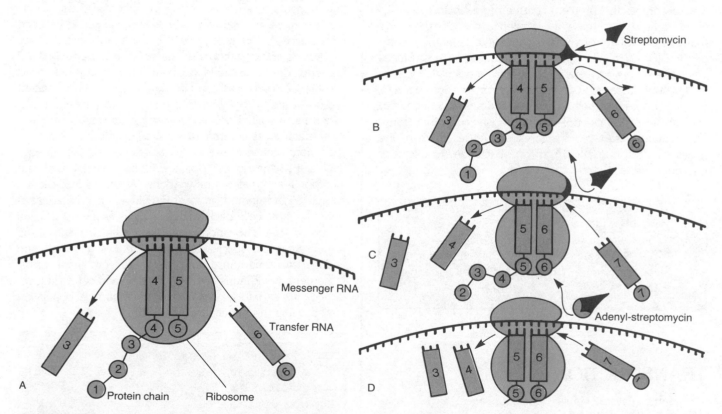

FIGURE 11.10
Mode of action is different in chromosomal and in plasmid-mediated antibiotic resistance. Streptomycin inhibits protein synthesis in bacteria. Ordinarily transfer RNA assembles amino acids on ribosomes according to instructions recorded in messenger RNA (A). Streptomycin attaches itself to a site on the ribosome and prevents protein synthesis, perhaps by blocking transfer-RNA attachment (B). Mutation of a chromosomal gene changes the ribosome's structure so that the streptomycin cannot bind (C), but R factor genes make an enzyme that attaches an adenine molecule to the streptomycin, which no longer fits the unchanged ribosome (D). (From R. C. Clowes. "The Molecule of Infection Drug Resistance." *Scientific American,* April, 1973.)

bacteria are found in highly concentrated populations and subjected to the artificial selective pressure of antibiotics. These circumstances are typically associated with the normal intestinal flora of humans and many domesticated animals. Hospitals and military organizations using large amounts of antibiotics in the treatment of Gram-negative infections have also demonstrated a high incidence of drug resistance. In one study, farm workers were found to harbor an abnormally large percentage of antibiotic-resistant, Gram-negative bacteria in their fecal material. These farmers regularly supplemented their pig feed with antibiotics. This same heavy selective pressure is placed on populations of Gram-negative pathogens in hospital environments. As more patients are treated with larger doses of broad spectrum antibiotics to control already resistant microbes, the frequency of resistance in pathogens such as *Proteus, Serratia, Pseudomonas,* and *Enterobacter* has drastically increased. The fact that patients and staff alike are confined to the hospital environment only serves to increase the possibility of exchanging drug-resistant bacteria and increases the severity of the problem. Furthermore, increased resistance may also increase the virulence of pathogens and make the treatment of Gram-negative infections more complicated (Box 19). Gram-positive pathogens such as *Staphylococcus aureus* have also experienced a dramatic increase in drug resistance. However, conjugation has not been observed, and a third method of plasmid transfer is needed to explain the high incidence of drug resistance in this group. The **transduction** process relies on viruses to carry the genes for drug resistance from one bacterium to others of the same species.

---

1   What does Hfr mean?
2   Where is a plasmid found within a cell?
3   Name two types of antibiotics that are sometimes ineffective due to chromosomal mutations in bacteria.

TRANSDUCTION

Many bacterial viruses (bacteriophages) have a **lytic life cycle** (Figure 11.11). In this cycle, the virus adsorbs to the surface of the host cell and injects its nucleic acid core through the outer covering. Once inside, the nucleic acid takes command of the host's metabolism to synthesize more virus particles. After the synthesis is complete, the host cell is ruptured to free new virions. The lytic cycle takes place very quickly (about 40 minutes) and there is no delay from the time of initial penetration to lysis of the host. During the lytic cycle, the DNA of the host is broken into small segments that are about the same size as virus nucleic acid. Occasion-

ally, in the case of certain types of bacteriophages, during the assembly of the complete virion a small segment of host DNA is incorporated into the virus protein coat in place of the phage genome. If the DNA found in this virus coat, for example, contains a gene for drug resistance, it may, on infection of another cell, be transferred to that cell and integrated into the endogenote, making the recipient cell drug resistant. This integration resembles the events that take place during transformation. Because any segment of host DNA may be transferred in this way, the process is called **generalized transduction.**

In the **lysogenic,** or **latent, life cycle,** lysis of the cell does not take place after the nucleic acid has penetrated the host. The virus integrates into the host chromosome, where it is called a **prophage** (previrus), and it replicates as part of the host chromosome (Figure 11.12). In this condition, the host cell is called a **lysogenic bacterium,** because it is capable of being destroyed at a later time if the virus prophage becomes active and reenters the lytic cycle (Figure 11.13). The delay of host destruction that is characteristic of the lysogenic cycle results in the formation of many generations of lysosgenic bacterial cells. During the period between virus infection and lysis, these virus-infected bacteria may function as active pathogens. The potential for **specialized transduction** of bacteria begins when the prophage shifts into the lytic cycle and accidentally incorporates an adjacent section of the host DNA into the virus core (Figure 11.14). When the modified prophage emerges from the host chromosome, it takes command of cell metabolism and synthesizes more protein coat and nucleic acid core. Since there is a different gene on the plasmid, it too is synthesized and incorporated into all the new virions. When the lytic cycle is completed, a hundred or so of the virions may be released from the host cell. Each of these contains the new gene (Figure 11.15). The virions may infect other bacteria of the same susceptible species and integrate this gene into the endogenote, thus transducing them to recombinant forms. If the gene being transferred by transduction is for drug resistance, it is easy to see how portions of a bacterial population may develop this genetic trait (Box 20).

---

1   What do viruses have to do with transduction?
2   A virus integrated into the host DNA loop is given what name?
3   How is specialized transduction related to drug resistance?

RELATEDNESS AND GENETIC ENGINEERING

The genetic material of a procaryotic cell is not a simple, single, static piece of DNA. Rather, this material is a complex

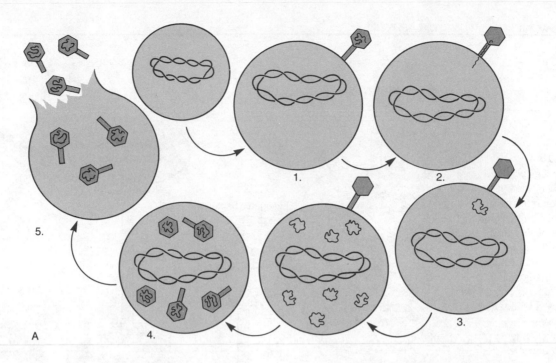

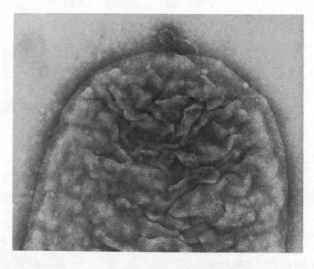

FIGURE 11.11

(A) There are five stages to the lytic life cycle of a virus. (1) The virion (virus particle) adsorbs to the surface of the susceptible host bacterial cell. (2) Penetration occurs when the nucleic acid core of the virion enters the cytoplasm of the host. (3) Synthesis of the protein coat and nucleic acid core of the virion takes place inside the host. (4) The entire virion is assembled. (5) The intact virions are released from the host cell in a burst. (B) The electron micrograph shows the attachment of the bacteriophage lambda to the surface of a host *E. coli* cell. Notice the long threadlike tail used to inject the virus DNA into the bacterium. (Electron micrograph courtesy of M. Schnoss, Stanford University Medical Center, Department of Biochemistry.)

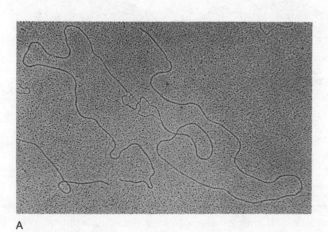

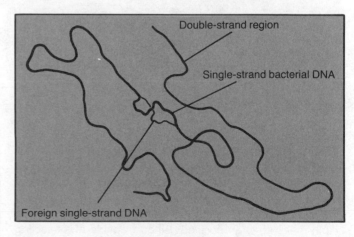

FIGURE 11.12

This illustrates how phage DNA is integrated into host DNA. The transducing phage here is φ80 psu₃ and contains approximately 2000 nucleotides. (Electron micrograph courtesy of N. C. Wu and N. Davidson, California Institute of Technology.)

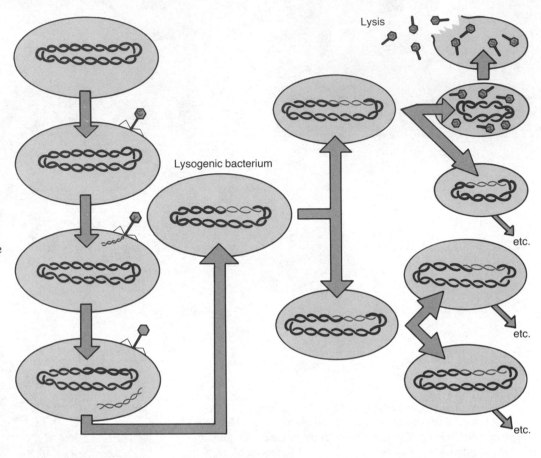

FIGURE 11.13

The lysogenic life cycle of a virus begins with adsorption and penetration. However, instead of continuing through lysis, the virus nucleic acid integrates into the host nucleoid. Since the host does not recognize the infection, it continues through its normal life cycle of repeated binary fission. Occasionally some environmental change stimulates some of the descendents of the infected cell to shift into the lytic cycle. Those that take this course are killed by the release of the virions. Bacteria that have prophage integrated into their DNA are referred to as lysogenic bacteria.

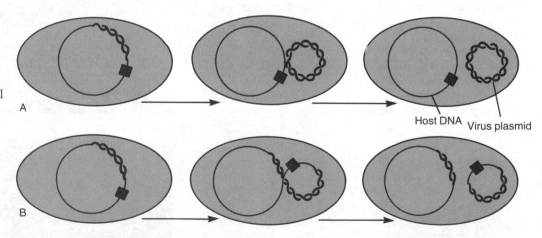

FIGURE 11.14

(A) This illustrates how the prophage normally leaves the endogenote DNA before entering the synthesis phase of the lytic cycle. (B) An additional gene for drug resistance may be integrated into the virus nucleic acid. Since the only gene adjacent to the prophage is incorporated into the virion, this process is known as specialized transduction (g = galactose gene).

of separate units, many of which may be dynamically interchanged among one another and exchanged between different cells. At one time or another, plasmids, prophages, and nucleoids have many common properties. They are circular double helices of DNA capable of self-replication, and they can produce proteins that affect the phenotype of the cell in which they are located. All are capable of interacting with one another.

$$\text{plasmid} \rightleftharpoons \text{prophage}$$
$$\text{nucleoid}$$

Many have been found to induce their own transfer by conjugation or the transfer of other genetic units. As more is learned about these various elements, the question of a

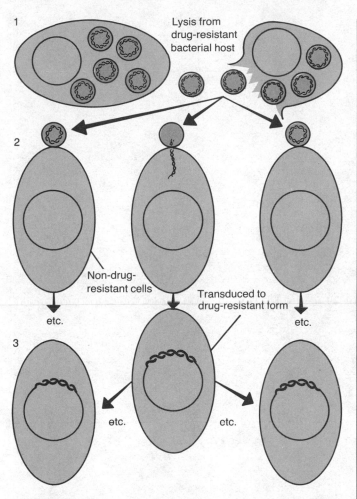

FIGURE 11.15
Specialized transduction takes place in stages. (1) A drug-resistant cell is infected with a lysogenic virus which forms a prophage. (2) When the prophage shifts into the lytic phase, it picks up the gene for drug resistance and incorporates it into all of the newly synthesized virions. (3) After bursting from the host cell, the drug-resistant gene is incorporated into the endogenote of susceptible cells by virus infection.

possible evolutionary relationship arises. It is possible that they all arose through a process of interconvertibility. If plasmids were to combine, resulting in an element essential for the functioning of a cell, such a cointegrated element could have become a nucleoid. If a prophage lost its ability to integrate into the host chromosome and become a protein-coated virus, it could have become a plasmid and possibly a plasmid might have gained the genes needed to form a protein coat and become a virus. The exact sequence of these proposed events may never be known; however, there is no doubt that a very close relationship exists among these genetic elements. The knowledge gained from research into these structural and functional relationships has in recent years

led to the development of tools and techniques for the controlled laboratory recombination of genes from a variety of different organisms. These processes are more commonly referred to as **gene manipulation,** or **genetic engineering.**

Plasmid-controlled conjugation is one of the most widely used gene transfer methods producing genetic recombinants in the laboratory. Detailed research into the structure and function of DNA has enabled microbiologists to link together specified gene sequences and existing plasmids (Figure 11.16). Plasmids produced in a test tube from the separate genetic elements of unrelated organisms are called **chimeras** (from Greek mythology, a creature with the head of a lion, body of a goat, and tail of a serpent). If the gene segments linked together are from related organisms, they are called **hybrid** plasmids. The term **composite** plasmid is used to refer to artificially constructed plasmids produced in a test tube. One of the first hybrid plasmids was produced in 1973 by Stanley Cohen and Annie Chang at Stanford University. This plasmid contained genes from two different *E. coli* cells. Chimeras have been produced by combining genes from *Staphylococcus aureus* and *E. coli;* and another combination has been made using the DNAs of *E. coli* and *Xenopus laevis,* a toad. After a composite plasmid is produced in a test tube, the methods of transformation are used to insert the plasmid into a host cell. This is accomplished by chemically altering the environment of the competent cell and stimulating it to take in the plasmid DNA. Once inside the host, the plasmid functions as any other normal plasmid, synthesizing proteins and replicating. In many instances, a plasmid located inside a host cell overreproduces itself. This results in the formation of a cell that contains numerous replicas of the plasmid. Scientists have taken advantage of this reproduction and cloned large amounts of genes specifically introduced into the cell on composite plasmids. These cells can be disrupted and the desired genes isolated in pure form. These may then serve as a further source of DNA for genetic manipulations in other cells, or they may be studied separately. Some of the genes that have already been cloned include histone genes from sea urchins, the gene for mouse mitochondrial DNA, the alpha and beta globin genes of rabbits, and genes for ribosomal RNA from a number of different eucaryotic cells.

The variety of genes that may be artificially introduced into bacteria appears to be endless, as long as essential procedures are followed. This means that the potential for permanently altering the genetic makeup of an organism is within the realm of possibility. Knowledge gained from the genetic engineering of microbes may also find application in eucaryotes. Mankind now has the tools and techniques to significantly modify many living organisms, including himself. This knowledge and ability places a great moral and ethical re-

CONJUGATION is a gene-transfer mechanism that is not limited to members of the same species or even to the same genus. Research has demonstrated that many types of Gram-negative bacteria may enter into this process.

TABLE 11.2
Efficiency of conjugation between different genera of Gram-negative bacteria.

F-lac Donor	Recipient	Frequency of F-lac Transfer*
Escherichia coli	*Escherichia coli*	$10^{-1}-10^{-3}$
Escherichia coli	*Salmonella typhi*	$10^{-4}-10^{-5}$
Escherichia coli	*Proteus mirabilis*	$10^{-5}-10^{-6}$
Salmonella typhi	*Escherichia coli*	$10^{-4}-10^{-5}$
Salmonella typhi	*Proteus mirabilis*	$10^{-4}-10^{-5}$
Salmonella typhi	*Serratia marcescens*	$10^{-7}-10^{-8}$
Salmonella typhi	*Vibrio comma*	$10^{-5}-10^{-6}$
Proteus mirabilis	*Escherichia coli*	$10^{-5}-10^{-6}$
Proteus mirabilis	*Proteus mirabilis*	$<10^{-10}$

SOURCE: Data supplied by L. S. Baron, Walter Reed Army Institute of Research.
*These frequencies do not reflect the fact that higher-frequency recipients can be obtained from other populations.

BOX 19
CROSSING AND CRISS-CROSSING

This finding has stimulated interest among many microbiologists for two reasons. First, biologists define a species based on its ability to sexually reproduce (produce viable genetic recombinants) *only* among members of the same group. If conjugation and genetic recombination take place between two bacteria such as *Salmonella typhi* and *Serratia marcescens*, the question may be asked, "Have these microbes been misclassified and should they be placed in the same genus?" Second, there is concern that during conjugation and recombination, genes from a pathogen such as *Salmonella typhi* may be transferred to a beneficial organism such as *E. coli*. If the genes transferred are those associated with the development of pathogenicity, might not *E. coli* itself become a pathogen? Both of these concerns are under investigation by microbial geneticists and taxonomists. The question of relatedness may never be answered completely; however, a more accurate evolutionary picture may result from such research. The question of naturally transferred pathogenicity may be unimportant when considering the rarity of this event and the fact that such recombinants may not be favored by natural or artificial selection forces. These "criss-crossed" recombinants may be "crossed out" by natural selection and competition.

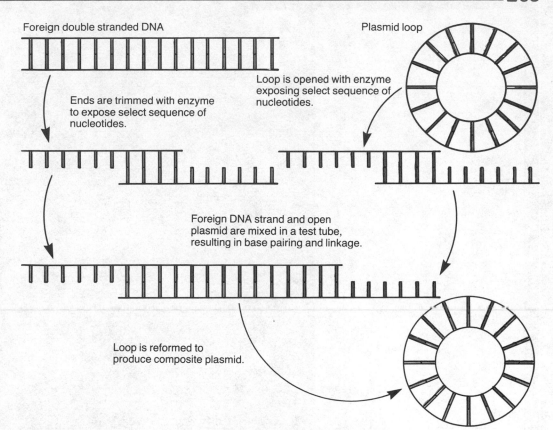

Foreign double stranded DNA

Plasmid loop

Ends are trimmed with enzyme to expose select sequence of nucleotides.

Loop is opened with enzyme exposing select sequence of nucleotides.

Foreign DNA strand and open plasmid are mixed in a test tube, resulting in base pairing and linkage.

Loop is reformed to produce composite plasmid.

FIGURE 11.16
The series of events shown here are all able to be carried out in a test tube. Composite plasmids produced by this method may be inserted into host cells by transformation.

sponsibility on those who might use these methods of genetic recombination. The areas of application are numerous and could be of great benefit. In the treatment of diseases, genes may be deliberately introduced into eucaryotic cells that lack the ability to synthesize essential molecules. For example, it may be possible in the future for insulin-producing genes replicated by cloning to be inserted into receptive human cells, resulting in a permanent cure of diabetes. Nitrogen-fixing genes now found only in a limited number of species may be cloned and inserted into the receptive cells of corn, wheat, rye, and other essential food plants. This would drastically reduce the amount of artificial nitrogen fertilizer necessary to cultivate these crops. Genes for the production of hormones, enzymes, vaccines, and drugs could be inserted into bacteria. These recombinant microbes could then be grown in large culture vessels and their metabolism products isolated at relatively low cost.

There was a concern within the scientific community about the possible misuse or indiscriminate application of recombinant DNA work. Without adequate guidance and control measures, scientists considered the possibility that new strains of disease-causing viruses could be produced that might be difficult, if not impossible, to control. Multiple-drug resistant plasmids could be produced in combination with toxigenic or other virulence factors that might be hazardous to the public. In order to safeguard scientists and the public from possible contamination with recombinant forms, however,

special containment procedures have been established. In addition, scientists have developed special strains of host bacteria that are extremely difficult to grow except in very selective environments. R. Curtiss has developed an *E. coli* strain that cannot synthesize the murein (peptidoglycan) layer of the cell wall except under special laboratory conditions; is sensitive to bile salts, detergents, and ultraviolet light; is unable to serve as a recipient in matings that have any of the five most common plasmids found in enteric bacteria; and is resistant to transduction by certain *E. coli* phage. With all these built-in growth limitations, this strain of *E. coli* could not likely survive outside the laboratory.

The successful development of genetic recombination techniques has enabled many scientists to manipulate the genes of a variety of cell types for the benefit of mankind. This has opened the door for the development of a new field known as **biotechnology** and spawned great interest on the part of the chemical-pharmaceutical industry. The range and variety of anticipated genetic processes and products may be limited only by the scope of the imagination. While most of the hoped-for products are 5 to 20 years in the future, a number of chemical-pharmaceutical companies have actively entered the area of biotechnology by employing or establishing research scientists skilled in the manipulation of genes. Their goal is to combine technological and marketing skills to produce gene products that are readily available to the public.

BOX 20

SOME UNPLEASANT FACTS

THE scope of the resistance problem posed by drug-resistant microbes can be seen by a review of statistics from around the world. Between 1936 and 1942 *Neisseria gonorrhoeae* developed almost total drug resistance to the sulfonamides. Fortunately, this event happened just as the "wonder drug" penicillin was being introduced. However, the frequency of penicillinase-producing *Neisseria gonorrhoeae* (PPNG) has increased steadily since the introduction of this antibiotic. In fact, some strains of *N. gonorrhoeae* have been identified that are not only resistant to the drug, but actually thrive in its presence! An increase in the frequency of these strains could have devastating effects for the estimated 3 million people that become infected each year in the United States. The use of alternative drugs to cope with this problem is effective, but there are drawbacks. These other antibiotics such as spectinomycin may be more costly, could also meet with resistance, and may not effectively eliminate *Treponema pallidum*, a coinvader in venereal disease cases. A near relative of *N. gonorrhoeae*, *N. meningitidis*, was also highly susceptible to sulfonamides prior to 1962. As with *N. gonorrhoeae*, this pathogen too has developed near total resistance. In addition, an estimated 65 to 85 percent of *Staphylococcus aureus* currently found in U.S. hospitals is penicillin-resistant due to the transduction of penicillinase genes. The United States is not the only country experiencing this problem. In Japan, the percentage of multiple-drug-resistant *Shigella* rose 60 percent between 1955 and 1968, and in England, *Salmonella typhimurium* increased in multiple-drug resistance by 21 percent between 1961 and 1964.

As a final note on this problem, the following appeared January 6, 1978, in *Morbidity and Mortality Weekly Report*, published by the Center for Disease Control in Atlanta.

FOLLOW-UP ON MULTIPLE-ANTIBIOTIC-RESISTANT PNEUMOCOCCI—SOUTH AFRICA

From May to November, 1977, type 19A pneumococci resistant to multiple antibiotics have been isolated from blood, cerebrospinal fluid (CSF), or pleural fluid of 15 South African children with meningitis (3 cases), pneumonia and bacteraemia, or pneumonia and empyema. The children were from Johannesburg and Durban, and all were less than three years of age. Eight of these children died, including the three with meningitis. Many other hospitalized children have had pharyngeal or sputum cultures positive for multiple-resistant pneumococci and had signs and symptoms of pneumonia. Some of these children died without documentation of the etiology of their pneumonia. Almost all infections occurred in children previously hospitalized for other medical conditions, notably malnutrition, measles, and pneumonia.

Culture surveys in several South African communities demonstrate that antibiotic-resistant pneumococci have at least five resistant patterns: (1) penicillin resistance only; (2) penicillin and tetracycline resistance; (3) penicillin, tetracycline, and chloramphenicol resistance; (4) penicillin and chloramphenicol resistance; and (5) penicillin, tetracycline, chloramphenicol, erythromycin, and clindamycin resistance. Strains in the last group are called multiply-resistant in this report. Several of the multiply-resistant pneumococci have also developed resistance to rifampicin.

TABLE 11.3

Prevalence of antibiotic-resistant pneumococci in nasopharyngeal cultures, South Africa, 1977.

	Number of persons cultured	Number of persons with pneumococci	Resistance Pattern (% Pneumococci)					
			Sensitive	Penicillin resistant	Penicillin, tetracycline resistant	Penicillin, chloramphenicol resistant	Penicillin, tetracycline, chloramphenicol resistant	Multiply-resistant
Durban								
Hospital 1	239	23	56.5	0	0	21.7	21.7	0
Hospital 2	408	38	21.1	7.9	0	52.6	18.4	0
Hospital 3	232	119	52.1	8.4	5.0	21.0	9.2	4.2
Other hospitals (N-4)	175	31	71.0	0	0	19.4	10.0	9.7
Community studies	472	245	98.0	0	0	1.6	0.4	0
Johannesburg								
Hospital 1	427	128	35.2	10.9	2.3	7.8	7.0	36.7
Hospital 2	116	81	2.5	3.7	0	2.5	0	91.4
Hospital 3	42	23	47.8	0	13.0	0	0	39.1
Hospital 4	51	12	75.0	0	0	0	0	25.0
Hospital 5	273	57	47.4	0	0	0	1.8	50.9
Hospital 6	72	14	57.1	0	0	0	0	42.8
Other hospitals (N-10)	382	57	94.7	1.8	1.8	0	0	1.8*
Community studies	902	236	91.9	4.2	0.4	0.4	0.4	2.5*
Cape Town								
Hospitals (N-5)	236	37	100.0	0	0**	0	0**	0
Total								
Hospitals (N-28)	2653	619	48.0	5.0	2.1	11.0	6.0	28.3
Communities	1374	481	95.0	2.1	0.2	1.0	0.4	1.2

*Multiply-resistant carriers could be traced epidemiologically to involved hospitals or other carriers.
**Two pneumococcal isolates, one resistant to penicillin and tetracycline and the other resistant to penicillin, tetracycline, and chloramphenicol have subsequently been identified.

The prevalence of resistant pneumococci found in community and hospital surveys in Durban, Cape Town, and the Johannesburg area are shown in Table 11.3. Carriers of multiply-resistant pneumococci were found in 8 of 28 hospitals surveyed and were largely in black children less than 3 years of age. Only 0.9 percent (4/434) of staff personnel caring for these children were found to be carriers. Spread among these hospitals appeared to have resulted from transfer of patients later found to be infected. Multiply-resistant pneumococci were found in 4.6 percent (7/152) of healthy family contacts of hospitalized carriers discharged home, which suggests that the spread into the community was slight. Community surveys were performed among factory workers and at day-care centers, orphanages, and health clinics; also included were surveys of new admissions to the hospitals involved. Carriers of multiply-resistant strains could be traced to previous contact with involved hospitals or other carriers.

Following these findings, hospital wards having carriers of multiply-resistant organisms were closed to new admissions. Carriers were placed in isolated wards or transferred to an isolation hospital. To evaluate control measures, subsequent surveys of patients and new admissions to the involved hospitals identified additional carriers who were then transferred to isolation hospitals. Erythromycin eradicated carriage of pneumococcal strains resistant only to penicillin, tetracycline, or chloramphenicol. A combination of rifampicin (30 mg/kg/day) and fusidic acid (30 mg/kg/day) given for 10 days was 63 percent effective in eradicating carriage of multiply-resistant pneumococci.

Various combinations and doses of novobiocin, cotrimoxazole, rifampicin, minocycline, and aerosol bacitracin were given with little success in eradicating carriage. Intravenous vancomycin (45 mg/kg/day) given twice daily for 5 days was effective in eradicating the organism. Nasopharyngeal colonization with a strain of *Streptococcus faecalis* which produces bacterocins and which inhibits the growth of the multiply-resistant pneumococci *in vitro* did not eradicate carriage of multiply-resistant organisms. No new cases of illness due to multiply-resistant organisms have been reported since these control measures were completed in November, 1977; however, reports of disease due to penicillin and chloramphenicol-resistant organisms continue in Durban, South Africa.

Gene products currently being produced include human insulin and growth hormones, foot-and-mouth vaccine for cattle, enzymes for the production of chemicals, and new bacterial strains useful in oil-recovery operations. It has been estimated that such biotechnological products will grow to a value of over $1.4 billion by the year 1985. A few companies currently involved in biotechnological research include Eli Lilly Pharmaceutical, Benentech, Shell Oil, Biogen, DuPont, Standard Oil, Abbott Laboratories, Dow Chemical, and Bristol-Myers.

SUMMARY

Genetics is the study of genes, how they produce characteristics, and how the characteristics are inherited. A gene is a portion of a DNA molecule composed of a specific series of nitrogenous bases that chemically codes for the production of a specific protein or RNA molecule, or serves as an operator in controlling the transcription of RNA within an operon unit. The catalog of an organism's genes, whether they are expressed or not, is called the genotype. The physical and chemical expression of the genotype is known as the phenotype. Alleles are different forms of a gene that are produced in cells as a result of mutation. Point mutation and other kinds of mutations may be caused spontaneously or by mutagenic agents. Because many mutagens are also carcinogens, it is possible to determine the carcinogenicity of a chemical agent by exposing microbes to the suspected agent and observing the mutation rate. This microbial test is known as the Ames test.

Natural and artificial selection increases the presence of mutations in populations by selecting for new gene combinations. Genetic recombination may be produced by transformation, conjugation, and transduction. These are processes that bring new combinations of genes together in a single cell. Many cells contain small circular pieces of extrachromosomal DNA that are self-replicating and contain a limited number of genes. Those controlling fertility are known as F factors while those containing genes for drug resistance are called R factors. Genes for drug resistance may occur on either the exogenote or endogenote of a cell. By transmitting these genes from one cell to another, they may increase in frequency in a population. Because of the close structural and functional resemblance of prophage, nucleoid, and plasmids, it is likely that an evolutionary relationship exists among them. The knowledge gained from research into these relationships has led to the development of the field of genetic engineering. Genetic engineering or manipulation holds great possibilities for the benefit or harm for mankind. Diseases may be cured or super-pathogens may be created.

WITH A LITTLE THOUGHT

On December 23, 1977, an outbreak of drug-resistant tuberculosis was described in a rural northern Mississippi county. Since then 5 more cases of tuberculosis due to organisms with confirmed primary resistance to isoniazid, para-aminosalicylic acid, and streptomycin (INH-PAS-SM) have been reported. This brings the total of such cases in the county to 19 since 1964.

Three of 5 new cases are known to be epidemiologically linked to cases previously reported with INH-PAS-SM resistance in the county. All 19 patients known to have this drug-resistant disease have been placed on alternative drug regimens, and all but one have had a good response. One patient was started on therapy but relapsed because of poor compliance. She is currently hospitalized in Illinois. Her 2-year-old son has been clinically diagnosed as having tuberculosis with negative bacteriology.*

As chief microbial geneticist of the investigation team looking into this case, describe the probable mechanism that was responsible for the development of this *Mycobacterium tuberculosis*.

---

*From "Follow-up on Drug-Resistant Tuberculosis—Mississippi." *Morbidity and Mortality Weekly Report*, vol. 27, no. 38, 1978. Center for Disease Control, Atlanta.

STUDY QUESTIONS

1 What three methods may result in the transfer of genes from one cell to another?

2 What is the difference among gene, genotype, and phenotype?

3 What is the difference among plasmid, prophage, endogenote, and extrachromosomal DNA?

4 Describe two kinds of mutations. How might each be produced?

5 What two methods may operate to repair a point mutation?

6 Why is the Ames test important to a microbiologist? a physician?

7 Describe the process of genetic recombination as it takes place inside a cell.

8 The increase in drug resistance among pathogens is related directly to gene transfer mechanisms. Describe how these operate.

9 What is multiple-drug resitance? How might it be generated in a population of pathogens?

10 Describe the construction of a composite plasmid and how it might be used to increase the genetic variety of microbes, plants, or animals.

SUGGESTED READINGS

AMES, B. N., J. MC CANN, and E. YAMASAKI, "Methods for Detecting Carcinogens and Mutagens in the *Salmonella* Mammalian Microsome Mutagenicity Test. *Mutagenicity Research,* 31:347–64, 1975.

CHAKRABARTY, A. M. "Plasmids in *Pseudomonas.*" *Ann. Rev. Genet.,* 10:7–30, 1976.

COHEN, S. N. "The Manipulation of Genes." *Scientific American,* 235:25–33, 1975.

COHEN, S. N. and J. A. SHAPIRO. "Transposable Genetic Elements." *Scientific American,* Feb., 1980.

CURTISS, R. "Genetic Manipulation of Microorganisms: Potential Benefits and Biohazards." *Ann. Rev. Microbiol.,* 30:507–33, 1976.

DEVORET, R. "Bacterial Tests for Potential Carcinogens." *Scientific American,* 241:40–49, 1979.

FALKOW, S. "Infectious Drug Resistance." *Pion Ltd.,* London, 1975.

FREIFELDER, D. "Recombinant DNA." *Readings From Scientific American,* W. H. Freeman and Company, San Francisco, 1979.

JACKSON, D. A., and S. P. STICH, editors. *The Recombinant DNA Debate.* Prentice-Hall, Englewood Cliffs, 1979.

NOVICK, R. P. "Plasmids." *Scientific American,* Dec., 1980.

WATSON, J. D. *Molecular Biology of the Gene,* 3rd ed. W. A. Benjamin, Menlo Park, Calif., 1976.

KEY TERMS

gene

mutagenic agent

photoreactivation

dark repair

carcinogenic agent

Ames test

genetic recombination

exogenote-endogenote-extrachromosomal DNA

transformation

plasmid

conjugation

F factors

generalized and specialized transduction

prophage

PPNG

genetic engineering

Pronunciation Guide for Organisms

Neurospora (nu-ros'por-ah)

PART 3
Host-Parasite Relationships

B

7µm

E

ACH microbe has unique life requirements, which are provided for by both the living and nonliving environment. However, microbes do not exist as isolated life forms but establish many different physical relationships with other organisms. Some of these symbiotic relationships are beneficial to humans, while others result in harm or death. The maintenance or reestablishment of a healthy balance between microbes and host requires a detailed understanding of the microbes, the organisms with which they interact, and the interactions themselves. The human body contains a number of systems that help resist both harm from microbes and damage from nonliving foreign materials. The efficient operation of these systems enables humans to establish good relationships with microorganisms and limit those that lead to disease and death. Many defense mechanisms are nonspecific in their actions. These represent a "first line of defense." The specific mechanism of host defense is known as the **immune system** and has the capability of recognizing the difference between "self" and "nonself." This system selectively and efficiently eliminates microbes and other specific material types that might harm the host. However, the immune system may at times function abnormally, causing harm to the body. When destructive antibody-antigen reactions occur, the host is more susceptible to microbial invasion and damage. In many cases, therapeutic control of abnormal reactions may be easily achieved, while in others the disease process still remains to be identified. If good health is to be maintained, there must be a balance between the host and parasite. One of the most important factors for preventing poor health or possibly host death is the development of immunity in an individual. However, in some individuals an effective immune system may be lacking or nonfunctional. In order to protect this type of person from infectious disease, a high level of public health must be established. By understanding the way infectious diseases move through a population of hosts and establishing good public health measures, these susceptible individuals will probably not become infected. This area of microbiology is known as **epidemiology.**

◀ *Bacteria (B) residing on the leg of the water flea* Daphnia. *Magnification* × *14,000. (Courtesy of R.A. Smucker and R.M. Pfister.)*

Symbiotic Relationships and Normal Flora

WE ARE NOT ALONE

SHIFTING RELATIONSHIPS

THE SKIN

THE ORAL CAVITY AND UPPER RESPIRATORY TRACT

THE GASTROINTESTINAL TRACT

THE GENITOURINARY TRACT

VIRULENCE: TOXINS AND INVASIVENESS

THE HEMOLYSINS AND OTHER SUBSTANCES

Learning Objectives

☐ Be familiar with the four possible symbiotic relationships that may exist among living organisms.

☐ Know the factors that affect a symbiotic relationship.

☐ Know the difference between resident and transient flora and give examples of each.

☐ Be familiar with microbes commonly found on the human skin and the nature of their relationship.

☐ Know which factors inhibit or foster microbial growth on the skin.

☐ Be familiar with microbes commonly found in the oral cavity and upper respiratory tract, and their relationship with these tissues.

☐ Know the current theory of caries formation.

☐ Be familiar with microbes commonly found in the gastrointestinal tract and their relationship with this tissue.

☐ Be familiar with microbes normally found in the genitourinary tract and their relationship with this tissue.

☐ Know the factors which contribute to the virulence of a bacterium.

☐ Understand the differences between exotoxins and endotoxins and give examples of each.

☐ Be able to define the term "invasiveness," give examples of invasive factors, and describe the nature of their activities.

☐ Understand the process of attenuation.

Each microbe has particular requirements for life and lives wherever the surroundings will meet these needs. This place, locality, or portion of an environment is defined by biologists as the habitat of an organism. In the habitat, each organism interacts with other organisms simply because they share the same living space. Each microbe will have its own niche, or functional role to perform, in its habitat. Just as the word "space" is the key to understanding habitat, the word "function" is the key to understanding the concept of niche. Included in a microbe's niche might be the kind of food that it needs, those organisms that feed on it, the temperature range and the pH range that it can tolerate, the kinds and

12

amounts of growth factors needed in the diet, and the amount of water that the microbe needs to stay alive. Some have very broad niches while others are narrow and have very specific requirements. The complete definition of a niche for an organism would include its interactions with other living things (biotic) as well as with the nonliving portions (abiotic) of the habitat.

Humans serve as suitable habitats for many microbes, which may be harmful, harmless, or beneficial. The skin, gastrointestinal tract, respiratory tract, and genitourinary tract all normally contain microbes. Whether any of these causes tissue damage depends on the susceptibility of the host and the pathogenicity and virulence of the microbe. The maintenance of good health depends on our understanding and controlling the relationships that exist between microbes and the human body.

WE ARE NOT ALONE

No organism lives alone even though it may have the ability to do so. Many autotrophs are able to grow in environments that contain only the most basic inorganic compounds; however, like heterotrophs in pure culture, cells in a population are constantly interacting with one another. In a natural setting, the interactions go beyond those that occur among members of the same species population and encompass other living things. Various close relationships may be established between two species of organisms which may or may not be beneficial to the organisms involved. A close physical relationship is known as **symbiosis.** The four fundamental symbiotic relationships are amensalism, commensalism, mutualism, and parasitism. **Amensalism** is a relationship between different species in which one species is harmed, but it is difficult to see how the other species is benefited. The classic example of an amensal relationship occurs between the mold *Penicillium* and the Gram-positive bacteria *Staphylococcus.* In this interaction, the mold releases the antibiotic penicillin, which has an inhibitory influence on the bacteria. The bacteria are obviously harmed, since they cannot grow and reproduce; but the presence or absence of bacteria seems to have little or no influence on the mold. If the mold is benefited, it must be very subtle.

In **commensalism,** one individual benefits but the other is not harmed. A commensalistic relationship exists between the bacterium *Streptococcus viridans* and humans. These bacteria are normally found in the human mouth and are not harmful. This environment is very favorable to them, and they may have all their nutritional needs met with little difficulty. Another example of commensalism is the relationship between the yeast *Candida albicans* and humans (Figure 12.1). The yeast lives on the surface of the skin and receives nutrients from glandular secretions. Humans receive no

benefit from this relationship; they simply serve as support for the yeast.

So far in these examples, only one individual has benefited from the association. However, there are situations in which individuals of two species live in close association and both benefit by the arrangement. This type of association is called **mutualism.** One interesting example of mutualism is seen in rabbits. Rabbits eat plant material that is high in cellulose even though they cannot produce the digestive enzymes capable of breaking down cellulose molecules into simple carbohydrates. While they cannot digest cellulose themselves, they manage to get energy out of these molecules with the help of special bacteria living in their digestive tract. These bacteria produce cellulose-digesting enzymes that break the cellulose molecules into smaller carbohydrate molecules

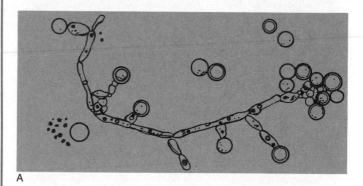
A

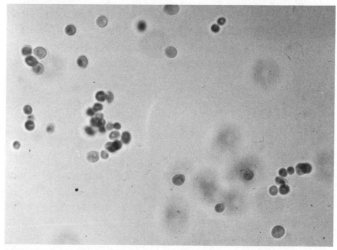
B

FIGURE 12.1

(A) Scrapings from the skin cultured on corn-meal agar will result in the growth of *Candida albicans.* This illustrates the filamentous hyphae of this microbe. Under normal circumstances, they are harmless; however, if conditions change, this microbe may cause the disease known as candidiasis. (B) *Candida* may also be found in the mouth, intestinal tract, and vagina. (Photo B courtesy of L. J. LeBeau, Department of Pathology, University of Illinois Medical Center, Chicago.)

which the rabbit's enzymes can break down further. The bacteria benefit because the rabbit's intestine provides them with a moist, warm, nourishing habitat. In return, the bacteria provide the rabbit with food molecules that the rabbit would otherwise be unable to digest.

Another mutualistic relationship exists between humans and *E. coli*. As already described in Chapter 10, we benefit greatly from our association with these intestinal bacteria. They provide us with vitamins, stimulate our immune systems, out-compete and produce antibiotics against pathogens, and release organic acids that optimize the movement of nutrients into the tissues. *Escherichia coli* also benefit greatly. As long as their host does not ingest excesses of antibiotics, these bacteria will live in their optimum growth environment.

In the last of the symbiotic relations, **parasitism,** the parasite lives in or on the host and benefits by obtaining nutrients and shelter, while the host is harmed. The relationship among bacteria, rats, fleas, and humans illustrates how complex parasitism can be. Fleas live in the fur of rats and feed on the rat's blood. The rat also serves as a host for other parasites, such as tapeworms, viruses, and bacteria. One bacterium, *Yersinia pestis,* does little harm to the rats, but if carried to humans, this bacterium may cause the disease known as "black death," or "plague." In this example, the flea can take in these disease organisms from one rat's blood and spread them to other rats as it moves from host to host. The rat's fleas may also hop onto humans and use them as hosts. Thus, the disease can be spread to people. The "black death" was spread in the mid-14th century in just this way, killing millions of people. In some countries of Western Europe, 50 percent of the population was killed by this disease (Figure 12.2).

SHIFTING RELATIONSHIPS

An interchange of relationships is also possible. Organisms that under normal circumstances are harmless or even beneficial can become deadly if the relationship shifts to parasitism. This shift depends on both organisms and the external environment. Commensalistic and mutualistic microbes are in a balance when their hosts are healthy. In fact, the term **health** may be defined from a microbiological point of view as a state of optimal well-being and balance between microbes and humans, and not merely the absence of disease. When nutritional imbalance, surgery, or an injury occurs, the body chemistry of the host is altered, **host susceptibility** increases, and microbes may shift into a **disease** relationship. For example, without minimal levels of essential amino acids in the diet, a person may develop health problems that could ultimately lead to death. The symptoms of a protein deficiency disease called **kwashiorkor** are clearly visible. A person with

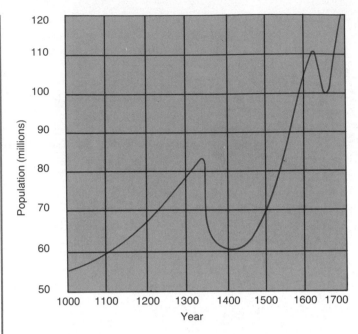

FIGURE 12.2
The sharp reduction in human population as a result of plague is difficult to appreciate today. However, the disease is still prevalant in the world and even the United States. In recent years, several cases of pneumonic plague (a lung infection of *Yersinia pestis*) were reported in the western United States.

this deficiency has a distended belly, slow growth and movement, emotional depression, and a high susceptibility to infection from normal flora or microbes that may be acquired from the environment. If not corrected, brain damage and possibly death may result. To correct this condition requires a change in diet to include complete protein found in fish, poultry, beef, and milk.

In a healthy individual, *Streptococcus viridans* found in the oral cavity is incapable of inducing disease. However, this microbe might be called an **opportunist** when it moves from a commensal to a parasitic relationship after a tooth extraction. *S. viridans* is often found in the circulatory system after this operation. When bacteria are carried passively in the circulating blood, the condition is known as a **bacteremia.*** This bacterium is also found in association with infections of the oral cavity due to gum (gingiva) damage. Commensalistic microbes are not the only microbes that can change relationship. *Escherichia coli,* usually a mutual friend, may also become parasitic. An injury to the intestine may result in the release of large amounts of these bacteria into the peritoneal cavity. When *E. coli* are out of their normal environment and uncontrolled by body defense mechanisms,

---

*Do not confuse bacteremia with the term "septicemia," which is a systemic disease with actively reproducing microbes and their products found in the circulatory system.

a severe case of peritonitis (inflammation of the peritoneum) may occur that could prove fatal.

---

1 What is commensalism?
2 Define "health" in terms of symbiosis.
3 What is meant by an opportunistic parasite?

In addition to host susceptibility, two other factors influence a shift in symbiosis and are related to the nature of the microbe. **Pathogenicity** refers to the ability of a species to cause disease. Within a pathogenic species, there may be strains that vary in their ability to cause harm. This characteristic is known as **virulence.** For example, some strains of *Staphylococcus aureus* may be weakly virulent while others are strongly virulent. Microbes that are not pathogenic or only weakly virulent will not in all likelihood shift from a commensal or mutualistic symbiosis. A major upset would be required in order for them to become parasites. Those that are pathogenic or strongly virulent have a great tendency to establish parasitic associations, and a great deal of effort may be needed to prevent them from causing disease.

Health care professionals must understand the nature of microbial symbiotic relationships if they are to take the best action to effectively control and prevent infectious disease. The disease process occurs when the body's defense mechanisms do not function efficiently enough to keep pathogens acquired from the outside in check or when microbes shift from harmless symbionts to parasites. Many believe that doctors, nurses, and other health care professionals must kill each and every microbe in order to bring patients back to good health. This is not necessarily true, since many infections are the result of symbioses shifts of normally beneficial microbes. The destruction of all normal flora by antimicrobial therapy may, in fact, only result in a different state of poor health—the exchange of one problem for another. Therefore, good therapy takes into account those normal, healthy symbioses and tries to reestablish them after the disease process has begun. Preventive medicine is also based on this concept. Maintaining good health by getting a proper balance of exercise and rest, eating a balanced diet, and avoiding disease microorganisms will encourage the maintenance of healthy symbiotic relationships (Figure 12.3).

All of the symbiotic microbes associated with the human body can be divided into two main groups. The **resident flora** are those regularly found inside the body (endosymbionts) or on its surfaces (ectosymbionts). If lost because of disease, chemical imbalance, or drug therapy, they will quickly return and reestablish the relationship. The resident flora are usually not pathogenic, but may be opportunistic, shifting into a parasitic relationship if the situation arises. Most residents enjoy a commensalistic association with humans and seldom invade the tissues due to effective body defense mechanisms. The **transient flora** are usually found in association with the skin and mucous membranes. They are temporary and may only be present for a matter of hours, days, or a few weeks. These microbes may or may not be pathogens and do not establish a permanent growing rela-

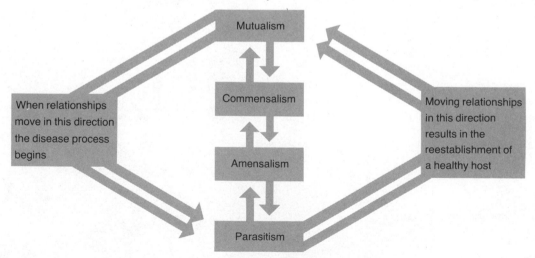

FIGURE 12.3
All symbiotic relationships are dynamic, and quick shifts among them can occur (inside double arrows). The most beneficial relationship for both microbes and humans is mutualism, and it is farthest from parasitism, which is the most destructive. Host susceptibility, pathogenicity, and virulence are factors that influence these relationships. Disease can result from a sudden shift from mutualism or commensalism to parasitism, and health may be regained by the reestablishment of mutualism or commensalism.

tionship with the body as do the residents. One of the main reasons for their temporary existence is the highly competitive nature of residents. As pointed out earlier, the competition between *E. coli* and transient pathogens found in the intestine does not allow these "foreigners" to gain a foothold. However, if the residents are eliminated, transients may also function as opportunists, becoming established as parasites and producing disease. Several areas of the body normally contain resident and transient flora. The skin, eyes and ears, mouth and upper respiratory tract, gastrointestinal tract and genitourinary tract all contain flora. The numbers and kinds of microbes in each area is limited and kept under control in many ways. Because of the uniqueness of the flora, each area may develop typical infections if these microbes get out of hand.

> 1 What is the difference between pathogenicity and virulence?
> 2 Why is it not beneficial to kill all microbes in the body at the time of infectious disease?
> 3 What type of relationship do most resident flora have with the body?

THE SKIN

Table 12.1 lists the variety of microbes typically associated with the skin. These are not equally distributed over the entire surface but concentrated in areas that provide the most favorable growth conditions. Populations are largest and most diverse in the axilla (underarm), groin, face, scalp, fingers and toes, palms, and other places where moisture and nutrients accumulate. Microbes are also found in association with sweat and sebaceous glands. The nature of the flora is greatly influenced by the chemical content and amount of secretions. Because of this, obese individuals and those that perspire a great deal will harbor more normal flora. Body odor is the result of microbial action and may be controlled by masking the odor (using deodorants), reducing the flow of perspiration (using an antiperspirant), destroying the microbes (with an antiseptic), or removing the microbes. In our culture, natural body odor is considered by most to be undesirable, and the problem is dealt with by antiseptic soaps, deodorants, and antiperspirants. The number of microbes on the surface may be greatly reduced by scrubbing with soap and water. This is the basic idea behind surgical scrubbing. However, this procedure never totally eliminates all of the microbes (hands are not sterile after a surgical scrub), since they are quickly replaced by resident flora associated with glands found deep within the skin (Figure 12.4).

Under normal circumstances, flora of the skin are kept under control as a result of the dryness and relatively low pH (4–6) of the secretions. Since many pathogens are unable to tolerate this lack of moisture and acid nature of the fluid

TABLE 12.1
Normal flora of the skin.

Diphtheroids, both anaerobic and aerobic (*Corynebacterium* sp., *Propionibacterium* sp.)
Staphylococci, nonhemolytic
Enterococci (*Streptococcus faecalis*)
Coliforms, Gram (−) bacilli (*E. coli, Enterobacter* sp.)
Acid-fast bacteria, bacilli (*Mycobacterium smegmatis*)
Proteus sp.
Pseudomonas sp.
Bacillus sp.
Candida albicans
Fungi
Numerous viruses

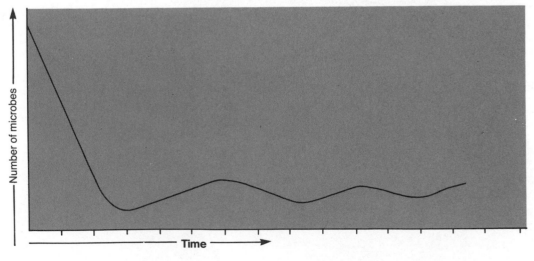

FIGURE 12.4
Surgical hand scrubbing will not remove all the microbes found on the skin. As the scrub proceeds, resident flora will be brought to the surface to replace the transients. Scrubbing, powdering, and gloving before surgery will help minimize the flora on the hands and reduce the danger of infection if the gloves should be cut during the operation.

that is present, they have difficulty growing. In addition, many of the secretions contain antimicrobial agents such as fatty acids and the enzyme lysozyme. This enzyme is found in tears, saliva, and many other body fluids and is able to hydrolyze the peptidoglycan layer in the cell walls of Gram-positive bacteria. Microbial penetration into the deeper portions of the skin is prevented by the keratinized epithelium. If these defense mechanisms should fail, a number of infections may develop. The nature of the infection will depend on host susceptibility, and the pathogenicity and virulence of the opportunist (Table 12.2).

One of the most common skin infections is acne, a disease of the sebaceous glands. It is impossible to estimate how many suffer from this infection since most people do not seek medical care. In the normal sebaceous follicle, the skin scales are shed and carried out in a continuous flow.

1 Name five bacteria common to the skin.
2 List two factors which keep normal flora in check on the skin.

In acne, the dead cells and oil do not move smoothly out of the pore but build up along the inside of the duct. Eventually, the buildup forms a plug, or **comedo** (blackhead). This prevents the normal flow of sebum, and the anaerobic diphtheroid *Corynebacterium acne* found in the tissue is able to hydrolyze the sebum into fatty acids which begin to destroy the wall of the gland. The swollen gland then ruptures under the surface of the skin. Phagocytes are mobilized to attack the bacteria, and pus is formed. The necrotic tissue accumulates to form the typical "whitehead," or pus-

TABLE 12.2
Diseases of the skin.

Disease	Cause	Gram Reaction	Nature of Microbe
Anthrax	*Bacillus anthracis*	(+)	Bacilli
Boils, carbuncles, furuncles	*Staphylococcus aureus*	(+)	Cocci
Cellulitis	*Staphylococcus aureus, Streptococcus pyogenes*	(+)	Cocci
Gas gangrene	*Clostridium perfringens, C. sporogenes*	(+)	Bacilli
Impetigo	*Staphylococcus aureus, Streptococcus pyogenes*	(+)	Cocci
Leprosy	*Mycobacterium leprae*	Not done acid-fast	Bacilli
Leptospirosis	*Leptospira icterohemorrhagiae*	Not done	Spirochetes
Nocardiosis	*Nocardia asteroides, Nocardia spp.*	(+)	Bacilli
Pemphigus neonatorum (impetigo of the newborn)	*Staphylococcus aureus*	(+)	Cocci
Pinta	*Treponema carateum*	Not done	Spirochetes
Pseudomonas infections including green nail and pyoderma	*Pseudomonas aeruginosa*	(−)	Bacilli
Rat-bite fever	*Spirillum minus*	Not done	Spirilla
	Streptobacillus moniliformis	(−)	Bacilli
Scarlet fever and other diseases caused by Group A beta hemolytic streptococci	*Streptococcus pyogenes*	(+)	Cocci
Tetanus	*Clostridium tetani*	(+)	Bacilli
Herpes simplex, cold sore, fever blister	Herpesvirus, herpes-zoster, shingles, varicella, chickenpox	Not useful	Viruses

Warts	Papillomavirus	Not useful	Viruses
Classic measles (rubeola or morbilli)	Paramyxovirus	Not useful	Viruses
German measles, rubella			
Smallpox	Poxvirus	Not useful	Viruses
Vaccinia			
Actinomycosis	*Actinomyces bovis, A. israelii* (F)	Not useful	Fungi
Candidiasis (moniliasis)	*Candida albicans* and *Candida* sp. (S)	Not useful	Fungi
Chromoblastomycosis	*Hormodendrum (Fonseccaea) compactum H. pedrosoi, Philophora* sp.	Not useful	Fungi
Cryptococcosis	*Cryptococcus neoformans* (D)	Not useful	Fungi
Mycetoma	*Nocardia* sp. *Streptomyces* sp. (F)	Not useful	Fungi
Sporotrichosis	*Sporotrichum schenckii* (D)	Not useful	Fungi

tule. Research indicates that growth of the bacteria and development of acne is correlated with hormonal changes during puberty. As these changes subside, the infections decrease in frequency and severity with little or no treatment. Avoiding foods such as chocolate, ice cream, or soft drinks is not necessary and probably has no effect. However, many dermatologists recommend that if a particular food appears to stimulate the disease, it should be avoided. Dozens of products are sold over the counter without prescription for the treatment of acne. They contain such ingredients as sulfur and salicylic acid, which peel off the top layers of skin and loosen hardened sebum. Other products, such as medicated soaps, various antiseptics, and face lotions, help keep the face clean, remove oil, and keep pores open. However, these preparations are effective for only about one in ten individuals. Specialists prescribe clindomycin hydrochloride, which applied to the skin delivers this antibiotic directly to the bacteria. Vitamin A capsules have been used to treat acne since the 1940s, but large doses of this natural vitamin can result in liver damage. More recently, a new drug known as 13-cisretinoic acid (a derivative of vitamin A) has been introduced. This drug reduces sebum production approximately 90 percent. Studies to date indicate the drug to be 75 percent effective against acne and produces few side effects; however, the carcinogenic effects of long-term use are unknown.

Another common skin infection is caused by the growth of *Staphylococcus aureus* in and around hair follicles. A boil (or furuncle) usually occurs on the neck, face, or buttocks. When *S. aureus* invades this tissue, it releases the enzyme **coagulase** which walls off the microbes from invading phagocytes. Chemicals known as **leukocidins** are also released by the microbes. These actively destroy phagocytes moving into the area (Figure 12.5). Two or three days after the initial infection, the accumulated necrotic tissue (pus) causes the boil to "point," or become yellow, and then rupture. When this occurs, the destroyed phagocytes and other dead tissue are released, leaving many *S. aureus* cells deep within the cavity. Never squeeze a boil. This makes good sense because the pathogens are deep within the lesion and may accidently be moved into the capillaries of the dermis. If this should occur when the resistance of the host is low, it is possible to develop **septicemia**, which is the presence of pathogens and their toxins in the blood. Septicemia may lead to more extensive systemic infections or staphylococcal endocarditis.

Fungal infections of the scalp, eyebrows, or eyelashes are known as **tinea capitis** ("tinea" means *worm*; "capitis" means *head*). One fungus that may be found on the skin is *Microsporum audouinii*, which is the cause of "grey-patch ringworm" (Figure 12.6). If conditions are favorable in prepuberty children, this microbe may infect the hair follicles and surrounding tissues. The hair becomes gray and brittle, and scalelike patches of temporary baldness are formed. In more severe cases, boggy areas of suppuration (pus production) may develop. The fungus may be picked up from cats and dogs and is easily spread from one child to another. In many cases, the infection resolves itself, but the antifungal drug griseofulvin and Whitfield's ointment (benzoic and salicylic acids) may be used to treat the infection.

1 Of what value is coagulase to the bacterium *Staphylococcus aureus*?
2 Why is the term "tinea capitis" a misnomer?
3 What does the term "suppuration" mean?

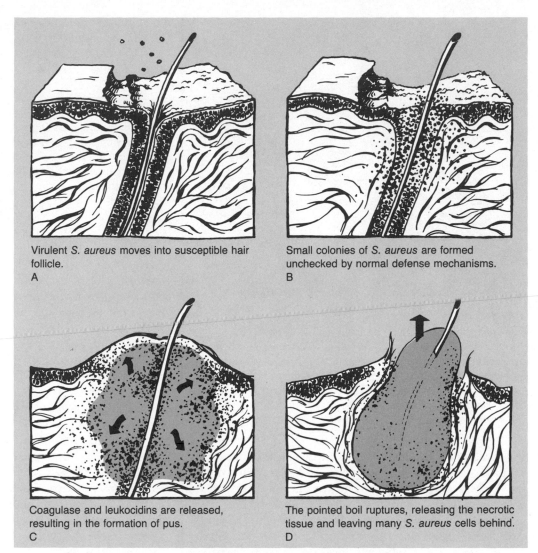

Virulent *S. aureus* moves into susceptible hair follicle.
A

Small colonies of *S. aureus* are formed unchecked by normal defense mechanisms.
B

Coagulase and leukocidins are released, resulting in the formation of pus.
C

The pointed boil ruptures, releasing the necrotic tissue and leaving many *S. aureus* cells behind.
D

FIGURE 12.5
The series of illustrations shows how *Staphylococcus aureus* normally found on the skin may enter into a parasitic relationship. Boils can form and rupture. (From M. Fisher, *Bacterial Infections of the Skin.* Abbott Laboratories.)

Many of the microbes associated with the eye and external ear are normal flora of the skin. The diphtheroids are the most frequently occurring bacteria of the eye and are located on the conjunctival membranes (Table 12.3). Lysozyme is the main agent of control in the eye, while ear infections are primarily controlled by a lack of moisture. Earwax (cerumen) restricts the movement of bacteria and other microbes into the canal. Table 12.4 lists infections associated with the eye, and Table 12.5 lists ear infections.

THE ORAL CAVITY AND UPPER RESPIRATORY TRACT

A brief look at Table 12.6 will give some idea of the complexity of the symbiotic relationships that exist in the oral cavity and upper respiratory tract. The list includes only microbes found in the pharynx, trachea, and nose since few bacteria are found in the bronchi or alveolar sacs of a healthy

TABLE 12.3
Normal flora of the *eye* and external ear.

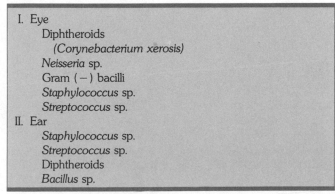

I. Eye
   Diphtheroids
      (*Corynebacterium xerosis*)
   Neisseria sp.
   Gram (−) bacilli
   Staphylococcus sp.
   Streptococcus sp.
II. Ear
   Staphylococcus sp.
   Streptococcus sp.
   Diphtheroids
   Bacillus sp.

person. Normal body defense mechanisms and the ever-decreasing diameters of the bronchi, bronchioles, and alveoli make it almost impossible for even transients to successfully move deep within the cavities of the lungs. The constant

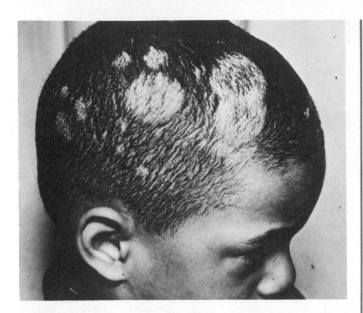

FIGURE 12.6
The gray patches on this child's head are the result of an infection known as grey-patch ringworm, which is caused by the fungus *Microsporum audouinii*. Itching may become severe as the fungus develops its typical ring formation on the scalp.(Courtesy of S. Lamberg. Reprinted from J. W. Rippon, *Medical Mycology, The Pathogenic Fungi and the Pathogenic Actinomycetes.* W. B. Saunders Co., 1974. Figure 7–6. Used by permission.

movement of the ciliated epithelium keeps contaminating microbes moving toward the mouth and nose, where they are finally coughed, sneezed, or blown out. Any interference with ciliary action inhibits this antimicrobial motion and may contribute to respiratory infection. Nicotine, found in tobacco smoke and many other chemicals, stuns the cilia and allows trapped microbes to settle into the lungs. If such inhibiting chemicals are kept in prolonged contact with the cilia, the motion may not be regained and the likelihood of lung disease increases. Bacterial diseases of the respiratory tract are listed in Table 12.7.

> I have had several gentlewomen in my house who were keen on seeing the little eels in vinegar; but some of 'em were so disgusted at the spectacle that they vowed they'd ne'er use vinegar again. But what if one should tell such people in the future that there are more animals living in the scum of the teeth in a man's mouth than there are men in a kingdom?

The quotation is from one of microbiology's famous forefathers, Anton van Leeuwenhoek (1632–1723). Our understanding of the microflora of the oral cavity has increased greatly since that time. The oral cavity is usually sterile at the time of birth, but within 12 hours a number of bacteria become established as normal flora. These microbes are derived from the vaginal tract of the mother or from microbes found on the infant's finger, in food, or on other materials placed in the mouth. *Streptococcus salivarius* is the most common and is usually found on the **gingiva** (gums) and soft tissues rather than on the surfaces of the teeth. *Streptococcus sanguis, S. mutans*, actinomycetes, and *Leptothrix* are all found attached to the enamel. The disease which causes cavities is known as **dental caries** and is influenced by the nature of the bacteria, tooth susceptibility, buffering capacity of saliva, and type of foods eaten. The warm, moist environment of the mouth promotes bacterial growth. *Streptococcus mutans* is a bacterium that is able to ferment sucrose (table sugar, candy) in this environment and produce an insoluble, sticky polysaccharide known as dextran. Dextran molecules adhere to the surface of the tooth and serve as a hold-fast for acid-producing bacteria responsible

TABLE 12.4
Diseases of the eye.

Disease	Cause	Gram Reaction	Nature of Microbe
Conjunctivitis (pink eye)	*Haemophilus aegypticus*	(−)	Bacteria, bacilli
Sty (a hordeolum)	*Staphylococcus* sp.	(+)	Cocci
Ophthalmia neonatoram	*Neisseria gonorrhoeae*	(−)	Bacteria, cocci
Trachoma and inclusion conjunctivitis	*Chlamydia trachomatis*	Not useful	Chlamydia
Mycotic keratitis	*Fusarium salani* *Aspergillus* and others	Not useful	Fungus
Endogenous oculomycosis	*Candida albicans*	Not useful	Fungus
Extension oculomycosis	*Rhizopus oryzae* *Mucor* sp.	Not useful	Fungus
Herpes	Herpesvirus	Not useful	Virus

TABLE 12.5
Diseases of the ear.

Disease	Cause	Gram Reaction	Nature of Microbe
Otitis externa (inflammation of the outer ear)	*Escherichia coli, Proteus* sp., *Pseudomonas* sp.	(−)	Bacteria, bacilli
	Hemolytic streptococci, *S. aureus*	(+)	Bacteria, cocci
Otitis media (middle ear infection)	*Streptococcus pneumoniae,* Beta-hemolytic streptococci, *Staphylococcus aureus*	(+)	Bacteria, cocci
	Haemophilus influenzae	(−)	Coccobacilli
Otomycosis (mycotic infection of the external ear and ear canal)	*Aspergillus niger*	Not useful	Mold
	Candida albicans		Yeast
Pharyngeal abscess	Beta-hemolytic streptococci, *S. aureus*	(+)	Bacteria, cocci

for decay. This sticky mat of bacteria and dextran is known as **dental plaque** (Figure 12.7). *Streptococcus mutans* and a number of lactobacilli lodged in dental plaque metabolize sucrose and other sugars, releasing lactic acid as an end product. Dental caries form in two stages. First, acid (pH 4.4) from the bacterial fermentation of food decalcifies the inorganic component of enamel. Second, enzymes from these same microbes hydrolyze enamel protein. This de-

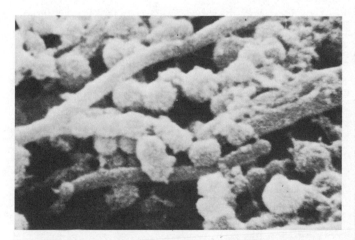

FIGURE 12.7
Bacterial dental plaque. (Copyright by the American Dental Association. Reprinted by permission.)

structive process can eventually reach the inner, living portion of the tooth and cause its death. It has been established that children fed high amounts of carbohydrates will be more susceptible to cavities than those on low sugar diets. Sucrose is the most **cariogenic** (cavity-producing) carbohydrate, while starch is the least. When sucrose is lodged in plaque in the fissures and crevices of the teeth, bacterial fermentation will produce enough acid to initiate decay in only ten to fifteen minutes.

Resistance to acid decalcification is greatest when fluoride is available and combines with the calcium phosphate portion of teeth. The addition of fluoride to water supplies enables the tooth structure to be strengthened while it is being formed and is, therefore, most effective in children. Adults do not incorporate the compounds from drinking water into tooth structure since growth has stopped, but they can receive benefits from topical treatments of fluoride in the form of fluoride toothpastes, mouthwashes, chewable tablets, and fluoride treatments. The chemical composition of saliva also helps to retard tooth decay by acting as a buffering agent to maintain saliva's pH of approximately 5.6–7.0. This action is primarily the result of bicarbonate ions. The old practice of brushing with baking soda (pH 8.3) introduces even more sodium bicarbonate into the mouth to neutralize acids. However, as many will affirm, the use of a flavored, abrasive dentifrice (toothpaste) is much more pleasant than the dry, gritty taste of baking soda. In addition, saliva contains the enzyme **lactoperoxidase,**

TABLE 12.6
Normal flora of the oval cavity and upper respiratory tract.

I. Mouth
  (a) Before Eruption of the Teeth
    Streptococci
      (S. viridans, S. salivarius)
    Staphylococci
    Diplococci, Gram (−)
      (Neisseria sp.)
    Diphtheroids
      (Corynebacterium sp.)
    Lactobacillus sp.
      (L. casei)
    Additional flora
  (b) After Eruption of the Teeth
    Spirochetes
      (Leptospira sp.)
    Bacteroides
      (Bacteroides melaninogenicus)
    Fusobacteria, Gram (−) anaerobes
      (Fusobacterium sp.)
    Vibrios, Gram (−)
      (Vibrio sp.)
    Actinomyces, Gram (+), branched
      (Actinomyces viscosus)
    Candida albicans
II. Pharynx and Trachea
    Streptococcus sp.
    Neisseria sp.
    Staphylococcus sp.
    Diphtheroids
    Haemophilus sp.
    Mycoplasma sp.
    Bacteroids
III. Nose
    Corynebacterium sp.
    Staphylococcus sp.
    Streptococcus sp.

which is able to destroy microbes by releasing singlet oxygen from hydrogen peroxide (H_2O_2).

Dr. Thomas Tomasi conducted studies at the Mayo Clinic in Rochester, Minnesota, to discover how secretions of the mouth respond to invading microbes. These studies revealed an increase in the presence of secreted antibodies in the saliva known as immunoglobulin A. The presence of these antibodies raised the question and hope of a possible vaccine against dental caries and other oral diseases. Researchers have found that an injection of dead *Streptococcus mutans*

1 What bacterial species is primarily associated with dental caries?
2 Define the term ''cariogenic.''
3 Of what value is lactoperoxidase?

can stimulate an increase in secreted immunoglobulin A in **gnotobiotic** (germ-free) laboratory rats and reduce the number of cavities formed in these test animals (Box 21). Another test group of five monkeys has also been used; and after five years of comparative studies, all the vaccinated monkeys were found to be cavity-free, while the control group developed sixty-four cavities. Enzymes produced by *S. mutans* and other cariogenic bacteria have also been used to prepare vaccines that have met with success. These preparations are now undergoing more detailed study and testing before being approved by the Food and Drug Administration for clinical tests with humans. The buffering action, antibody response, and enzymatic reactions found in saliva all help to limit oral diseases; however, many microbes may cause infections of the oral cavity. Table 12.8 lists some of the more common diseases and their causes.

THE GASTROINTESTINAL TRACT

The mouth is the gateway to the intestinal tract. Many microbes on food, dust, and other materials move through the gastrointestinal tract after being swallowed. The first portion of the tract, the esophagus, contains only transient flora on their way to the stomach. When microbes enter the stomach, many are destroyed by the high acidity (pH 2) of the contents. However, there are approximately 10^3 to 10^5 bacteria per gram of contents in persons on a regular diet. This number increases after a large meal since an open pyloric valve allows the mass of food to be neutralized more quickly. A variety of resident flora exists among the cells that make up the lining of the stomach. These bacteria are primarily aciduric streptococci and lactobacilli. They are not destroyed by the acid because they are protected by the mucous lining of the stomach.

As food and microbes move into the small intestine, the pH becomes more alkaline and the number of microbes increases. The adult duodenum contains 10^6 bacteria per gram of contents, and because the duodenum is close to the stomach, the bacteria found there are similar to those in the stomach. The environment becomes more favorable for growth in the jejunum (the middle portion of the small intestine) and ileum (the final portion of the small intestine), and the flora continue to increase (10^7). Once inside the large intestine or colon, the adult flora reach about 10^{10} bacteria per gram and contain a variety of microbes (Table 12.9). Most of these (95 percent) are anaerobes such as *Bacteroides* and *Clostridium,* while others are facultative (e.g., *E. coli*). This imbalance is due to the fact that the oxygen concentration has been reduced by chemical and microbial activities in the intestine.

Bacteria within the intestine carry out a variety of activities. As described earlier, many synthesize vitamins B_{12} and

TABLE 12.7
Bacterial diseases of the respiratory tract.

Disease	Cause	Gram Reaction	Morphology
Diphtheria	Corynebacterium diphtheriae	(+)	Bacilli
Leprosy	Mycobacterium leprae	*	Bacilli
Melioidosis	Pseudomonas pseudomallei	(−)	Bacilli
Nasopharyngitis**	Neisseria meningitidis	(−)	Cocci
Ornithosis	Chlamydia ornithosis	(−)	Bacilli
Parapertussis	Bordetella parapertussis	(−)	Bacilli
Pneumonia	Diplococcus pneumoniae	(+)	Cocci
	Haemophilus influenzae	(−)	Bacilli
	Klebsiella pneumoniae	(−)	Bacilli
	Staphylococcus aureus	(+)	Cocci
	Streptococcus pyogenes	(+)	Cocci
Pneumonic plague	Yersinia (Pasteurella) pestis	(−)	Bacilli
Primary atypical pneumonia	Mycoplasma pneumoniae	†	Pleomorphic
Psittacosis	Chlamydia psittaci	(−)	Bacilli
Streptococcal lymphoiditis††	Streptococcus pyogenes	(+)	Cocci
Tuberculosis	Mycobacterium tuberculosis	*	Bacilli
Whooping cough (pertussis)	Bordetella pertussis	(−)	Bacilli

*The Gram-stain reaction is not used diagnostically. The acid-fast staining procedure is used instead.
**Note that this is the most common manifestation of the bacterial pathogen.
†Gram reactions are of little value with this pathogen.
††Grouped here because they are characterized by involvement of the lymphoid tissue of the upper respiratory passages.
SOURCE: Modified from G. A. Wistreich and M. D. Lechtman, *Microbiology and Human Disease*, 3rd ed. Copyright © 1973 by Benziger Bruce & Glencoe, Inc. Reprinted with permission of Macmillan Publishing Co., Inc.

TABLE 12.8
Diseases of the oral cavity and related tissues.

Disease	Cause	Gram Reaction	Nature of Microbe
Stomatitis	Hemolytic streptococci	(+)	Bacteria, cocci
Necrotizing ulcerative gingivitis (trenchmouth)	"Fusiform" bacillus and Treponema vincentii	(−)	Bacilli, spirochetes
Canker sore	Unknown—*not* a virus, possibly an L-form Streptococcus	(+)	Bacteria, cocci
Cold sore	Herpes simplex	None	Virus
Thrush	Candida albicans	None	Yeast
Chancre (syphilis)	Treponema pallidum	No reaction	Bacteria, spirochetes

K. They also produce thiamine, pyridoxine (B_6), and riboflavin. The normal flora outcompete pathogens and stimulate the immune response when they accidentally move into the tissues. Many produce organic acids (acetic, butyric, and propionic) that inhibit pathogens and optimize nutrient absorption. Occasionally these benefits may be outweighed by the production and elimination of the gases and foul odor. These bacterial products become concentrated as water is resorbed through the wall of the rectum, and the fecal material becomes less fluid. The microbial population reaches its greatest concentration (10^{11}) in discharged feces. In cases of intestinal upset that result in diarrhea, the numbers of microbes may decrease drastically. A number of factors may affect the balance of normal flora in the intestine. Change in diet, antibiotic therapy, or emotional trauma may cause disruptions that result in common diarrhea. Once the disruptive influence

BOX 21

GERM-FREE
ANIMALS

GNOTOBIOTIC laboratory animals are used by many research labs to determine the exact nature of microbe-animal symbiosis. The term refers to the fact that researchers know exactly which microbe is in contact with the animal and that all others have been excluded from the relationship. Gnotobiotic animals are bred in sterile environments and their young are surgically removed from the uterus to limit contact with flora already present on the mother's body. These offspring are then maintained and bred under the same sterile environmental conditions for generations to guarantee the exclusion of all microbes from the animals' bodies. All food, water, and other materials placed in the rearing compartments are sterilized. These animals may then be selectively infected with a pure culture of a known microbe and observations made to determine the nature of the interaction. A number of gnotobiotic animals have been bred in this way. Birds, rabbits, mice, hamsters, rats, and even monkeys are available for research work. Because gnotobiotic animals are not "normal," they must be handled with special care. In many cases, their defenses are so low that a minimal exposure to an otherwise harmless microbe, such as *Micrococcus luteus,* will result in the animal's quick death. It is also important in such research to know that these animals demonstrate anatomical and physiological differences compared to their counterparts that have normal flora. For example, the entire intestinal wall of germ-free rats is much thinner than normal animals, and they lack adequate levels of vitamin K, which must be supplied to them in their special diet. Despite the handicaps involved with research using these animals, the information gained has been of great value.

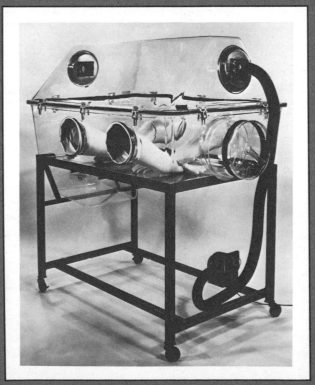

A germ-free isolator. (Courtesy of the GERMFREE Laboratories, Inc.)

TABLE 12.9
Normal flora of the gastrointestinal tract.

I. Intestine of Breast-Fed Infants
  Lactobacillus bifidus (*Bifidobacterium*)
  Lactic streptococci
II. Normal Adult Flora (Bottle-Fed Infants)
  Coliforms
    (*E. coli, Serratia, Enterobacter*)
  Enterococci
    (*Streptococcus faecalis*)
  Bacteroides
    (*Bacteroides fragilis*)
  Lactobacilli
    (*Lactobacillus acidophilus*)
  Clostridia
    (*Clostridium perfringens*)
  Pseudomonas sp.
  Proteus sp.
  Alcaligenes faecalis
  Yeast
  Fungi
    (*Penicillium* sp.)
  Enteroviruses

has been brought under control or eliminated, the intestinal tract regains its balance and the symptoms disappear. If the fecal material is analyzed for microbial content and pH during the illness, the acid-producing microbes, especially the resident lactobacilli, decrease in numbers and the pH becomes more alkaline. When the symptoms are relieved, the pH again becomes more acid and the lactobacilli become reestablished. In some cases, outside help is necessary to bring the problem under control. It has been noted that eating more acidic foods, such as yogurt and cultured buttermilk, more quickly resolves the problem. Eating fresh fruits and uncooked vegetables also seem to help. Today this is interpreted to mean that these foods encourage the reestablishment of a more acid intestinal environment and serve as a source of acid-producing bacteria. Whether this actually works is questionable; however, the concept of microbial antagonism is valid and has been adopted by the medical profession for a number of years. **Antagonism** takes place between microbes if one releases a substance that kills or inhibits the activities of the other, but is completely harmless against the species producing the material. By encouraging the maintenance of microbes such as *E. coli* and *Lactobacillus acidophilus* in the intestine, their antagonistic characteristics help maintain good health. As early as 1908, I. Mechnikov came to the conclusion that the most important factor contributing to "senility and premature death" was self-intoxication caused by the "putrid material" in the colon. (His conclusion was based on observations made in Bulgaria, where yogurt played an important part in the nutrition of many people claiming to be over 100 years old.) He conceived the idea of changing the bacterial

flora of the lower intestine from this putrefactive state to acid-forming flora. This was done by feeding to people the acid-producing bacterium commonly found in yogurt, *Lactobacillus bulgaricus*. However, Mechnikov did not know that this microbe will not implant in the intestine. Later in 1921, it was shown that a diet of lactose sugar markedly stimulated the resident intestinal bacterium *L. acidophilus* within only four to six days and resulted in the suppression of other bacterial species by antagonism.

With the introduction of oral antibiotics, patients frequently develop gastrointestinal imbalance in normal flora and a loss of *L. acidophilus* and other antagonists. Yeast and mold infections are often diagnosed in these cases and standard therapy has become the administration of concentrates of *L. acidophilus* in a lyophilized tablet form known as **Lactinex.** These tablets are sold at drug stores and must be refrigerated to maintain their viability. The successful treatment of post-radiologic diarrhea by the use of *L. acidophilus* was reported as early as 1973. These tablets are also effective against other infections and illnesses, including cold sores, staphylococcal enterotoxemia, and infantile diarrhea (*E. coli* infection). More recently, another form of *L. acidophilus* has been introduced. Sweet acidophilus milk is made by adding a high concentration of viable *L. acidophilus* to pasteurized, chilled milk just before packaging. Sweet acidophilus tastes like regular milk, and its spoilage time does not vary significantly from regular milk. When it does sour, it changes to a product known as acidophilus milk. It has a very bitter taste and consistency that resembles yogurt.

Table 12.9 showed the difference between the normal intestinal flora of breast-fed infants and adults. This difference also extends to bottle-fed infants. Human milk is very unlike the milk of other mammals (Table 12.10) and contains a disaccharide amino sugar and possibly other agents that encourage the growth of the bacterium *Bifidobacterium bifidus* (*Lactobacillus bifidus*). This bacterium functions as the predominant acidogenic bacteria in the intestinal tract of breast-fed infants until the baby is weaned. When cow's milk and solid foods replace mother's milk, the baby may experience cramping, emotional upset, and changes in stool color, consistency, and odor. At this time, the intestinal flora shift from the typical breast-fed, *Bifidobacterium*-predominating flora to the normal flora of a bottle-fed infant. This period of change can be very difficult for the baby. Many mothers have found that supplementing the diet with sweet acidophilus milk or Lactinex tablets reduce the trauma of weaning.

---

1 Of what value are *E. coli* and *Lactobacillus acidophilus* to maintaining overall health?
2 Where in the body and at what time in life is *Bifidobacterium bifidus* found?

TABLE 12.10

Average composition of the milk of certain mammals.

	Water%	Protein%	Fat%	Lactose%	Ash%
Cow	87.29	3.42	3.66	4.92	0.71
Goat	87.33	3.50	4.15	4.20	0.82
Human	87.60	1.20	3.80	7.00	0.21
Ass	89.08	1.98	2.45	6.04	0.45
Buffalo	82.44	4.74	7.40	4.64	0.78
Camel	87.67	3.45	3.02	5.15	0.71
Cat	83.05	7.00	4.50	4.85	0.60
Dog	74.55	3.15	10.20	11.30	0.80
Elephant	85.63	3.20	3.12	7.42	0.63
Fox	81.86	6.35	6.25	4.23	1.31
Horse	89.18	2.60	1.59	6.14	0.49
Llama	86.55	3.90	3.15	5.60	0.80
Pig	80.63	6.15	7.60	4.70	0.92
Porpoise	41.28	11.20	45.80	1.15(?)	0.57
Rabbit	68.50	12.95	13.60	2.40	2.55
Reindeer	63.38	10.25	22.42	2.50	1.45
Seal	34.00	12.00	54.00	none	0.53
Whale	69.80	9.43	19.40	?	0.99
Zebu	86.20	3.00	4.80	5.30	0.70

SOURCE: L. M. Lampert, *Modern Dairy Products*. Chemical Publishing Company, 1975.

Two other problems associated with the maintenance of normal flora are also being investigated. Colic in bottle-fed infants (constipation, intestinal cramping, and pain) may result from an imbalance in normal flora due to a lack of stabilizing antagonists. The symptoms resemble those occurring at weaning. If this is true, it may account for the fact that breast-fed infants have fewer cases of constipation and are less likely to have serious digestive upsets. Supplementing a bottle-fed infant's diet with expectorated milk (donated from another person), Lactinex granules, or sweet acidophilus milk (room temperature) may serve to stabilize the intestinal flora and relieve the problem. These kinds of disruptions can have effects that go beyond the abdominal region. Thrush was mentioned earlier as an oral yeast infection that could result from excess antibiotic therapy and loss of normal flora. Other physiological changes may also occur when the normal flora are disrupted. Statistics indicate that bottle-fed infants are more likely to develop respiratory infections, such as bronchitis and pneumonia, in comparison to breast-fed infants with more stable flora.

Another disease that may be microbial in nature and related to intestinal imbalance is **sudden infant death syndrome** (SID, or "crib death"). There is evidence that this fatal disease could be the result of extremely small amounts of *Clostridium botulinum* toxin that go undetected. A neurological problem is also suspected and may be the reason for the disease state being established. A number of infant botulism and *Clostridium perfringens* food poisoning cases have been recognized. Honey is not recommended for infants because it contains *Clostridium* spores as a normal contaminant. Under normal circumstances, the presence of *C. botulinum* should pose little danger due to the acid nature of the contents of the intestine. However, if the acidogenic bacteria have been lost or are for some reason unable to maintain acid production, microbes such as *Clostridium* may be able to produce enough toxin to cause death. Studies are now being conducted in this area, and in a few cases, *C. botulinum* has been isolated from the fecal material of SID victims. A number of other microbial diseases may occur in the gastrointestinal tract. A listing of these are in Table 12.11.

THE GENITOURINARY TRACT

The microflora of the adult urinary tract resemble that found on the surface of the skin, but the population is reduced in number. In both males and females, the lower one-third of the urethra contains mixed populations; however, in healthy individuals, the remaining two-thirds has few if any microbes. Urinary tract infections in females are much more likely to occur than in males because of the shorter length of the duct and the close anatomical relationship between the anus and the urethral meatus (opening). Some of the most common causes of urinary tract infections include *E. coli, Proteus, Pseudomonas,* and *Streptococcus faecalis* derived from contaminating fecal material. In voided urine collected from healthy patients, there are approximately 10,000 bacteria per ml of fluid. These are normal flora from the lower portion of the urinary tract. Counts in excess of 100,000 per ml are considered to be indicative of a urinary tract infection.

The flora of the adult vagina are regulated in much the same way as that found in the intestine. Acidogenic bacteria

TABLE 12.11
Diseases of the gastrointestinal tract.

Disease	Cause	Gram Reaction	Nature of Microbe
Gastroenteritis	E. coli	(−)	Bacteria, bacilli
	Shigella spp.		
	Salmonella spp.		
Cholera	Vibrio cholerae		Bacteria, vibrio
Amoebic dysentery	Entamoeba histolytica	Not useful	Protozoan, amoeba
Giardiasis	Giardia lamblia	Not useful	Protozoan, flagellate
Balantidiasis	Balantidium coli	Not useful	Protozoan, ciliate
Food poisoning	Clostridium perfringens	(+)	Bacteria, bacilli
	Staphylococcus aureus	(+)	cocci
	Bacillus cereus	(+)	bacilli
Aflatoxin poisoning	Aspergillus niger	Not useful	Fungus
Infectious hepatitis	Virus A	Not useful	Virus
Epidemic infant diarrhea	Echovirus	Not useful	Virus

create an environment that is antagonistic to the growth of potential pathogens. Their loss from the fluid lining of the vagina results in a more favorable environment for pathogenic microbes. Infection of the vagina is generally referred to as **vaginitis** and may be caused by the growth of such microbes as *Candida albicans* (a yeast infection), *Trichomonas vaginalis* (a flagellated protozoan), or a variety of bacteria.

The vaginal flora undergo a sequence of changes beginning at birth. **Döderlein's bacillus** (an acidogenic lactobacillus) appears in the vagina soon after birth. Many suspect that this microbe originates from the vagina of the mother and is transferred to the infant by contact. These bacteria remain in the vaginal tract for only a period of several weeks. When the secretions change to a more neutral pH as a result of hormonal changes in the infant, a mixed flora become established and remain until puberty (Table 12.12). At this time the pH again changes and these acidogenic bacteria reappear to remain as the predominant acid producers, until menopause when once more they are lost and replaced by a mixed population. If at any time during the reproductive years Döderlein's bacillus is suppressed by a change in body chemistry or antibiotic therapy, the likelihood of infection increases. And likewise, as the pH of the vagina becomes more acidic, pathogens are suppressed in favor of Döderlein's bacillus. The composition of therapeutic vaginal creams is founded on this basis. Most of these ointments are buffered to an acidic pH (4–5) normally found in the fluid secretions of the vagina. When these creams are applied, the optimum growth environment for Döderlein's bacillus is established and pathogens are inhibited.

Because of the close physiological, morphological, and ecological relationship among *L. acidophilus* in the adult intestine, *Bifidobacterium* in the breast-fed infant's intestine, and Döderlein's bacillus in the adult vagina, speculation exists

TABLE 12.12
Normal flora of the genital tract.

I. Anterior Urethra:
   Similar to skin but in lesser numbers.
II. Vagina:
   (a) After birth:
      Döderlein's bacillus, a *Lactobacillus* sp.,
         predominates for several weeks only, later
         changing to mixed flora until puberty is reached.
   (b) Normal adult:
      Döderlein's bacillus returns.
      Clostridium sp.
      Peptostreptococcus sp.
      Streptococcus sp.
      Coliforms
      Spirochetes
      Staphylococcus sp.
      Diphtheroids
      Yeast
         (*Candida albicans*)
   (c) After menopause:
      Döderlein's bacillus is lost, mixed flora return.

that these three bacteria are, in fact, the same microbe. The difference among them may be the result of modified gene expression in the different environments. If this is true, the potential exists for resolving infections that result from an imbalance of normal flora in a number of different anatomical locations.

---

1 Why should honey not be fed to infants?

2 An infection of the vagina is known by what medical term?

3 What happens to the pH of the vagina as an infection develops? As the infection is brought under control?

Other infections of the genital tract include puerperal fever and various types of venereal diseases. As a result of antibiotic therapy and good aseptic technique in hospital delivery rooms, puerperal fever is an uncommon disease in the United States. (The term "puerperal" refers to the puerperium, or period of confinement after labor.) The disease may be caused by a hemolytic *Streptococcus* that enters the uterus during childbirth; however, infections caused by other bacteria may also result in the same symptoms (e.g., *Staphylococcus, E. coli, Bacteroides,* and *Clostridium*). In the past, this infection was a major cause of maternal death after delivery, but today it is rare and easily controlled. Venereal diseases of the vagina may be caused by *Neisseria gonorrhoeae, Treponema* (a number of species) and herpesvirus. These will be described in later chapters.

VIRULENCE: TOXINS AND INVASIVENESS

All symbiotic relationships (commensalism, mutualism, amensalism, and parasitism) between microbes and the human body are dynamic. However, when the harmless forms shift and become parasitic, the interaction between host and parasite results in the slow weakening and possible death of the host. Many of the relationships mentioned previously begin on the mucous membranes of the respiratory, genitourinary, and gastrointestinal tracts. In order to be successful, parasites must first **adhere** to the appropriate host cells. This adherence is a specific characteristic of the relationship rather than a casual or random feature. *Streptococcus sanguis* has the ability to adhere to the gingiva but not to the enamel of the teeth; while *S. mutans* will adhere to the enamel when sucrose is available for the production of sticky dextran molecules. *Lactobacillus acidophilus* may adhere to the cells lining the large intestine of humans but not to these cells in other animals. The success of a parasite, in many cases, depends on the ability of the microbe to set up this first **focus of infection.** Many pathogens, such as *Corynebacterium diphtheriae, Vibrio cholerae,* and *Bordetella pertusis* remain on surfaces and cause disease symptoms by releasing poisons that cause tissue damage deep within the host. Others penetrate the surface through small, naturally occurring crevices or cracks to become established as deep tissue parasites. Some microbes are not as opportunistic and are capable of releasing a variety of molecules that allows them to enter, multiply, and spread within the host.

In most cases, the initial infecting population is too small to cause enough damage to be visible, or the microbes may be latent. Latent infections are the result of a balance between parasite and host. This balance occurs because either the host defense mechanisms are keeping the parasite in check or the genes responsible for host damage are not being expressed. Latency may continue for a considerable period before converting to an active, symptomatic infection. This is the case with many strep throat infections and also with infections of *Mycobacterium tuberculosis* that have been latent in a host for several years. The degree to which *M. tuberculosis* or any other pathogen causes host harm is called virulence and is the result of substances produced by or associated with the microbe during the course of the infection. Once the microbes are established within the host, symptoms of microbial virulence may be expressed in two characteristic ways: **toxin production** or **invasiveness.**

Exotoxins are protein poisons produced in the cytoplasm of Gram-positive and some Gram-negative bacteria. These molecules are actively released from the cell during its growth phase and each has unique properties (Table 12.13). Most exotoxins are heat labile (easily destroyed). For example, the exotoxin of *Clostridium botulinum* responsible for botulism poisoning is destroyed by heating the molecule at 60°C for only 30 minutes. However, the exotoxin produced by *Staphylococcus aureus* known as **enterotoxin** (causing food poisoning) is heat stable and will not be destroyed even after 30 minutes of boiling. Exotoxins are also strong antigens; that is, they readily stimulate the immune system to produce antibodies with which they react. This antigenic character is the basis for the DPT vaccination and the production of other toxoids. A **toxoid** is a toxin that has been chemically or physically altered so that it is no longer harmful but is still able to stimulate the immune system. Toxoids are manufactured by heating the toxins or exposing them to formaldehyde. DPT vaccinations are made by altering the toxins of *Corynebacterium diphtheria, Bordetella pertusis,* and *Clostridium tetani* and combining them in a trivalent (three in one) vaccine.

The symptoms of exotoxin poisoning usually begin just after the bacteria have become established, and they are progressive. Damage increases as the microbes grow in number, and tissues located at a distance from the focus of infection may be affected. This is the typical situation in diphtheria. The bacteria remain localized in the pharynx while

TABLE 12.13
Exotoxins released by pathogenic bacteria.

Exotoxin	Example of Bacterial Source	Gram Reaction	Effected Structure	Disease
Neurotoxin	*Clostridium botulinum*	(+)	Myoneural junctions	Botulism
Neurotoxin (tetanospasmin)	*Clostridium tetani*	(+)	Central nervous system	Tetanus (lockjaw)
a toxin	*Clostridium perfringens*	(+)	General	Gas gangrene
Diphtheritic toxin	*Corynebacterium diphtheriae*	(+)	General	Diphtheria
a toxin	*Staphylococcus aureus*	(+)	General	Pyogenic infections
"Enterotoxin"	*Staphylococcus aureus*	(+)	Nerve cells	Food poisoning
Enterotoxin	*Shigella dysenteriae*	(−)	Intestinal epithelium	Bacterial dysentery
Enterotoxin	*Vibrio cholerae*	(−)	Intestinal epithelium	Cholera
"Guinea pig toxin"	*Yersinia pestis*	(−)	General	Bubonic plague
Hemolysins	Streptococci staphylococci	(+)	Lyse red blood cells	Septicemia

the toxin spreads throughout the body resulting in a severely inflamed throat, nerve paralysis, fever, and pain. Treatment for this infection includes both an antibiotic to destroy the bacteria and an antitoxin to destroy the circulating toxin molecules.

---

1 What is meant by a focus of infection?
2 What is the difference between an exotoxin and an enterotoxin?
3 Of what value is a toxoid? Name one.

THE HEMOLYSINS AND OTHER SUBSTANCES

Members of an important category of exotoxins are able to destroy host cells by disrupting host cell membranes. These molecules either attack the phospholipid layer or affect the sterols in host cells. Since the production of these exotoxins is easily demonstrated by culturing bacteria on blood agar, this group of toxins has become known as the **hemolysins.** *Streptococcus, Staphylococcus,* and *Clostridium* are a few genera of bacteria that may have the ability to produce these molecules. Hemolysins that completely destroy red cells are known as **beta hemolysins.** When beta-hemolytic bacteria are grown on blood agar, they produce a clearing around the colonies (Figure 12.8). **Streptolysin S**

is one kind of beta hemolysin produced by some pathogenic streptococci. Endocarditis caused by a streptococci-releasing streptolysin S is an acute disease showing severe heart valve ulcerations and destruction, anemia, and death within only a few days if left untreated. The second form of hemolysin will cause only partial destruction of red cells and is known as **alpha hemolysin.** These toxins produce a green zone around the colonies when cultured on blood agar. Streptococci that release **streptolysin O** are alpha hemolytic and an endocarditis resulting from such an infection is much less severe (subacute) than one caused by streptococci-producing streptolysin S. Subacute endocarditis symptoms include fever, anemia, weakness, gradual valve scarring (with the development of a heart murmur), and enlargement of the spleen.

Endotoxins are more typically found in Gram-negative bacteria and differ significantly from the exotoxins. Bacteria such as *E. coli, Pseudomonas, Salmonella,* and *Shigella* all contain endotoxins. In addition, many fungi contain molecules that act as endotoxins. These are collectively known as mycotoxins (fungal poisons). One in particular, aflatoxin, has been implicated in a number of poisoning incidents involving *Aspergillus*-molded grain fed to sheep and trout. Endotoxins are a part of the cell structure and are not released from the cell during its life. These molecules are lipopolysaccharide-protein molecules of the cell walls and are heat stable. They are only released after the wall has been damaged or destroyed, and these components cause fever, inflammation, and tissue damage. It is the endotoxin molecule that is able to cause fever by stimulating the release of a molecule known

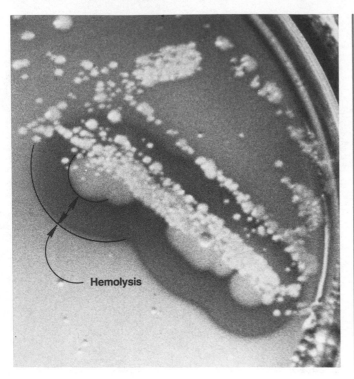

FIGURE 12.8
The destructive effects of hemolysins can be seen as they diffuse from the bacterial colony and spread into the blood-containing nutrient agar medium.

as **endogenous pyrogen** from polymorphonuclear leukocytes (PMN cells have deeply lobed nuclei). This molecule operates directly on the thermoregulatory center in the brain to cause fever.

Endotoxins are weak antigens and toxins. A much larger dose is required to stimulate tissue damage and antibody formation. However, if an infection contains a large number of actively growing cells, their death will result in the release of enough endotoxin to produce symptoms such as fever, localized tissue shock, or diarrhea. As weak antigens, they have limited ability to form toxoids; however, since humans are constantly exposed to low levels of endotoxins, they do stimulate a nonspecific antibody resistance to endotoxins in general. However, they are of little protective value.

In addition to exotoxins and endotoxins, many pathogens release other substances that increase their virulence. In general, these chemicals function as enzymes and enhance the pathogen's **invasiveness**—its ability to enter, spread, and multiply. Since a number of nonpathogenic bacteria also release these products, it is suspected that they are not responsible for major disease symptoms but do contribute to the course of the infection. One of these invasive factors is known as **collagenase**. This enzyme is able to hydrolyze the protein collagen found in many connective tissues. *Clostridium perfringens* is capable of releasing collagenase and

disrupting the continuity of the tissue in which it has become established. When this molecule is released, it disrupts the normal circulatory pattern and further decreases an already low oxygen supply, encouraging the spread of this anaerobic pathogen.

Coagulase is released by pathogenic *Staphylococcus aureus* and causes clotting in plasma. This is an advantage to the pathogen since it walls itself off from invading phagocytes. The production of coagulase is one of the key identification characteristics for pathogenic staphylococci used in the lab (refer to Chapter 9). Another enzyme that can be released by staphylococci is **hyaluronidase.** Hyaluronic acid is an organic molecule found between cells and acts as a cell adhesive. The enzyme hyaluronidase digests this "tissue glue" and allows the bacteria to spread deeper into the host. **Leukocidins** and **leukostatins** are able to kill or inhibit white blood cells. Leukocidins are usually most effective after the phagocytes have engulfed streptococci or staphylococci and the chemical comes in direct contact with the interior of the phagocyte. When the leukocidins are released, the phagocyte "rounds up" and bursts. The release of a leukostatin interferes with the phagocyte's ability to engulf the pathogen. Many will simply move around the pathogen as if it did not exist. The streptococci are also able to reproduce molecules known as **fibrinolysins**. These substances dissolve human fibrin blood clots that have been formed around the invading bacteria. By breaking the clot, the pathogen can spread and reproduce.

---

1 Endotoxins are typically found in association with Gram-positive or Gram-negative bacteria?
2 What is a pyrogen?
3 Name three types of invasive factors.

The ability of a microbe to produce toxins or invasive chemicals is genetically controlled. The genes responsible for toxin production are located on either the exogenote or endogenote DNA and may be acquired by mutation, conjugation, transformation, or transduction. Pathogens which demonstrate these inherited features undergo natural and artificial selection, and the frequency of pathogens in a population will depend on how favorable the environment is. When pathogens are transferred from one susceptible host to another in sequence, the environment becomes increasingly favorable for the expression and maintenance of virulence genes in the population. After several passages, the virulence of the pathogen increases. Likewise, a pathogen may lose its virulence if maintained outside a host in pure culture (Figure 12.9). The loss of virulence in a population of pathogens as a result of laboratory culturing is known as **attenuation.** Attenuation is thought to occur because of

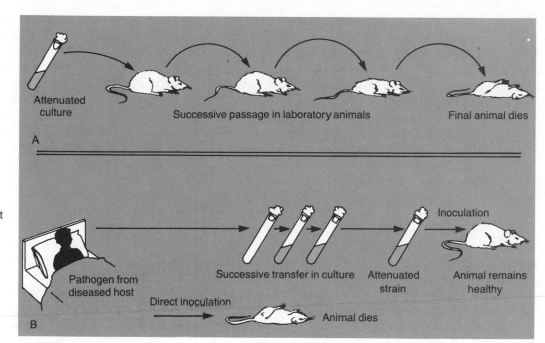

FIGURE 12.9

(A) The transfer of nonvirulent pathogens through a series of susceptible hosts reestablishes the virulence of the microbe. (B) The transfer of a virulent pathogen through a series of pure culture results in the loss of virulence, a process known as attenuation.

more favorable selection pressures for the nonvirulent forms or because of the actual loss of the gene responsible for virulence. This second case applies to virulence genes carried on an exogenote that is not self-replicating (Figure 12.10). The laboratory attenuation of pathogens has great practical value in vaccine production. Killed cells are not strong antigens in comparison to live cells. In some situations, therefore, it is advantageous to use live, attenuated pathogens to better stimulate the immune system. Some of the most common vaccines formed in this way are BCG (bacille de Calmette et Guérin) for tuberculosis, and the viruses for smallpox, polio, and measles.

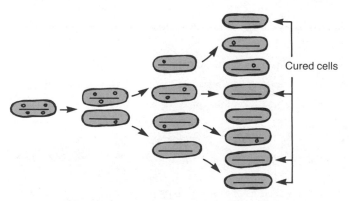

FIGURE 12.10

The sequence shows the sequential loss of exogenote DNA from the descendents of virulent cells. This occurs because the plasmid is not self-replicating. If this plasmid contains a gene for toxin production, the resulting population will contain a very low frequency of virulent cells. This is attenuation due to curing, and the cells lacking the plasmid are known as cured cells.

Antibody that is produced by artificial stimulation with vaccines or by contact with normal flora helps to keep parasites in check and maintain healthy symbiotic relationships. The relationships that exist between the human body and normal flora are complex and everchanging. As the body chemistry changes due to hormonal imbalances, antibiotic therapy, or other factors, resistant and transient flora may shift from harmless, nonvirulent forms to more virulent parasites. However, many pathogens are unable to adhere to the body long enough to establish anything but a transient relationship, or they find the human an unsuitable host. Those that do find a niche must be able to successfully cope with many antimicrobial mechanisms, including the competitive forces of resident flora and body defense mechanisms, such as dryness, antibodies, acids, and enzymes. The success of adaptations between human and microbe is evident in the many offensive and defensive mechanisms found in both, and in the fact that there are many symbiotic relationships maintained as either mutualism or commensalism.

SUMMARY

Many different kinds of physical relationships exist among organisms. These symbiotic relationships take four major forms known as amensalism, commensalism, mutualism, and parasitism. In an amensalistic relationship, one species is harmed while the other appears to be unaffected. In a commensal relationship, one species benefits and the other is neither harmed nor helped. Both interacting species in a mutual relationship benefit from the arrangement. In parasitism, the parasite lives in or on a host and benefits by obtaining nutrients and shelter, but the host is harmed. In a healthy individual, most microbes exist in either mutual or commensal symbiosis. However, a disruption in the host metabolism can easily result in a shift to a parasitic interaction. Microbes that are able to take advantage of this increased host susceptibility are known as opportunists and cause host harm if they are pathogenic and virulent. The maintenance or reestablishment of good health after a parasitic symbiosis requires treatment based on a knowledge of normal microbial symbiotic relationships.

Microbes normally occurring on the body are classified into two main groups. Resident flora are those that are regularly found inside the body or on its surface. They are mutual or commensal and will quickly return if removed. The transient flora are only temporary and do not establish a permanent growing relationship with the body as do the residents. Each major portion of the body carries with it a normal and unique complement of resident and transient flora. These are kept in check by a variety of defense mechanisms, including dryness, pH, enzymes, and competition within the floral populations. Occasionally, residents or transients shift from beneficial or harmless forms to parasites, but they can be brought back into more normal relationships with proper treatment and therapy. All symbiotic relationships that make the shift to parasitism depend on the susceptibility of the host and the virulence of the microbes. Virulence is the degree to which a pathogen is able to cause host harm and is the result of substances produced by or associated with the microbe during the course of the infection. The two major factors contributing to virulence are toxigenicity and invasiveness. Toxins may be exotoxins released from an actively growing cell or endotoxins that are components of bacterial cell walls. Endotoxins are not released until the parasite is disrupted. Invasive factors are substances that function as enzymes and increase the pathogen's ability to enter, spread, and multiply in the host. Invasive factors include such agents as collagenase, coagulase, and hyaluronidase. In laboratory-pure cultures, the virulence of a pathogen may be decreased or lost by the process known as attenuation. Virulence may be increased by returning the pathogen to a suitable living host and transmitting the microbe from host to host.

WITH A LITTLE THOUGHT

A man is admitted to the hospital with candidiasis (*Candida albicans* yeast infection) of the thumbs and nail beds.

The patient is 40 years old, malnourished, diabetic, and alcoholic. He works 12 hours a day, 6 days each week at a downtown car wash, wears a protective plastic rain coat and wading pants and his hands are constantly in water.

Candida is a budding, yeastlike fungus found throughout the world and is normally found in the surface of the skin. This microbe grows well in the lab on Sabouraud's dextrose agar (dextrose, peptone, and water) at 36°C. The physician in charge is fearful that the yeast may become systemic and establish foci of infections in the kidneys, spleen, lungs, liver, meninges, or endocardium.

What factors influenced this microbe to shift from its normally harmless, symbiotic relationship to this parasitic stage? How might this infection be brought under control? What recommendations might be made to the patient in order to prevent a reoccurrence of the infection?

STUDY QUESTIONS

1   Define symbiosis and give four examples.

2   What is an opportunistic microbe? What is an antagonistic microbe?

3   What is the difference between bacteremia and septicemia?

4   What factors control the normal flora found on the surface of the skin?

5   Describe the formation of a cavity in the enamel of a tooth. What role do microbes play in this disease process?

6   What are gnotobiotic animals? Of what value are they in understanding the role of normal flora?

7   Identify these microbes and describe their roles in maintaining good health: *Lactobacillus acidophilus, Bifidobacterium bifidus,* and Döderlein's bacillus.

8   What is the difference between pathogenicity and virulence?

9   What are the major differences between exotoxins and endotoxins?

10   List four chemicals associated with the invasiveness of pathogens and describe their functional advantage to the microbes that produce them.

SUGGESTED READINGS

ANDREWS, M. *The Life That Lives on Man.* Taplinger Publishing, New York, 1978.

BURNETT, G. W., and G. SCHUSTER. *Oral Microbiology & Infectious Disease.* Williams & Wilkins, Baltimore, 1976.

CALDWELL, R. C., and R. E. STALLARD. *A Textbook of Preventive Dentistry.* W. B. Saunders, Philadelphia, 1977.

CURWEN, M. D., F. WILLIAM, S. ROSENBERG, and W. SHALITA. "Logical Lines to Acne Control." *Patient Care, The Journal of Practical Family Medicine,* Vol. IV, no. 8 ed. Miller and Fink Corp., April, 1975.

GIBBONS, R. J., and J. VAN HOUTE. "Bacterial Adherence in Oral Microbial Ecology." *Ann. Rev. Microbiol.,* 29:19–44, 1975.

"Ingestion of *Streptococcus mutans* Induces Secretory Immunoglobulin A and Caries Immunity." *Science,* 192:1238–40, June, 1976.

JENKINS, G. N. *The Physiology and Biochemistry of the Mouth,* 4th ed. Blackwell Scientific Publications, 1978.

MECHNIKOV, I. *The Prolongation of Life, V. Lactic Acid as Inhibiting Intestinal Putrefaction.* G. P. Putnam's Sons, New York, 1908.

METTLER, L., A. REMEVKE, and G. BRIELER. "The Effect of Bifidus Bacteria Replacement Therapy on Pararadiological and Post-irradiation Dysbacteria and Intestinal Radio-reaction." *Strahletherapie,* 145:588–99, 1973.

O'BRIEN, T. C. "Immunologic Aspects of Dental Caries." Special Supplement to *Immunology Abstracts,* 1976.

SKINNER, F. A., and J. G. CARR, editors. *The Normal Microbial Flora of Man.* Academic Press, London, 1974.

KEY TERMS

symbiosis (sim″bi-o′sis)	resident and transient flora	hemolysis
amensalism		endotoxin
commensalism	coagulase	invasiveness
mutualism	leukocidin	collagenase
parasitism	septicemia	hyaluronidase
bacteremia	exotoxin	attenuation

Pronunciation Guide for Organisms

Microsporum (mi-kros′po-rum)

Leptothrix (lep′to-thriks)

Micrococcus (mi″kro-kok′us)
   luteus (lu′te-us)

Vibrio (vib′re-o)
   cholerae (kol′er-ee)

Immunology I: Host Defense Mechanisms and Immunity

NONSPECIFIC HOST DEFENSE MECHANISMS

THE RETICULO-ENDOTHELIAL SYSTEM AND INFLAMMATION

IMMUNITY—A SPECIFIC HOST DEFENSE

BASICS OF THE IMMUNE SYSTEM

IMMUNOGLOBULINS AND ANTIGENS

LYMPHOCYTES AND THE IMMUNE SYSTEM

ANTIBODY-ANTIGEN REACTIONS
Agglutination
Precipitation
Lysis
Opsonization
Neutralization

OTHER SEROLOGICAL TESTS

CELL-MEDIATED IMMUNITY (CMI)

THE ANAMNESTIC RESPONSE AND IMMUNIZATION

Learning Objectives

☐ Name the major nonspecific host defense mechanisms of the human body.

☐ List the leukocytes capable of phagocytosis and the roles played by prostaglandins, histamines, serotonin, and heparin.

☐ Be familiar with the nature of the reticulo-endothelial system and its two major components.

☐ Be able to explain the inflammatory response.

☐ Explain the nature of the immune system and how its failure may result in immunodeficiency disease or hypersensitivities.

☐ Be able to define innate, acquired, natural, artificial, and passive immunity, and give examples of each.

☐ Be able to describe the nature of the five classes of immunoglobulins and the basic features unique to each class.

☐ List the major difference between antibody-mediated immunity and cell-mediated immunity.

☐ Describe the five major types of antibody-antigen reactions.

☐ Know the components and nature of the complement system and relate it to opsonization and the properdin system.

☐ Explain complement fixation.

☐ Be familiar with the basic serological tests such as flocculation, ASTO, fluorescent antibody, electrophoresis, RIA, and serotyping.

☐ Be able to describe the basics of cell-mediated immunity and the roles played by the various T-cells, macrophages, and B-cells.

☐ Draw and explain a graph illustrating the primary and secondary (anamnestic) responses following contact with an antigen.

☐ Know the basic immunization schedule for humans.

13

The body contains a number of different systems that help humans resist harm from microbes or damage from nonliving foreign materials. The efficient operation of these systems enables individuals to establish good host-parasite relationships and limit those that lead to disease and death. Many of these defense mechanisms are unable to discriminate between various damaging materials in the environment. Their general nature provides the body with an outer "first line of defense" against mechanical and chemical harm. Without these nonspecific host defense mechanisms, tissue damage could become extensive enough to allow resident or transient flora to establish a parasitic relationship. Other defense mechanisms in the body are highly specific in their action and recognize the difference between one kind of molecule and another as well as the difference between "self" and "nonself." This system, known as the immune system, is able to selectively and efficiently eliminate microbes and other foreign agents that might cause host harm. The system is also able to "learn" from these experiences and "remember" how to defend the body from further attacks of the same agent. In order to better understand the operation of both nonspecific and specific host defense mechanisms, a great deal of research has been done in the areas of anatomy, physiology, biochemistry, immunology, and serology. The benefits have resulted in the development of laboratory tests that aid in the identification of malfunctions in host defense, the diagnosis of specific diseases, and the prevention of many diseases. This chapter will review the nature of these defense mechanisms, describe a number of important serological tests performed in the laboratory, and discuss the stimulation of the immune system that provides specific immunity.

NONSPECIFIC HOST DEFENSE MECHANISMS

The human body has a variety of defense mechanisms that enables it to resist infection and harm from outside agents. Some of these barriers are able to restrict both living and nonliving materials and show no specificity in their actions. **Nonspecific host defense mechanisms** include features of the skin and mucous membranes, phagocytosis, and the inflammatory response. The skin presents an outer barrier of **keratinized epithelium** to microbes and nonliving agents. This dry layer of insoluble protein forms a shield that protects underlying living tissue. Moist mucous membranes lack this thickened protein layer but have other features that also prevent infections and tissue harm. **Mucus** is a freely moving liquid produced by goblet cells and contains inorganic salts, various organic molecules, loose epithelial cells, and leukocytes. **Lysozyme** and **lactoperoxidase** found in this fluid are enzymes active in the destruction of many types

of microbes as well as organic molecules. Foreign materials caught in the mucus layer on the **ciliated epithelium** of the upper respiratory tract are moved in escalator fashion toward the nose and mouth where they may be forcibly expelled or killed by sneezing, coughing, or swallowing. The turbulence of the air moving through the complex arrangement of turbinate bones in the nose and through the tubular network of the upper respiratory tract limits the penetration of foreign materials into the bronchi, bronchioles, and alveoli. This action prevents microbes or small foreign objects from becoming lodged in the tissue and causing harm. The low pH of both mucus and stomach fluids also aids in destruction. The hydrolytic action of digestive enzymes is enhanced by the emulsifying effects of **bile** as materials are moved by **peristalsis** through the gastrointestinal tract.

Phagocytosis is one of the most important nonspecific host defense mechanisms and involves the engulfment of foreign materials and microbes by a variety of specialized cells known as **phagocytes.** Certain members of a group of white blood cells known as leukocytes have the ability to phagocytose dangerous materials. Leukocytes are classified according to shape, size, nuclear structure, and cytoplasmic content. **Granulocytes** are leukocytes that contain granules in their cytoplasm and are about twice the size of a red blood cell. The three types of granulocytes that have been identified are neutrophiles, eosinophiles, and basophiles. Leukocytes that lack cytoplasmic granules are known as **agranulocytes** and may be either monocytes or lymphocytes (Figure 13.1).

The most abundant granulocytes are **neutrophiles** (approximately 60 percent of the white cells in blood) and are so named because of their pink staining reaction in neutral dyes. These cells are also known as **polymorphonuclear leukocytes** (PMNs) since their lobed nucleus may take on a number of shapes. PMNs are manufactured in the bone marrow and live only a few days after their release into the circulatory system. They are capable of amoeboid motion and may pass through the endothelial lining of blood vessels into the surrounding tissue, a process known as **diapedesis** (Figure 13.2). Polymorphonuclear leukocytes are important factors in inflammation and are able to release chemicals known as **prostaglandins.** These compounds are responsible for increasing vascular permeability, stimulating smooth muscle contraction, and lowering the threshold of pain.

Eosinophiles contain granules that stain deep red in acid dye. These leukocytes may also have a lobed nucleus but only make up about three percent of all white cells in circulation. Eosinophiles may be a factor in allergies by breaking down antibody-antigen complexes into products that cause tissue damage and blood vessel constriction. Eosinophiles are also able to release profibrinolysin that later becomes activated to fibrinolysin, an enzyme that is able to

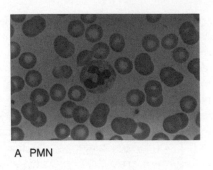

A PMN

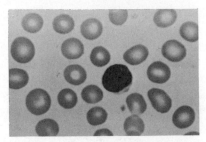

D Monocyte

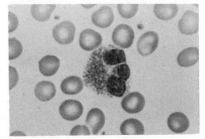

B Acidophile

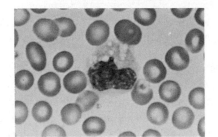

E Lymphocyte

FIGURE 13.1
Human white blood cells, leukocytes, are classified according to their size, shape, and cellular content. The two general groups shown here are the granulocytes (A, B, C) and agranulocytes (D, E).

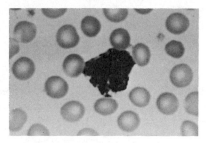

C Basophile

destroy blood clots. **Basophiles** also contain granules but stain deep blue in basic dyes. They are similar in size and shape to eosinophiles and also play a role in allergic responses. Basophiles are able to move by diapedesis through capillary walls. Cells that may be basophiles, or are at least physiologically identical to them, are located immediately outside the capillary wall and in other tissues. These granule-containing cells are known as **mast cells.** Mast cells can release the chemicals histamine, serotonin, and heparin when properly stimulated by a particular class of antibody (IgE). **Histamine** is a small, nitrogen-containing molecule that is able to stimulate the contraction of smooth muscle found around blood vessels and bronchioles, increase the permeability of blood vessels, and stimulate the release of nasal and bronchial mucus secretions. **Serotonin** stimulates vasoconstriction, while **heparin** is a natural anticoagulant that prevents the formation of blood clots. The total number of basophiles in the blood (approximately one percent of the leukocytes) increases slightly during infection and may be correlated with an increase in heparin release that prevents clot formation and the adherence of red blood cells to vessel walls during inflammation.

The agranulocytes and granulocytes are formed from special cells within red bone marrow known as **stem cells.** When agranulocytes are differentiated and released into the blood, they are either monocytes or lymphocytes. **Monocytes** make up between three to nine percent of the leukocytes, have a nucleus with a variable shape, and are generally larger than the lymphocytes. These cells respond to the release of chemicals from damaged tissue by moving toward the disrupted area by positive chemotaxis. Monocytes are capable of moving through capillary walls and changing into highly active tissue phagocytes known as **macrophages.** Macrophages are about five times larger than monocytes, contain more lysosomes, and are able to phagocytose red blood cells and necrotic tissue. The **lymphocytes** are smaller than macrophages and contain relatively large, rounded nuclei. These cells account for approximately 30 percent of circulating leukocytes and are vital in the immune response system. Further details on their function will be described later.

Destruction of foreign microbes and nonliving material by neutrophiles and macrophages involves the production of singlet oxygen, hydrogen peroxide, proteolytic enzymes,

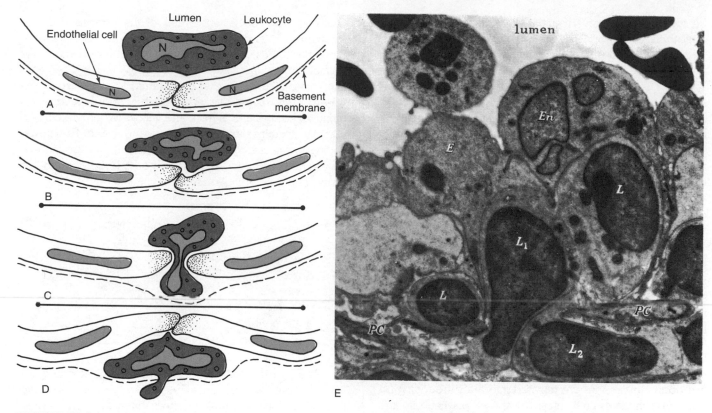

FIGURE 13.2

Many types of leukocytes are capable of moving from the lumen (inner space) of blood vessels through the endothelial lining into the surrounding tissue. This process is called diapedesis. In many cases, the cells stick to the interior of the blood vessel before moving. (In Part E, L_1 and L_2 are lymphocytes, PC is the periendothelial sheath, E is the endothelium, and En is the endothelial nucleus. From V. T. Marchesi and J. L. Gowans. "The Migration of Venules in Lymph Nodes: An Electron Microscope Study." *Proceedings of the Royal Society,* B159, plate 14, 1964.)

and lipases. Neutrophiles are capable of phagocytosing about twenty-five bacteria, while macrophages may ingest as many as one hundred cells. Nonspecific phagocytosis will continue until the cell reaches its maximum holding capacity and begins to self-destruct from the inside as a result of the release of hydrolytic enzymes. The neutrophiles and macrophages eventually die and contribute to the formation of pus.

---

1 What is the dry layer of insoluble protein on the body's surface which acts as a microbial barrier?

2 Name examples of granulocytes and agranulocytes.

3 What class of compounds is responsible for increasing vascular permeability, stimulating smooth muscle contraction, and lowering the threshold of pain?

THE RETICULO-ENDOTHELIAL SYSTEM AND INFLAMMATION

Many monocytes converted to macrophages become attached to tissues rather than continually wandering throughout the body. These cells continue to phagocytose foreign microbes and nonliving materials from relatively fixed positions. All of these fixed macrophages form a reticulum (network) of phagocytes throughout the body that operates as a functional system. Since many of these cells were first identified with the endothelium of blood vessels, the system has become known as the **reticulo-endothelial,** or **fixed macrophage, system.** The system is commonly divided into two major components: (1) tissue macrophages that are fixed to the walls of blood vessels and various other tissues, and (2) lymphocytic cells which are largely fixed in lymphoid tissues, such as lymph nodes, spleen, and bone marrow. Tissue macrophages often referred to as histiocytes may be found in connective tissue. Other special terms used to refer to fixed macrophages are alveolar cells, or "dust cells," found in the lungs, Kupffer cells in liver sinuses, Peyer's patches in the intestine, and microglia of the central nervous system. These cells are concentrated in positions that enable them to quickly filter foreign material as it enters the body. For example, the liver macrophage system serves to clear blood of microbes that may have entered the circulatory system from food taken in from the intestinal tract. The spleen macrophages remove damaged platelets and other cell debris found in the blood. In some cases, both the wandering and fixed macrophages work together during tissue injury or invasion by foreign objects in a nonspecific response known as inflammation.

The **inflammatory response** is a sequence of tissue events that results in the elimination of damaging foreign material or microbes and the repair of injured tissue. This sequence probably occurs throughout the body on a small, unnoticed scale everyday, but is most easily recognized when it occurs on the surface of the skin. The visible signs of inflammation include redness, swelling, heat, and pain. These symptoms may be caused by irritants such as bacteria, viruses, heat, chemicals, or physical damage. When damage first occurs, histamines are released into the surrounding tissue. This stimulates an increase in capillary permeability and allows fluid **(plasma)** to move from the vessels into the tissue to cause swelling **(edema).** Capillaries dilate, and an increase in blood flow reddens and warms the area. Fibrinogen moves from the blood into the tissue and begins to form a fibrin clot to wall off the site of irritation. This isolating mechanism reduces the chance of microbes spreading from the initial site of infection. Microbes that produce coagulase (e.g., *Staphylococcus aureus*) stimulate this process more rapidly than pathogens not producing toxins. As a result, staphylococcal infections may quickly become localized and nodular (e.g., in the formation of a boil).

At this same time, PMNs (neutrophiles) move by diapedesis into the tissue spaces and begin phagocytosis. Following the PMNs, agranulocytes (lymphocytes and macrophages) enter the inflammatory site. These cells release a number of chemical agents that attract other macrophages, kill foreign cells, help keep leukocytes localized, and stimulate phagocytosis. Other chemicals, **kinins** and prostaglandins, are released during the response. While both are thought to aid in continuing the inflammation process of vasodilation, only the prostaglandins may be responsible for pain and edema. As the response continues, surrounding host tissue is enzymatically destroyed. Macrophages are able to engulf this necrotic tissue until the macrophages themselves are killed by the release of hydrolytic enzymes. The inflammatory site becomes depleted of oxygen, lactic acid is formed, and the temperature rises. These changes contribute to further tissue destruction, increased macrophage activity, and

death of the less efficient PMNs. If the necrosis becomes extensive, a pus cavity may be formed and rupture from the surface. Pus produced during inflammation may take several forms depending on its composition. If the pus contains more fluid than cells, it is referred to as a **serous exudate.** A **purulent,** or **suppurative, exudate** contains many leukocytes and is thick and yellow. When the erupted material contains fibrin, the exudate is called **fibrinous.** During the final stages of inflammation, the destroyed leukocytes and necrotic tissue are absorbed into surrounding tissues and the damaged area is repaired.

---

1 Monocytes that move through capillary walls become what type of phagocytic cells?

2 Name two specialized terms used to refer to fixed macrophages other than reticulo-endothelial macrophages.

3 What role do the kinins serve in the inflammatory response?

The inflammatory response is not always able to effectively control the invasion of pathogens. As discussed earlier, many pathogens have capsules that allow them to resist phagocytosis and the killing effects of lysosomal enzymes. *Streptococcus pneumoniae, Salmonella typhi,* and *Brucella abortus* are all able to resist the destructive effects of macrophages. They may actually be carried inside these cells after engulfment to other parts of the body where they may establish other foci of infection. In some situations, the inflammatory response does not destroy invading pathogens but may become so exaggerated as to actually cause harm. In many upper respiratory infections, the host responds to a minimal infection with an inflammatory response that fills alveoli with edema fluid (Table 13.1). This is an excellent medium for bacterial and fungal growth. Normal breathing movements and coughing cause this infected fluid to be spread from one sac to another through the pores of Kohn (interalveolar pores), stimulating further inflammation (Figure 13.3). The invasion of granulocytes causes the tissue to take on a red color followed by a change to gray as the PMNs become concentrated and attempt to engulf infecting microbes. The fluid-filled sacs may restrict gas exchange and cause death, if congestion becomes extensive. Treatment of bacterial pneumonia with antibiotics is of great help since the quick elimination of the pathogen will shorten the period of spreading inflammation, edema, and congestion.

The inflammatory response and activity of phagocytes may be decreased by the use of certain drugs. Many common anti-inflammatory drugs are used to mildly suppress the pain, heat, and swelling associated with prostaglandin release. A number of these medications belong to a class of

TABLE 13.1

Bacterial Pneumonias
Klebsiella-enterobacterial
Staphylococcal
Streptococcal
Haemophilus
Tuberculous
Pseudomonas
Escherichia coli
Bacteroides
Proteus
Psittacosis
Q fever
Mycoplasmal
Nonbacterial Pneumonias
Viral

compounds known as salicylates (for example, aspirin), while many others are organic acids. More powerful anti-inflammatory drugs belong to the class of lipids known as steroids. Corticosteroids may be administered to patients with rheumatoid arthritis, asthma, bursitis, and tissue transplants. These drugs inhibit phagocytosis by decreasing capillary permeability and histamine action, stabilizing lysosomal membranes, and interfering with kinin production. Persons being given large amounts of these drugs may also be placed on antibiotic therapy to compensate for the reduced activity of

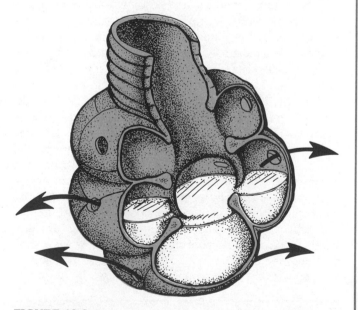

FIGURE 13.3
Movement of fluid during normal breathing is illustrated in a portion of lung tissue filled with edema fluid. Normal breathing movements force the pathogen-containing fluid through the smaller pores of Kohn into surrounding alveoli. This stimulates further inflammation and fluid release.

their naturally protective inflammatory response mechanism.

IMMUNITY—A SPECIFIC HOST DEFENSE

Inflammation is a nonspecific host defense mechanism that occurs in response to a range of stimuli. However, there is another defense mechanism that provides the host with resistance to specific agents. **Immunity** ("immunis" means *exempt*) has been recognized as a naturally occurring protective mechanism since before Jenner (1749–1823) developed the smallpox vaccine. During the mid-1600s, many people recognized that an individual who had become ill with smallpox would also become protected against a second case of the disease. It was common practice in some villages in China, Africa, Scandinavia, and Europe for a mother to "buy the pox." She might send her child to the home of an infected person for a week or obtain scrapings from the skin pox and "scratch" this material into her own child's skin. Her hope was to induce a mild form of the disease and make her child insusceptible to severe scarring and death (Figure 13.4). Records indicate that few children died from this practice, but those that had undergone treatment were contagious during the time of their mild illness and no doubt contributed to the spread of the viral disease. This folk medicine became known as "variolation" ("varus" means *pimple* or *pox*) or "inoculation" ("inoculare" means *graft*) and later was a clue that ultimately led to the development of vaccinations (refer to the Introduction). Since this early period, vaccines have been produced to prevent such diseases as anthrax (late 1800s), rabies (late 1800s), diphtheria (early 1900s), and tetanus (early 1900s). Currently, research efforts in the field of immunology have stimulated interest in the possibility of developing vaccines against cancer. The control of infectious diseases by the immune system is not always successful, and in some individuals, this complex defense mechanism may be ineffective and result in **immunodeficiency diseases**. In others, the system may overreact and cause harm in the form of **hypersensitivities (allergies)**. In addition, the medical profession has long been concerned with the role of the immune system in the transplantation of tissues from donor to recipient. Successful work in this area first began in 1902 when Karl Landsteiner first identified the A, B, O blood groups and established the

---

1 Why does the use of anti-inflammatory drugs increase the chance of infectious disease?
2 What is meant by "immunodeficiency disease"?

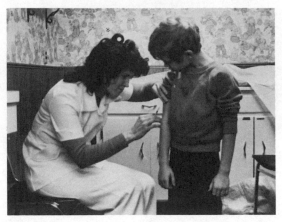

A

B

FIGURE 13.4
(A) This eighteenth century drawing encouraged public inoculation against smallpox. (B) More recently, health officials have reminded the public that many infectious diseases still exist and to prevent outbreaks, vaccinations must be obtained. (Part A courtesy of Yale Medical Library, Clements C. Fry Collection. Photo B courtesy of K. Knack and David Ross.)

basic principles behind blood transfusions. Today the transplantation of heart, kidney, skin, and corneal tissues and other organs is finding greater success as more is learned of the immune system.

BASICS OF THE IMMUNE SYSTEM

The immune system of higher vertebrates, including humans, responds to foreign agents in two ways: innate and acquired immunity. The **innate**, or **inborn, immune system** of these animals allows each species to resist particular infections without having been previously exposed to the infectious agent. No previous experience or contact with the agent is required since the organism is born with these protective mechanisms. No antibody production is involved. Basic physiological, genetic, biochemical, and anatomical differences in species are responsible for defending the organism against pathogens. For example, certain populations of humans in Africa are resistant to malaria infections because they lack the genes necessary to produce cell-surface receptor sites for these protozoan parasites. Since the receptor site is not produced, the microbes cannot attach and invade red blood cells to establish the disease state. Other factors which are innate and provide immunity include the pH of body fluids, body temperature*, and metabolic pathways that are selectively harmful to certain pathogens.

---

*The reason chickens are immune to anthrax is their high body temperature. Experiments have shown that if their body temperature is lowered, they succumb to the infection.

Acquired immunity may be gained when an individual becomes relatively and temporarily insusceptible to an infectious agent either by natural or artificial means. **Naturally acquired active immunity** develops after an attack of a disease and antibodies to that microbe have been produced. The protection provided by this exposure is usually long lasting and specific as in the case of many childhood diseases (mumps, measles, and polio). **Naturally acquired passive immunity** is provided to infants by the transmission of maternal antibodies through the placenta or in the colostrum of breast milk (Figure 13.5). Protection provided by these antibodies is only temporary (4–6 months), since these protein molecules may either react with an antigen or be naturally metabolized to other forms in the infant's body. Immunity may also be acquired artificially in active or passive ways. Vaccination with live, attenuated, or dead microbes, or with toxoids stimulates the immune system and provides the individual with a long-lasting, specific defense. Vaccinations are examples of **artificially acquired active immunity. Artificially acquired passive immunity** may be obtained by the injection of antibody previously formed in another organism. The injection of this **antiserum** provides immediate protection but is only valuable for a short time since the body has not been stimulated to manufacture its own antibody. A common antiserum used by physicians

1 What form of immunity does not involve the action of antibodies?
2 Why is the injection of antiserum a passive form of immunization?

A

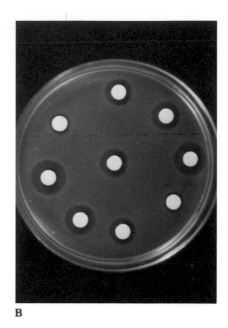

B

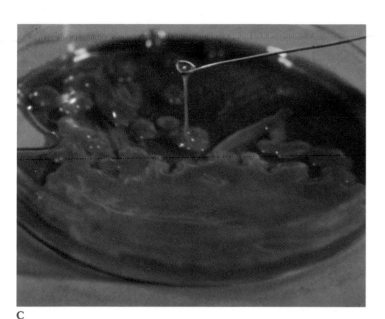

C

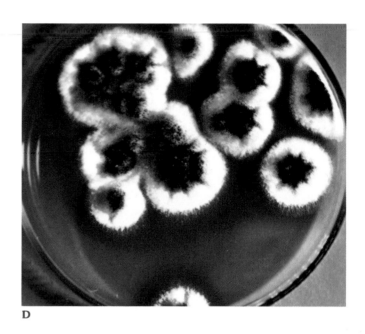

D

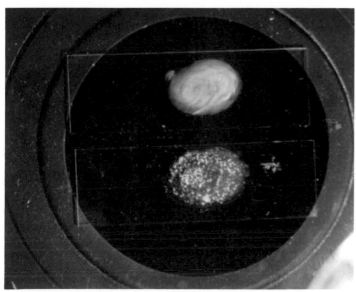

E

(A) *Pseudomonas aeruginosa* growth in nutrient broth. Left photo: Production of blue-green pigment and pellicle. Right photo: Characteristic blue-green pigment may be elicited or enhanced by vigorous shaking of the tube (right tube). (B) antibiotic sensitivity test. (C) *Klebsiella pneumoniae* cultured on Endo agar. (D) *Aspergillus niger* cultured on mannite agar. (E) *Staphylococcus aureus* coagulase slide test. Upper: negative test with *S. epidermis* strain; lower: positive reaction with coagulase-positive *S. aureus* strain. (Courtesy of Abbott Laboratories.)

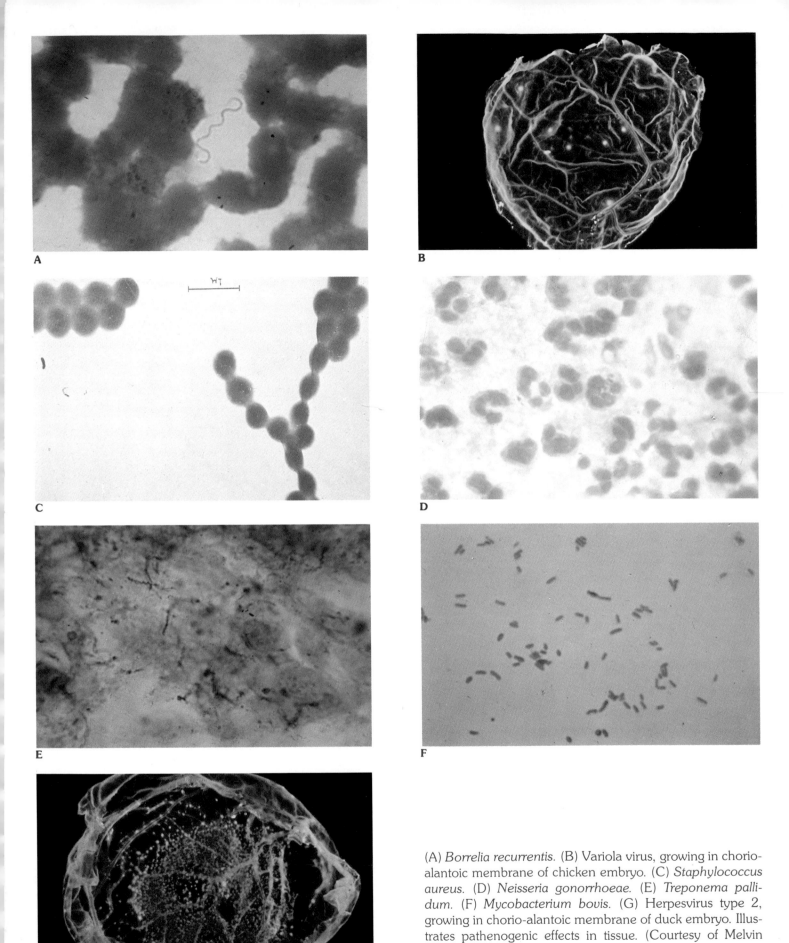

(A) *Borrelia recurrentis*. (B) Variola virus, growing in chorio-alantoic membrane of chicken embryo. (C) *Staphylococcus aureus*. (D) *Neisseria gonorrhoeae*. (E) *Treponema pallidum*. (F) *Mycobacterium bovis*. (G) Herpesvirus type 2, growing in chorio-alantoic membrane of duck embryo. Illustrates pathenogenic effects in tissue. (Courtesy of Melvin Rheins.)

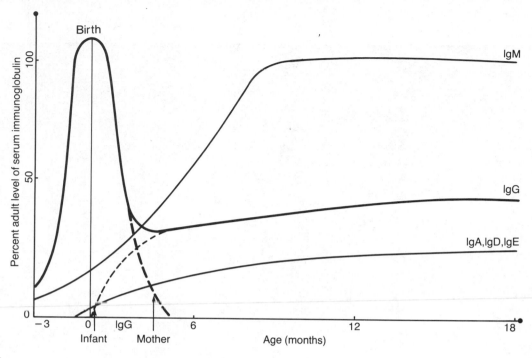

FIGURE 13.5

At birth an infant does not form antibodies even though it has the ability to produce them. Only one type of antibody (IgG) is able to cross the placenta; however, soon after birth the infant begins the formation of all types of antibodies (IgM, IgG, IgA, IgD, and IgE). Adult levels of antibody type IgM are reached by nine months of age; however, the others require a longer period. Not only are high levels of antibody important to the infant, but the developing child must have a variety of antibodies of each type. This variety is achieved by repeated contact with antigens that stimulate a broad range of active acquired immunity. Most outgrow "childhood" illnesses between four and six years of age. (From M. Adinolfi, editor. *Immunology and Development.* Spastics International Medical Publications.)

for the treatment of rubella (German measles) is gamma globulin. Antisera are obtained from the blood of vaccinated or previously infected horses, cows, rabbits, or humans. Table 13.2 summarizes the main differences among the various forms of immunity.

IMMUNOGLOBULINS AND ANTIGENS

All immune responses involve highly specific reactions between antibodies and antigens. An **antigen (Ag),** or **immunogen,** is a large organic molecule that is able to stimulate production of a specific antibody with which it may chemically combine. An **antibody (Ab)** is always a protein molecule manufactured by the body in response to the presence of a foreign molecule known as an antigen. Proteins that function as antibodies are known as **immunoglobulins (Ig).** The basic structure of most antibodies is that of a

"slingshot" or "Y" configuration (Figure 13.6). These proteins are composed of two, long "heavy" chains, and two, short "light" chains attached to one another by covalent bonds. The unit is able to spread apart, or flex, at the "hinge," making the space between the top of the "Y" larger or smaller. All antibodies are coded for by DNA and manufactured by protein synthesis. The portion of an antibody that chemically bonds to a specific antigen is known as the antigenic bonding site and is located at the top of each arm of the "Y." The bonds that hold antibody and antigen together are usually weak hydrogen bonds or ionic bonds and are easily broken. The size and shape of the antigenic bonding site governs the specificity of the antibody-antigen reaction. Any alteration in this site by heat or chemicals will interfere with the combining of the two. The number of bonding sites on an antibody is referred to as the **valency** of the molecule and will influence the way in which antigen-antibody combinations interact. Antibodies with two antigenic bonding sites are bivalent and those with more than two are polyvalent. About 85 percent of all human antibodies are bivalent.

TABLE 13.2
The immune system.

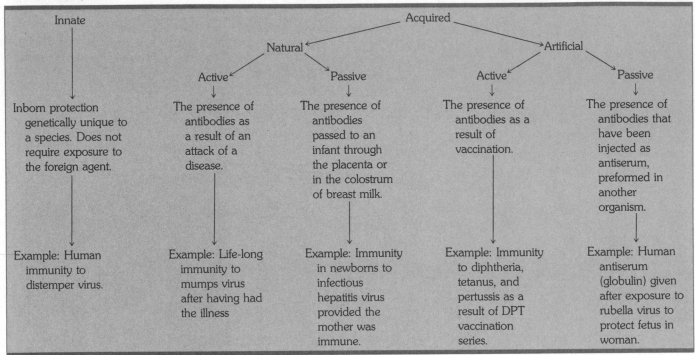

	Innate	Acquired			

		Natural		Artificial	
		Active	Passive	Active	Passive
Inborn protection genetically unique to a species. Does not require exposure to the foreign agent.		The presence of antibodies as a result of an attack of a disease.	The presence of antibodies passed to an infant through the placenta or in the colostrum of breast milk.	The presence of antibodies as a result of vaccination.	The presence of antibodies that have been injected as antiserum, preformed in another organism.
Example: Human immunity to distemper virus.		Example: Life-long immunity to mumps virus after having had the illness	Example: Immunity in newborns to infectious hepatitis virus provided the mother was immune.	Example: Immunity to diphtheria, tetanus, and pertussis as a result of DPT vaccination series.	Example: Human antiserum (globulin) given after exposure to rubella virus to protect fetus in woman.

FIGURE 13.6
An antibody molecule is "Y"-shaped and composed of several polypeptide chains. The variable portion of the molecule differs among antibodies and contains the antigenic bonding site. In some antibodies, the constant portion of the molecule may be attached to a cell surface.

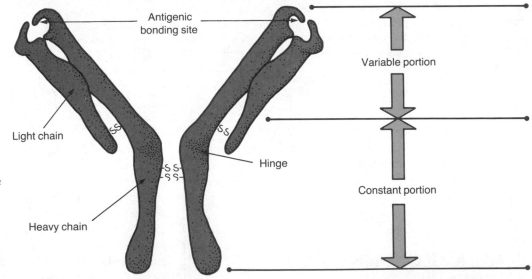

Antigenic bonding site

Light chain

Heavy chain

Hinge

Variable portion

Constant portion

Antibodies produced by the body can be divided into several classes based on the structure of the heavy chain. The light chains are more consistent in composition. Within any one class, the antibodies all demonstrate a "constant" amino acid sequence for a portion of the molecule but vary in the nature of the amino acid sequence of their antigenic bonding sites. This variation accounts for the specificity of different antibody-antigen reactions. The classes of antibodies have been named with letters of the Greek alphabet and are typically referred to in "shorthand notation" as IgM, IgG,

IgA, IgE, and IgD.* Notice in Table 13.3 that IgM and IgA antibodies are joined together in polyvalent forms. The joining is due to the presence of additional chains known as "J" chains.

A great many complex substances can serve as antigens. Table 13.4 lists a number of antigenic materials composed of large protein or polysaccharide molecules. Lipids

*M, G, A, E, and D are mu, gamma, alpha, epsilon, and delta, respectively.

TABLE 13.3
Serum immunoglobulins.

Immunoglobulin	Heavy Chains	Other Chains	Percent of Total Ig	Complement Fixation	Ability to Cross Placenta	Location	Valency	Concentration Range in Serum (g/ml)	Major Characteristic
IgM	Mu	J	6	+	−	Blood serum	5–10	0.5–2	Very effective agglutinator; produced early in immune response—effective first line defense against bacteraemia.
IgG	Gamma$_1$ Gamma$_2$ Gamma$_3$ Gamma$_4$	−	80	+	+	Blood serum, extra-cellular body fluids, and on killer lymphocytes	2	8–16	Most abundant Ig of internal body fluids, particularly extravascular where it combats microorganisms and their toxins.
IgA	Alpha$_1$ Alpha$_2$	J	13	−	−	Blood serum and secretions, and on helper lymphocytes	2 and polyvalent	1.4–4	Major Ig in seromucous secretions where it defends external body surfaces.
IgE	Epsilon	−	0.002	−	−	Blood serum and fixed to basophiles and mast cells	2	$17–450 \times 10^{-9}$	Raised in infections. Responsible for symptoms of allergy.
IgD	Delta	−	1	−	−	Blood serum and on lymphocytes in newborns	2	0–0.4	Transient: present on lymphocyte surface.

TABLE 13.4
Substances that can serve as antigens.

Toxoids of diphtheria, pertussis, and tetanus
Killed *Salmonella, E. coli, Klebsiella,* streptococci,
    Enterobacter, and others
Attenuated polio virus (Sabin vaccine)
Live cowpox virus
Isoagglutinins (on RBCs)
Bacterial exotoxins
Capsular material of bacteria
Flagellar protein
Mold spores
Pollens
Dust
Plant oils (poison ivy)
Foods (chocolate, banana, and others)
Insect venoms

and nucleic acids may also occasionally serve as antigens. The ability to recognize the difference between "self" and "foreign" antigenic materials develops early in life. The presence of high concentrations of an antigen in the fetus or newborn before the immune system has become fully developed (competent) induces a state of immunologic tolerance to the material. **Immunologic tolerance** is the inability of the immune system to produce antibodies against a known antigen. The specificity of antibody-antigen reactions is important in preventing the development of autoimmune disease that results from the loss of immunologic tolerance to "self." **Autoimmunity** occurs when antibodies are produced against molecules native to the body (autoantigens). If this reaction should occur, severe tissue damage, pain, and deformities may result. Rheumatoid arthritis is an autoimmune disease that affects the joints, causing inflammation, damage to synovial membranes, and atrophy (wasting away) of the bones.

1 What is meant by the term "valency"?
2 Which immunoglobulin occurs in the blood at the highest level?
3 How does immunologic tolerance relate to autoimmunity?

In order to serve as an antigen, a molecule must be able to stimulate the formation of antibody and chemically combine with it. The portion of antigen molecule that determines the specificity of an antibody-antigen reaction and combines with the antibody's antigenic bonding site is known as the antigenic determinant. Smaller molecules which are able to react with the antigenic bonding site but do not stimulate the formation of antibody are known as **haptens.** Haptens may be antigenic determinants, but the reverse is not necessarily true. Haptens have been used to study an-

tibody-antigen reactions without stimulating the production of more antibody.

LYMPHOCYTES AND THE IMMUNE SYSTEM

The cells primarily responsible for the operation of the immune system are the lymphocytes. As described earlier, these leukocytes are derived from stem cells in the bone marrow. When the lymphocytes leave the marrow and enter the lymph system (lymph node, spleen, tonsils, thymus, etc.), some fall under the influence of the thymus gland and become differentiated into cells called **T-lymphocytes (T-cells).** The differentiation of other lymphocytes is not controlled by the thymus and these are known as bursa-equivalent cells, or **B-lymphocytes (B-cells).** This term comes from the fact that chickens produce equivalent types of lymphocytes in a gland known as the bursa of Fabricius located near their intestinal tract. T-cells formed from the thymus do not secrete antibody but aid the B-cells in regulating the production of antibodies and play a vital role in a process known as **cell-mediated immunity, (CMI).** During cell-mediated immunity, the T-cells respond to the presence of specific antigen by excreting a range of other chemicals, **lymphokines,** able to kill foreign cells, or they may function as T-killer cells with macrophages to directly kill tumor cells or transplanted tissues. This form of host defense mechanism is especially effective against intracellular pathogens, fungal infections, chronic bacterial infections, and certain viruses. Cell-mediated immunity will be described in more detail later.

B-cells play a vital role in the process known as **humoral,** or **antibody-mediated, immunity (AMI).** These differentiated lymphocytes secrete antibody into body fluids (humor) in response to antigenic stimulation (Table 13.5). After release from the marrow, B-cells that come in contact with antigen are stimulated to enlarge and become blast cells. These divide by mitosis into numerous smaller lymphocytes capable of responding to the same antigen by producing small amounts of antibody. Cells that are able to respond in this manner are said to be **immunocompetent.** This repeated stimulation and division results in the formation of a clone of cells known as plasma cells. The **plasma cells** are the primary antibody-producing cells of the body and are responsible for humoral immunity. The population of plasma cells in an individual is composed of many subpopulations that are responsive to different types of antigens because of the presence of specific antigenic-bonding sites on the cell surface. Each plasma cell produces antibody against only one antigen.

The antigenic-bonding sites, or receptors, are antibodies produced by the cell. When the appropriate antigen makes contact with the receptor, the plasma cell begins the production and release of antibody. If the cell does not have a

TABLE 13.5
Antibody-mediated and cell-mediated immunity.

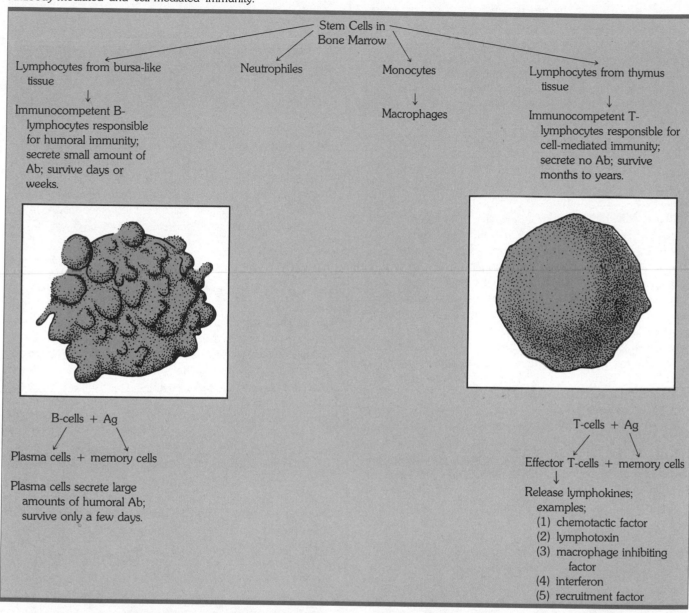

Stem Cells in Bone Marrow

Lymphocytes from bursa-like tissue
↓
Immunocompetent B-lymphocytes responsible for humoral immunity; secrete small amount of Ab; survive days or weeks.

Neutrophiles

Monocytes
↓
Macrophages

Lymphocytes from thymus tissue
↓
Immunocompetent T-lymphocytes responsible for cell-mediated immunity; secrete no Ab; survive months to years.

B-cells + Ag
↓
Plasma cells + memory cells

Plasma cells secrete large amounts of humoral Ab; survive only a few days.

T-cells + Ag
↓
Effector T-cells + memory cells
↓
Release lymphokines; examples;
(1) chemotactic factor
(2) lymphotoxin
(3) macrophage inhibiting factor
(4) interferon
(5) recruitment factor

particular receptor site (immunoincompetent), it will not be able to form antibody against that antigen. Any one plasma cell can respond to a particular antigen; however, not all plasma cells are able to respond to all antigens. This fact means that there are thousands of overlapping subpopulations of immunocompetent cells in the body able to respond to the presence of a great many different antigens and establish a high and diverse antibody titer. The production of antibodies and their specific reactions with antigens provide the body with an effective, selective defense mechanism against invading foreign materials and microbes.

1 What are lymphokines?
2 Which cells are primarily responsible for antibody production?
3 What do receptor sites have to do with immunocompetence?

ANTIBODY-ANTIGEN REACTIONS

Five general categories of antibody-antigen reactions have been recognized: agglutination, precipitation, lysis, opsonization, and neutralization.

AGGLUTINATION

An **agglutination** reaction occurs when an antibody combines with an antigen that is a part of a large, insoluble particle (bacterium, RBC, etc.) to form an insoluble aggregate of material. Many of these reactions involve bivalent or polyvalent antibodies which may interlink to form complex arrangements. These complexes are more easily recognized by phagocytes than the smaller antigens and are quickly engulfed and destroyed. The identification of an ABO blood type is probably one of the most familiar agglutination reactions. In this reaction, a sample of blood is placed on a slide and mixed with a solution of antiserum containing a single known antibody. If the surface of the red blood cells contains the antigen (agglutinogen) which will specifically combine with the antibody (agglutinin), the cells will agglutinate and fall out of solution as a result of becoming interconnected by a **latticelike network.** If the antiserum contains antibody that will not combine with the surface antigens, no reaction will be observed. By mixing the several types of antisera with separate samples of an individual's blood, the exact antigenic nature may be determined easily (Figure 13.7).

The agglutination reaction has been adapted for other lab tests. One indirect, or passive, agglutination reaction is used in the identification of rheumatoid factor (RF). This molecule is an immunoglobulin found in the synovial fluid and blood of patients with rheumatoid arthritis (RA) and a number of other collagen inflammatory diseases (for example, systemic lupus erythematosus, or SLE). This molecule probably does not start the inflammation typical of this disease but may exaggerate the condition. The **RF test** for the identification of rheumatoid factor is performed on the surface of a black card. A sample of blood thought to contain the factor is mixed with small, gamma-globulin-coated latex particles. If the factor is present, the white latex particles will agglutinate on the dark surface. This test is very sensitive but not as specific as the Rose-Waaler test, which is based on a specific antibody-antigen reaction involving sheep red blood cells. A positive Rose-Waaler test result is the agglutination of the sheep cells when they are in the presence of rheumatoid factor. Another common agglutination test identifies the presence of the **heterophile antibody** characteristically found in sera of individuals with infectious mononucleosis caused by Epstein-Barr virus. A slide spot test, the MONOSPOT®, is used to screen suspected cases of mononucleosis and is similar to the RF test. Confirmation is made by performing the Paul-Bunnell test which, like the Rose-Waaler test, relies on the agglutination of specially treated sheep red blood cells (Figure 13.8).

PRECIPITATION

Unlike agglutination reactions, **precipitin reactions** involve a small, soluble antigen. When the antibody-antigen reaction occurs, a few small, soluble lattice complexes form followed by a long period of increased lattice development and finally precipitation. Precipitin reactions take place over a long period and are less visible than agglutination. Maintaining an optimum balance between antibody and antigen is important to the efficient operation of this immune system, since phagocytosis is most efficient when the lattice is completely developed. If either the antibody or antigen is found in excess, the lattice structure will be incomplete and precipitation does not occur (Figure 13.9). If the antigen is in excess, an inflammatory response may occur that would damage tissues. Maintenance of the optimum ratio between antibody and antigen is accomplished by a particular subpopulation of T-cells (T-suppressor cells) that suppresses both the macrophages and the B-cells that reduce antibody formation.

Detection of a precipitin reaction between antibody and antigen is of great value in the diagnosis of many diseases.

FIGURE 13.7
In ABO blood typing, antibody and antigen are mixed on a glass slide to determine a patient's blood type.

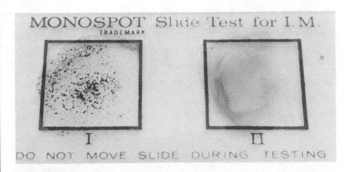

FIGURE 13.8
The MONOSPOT® slide test for infectious mononucleosis is commonly performed in medical laboratories. The agglutination reaction is seen on the left.

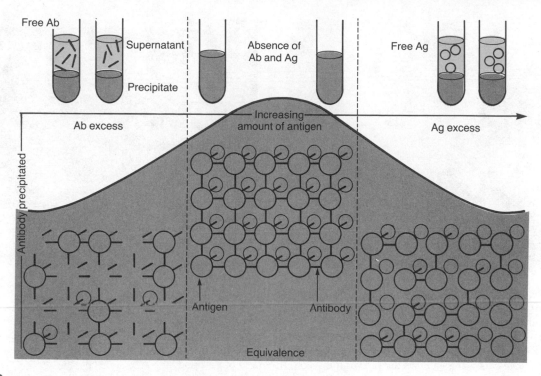

FIGURE 13.9

The relative concentrations of antigen and antibody influences the formation of the lattice network. An excess of antibody (left side of graph) results in poor precipitation, leaving antibody free in the fluid; an excess of antigen (right side of graph) complexes all the antibody but leaves extra antigen in the fluid. The optimum ratio between antigen and antibody (middle portion of graph) allows the formation of a complete lattice network and shows the greatest amount of precipitate. T-helper cells and T-suppressor cells in the body help maintain the optimum level of serum antibody so that the optimum lattice network can be formed with incoming antigen. The result is more efficient phagocytosis of the foreign material. (From J. A. Bellanti, *Immunology.* W. B. Saunders, 1971.)

Because they are more difficult to balance and require more time than agglutinations, precipitin reactions are identified by using a special method known as **agar diffusion.** This technique is based on the principle that antibodies and antigens will diffuse through a semisolid agar medium and form a precipitate when they make contact with one another. The test is performed on special immunodiffusion plates that have incorporated into the medium a known class of anti-immunoglobulin (IgG, IgM, IgE, etc.) or a specific antigen (e.g., antigen against *Salmonella typhi* antibody). A sample of solution suspected of containing an antibody is dropped into a small well punched through the medium. When the diffusing antibody (if present) meets the antigen, a ring will be formed around the well (Figure 13.10). By using a serial dilution of the sample, rings of different sizes will be formed and a quantitative measure of the amount of antibody may be determined by comparing the test results to a known standard. The determination of serum immunoglobulin levels may be used as good indicators of many systemic diseases (Table 13.6). For example, patients with an elevated

IgG and leukopenia (a reduction in leukocytes) may be diagnosed as having a systemic autoimmune disease. Those with tuberculosis are likely to have a slight to moderate elevation of IgG, a normal to marked elevation of IgA, and a slight or decreased amount of IgM.

1 What is a rheumatoid factor?
2 The presence of the heterophile antibody can be indicative of what type of infection?
3 What role does a T-suppressor cell play in antibody-antigen reactions?

LYSIS

The third type of antibody-antigen reaction results in the **lysis** of microbial cells or red blood cells. The series of reactions that ultimately forms holes in the plasma membrane begins with the attachment of free antibody to cell surface

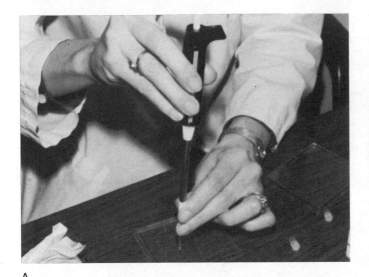

A   B

FIGURE 13.10
The radial immunodiffusion plate is used to detect and measure the amount of immunoglobulin in a sample. (A) The sample is placed in the wells of the diffusion plate, allowed to diffuse at 23°C for 12 to 24 hours, then the sizes of the rings are measured. (B) A close-up of an immunodiffusion plate containing IgG and the precipitin reaction. (Courtesy of Kallestad Laboratories, Inc.)

antigens and is followed by a complex series of reactions involving 11 serum proteins collectively known as the **complement system** (C system). Components of the C system are identified by numbered letters, C1 through C9. The additional components that account for the 11 elements are subcomponents of C1 and are designated as C1q, C1r, and C1s (C1 = C1q + C1r + C1s). After a foreign microbe enters the body, it may react with antibodies IgG or IgM, which form a coating on the cell. This sticky antibody coating enables C1 to attach and become activated. Activated C1 is then able to attach to and activate C4, which in turn attaches to and activates C2, etc. (Figure 13.11). This cascading series of attachments and activations results in the formation of many combinations of complement proteins that have significant biological activity. The final complex formed by the attachment and activation of C8 and C9 to the cell surface results in the formation of a hole in the plasma membrane through which essential cell components may leak (Figure 13.12).

OPSONIZATION

Another biologically active complex in the complement series is the product formed after C3 and C5 are attached to one another and activated. The presence of this molecule on the cell surface makes the microbe more recognizable to phagocytes, stimulating the movement of phagocytes toward the microbe and increasing their phagocytic ability. The adsorption of certain antibodies or complement (C3C5 complex) onto the surface of foreign material that results in the enhancement of phagocytosis is known as **opsonization.** Opsonization is the fourth main antibody-antigen reaction associated with humoral antibodies. Complement and certain antibodies are the two main **opsonins** that stimulate phagocytosis. C3C5 complement also stimulates T-cells to move into the area (positive chemotaxis), which begins the process of cell-mediated immunity and causes the release of histamines from leukocytes. The histamine release increases capillary permeability, diapedesis, and smooth muscle contraction. Taken collectively, these reactions result in local inflammation. The activation of this complement factor need not follow the pathway beginning with the antibody-antigen reaction. A bypass system present in humans called the **properdin system** is able to activate the C3C5 complex, initiating these same protective responses. The important role played by this system is illustrated by individuals who lack an efficient complement system. Detailed serological tests

1 The complement system is composed of what kind of organic molecules?
2 How does the complement complex affect a bacterial cell?
3 Name two types of opsonins.
4 What is the properdin system?

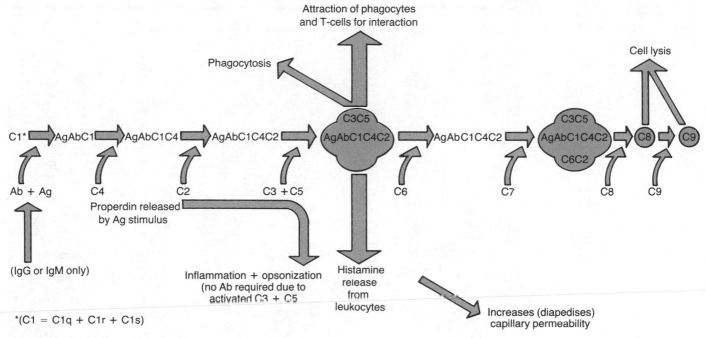

FIGURE 13.11

Immunoglobulins IgG and IgM are able to simulate the cascading action of the complement system after they react with antigens on the surface of bacteria or other foreign agents.

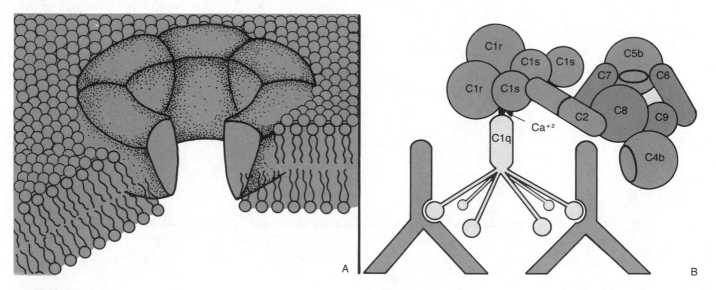

FIGURE 13.12

Antibody combines with cell membrane antigens to initiate cell lysis. (B) The antibody-antigen combination then binds with C1q, and the reaction sequence begins. The result is the formation of the C8C9 complex that forms a large pore (A) in the cell membrane. (From M. Mayer. "The Complement System." *Scientific American*, Nov., 1973.)

have identified that individuals suffering from severe or recurrent infections of *Neisseria meningiditis* lack complement components C5 through C9. In addition to meningitis, a number of other systemic and cutaneous skin diseases are associated with complement deficiencies (Table 13.7).

Lysis of cells by antigen-antibody-complement reactions has also been applied to the laboratory analysis of diseases. This serological test is known as complement fixation and is used in the identification of diseases such as leptospirosis, mycoplasmal pneumonia, Q fever, polio, rubella,

TABLE 13.6
Serum immunoglobulin levels in disease.

Diseases	IgG	IgA	IgM
Immunodeficiency disorders			
Combined immunodeficiency	↓↓↔↓↓↓	↓↓↔↓↓↓	↓↓↔↓↓↓
X-linked hypogammaglobulinemia	↓↓↔↓↓↓	↓↓↔↓↓↓	↓↓↔↓↓↓
Variable common immunodeficiency	↓↔↓↓↓	↓↔↓↓↓	↓↔↓↓↓
Selective IgA deficiency	N	↓↓↓	N
Protein-losing gastroenteropathies	N↔↓↓↓	N↔↓↓↓	N↔↓↓↓
Acute thermal burns	N↔↓↓↓	N↔↓↓↓	N↔↓↓↓
Nephrotic syndrome	N↔↓↓↓	N↔↓↓↓	N↔↓↓↓
Monoclonal gammopathies (MG)			
IgG (e.g., G-myeloma)	N↔↑↑↑	N↔↓↓↓	N↔↓↓↓
IgA (e.g., A-myeloma)	N↔↓↓↓	N↔↑↑↑	N↔↓↓↓
IgM (e.g., M-macroglobulinemia)	N↔↓↓↓	N↔↓↓↓	N↔↑↑↑
L chain disease (i.e., Bence Jones myeloma)	N↔↓↓↓	N↔↓↓↓	N↔↓↓↓
Chronic lymphocytic leukemia			
Infections			
Infectious mononucleosis	↑↔↑↑	N↔↑	↑↔↑↑
Subacute bacterial endocarditis	↑↔↑↑	↓↔N	↑↔↑↑
Tuberculosis	↑↔↑↑	N↔↑↑↑	↓↔N
Actinomycosis	↑↑↑	↑↑	↑↑↑
Deep fungus diseases	N	N↔↑	N
Bartonellosis	↑	↓↔N	↑↑↔↑↑↑
Liver diseases			
Infectious hepatitis	↑↔↑↑	N↔↑	N↔↑↑
Laennec's cirrhosis	↑↔↑↑↑	↑↔↑↑↑	N↔↑↑
Biliary cirrhosis	N	N	↑↔↑↑
Chronic active hepatitis	↑↑↑	↑	N↔↑↑
Collagen disorders			
Lupus erythematosus	↑↔↑↑	N↔↑	N↔↑↑
Rheumatoid arthritis	N↔↑↑↑	↑↔↑↑↑	N↔↑↑
Sjögren's syndrome	N↔↑	N↔↑	N↔↑↑
Scleroderma	N↔↑	N	N↔↑
Miscellaneous			
Sarcoidosis	N↔↑↑	N↔↑↑	N↔↑
Hodgkin's disease	↓↔↑↑	↓↔↑	↓↔↑↑
Monocytic leukemia	N↔↑	N↔↑	N↔↑
Cystic fibrosis	↑↔↑↑	↑↔↑↑	N↔↑↑

N = normal, ↑ = slight increase, ↑↑ = moderate increase, ↑↑↑ = marked increase, ↓ = slight decrease, ↓↓ = moderate decrease, ↓↓↓ = marked decrease, ↔ = range.

SOURCE: Modified and reproduced with permission from S. E. Ritzmann and J. C. Daniels, editors. *Serum Protein Abnormalities: Diagnostic and Clinical Aspects*. Little, Brown, 1975.

histoplasmosis, coccidioidomycosis, and streptococcal infections.

The presence of a symptomatic or an asymptomatic infection causes a patient's immune system to produce serum antibody. The identification of the presence of this antibody in the serum may be used to confirm a diagnosis and determine a patient's clinical stage of infection (refer to Chapter 9). The **complement fixation** test utilizes a patient's serum, the test antigen, complement from guinea pigs, and antibodies to sheep red blood cells to determine whether or not the sheep red cells may be lysed by guinea pig comple-

ment. If no hemolysis occurs, the antibody to the suspected disease is present and the patient has the illness; if hemolysis occurs, the antibody to the suspected disease is absent and the patient does not have the disease (Table 13.8).

To perform complement fixation, a blood sample is drawn and the serum is isolated for testing. Heat treating destroys all the naturally occurring complement but does not harm the suspected antibody. The serum is mixed with guinea pig complement and "test" antigen associated with the disease in question (e.g., streptococcal antigen). If the patient has produced antibody against the disease (e.g.,

TABLE 13.7
Systemic and cutaneous difficulties associated with complement system deficiencies.

Component	Systemic Disease	Cutaneous Disease
Clq	Bacterial infections	?
Clr	LE-like glomerulonephritis	DLE-like lesions
C2	SLE, glomerulonephritis, vasculitis, dermatomyositis. May be normal.	"Butterfly" rash, heliotrope, anaphylactoid purpura.
C3	Recurrent pyogenic infections	Pyoderma
C6	Disseminated gonococcal disease	Gonococcal, cutaneous, and arthritis syndromes
C8	Disseminated gonococcal disease	?
C5 dysfunction	Recurrent Gram-negative infections	Leiner's syndrome
Opsonic defect in sickle cell disease	Pneumococcal meningitis, osteomyelitis	None

LE = lupus erythematosus; SLE = systemic lupus erythematosus; DLE = discoid lupus erythematosus.

streptococci), the test antigen will react with the antibody and initiate the complement sequence. This reaction will destroy all the guinea pig complement and make it unable to lyse red blood cells. Sheep RBCs plus antibodies to sheep RBCs are added to this mixture to visibly detect whether guinea pig complement neutralization has occurred. If no hemolysis occurs, antibody originally present in the patient's blood already complexed the complement and made it un-

TABLE 13.8
Complement fixation test.

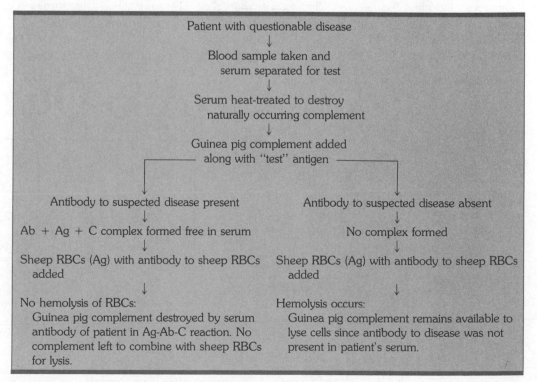

Patient with questionable disease
↓
Blood sample taken and serum separated for test
↓
Serum heat-treated to destroy naturally occurring complement
↓
Guinea pig complement added along with "test" antigen

Antibody to suspected disease present
↓
Ab + Ag + C complex formed free in serum
↓
Sheep RBCs (Ag) with antibody to sheep RBCs added
↓
No hemolysis of RBCs:
Guinea pig complement destroyed by serum antibody of patient in Ag-Ab-C reaction. No complement left to combine with sheep RBCs for lysis.

Antibody to suspected disease absent
↓
No complex formed
↓
Sheep RBCs (Ag) with antibody to sheep RBCs added
↓
Hemolysis occurs:
Guinea pig complement remains available to lyse cells since antibody to disease was not present in patient's serum.

available to destroy the RBCs. This indicates that the patient had been in contact with the pathogen. If hemolysis occurs, antibody to the suspected disease was not present, and guinea pig complement remained available to react with the sheep blood cells and cause their destruction. This indicates that the patient had not been infected with the pathogen. Table 13.8 summarizes the sequence of events in the complement fixation test.

NEUTRALIZATION

The final type of antibody-antigen reaction associated with humoral immunity is known as **neutralization.** Antibodies that participate in these reactions are known as **antitoxins.** Neutralization reactions may occur between soluble toxin molecules released by pathogens, or they may occur with individual viruses. Many of the antibodies involved act as monovalent units and, therefore, seldom form lattice networks. The antibody-antigen complex blocks the reaction site of a toxin or the attachment site of the virus. Normal body metabolism or phagocytosis is responsible for the final elimination of the foreign agent.

OTHER SEROLOGICAL TESTS

The identification of antitoxins, other antibodies, and specific antigens may be accomplished through a number of techniques, including flocculation, fluorescent antibodies, serotyping, and automated immunodiagnostic systems. A **flocculation** reaction occurs when the antibody-antigen combination separates from solution in clearly visible parti-

cles rather than in the more solid mass typical of precipitation or agglutination. Flocculation tests are usually run to identify the presence of serum antibodies associated with diseases such as syphilis, hepatitis, infectious mononucleosis, and trichinosis (a parasitic worm infection). One of the most common flocculation tests is known as the venereal disease research laboratory test, or VDRL. A positive VDRL test shows a high or increasing titer of a serum antibody known as the **Wassermann,** or **reaginic, antibody** and is characteristic of most individuals with syphilis. Syphilitic patients release a compound called cardiolipin, which stimulates their immune system to produce reaginic antibody. Since cardiolipin is such a common antigen, many other diseases may also show positive VDRL tests (for example, infectious mononucleosis, systemic lupus erythematosus). Even though the VDRL is not specific for syphilis, it is a valuable screening test. Other flocculation tests for reaginic antibody are the rapid plasma reagin (RPR) and the unheated serum reagin (USR).

Another important serological test called the ASTO (antistreptolysin O) test is used to identify the presence of a serum antibody produced against the strongly antigenic streptococcal product streptolysin O. This is the most commonly performed serological test for the identification of rheumatic fever (Figure 13.13). The release of streptolysin O during some streptococcal infections (e.g., septic sore throat, scarlet fever) stimulates the production of antistreptolysin O antibody and a second antibody (an autoantibody) that is able to react with cardiac tissue. It is the presence of the autoantibody that results in rheumatic fever symptoms, such as heart muscle and valve damage, fever, inflammation, scarring, and anemia. A high ASTO may indicate the presence of damaging autoantibody and helps to differentiate

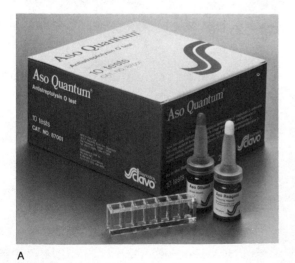

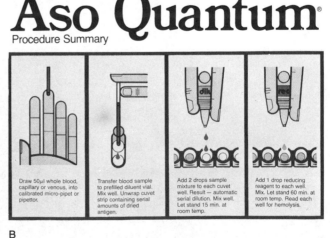

A   B

FIGURE 13.13
The Aso (ASTO) test system is a unique method for the quantitative determination of antistreptolysin O. (Courtesy of Sclavo, Inc.)

between streptococcal infections and other diseases with similar symptoms.

The **fluorescent antibody technique (FA)** is a rapid and accurate method for demonstrating antibody-antigen reactions. For this procedure, the antigen (tissue, bacteria, or serum) is fixed to a slide and stained with the appropriate antibody that has been chemically combined with a fluorescent dye. If the antigen is present, a reaction will occur and fix the dye directly to the antigen. If the antigen is missing, the dye can easily be washed off. When viewed through a dark-field microscope lighted with a special wavelength of ultraviolet light, the fixed fluorescent antigen-antibody complex will glow. This direct FA method is usually used to identify such antigens as those found on the surface of pathogenic streptococci; however, many other direct FA stains have been developed for the diagnosis of enteropathogenic *E. coli, Neisseria meningitidis, Salmonella typhi, Shigella sonnei, Listeria monocytogenes,* and *Haemophilus influenzae.* Indirect FA methods are also used to identify antibodies. One of the most successful indirect methods is known as **FTA-ABS** (fluorescent treponemal antibody absorption) test and is used to identify the presence of *Treponema pallidum* antibody in human serum. To carry out this test, virulent *T. pallidum* from an infected rabbit is mixed with the patient's serum. If the patient has syphilis and antibody to *T. pallidum* is present, an antibody-antigen reaction will take place on the slide. Next, a fluorescent antihuman antibody is added to the slide as a dye that will adsorb to the surface of the cells containing antibody. The presence of glowing spirochetes is a positive FTS-ABS test for syphilis. A number of other antibodies may be identified by using the indirect method. New automated methods for more rapid determination have been developed and are currently available for use in diagnosis (Box 22).

Two other techniques that have been of great value in the identification of antibodies are **electrophoresis** and **radioimmunoassay (RIA).** Electrophoresis is the separation of molecules based on differences in their electrical charge. Molecules with a strong positive charge move toward a negative pole (electrode) faster than those with a weak positive charge; negatively charged molecules move toward a positive electrode faster than those with a weak positive charge. Zone electrophoresis works on this principle and is performed on a thin sheet of transparent cellulose acetate (Fig-

ure 13.14). The sample to be analyzed is placed between the positive and negative electrodes, and the current is turned on for about ninety minutes. After this period, different proteins have moved across the acetate surface and separated. Each type of protein may then be identified with a stain and its concentration determined. Immunoelectrophoresis (IEP) is a combination of the agar diffusion method described earlier and electrophoresis. Immunoelectrophoresis is used for the identification of immunoglobulins that may be found in serum, urine, or other body fluids.

Radioimmunoassay is highly specific and sensitive, and may be done with any type of molecule that is able to stimulate antibody formation. This method may be used to identify the presence and amount of antibodies that differ in only a single amino acid in tissues and fluids. Radioimmunoassay operates much like direct and indirect fluorescent antibody methods and has been successful as an identification tool for new molecules involved in antibody-antigen re-

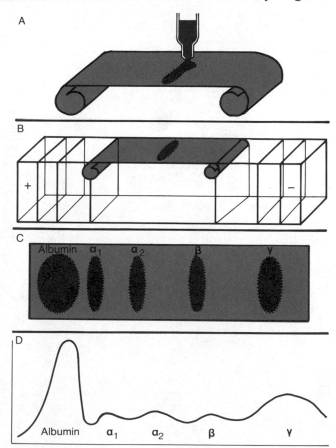

FIGURE 13.14

In zone electrophoresis, immunoglobulins on a thin, transparent acetate sheet migrate toward a negatively or positively charged electrode (A, B). The result is a series of separate spots on the sheet that are easily seen after they have been sprayed with a dye (C). The amount of immunoglobulin may then be determined and aids in diagnosing a particular disease (D).

THE use of the indirect immunofluorescence technique for detection of antirubella antibodies was reported by Brown, Maassab, Veronelli, and Francis; and Schaeffer, Orsi, and Widelock, who reacted cells which were infected with rubella virus and fixed onto slides with diluted serum samples. Antirubella antibodies, when present in the serum, combined with the intracellular viral antigen. The FIAX® test also employs the indirect immunofluorescence technique. Instead of infected cells affixed to glass slides, soluble viral antigen is immobilized on the surface of StiQ™ Samplers. Also, instead of subjective microscopic evaluation, objective measurement is made using the FIAX® Fluorometer.

The FIAX® 100 digital fluorometer and HP85 computer perform all IDT's assays by simply inserting a test-dedicated cassette into the computer. (Courtesy of International Diagnostic Technology, Inc.)

The initial step in the test procedure involves the reaction of viral antigen on one side of the StiQ™ Sampler with the diluted serum sample. If antirubella antibodies are present, they become bound to antigenic sites on the Sampler surface. Unreacted constituents lodging on the Sampler are washed free of the surface by a brief buffer wash. The specifically bound antibodies are next reacted with fluorescein isothiocyanate-labeled antibodies to human immunoglobulins. Finally, unreacted, labeled antibodies are removed from the Sampler surface by another buffer wash. These four steps are carried out in a 100-minute incubation sequence, before reading.

The labeled antibodies remaining on the surface are measured by inserting the StiQ™ Sampler in the FIAX® Fluorometer. The second side of the StiQ™ Sampler has a surface not coated with viral antigen. Fluorescence accumulating on it, due to other factors, is also measured and subtracted from the fluorescence on the antigenic side to obtain the delta fluorescence signal units, which represent the net amount of fluorescence due to antirubella antibodies.

Besides antirubella antibodies, a number of other test kits are available*:

Test Kit	
Immunoglobulin G	Ultra low level immunoglobulin G
Immunoglobulin A	Low level albumin
Immunoglobulin M	AntiDNA antibodies
Complement C3	Antitoxoplasma antibodies
Complement C4	Antinuclear antibodies (ANA)

*International Diagnostic Technology, 2551 Walsh Avenue, Santa Clara, Calif., 95050.

BOX 22

THE AUTOMATED IMMUNO-DIAGNOSTIC FIAX®/StiQ™ METHOD FOR ANTIRUBELLA ANTIBODIES

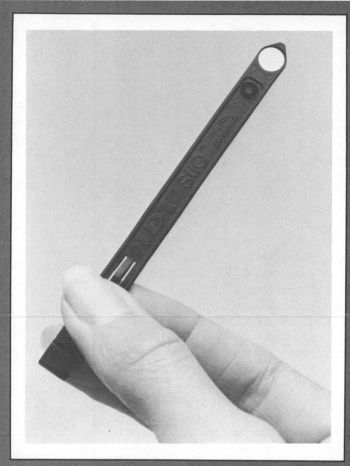

IDT's unique STiQ™ Sampler extends its performance range to include rubella-G and ANA with pattern recognition capabilities. (Courtesy of International Diagnostic Technology, Inc.)

Anti-Rubella Antibodies Assay

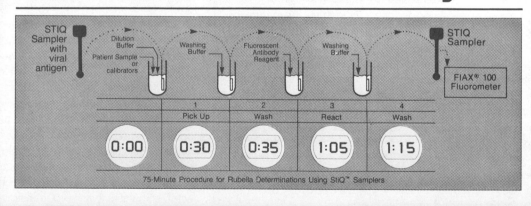

Courtesy of International Diagnostic Technology, Inc.

actions and in detecting a specific IgE antibody in an allergic response. The presence of a specific antibody in tissue is identified by using radioactive iodine (^{125}I) coupled to an antigen as a marker. If the molecule in question is present, radioactivity may be detected by placing a photographic film over the surface of the test slide and "taking a picture" of the molecules. The presence of white spots on the film indicates the location of the radio-labeled molecules, while the intensity of the spots may be used to determine the amount of labeled antibody in the sample.

The serological identification of a specific strain of a pathogen also has diagnostic value. In many cases, the symptoms of an infection depend on the nature of cell products released by a clone of pathogens. Genes for virulence and toxigenicity occur together in the same clone with genes for antigenic cell wall material. Therefore, it is possible to identify a pathogen by serologically testing for cell wall antigens. In the early 1930s, Rebecca Lancefield recognized the importance of such serological tests and developed a classification system for streptococci based on the antigenic nature of cell wall carbohydrates (C compounds). This system has become known as the Lancefield system and each different **serotype** (serological kind) is identified by letter, A through O (Table 13.9). *E. coli* and other bacteria have also been serotyped using specific antigen-antibody reactions involving flagella antigens (serotype H), capsular antigens (serotype K), and cytoplasmic antigens (serotype O). Within *E. coli* serotype group O, there have been over 137 different subgroups identified. The value of performing such specific diagnostic tests may be seen in the fact that *E. coli* O-55, O-111, and O-127 serotypes are those most frequently associated with infantile diarrhea. The serotyping of *E. coli* from fecal samples of infants with this disease is of diagnostic value and aids in identifying an infection's source.

Serological tests are designed to identify the nature of antibodies, antigens, and their reactions. Performing these reactions *in vitro* (in glass) has aided in developing a more complete understanding of how the specific host defense mechanism of humoral immunity operates *in vivo* (in the body). The union of antigen (protein, polysaccharide, lipid, or nucleic acid) with an antibody (released from plasma cells) occurs within minutes, but time may be required to show precipitation or agglutination and may involve other substances (e.g., complement). The second specific host defense mechanism, cell-mediated immunity, does not operate in this "immediate" fashion but is "delayed," occurring over a more extended period.

1 Radioimmunoassay (RIA) has been useful in identifying what types of immunological problems?
2 Why is serotyping of such great importance?
3 What does the term *in vitro* refer to?

CELL-MEDIATED IMMUNITY (CMI)

"Delayed," or cell-mediated, immunity is an acquired specific host defense mechanism involving antigen-sensitized T-lymphocytes that are able to destroy microbes, transplanted tissues, or cancer cells. The destruction may be caused by the T-cells directly, by chemicals released from the T-cells, or by macrophages stimulated to enter the reaction site. In most cases, the antigens which stimulate these reactions are an integral part of the harmful agent. This form of immunity is highly effective against certain virus infections and chronic fungal and bacterial infections. As the name implies, cells act as intermediary agents in eliminating specific foreign agents during CMI. Thymus-derived lymphocytes (T-cells) that contain specific receptor sites on their surfaces are responsible for CMI. These T-cells, like the B-cells, receive their committment during processing. When they emerge from the thymus, they are immunocompetent and committed to one antigen. When a foreign antigen binds to the receptor sites, a complex series of events occurs in 24–48 hours and reaches a peak in about 72 hours. The tissue develops a red bump or "mosquito-bite" appearance during this period. The TB skin test is an example of this reaction. The extent of the **induration** (hard spot) depends on whether or not T-cells have become antigen-sensitized (immunocompetent) and to what degree sensitization has occurred. The reactions that result in the elimination of the foreign agent begin with the antigen binding to the surface receptors. This stimulates the T-cells to change into a greatly enlarged cell known as a blast cell—a process referred to as blast transformation. Blast cells are about five times larger than T-cells, undergo mitosis, and produce several different lymphokines responsible for the induration associated with CMI (Table 13.10). Because of the release of the lymphokine known as chemotactic factor, macrophages are stimulated to move from surrounding tissues into the reaction area. When they arrive, they are transformed into "angry," or "killer," macrophages which are much better able to ingest and decompose foreign materials. Also note that although the T-cell response is specific for a particular antigen, once they are formed, the killer macrophages are equally effective against any kind of antigen. In addition to "killer" cell transformation, migration-inhibiting factor prevents them from leaving the area and encourages phagocytosis. Recruitment factor stimulates blast cell transformation, which magnifies the entire series of events. The release of lymphotoxins from T-cells results in the destruction of target cells, while interferon inhibits the replication of infecting viruses. Because cell-mediated immunity requires the presence of living, sensitized T-cells, passive immunity of this form can only be acquired by the transference of whole lymphocytes and cannot be accomplished by simply transferring serum. Table

TABLE 13.9
Lancefield classification of selected streptococci.

Lancefield Group	Species Example	Group Antigen (Hapten)			Cellular Products	Extracellular Products	Clinical Diseases
		Chemical Nature	Cellular Location	Hemolysis			
A	S. pyogenes	Rhamnose-N-acetylglucosamine polysaccharide	Wall	Beta	Hyaluronic acid capsule Group A carbohydrate M protein T protein	Streptolysins Hyaluronidase Streptokinase Leukocidin Erythrogenic toxin	Pharyngitis-tonsillitis Skin infections (impetigo) Erysipelas Scarlet fever Puerperal fever Rheumatic fever Glomerulonephritis Endocarditis
B	S. agalactiae	Rhamnose-glucosamine polysaccharide	Wall	Beta	Group B carbohydrate	Streptolysins	Neonatal sepsis and meningitis Puerperal fever
C	S. equi	Rhamnose-N-acetylgalatosamine polysaccharide	Wall	Beta	Group C carbohydrate	Streptolysins	Endocarditis
D	Enterococci (S. faecalis) Nonenterococci (S. bovis)*	Glycerol-teichoic acid containing D-alanine-glucose	"Intracellular" between wall and membrane	Gamma	Group D carbohydrate amino acid complex		Endocarditis Urinary tract infections
K	S. salivarius	Rhamnose polysaccharide	Wall	Alpha	Group K carbohydrate	Glucanlike substance	Endocarditis Dental caries
H	S. sanguis	Rhamnose polysaccharide	Wall	Alpha	Group H carbohydrate	Glucanlike substance	Endocarditis Dental caries
Not grouped	S. mitis			Alpha			Endocarditis Dental caries
Not grouped	S. mutans			Alpha		Glucan	Endocarditis Dental caries
Not grouped (anaerobic strains)	Peptostreptococcus spp.			Gamma			Brain abscess Lung abscess Puerperal fever

*Other species of group D streptococci occur that may be alpha, beta, or gamma hemolytic.
SOURCE: Modified from M. Wilson, H. Mizer, and J. Morella, *Microbiology in Patient Care*, 3rd ed. McMillan Publishing, Inc. Copyright © 1979.

TABLE 13.10
Lymphokines associated with CMI.

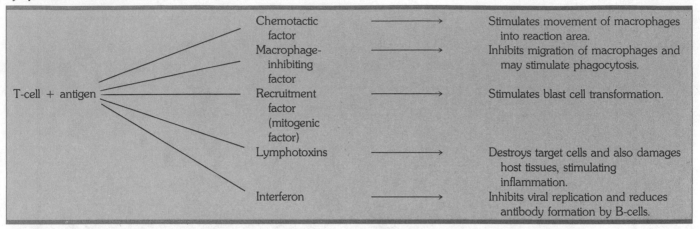

T-cell + antigen		
Chemotactic factor	→	Stimulates movement of macrophages into reaction area.
Macrophage-inhibiting factor	→	Inhibits migration of macrophages and may stimulate phagocytosis.
Recruitment factor (mitogenic factor)	→	Stimulates blast cell transformation.
Lymphotoxins	→	Destroys target cells and also damages host tissues, stimulating inflammation.
Interferon	→	Inhibits viral replication and reduces antibody formation by B-cells.

13.11 gives a brief comparison between CMI and antibody-mediated reactions, or AMI.

CMI, humoral immunity, and inflammation have been presented as separate defense mechanisms; however, they all work together to maintain a healthy body. For example, a response that begins as CMI may change to inflammation as phagocytic macrophages self-destruct, neutrophiles move into the reaction area, and kinins and prostaglandins are released. The destruction of invading microbes and host tissue releases antigens into body fluids that stimulate the humoral immune system, and antibody is released.

---

1 CMI is highly effective against what types of infections?
2 Which type of cell, T-cell or macrophage, has a "killer" response to the presence of a specific antigen?
3 Why is passive immunization for CMI impractical?

THE ANAMNESTIC RESPONSE AND IMMUNIZATION

Whether the humoral or cell-mediated immune systems are stimulated into action depends on the location, amount, and length of time antigen is present in the body. Simply having the foreign material in the body may not be enough to stimulate these special cells and provide the body with a form of specific, long-term protection. A **minimal effective exposure** to the antigen is necessary to cause lymphocytes (B and T) to be transformed into immunocompetent cells. In fact, the first minimal effective exposure of lymphocytes to an antigen only results in the development of a minimal response (Figure 13.15). This first exposure notifies, or **sen-**

sitizes, a limited number of lymphocytes to the antigen's presence and converts them to immunocompetent cells that produce only a small amount of antibody. The period immediately following sensitization is known as the **primary response phase** and is identified by a slight increase in antibody produced against the specific sensitizing antigen. During the primary phase, immunocompetent cells undergo mitosis. Some become plasma cells while others serve as "memory cells," since they now "remember" how to manufacture the antibody.

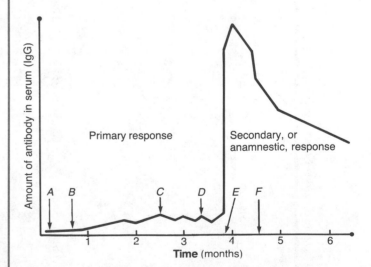

FIGURE 13.15
In an anamnestic response test, the first exposure to antigen (A) results in sensitization. (B) Antibody may be detected in serum within a few days. (C) Titer slowly rises to a low peak and falls slowly. (D) With a second exposure there is an immediate drop in titer due to antibody-antigen reactions. (E) A rapid rise in antibody titer follows and is known as the anamnestic response. (F) After the second and all further exposures, antibodies disappear more slowly.

TABLE 13.11
A comparison between antibody-mediated (humoral) and cell-mediated (cellular) immunity.

	Antibody-mediated	Cell-mediated
Sensitizing Material	Proteins, polysaccharides, lipids	Proteins or proteinhapten
Reaction Time	"Immediate," minutes to hours	"Delayed," 18–24 hours
Initiating Event	Union of antibody with antigen	Reaction of "sensitized" lymphocytes with antigen
Transfer	Circulating serum antibody	Cells
Effector Mediators	Complement, vasoactive amines	Lymphokines
Benefits	Antimicrobial immunity to many microbes and extracellular products (e.g., bacterial toxins)	(1) Antimicrobial immunity (viruses, bacteria, fungi, protozoans), "intracellular" parasites (2) Transplantation immunity (3) Tumor immunity
Diseases	(1) Autoimmunity (2) Rheumatoid arthritis (3) Allergies	(1) Autoimmunity (2) Contact hypersensitivity (3) Some drug reactions (4) Malignancy

SOURCE: Modified from J. A. Bellanti, *Immunology: Basic Process*, 2nd ed. W. B. Saunders Company, 1979.

The clone* of memory cells (sensitized, immunocompetent lymphocytes) produced during this period is able to produce antibody more rapidly than those that were first sensitized to the antigen. Therefore, when the antigen is reintroduced to the body, large amounts of antibody are produced very rapidly by this increased number of lymphocytes. The rapid increase in antibody titer following second exposure to an antigen is known as the **secondary,** or **anamnestic, response phase** of antibody production. As a result of B-cell activity, the serum antibody level reaches a maximum and begins to decline as the antigen is removed from the body and antibody production rates fall. The rate at which the titer drops varies and is characteristic of a particular antibody. If the titer falls too far, there may not be enough antibody or sensitized lymphocytes to provide protection if the body comes in contact with a large quantity of antigen. However, this usually does not happen since all minimal effective exposures to the antigen after the second exposure will result in further anamnestic responses, and the antibody titer will be re-elevated or "boosted" to its maximum protective level.

Understanding antibody formation (see Figure 13.15) helps explain the need for "booster shots." For example, if an individual lives or works in an area where tetanus toxin can regularly stimulate an anamnestic response, for example, by stepping on a nail or being scratched by a thorn, a constant high level of immunity to tetanus toxin will be naturally maintained. Each time the memory cells for antitetanus antibody come in contact with a minimal effective level of tetanus toxin, the antibody titer will be boosted to its maximum protective level. If natural exposure does not occur, the antibody titer against this antigen will fall slowly and could reach a dangerously low, unprotective level. When an individual comes into a hospital emergency room and the physician suspects that *Clostridium tetani* may have entered a wound, there is no quick way of determining how "naturally" protected the patient may be. Therefore, a booster injection of tetanus toxoid is given to artificially stimulate antibody production and ensure a high level of protection. This procedure is effective since the antibody titer will rise faster than invading *C. tetani* can produce toxin.

The controlled stimulation of immunity to a particular antigen is known as **immunization,** or **vaccination. Passive immunization** is done by administering antibodies from one individual to another, while **active immunization** is accomplished by injecting an appropriate antigen to stimulate the individual's own immune system. Passive immunization is mainly used to protect individuals who might develop disease symptoms before their own immune response can form protective antibodies (about 7–10 days) and for individuals who have an immunodeficiency disease. The most common antiserum used is **gamma globulin** (mainly IgG) obtained from humans or animals. Immunoglobulin may be given to protect individuals from

---

*Refer to Chapter 11.

infections, such as varicella-zoster (chickenpox and shingles), measles, hepatitis A virus, and rubella virus (Table 13.12). Caution must be used before passive immunization to guard against a possible allergic reaction, **serum sickness,** to antigens that may remain in the antiserum after it has been prepared. Passive immunization by the transfer of humoral antibodies has been very successful; however, attempts at transferring immune lymphocytes (T-cells) and

TABLE 13.12
Passive immunization materials.

Disease	Product
Black widow spider bite	Black widow spider antivenom, horse
Botulism	ABE polyvalent antitoxin, horse
Diphtheria	Diphtheria antitoxin, horse
Gas gangrene	Gas gangrene antitoxin, polyvalent, horse
Hepatitis A (infectious)	Immune serum globulin, human
Hepatitis B (serum)	Hepatitis B immune globulin, human
Hypogammaglobu-linemia	Immune serum globulin, human
Measles	Immune serum globulin, human
Mumps	Mumps immune globulin, human
Pertussis	Pertussis immune globulin, human
Rabies	Rabies immune globulin, human (Horse antirabies serum may be available but is much less desirable.)
Rh isoimmunization (erythroblastosis fetalis)	Rh$_o$ (D) immune globulin, human
Snakebite	Coral snake antivenom, equine crotalid antivenom, polyvalent, horse
Tetanus	Tetanus immune globulin, human. (Cow and horse antitoxins may be available but are not recommended. They are used at 10 times the dose of tetanus immune globulin.)
Vaccinia	Vaccinia immune globulin (VIG), human. (May be obtained from CDC.)
Varicella	Varicella-zoster immune globulin (VZIG), human, or zoster immune globulin (ZIG), human

SOURCE: Modified from H. H. Fudenberg et al, *Basic & Clinical Immunology,* 2nd ed. Lange Medical Publications, 1978.

macrophages to provide passive cell-mediated immunity has not yet been of value. Active immunization has been the most successful method. In this procedure, a first injection of antigen is given to sensitize the B-lymphocytes and initiate the primary response phase. One to several weeks later, a second injection is given to initiate the anamnestic response and provide the maximum level of protection. In certain cases, such as DPT, a third injection may be required to ensure that antibody levels for all antigens in the polyvalent vaccine have been elevated. Side effects to active immunization can also occur and may include dizziness, soreness at the site of injection, fever, or diarrhea. Table 13.13 lists a number of diseases for which active immunization is available, the type of agent used, and an immunization schedule for each disease. Because of highly effective immunization programs and other control measures used throughout the world, the incidence of many deadly diseases has been effectively reduced. As a result, many immunization campaigns conducted in the United States and other countries have been discontinued or limited. For example, vaccinations against plague and rabies are not a regular part of the immunization program for infants and children (Table 13.14). The use of antibiotics has successfully controlled plague, while an outstanding program of domestic animal immunization has greatly reduced the incidence of human rabies in the United States. The introduction of the measles vaccine in 1960 and its widespread use also demonstrates the effectiveness of immunization (Figure 13.16). The success of immunization programs in other countries varies with social, economic, and medical conditions. Before venturing off to foreign lands, it is best to check with a doctor to determine whether a special immunization might be legally required to enter the country. The *Medical Letter* (a nonprofit newsletter for doctors) recommends the following:

Tetanus and diphtheria   Everybody, traveling or not, should receive a tetanus-diphtheria booster injection every ten years.
Polio   Travelers not previously immunized against polio should receive polio vaccine.
Measles   People born after 1957 who have not received measles vaccine and who have not had the disease probably should receive the vaccine before traveling.
Hepatitis   Immunoglobulin to prevent hepatitis A should be taken only by travelers to developing countries who are going outside the usual tourist routes or are planning to stay three months or longer.
Yellow fever   Vaccine should be taken for travel to infected areas.
Typhoid   Vaccine recommended for travelers to rural areas of tropical countries and for any area where an outbreak is occurring.
Plague   Vaccination only for travelers to interior regions of Vietnam, Kampuchea, and Laos.

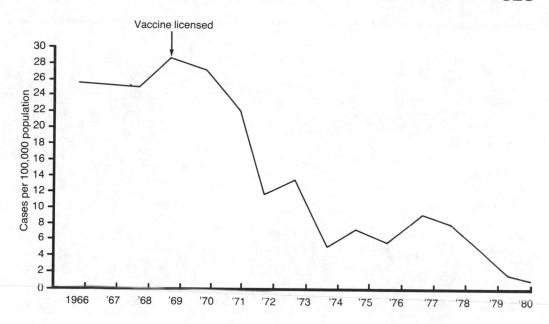

FIGURE 13.16
Reported cases of rubella by year in the United States. (Courtesy of the Center for Disease Control, Atlanta.)

Rabies   Vaccination recommended only for travelers anticipating unusual contact with animals having rabies.
Cholera   The risk is low and vaccines have limited effectiveness.
Smallpox   There no longer is a need for smallpox vaccination.

Advances in immunology and the development of vaccines have been very successful in the past decade. Research is currently being conducted in the development of human fibroblast vaccines against rabies and hepatitis B viruses. Work is also being done to produce a vaccine against *Streptococcus mutans* for the prevention of tooth decay and periodontal disease. In addition, there is strong speculation that within the next few years a polyvalent vaccine will be available to protect children under the age of one year against bacterial meningitis caused by *Haemophilus influenzae, Neisseria meningitidis,* and *Streptococcus pneumoniae.* Detailed research into the antigenic differences between normal human cells and cancer cells has raised the possibility that the immune system is able to recognize such differences and destroy cancer cells through cell-mediated immunity. Should this turn out to be the case, immunization against certain forms of cancer may also be on the horizon.

SUMMARY

Nonspecific and specific host defense mechanisms found in the human body prevent microbes and many nonliving agents from causing harm. Nonspecific mechanisms include phagocytosis, the inflammatory response, digestive enzymes, reflexes (for example, coughing, sneezing, swallowing), peristalsis, and special features of the mucous membranes (for example, lower pH, enzymes, and cilia). Phagocytosis is accomplished by many cell types throughout the body. Those located in the blood are known as leukocytes and may be neutrophiles (PMNs), eosinophiles, basophiles, and monocytes. Monocytes are able to move by diapedesis through the endothelial lining of the blood vessels and differentiate into macrophages. The macrophages are important in the inflammatory response and cell-mediated immunity. The inflammatory response is a generalized tissue reponse to damage that results in the movement of macrophages, PMNs, and certain chemicals such as histamines and prostaglandins into the damaged area to phagocytose the foreign agent and repair host damage. Individuals that lack the ability to respond via the inflammatory response have limited protection. In other cases, individuals suffer from diseases that result from an abnormal reaction. Examples include rheumatoid arthritis and systemic lupus. Many drugs, both prescription and nonprescription, are available to aid in reduc-

1   What is meant by a minimal effective exposure?
2   What enables the anamnestic response to occur?
3   Why are boosters to tetanus given in an emergency situation?
4   Why is serum sickness only associated with passive immunization?

TABLE 13.13
Active immunization.

Disease	Product (Source)	Type of Agent	Primary Immunization	Duration
Cholera	Cholera vaccine	Killed bacteria	2 doses, 1 week or more apart	6 months*
Diphtheria	DTP, DT (adsorbed) for child under 6; Td (adsorbed) for all others	Toxoid	3 doses, 4 weeks or more apart, with an additional dose 1 year later for a child under 6. (Can be given at the same time as polio vaccine, if doses at least 8 weeks apart.)	10 years**
Influenza	Influenza virus vaccine, monovalent or bivalent (chick embryo). Composition of the vaccine is varied depending upon epidemiologic circumstances.	Killed whole or split virus A and/or B	1 dose. (Two doses, 4 weeks apart may be preferable when a major new antigenic component is first incorporated into the vaccine. Two doses are recommended for recipients under 6 years of age. Split virus products should be used in persons under 18 years because of a lower incidence of side effects.)	1 year**
Measles†	Measles virus vaccine, live, attenuated (chick embryo)	Live virus	1 dose at age 15 months	Permanent
Meningococcus	Meningococcal polysaccharide vaccine, group A or group C	Polysaccharide	1 dose. Since primary antibody response requires at least 5 days, antibiotic prophylaxis with rifampicin (600 mg or 10 mg/kg every 12 hours for 4 doses) should be given to household contacts.	Permanent (?)
Mumps†	Mumps virus vaccine, live (chick embryo)	Live virus	1 dose	Permanent
Pertussis	DTP	Killed bacteria	Same as DTP	See **
Plague	Plague vaccine	Killed bacteria	3 doses, 4 weeks or more apart	6 months**
Pneumococcus	Pneumococcal polysaccharide vaccine, polyvalent, 14 strains	Polysaccharide	0.5 ml	Uncertain—probably at least 3 years**
Poliomyelitis	Poliovirus vaccine, live, oral, trivalent (monkey kidney, human diploid)	Live virus types I, II, and III	2 doses, 6–8 weeks or more apart, followed by a third dose 8–12 months later. (Can be given at the same time as primary DPT immunization.)	Permanent
	Poliomyelitis vaccine	Killed virus types I, II, and III	3 doses, 1–2 months apart, followed by a fourth dose 6–12 months later**	2–3 years, perhaps longer
Rabies	Rabies vaccine (duck embryo). (Vaccine derived from rabbit neural tissue may still be available but is not preferred because of the higher rate of neurologic	Killed virus	Preexposure: 2 doses, 1 month apart, followed by a third dose, 6–7 months later. Or 3 doses, 1 week apart, followed by a fourth dose 3 months later.	2 years**

side-reactions.) An experimental diploid vaccine is available through state health departments for postexposure immunization of patients with demonstrated allergy to duck embryo vaccine or as a booster for those who have received the full course of rabies immune globulin and duck embryo vaccine and have failed to develop antibody titers > 1.8.

Postexposure:
Always give rabies immune globulin as well. If not previously immunized, give a total of 23 doses. Give 2 injections per day for the first 7 days, then 7 daily doses, and boosters on days 24 and 34 following start of treatment. Alternatively give 21 daily doses and boosters on days 24 and 34. If previously immunized and an antibody response demonstrated, do not give serum therapy. For nonbite exposure, give 1 booster dose. For the bite of a rabid animal, give 5 daily boosters followed by 1 dose on day 25. If an antibody response was not previously demonstrated, treat as unimmunized.

Rocky Mountain spotted fever	RMSF vaccine (chick embryo)	Killed bacteria	3 doses at 7–10 day intervals	1 year**
Rubella†	Rubella virus vaccine, live (duck embryo, rabbit kidney, human diploid)	Live virus	1 dose	Permanent

*Revaccination interval required by international regulations.
**A single dose is a sufficient booster at any time after the effective duration of primary immunization has passed.
†Combination vaccines available.
SOURCE: Modified from H. H. Fudenberg et al, *Basic & Clinical Immunology,* 2nd ed. Lange Medical Publications, 1978.

TABLE 13.14
Immunization of normal infants and children.

D	diphtheria	
P	pertusis	
T	tetanus	Given at 3 months, 4 months, 5 months, and 6 months to 1 year of age
P	polio	
D	diphtheria	
T	tetanus	
P	polio	Given at 6 to 13 years of age every 5 years
T	tetanus	
P	polio	Given at 14 years of age and over every 5 years
Tetanus toxoid		If booster over one year (depends upon policy—could be from 6 months to 3 years). Give for laceration, burn, puncture wound, dog bite, and cat scratch. (Tetanus antioxin given if no previous immunization.)
M	red measles (rubeola)	
M	mumps (epidemic parotitis)	Given at 12 months of age in a single injection
R	German measles (rubella)	
	tuberculin	Given at one year of age, then once yearly

ing the redness, swelling, heat, and pain associated with inflammation.

The specific host defense mechanism is known as the immune system and is subdivided into two categories, humoral and cell-mediated. Immunity may be innate or acquired and is further labeled as being naturally or artificially derived. Active immunity results from the presence of antigen in the body which stimulates the formation of antibody. Passive immunity results from the administration of preformed antibody and only temporarily helps resist infection. Antibodies, or immunoglobulins, are protein molecules formed by sensitized lymphocytes and plasma cells in response to the presence of a specific antigen (usually a protein). The antibodies are classified into five major groups based on their differences in the constant portion of their heavy polypeptide chains: IgG, IgM, IgD, IgA, IgE. Serum antibody responsible for humoral immunity is produced by plasma cells derived from B-lymphocytes, while T-cells are involved in cell-mediated immunity. Cells that have been sensitized to a particular antigen and are able to produce antibody are referred to as immunocompetent.

Antibody-antigen reactions may occur as agglutination, precipitation, lysis, opsonization, or neutralization. Many of these reactions have been reproduced in the lab and adapted for use as serological tests, including RF (rheumatoid factor), ASTO (antistreptolysin O), complement fixation, serotyping, fluorescent antibody, and immunodiffusion. Cell-mediated immunity results in the destruction of pathogens, cancer cells, and tissue transplants. This takes place either directly by the release of interferon or lymphotoxins, or indirectly by macrophages stimulated to enter the reaction site of other lymphokines. The cell primarily responsible for CMI is the T-cell. The development of the immune state occurs in two phases, the primary phase and the anamnestic response. Knowing how these stages occur has made it possible to artificially stimulate the formation of high antibody titer in serum and activate protective humoral and cell-mediated systems by the process of immunization.

WITH A LITTLE THOUGHT

Since "Dirty Dave" was able to crawl, he has been into everything. The garbage can, sandbox, mud puddles, garden, and wooded lot next door have been his playgrounds. He is constantly dirty and runs around with a "runny nose," cuts, and bruises. It seems that he is always sick. Dirty Dave never plays with "Finicky Fred" because Fred is never permitted outside. He never gets dirty and certainly never plays in the sandbox or any other "dirty places" that are so much fun for Dirty Dave. Every time Fred gets a runny nose or a scratch, his parents rush him to the doctor's office where he *always* gets an antibiotic to cure his ill. Fred has never been sick a day in his six-year-old life.

But Dirty Dave is not really in such bad shape. In fact, he may be better off in the long run in comparison to Finicky Fred. Why?

STUDY QUESTIONS

1 What is the difference between a nonspecific host defense mechanism and a specific host defense mechanism? Give examples.

2 What is the relationship among PMNs, diapedesis, prostaglandins, histamine, mast cells, and the inflammatory response?

3 Correlate the symptoms of the inflammatory response with specific tissue changes.

4 What are the advantages of anti-inflammatory drugs? What natural host defense mechanisms are inhibited by these drugs?

5 Why is immunization safer and more effective than variolation?

6 Describe the differences between innate and acquired immunity; natural active and natural passive immunity; and active artificial and passive artificial immunity.

7 Briefly explain each of the following serological tests and its importance: RF, complement fixation, immunoelectrophoresis, immunodiffusion, VDRL, and ASTO.

8 What is the difference between humoral and cell-mediated immunity?

9 Define the following terms: immunocompetent, sensitization, lymphokines, lattice network, fluorescent antibody, serotyping.

10 Describe the development of immunity, both humoral and cell-mediated, as it occurs from just before birth through adulthood.

SUGGESTED READINGS

BELLANTI, J. A. *Immunology: Basic Process,* 2nd ed. W. B. Saunders Company, Philadelphia, 1979.

COOPER, M. D., and A. R. LAWTON. "The Development of the Immune System." *Scientific American,* Nov., 1974.

EISEN, H. N. *Immunology.* Harper & Row, Hagerstown, Md., 1980.

FUDENBERG, H. H. et al. *Basic & Clinical Immunology,* 3rd ed. Lange Medical Publications, Los Altos, Calif., 1980.

LEDER, P. "The Genetics of Antibody Diversity." *Scientific American,* May, 1982.

"Hepatitis B Vaccine Passes First Major Test." *Science,* vol. 210, Nov., 1980.

MAYER, M. M. "The Complement System." *Scientific American,* Nov., 1973.

MILSTEIN, C. "Monoclonal Antibodies." *Scientific American,* Oct., 1980.

ROITT, I. M. *Essential Immunology,* C. V. Mosby, St. Louis, 1981.

ROSE, N. R., and H. FRIEDMAN. *Manual of Clinical Immunology.* American Society for Microbiology, Washington, D.C., 1980.

SELL, S. *Immunology, Immunopathology, and Immunity,* 3rd ed. Harper & Row, Hagerstown, Md., 1980.

KEY TERMS

phagocytosis

diapedesis (di″ah-pe-de′sis)

reticulo-endothelial system

inflammation

kinins

innate and acquired immunity

antigen

antibody

immunologic tolerance

antigenic determinant

B-cells

T-cells

cell-mediated immunity

antibody-mediated immunity

complement system

opsonization

serotype

anamnestic, or secondary, response (an-am-nes′tik)

Immunology II: Hypersensitivity, Transplants, and Tumors

HYPERSENSITIVITY

TYPE I HYPERSENSITIVITY

SENSITIVITY TESTING AND TREATMENT

TYPE II HYPERSENSITIVITY

TYPE III HYPERSENSITIVITY

TYPE IV HYPERSENSITIVITY
TB Skin Test Reaction
Contact Dermatitis

TRANSPLANTATION IMMUNOLOGY

THE ABO AND Rh BLOOD SYSTEMS

TRANSFUSION REACTIONS

HISTOCOMPATIBILITY ANTIGENS

CANCER AND IMMUNOLOGY

IMMUNOTHERAPY AND CANCER CONTROL

Learning Objectives

☐ Define the terms "allergy," "immunologic hypersensitivity," and "allergen."

☐ Describe the events which occur in a Type I hypersensitivity and give examples of the reactions.

☐ Know the differences between generalized and localized anaphylaxis.

☐ Be able to explain skin sensitivity testing and the advantages of desensitization shots.

☐ Describe the events which occur in a Type II hypersensitivity and relate these events to immunosurveillance and neoplasia.

☐ Describe the events which occur in a Type III hypersensitivity and give examples of this reaction.

☐ Describe the events which occur in a Type IV hypersensitivity.

☐ Be able to explain a TB skin test and how it is interpreted.

☐ Be familiar with the nature of contact dermatitis and some of its causes.

☐ Be able to describe a tissue rejection response.

☐ Know the ABO and Rh classification systems for human blood typing.

☐ Describe the symptoms of erythroblastosis fetalis and the role RhoGAM plays in the control of this immunological problem.

☐ Recognize the basis of immune transfusion reactions.

☐ Define the terms "histocompatability antigen," "HLA," and "major histocompatability complex."

☐ Know the relationship between HLA antigens and certain immunological problems.

☐ Recognize the proposed relationship between cancer and immunology.

☐ Be able to define CEA and TSA, and correlate them with the immunosurveillance system.

☐ Recognize the proposed effects of immunotherapy on cancer control.

14

The immune system plays many vital roles in maintaining the health of the body. One very important function is the control of foreign microbes. Infectious microbes are quickly recognized as foreign, brought under control, and eliminated before they cause significant harm. However, the immune system may also malfunction. Under certain conditions, antibody-antigen reactions can result in responses known as hypersensitivities, or allergies. Possibly the most familiar are allergies to foods, pollens, and drugs; however, the range of tissue damage may extend beyond these typical examples. A number of diseases trigger abnormal immune responses that contribute to tissue destruction. Rheumatic fever, glomerulonephritis, and many other diseases result from harmful antibody-antigen reactions. Systemic lupus and rheumatoid arthritis involve chronic responses of both the immune and inflammatory systems, and may cripple or kill. In many cases, therapeutic control of abnormal reactions may come easily, while in others the disease process still remains to be identified. Research efforts have led immunologists from the study of infectious disease control to the fields of tissue transplantation and tumor immunology. The extension of this research into these areas results from detailed investigations into the nature of "self" and "nonself," since a pathogenic microbe is as foreign to the human body as a tumor or a transplanted organ. The immune system protects by limiting the spread of cancer cells as well as preventing microbial infections. The fundamental protective immune response to a foreign object may be supplemented through the use of vaccines and protective antibodies. A complete understanding of the system is essential if a physician is to help a patient requiring a tissue transplant, experiencing immunodeficiency, or having autoimmune diseases.

HYPERSENSITIVITY

The immune system is an extremely complex defense mechanism. Under normal circumstances, this complexity serves to increase the efficiency and accuracy of eliminating harmful foreign agents. However, there are times when the immune system ceases to be beneficial, and reactions occur that cause more harm than good. Harm may result after an individual develops the ability to produce certain antibodies against a normally harmless molecule. These reactions are commonly referred to as **allergies;** however, the more appropriate term used today is **immunologic hypersensitivity.** For example, allergic rhinitis (hay fever) is a set of symptoms initiated by an antibody (IgE) reaction with ragweed and other pollen antigens. All hypersensitivities involve the interaction of an antigen with either a humoral antibody or with specially sensitized lymphocytes. If the antigens are such things as dust, pollen, tobacco, fungi, eggs, aspirin, penicillin, etc., they are called **allergens** and may stimulate such allergic reactions as asthma, eczema, migraine headache, hives, itching throat, serum sickness, or skin rashes. If the antigens are modified cell-surface antigens, they may stimulate such immunologic disorders as glomerulonephritis, rheumatic fever, colitis, pernicious anemia, systemic lupus, or rheumatoid arthritis.

As knowledge in this area has increased, there have been many attempts to classify hypersensitivity diseases. In 1963, P. G. H. Gell and R. R. A. Coombs developed a very useful classification system for reactions responsible for hypersensitivities. This system correlates clinical symptoms with information about the immunlogic events that occur during hypersensitivity reactions (Table 14.1).

TYPE I HYPERSENSITIVITY

This type of hypersensitivity is also known by such names as **immediate, reaginic, atopic,** and **anaphylactic hypersensitivity.** Each name provides some insight into the nature of this abnormal antibody-antigen reaction. The symptoms of this reaction (anaphylaxis) are associated primarily with the release of immunoglobulin E (IgE), also known as **reagin.** IgE is a bivalent antibody. Unlike most other immunoglobulins, it has the ability to bind to special receptor sites on the surface of mast cells and basophiles (Figure 14.1). The sticky portion of IgE molecules is at the base of the "Y" configuration and is known as the **Fc** portion of the antibody. B-cells capable of producing IgE become sensitized to the allergen when it first enters the body through the skin, respiratory tract, or gastrointestinal tract. Upon second exposure, an anamnestic response occurs, releasing additional IgE, which becomes attached to more mast cells and basophiles. Allergen that reacts with this bound antibody stimulates mast cells and basophiles to release chemicals known as **vasoactive amines** (Figure 14.2). This group includes histamine, slow-reacting substance of anaphylaxis (SRS-A), eosinophil chemotactic factor of anaphylaxis (ECF-A), heparin, serotonin, kinins, and prostaglandins (Table 14.2). The release of these amines is responsible for the symptoms of both **generalized** and **localized anaphylaxis.**

Generalized anaphylaxis is a shock reaction that occurs after second exposure to an allergen. The symptoms occur throughout the body and vary from a mild reaction to a massive response that may be fatal within minutes. The release of vasoamines may cause a skin reaction and the formation of hives, technically known as **urticaria.** This is the temporary formation of a smooth, raised patch of skin (wheal) which begins as a reddened area and changes to a white center. As the reaction continues, it may spread, forming "feelers" or "pseudopods," which itch severely but rarely last more than a few days. The more deadly form of gen-

TABLE 14.1

Gell-Coombs classification for hypersensitivity.

Class	Name	Example	Problem	Antigen	Response	Hypersensitivity Transferred to Nonsensitized Person by
Type I	Anaphylactic, or immediate, hypersensitivity	Bronchial asthma, Systemic anaphylaxis, Bee stings, drug reaction, Skin allergies, tissue transplant, Food allergies, Periodontal disease	IgE-mast cell binding	Usually from outside the body (e.g., grass pollen, mold spores)	5–30 minutes, Wheal formation, Kinin release, Edema fluid	Serum antibody
Type II	Cytotoxic	Glomerulonephritis, Rh disease in newborn, drug reaction	IgG or IgM binding with tissue; Cell lysis by complement	Cell surface	Cell lysis	Serum antibody
Type III	Arthus (not found in humans), Serum sickness, Immune complex disorders	Tertiary syphilis, Lepromatous leprosy, Guillain-Barré syndrome, Rheumatoid arthritis, Transplanted tissues, Systemic lupus erythematosus, Hodgkin's disease, Crohn's disease, drug reaction, Lymphoblastic leukemia, Ulcerative colitis	Soluble IgG, IgM Antigen-antibody complexes, complement, platelets, PMNs.	Outside the body	4–8 hours, Acute inflammation, Edema, Erythema, Cell lysis	Lymphoid cells
Type IV	Delayed hypersensitivity	Drug allergies, TB skin test response, Contact dermatitis: fabric, fur, cosmetic, poison ivy, metal, Food allergies, Transplanted tissues, Chronic microbial infections	T-cells, lymphokines, macrophages	Cell surface or outside the body	24–48 hours; Induration, Erythema, Cytotoxic and lymphokine necrosis	Serum antibody

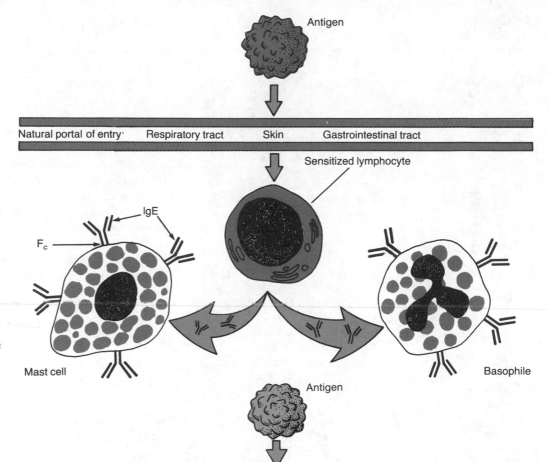

FIGURE 14.1

Antigens that stimulate the release of IgE are known as allergens. The IgE molecules released from sensitized lymphocytes attach to the surface of mast cells and basophiles at the Fc end of the antibody molecule.

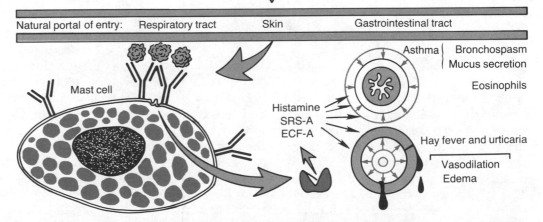

FIGURE 14.2

The allergic response classified as Type I results when a cell-bound antibody, IgE, reacts with an allergen and stimulates the release of vasoactive amines. These molecules stimulate a wide range of tissue changes. (Modified from O. L. Frick. "Immediate Hypersensitivity." *Basic and Clinical Immunology.* Lange Medical Publications, 1978.)

eralized anaphylaxis progresses from mild urticaria to (1) extensive skin reddening and itching; (2) capillary breakage in the skin, eyes, and internal organs; (3) a burning sensation in the rectum, mouth, and possibly the vagina; (4) severe headache; (5) hot "flashes"; (6) a quick drop in blood pressure; (7) constriction of bronchial smooth muscle; and (8) death (Box 23). Localized anaphylaxis is characterized by many of these same symptoms; however, they only occur in association with a specific organ (for example, skin, nasal mucosa, intestinal lining) and are rarely fatal. Two of the most familiar examples of localized anaphylaxis are food allergies and bronchial asthma.

> 1 Give two common examples of hypersensitivities.
> 2 What is the relationship between IgE and vasoactive amines?
> 3 What does "urticaria" mean?

As food moves through the digestive tract, secretory IgA combines with food antigens and prevents them from making contact with lymphocytes. In cases of food allergies, the individuals suffer from a lack of IgA, which allows allergens to enter the circulatory system. Once inside, the aller-

TABLE 14.2
Vasoactive amines and their actions.

Amine	Action
Histamine	Stimulates smooth muscle contraction around small blood vessels and bronchioles; increases secretion of mucus in nose and bronchioles; antihistamines block their action.
Slow-reacting substance of anaphylaxis (SRS-A)	Produced during Type I response; stimulates smooth muscle contraction; increases capillary permeability; antihistamines have no effect on SRS-A.
Eosinophile chemotactic factor of anaphylaxis (ECF-A)	Stimulates migration of eosinophiles into reaction area; these cells are suspected of phagocytosing allergic antibody-antigen complexes to limit the reaction.
Serotonin	Simulates capillary dilation; increases permeability; stimulates smooth muscle contraction; of little importance in Type I reaction.
Heparin	Prevents clot formation.
Kinins	Stimulate chemotaxins, smooth muscle contraction, dilation of arterioles, and increase capillary permeability.
Prostaglandins	Stimulate smooth muscle contraction, increase capillary permeability, and lower the threshold of pain.

gen sensitizes lymphocytes which later produce IgE that attaches to mast cells and basophiles. Upon second and all subsequent exposures to the allergen, an immediate hypersensitivity Type I reaction will occur provided the allergen reenters body tissue. The reaction occurs within one to sixty minutes after ingestion and results in a shock reaction, usually involving only one organ. Once established, Type I food allergies are usually permanent but may be controlled with antihistamines or by avoiding the allergen. Some symptoms of food allergies include bed wetting, headaches, and minor personality changes. These symptoms have been noted in children allergic to foods such as milk, chocolate, wheat or corn grain cereals, and eggs. The severity of some of these reactions is predictable and may vary from a mild localized to a fatal systemic anaphylaxis.

Bronchial asthma ("asthma" means *panting*) symptoms primarily include contraction of bronchiole smooth muscle, itching of the eyes and skin, and excessive mucus secretion that severely impairs breathing. The intensity of this reaction may be severe compared to hay fever, and it is the result of vasodilation and edema. The constriction of the bronchi results in a wheezing or whistling sound during exhalation. The accumulation of mucus in the nose and throat causes the asthmatic person to cough, have a large amount of nasal discharge, and become very tired. Asthma "attacks" caused by allergens occur when the individual is in contact with the antigen. Dust, tree and grass pollens, animal dander (dandruff), molds, and other agents that may be inhaled can serve as allergens. These may become concentrated on windy days, during certain seasons, or in certain parts of the country. Unless diagnosed and properly treated, asthma may lead to serious physical and psychological handicaps. The swelling of nasal passageways may close the eustachian tubes and lead to ear infections. The increase in mucus and restricted breathing may quickly lead to bronchial pneumonia in children and emphysema in older adults. The physical inability of a child to participate in games and sports can cause pyschological scarring and may serve as a nonallergenic (psychological) stimulation for further asthma attacks (intrinsic asthma).

SENSITIVITY TESTING AND TREATMENT

The **skin sensitivity test** and the **radio allergo sorbant test (RAST)** are used in the diagnosis of Type I hypersensitivity. Since skin tests for the determination of food allergies may be misleading or inaccurate, the RAST test has become a widely used diagnostic tool. RAST is a radiological assay capable of detecting and determining the level of IgE antibodies formed against specific allergens. The skin sensitivity test is used to diagnose Type I allergies to inhalants and some foods. A complete history of the patient's difficulties is taken to help determine the most likely cause of the allergy and which antigens will be used in the test. Purified samples of possible allergens are placed on the surface of the patient's skin and each is "picked" with a lancet to inject the antigen (Figure 14.3). After a twenty-minute period, antigens which have been able to react with IgE-bound antibody and stimulate the release of vasoamines will form wheals on the skin. The patient is allergic to antigens that form wheals. "Feelers" or "pseudopods" that extend from the wheals indicate a stronger hypersensitivity. Several steps may be taken to help control the reaction (Figure 14.4). The simplest control measure is to avoid contact with the allergen. This may mean not eating certain foods, staying away from dogs and cats, not going out on windy days, or even moving to another part of the country. The use of air conditioners and air purifiers may be necessary to control

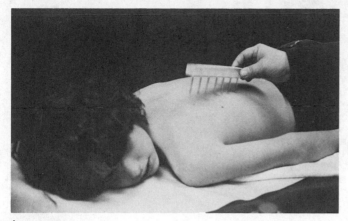

A

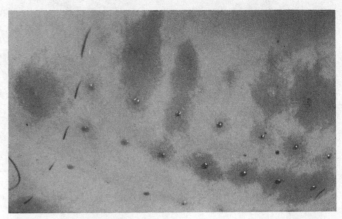

B

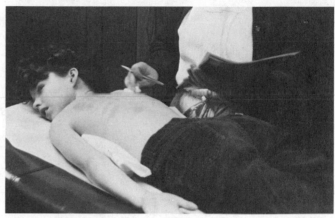

C

FIGURE 14.3

Skin sensitivity testing determines hypersensitivity to inhalants and certain foods. (A) A series of droplets containing allergen is placed on the patient's back, and each site is "picked." As the allergens stimulate the IgE-bound mast cells, inflammation and wheals develop on the skin. (B) Extensive inflammation and wheals have developed from many of the test substances, along with the formation of "pseudopods" or "feelers" on test sites in row one. (C) After a twenty-minute reaction period, the test is read. Allergens stimulating maximum wheal formation are recorded + + + +. (Courtesy of T. Smith.)

FIGURE 14.4

Four approaches to control Type I reactions are (1) limiting contact with the allergenic substances; (2) counteracting or interfering with the release of vasoactive amines; (3) reversing the tissue response; and (4) interfering with the sensitization of restimulation by providing immunotherapy.

Antigen

Environmental control
Food elimination diet   (1)

Natural portal of entry   Respiratory tract   Skin   Gastrointestinal tract

(4) Immunotherapy

(2)

Diethylcarbamazine

Antihistamines

Mast cell

Histamine
SRS-A
ECF-A

Cromolyn sodium   (2)

Asthma { Bronchospasm
          { Mucus secretion

Eosinophils

Epinephrine-isoproterenol }
Corticosteroids   } (3)
Theophylline   }

Epinephrine   } (3)
Phenylephrine   }

Hay fever and urticaria

Vasodilation
Edema

333

ONE of the most frequent Type I allergic responses occurring in the United States is anaphylactic sensitivity to the venom of stinging insects. Approximately 50 people die each year in the United States from generalized systemic anaphylactic shock after being stung or bitten by insects in the group Hymenoptera (bees, wasps, hornets, yellow jackets, fire ants, etc.).

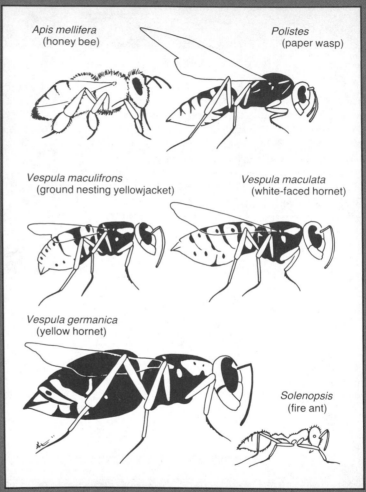

Apis mellifera
(honey bee)

Polistes
(paper wasp)

Vespula maculifrons
(ground nesting yellowjacket)

Vespula maculata
(white-faced hornet)

Vespula germanica
(yellow hornet)

Solenopsis
(fire ant)

Courtesy of Pharmacia Diagnostics.

Hypersensitivity develops after being stung by any member of this group. A person known to be allergic to honey bee stings is most likely allergic to the stings of all other insects in this group. Nonsensitized individuals only experience a sting as a localized soreness and irritation; however, those who have become moderately sensitized may respond to the venom antigen with such symptoms as massive swelling, itching and hives, sneezing, asthma attack, dizziness, vomiting, abdominal cramps, or diarrhea. If a highly sensitized person (due to a high antibody titer against the venom) receives several stings at the same time, the allergic response is hastened and may induce anaphylaxis that results in death. Individuals known to be strongly hypersensitive are advised to be sure to remove the venom sac and stinger of honey bees with a knife or fingernails and carry an anaphylaxis kit with them at all times. This kit contains an antihistamine tablet and a preloaded syringe or "pencil" of epinephrine or Sus-Phrine. These drugs will quickly help to counteract the symptoms of anaphylaxis. Sensitivity may be reduced by administering desensitization injections of dilute mixed extracts of Hymenoptera insect venom. These injections are given in increasing dosages until the risk of anaphylaxis is reduced and the shots may be continued at a maintenance level every two to six weeks for the rest of the patient's life.

asthma and other respiratory hypersensitivities. Drugs can also be used to block the release of vasoamines (e.g., cromolyn sodium), interfere with vasoamine action (e.g., antihistamines), or counteract symptoms as they occur (e.g., epinephrine, theophylline). Immunotherapy is another control method that is effective in many individuals. This method is also known to many as **desensitizing "shots."** After a skin sensitivity test, an antigenic serum is prepared that contains a very low concentration of the antigens known to be responsible for the hypersensitivity. This serum is injected intramuscularly on a regular basis, and the dosage is slowly increased over a period of months (Figure 14.5). How frequently individuals receive desensitization shots depends on their degree of hypersensitivity to the allergens. During the peak pollen season, some people may require shots every two or three days, while others respond very positively to shots given only every three weeks. Because of the possibility of inducing a severe anaphylaxis, many states require that the injection only be given in the presence of a physician who must remain available for a period of about twenty minutes. After this time, the likelihood that the injection has induced a dangerous Type I anaphylactic shock will have

passed. The injection site is also checked for the severity of localized anaphylaxis to help determine future antigen serum concentrations. It is hoped that this low-level contact with the allergens will not cause a hypersensitivity reaction but will stimulate the formation of protective IgG antibody. The increasing titer of IgG antibody serves as a blocking antibody by reacting with the allergen before it has the opportunity to contact the IgE-mast, cell-bound antibody.

In recent years, another method of desensitization has undergone extensive research. The process known as **drop therapy** consists of placing droplets of the suspected allergen under the patient's tongue. To determine the proper level for desensitization, the drops are given initially at a high dosage and allergic reactions are noted within ten minutes. The strength of the allergen is steadily decreased until a level is reached at which no reaction occurs. This is used as the level of desensitization. The patient is given a vial of droplets to use daily to prevent allergic reactions. Dr. Richard Mackarness is one of the main promoters of this system and his work in the United Kingdom is showing remarkable success. Slowly this method of desensitization is receiving interest and may prove to be a major method of treatment for allergies.

---

1 What type of allergens might provoke an asthma attack?

2 What type of material is introduced into the skin to perform a skin sensitivity test?

3 What type of material is injected into a hypersensitive person during the desensitization procedure?

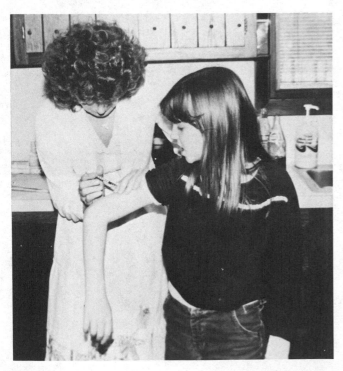

FIGURE 14.5
Immunotherapy by means of desensitizing injections may be needed every two or three days. For less severe cases of Type I hypersensitivity, injections may only be needed every two or three weeks during the peak allergy season. With each injection, IgG formation is stimulated. The patient must be observed for twenty minutes after an injection as a safeguard against possible anaphylactic reaction. (Courtesy of B. Young and Amy Ross.)

TYPE II HYPERSENSITIVITY

Type II hypersensitivities are generally referred to as **cytotoxic reactions,** since they result in the destruction of host tissue by lysis or by interfering with the elimination of harmful, antigenically different, human cells. The development of immunologic tolerance (refer to Chapter 13) occurs early in development. During this period, the immune system "learns" to recognize the difference between "self" and "nonself." From that time on, human cells and other materials with antigenic surfaces are continuously "looked over," or surveyed, by the immune system. Those recognized as "nonself" are eliminated by either humoral or cell-mediated immunity. "Nonself" antigens include foreign agents (e.g., microbes), as well as damaged "self" tissue (e.g., RBCs), and altered "self" tissues. This process is known as **immunosurveillance** and is of great benefit in eliminating cells which may form neoplasms (tumors). A **neoplasm** is a growth of new cells that have surface antigens which differ from other cells of the body. A neoplasm that spreads through the body (metastasis) and causes changes that could result in death is referred to **malignant cancer.** Normal body cells may change to neoplastic cells as a result of viral infections or contact with carcinogenic agents. In most people, immunosurveillance recognizes these abnormal cells and eliminates them. However, one kind of Type II hypersensitivity results from the formation of blocking antibodies that interfere with immunosurveillance antibodies. Blocking antibodies combine with surface antigens of cancer cells and "mask" these cells. This allows cancer cells to go undetected, and they are free to multiply and metastasize.

Another Type II hypersensitivity involves the complement system and is known as complement-dependent lysis (Figure 14.6). This series of destructive reactions begins with an antibody-antigen reaction on the surface of human tissue. The antigen may have come from outside the body (e.g., penicillin, tetracycline), or may be due to a biochemical change on the host cell surface. The most likely antibodies able to combine with these antigens belong to the IgM and IgG classes, and contain complement binding sites on the Fc portion of the molecules. The binding of C1q complement to the antibody begins the complement-dependent sequence, and ends with lysis and inflammation of host tissue. Cells undergoing complement-dependent lysis include red blood cells during autoimmune hemolytic anemia, brain

cells, T-cells during systemic lupus, and dermal-epidermal junction cells during skin (pemphigoid) diseases. Type II hypersensitivity also occurs in conjunction with drugs but differs from Type I allergies. During Type II reactions, normally nonantigenic drug molecules attach to the surface of red blood cells and form strong antigens that stimulate the release of antibody and lysis by complement fixation. This kind of drug allergy may occur after a patient is given penicillin, tetracycline, para-aminosalacylic acid, insulin, antihistamines, or numerous other drugs. A modification of this lytic reaction occurs in the disease glomerulonephritis stimulated by a previous Group A beta-hemolytic streptococcal infection. In about 5 percent of human glomerulonephritis cases, antibodies released as a result of the streptococcal infection cross-react with antigenic material normally found on the basement membrane of the glomeruli. As these antibodies accumulate over the membrane surfaces, comple-

---

1 What is meant by immunosurveillance?

2 How is the complement system involved in Type II hypersensitivities?

3 Give two examples of Type II complement-dependent hypersensitivities.

A Antigen on host tissue either (1) derived from outside the host or (2) due to biochemical change in host tissue.

B Humoral antibody (IgM or IgG) attaches to antigen on surface of tissue.

C Complement (C) series begins by attaching to antibody-antigen complex.

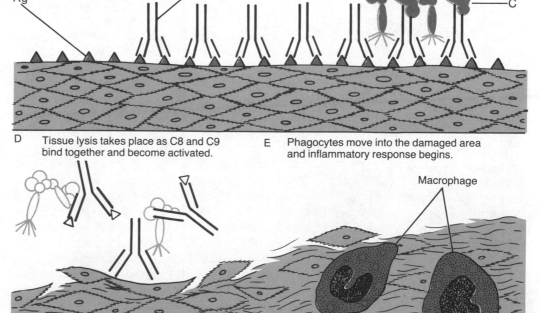

D Tissue lysis takes place as C8 and C9 bind together and become activated.

E Phagocytes move into the damaged area and inflammatory response begins.

FIGURE 14.6
In a Type II hypersensitivity reaction sequence, complement-dependent lysis occurs on the surface of tissues and cells which contain antigen that has been acquired from outside the body or as a result of cell-surface changes. Lysis by antibody-dependent, cell-mediated cytotoxicity occurs in this same basic manner; however, the antibody that combines with antigenic tissue is fixed to the surface of special cells known as killer, or null, cells.

ment-dependent lysis disrupts the area and stimulates an inflammatory reaction. The remaining 95 percent of nephritis cases result from Type III hypersensitivity and will be explained later in this chapter.

TYPE III HYPERSENSITIVITY

The third kind of hypersensitivity is known as the **immune-complex reaction.** The symptoms include an acute inflammatory response and tissue destruction typically associated with basement membranes and blood vessels. This reaction is caused by soluble antibody-antigen complexes that accumulate on target tissues. The formation and action of these antigen-antibody complexes are explained by the lattice-network precipitation curve presented in Chapter 13 (Figure 13.9). Precipitin reactions that take place in the presence of excess or equal amounts of antibody form insoluble antibody-antigen lattice complexes that are easily engulfed by phagocytes and removed from the body. However, the introduction of excessive amounts of antigen results in the formation of small, soluble immune-complexes that escape phagocytosis and may accumulate. In most cases, the immunoglobulins involved in this type of hypersensitivity are IgG and IgM. The concentration of immune-complexes stimulates the complement system, the release of kinins, polymorphonuclear leukocyte (PMN) migration, an increase in capillary permeability, and the release of histamines. Many diseases have Type III hypersensitivity associated with their major symptoms (refer to Table 14.1).

One of the most typical and well-studied Type III reactions was first identified in 1903 by Maurice Arthus. What is now known as the **Arthus reaction** has not been found to naturally occur in humans but may be experimentally produced in laboratory animals. After a series of horse serum injections, the skin of the animal becomes sensitized to antigenic horse protein. Further injections of serum result in an Arthus reaction at the injection site. The excess antigen binds with antibody, and immune complexes accumulate on cells which form the boundary between dermis and epidermis. The complement system stimulates a massive migration of PMNs into the area, which plugs the vessel and releases vasoactive amines that cause vasoconstriction and increase permeability. The endothelium begins to undergo necrosis as hydrolytic enzymes are released from dying PMNs. The injection site swells with edema fluid and hemolyzed red blood cells as inflammation progresses (Figure 14.7). In severe cases, the limited oxygen supply stimulates contaminating clostridia and ultimately leads to the development of gangrenous tissue.

Human Type III hypersensitivities that most closely resemble the Arthus reaction are known as **serum sickness** and **hypersensitive pneumonitis.** Pneumonitis (inflammation of the lungs) may be caused by a wide variety of

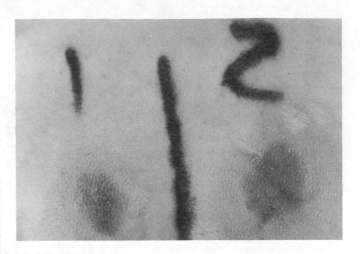

FIGURE 14.7
These two skin changes occurred eight hours after the intradermal injection of the antigen. Note the edema and necrosis in the central area. Serum sickness and hypersensitive pneumonitis are similar responses in humans. (Courtesy of L. J. LeBeau, Department of Pathology, University of Illinois Medical Center, Chicago.)

antigens and each has been associated with an unusual disease name such as "farmer's lung" and "bird fancier's disease" (Table 14.3). Serum sickness is a response that may occur in any number of tissues after repeated injections of horse antiserum used to provide acquired passive immunity. Signs of serum sickness appear between three days and three weeks after injection of the foreign antigen. Symptoms that occurred after DPT horse serum injections in the early 1900s included fever, malaise, urticaria (skin rashes), and enlargement of the lymph nodes and spleen. Because this form of protection is used less frequently today, the incidence of serum sickness in the United States is very low. However, caution is still exercised when immunizing against viral infections (for example, the "flu"), since many of these vaccines are attenuated or killed viruses that have been cultured in duck eggs and contain duck protein (Figure 14.8). Vaccinations for rabies virus from a nonhuman source cause serum sickness in many people. The incidence of this hypersensitivity is so high that physicians must weigh the risk of inducing serum sickness against the risk that the patient has actually contracted the viral infection. Today "serum sickness" Type III hypersensitivity may also be produced in some individuals in response to antibiotic therapy or kidney transplants.

All of the disorders mentioned so far result primarily from Type III hypersensitivity. Immune-complex disease symptoms have also been identified that are secondary characteristics of a number of other illnesses. For example, Type III hypersensitivity is found in some individuals after a Group A beta-hemolytic streptococcal infection and is responsible for about 95 percent of glomerulonephritis cases.

TABLE 14.3

Hypersensitive pneumonitis.

Disease	Antigen	Antigen Source
Farmer's lung	Thermophilic actinomycetes	Moldy hay
Mushroom grower's lung	*Trichophyton vulgaris* and *Micropolyspora faeni*	Mushroom beds
Thatched roof disease	Thatch proteins, fungi (?)	Thatched roof
Malt worker's disease	*Aspergillus clavatus* and *A. fumigatus*	Moldy barley and malt dust
Maple bark stripper's disease	*Cryptostroma corticale* spores	Tree bark
Bagassosis	*Thermoactinomyces vulgaris* and *T. sacchari*	Sugarcane stalks
Sequoiosis	*Pullularia pullulans* spores	Redwood trees
Cheese washer's disease	*Penicillium casei* spores	Cheese casings
Bird fancier's disease	Avian serum and gut proteins (?)	Bird excreta
Fish meal lung	Fish proteins	Fish meal in pet food
Furrier lung	Animal protein (?)	Fox fur
Air conditioning lung	Thermophilic and aspergillus organisms	Fungi present in water in air conditioning system
Meat wrapper's disease	Unknown	Found in fumes of hot, melted plastic wrap

SOURCE: Modified from J. Caldwell and B. Kaltreider. "Pulmonary and Cardiac Disease." *Basic and Clinical Immunology.* Lange Medical Publications, 1978.

**IMPORTANT INFORMATION
ABOUT SWINE INFLUENZA (FLU) VACCINE
(MONOVALENT)**

July 15, 1976

The Disease

Influenza (flu) is caused by viruses. When people get flu they may have fever, chills, headache, dry cough or muscle aches. Illness may last several days or a week or more, and complete recovery is usual. However, complications may lead to pneumonia or death in some people. For the elderly and people with diabetes or heart, lung, or kidney diseases, flu may be especially serious.

It is unlikely that you have adequate natural protection against swine flu, since it has not caused widespread human outbreaks in 45 years.

The Vaccine

The vaccine will not give you flu because it is made from killed viruses. Today's flu vaccines cause fewer side effects than those used in the past. In contrast with some other vaccines, flu vaccine can be taken safely during pregnancy.

One shot will protect most people from swine flu during the next flu season; however, either a second shot or a different dosage may be required for persons under age 25. If you are under 25 and a notice regarding such information is not attached, this information will be provided to you wherever you receive the vaccine.

Possible Vaccine Side Effects

Most people will have no side effects from the vaccine. However, tenderness at the site of the shot may occur and last for several days. Some people will also have fever, chills, headache, or muscle aches within the first 48 hours.

Special Precautions

As with any vaccine or drug, the possibility of severe or potentially fatal reactions exists. However, flu vaccine has rarely been associated with severe or fatal reactions. In some instances people receiving vaccine have had allergic reactions. You should note very carefully the following precautions:

- Children under a certain age should not routinely receive flu vaccine. Please ask about age limitations if this information is not attached.
- People with known allergy to eggs should receive the vaccine only under special medical supervision.
- People with fever should delay getting vaccinated until the fever is gone.
- People who have received another type of vaccine in the past 14 days should consult a physician before taking the flu vaccine.

If you have any questions about flu or flu vaccine, please ask.

FIGURE 14.8

This serum sickness warning was given to everyone receiving "swine flu" immunizations in 1976. Similar warnings are still issued with many other vaccinations.

Guillian-Barré syndrome may occur after viral infections and is a form of acute idiopathic polyneuritis. The onset of Guillian-Barré syndrome symptoms is very rapid (acute), results in damage to "self" (idiopathic), and produces inflammation of numerous peripheral nerves (polyneuritis). Systemic lupus erythematosus (SLE) results from the formation of immune-complexes of antinuclear antibody and nucleic acid antigen. Because SLE cases have been reported to occur in the same family over a number of generations, many suspect that the disease may in some way be influenced by genetic factors. It is also suspected that certain drugs may stimulate the formation of immune-complexes. A number of cases have reportedly been triggered by sulfa drugs, penicillin, and birth control pills. Many other diseases also demonstrate immune-complex hypersensitivity responses; however, the antigens responsible for the reactions and the dynamic nature of complex-tissue interaction have not been identified in all cases.

---

1 What symptoms typically occur in conjunction with a Type III hypersensitivity?
2 How does serum sickness relate to allergies and "flu" shots?
3 What is Guillian-Barré syndrome?

TYPE IV HYPERSENSITIVITY

Delayed cell-mediated hypersensitivity is very similar to cell-mediated immunity (CMI). However, the tissue response to a foreign antigen goes beyond protective levels and results in damage such as ulceration and necrosis. Some of the most familiar Type IV allergies result in skin damage, and possibly scarring, due to antigens released during the infectious diseases smallpox, measles, cold sores, and leprosy. Other skin allergies are seen as skin graft rejections, TB tine test, and contact dermatitis due to such materials as plant oils (e.g., poison ivy), fabrics, metals, chemicals, cosmetics, and furs. Internal tissue damage triggered by delayed hypersensitivity occurs in cases of food and drug antigens, foreign transplanted tissues, and chronic microbial infections (for instance, certain viruses, bacteria, and fungi).

Delayed hypersensitivity begins in much the same way as cell-mediated immunity. The allergen combines with special surface receptors on sensitized T-cells, triggering them to enter a period of blast transformation after which lymphokines and interferon are released into surrounding tissue. No serum antibody or complement are involved in these reactions. A delay of about 12 hours occurs before the effects of these processes are seen. Induration (a hard spot) and erythema (reddening) are the typical symptoms of de-

layed hypersensitivity. They reach a maximum between 24 to 48 hours and may completely disappear after about 72 hours. During the response, PMNs are stimulated to enter the reaction area but are quickly replaced by very large numbers of mononuclear phagocytes and lymphocytes. Induration is caused by the massive infiltration of these cells and not by the accumulation of edema fluid as in Type I hypersensitivity. In tissues that have received a large dose of allergen, the activity of T-cells and lymphokines becomes magnified, resulting in tissue necrosis and ulceration at the center of the reaction site. Allergen stimulation of delayed hypersensitivity on the skin may also stimulate this same reaction elsewhere in the body. For example, it is possible for a TB skin test to stimulate delayed hypersensitivity around foci of infections in individuals with active cases of tuberculosis. Delayed hypersensitivity is also known as cell-mediated hypersensitivity because a subpopulation of lymphocytes known as **null,** or **killer, T-cells,** becomes active in the killing of target cells. Killer T-cells are able to recognize other cells that have the attached allergen and cells with surface changes caused by certain viral infections. The killing of these abnormal cells is characteristic of delayed hypersensitivity and is known as **direct cell-mediated cytotoxicity.**

TB SKIN TEST REACTION

Perhaps the most familiar delayed response is the TB skin test to detect hypersensitivity to *Mycobacterium tuberculosis.* The test is performed with either OT (old tuberculin) or PPD (purified protein derivative.) Old tuberculin is a concentrated, filtered solution of old *M. tuberculosis* broth culture. It contains the allergen, tuberculin protein, as well as many other antigenic molecules. These extra antigens are able to stimulate other forms of hypersensitivity and may induce a false positive test. To reduce the chance of this nonspecific reaction, OT may be chemically purified and administered as purified protein derivative. A number of methods have been developed to introduce the tuberculin protein (in the form of OT or PPD) to sensitized skin cells. The **tine test** uses a small, disposable plastic applicator with several small, metal tines (points) coated with dried OT. The tine is pressed to the skin on the inside of the forearm to "inject" the antigen into the epidermis. The **Mantoux test** is performed by administering either OT or PPD intradermally with a hypodermic syringe. The needle is inserted just under the surface and the skin is gently lifted to inject the antigen into the epidermal layer.

Sensitivity to *M. tuberculosis* usually takes place within 4–7 days after an individual has first come in contact with the antigen. If the tine test is given in this period, no delayed hypersensitivity will be detected. After this period, however, the introduction of OT will initiate the delayed response and will require approximately 12 hours before the skin reaction can be seen. Induration and erythema develop gradually and reach a peak between 24 to 48 hours and then slowly disappear. The test should only be "read" during the peak response period. A positive TB skin test shows an area of induration of at least 10 millimeters in diameter with surrounding erythema (Figure 14.9). A positive skin test indicates that the patient has been exposed and sensitized to the mycobacterial antigen but does not necessarily mean that the patient has an active infection. A confirmed case of tuberculosis requires several tests including the tuberculin skin test, chest x-ray, and sputum samples for acid-fast staining and culturing. A negative skin test may occur during a se-

1 What subpopulation of lymphocytes becomes activated in delayed hypersensitivity?
2 What type of material is introduced under the skin during a tine test?
3 What does a positive tine test mean?

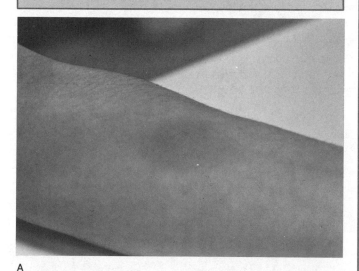

A

B

FIGURE 14.9
(A) This positive tuberculosis skin test (Mantoux test) shows erythema at the site of injection. (B) Note the induration. A Mantoux reaction is at its peak from 48 to 72 hours after the injection.

vere case of tuberculosis, measles, Hodgkin's disease, or after immunization with BCG vaccine. These symptoms and others are familiar to many who experience Type IV hypersensitivities to something they touch.

CONTACT DERMATITIS

Contact dermatitis is a Type IV hypersensitivity that occurs in response to something a person touches. Although many substances may harm the skin, not all dermatitis (inflammation of the skin) reactions are true allergies, since the immune system plays no natural part in defending the body against the offending material. Substances that are not selective in their action and affect all people in a harmful way are known as **primary irritants.** Many household cleansers, detergents, acids, and alkalis may cause dermatitis (e.g., "detergent hands" or eczema). These are primary irritants since contact with the chemicals does not stimulate antibody formation. Table 14.4 classifies and gives examples of a number of materials shown to be responsible for true, allergic contact dermatitis. Of all those listed, the most common form of contact dermatitis results from contact with the chemical urushiol found in poison ivy. After first contact and sensitization to this allergen, all second and subsequent contacts result in delayed hypersensitivity in most people. Touching the plant or coming in contact with smoke from burning these leaves will cause a rash, itching, lumps, blisters (vesiculation), and finally crusting and scaling. More commonly these symptoms are known as **eczema,** and they vary in severity from one individual to another and from one experience to another.

Two other very common hypersensitivities occur in response to metals and small allergens found in cosmetics. Although chromium and nickel are unable to stimulate antibody production themselves, these metals join with more complex proteins and polysaccharides that, in combination, are able to sensitize lymphocytes. Both metals are frequently used in watches, zippers, clasps, needles, pins, and costume jewelry. A cosmetic is defined by the Committee on Cutaneous Health and Cosmetics (American Medical Association) as "any substance that is applied to the skin, hair, nails, or teeth for lending attractiveness, for makeup, for cleansing, or for 'conditioning'." Even though the severity of responses to cosmetics are for the most part only minor rashes and itching, there are an estimated 3,000,000 people in the United States and Canada who annually suffer from cosmetic allergies of varying types. Many become sensitized by direct application of nail polishes, deodorants, hair dyes and sprays, shaving lotions, and perfumes. While many others are indirectly sensitized, according to Dr. Alice Mills and Lou Joseph of the American Academy of Allergy, Northwestern University: "The wily sprees of cosmetic allergy can often assume a grotesque turn when a romance sours be-

TABLE 14.4
Common contact dermatitis allergens.

Plants	Metals	Natural Materials	Synthetic Materials
Rhus genus poison ivy, poison oak, and poison sumac. Pine trees Celery Tulips Daisies Tumbleweed Potato	Chromium: jewelry, belt buckles, needles, scissors, watches. Nickel: coins, zippers, pins, hairpins, curlers. Copper: plumbing, jewelry. Zinc: gutters, metal roofs, galvanized steel (garbage cans).	Leather: shoes and gloves, watch bands. Rubber: baby pants, dress shields. Fish: cod, tuna (during preparation). Meat: beef, pork (during preparation). Fowl: chicken, turkey (during preparation). Furs: rabbit, fox. UV light.	Plastics: raincoats, buttons, toys, handles. Dyes: fabrics for cloths and furniture. Chemicals found in: cosmetics, clothing, paint, deodorants, building materials, perfumes, hair dyes, nail polish.

cause the boyfriend develops swollen lips every time he kisses his girl. And many husbands have complained about being allergic to their spouses' perfumes, hair spray, lipstick, or facial powder.''

Contact dermatitis rarely results in severe, permanent tissue damage. However, delayed hypersensitivity occurring after tissue transplants may result in direct cell-mediated cytotoxicity. Whether the tissue being transplanted is blood, bone marrow, kidney, heart, etc., the antigenic components found on the donor cell surfaces will sensitize recipient lymphocytes and eventually lead to rejection.

> **1** How does contact dermatitis occur?
> **2** What types of materials might serve as primary irritants?

TRANSPLANTATION IMMUNOLOGY

Attempts to restore health by the transplantation of tissues from a donor to a recipient have long been attempted by physicians. In early attempts, whole blood, limbs, and internal organs of animals and humans were transferred to recipients with the hope that the new and old tissues would "grow together" (Figure 14.10). However, very little success was achieved even though the tissues "looked alike." Within a matter of days, the transplanted organ would wither and die, and in countless cases, so did the recipient. The degree of resemblance of donor tissue to recipient tissue is very important to the success of transplantation operations but not on the superficial level conceived by early physicians. In order for a donor tissue to be successfully transplanted, it must antigenically resemble the recipient as closely as possible. In addition, the immune system of the recipient must be temporarily suppressed to give the new tissue a chance to "fit in" and be accepted as "self."

The immune responses that result in the destruction of transplanted tissues are known collectively as **rejection responses** and are triggered by naturally occurring antigens on the surface of transplanted cells. Cell-surface antigens of a different species are seen by the immune system of the

FIGURE 14.10
This illustration (circa 1693) shows a primitive attempt at whole blood transfusion. The patient's blood is being let from his right arm, while the blood from a dog is being transfused into his left arm. (Courtesy of the National Library of Medicine.)

human recipient as "nonself" and rejected very quickly. Antigens on the surface of cells from a genetically related individual of the same species stimulate the rejection response less rapidly, and, with a little help, may be successfully incorporated into the recipient's body. Glycoproteins found within the same species that function as cell-surface antigens are known as **isoantigens.** Their structure is genetically controlled, and they may be found on the surfaces of blood cells (e.g., red and white cells, and platelets) and on fixed tissue cells (e.g., skin, heart, kidney, etc.). Cells have a variety of isoantigens on their surfaces; however, the most clinically important are those on red blood cells associated with the ABO and Rh blood systems, and the histocompatibility antigens found on the surface of white blood cells and most fixed tissues.

THE ABO AND Rh BLOOD SYSTEMS

The surface of red blood cells contains genetically determined isoantigens. A person having two genes for isoantigen A (genotype AA) or one gene for A and a second for isoantigen H (genotype AO) belongs to the phenotype blood group A. A person with two genes for isoantigen B (genotype BB) or one gene for B and a second for isoantigen H (genotype BO) belongs to phenotype blood group B. Those that have genes for isoantigen A and B belong to blood group AB; and individuals who have a pair of alleles for isoantigen H (genotype OO) belong to blood group O (Table 14.5). Individuals in a particular blood group have nat-

urally occurring serum antibodies against isoantigens not found on their own red blood cells. These antibodies circulate freely in the serum and are known as isoantibodies, or **isoagglutinins.**

The second important isoantigen system was first discovered by Levine and Stetson while investigating a case of erythroblastosis fetalis (1939) at the same time that Landsteiner and Wiener were researching the red blood cell antigens of Rhesus monkeys. The genetics of the Rh (Rhesus) isoantigen system has not been as clearly established as that of the ABO system; however, it is suspected that three different alleles (C, D, and E) may be responsible for the two phenotypic blood groups, Rh(+) and Rh(−) (refer to Table 14.5). Rh antibodies are not found naturally in blood serum as are the ABO isoagglutinins. The antibodies are only formed when the immune system is directly stimulated by transfusion or contact with antigen during pregnancy. Having a working knowledge of both the ABO and Rh systems is essential in understanding and performing blood transfusions (short-term transplants), and in preventing hemolytic disease of the newborn.

Hemolytic disease of the newborn is the destruction of red blood cells by maternal antibodies that have formed against fetal isoantigens, crossed the placenta, and entered the circulatory system of the infant. This disease is more commonly known as **erythroblastosis fetalis.** Stimulation of antibody formation most often takes place within 24 hours of birth when red cells from the newborn enter the mother's circulatory system through the uterus. All subsequent pregnancies involving a fetus with the same red cell antigen as the first fetus will serve to stimulate formation of

TABLE 14.5
ABO and Rh blood systems.

Genotype	Phenotype	Red Cell Antigen	Serum Isoagglutinins
AO AA	A	A	Anti-B
BO BB	B	B	Anti-A
OO	O	H	Anti-A Anti-B
AB	AB	A, B	None

| | Genes | | Expressed Antigens | |
	Fisher-Race	Wiener	Fisher-Race	Wiener
Rh(+)	cDe	R^0	c, D, e	Rh_0, hr', hr"
	CDe	R^1	C, D, e	Rh_0, rh', hr"
	cDE	R^2	c, D, E	Rh_0, rh", hr'
	CDE	R^Z	C, D, E	Rh_0, rh', rh"
Rh(−)	cde	r	c, e	hr', hr"
	Cde	r'	C, e	rh', hr"
	cdE	r"	c, E	rh", hr'
	CdE	r^y	C, E	rh', rh"

these antibodies. After crossing the placenta, these antibodies lead to anemia, jaundice, erythroblastosis (an abnormal increase in red cell production), and possibly fetal death. The most frequent hemolytic disease occurs between mothers that are blood type O and infants that are blood type A; however, the symptoms of this hemolytic disease are usually very mild compared to those caused by Rh incompatibility. If an Rh(−) woman is pregnant with an Rh(+) fetus for the first time [the father must be Rh(+)], the pregnancy and delivery will in all likelihood result in a baby without erythroblastosis fetalis. However, as placenta membranes are broken, Rh(+) red cells from the baby may enter the mother's circulatory system and stimulate the immune system to produce anti-Rh(+) antibodies. On second and all subsequent pregnancies involving Rh(+) infants, the mother's antibody titer can rise, and antibodies will cross the placenta into the fetus, resulting in hemolytic disease (Figure 14.11).

In the past, erythroblastosis fetalis was a serious problem and was treated by immediate transfusion of the infant's blood with blood lacking harmful maternal antibodies. Today the condition can be treated in advance of an Rh-incompatible pregnancy by the use of RhoGAM. **RhoGAM** is a solution of specific antibodies to the D antigen found on Rh(+) red cells. After an Rh(−) mother gives birth to her first Rh(+) baby, RhoGAM is administered within 48 hours of delivery to combine with any Rh(+) infant red cells that may have entered the mother's system. This prevents them from stimulating the immune response. If a second Rh(+) pregnancy occurs, no anti-Rh antibodies will have been manufactured by the mother, and the baby will not develop the hemolytic disease.

A quick and accurate determination of blood type is essential in preventing hemolytic diseases of the newborn and assuring that blood transfusions will be successful. The **Coombs test** is an agglutination test for the identification of cell-surface antigens that under normal circumstances do not form a clearly visible, antibody-antigen lattice network. By adding antibodies formed in other animals (Coombs antiglobulin) to antibody-antigen reacted blood, larger, more visible lattice networks are formed (Figure 14.12).

---

1 Isoantigens involved in the rejection response are what kind of cell-surface molecules?
2 What is an isoagglutinin?
3 Why is RhoGAM used? How does it work?

TRANSFUSION REACTIONS

The antibody-antigen reactions occurring in an individual after blood has been transplanted are known as **immune transfusion reactions.** The two types of reactions which have been identified are **immediate incompatibility** and **delayed incompatibility.** The immediate reaction results from the presence of antibody in the blood that is able to react with isoantigens on the surface of red cells. When the antibody is in the recipient's blood and the antigen on the donor's cells, a major or severe reaction occurs. This type of reaction results in chills, fever, shock, and hemolysis of transfused red cells. A person with type A blood who receives a transfusion of type B blood will experience this im-

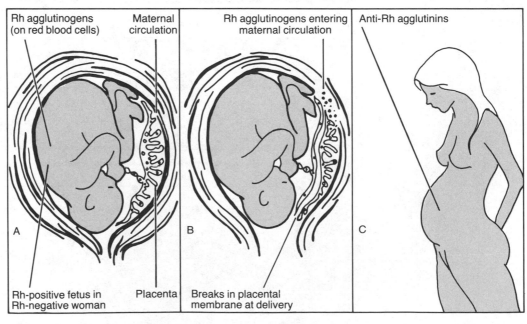

FIGURE 14.11
(A) An Rh(−) woman is pregnant with an Rh(+) fetus. (B) Some of the fetal red blood cells with Rh agglutinogens may enter the maternal blood at birth. (C) As a result, the woman's cells may produce anti-Rh agglutinins.

Rh agglutinogens (on red blood cells)   Maternal circulation

Rh-positive fetus in Rh-negative woman   Placenta

Rh agglutinogens entering maternal circulation

Breaks in placental membrane at delivery

Anti-Rh agglutinins

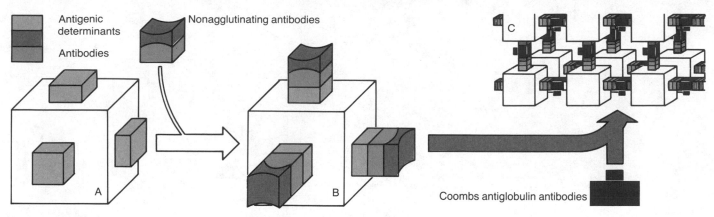

FIGURE 14.12

In the Coombs test, once the antibody-antigen reaction occurs (A) lattice networks may be formed by the addition of Coombs antiglobulin antibody (B). (C) The result is a clearly visible and accurate antibody-antigen reaction. (Modified from E. W. Nester, C. E. Roberts, B. J. McCarthy, and N. N. Pearsall, *Microbiology and Molecules, Microbes, and Man,* 2nd ed. Copyright © 1978, 1973 by Holt, Rinehart and Winston. Reprinted by permission.)

mediate major incompatibility reaction, since the recipient's blood contains large quantities of anti-B antibody. In an emergency, however, a type A recipient may be transfused with type O blood and only experience a minor incompatibility reaction. This mild response is due to the comparatively small amount of antibody in the donor's blood (anti-A and anti-B) that will react with the isoantigens (type A) on the surface of a small number of the recipient's red cells. Immediate minor incompatibility reactions occur when the antibody is in the donor's blood and the antigen is on the recipient's red cells. Because type O blood lacks the surface isoantigens A and B, this blood may be transfused into most people with only a minor incompatibility reaction in recipients with types A, B, and AB blood. Persons with type O blood are known as universal donors. Table 14.6 summarizes the ABO blood types, and the nature of the isoantigens and antibodies in transfusion reactions.

Delayed immune transfusion reactions result in hemolytic anemia after the recipient's immune system has been stimulated to produce antibodies by the transfused isoanti-

gens. Since the average life span of a red cell is only 120 days, the response is not usually severe, but the presence of the antibody will prevent the patient from receiving any future transfusions of the same type blood.

Nonimmune reactions may also create problems for the recipient, unless donor blood is screened before it is used. Blood donation agencies throughout the country regularly review a potential donor's health history to reduce the incidence of nonimmune complications such as hepatitis, malaria, passively transferred hypersensitivities, and others (Figure 14.13). Blood banks also perform a procedure known as **leukophoresis.** This technique separates granulocytes from the circulating blood of a donor. The other components of the blood (red cells, platelets, plasma, and agranulocytes) are returned to the donor during the process. Separation requires that the donor's blood be passed through the leukophoresis machine for three to four hours, and may only be performed approximately every two months. Granulocytes extracted by this process are used for patients with leukemia, certain forms of chemotherapy, and some trans-

TABLE 14.6

The ABO system and blood transfusions.

Blood Type	Isoantigen (Agglutinogen) on the Red Cell	Antibody (Agglutinin) in the Plasma	Preferred Blood Type for Transfusion	Permissible Blood Type for Emergency Transfusion
A	A	Anti-B	A	O
B	B	Anti-A	B	O
AB	A and B	Neither anti-A nor anti-B	AB	A, B, O
O	Neither A nor B	Both anti-A and anti-B	O	O

PLEASE PRINT YOUR

LAST NAME   FIRST NAME   INITIAL

SOCIAL SECURITY NUMBER   AGE   BIRTHDATE

COMMUNITY BLOOD CENTER

DO NOT MARK BELOW THIS LINE

MAY WE CALL YOU SHOULD YOUR BLOOD TYPE BE NEEDED?

YES _____ NO _____ IN HOW MANY MONTHS? 2 _____ 3 _____ 4 _____ OTHER _____

SHOULD WE CALL HOME? _____ AT WORK? _____

NUMBER & STREET   CITY

STATE   ZIP   HOME PHONE   BUSINESS PHONE

BADGE # _____

PLACE OF EMPLOYMENT

OCCUPATION

DO NOT MARK BELOW THIS LINE

I.D. ___ - __ - ____

SEG _____ LOT _____

PLACE UNIT CONTROL # HERE

SEG #, LOT #, NAME & UNIT #
VERIFIED BY _____

NAME CHECKED, CONTROL # ON BAGS AND THIS FORM
VERIFIED SAME & DRAWN BY _____ @ _____ A.M. P.M.

ABO	Rh	COMMENTS

BC   Midl   Mobile

REVIEW BY _____ DATE ___/___/___
COMMENTS _____

DATE	M F	Gp-Rh.	B.P.	TEMP	HB	ARMS CK'ED?
	Ca I					
	O B	WT.		PULSE	HCT	EXAM'R

DEFERRED BY _____ ___/___/___
UNTIL _____
WHY _____
CODE _____

FIGURE 14.13

Blood transfusion problems can be greatly reduced by surveying the donor's medical history before blood is drawn. Note that both infectious diseases and immuno-related problems are monitored.

PLEASE READ AND ANSWER ALL QUESTIONS   YES   NO

1. Are you in good health?.................................... 1 ☐ ☐_____
2. Are you taking medication, drugs or pills? .... 2 ☐ ☐_____
3. Do you have a rash, hives, boils, cold sore, burns, bruises, or skin infection? .... 3 ☐ ☐_____
4. Do you have a history of asthma, hay fever, or allergy?........... 4 ☐ ☐_____

Have you . . . .

5. seen a physician or doctor in the past year? .... 5 ☐ ☐_____
6. had symptoms of cold, flu or cough within the last week? .... 6 ☐ ☐_____
7. had dental work within the last 3 days? .... 7 ☐ ☐_____
8. ever had surgery or serious illness? .... 8 ☐ ☐_____
   within last six months? .... ☐ ☐_____
9. been exposed to a contagious disease, mumps, measles, or chickenpox within last month? .... 9 ☐ ☐_____
10. ever received blood and / or plasma transfusion? Date_____ 10 ☐ ☐_____
11. been pregnant within the last six months? .... 11 ☐ ☐_____
12. had recent vaccinations or immunizations?.................... 12 ☐ ☐_____
    Date _____ Type of Injection _____
13. had any of the following diseases? (Circle, if yes)   13 ☐ ☐_____
    Heart   Kidney   Lung   Liver   Ulcers   Cancer
    Infectious Mono   Tuberculosis   Rheumatic Fever   Diabetes
    Bleeding Tendencies   Arthritis   Malaria   Heart Murmur
14. had convulsions, epilepsy, or fainting spells?................14 ☐ ☐_____
15. taken nonprescription drugs or narcotics by mouth or injection? ....15 ☐ ☐_____
16. traveled or resided outside the U.S.A. in the last 3 years? ....16 ☐ ☐_____
    Date _____ Location _____
17. ever had yellow jaundice or hepatitis, or contact with hepatitis?.....17 ☐ ☐_____
    Date _____
18. had tattoo or ears pierced within the last six months?.............18 ☐ ☐_____
19. been about the same weight for the last 6 months?..............19 ☐ ☐_____
20. any venereal diseases in the last 5 years? Date _____20 ☐ ☐_____
21. ever been told not to donate blood or plasma?...................21 ☐ ☐_____
22. What was the date of your last blood or plasma donation? _____

I certify that I have, to the best of my knowledge, truthfully answered the above questions and understand that this information is important in determining whether I should be a blood donor at this time.

I voluntarily agree to donate my blood to be used as decided by COMMUNITY BLOOD CENTER. Further, I give my permission to the blood center to do detail typing on my blood.

_____ DATE   _____ SIGNATURE

_____ WITNESS

plant operations. The success of leukocyte transplantation and the transplantation of fixed tissues also depends on a close antigenic match between recipient and donor. The most important cell-surface antigens responsible for the rejection of tissue grafts and transplants from one person to another are the histocompatibility antigens.

HISTOCOMPATIBILITY ANTIGENS

Histocompatibility antigens are complex glycoproteins that are components of cell membranes. These molecules are embedded in the phospholipid bilayer with one end located inside the cell and the other extended from the membrane surface (Figure 14.14). The synthesis of these antigens is under the direct control of special **human leukocyte antigen (HLA)** genes found in all cells of the body. These antigens are so named because they were first identified on the surface of human leukocytes. Since their discovery, approximately 77 different surface antigens have been identified, and labeled with letters and numbers. Because of the many possible HLA gene combinations in the human population, each person has the ability to synthesize a unique combination of histocompatibility antigens known as the **major histocompatibility complex (MHC)**. The MHC found on the surface of leukocytes and fixed tissue cells is unique to each person and may be used to identify a specific tissue type. The presence of these antigens early in fetal development is responsible for establishment of the immune tolerance system (refer to Chapter 13). The body "learns" to recognize and accept this combination of antigens as "self," and does not produce antibody against them. The recognition of "self" as opposed to "nonself" is the basis of the immune surveillance system. If a tissue from another person is transplanted, the recipient's immune system will respond to the presence of the "nonself" MHC by producing antibodies against the tissue.

The destruction of transplanted tissue may take place as hyperacute, acute, or chronic rejections. **Hyperacute (immediate) rejection** begins within minutes after the tissue enters the body and is similar to immune-complex Type III hypersensitivity. **Acute (delayed) rejection** begins about six days after surgery and is the result of cell-mediated rejections (Type IV). **Chronic rejection** is the slowest destruction process and appears to be cell mediated; however, humoral antibodies contribute to the slow loss of tissue function over an extended period.

The likelihood of tissue acceptance can be increased by obtaining a graft from another individual of the same species. These tissues are known as **allografts** ("allo" means *other*). Individuals with a great degree of genetic similarity are more likely to have similar major histocompatibility complexes since all 77 histocompatibility alleles are determined by just 5 HLA genes (HLA-A, HLA-B, HLA-C, HLA-D, and HLA-DR). Based on studies of heredity, it has been demonstrated that these genes are transmitted from parents to offspring according to Mendelian principles. During sexual reproduction, each parent contributes to the child a unique combination of HLA genes. Because of the inheritance patterns involved, two siblings have approximately a 25-percent chance of being HLA-A identical, 50-percent chance of sharing one of the HLA-A genes, and only 25-percent chance of having completely different HLA-A genes. Therefore, a deliberate attempt is made to use tissues from siblings (brothers and sisters) or other genetically related, his-

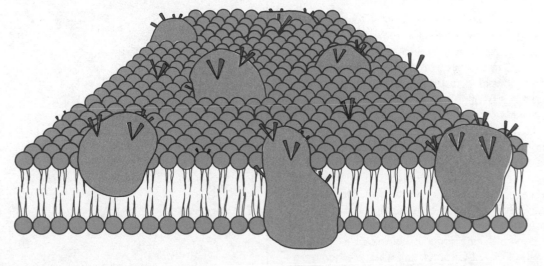

FIGURE 14.14
Histocompatibility antigens are glycoproteins that pass from one side of the phospholipid bilayer to the other. Five genes are associated with the formation of these antigens and a total of 77 molecules have been identified to date. Because of the many possible combinations, each person has a unique histocompatibility antigen combination (MHC). (From B. Cunningham. "The Structure and Function of Histocompatibility Antigens." *Scientific American,* Oct., 1977.)

tocompatible individuals. The selection of an appropriate donor for transplant is basically the same procedure required for blood transfusions. Various tests are run to determine whether antibody-antigen reactions will occur after surgery and result in the rejection of the graft. *In vitro* studies are performed to determine how closely HLA antigens match. The greater the degree of similarity, the more successful the transplant will be. The lymphocytotoxic cross-match test is performed by mixing the recipient's serum with donor lymphocytes to determine whether antibodies are present which will initiate rejection. The cell-mediated lymphocytosis test is used to detect possible acute rejection by mixing recipient leukocytes with donor serum. After an extensive incubation period, the presence of newly activated killer T-cells is an indication that rejection is likely. Even though tissue cross-matching shows a high degree of histocompatibility (for example, between identical twins), there remains the possibility that some form of rejection may occur. The detailed identification of HLA antigens on the surface of human cells is vital prior to transplantation surgery and has also become valuable as an early warning signal for certain genetically related diseases (Box 24).

---

1 What factor limits the severity of a delayed immune transfusion reaction?
2 What is the advantage of leukophoresis?
3 What is the correlation between MHC and tissue typing?
4 What is an allograft?

CANCER AND IMMUNOLOGY

Changes in a normal cell resulting from altered genes or abnormal gene action may cause the formation of a cancer cell. This process is called **neoplasia** and the resultant cell mass is a tumor. Cancer cells in humans do not reproduce any faster than normal surrounding cells but continue to reproduce by mitosis without stopping. Their nonstop mitosis brings about a logarithmic increase in the number of cancer cells which, in humans, may move from their original location to other parts of the body by metastasis. The presence of these cells may cause little or no harm (benign tumor), or they may severely interfere with normal tissue functions and cause death (malignant tumor). Many cancers arise spontaneously as a result of naturally occurring mutations; however, the incidence of cancer known to be caused by viral infection and carcinogenic chemicals is increasing rapidly. Dr. Albert Szent-Gyorgyi and many other prominent scientists suspect that neoplasia is the failure of a naturally occurring "biochemical mitosis braking system" in cells.

In the normal process of development, cells of the embryo undergo differentiation and specialization to form tissues. During this time, some form of biochemical brake is applied to the genetically controlled cell division process. This braking system may be the gene repression system (refer to Box 13) or some other, as yet unknown, mechanisms. The brake prevents the differentiated cell from dividing, or slows the division to a rate that is typical for the tissue in which the cell is located. When the mitotic brake is removed, the cell returns to a division cycle resembling that of embryonic cells. This shift has been confirmed by comparing the HLA type of normal cells to cancer cells in patients with carcinoma of the colon, liver, and other organs. These studies have shown that cell-surface antigens normally found on embryonic cells (embryonic antigens) disappear after the differentiation process but reappear as **carcinoembryonic antigen (CEA)** on the surface of these cancer cells (Figure 14.15) and in the serum of adults with certain forms of cancer (Table 14.7)

TABLE 14.7
Circulating CEA levels in various conditions.

Clinical Status	Percent Elevated CEA
Normal	
Healthy, unselected	11
Healthy, nonsmokers	3
Healthy, smokers	19
Healthy, pregnant	3
Nonmalignant diseases	
Alcoholic cirrhosis of liver	70
Alcohol addiction	65
Pulmonary emphysema	57
Kidney transplant	56
Pancreatitis	53
Granulomatous colitis	47
Pneumonia	46
Gastric ulcer	45
Ulcerative colitis	31
Malignant diseases	
Colorectum, all stages	72–81
Dukes' A	38–44
Dukes' B	60–76
Dukes' C	60–75
Metastasized	80–89
Stomach	61
Pancreas	91
Breast	47
Lung	76
Prostate	40
Bladder	42
Gynecologic	65
Lymphomas	35
Acute and chronic leukemias	37

SOURCE: D. M. Greenberg. "Oncofetal and other Tumor-Associated Antigens of the Human Digestive System." *Pathology of the Gastro-Intestinal Tract.* B. Morson (editor). *Current Topics in Pathology* vol. 63. G. Kirsten WH (editors). Springer-Verlag, 1976.

THE genes responsible for HLA antigens are located on human chromosome 6, very close to other genes which have medical significance. Because of their close relationship with one another, the genes that make an individual inclined toward a particular disease will be inherited along with a particular HLA gene. Therefore, the presence of specific HLA gene products (histocompatibility antigens) on human leukocytes may be used to predict the presence of the closely linked genes. These genes influence such diseases as arthritis, multiple sclerosis, rheumatic fever, juvenile diabetes, chronic active hepatitis, systemic lupus, psoriasis, myasthenia gravis, and others. By performing the HLA blood tests noted earlier, it is possible to predict the likelihood that an individual will acquire one of these genetically related diseases. For example, 95 percent of the individuals with a form of arthritis known as **ankylosing spondylitis** (arthritis of the spine) will have histocompatibility antigen type B27. This disease is characterized by lower back pain due to the inflammation of the sacroiliac joints, spine, and other larger joints. No rheumatoid factor is found in these patients and erosion of vertebrae occurs early in the disease. Later the bones of the sacroiliac joints may become fused. The diagnosis of ankylosing spondylitis may be made more accurate by correlating symptoms with an HLA blood type. This is of great value in avoiding a misdiagnosis since rheumatoid arthritis and other lower back problems can show similar symptoms. **Myasthenia gravis** is another HLA-associated disease. This is an autoimmune disease in which antibodies are formed (possibly as a result of viral infection) that attach to acetylcholine receptors on nerve cells. The antibodies block the receptors and interfere with impulse action, resulting in fatigue, muscular exhaustion, and progressive weakness. Muscles of the face, lips, tongue, throat, and neck are frequently affected. HLA blood-typing research has shown that persons with histocompatibility antigen B8 are more inclined to carry the genes associated with myasthenia gravis than others in the general population. The ability to forecast and diagnose myasthenia gravis, ankylosing spondylitis, and other HLA gene-associated diseases by HLA typing is only now beginning to be used in the clinical setting.

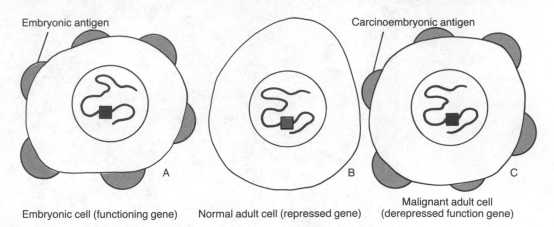

Embryonic cell (functioning gene)   Normal adult cell (repressed gene)   Malignant adult cell (derepressed function gene)

FIGURE 14.15

(A) During embryonic development, the cell contains operating genes (blackened area) that control the synthesis of embryonic antigen. (B) Differentiation and specialization result in the "turning off," or repression, of the genes. Therefore, no embryonic antigen is produced. (C) During neoplasia, some cells have this gene "turned on," or derepressed, and the cancer cells again show this surface antigen; however, it is now called carcinoembryonic antigen.

Stimulation of cells by carcinogenic agents also induces mutations and gene changes, resulting in the production of other types of surface antigens. These newly synthesized glycoproteins are known as **tumor-specific antigens (TSA).** Chemical carcinogens may stimulate tumor-specific antigen production by derepression, as in the case of carcinoembryonic antigen, or a viral infection (only studied in animals) may cause TSA to be formed (Figure 14.16). Studies have shown that genetically identical cells exposed to the same carcinogenic agent will each develop its own specific surface antigen. If each mutated cell develops into a tumor, each tumor will have its own specific antigenic makeup. Many

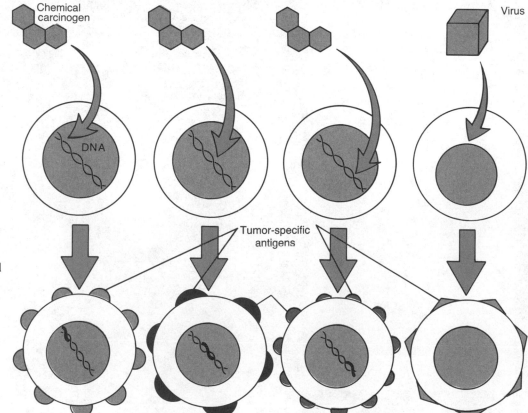

FIGURE 14.16

Whether stimulated by chemical carcinogen or viral infection, each cell undergoes unique changes that may result in the formation of specific surface antigens. Since they are unique to cancer cells, they have been called tumor-specific antigens. (From J. Bellanti, *Immunology II.* W. B. Saunders Company, 1978.)

researchers are convinced that low levels of naturally occurring carcinogens in the environment constantly stimulate the release of the biochemical braking system. They contend that body cells regularly undergo neoplasia. It is thought that the reason more people do not contract cancer is the result of both the efficient operation of DNA repair mechanisms and the successful operation of the immune system (Figure 14.17). If DNA repair mechanisms (refer to Chapter 11) are operating efficiently, mutated genes will quickly be restored to their original code and the mitotic brake reapplied, before TSA is produced and a tumor develops. Should the systems not be able to function, the cell releases the brake and slips into uncontrolled, continuous mitosis. Destruction of the abnormal cell and control of the tumor then becomes the responsibility of the immune system.

Several lines of evidence point to the involvement of the immune system in controlling cancer. First, patients with cancers that have undergone remission (improvement) are correlated with improved immune responses. Second, children with demonstrated immunodeficiency diseases have a higher incidence of cancer than those without such illnesses. Third, patients on a long-term immunosuppressive therapy (for example, transplant patients) show a higher incidence

1 What cell-surface antigens disappear after differentiation?
2 What cell-surface antigens appear on the cell after cancer cell formation?
3 How does the immunosurveillance system work with respect to cancer cell formation?

of cancer. Fourth, the elderly show a normal but progressive loss in the responsiveness of their immune system and a progressive increase in the incidence of cancer. However, the immune system can only respond to the presence of these abnormal cells if the TSA is "recognizable." Should the mutated cancer cell not produce any TSA at all (as is the case with certain animal-virus induced cancers), tumor growth and metastasis may result. If the TSA is a weak antigen and not strong enough to sensitize lymphocytes to its presence (as many human cancers) or the immune system fails to respond, the same result will occur. When TSA is strong and recognizable, both cell-mediated and antibody-mediated immune responses are activated to destroy the abnormal cells. Research has shown that activated macro-

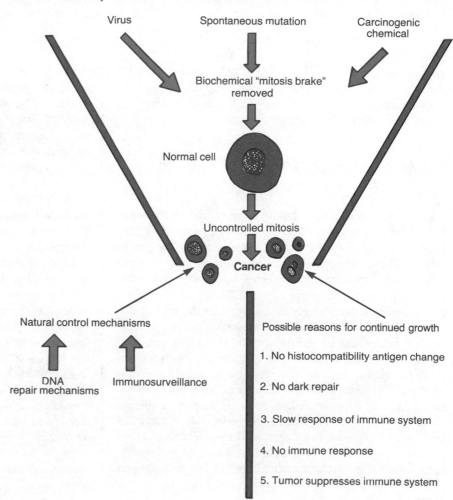

FIGURE 14.17
Cancer cells result from altered genes or abnormal gene activity. If the immune system of the DNA dark-repair mechanism is operating, these abnormal cells may be controlled. If unchecked, the cancer cells will grow to form a tumor, may metastasize, and can result in death of the patient.

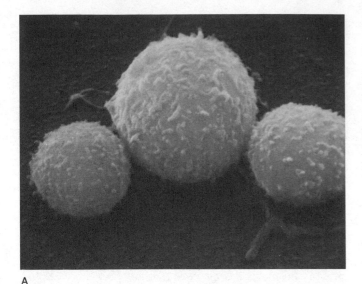

A

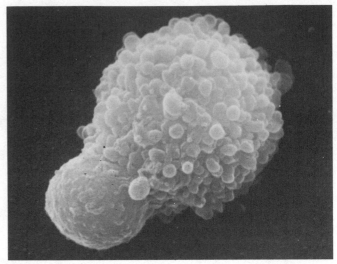

B

FIGURE 14.18
(A) In this scanning electron micrograph, lymphocytes attack a cancer cell. The cancer cell (large spheroid, center) has molecular "labels" (antigens) on its surface that enable one of the sensitized lymphocytes (smaller cells) to selectively attack and kill it. (B) Death of the cancer cell is indicated by the blebs, or deep folds, appearing on its surface membrane. The mechanism by which lymphocytes kill tumor cells is unclear but probably involves the release of a toxic factor that disrupts the cell membrane. (Courtesy of A. Liepins, Sloan-Kettering Institute for Cancer Research.)

phages have a special ability to recognize, attack, and kill malignant cells (Figure 14.18). In addition, antibodies produced against TSA are involved in tumor destruction by complement-dependent cytotoxic reactions.

IMMUNOTHERAPY AND CANCER CONTROL

The important role played by the immune system in cancer control has stimulated interest in possibly developing immunizations against cancer and immunotherapy for cancer victims. Until recently, the major approach to cancer control was to reduce or eliminate the tumor by surgically removing the mass, or destroying the abnormal tissue with radiation and chemotherapy. An immunotherapeutic alternative designed to supplement the body's natural immunosurveillance defenses may be more successful in the long run. One of the most recent methods being explored involves the use of BCG vaccine as an immunotherapeutic agent for cancer victims. BCG vaccine (Bacille Calmette-Guérin) is an attenuated form of *Mycobacterium bovis* developed as an immunization against tuberculosis. However, this vaccine has been found to stimulate a heightened responsiveness in the immune system (Refer to Chapter 20). Research with malignant melanoma (a pigmented tumor) in lab animals and select human patients has shown that the injection of the BCG vaccine directly into the mass can result in better nonspecific cell-mediated immunity (i.e., increased macrophage activity). The response is rapid and complete enough in certain situations to cause the complete destruction of the melanoma. Individuals with acute lymphoblastic leukemia, acute myelogenous leukemia, and acute myeloblastic leukemia who are given BCG as an immune system stimulator have responded well. Many have shown prolonged remission and increased survival compared to those only treated chemotherapeutically. However, no direct evidence at this time indicates that BCG was the only factor responsible for these reported successes. How the BCG produces this heightened immune response is not yet known, but several ideas have been proposed. One suggests that the cancer cells are killed accidently by macrophages (the "innocent bystander theory"); another proposes that the vaccine magnifies anticancer, cell-immune responses already taking place; while a third is based on the idea that BCG stimulates the replacement of stem cells lost due to chemotherapy. Four other agents currently being investigated for their immunotherapeutic value are the bacterium *Corynebacterium parvum* and the chemicals glucan, levamisole, and dinitrochlorobenzene (DNCB). Only continued research will reveal the exact mechanisms operating in these nonspecific methods that heighten the immune response to cancer.

Another option to cancer control centers around increasing the specific immunity of an individual. One approach is the use of "unblocking sera" that contain a large amount of tumor-specific antibody. It is thought that this antibody will enhance TSA and make it more "recognizable" to the immune system. To date, laboratory tests show little success, and some danger, with this method. Another option is the immunization of a patient with TSA from the tumor itself. This approach is based on the idea that the injection of concentrated antigen will stimulate the immune system before tissues are damaged too extensively. The weak antigenicity of human TSA has been a stumbling block in this area of therapy.

More recently another innovative method of cancer control has found initial success. This technique utilizes **monoclonal antibody** from **hybridoma cells** to provide the cancer patient with artifically acquired passive immunity against the cancer cells. Hybridomas are cells resulting from the controlled fusion of two cell types: B-lymphocytes previously sensitized to a specific antigen and myeloma cancer cells from a bone marrow tumor. These two cell types are used because of their unique properties. While the B-cells do not grow well as a continuous cell culture, they can be sensitized to produce a specific antibody. The myeloma cells grow well in culture but only produce nonspecific antibody. To obtain hybridoma cells which produce monoclonal antibody against cancers such as leukemias, normal mice or humans are immunized with cell-surface proteins unique to the cancer cells (i.e., histocompatibility antigens, TSA, etc.). The mouse's or human's spleen is then removed and the sensitized B-lymphocytes isolated. The hybridoma cells are produced when these B-cells are caused to fuse with myeloma cells. By carefully conducted selection techniques, the hybridoma cells that produce the desired antibody are identified, separated, and grown in pure culture as clones. These hybrid cells are the "seed cells" of a continuous cell culture that is committed to the production of a single specific antibody—the monoclonal antibody. While this immunotherapy method of cancer control is in its infancy, there has been at least one case in which this method effectively controlled one form of leukemia by providing artifically acquired passive immunity against cancer.

A final control that has some appeal is the "adoptive" approach. In this method, the immunodeficient cancer patient receives anticancer chemicals (e.g., immune RNA, lymphokines,) or whole cells (e.g., thymus or leukocytes) from a healthy donor to heighten immunosurveillance and tumor destruction. All of these methods are only in the research stage at this time, and many problems exist with each approach. Although the concept of immunotherapy for cancer control is well founded, the practical application of such methods has not as yet been proved. A great deal of work in the areas of immunology, virology, cytogenetics, and ser-

ology remains, before enough data will be available to more clearly direct researchers to the formulation and production of a clinically successful, immunological method to cancer control.

SUMMARY

Under normal circumstances, the immune system provides an efficient basic defense mechanism against specific agents recognized as foreign to the body. However, some individuals experience harmful antibody-antigen reactions known as allergies, or more technically, hypersensitivities. Type I hypersensitivities are called anaphylactic because they are characterized by the release of vasoactive amines (for example, histamines, SRS-A serotonin, kinins, and prostaglandins) from IgE-bound mast cells and basophiles. Symptoms may be localized or generalized (systemic) and include cell lysis, capillary damage, reddening, edema, burning sensations, and smooth muscle contraction. Type I hypersensitivities include bronchial asthma, anaphylactic shock, drug reactions, food allergies, and tissue transplant rejection reactions. Type I hypersensitivies may be determined by a skin sensitivity test, and control may be achieved by the use of drugs or immunotherapy known as desensitization.

Type II hypersensitivity is known as cytotoxic because these antibody-antigen reactions result in complement-dependent lysis of cells. Many cases of glomerulonephritis and drug allergies are due to Type II reactions. Type III reactions involve the formation of immune-complexes that become attached to basement membranes. They result in acute inflammation, edema, erythema, and cell lysis. Chronic inflammatory diseases, such as rheumatoid arthritis, systemic lupus, tertiary syphilis, and Crohn's disease, are Type III hypersensitivities. Serum sickness and hypersensitive pneumonitis are also Type III. Delayed cell-mediated Type IV reactions involve the reaction of T-cells, lymphokines, and macrophages. The most classical and familiar delayed reactions are the TB skin test response and contact dermatitis eczema.

Both Type IV and Type I hypersensitivities are involved in tissue rejection after transplantation surgery. Antigens responsible for this reaction are located on the surface of transplanted cells and may be isoantigens, for example, those associated with the ABO and Rh systems, or histocompatibility antigens. The Coombs test is a selective, sensitive test for the identification of many of these antigens. It is used as a diagnostic tool to prevent transfusion reactions and hemolytic diseases of the newborn. The identification of the histocompatibility antigens known as HLA are important in tissue cross-matching tests to determine the acceptability of donor tissue in transplantation operations. A high degree of histocompatibility usually means a greater degree of transplant success. The presence of these antigens on leukocytes can be used to indicate HLA gene-associated diseases such as rheumatoid arthritis, ankylosing spondylitis, myasthenia gravis, and others.

The immunosurveillance system responsible for rejection reactions also plays an important role in eliminating cancer cells. These cells may have formed in the body as a result of virus infection or the mutagenic action of carcinogenic chemicals. Neoplasia results in the formation of cancer cells with new, specific cell-surface antigens known as carcinoembryonic antigens or tumor-specific antigens. The presence of these antigens is not only indicative of cancer cells, but they also stimulate the immune system to eliminate the abnormal cells as unwanted foreigners. A growing, metastisizing cancer may result from no dark repair of mutations, the lack of TSA on the cell surface, or a faulty immune system. Laboratory attempts at cancer control through immunotherapy have had only limited success, but it is hoped further work in the area be more fruitful.

WITH A LITTLE THOUGHT

Periodontal disease is a chronic inflammatory disease caused by the accumulation of oral bacteria in a dental plaque that forms around the gingival margin. Specific microbial products have not been identified as etiologic (causitive) agents nor has the specific nature of the periodontal disease process been fully explained. However, there is considerable evidence that these microbial products act as antigens and initiate periodontal inflammation by activating the immune responses of the host.

This extract was taken from D. Rowe. "Immune Reactions and Periodontal Disease." *Dental Hygiene,* vol. 53, April, 1979. Speculate what the author might have included in the remaining part of the text under the headings: Basic Immunological Concepts; Tissue-Destructive Mechanisms in Periodontal Disease; Anaphylaxis Hypersensitivity Reaction; Cytotoxic Reaction; Immune-Complex Reaction; Cell-Mediated Hypersensitivity Reaction; and Conclusion.

STUDY QUESTIONS

1 What is immunologic hypersensitivity? List the four types and give examples of each.

2 Explain the difference between immediate anaphylactic hypersensitivity and delayed cell-mediated hypersensitivity.

3 What is anaphylaxis? How might this response be induced in a sensitized individual? How might the problem be reduced?

4 Some commercially available aerosol inhalers contain adreneline (epinephrine). Why is this form of medication of value to a person with upper respiratory hypersensitivity such as asthma?

5 What are the differences between Type II and Type III hypersensitivities? What function do null, or killer, T-cells play in the process?

6 Explain the ABO blood system and how transfusion reactions occur. How are these rejections eliminated?

7 What is the difference between contact dermatitis and skin reactions due to primary irritants?

8 What types of hypersensitivity are involved in transplant rejections?

9 What are histocompatibility antigens and how are they related to recognition of "self," recognition of transplants, tumor immunology, and tissue cross-matching?

10 Describe how immunosurveillance is thought to control cancer. What evidence supports this theory? What evidence disputes this theory?

SUGGESTED READINGS

ALLISON, A. C., and J. FERLUGA "How Lymphocytes Kill Tumor Cells." *New England Journal of Medicine,* 295:165–67, 1976.

Asthma and Allergy Foundation of America, 19 W. 44th Street, New York, 10036. Several pamplets on allergies.

BROWN-SKEERS, V. *"How the Nurse Practitioner Manages the Rheumatoid Arthritis Patient." Nursing 79,* June, 1979.

CRAVEN, R. "I Died for a Few Seconds." *American Journal of Nursing,* April, 1972.

CUNNINGHAM, B. A. "The Structure and Function of Histocompatibility Antigens." *Scientific American,* Oct., 1977.

GELL, P. G. H., R. R. A. COOMBS, and P. J. LACHMANN, editors. *Clinical Aspects of Immunology,* 3rd ed. F. A. Davis, Philadelphia, 1975.

HUNTLEY, C. "Atopic Dermatitis and Contact Dermatitis in Children." *American Family Physician,* Aug., 1977.

KOFFLER, D. "Systemic Lupus Erythmatosus." *Scientific American,* July, 1980.

LISTER, J. "Nursing Intervention in Anaphylactic Shock." *American Journal of Nursing,* April, 1972.

OLD, L. J. "Cancer Immunology" *Scientific American,* May, 1977.

PATTERSON, R., ed. *Allergic Diseases: Diagnosis and Management* J. B. Lippincott, Philadelphia, 1980.

ROITT, I. *Essential Immunology.* C. V. Mosby, St. Louis, 1981.

ROSE, N. R. "Autoimmune Diseases" *Scientific American,* Feb., 1981.

ROWE, D. J. "Immune Reactions and Periodontal Disease." *Dental Hygiene,* vol. 53, April, 1979.

WHITE, J. "Systemic Lupus Erythematosus." *Nursing 78,* Sept., 1978.

KEY TERMS

immunologic hypersensitivity

allergen

vasoactive amines

anaphylaxis (an"ah-fi-lak'sis)

cytotoxic reaction

immunosurveillance

neoplasm

immune complex reaction

Guillian-Barré syndrome

delayed hypersensitivity

eczema (ek'ze-mah)

histocompatibility antigens

carcinoembryonic antigen

tumor-specific antigen

Epidemiology

EPIDEMIOLOGY: SCOPE AND TERMINOLOGY

RESERVOIRS AND EPIDEMICS

CARRIERS AND VECTORS

AN ORGANIZATIONAL APPROACH

CLASSIFICATION OF DISEASES

HOW DISEASES SPREAD
Airborne Transmission
Waterborne Transmission
Foodborne Transmission
Zoonotic Transmission

TRANSMISSION ROUTES

HERD IMMUNITY AND DISEASE CYCLES

THE CONTROL OF COMMUNICABLE DISEASES
Active Immunization
Vector Control
Patient Isolation
Quarantine
Special Precautions

NOSOCOMIAL INFECTION

Learning Objectives

☐ Be familiar with the scope of epidemiology and know basic terminology including "morbidity," "mortality," and "prevalence."

☐ Know the major reservoirs of infectious microbes and the difference among the terms "endemic," "epidemic," and "pandemic."

☐ Recognize the importance of clinical cases, subclinical cases, carriers, and vectors as sources of infectious agents.

☐ Be familiar with the local, state, national, and international public health agencies and the roles each plays in controlling infectious disease.

☐ Name the five classes of reportable diseases and cite examples of each.

☐ Know the four ways infectious diseases can spread through a population.

☐ Define and give examples of portals of entry and portals of exit.

☐ Explain the concept of herd immunity in terms of understanding disease cycles and the need for "public" health.

☐ Name and explain the major methods used to control a communicable disease.

☐ Understand the source, nature, cause, and importance of nosocomial infections.

The development and efficient functioning of the immune system contributes to an individual's resistance to infection and the maintenance of good health. Good health results when there is a balance between host and parasite, and both cope with a changing environment without disturbing the relationship (Figure 15.1). Should any of these elements (host, parasite, or environment) change, the delicate balance will shift and could be fatal to host, parasite, or both. Even though both host and parasite have a natural tendency to maintain this balance, there are instances when changes disrupt the equilibrium.

One of the most important factors responsible for preventing poor health or possibly host death is the develop-

15

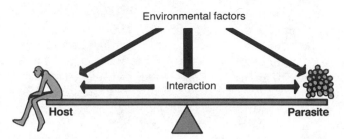

Environmental factors

Interaction

Host   **Parasite**

FIGURE 15.1

The balance necessary for good health depends on the interactions between host and parasite. In many ways, both are affected by environmental factors. Poor health, and possibly death, of the host may result from parasite activities, poor host-defense mechanisms, or an environmental imbalance.

ment of an individual's immune state. As an increasing number of the human host population become immune to a particular disease, a level of population resistance develops that contributes to the health of the individual. This results from a reduced number of unhealthy, susceptible persons in the population who would otherwise act as sources of further infection. Therefore, maintaining good health goes beyond establishing a balanced host-parasite relationship for an individual and must include measures resulting in good health for the general public. To maintain a high level of public health, a working knowledge is required in such areas as (1) human and parasite anatomy and physiology, (2) the route by which a parasite enters and leaves the body, (3) defensive and offensive mechanisms available to both host and parasite, (4) the life cycle of the parasite, and (5) the relationship between initial symptoms and the communicable period of infection. Many local, state, national, and international organizations are involved in this special area of study known as epidemiology, and each attempts to raise the general public health to a higher level. Their investigations have resulted in the successful control of many diseases and the identification of special epidemiological problems such as those associated with hospital-acquired infections.

EPIDEMIOLOGY: SCOPE AND TERMINOLOGY

Epidemiology is the study of environmental factors and others that affect host-parasite relationships and determine the frequency and spread of disease through a host population. The term "epidemiology" is derived from the Greek *epidemios*, which means *prevalence*, or *frequency*. The frequency of infection in a population is expressed in terms of both morbidity and prevalence. The **morbidity,** or **incidence rate,** of a disease describes the number of individ-

uals infected with a particular microbe in a given unit of time. This is usually expressed as the number of cases per 1000 or 100,000 individuals each year. For example, if there were 754 cases of influenza per 100,000 members of the population the morbidity would be expressed 754/100,000. The **prevalence rate** refers to the total number infected at any one time regardless of when the disease began and is more specific than the morbidity rate. For example, if 21 women were found to have *Streptococcus pyogenes* throat infections in a group of 100 during one week, the prevalence rate for that week was 21/100. Both morbidity and prevalence express a particular disease's frequency in a population; however, neither describes how many of the infected individuals may have died as a result of the infections. The **mortality rate** is the number of deaths occurring in a population during a stated period of time.

The determination of morbidity, mortality, and prevalence rates helps public health agencies direct health care personnel to control the spread of infectious diseases. A sudden increase in a disease's morbidity rate may forecast a need for increased medical and public health action to prevent a future increase in mortality. Even though a particular infectious disease may rarely be fatal (i.e., has a low mortality rate), maintaining low morbidity rates reduces human suffering and financial loss due to employees' sick leave.

RESERVOIRS AND EPIDEMICS

Since many pathogenic microbes are opportunistic and do not rely exclusively on the human body for residence, they will not likely ever be extinguished. The goal of public health offices and epidemiologists is to maintain low morbidity rates for these diseases. *Clostridium tetani, C. perfringens, Staphylococcus aureus,* and many other microbes survive very well outside the human body. Infection by these microbes may be controlled but never eliminated since there will always be a large reservoir of these infectious microbes in the soil. However, microbes that rely on the human body as their **reservoir of infection** may be eliminated through public health measures aimed at destroying these pathogens and preventing their spread and multiplication. Viral infections such as smallpox and measles, and the bacterium *Neisseria gonorrhoeae* are examples of pathogens that can be eliminated, if appropriate control measures are available and put into use.

A disease which is constantly in a given geographical area at a low morbidity rate acceptable to public health authorities is called an **endemic disease.** If the disease's morbidity rate rises above a normal, expected level and results in high mortality or great public harm, the disease is reclassified as an **epidemic.** The morbidity rate at which a disease is classified as epidemic and not endemic depends on the nature of the microbe (virulence), the density of the host population, the level of immunity to the disease (susceptibility), and the place and time of the increase. For example, an outbreak of three or four cases of spinal meningitis caused by *Neisseria meningitidis* would constitute an epidemic in most large cities in the United States; however, ten times this number might occur in an underdeveloped African or east Asian country before health care officials would classify the disease as an epidemic and mobilize public health care personnel. The term **pandemic** may be used to describe an epidemic that has grown to involve an exceptionally large geographical area or a situation in which several major epidemics have developed at the same time in many different parts of the world. The highly effective medical and public health care facilities of developed countries, such as the United States, allow maintenance of very low morbidity levels, while the underdeveloped nations must accept higher morbidity rates as they strive to obtain better facilities and trained personnel.

---

1. What is the difference between morbidity rate and prevalence rate?
2. Why will diseases like tetanus and gas gangrene never be eliminated but only controlled?
3. What factors determine whether a disease is classified as endemic or epidemic?

An epidemic may arise from a common source, such as contaminated food or water, and spread through the population as individuals come in contact with the contaminated material. This is known as a **common-source epidemic,** which appears suddenly in the population, reaches a peak very quickly, and subsides equally as rapidly, once the source of the microbe has been identified. Another form of epidemic known as a **propagated epidemic** results from person-to-person transmission of the infectious agent. Because each individual who becomes ill must acquire the microbe from someone else, the number of cases rises slowly. As the number of susceptible individuals in the population declines due to the development of post-infection immunity and a decrease in the frequency of contact with infected individuals, the epidemic slowly declines to the previous endemic level (Figure 15.2). Person-to-person, or propagated, epidemics usually last longer than the common-source epidemic.

CARRIERS AND VECTORS

The occurrence of disease in a population is, in some instances, difficult to determine since not all individuals demonstrate clinical symptoms. Patients that show the typical symptoms of a disease are known as **clinical cases.** They are the most likely to seek medical attention, and their cases are primarily used in determining morbidity rates. Because clinical cases are easily identified, they can be quickly treated and removed from the reservoir of infectious individuals. This prompt action helps prevent epidemics. Patients displaying only minor symptoms of the infection are referred to as **subclinical cases,** and they tend not to seek medical help as readily as clinical cases. The presence of subclinical cases in the population encourages person-to-person transmission and maintenance of endemic levels of disease. These individuals also represent a potential source of propagation epidemics. Another type of individual that represents an even more important threat to the public health is known as a carrier. **Carriers** are those who have a particular pathogenic microbe but fail to show clinical or subclinical signs of infection. Both healthy and chronic carriers can transmit pathogenic microbes throughout their entire lives. For example, people who have acquired *Salmonella typhi* and have the pathogen established in their gallbladders may never experience the symptoms of typhoid fever, but they are capable of transmitting the pathogen through fecal material. Chronic carrier states also exist for diphtheria, certain streptococcal infections, and enteropathogenic *E. coli*. Casual carriers are those who only serve as temporary transmitters of the pathogen, since the microbe never establishes a damaging host-parasite relationship. The pathogens are only

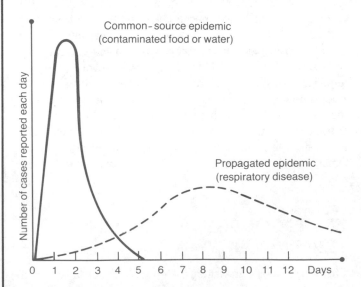

FIGURE 15.2

These curves illustrate the differences between common-source and propagated epidemics. The common-source epidemic decreases rapidly due to identification and avoidance of the contaminated material. The slow increase in propagated epidemics is controlled by the low rate of transmission among people having the disease-causing agent.

transient flora in casual carriers. The chronic and casual carrier states are the most difficult to identify and have major epidemiological significance, since these individuals serve to maintain a pathogen in the population despite control measures. A third type of carrier, known as the incubatory, or convalescent, carrier, shows no symptoms during the communicable period since these carriers are either in the incubation period of the disease, or they have overcome the disease but are still able to serve as a source of infection. Since these individuals either have symptoms or will develop them, they may be more easily identified and eliminated from the reservoir of infection.

> **1** What kind of epidemic appears suddenly, reaches a peak quickly, and subsides rapidly?
> **2** Which individuals encourage the maintenance of a disease in a population at endemic levels?
> **3** What name is given to an individual who shows no symptoms of infection during the communicable period?

Another type of reservoir is infected animals that may serve as vectors, or transmitters, of disease. **Vectors** are usually insects such as lice, ticks, fleas, cockroaches, and flies that carry pathogens from one human to another (Table 15.1). In some cases, the pathogen has established a separate reproductive phase in the insect vector. For example, the protozoan *Plasmodium vivax* is carried from person to person by the *Anopheles* mosquito. While in the mosquito, the protozoan increases in number by reproducing a form known as the sporozoite, which becomes concentrated in the salivary glands. Sporozoites are injected into a human host along with anticoagulant saliva while the insect is withdrawing blood. Once inside the human, *P. vivax* begins its life cycle that results in malaria. The identification and elimination of this reservoir of infection is sometimes difficult, and may take years of investigation and effort. For example, the effort to eliminate Rocky Mountain spotted fever was started in the 1920s and continues today. This disease is probably the most important rickettsial infection in the United States today, and the number of cases reported each year is growing. Originally localized in the Wyoming, Oregon, Idaho, and Washington areas, Rocky Mountain spotted fever has spread throughout the country and recent cases have been identified as far east as Maryland, Maine, West Virginia, and North Carolina. The wood tick, *Dermacentor andersoni*, is the vector of this pathogen (*Rickettsia rickettsii*) in the west; while another species, *D. variabilis* (the eastern dog tick), is the vector in the eastern United States (Figure 15.3). Transmission of the rickettsia occurs when the tick "burrows" into the skin to feed on a

blood meal. This usually happens wherever clothes fit tightly, for example, at the belt line, around collars, and around boot tops. If the insect vector is not removed before it starts to feed (between 4 to 6 hours), the pathogen will be transmitted to the human host. Once inside, the pathogen causes the sudden onset of a moderate fever which may last 2–3 weeks. The patient will experience headache, chills, conjunctival infection, and a rash on the arms and legs that will eventually move to the palms and soles. As the rash spreads over the entire body, small, pinpointed, purplish spots, known as **petechiae,** form as a result of submucosal hemorrhaging. The fatality rate is about 20 percent in untreated cases, but treatment with tetracycline or chloramphenicol makes fatalities rare. Research and surveillance of Rocky Mountain spotted fever and many other infectious diseases occur at local, state, national, and international levels. By coordinating efforts, public health organizations have been able to control many diseases such as typhoid, cholera, and ma-

TABLE 15.1
Vectors of disease.

Vector	Diseases
Dogs and cats	Rabies virus
	Leptospirosis
	Salmonellosis
	Ringworm
	Anthrax
	Tularemia
	Tuberculosis
	Toxoplasmosis
Birds	Psittacosis
	Encephalitis
	Salmonellosis
	Leptospirosis
	Numerous fungal infections
Rodents (mice, rats, etc.)	Salmonellosis
	Brucellosis
	Tularemia
	Leptospirosis
	Ringworm
Insects (fleas, ticks, flies, mosquitos)	Rocky Mountain spotted fever
	Leishmaniasis
	Chagas disease
	Epidemic typhus
	Trench fever
	Endemic typhus
	Scrub typhus
	Rickettsial pox
	Plague
	Tularemia
	African sleeping sickness
	Yellow fever
	Malaria
	Dengue fever

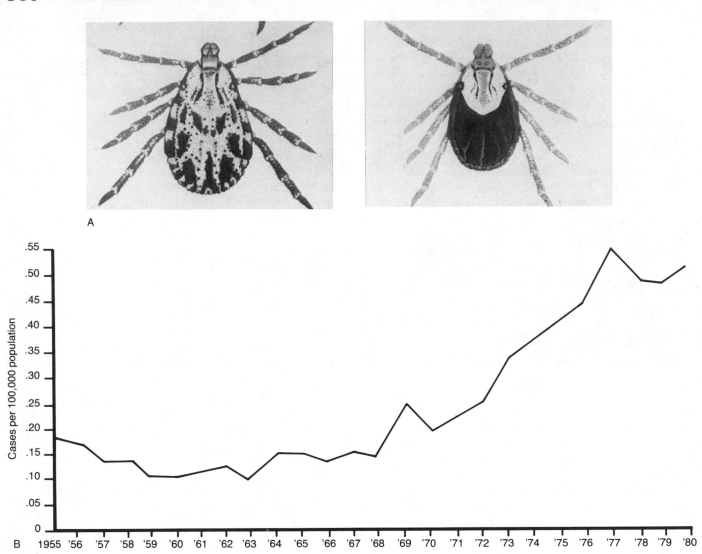

FIGURE 15.3
The vector of Rocky Mountain spotted fever is the tick. (A) The two species responsible are
Dermacentor andersoni (left) and *D. variabilis* (right). (B) Reported cases of Rocky Mountain
spotted fever in the United States. (Courtesy of the Center for Disease Control, Atlanta.)

laria, which have in the past been responsible for both high
morbidity and mortality rates.

AN ORGANIZATIONAL APPROACH

Since the 1700s, the United States and many other
countries have publicly acknowledged the importance of
maintaining good public health. Many organizations directed
toward achieving this goal have been established. At the
local level, many city and county governments have health
departments responsible for implementing and maintaining
local public health regulations. As problems arise, these
agencies provide counsel and services to aid in resolving
problems in county and local hospitals. County health departments maintain surveillance records on infectious diseases and illnesses that are of a public concern, and provide
immunizations and referrals to those seeking help. When
problems extend beyond the geographical boundaries or
scope of these local departments, state health departments
are contacted. These agencies not only advise and make
recommendations for the establishment of state health laws
and regulations, but their size enables them to provide services not available at the local level. Many state health departments have research and diagnostic laboratories designed to accurately identify pathogens and toxins, such as
rabies virus, enteropathogenic *Escherichia coli,* drug-resistant staphylococcal strains, and *Neisseria gonorrhoeae.*

The U.S. Department of Health, Education, and Welfare was created at the national level in 1953 and includes as a major component the **United States Public Health Service (USPHS).** This cabinet-level organization dates back to the late 1700s, when it was originally known as the Marine Hospital for American Navy Men. The USPHS has several subgroups which make up a large, complex network of public health teams throughout the country. One of the most important is the **Center for Disease Control (CDC)** centered in Atlanta, Georgia, established in 1973. As noted in the *U.S. Government Manual:*

> The CDC is the Federal agency charged with protecting the public health of the Nation by providing leadership and direction in the prevention and control of diseases and other preventable conditions. It is comprised of eight major operating components: National Institute for Occupational Safety and Health, Bureau of Epidemiology, Bureau of Health Education, Bureau of Laboratories, Bureau of Smallpox Eradication, Bureau of State Services, Bureau of Training, and Bureau of Tropical Diseases.
>
> The Center administers national programs for the prevention and control of communicable and vector-borne diseases and other preventable conditions, including the control of childhood lead-based paint poisoning and urban rat control. The Center directs and enforces foreign quarantine activities and regulations; provides consultation and assistance in upgrading the performance of clinical laboratories, and evaluates and licenses clinical laboratories engaged in interstate commerce; and administers a nationwide program of research, information, and education in the field of smoking and health.
>
> To assure safe and healthful working conditions for all working people, occupational safety and health standards are developed, and research and other activities are carried out, through the Center's National Institute for Occupational Safety and Health.
>
> The Center also provides consultation to other nations in the control of preventable diseases, and participates with national and international agencies in the eradication or control of communicable diseases and other preventable conditions.

At the international level, the **World Health Organization (WHO)** serves to distribute technical data and health information, monitors potential epidemic situations, and aids in the control of contagious diseases. One of the most successful efforts of WHO has been the smallpox eradication program. Because of a highly effective worldwide immunization and virus control program, smallpox is now the first infectious disease eliminated by the efforts of mankind (Figure 15.4). WHO, in conjunction with other international organizations such as the Pan American Health Organization and the American Public Health Association, has also embarked on a measles eradication program. The World Health Organization also provides nations with such services as assistance in developing health standards, consultation, collection and analysis of health data, collection or production of

standard drug materials, and training programs for medical personnel and research. Special research centers are currently maintained for the study of gonococcal, meningococcal, staphylococcal, and streptococcal infections; whooping cough, leprosy, plague, and many other diseases.

---

1 What animal is the vector of Rocky Mountain spotted fever?
2 What role does the CDC play in epidemiology?
3 What efforts are needed to eliminate a disease like smallpox?

CLASSIFICATION OF DISEASES

The effective control of any infectious disease depends on the accurate identification and reporting of the disease to health care authorities. Five classes of reportable diseases have been established to provide consistency in the reporting of communicable diseases:

Class 1 *Case report universally required by international health regulations.* This class is limited to the diseases subject to the International Health Regulations (quarantinable diseases): cholera, plague, smallpox, and yellow fever; and to the diseases under surveillance by WHO: louse-borne typhus fever and relapsing fever, paralytic poliomyelitis, influenza, and malaria.

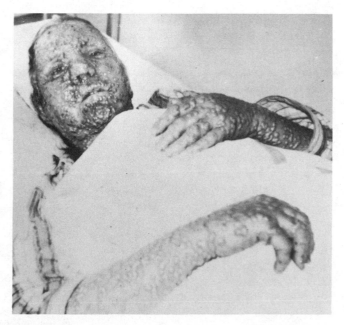

FIGURE 15.4
A rare sight—smallpox victim. (Courtesy of the Center for Disease Control, Atlanta.)

Class 2   *Case report regularly required wherever the disease occurs.* Two subclasses are recognized, based on the relative urgency for investigating contacts and sources of infection, or for starting control measures.
>   A. Case report to local health authority by telephone, telegraph, or other rapid means. (Examples: typhoid fever and diphtheria.)
>   B. Case report by most practicable means. (Examples: brucellosis and leprosy.)

Class 3   *Selectively reportable in recognized endemic areas.* In many states and countries, diseases of this class are not reportable. Reporting may be prescribed in particular regions, states, or countries by reason of undue frequency or severity. Three subclasses are recognized:
>   A. Case report by telephone, telegraph, or other rapid means in specified areas where the disease ranks in importance with Class 2A; not reportable in many countries. (Examples: tularemia and scrub typhus.)
>   B. Case report by most practicable means. (Examples: bartonellosis and coccidioidomycosis.)
>   C. Collective report weekly by mail to local health authority. (Examples: clonorchiasis and sandfly fever.)

Class 4   *Obligatory report of epidemics—no case report required.* Prompt report of outbreaks of particular public health importance by telephone, telegraph, or other rapid means; forwarded to next superior jurisdiction by telephone or telegraph. Pertinent data includes number of cases, within what time, approximate population involved, and apparent mode of spread. (Examples: food poisoning and infectious keratoconjunctivitis.)

Class 5   *Official report not ordinarily justifiable.* Diseases of this class are of two general kinds: those typically sporadic and uncommon, often not directly transmissible among humans (e.g., chromoblastomycosis); or of such epidemiological nature as to offer no practical measures for control (For example, the common cold.)*

The identification of the **etiologic agent** (cause of the disease) and its prevalence is only the initial step of communicable disease control. A clear understanding of how the agent moves through the environment and interacts with the host population is also essential.

---

*Modified from A. S. Benenson, ed., *Control of Communicable Diseases in Man,* 12th ed. An official report of the American Public Health Association, 1975.)

HOW DISEASES SPREAD

Microbes can be transmitted through a population either by direct or indirect means. **Direct transmission** occurs when the microbes move from person to person either by direct body contact such as in sexual activity or handshaking, or by other close personal association, for example, sneezing or coughing. **Indirect transmission** occurs when the infectious microbes are carried from one person to another on some other object, such as food, water, dust, soil, animals, or inanimate objects.

AIRBORNE TRANSMISSION

Microbes do not grow or reproduce in the air to any great extent but are found there primarily as transients. They come from the soil, water, plants, and animals as wind and rain move over these objects. Natural antimicrobial factors, such as ultraviolet radiation, photooxidation, temperature extremes, and lack of water all reduce the likelihood of microbes' survival and transmission to new hosts or other suitable growth environments. The microbial density, or load, of air is variable and depends on many factors. The dry, warm wind of a summer day fills the air with a greater number and variety of microbes than is found after a rain shower or during the winter. Air that is recirculated by a forced air furnace or an air conditioner contains a large number of suspended particles and microbes. The presence of these microbes becomes of particular importance during winter months as harsh weather conditions force people inside. The longer periods of close contact inside buildings increases the chance that airborne pathogens will be able to move quickly from one susceptible host to another.

Microbes that leave the upper respiratory tract vary in size. Those that are large usually fall quickly and are more likely to be transmitted to another host in dust rather than through the air. The smaller particles, however, may remain suspended in the air for many hours and serve as sources of infection. Sneezing, coughing, or talking results in the explosive release of microbes into the air (Figure 15.5). Many of the microbes are in small clusters surrounded by mucus that quickly dries, forming particles known as **droplet nuclei.** The layer of mucous helps protect the microbes from drying and enables them to remain infective for a longer period. In general, the Gram-positive bacteria are naturally more resistant to drying than the Gram-negative bacteria because of their thicker, more protective cell wall. Both streptococci and staphylococci can survive two or three months in the dry dust of a house or hospital. *Mycobacterium* is also very resistant to drying but less likely to survive in the air, since this pathogen is highly sensitive to photooxidation. This discovery led to the addition of sun porches and the removal of window shades in many tuberculosis sanitoriums.

1 Of what value is the classification of diseases by WHO?
2 What does the term "etiological agent" mean?
3 Small clusters of microbes surrounded by dried mucous have what name?

The suspension of microbe-containing particles in the air, **aerosol formation,** plays a key role in the transmission of many diseases (Table 15.2). Aerosols may be formed by sneezing, talking, coughing, dry mopping floors, dusting, making beds, or simply removing clothing. In order to decrease aerosol formation and airborne transmission of pathogens in the hospital, several important procedures are followed. Beds are made by carefully unfolding linens and not shaking the sheets; floors and counter tops are wet mopped and dusted; and special protective hair coverings and masks are worn in areas where patients are most sus-

ceptible to airborne pathogens; (for example, surgery, nursery, burn units). Since the complete elimination of aerosols is impossible, other procedures and equipment have been developed to reduce the microbial load in the air. Ultraviolet lights are installed in cabinets used to transfer pathogens and in the duct work of heating systems to destroy airborne pathogens. Special hospital rooms used for highly susceptible patients, such as burn victims or patients on immunosuppressive drugs, may have a positive air pressure so that airborne microbes in the room will move out the door when it is opened and prevent microbes from entering the room from the hall. Individuals with highly contagious diseases may be isolated in a room under negative pressure. The air in these special rooms is removed through the ducts very rapidly, causing the air from the hall to move through the door into the room. This restricts the movement of airborne pathogens from the patient into the rest of the hospital. In some cases, electrostatic precipitators may be installed to electrically remove and destroy small particles suspended in

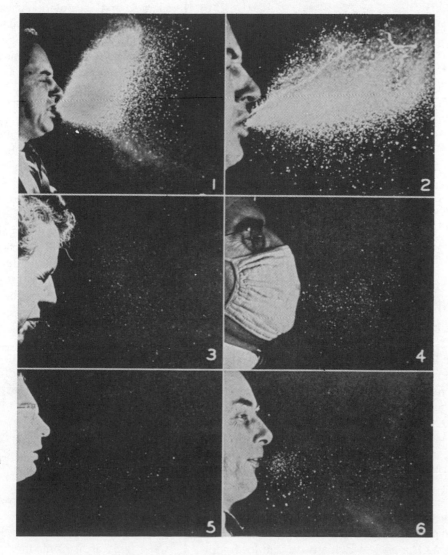

FIGURE 15.5
These high-speed photographs show the release mucus contaminated by microbes.
(1) Violent sneeze in a normal subject; (2) Head cold sneeze; (3) Stifled sneeze; (4) Sneeze through a dense face mask; (5) Cough; and (6) Enunciation of the letter "f." (From M. W. Jennison. "Aerobiology." *American Association for the Advancement of Science,* no. 17, 1942.)

TABLE 15.2
Human airborne infections.

Organism	Disease
Bacteria	
Corynebacterium diphtheriae	Diphtheria
Streptococci	Pneumonia, scarlet fever, septic throat
Mycobacterium tuberculosis	Tuberculosis
Neisseriae spp.	Meningitis
Bordetella pertussis	Whooping cough
Mycoplasma pneumoniae	Pneumonia
Chlamydia spp.	Psittacosis, legionnaire's disease
Viruses	
Smallpox	Smallpox
Chickenpox	Chickenpox
Measles	Hard (rubeola) and German (rubella) measles
Mumps	Mumps
Influenza	Flu
Poliomyelitis	Polio
Fungi	
Actinomyces spp.	Lung infection
Candida spp.	Infections of any body tissue
Blastomyces spp.	Resembles tuberculosis
Histoplasma capsulatum	Histoplasmosis; lung infection resembles TB or cancer
Coccidioides spp.	Coccidiodomycosis; also known as San Joaquin fever lung infection

the air. Precipitators were once very large, expensive units and not affordable to most households; however, technological advances have reduced their size and cost, making them more readily available.

WATERBORNE TRANSMISSION

Many microbes incapable of surviving airborne transmission can, nevertheless, be successfully transmitted from one person to another through water. This environment provides greater protection for microbes, enabling them to survive longer and be moved great distances. Although many waterborne pathogens seldom grow or reproduce in this environment, the presence of dissolved nutrients helps protect them and enhance their survival. For example, water in a shallow, warm, slow-moving lake or river that is heavily contaminated with sediments and organic material is a better environment, and more likely to harbor pathogens, than the deep, clear, cold water of a fast-moving mountain stream. The churning action of a rushing mountain stream mixes atmospheric oxygen into the water, reducing the microbial load by oxidation. The ancient Romans realized the impor-

tance of this action and cascaded water from the mountains into the city in open aquaducts. Because large amounts of water are used by all people for food preparation, drinking, canning, industrial processes, and recreation, a small number of waterborne pathogens may increase in concentration to become a public health hazard. The source of most waterborne pathogens is usually human or animal excrement that has accidently entered the water from damaged sewer pipes, septic field overflow, or sewage plants that have released untreated or poorly treated wastes. Once the source of the pathogens has been identified and eliminated, the concentration of waterborne pathogens quickly falls. The sudden occurrence and loss of pathogens results in common-source or minor sporadic outbreaks of illness. An **outbreak** is defined as an incident in which either two or more persons become ill from the same type of pathogen after water consumption, or evidence implicates water as the source of illness. Table 15.3 lists the major types of waterborne outbreaks by etiological agent and type of water involved in the transmission process. Because most of the illnesses are sporadic and accurate identification of the etiological agent may be impossible, the disease may have to be identified according to its symptoms; for example, acute gastrointestinal illness. A complete identification of the pathogen and its source can only be made if symptomatic cases are quickly identified, and specimens are collected and analyzed (Box 25). In many cases, however, the etiological agent makes only a fleeting appearance, and there is no opportunity to make a complete and accurate identification. Such was the case in 1979 when an outbreak of gastrointestinal illness involved at least 239 persons who had visited a county park in Michigan. The illness was suspected to be caused by a virus and transmitted by water to those swimming in the lake. Public health authorities were notified that several people had become ill two days after visiting the park, and attempts to identify other victims by radio, television, and newspaper were initiated. By the time symptomatic cases were identified and stool samples were collected, the etiological agent was unable to be accurately identified. No further cases were reported after the initial outbreak.

> 1 Aerosols play a key role in disease transmission. How might they be produced?
> 2 What factors aid in reducing airborne disease transmission?
> 3 What type of factors help protect pathogens that are waterborne?

The control of waterborne diseases is primarily accomplished by water and wastewater treatment processes. The proper installation and operation of sewage treatment and water purification plants throughout the country has signifi-

TABLE 15.3
Waterborne disease outbreaks by etiology and type of water system.

	Municipal*		Semipublic		Individual		Total	
	Outbreaks	Cases	Outbreaks	Cases	Outbreaks	Cases	Outbreaks	Cases
Acute gastrointestinal illness	5	518	13	1396	2	24	20	1938
Chemical poisoning	4	612	1	11	1	10	6	633
Giardiasis	2	950	2	62	0	0	4	1012
Salmonellosis	1	206	1	7	0	0	2	213
Hepatitis	0	0	1	47	0	0	1	47
Shigellosis	0	0	1	17	0	0	1	17
Total	12	2286	19	1540	3	34	34	3860

*****NOTE: Municipal systems** are defined as public or investor-owned water supplies that serve large or small communities, subdivisions, and trailer parks of at least 15 service connections or 25 year-round residents. **Semipublic water systems** are those in institutions, industries, camps, parks, hotels, service stations, etc., which have their own water system available for use by the general public. **Individual water systems,** generally wells and springs, are those used by single or several residences, or by persons traveling outside populated areas. (From *Foodborne & Waterborne Disease Surveillance,* Annual Summary, 1977. Issued August, 1979. Center for Disease Control, Atlanta.)

cantly reduced the incidence of waterborne infections. Diseases such as typhoid fever and cholera, which were prominent killers just 100 years ago, have been brought under control. The success has reached such high levels that epidemiologists can now concentrate on the identification of other, rarer waterborne pathogens and improve the accuracy of reporting procedures. The graph in Figure 15.6 illustrates a sharp downturn in the average annual number of waterborn diseases from 1938 to 1955, followed by a slow increase. The increase is due to more accurate identification and reporting practices, and not to a recurrence of major pathogens.

FOODBORNE TRANSMISSION

A general rule is that the food we find nutritious is probably nutritious for microbes as well. Microbes contaminating food may grow there and release toxins. Foodborne diseases developing from the ingestion of live microbes which continue to grow in the host are called **food infections.** Illnesses resulting from the ingestion of toxic by-products of microbes that are no longer alive are known as **food poisonings.** The most common food infection is caused by the ingestion of a variety of *Salmonella* species. The disease symptoms are known as gastroenteritis. The most common food poisoning is probably caused by the ingestion of a toxin of *Staphylyococcus aureus* known as **enterotoxin.** The exact number of individuals that suffer from this form of poisoning is not known because this is not a reportable disease (Table 15.4). In the past, this illness was known as **ptomaine food poisoning,** since those suffering from the illness were shown to have an increase in ptomaine molecules in their fecal material. However, ptomaines are not responsible for the sometimes abrupt and violent onset of cramps, nausea, vomiting, and diarrhea. More accurate research has identi-

fied the cause as the heat-stable enterotoxin released by *Staphylococcus aureus* growing in such high carbohydrate foods as potato salad, cream-filled pastries, gravy, and dressings. The most likely source of these pathogens is from persons with clinical infections (for example, infected fingers, upper respiratory infections) and others in the carrier state.

TABLE 15.4
Confirmed foodborne disease outbreaks and cases for a recent year.

Type	Outbreaks	Cases	
Bacterial			
Arizona hinshawii	1	13	
Bacillus cereus	0	0	
Clostridium botulinum	20	75	
C. perfringens	6	568	
Salmonella spp.	41	1706	
Shigella spp.	5	67	
Staphylococcus spp.	25	905	
Vibrio cholerae (not 01)	1	2	
V. parahaemolyticus	2	118	
Suspect group D *Streptococcus*	0	0	
Yersinia enterocolitica	0	0	
	101	3454	**Total**
Viral			
Hepatitis A	4	72	
Chemical			
Heavy metals	8	326	
Ciguatoxin	3	22	
Scombrotoxin	13	71	
Monosodium glutamate	2	11	
Mushroom poison	5	14	
Miscellaneous	6	11	
	37	455	**Total**

AN outbreak investigation is intended to gather information about the hosts and their environment, and the disease-causing agent and its environment. Questions about the hosts and their environment begin with *what, when, where, who,* and *why;* questions about the agent and its environment begin with *what, where, how,* and *why.* Table 15.5 lists many factors that have to be identified in a successful foodborne disease investigation.

An investigation starts with a case report or an examination of surveillance data. Investigators then interview patients and controls, search for additional cases, identify the vehicle and agent, seek sources of contamination, and determine specific factors that contributed to the outbreak. An investigation should be initiated as soon as the possibility of an outbreak or epidemic is suggested. Any delay may defeat the primary purpose of the investigation—to gather information and formulate control measures.

BOX 25

INVESTIGATING FOODBORNE ILLNESSES

TABLE 15.5
Factors in a foodborne disease investigation.

Agent	Agent Environment	Host Environment	Host
Type *(What)*	Source *(Where)*	Human ecology *(Why)*	Verify diagnosis *(What)*
Bacteria	Human	Biological	Symptoms
Rickettsias	Animal	Chemical	Duration
Viruses	Fomites	Physical	Severity
Protozoa		Cultural	Incubation periods
Fungi	Contamination *(How)*	Socio-economic	Specimens
Nematodes	Infected animal		Infectivity
Cestodes	Soil	Control *(What to do)*	
Trematodes	Feces to carcasses	Agent	Descriptive epidemiology
Algae	Raw ingredients	Refrigeration	Time *(When)* (of
Toxicants in	Infected worker (feces, nose,	Thorough cooking or	onset) (of eating)
plants and	throat, infection, skin)	thermal processing	(of year)
animals	Preparation or processing	Hot holding	Place *(Where)* (of
Chemicals	method	Sanitation	residence) (of
	Equipment	Safe storage	eating) (of food
Nature *(Why)*		People	source)
Numbers	Vehicle *(What)*	Personal hygiene	Person *(Who)*
Toxigenicity	Food history attack rates	Carrier control	Age
Spore former	Samples	Immunization (typhoid)	Sex
Thermostabile	History of preparation	Surveillance	Foods eaten
Growth range	Source	Supervision	Food preferences
Optimum		Training	Occupation
growth	Survival, growth, and multiplication	Equipment	Ethnic group
Aerobic	Times, temperatures, procedures	Construction	Socio-economic group
Anaerobic	Storage upon arrival	Maintenance	Other factors
Facultative	Preparation	Different for raw and	Compare with
Nutrients	Cooking or thermal processing	prepared foods	controls
pH range	Storage during serving		
State of maturity	Chilling		
Concentration	Reheating		
Resistance	Storage after reheating		
Water activity	Storage before and after		
	sampling		

SOURCE: *Guide for Investigating Foodborne Disease Outbreaks and Analyzing Surveillance Data.* Center for Disease Control, Atlanta, 1980.

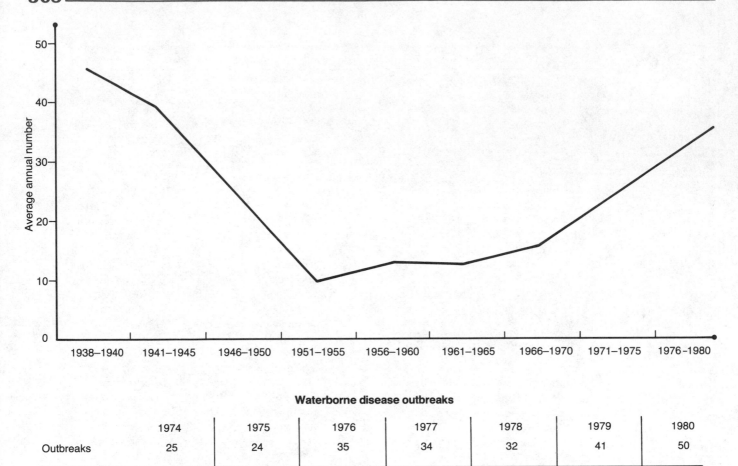

Waterborne disease outbreaks

	1974	1975	1976	1977	1978	1979	1980
Outbreaks	25	24	35	34	32	41	50
Cases	8363	10,879	5068	3860	11,435	9720	20,008

FIGURE 15.6
Average annual number of waterborne disease outbreaks. (Courtesy of the Center for Disease Control, Atlanta.)

Milk from infected cows is also an important source of enterotoxin-producing *S. aureus.*

The control of foodborne illnesses must begin at the initial stage of food production, such as the harvesting of grains, and should not end until the food has either been eaten or properly disposed. Contamination may occur during harvesting, processing, packaging, preparation, and consuming. Advances in commercial food handling and processing have greatly reduced the risk of food infection or poisoning. Should an outbreak result from improper food packaging, an investigation is launched to identify and eliminate the source of the problem. The success of such an investigation depends on the skill of the investigators and their knowledge of microbiology, epidemiology, psychology, and many other fields. The stage during which food contamination most likely occurs is in the home, when the food is being prepared and eaten. Consumer education programs are designed to instruct the public in the safe preparation of foods to reduce the likelihood of poisoning and infection. Public service announcements warning of the dangers of *Salmonellosis* frequently appear during holidays, when many people prepare especially large meals (Figure 15.7). Control of foodborne illness begins with proper food preservation techniques and the preparation of meals in a manner that does not cause contamination. Foods must be cooked at the proper temperature for the required period to ensure that possible contaminants are destroyed, not incubated. Cooling leftover food before putting it into the refrigerator is a potentially dangerous practice that dates back to the time of the old "icebox." If the food was placed in the box before it had cooled, the block of ice would melt and have to be replaced too quickly. Today's refrigerators elim-

FIGURE 15.7
Public service announcements warn of the dangers of food poisoning due to bacterial contamination. These appear more frequently during the holiday season when foods such as turkey, dressing, and cream-filled desserts are prepared.

The Cost of One Rabid Dog — California

On May 10, 1980, a dog in Yuba County, California, was placed under observation after it bit 3 persons in a parking lot in the Olivehurst area. Because the dog appeared ill, it was killed, and tissues were tested and found positive for rabies on May 12. The subsequent investigation by Sutter-Yuba County Health Department personnel eventually resulted in the identification of 70 persons, who received antirabies prophylaxis because of known or probable exposure to the dog. Because investigators found that only 20% of the dogs and cats in the area had up-to-date vaccinations, special clinics were held in which 2,000 dogs were vaccinated; over 300 unclaimed dogs and cats were destroyed.

No persons or other animals were known to develop rabies as a result of this episode. However, the costs generated by this single rabid dog were estimated as: $92,650 for human antirabies treatment, $4,190 for animal vaccination and veterinary services, and $8,950 for health-department and animal-control programs. The total cost of the episode was $105,790, or over $1,500 per person treated, not including lost work time, patient travel time, and costs of the 6 months' quarantine imposed on animals exposed to the rabid dog.

Reported by MK Cusick, MD, Sutter-Yuba County Health Dept, G Humphrey, DVM, Dept of Health Svcs, California; Viral Diseases Div, Center for Infectious Diseases, CDC.

Editorial Note: There have been few reports in the literature in which costs could be assigned to 1 specific case of rabies. Although human rabies has become a rarity in the United States — 0-5 cases per year — the cost of controlling the disease is an increasing burden on health budgets.

FIGURE 15.8
(From *Morbidity and Mortality Weekly Report*, vol. 30, 1981. Center for Disease Control, Atlanta.)

inate this problem, but many people continue the practice without realizing the consequences.

ZOONOTIC TRANSMISSION

Zoonoses are diseases of animals that on occasion are transferred to humans (Table 15.6). They can be caused by bacteria, viruses, fungi, or protozoans. Transfer from the infected animals often occurs directly. For example, rabies virus can be transferred to humans bitten by an infected skunk. Transfer may also occur indirectly through a vector. Equine encephalitis virus is carried from infected horses to humans by mosquitos. Because these pathogens have a second living reservoir of infection, they maintain a high degree of virulence and large infecting population. Recognition of the importance of zoonotic diseases dates back to the time of Hippocrates, when attempts were first made to control their spread. Today control measures include laws and regulations ordering the destruction of infected animals, animal immunization to prevent infection, treatment of infected animals, and elimination of vectors responsible for indirect transmission (Figure 15.8). These control measures interfere with the pathogen's movement at different points. The successful transfer of a zoonotic pathogen depends on several important factors.

1 What is the most common food infection?
2 How is foodborne illness controlled?
3 Name three zoonotic diseases.

TRANSMISSION ROUTES

For a pathogen to be transferred successfully from one host to another, five difficulties must be overcome. First, the pathogen must be able to abandon its current host. The place from which the microbe leaves the host is known as the **portal of exit.** In many diseases, this is the upper respiratory tract or gastrointestinal tract. Second, the microbe must be able to survive the movement from one host to another. This trait is genetically determined and influences exactly how the microbe will best be transferred. Those that are able to survive outside the body for long periods (for example, *Staphylococcus*) may be transferred by air, water, or food. Those that are more intolerant of such drastic changes (for example, *Treponema*) must be transferred in a more direct manner such as occurs during sexual intercourse. Third, successful transmission may only take place if the microbe finds a suitable host. Pathogens are opportunistic and do not select their host before they make their move. If by chance the microbe falls onto an unsuitable species, a host-parasite relationship will not be established, and the pathogen will die. Death may also occur if a suitable host is found but is not susceptible at the time of contact. Host susceptibility is the fourth essential for successful transmission. Finally, the microbe must be able to enter the suitable, susceptible host. The place through which the microbe enters the host is known as the **portal of entry.** Table 15.7 lists several important human diseases, their usual portals of exit and entry, and how they may be spread through the population. Epidemics can be brought under control at any of these five points. However, in order to control disease, a basic understanding of population dynamics and disease cycles is needed.

TABLE 15.6
Zoonoses.

Disease	Etiological Agent	Animal Reservoirs	Vectors, Mode of Transmission
Bacterial			
Anthrax	*Bacillus anthracis*	Cattle, horses, sheep, goats, swine, wild animals, birds	Contaminated wool, hair, hides, air, food, water
Brucellosis	*Brucella* spp.	Cattle, swine, goats, sheep, horses, mules, fowl, dogs, cats, deer, rabbits	Milk, meat, other contaminated food; contact
Glanders	*Pseudomonas mallei*	Horses	Contact; fomites, inhalation
Leptospirosis	*Leptospira interrogans (icterohaemorrhagiae)*	Dogs, cattle, swine, rodents	Contact with infected reservoir animals; inhalation of contaminated dust
Plague	*Yersinia pestis*	Rodents	Fleabite
Salmonellosis	*Salmonella* spp.	Fowl, birds, cattle, swine, rodents, cats, dogs, turtles	Ingestion of contaminated foods
Tularemia	*Francisella tularensis*	Wild animals and birds	Contact; tick bites
Q fever	*Coxiella burnetii*	Rats, birds, domestic animals	Tick bite; inhalation of aerosolized rickettsias
Rocky Mountain spotted fever	*Rickettsia rickettsii*	Wild rodents	Tick bites
Murine typhus fever	*R. typhi*	Rats	Flea bites
Psittacosis and ornithosis	*Chlamydia* spp.	Birds (especially psittacine varieties), domestic fowl	Contact; inhalation
Viral			
Equine encephalitis	Arbovirus	Horses, mules, birds	Mosquito bites
Rabies	Rabies virus (rhabdovirus)	Dogs and other canines, cats, bats, skunks	Animal bites
Jungle yellow fever	Yellow fever virus	Monkeys, marmosets, lemurs	Mosquito bites
Protozoan			
Malaria	*Plasmodium vivax*	Monkeys and humans	Mosquito bites
Leishmaniasis	*Leishmania* spp.		Insect bites (sandfly)
Fungal			
Ringworm	Several species of pathogenic dermatophytes	Cats, dogs, other domestic animals	Contact
Histoplasmosis	*Histoplasma capsulatum*	Birds and domestic fowl	Inhalation of spores

HERD IMMUNITY AND DISEASE CYCLES

As a disease travels through a population, surviving individuals develop an active acquired immunity which makes them unsusceptible to reinfection. This also removes them as a possible source of infection to others. In addition to the development of individual immunity, the population as a whole develops what is known as **collective,** or **herd, immunity.** The concept of herd immunity is based on the relative proportions of immune and susceptible individuals in the population. Research has indicated that there is a point (characteristic of each type of disease) at which the collective immunity of the population serves as protection for susceptible individuals. This takes place because there is a reduced probability of contact among the few remaining infected individuals and those that are still susceptible. Therefore, even though some in the population may not have had the disease or not been immunized against it, they will probably not become infected, since they will not likely come in contact with an infected individual. For this reason, public health officials immunize large portions of the susceptible population and attempt to maintain a high level of herd immunity. Any increase in the number of susceptibles may

TABLE 15.7
Transmission routes of various human diseases.

Disease	Portal of Exit	Portal of Entry	Mode of Transmission
Bacillary dysentery	Intestinal/ urinary tract	Gastrointestinal tract	Patients and carriers; fecal-oral route
Brucellosis	Gastrointestinal tract	Gastrointestinal tract	Oral ingestion of infective material
Diphtheria	Oral/respiratory tract	Respiratory tract	Nasal and oral secretions; respiratory droplets
Gonorrhea	Genital tract	Genital tract	Sexual intercourse
Hepatitis (epidemic)	Gastrointestinal tract	Gastrointestinal tract	Fecal-oral route
Hepatitis (serum)	Blood	Blood	Injections, transfusions
Influenza	Oral/respiratory tract	Respiratory tract	Respiratory
Pneumonia (streptococcal)	Oral/respiratory tract	Respiratory tract	Respiratory droplets
Polio	Gastrointestinal tract	Oral/respiratory tract	Infected feces and pharyngeal secretions
Syphilis	Genital tract	Genital tract	Sexual intercourse, contact with open lesions; blood transfusion; transplacental inoculation
Tuberculosis	Oral/respiratory tract	Respiratory/ gastrointestinal tract	Sputum; respiratory droplets; infected milk
Typhoid	Intestinal/ urinary tract	Gastrointestinal tract	Infected urine and feces
Rubella (German measles)	Oral/respiratory tract	Respiratory tract	Probably respiratory droplets

result in an endemic disease becoming epidemic. The proportion of immunes to susceptibles must be constantly monitored, because new individuals continuously enter the population through migration, birth, and antigenic drift. **Antigenic drift** is the spontaneous change in the antigenic character of a pathogenic microbe which converts it to a strain previously unrecognized by host immune mechanisms. This kind of mutation affects influenza viruses frequently, changing them from one antigenic form to another. Antigenic drift is also thought to be caused by the hybridization of different subtypes of influenza viruses. The "hybridization theory" states that two subtypes of virus can intermingle to form a new antigenic type. This may occur between an animal strain of the virus and the human strain. This is thought to happen when antibody levels in the human population become so high that the virus can no longer infect humans. The virus is transmitted to animals where the hybridization takes place. An animal influenza type A virus may rapidly adapt to infect humans. Genetic reassortment during coinfection with influenza A viruses derived from humans and animals can also lead to hybridization. The close contact of humans with animals occurs frequently in Asia, and it is of interest that most great pandemics have originated in China, where contact between humans and swine are frequent and close. The pandemic of 1918 was probably the result of such a hybridization as indicated by the fact that survivors of that pandemic have antibodies to the animal form of swine flu.

Whenever antigenic drift occurs, the population of susceptibles increases since the immune system has not been sensitized to the new mutant strain. If the percent of susceptibles rises above the threshold level of protection provided by herd immunity, the morbidity rate will also rise. The morbidity rate of diphtheria among school children may reach epidemic levels, if the number of susceptibles is allowed to rise above 30 percent. The goal of public health agencies is to ensure that at least 70 percent of the population is immunized against this disease in order to provide herd immunity to those that are not immunized.

The concept of herd immunity also helps to explain why some diseases run in cycles. When a disease strikes a population, a high proportion of individuals shift from sus-

ceptibles to immunes. The high proportion of immunes in the group provides a level of protection for uninfected, unimmunized individuals who enter the population through birth. This level of protection decreases through time as the proportion of immunes again rises to an unsafe level (Figure 15.9). After the group crosses the threshold of protection provided by herd immunity, the morbidity rate increases to epidemic proportions. Post-epidemic infection and artificial immunity again establish a high level of herd immunity and the epidemic passes, but only temporarily. Another epidemic occurs, and a disease cycle is established when the percentage of susceptibles again increases due to the influx of susceptibles. Disease cycles have been identified for measles, chickenpox, diphtheria, and many others (Figure 15.10). Breaking the cycle and maintaining herd immunity can be accomplished in a number of ways.

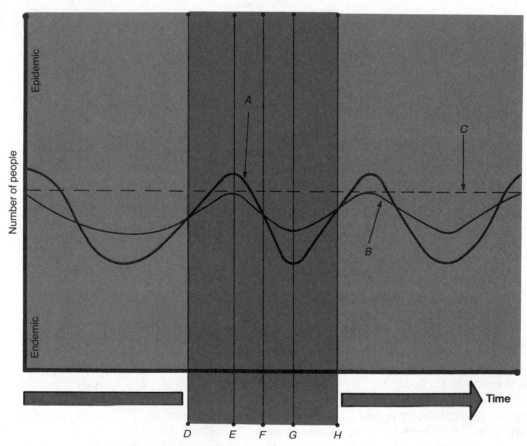

FIGURE 15.9
Herd immunity.

A = number of infected individuals
B = number of immune individuals
C = herd immunity (usually given as percent)
D = proportion of immunes to susceptibles at herd immunity level
D–E = number of infected rises to peak of epidemic (E); number of immunes increase slowly
E–F = number of infected falls as epidemic is brought under control as herd immunity is reached
F = endemic; controlled by low number of infected individuals
G–H = infected increase as number of immunes remains below safe herd immunity level

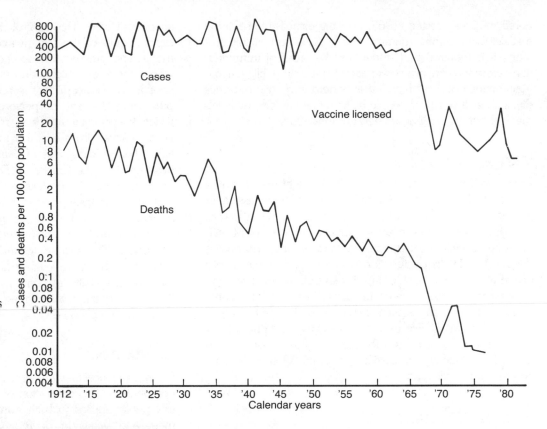

FIGURE 15.10
The upper graph line shows how measle cases cycle from periods of low morbidity to high. The lower line represents the decline in mortality. This drastic decrease has continued in the United States following the introduction of attenuated measles vaccine in 1963. (Courtesy of the Center for Disease Control, Atlanta.)

---

1   How does the concept of herd immunity explain the fact that susceptible members of a community may not become ill with an infectious disease?
2   What term is used to refer to the spontaneous change in the antigenic character of a pathogenic microbe?
3   How does the concept of herd immunity explain why some diseases run in cycles?

THE CONTROL OF COMMUNICABLE DISEASES

Tracing and controlling communicable diseases depends on the precise identification of the etiological agent. It is incorrect to assume that a particular epidemic comes from a common source or results from person-to-person transmission. This information can only be determined after the specific microbe has been identified. Simple identification of the genus and species of the pathogen is not sufficient. Specimens must be collected and biochemical tests run to determine the *type* or *strain* of microbe involved in the epidemic. Based on this knowledge, decisions can be made to deter-

mine the actual source of the pathogen, what methods are required to stop quickly the spread of disease, and what course of action will best prevent a recurrence of the epidemic. Several methods of control are available and include: (1) active immunization, (2) vector control, (3) patient isolation, (4) quarantine, and (5) special precautions for health care personnel.

ACTIVE IMMUNIZATION

The injection of toxoids (e.g., DPT vaccine), attenuated pathogens (e.g., polio virus vaccine), and live microbes (e.g., cowpox virus) stimulates the immune system into action and provides a high level of individual protection. The establishment and maintenance of active immunization programs that begin early in life has enabled the United States and many other countries to increase the level of host resistance to many communicable diseases. In the United States alone, approximately 73 percent of all children entering kindergarten have been immunized against poliomyelitis, 85 percent against diphtheria, pertussis, and tetanus, and more than 90 percent against measles. The level of protection has become so high that morbidity rates are their lowest in history. For example, at the end of the first 20 weeks in 1979, there were only 7520 cases of measles reported in the United States compared to 14,625 for the same period in 1978. This decrease stems from the development of the live attenuated measles vaccine (1963) and the establishment of the measles eradication program (1977). As a part of this program, it is required that students show proof of immunization before they are allowed to attend classes in public schools. Maintenance of these high levels of herd immunity becomes difficult because many people believe that low disease levels mean that they no longer need immunization.

VECTOR CONTROL

Communicable diseases may also be brought under control by eliminating vectors responsible for the transmission of pathogens. Mosquito control programs are directed toward the elimination of this insect vector and the control of such diseases as yellow fever, malaria, and equine encephalitis (Figure 15.11). In conjunction with such programs, public health workers also seek out and treat both active cases and carriers of infectious disease to further reduce vector transmission. This often requires close serological screening of suspect individuals. For example, a suspected typhoid fever carrier is given a blood test for the identification of a serum antibody formed by the carrier against the Vi capsular antigen of *Salmonella typhi*.

PATIENT ISOLATION

The purpose of patient isolation is to remove communicable individuals from the general population. This isolation helps maintain a high proportion of immunes, reduces the size of the reservoir of infection, and enables the patient to receive specialized care. This control measure makes possible the rapid reentry of the patient into the population as an immune, and prevents the spread of disease in the hospital, where there is a high concentration of highly susceptible people. This susceptibility is primarily due to patients with preexisting diseases or persons receiving immunotherapy which lowers their ability to resist infection. Isolation prevents hospitals from becoming reservoirs of epidemic infection. The five categories of patient isolation that have been designated by the Center for Disease Control are (1) strict, (2) respiratory, (3) protective, (4) enteric, and (5) wound and skin. The specifications for each type of isolation cover six major areas: (1) rooms, (2) gowns, (3) masks, (4) hands, (5) gloves, and (6) articles such as thermometers, dishes, and books. The isolation technique used for a particular disease depends on the level of herd immunity in the population and the ability to control the disease. Table 15.8 lists the patient isolation techniques, their specifications, and diseases requiring isolation.

QUARANTINE

Quarantine measures are similar to isolation procedures but not identical. This separation of an individual from the general public applies to both humans and domesticated animals, and varies in degree. Complete quarantine limits movement of healthy individuals who have been exposed to a communicable disease for the incubation period of the disease. A modified quarantine is less restrictive, used in suspected cases, and may precede complete quarantine. In

A

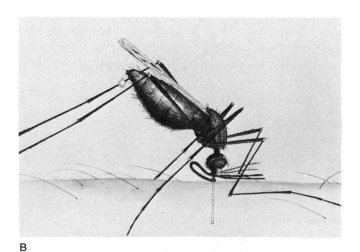

B

FIGURE 15.11
(A) Many cities regularly use chemicals to control mosquitos. This vector control has helped limit the spread of disease. (B) The *Anophiles* mosquito serves as a vector for the protozoan *Plasmodium,* the cause of malaria. (Part A courtesy of Microchem Inc. Part B courtesy of the Center for Disease Control, Atlanta.)

TABLE 15.8
Patient isolation.

Type	Specifications	Diseases Requiring Isolation*
Strict isolation	Visitors report to nurses' station before entering room. Private room necessary; door must be kept closed. Gowns must be worn by all persons entering room. Masks must be worn by all persons entering room. Hands must be washed on entering and leaving room. Gloves must be worn by all persons entering room. Articles must be discarded or wrapped before being sent to Central Supply for disinfection or sterilization.	Anthrax, inhalation Burn wound, major, infected with *Staphylococcus aureus* or group A *Streptococcus* Congenital rubella syndrome Diphtheria (pharyngeal or cutaneous) Disseminated neonatal *Herpesvirus hominis* (herpes simplex) Herpes zoster, disseminated Lassa fever Marburg virus disease Plague, pneumonic Pneumonia, *Staphylococcus aureus* or group A *Streptococcus* Rabies Skin infection, major, infected with *Staphylococcus aureus* Smallpox Vaccinia (generalized and progressive, and eczema vaccinatum) Varicella (chickenpox)
Respiratory isolation	Visitors report to nurses' station before entering room. Private room necessary; door must be kept closed. Gowns not necessary. Masks must be worn by any person entering room unless that person is not susceptible to the disease. Hands must be washed on entering and leaving room. Gloves not necessary. Articles: those contaminated with secretions must be disinfected.	Measles (rubeola) Meningococcal meningitis Meningococcemia Mumps Pertussis (whooping cough) Rubella (German measles) Tuberculosis, pulmonary (including tuberculosis of the respiratory tract)—suspected or sputum-positive (smear)
Protective isolation	Visitors report to nurses' station before entering room. Private room necessary; door must be kept closed. Gowns must be worn by all persons entering room. Masks must be worn by all persons entering room. Hands must be washed on entering and leaving room. Gloves must be worn by all persons having direct contact with patient. Articles: see manual text (CDC).	Agranulocytosis Dermatitis, noninfected vesicular, bullous, or eczematous disease when severe and extensive Extensive, noninfected burns in certain patients Lymphomas and leukemia in certain patients (especially in the late stages of Hodgkin's disease and acute leukemia)

Enteric precautions	Visitors report to nurses' station before entering room. Private room necessary for children only. Gowns must be worn by all persons having direct contact with patient. Masks not necessary. Hands must be washed on entering and leaving room. Gloves must be worn by all persons having direct contact with patient or with articles contaminated with fecal material. Articles: special precautions necessary for articles contaminated with urine and feces. Articles must be disinfected or discarded.	Cholera Diarrhea, acute illness with suspected infectious etiology Enterocolitis, staphylococcal Gastroenteritis caused by enteropathogenic or enterotoxic *Escherichia coli*, *Salmonella* spp., *Shigella* spp., *Yersinia enterocolitica* Hepatitis, viral, type A, B, or unspecified Typhoid fever *(Salmonella typhi)*
Wound and skin precautions	Visitors report to nurses' station before entering room. Private room desirable. Gowns must be worn by all persons having direct contact with patient. Masks not necessary except during dressing changes. Hands must be washed on entering and leaving room. Gloves must be worn by all persons having direct contact with infected area. Articles: special precautions necessary for instruments, dressings, and linen. Note: see manual for Special Dressing Techniques to be used when changing dressings.	Burns that are infected, except those infected with *Staphylococcus aureus* or group A *Streptococcus* that are not covered or not adequately contained by dressings (see Strict Isolation) Gas gangrene (due to *Clostridium perfringens)* Herpes zoster, localized Melioidosis, extrapulmonary with draining sinuses Plague, bubonic Puerperal sepsis—group A *Streptococcus*, vaginal discharge Wound and skin infections that are not covered by dressings or that have copious purulent drainage that is not contained by dressings, except those infected with *Staphylococcus aureus* or group A *Streptococcus*, which require Strict Isolation Wound and skin infections that are covered by dressings and the discharge is adequately contained, including those infected with *Staphylococcus aureus* or group A *Streptococcus*. Minor wound infections, such as stitch abscesses, need only Secretion Precautions.

*See "Isolation Techniques for Use in Hospitals" for details and recommended duration of isolation (Center for Disease Control, Atlanta.)

less severe cases of communicable disease, susceptibles are separated from contact with communicable individuals by a sanitary boundary, as disease control measures are put into effect. While this form of epidemic control is probably the oldest form of control, it has not been found to be particularly effective and is no longer extensively used. However, quarantine is still recommended for use by the World Health Organization to control the spread of smallpox, cholera, and yellow fever. The greatest difficulty preventing quarantine from being effective is the ease with which prodromal carriers may move through the population or world. An individual may become infected in one country, board a jet for travel to another country, and not develop symptoms of the disease for several days. During this prodromal period, the person may serve as a traveling source of the infectious microbe.

---

1 Which methods are used to control communicable diseases?
2 What are the five categories of patient isolation?
3 Explain the difference between isolation and quarantine?

SPECIAL PRECAUTIONS

Allied health personnel voluntarily place themselves in a position to acquire and transmit infectious diseases as they provide care and treatment for others. This is a special problem in hospitals for a number of reasons. First, patients may have entered the hospital with a virulent infection that may be easily spread during treatment. Second, the concentration of people with weakened immunity increases the likelihood that infectious diseases may be acquired by them and transmitted to others by hospital personnel. Third, the intensive use of antibiotics in hospitals to control and prevent microbial infections has resulted in the development and concentration of drug-resistant microbes. Their greater resistance may prolong illness and increase the communicable period while culture and susceptibility tests are being performed. Fourth, many medical and surgical procedures, such as catheterization and biopsy, increase the risk of infection above normal. Surgical patients require special surveillance and treatment. Finally, many drugs used to treat patients lower their resistance to infection, and care must be taken to assure that they do not become infected.

Special precautions to deal with the spread of infections in hospitals include the use of proper housekeeping procedures. It is also important to use proper aseptic technique when dealing with patients. Dressings on wounds must be carefully handled in order to prevent contamination when they are changed (Figure 15.12). **Fomites** (equipment, material, and other inanimate objects), which could serve as sources of infection, must be properly disinfected or sterilized. These items include linens, bedpans, thermometers, catheters, hypodermic needles, and eating utensils. Finally, proper handwashing technique is essential in limiting the spread of infectious disease.

NOSOCOMIAL INFECTION

Even though detailed attention is given to safety procedures, many patients entering the hospital acquire infections during their stay. The American Public Health Association defines such **nosocomial,** or hospital-acquired, **infections** as:

> An infection originating in a medical facility, e.g., occurring in a hospitalized patient in whom it was not present or incubating at the time of admission, or is the residual of an infection acquired during a previous admission. Includes infections acquired in the hospital but appearing after discharge; it also includes infections among staff.

The need to prevent nosocomial infections and cut health care costs make it imperative that the period of hospitalization be reduced as much as possible. Approximately three to five percent of all patients admitted to hospitals annually in the United States develop nosocomial infections which either lengthen their confinement or result in additional treatment. The additional expenses amount to over one billion dollars each year. In order to reduce hospitalization time and costs, infection-control committees are established which survey and identify problem areas, and institute appropriate control measures. These committees are composed of physicians, nurses, administrators, and other hospital personnel. They meet regularly to establish infection-control policies and mechanisms by which they may be successfully imple-

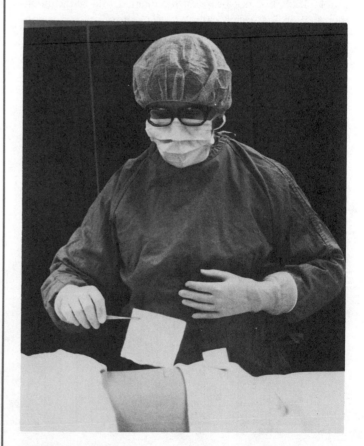

FIGURE 15.12
Special aseptic techniques are used in changing a sterile dressing. This limits the spread of infectious microbes and prevents further contamination of the wound. (Courtesy of M. Mahan.)

mented. Infection-control policies generally cover seven main areas: (1) isolation, (2) aseptic technique, (3) employee health programs, (4) in-service training to help reduce infections, (5) catheterization, intravenous procedures, and other closed system care, (6) physician review for appropriate antibiotic therapy, and (7) methods of providing a clean environment.

One factor which helps reduce the nosocomial infection rate is the establishment of hospital units which provide specialized care. The separation of patients into medical, surgical, obstetric, and other areas enables the hospital to better control the transmission of infectious microbes by using special control measures unique to each area. The incidence of nosocomial infections is greatest in surgical patients, since they have experienced severe trauma as a result of their operations and have gained a new portal of entry for infectious microbes (Table 15.9). Surgery involving the abdominal cavity and bacteria-filled intestinal tract may result in infection. Special precautions to avoid these infections include the prophylactic administration of an antibiotic, the clearing of the intestinal tract by enema before the operation, and the use of appropriate aseptic technique during and after surgery. However, the highest incidence of nosocomial infections occurs in association with the urinary tract (Table 15.10). If not properly maintained, the slight movements of an indwelling catheter will "pump" microbes con-

taminating the external skin up the urethra between the catheter and the urethral mucosa into the bladder. Microbes may also enter on small air bubbles that may move up the inside of the catheter tube after the specimen has been taken. The four pathogens most frequently involved in nosocomial infections are *E. coli, Staphylococcus aureus, Streptococcus* (group D), and *Pseudomonas* (Table 15.11). The primary reasons for this predominance are their wide occurrence, ability to survive outside the human body for long periods, and resistance to antibiotics.

---

1 Name three fomites.
2 What seven main areas are covered by an infection control policy?
3 What are the four pathogens most frequently involved in nosocomial infections?

There are several factors which influence the establishment of nosocomial infections, and each must be taken into account when attempting to reduce these infections. The **virulence** of a microbe is an important consideration and varies widely in any one species. Not only are highly virulent strains able to cause disease but mildly virulent forms may

TABLE 15.9
Relative incidence of nosocomial infections by service and quarter.

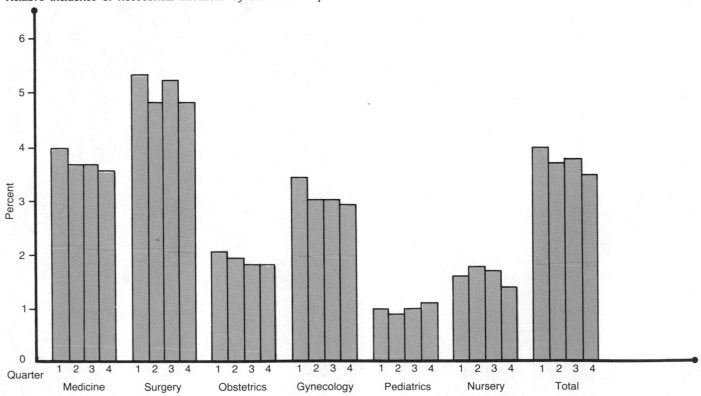

SOURCE: Center for Disease Control, Atlanta.

TABLE 15.10
Relative incidence of nosocomial infection by site.

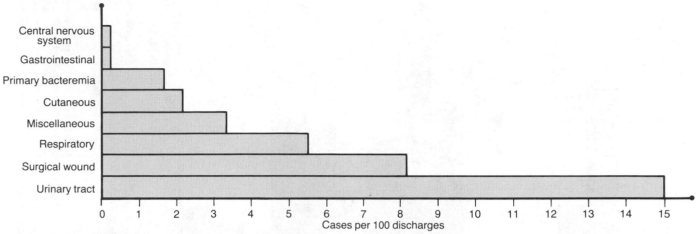

Cases per 100 discharges

SOURCE: Center for Disease Control, Atlanta.

also cause harm when normal body defense mechanisms are impaired. Once considered to be nonpathogenic, *Serratia marcescens,* has proved to be the cause of numerous hospital-acquired infections, when introduced into a debilitated patient. Another important factor is the number of microbes that are available to establish an infection. The higher the concentration, the more likely it is that the microbes will cause disease. Prolonged contact with the microbes at an increased number of suitable sites also encourages infection. Although in many cases coincidence has a great deal to do with acquiring an infection, it has been demonstrated that patients who have previously received, or are currently receiving, antimicrobial therapy are highly susceptible to infections caused by such Gram-negative bacteria as *Serratia, Klebsiella, Pseudomonas,* and *Acinetobacter.* This occurs because the normal flora of the body is eliminated by the antibiotic and replaced by foreign, drug-resistant pathogens. Therefore, special care must be taken when dealing with patients who are receiving antibiotics on a regular basis. Care must also be exercised to ensure that all sources of pathogens are properly handled and disinfected (refer to Chapter 10). Good aseptic technique should always be followed when using intravenous devices, catheters, inhalation therapy equipment, tracheotomy tubes, and sterile dressings.

SUMMARY

Host-parasite relationships are dynamic and influenced by many environmental factors. Epidemiology is the study of the factors which influence host-parasite relationships and lead to an increase or decrease in a population's disease. Following a disease as it moves through a population involves monitoring morbidity, mortality, and prevalence rates. This data aids public health agencies in determining whether or not health care forces should be mobilized in order to lower the morbidity rate to an acceptable level. Diseases that occur in a localized area at rates that are acceptable to public health agencies are known as endemic diseases. Those that have exceeded this level or spread to encompass a larger geographical area are known as epidemic diseases. Pandemic diseases occur when large numbers of individuals become infected on a worldwide basis. An epidemic moves through a population as either a common-source epidemic or a person-to-person propagated epidemic. Most common-source epidemics have their origin in soil, water, or other inanimate objects that serve as reservoirs of infection. Humans may also serve as reservoirs of infection as clinical, subclinical, or carrier cases. On occasion, the disease-causing microbe may be moved from one host to another by insect vectors, such as mosquitos, ticks, fleas, or flies. Through the efforts of local, state, national, and international public health organizations, many communicable diseases, such as smallpox and typhoid fever, have been brought under control. This has been made possible by a high degree of cooperation, the establishment of a disease classification system, and research in the areas of transmission (airborne, waterborne, foodborne, and animalborne), disease cycles, and herd immunity. Disease cycles have been broken and brought under control in the general population by eliminating or controlling the transmission of microbes from one individual to another, increasing the number of immune individuals in the population by immunization, and reducing reservoirs of infection. The adoption of special precautions and procedures is essential to preventing both staff and patients from acquiring infections in the hospital environment. Hospital-acquired infections are known as nosocomial infections. Careful surveillance and the use of good aseptic techniques are essential to reducing hospitalization time and costs.

TABLE 15.11

Incidence* and relative frequency** of selected pathogens causing nosocomial infections, by site, January–December, 1976.

	Primary bacteremia	Surgical wound	Lower respiratory	Urinary tract	Cutaneous	Other	All sites
S. aureus	2.1 (12.6%)	18.9 (15.5%)	7.4 (10.4%)	2.9 (1.7%)	8.7 (33.7%)	5.8 (12.6%)	45.8 (10.0%)
S. epidermidis	1.3 (7.8%)	5.5 (4.5%)	0.4 (0.5%)	5.1 (2.9%)	1.1 (4.3%)	1.6 (3.5%)	15.0 (3.3%)
S. pneumoniae	0.2 (1.2%)	0.1 (0.1%)	2.7 (3.7%)	— (—)	0.1 (0.2%)	0.2 (0.4%)	3.3 (0.7%)
Streptococcus, Group A	0.1 (0.6%)	0.9 (0.8%)	0.3 (0.4%)	0.1 (0.1%)	0.2 (0.6%)	0.5 (1.1%)	2.1 (0.5%)
Streptococcus, Group B	0.3 (1.9%)	1.8 (1.5%)	0.3 (0.5%)	1.5 (0.9%)	0.3 (1.3%)	0.8 (1.7%)	5.0 (1.1%)
Streptococcus, Group D	1.2 (7.4%)	12.7 (10.5%)	0.9 (1.2%)	24.4 (13.9%)	1.6 (6.1%)	2.7 (5.9%)	43.5 (9.6%)
E. coli	2.4 (14.4%)	19.1 (15.8%)	5.2 (7.2%)	55.3 (31.7%)	2.2 (8.5%)	4.2 (9.3%)	88.4 (19.4%)
Klebsiella spp.	1.8 (10.6%)	6.4 (5.2%)	7.9 (11.1%)	15.3 (8.8%)	1.1 (4.3%)	2.1 (4.6%)	34.6 (7.6%)
Enterobacter spp.	0.9 (5.4%)	4.9 (4.0%)	4.7 (6.5%)	7.2 (4.1%)	0.8 (3.1%)	1.1 (2.4%)	19.6 (4.3%)
Proteus-Providentia spp.	0.6 (3.6%)	8.6 (8.1%)	4.0 (5.6%)	18.2 (10.4%)	1.4 (5.5%)	2.6 (5.7%)	35.4 (7.8%)
Pseudomonas	1.2 (6.9%)	7.2 (5.9%)	7.1 (10%)	20.7 (11.9%)	1.6 (6.2%)	3.1 (6.8%)	40.9 (9%)
Serratia spp.	0.5 (3.0%)	1.4 (1.3%)	2.1 (2.9%)	3.7 (2.1%)	0.4 (1.4%)	0.6 (1.3%)	8.7 (1.9%)
Bacteroides fragilis	1.0 (6.1%)	7.7 (6.3%)	0.1 (0.2%)	0.1 (–)	0.5 (1.8%)	1.3 (2.8%)	10.7 (2.3%)
Candida spp.	0.5 (3.3%)	1.0 (0.8%)	2.5 (3.5%)	7.3 (4.2%)	0.7 (2.9%)	3.8 (8.3%)	15.8 (3.5%)
Other fungi	0.2 (1.4%)	0.3 (0.3%)	0.5 (0.7%)	2.9 (1.7%)	0.2 (0.8%)	0.7 (1.5%)	4.8 (1.0%)
Other pathogens	2.0 (12.0%)	11.7 (9.6%)	6.1 (8.5%)	5.9 (3.4%)	1.3 (5.1%)	4.1 (8.9%)	31.2 (6.8%)
No culture; No pathogen	0.3 (1.8%)	13.7 (11.2%)	19.1 (26.8%)	4.0 (2.3%)	3.5 (13.6%)	10.7 (23.3%)	51.3 (11.2%)
ALL PATHOGENS ***	16.5 (100.0%)	121.9 (100.0%)	71.4 (100.0%)	174.6 (100.0%)	25.7 (100.0%)	45.9 (100.0%)	456.1 (100.0%)
Secondary bacteremia	NA —	3.8 (3.1%)	3.0 (4.1%)	4.5 (2.6%)	1.3 (5.0%)	3.7 (7.9%)	16.3 (3.2%)

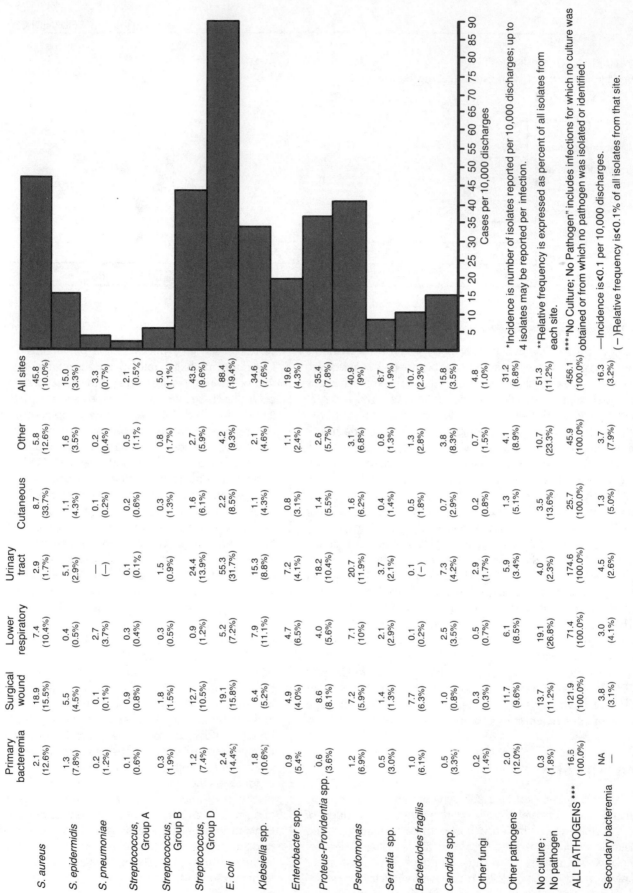

Cases per 10,000 discharges

*Incidence is number of isolates reported per 10,000 discharges; up to 4 isolates may be reported per infection.

**Relative frequency is expressed as percent of all isolates from each site.

***"No Culture; No Pathogen" includes infections for which no culture was obtained or from which no pathogen was isolated or identified.

—Incidence is <0.1 per 10,000 discharges.

(—)Relative frequency is ≤0.1% of all isolates from that site.

SOURCE: *National Nosocomial Infection Study Report.* Annual Summary, 1976. Issued Feb., 1978. Center for Disease Control, Atlanta.

WITH A LITTLE THOUGHT

An outbreak of gastroenteritis apparently caused by *Salmonella* has occurred throughout a hospital. Not one unit seems to be unaffected. As the head of the infection control committee, arrange a meeting of your group and outline a plan of action which will result in the identification of the responsible microbe, determine whether the epidemic is common-source or propagated, determine its most likely transmission route, and suggest measures to control the spread of the disease.

STUDY QUESTIONS

1 What factors might influence a change in the morbidity rate of a disease? the mortality rate?

2 Describe two reservoirs of infection of common human pathogens.

3 What is the difference among endemic, epidemic, and pandemic?

4 Describe how a common-source epidemic moves through a population. Why does the morbidity rate change so quickly in comparison to a propagated epidemic?

5 What are the three main health states of people who serve as reservoirs of infection?

6 What is a vector? What are zoonotic infections?

7 What are the basic differences among the five main classes of reportable diseases? How does classifying diseases aid in their control?

8 Name three transmission routes of infection. How might each be controlled?

9 Explain the concept of herd immunity. What methods might be used to maintain this form of protection?

10 What special problems occur in preventing the transmission of infectious diseases in hospitals?

SUGGESTED READINGS

BENENSON, A. S., ed. *Control of Communicable Diseases in Man,* 13th ed. The American Public Health Association, Washington, D.C., 1981.

BURNET, M., and D. O. WHITE. *Natural History of Infectious Disease,* 4th ed. Cambridge University Press, New York, 1972.

Center for Disease Control, *Dateline: CDC 30th Anniversary Issue,* vol. 8, no. 7. July, 1976.

GREENBERG, B. "Flies and Disease." *Scientific American,* Jan., 1965.

HENDERSON, D. A. "The Eradication of Smallpox." *Scientific American,* 235:25–33.

KAPLAN, M. M., and R. G. WEBSTER. "The Epidemiology of Influenza." *Scientific American,* 237:88–106.

LINIENFELD, A. M. *Foundations of Epidemiology.* Oxford University Press, New York, 1976.

MAUSNER, J. S., and A. K. BAHN. *Epidemiology, An Introductory Text.* W. B. Saunders, Philadelphia, 1974.

WISHNOW, R. M., and J. L. STEINFELD. "The Conquest of the Major Infectious Diseases in the U.S.: A Bicentennial Retrospect." *Annual Review of Microbiology,* 30:427–50, 1976.

KEY TERMS

morbidity

mortality

reservoir of infection

epidemic

endemic

pandemic

common source and
   propogated epidemic

clinical

subclinical

carrier

vector

petechiae (pe-te′ke-eye)

etiological agent
   (e″te-o-loj′ek-al)

zoonosis (zo-o-no′sis)

herd immunity

fomite

nosocomial

Pronunciation Guide for Organisms

Anopheles (an-of′e-lez)

Dermacentor (der″mah-sen′tor)

Acinetobacter (a-si-ne″to-bak′ter)

PART 4
Medical
Microbiology

THOUGH only a small portion of all microbes are human pathogens, they have a significant impact on our health, society, and economy. Each year millions of people fall victim to various types of infectious diseases caused by bacteria, fungi, protozoans, and viruses. The successful control of many pathogens has resulted from exhaustive research into microbial physiology, genetics, immunity, and host-parasite relationships. The field of microbiology that specializes in the study of microbes that cause disease in humans, animals, and many plants is known as medical microbiology. Investigators in this field have utilized information from many other areas in their attempts to better understand and control pathogens. For convenience, medical microbiologists have classified microbial pathogens into several groups. Pathogenic bacteria are probably the most familiar, and their subgroups include pyogenic cocci (e.g., staphylococci); Gram-positive bacilli (e.g., clostridia); the enteric Gram-negative bacilli (e.g., *E. coli*); small, Gram-negative bacilli (e.g., the plague bacillus); corynebacteria; acid-fast bacteria; spirochetes (e.g., *Treponema*); and the related forms *Mycoplasma*, rickettsia, and chlamydia. A second important group is the pathogenic fungi. These may be subgrouped according to their morphological type as either yeastlike or filamentous. Diseases caused by the fungi are known as **mycoses.** Those that involve the skin and mucous membranes are called superficial mycoses, while those that become established deep within body tissues are systemic mycoses. Other eucaryotic pathogens belong to the protozoan group. Though there are relatively few pathogens in this group, millions have experienced the symptoms of their infections. Diseases such as malaria, African sleeping sickness, vaginitis, and amoebic dysentery are common throughout the world. The last major class of pathogens is the viruses. These are noncellular and classified according to such features as nucleic acid type, size and morphology, susceptibility to physical and chemical agents, and immunological properties. Viral diseases range from the common cold and cold sores to rabies and some forms of cancer. They are known to infect all cell types including those of bacteria, animals, and plants.

All pathogens are capable of establishing host-parasite relationships; however, the type of disease symptoms that result may vary from person-to-person and differ

◄ *Agglutination of* Shigella flexneri *and* flexneri *polyvalent antisera. Magnification:* ×38,000. *(Electron micrograph by F. Siegel. Courtesy of the Burroughs Wellcome Co.)*

depending on the focus of infection. To bring a pathogen under control, it is essential to accurately identify the pathogen. Once identification is confirmed, other information pertaining to a pathogen's occurrence in the population, mode of transmission, incubation and communicability periods, antibiotic susceptibility, and epidemiology may be learned. The following series of chapters will present representative pathogens and describe their basic characteristics. The most common disease symptoms caused by these pathogens will also be discussed.

Bacterial Pathogens and Diseases (I)

PYOGENIC COCCI

Staphylococci
Streptococci
Neisseria

GRAM-POSITIVE, SPORE-FORMING BACILLI
Bacilli
Clostridia

ENTERIC GRAM-NEGATIVE BACILLI
Coliforms
Proteus
Salmonella
Shigella
Vibrio
Pseudomonas

Learning Objectives

☐ Know the three major groups of pathogenic microbes described in this chapter.

☐ Be familiar with the seven basic characteristics of the representative bacteria presented in the chapter, including morphology and staining, respiration-fermentation, motility, reservoir of infection, disease, pathogenic relatives, and features.

☐ Describe the symptoms, mode of transmission, virulence characteristics, and chemotherapeutic agents used in the treatment of the disease-causing microbes presented.

☐ Understand the medical terms presented in the Key Terms.

Pathogenic bacteria cause a variety of infectious diseases that have been recorded throughout history. Many have been described in the Bible and by such notable writers as Aristotle (384–322 B.C.) and Hippocrates (460?–377 B.C.). Fear and misunderstanding surrounding the causes and cures of these diseases encouraged the use of many disease-related terms and phrases in everyday vocabulary. People may be "plagued" with a problem, have a "germ" of an idea, or become "inflamed" with anger. While only a few pathogens are responsible for the majority of bacterial diseases, there is great variety among pathogenic bacteria. Structural, physiological, and symptomatic differences are used by microbiologists to classify pathogenic bacteria into several distinct groups. The pyogenic cocci include the staphylococci, streptococci, and members of the genus *Neisseria*. Infections caused by these pathogens typically result in the formation of pus ("pyo" means *pus,* "gen" means *forming*). The Gram-positive, spore-forming bacilli are in the genera *Bacillus* and *Clostridium.* While *B. anthracis* has been a significant problem in the past, effective control measures and antibiotic therapy have brought this zoonotic infection under control. The clostridia responsible for diseases such as gangrene, tetanus, and botulism have also been controlled, but cases continue to appear occasionally throughout the country. The enteric Gram-negative bacilli are subdivided into the coliforms, *Proteus, Salmonella, Shigella, Vibrio,* and *Pseudomonas.* These are all known as enteric bacilli because they

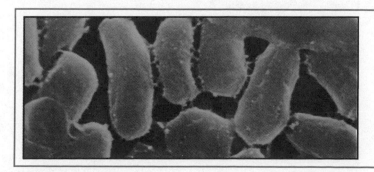

16

are typically found in the intestinal tract of humans and domesticated animals such as cattle, horses, and swine. The coliforms are microbes that resemble *E. coli* and are also subdivided into two related groupings of *Escherichia, Enterobacter, Klebseilla* and *Serratia;* and *Arizona, Edwardsiella* and *Citrobacter.* Much information has already been presented in previous chapters about some of these microbes; however, it is the intention of this chapter to focus on representative pathogens from these groups. The morphology, physiology, and epidemiology are presented, as well as background on certain diseases associated with infections caused by some of these microbes. There are seven other major categories of pathogenic bacteria and related forms that will be described in a similar manner in Chapter 17.

PYOGENIC COCCI

STAPHYLOCOCCI: *STAPHYLOCOCCUS AUREUS*

1 Morphology and Staining Reaction: Gram-positive clusters of cocci.

2 Respiration-Fermentation: Facultative anaerobes; ferment lactic acid from glucose.

3 Motility: Nonmotile.

4 Source of Infection-Reservoir: Soil and carriers (70–90 percent in two-week-old infants).

5 Disease: Boils, carbuncles, impetigo neonatorum, enterocolitis, pneumonia, scalded skin reaction, septicemia, meningitis, osteomyelitis, otitis, sinusitis, toxic shock syndrome.

6 Pathogenic Relatives: *S. epidermidis* is an opportunistic species found in normal flora on the skin.

7 Features: Heat resistant (TDT = 30 minutes/80°C), pathogenic strains are usually pigmented, antibiotic resistance develops rapidly, will grow in salt concentrations as high as 15 percent.

    Staphylococcus aureus are Gram-positive, spherical bacteria that typically appear as grapelike clusters when viewed through the microscope. Nutrient agar colonies range from white to golden yellow, depending on the amount of pigment produced. *S. aureus* may produce several types of hemolysins known as staphylolysins. The four antigenic types are designated as alpha, beta, gamma, and delta. Each type differs in its ability to destroy different blood types. For example, alpha staphylolysin destroys rabbit and sheep red blood cells, while the delta lysin destroys human, sheep, rabbit, horse, rat, guinea pig, and mouse red cells. Staphylococcal strains that produce the alpha and delta hemolysins are pathogenic for humans. This combination of alpha and

delta action also destroys leukocytes and skin tissue. When *S. aureus* is grown on a medium containing the appropriate blood type, characteristic discoloration and zones of clearing are identifiable as a result of the production and action of these hemolysins (Figure 16.1). Alpha and delta lysin production occurs in about 90 percent of all human staphylococcal infections and is found in many of these cases in conjunction with the release of coagulase. This enzyme is associated with virulence and may be responsible for stimulating the formation of fibrin clots, which serve to protect and isolate invading pathogens from phagocytosis. The presence of these two characteristics are determined by laboratory analysis. Laboratory identification of staphylococci is also based on their ability to grow in a high salt concentration and ferment the carbohydrate mannitol. When placed on mannitol salt agar medium (7.5 percent NaCl), *Staphylococcus aureus* will ferment the mannitol and release acids which change the pH indicator in the medium from red to yellow. More specific identification of the type of *S. aureus* is done by a process known as **phage typing.** The test is based on the ability of known bacterial viruses (phage) to lyse strains of *S. aureus* due to the presence of different phage receptor sites on the cell surface. Phage typing is a valuable identification procedure used in locating and tracing nosocomial staph infections. *Staphylococcus aureus* produces disease by multiplying and spreading in tissues and producing extracellular products (Table 16.1). Staph is responsible for a variety of purulent (pus-containing) infections that may be caused by opportunistic cells. About one-third of hospital-acquired infections are caused by staph originally on the skin of the patient acquiring the disease. One such disease is known by such names as Ritter's disease, exfoliative dermatitis, and bullous impetigo. The more

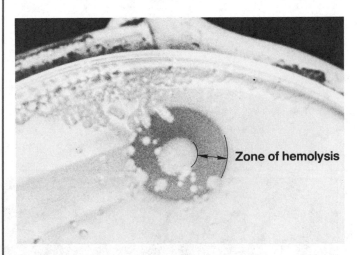

FIGURE 16.1
Hemolysins are just one of the many extracellular products produced by *Staphylococcus aureus*. When these molecules make contact with red blood cells in the blood agar medium, the cells are destroyed.

TABLE 16.1
Staphylococcus aureus extracellular products and their actions.

Product	Action
Hemolysins	
alpha	Destroys red blood cells, dermonecrotic (skin destruction), leukotoxic.
delta	Destroys red blood cells.
Leukocidin	Inhibits phagocytosis by granulocytes; causes loss of mobility, swelling, and bursting.
Enterotoxin	Heat-stable toxins of five known types. Responsible for gastrointestinal upset typical of food poisoning.
Exfoliatin	Toxin responsible for scalded skin dermatitis; causes the loss of surface layers of skin.
Coagulase	Reacts with prothrombin to form a complex which has the ability to cleave fibrinogen and cause the formation of fibrin clot.
Hyaluronidase	Breaks down hyaluronic acid located between tissue cells, allowing for penetration.
Staphylokinase	A fibrin clot-splitting enzyme; may be released to allow the pathogen to spread through tissue.

favored name is **scalded-skin syndrome** (Figure 16.2). This infection has its highest incidence in neonates (infants to the age of four weeks) and is usually found at the umbilical stump and circumcision site. If the infection spreads to include a more extensive area, it may be referred to as **pemphigus (impetigo) neonatorum.** Pemphigus neonatorum is the most frequent nursery-acquired staph infection.

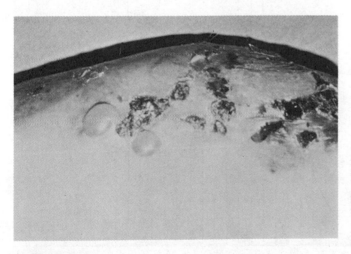

FIGURE 16.2
Scalded-skin syndrome is also known as bullous impetigo, and this *Staphylococcus aureus* skin infection typically occurs in neonates. (From *Bacterial Infections of the Skin,* Abbott Laboratories.)

This disease is typically spread through the nursery by hand contact. The infection progresses from the formation of small vesicles on the skin surface to fluid and bacteria-filled bullae which eventually rupture to form a red, thin varnishlike crust. The final form may resemble a ringworm infection. Topical antibiotics may be used in the treatment; however, the high incidence of plasmid-transferred, drug-resistant *S. aureus* in hospitals may require the use of semisynthetic drugs such as oxacillin or nafcillin.

> **1** What are the four antigenic types of staphylolysin?
> **2** Of what value is phage typing?
> **3** Define the term "purulent."

Staphylococcus aureus may also be responsible for deep-tissue diseases. The type of disease depends on the portal of entry and where the pathogens become localized. If the bacteria enter through the upper respiratory tract, sinusitis, middle ear infections, or pneumonia may result. Staphylococcal pneumonia is a very serious disease due to the release of many necrotizing toxins and hemolysins. Treatment of such an infection is especially difficult when the strain is drug resistant. If bacteria enter the blood through a break in the skin, either accidentally or as a result of surgery, the pathogens may cause septicemia. Localization of these pathogens in the bone marrow will then result in osteomyelitis. If the infection becomes established in the periosteum (sheath of tissue surrounding the bone), the disease is known as periostitis. Any tissue or organ of the body can be infected by *S. aureus*.

The primary reservoir of infectious pathogenic staph is humans. The pathogen may be carried in the nasal passageways, as well as on the skin. It has been estimated that between 30 and 40 percent of the general population carry coagulase-positive staph in the anterior part of their noses, and people with "runny noses" are responsible for transmission by hand contact. Once transmitted, *S. aureus* has a variable incubation period which may be as long as ten days, depending on host susceptibility. Newborns, the elderly, and debilitated people are most susceptible to infection and disease. Because of the difference in drug resistance between "hospital staph" and that found outside the hospital, infections acquired outside the hospital will generally respond more quickly to antibiotic therapy.

Another staphylococcal-induced disease came into national prominence during the late 1970s. **Toxic shock syndrome (TSS)** is an acute disease associated with strains of *S. aureus* that produce a unique epidermal exotoxin. Confirmed cases of TSS have been found to occur more commonly in caucasian women under 30 years of age who are menstruating and using tampons regularly.

The association between tampon use and TSS is well-known; however, the disease has also occurred in males and preadolescent females with *S. aureus* infections localized in areas other than the vagina. Although the overall risk of TSS is low (1.2 to 3/100,000 women/year), the documented correlation with tampon use has resulted in the removal of one brand of tampon from the market and the use of warning labels on other brands (Figure 16.3). It is suspected that *S. aureus* is carried from the fingers or the introitus into the vagina in the process of tampon insertion. While many women make it a practice to wash their hands after insertion, few apparently wash before insertion. Once inside, the bacteria may find the environment suitable for the elaboration of the toxin which enters the body through microulcerations of the vaginal wall. Although conventional and superabsorbent tampons may produce mucosal drying and epithelial changes, microulcerations occur more often after the use of superabsorbent tampons. The disease typically begins abruptly with fever, vomiting, diarrhea, occasionally a chill, and abdominal pain or cramps. Hypotension develops within 72 hours. There is a rash which is often attributed to the fever but may go unnoticed. A thick, odorless vaginal discharge may be present but go unnoticed in a menstruating patient. Many women complain of sensitive skin, sore throat, or muscle tenderness. A confirmed case of TSS includes fever, hypotension, rash, desquamation, involvement of organ systems, and the absence of evidence for other causes.

1 Why are some forms of *S. aureus* known as "hospital staph"?

2 How might a woman prevent toxic shock syndrome?

STREPTOCOCCI: *STREPTOCOCCUS PYOGENES*

1 Morphology and Staining Reaction: Gram-positive cocci in chains or pairs.

2 Respiration-Fermentation: Facultative anaerobes; ferment acids from glucose and maltose.

Important Information About Toxic Shock Syndrome Based On Food and Drug Administration (FDA) Advice

Tampons have been associated with a rare but serious disease called Toxic Shock Syndrome (TSS) which sometimes can be fatal. Toxic Shock Syndrome occurs mainly in girls and women using tampons during their period.

WARNING SIGNS OF TSS ARE:

(1) SUDDEN FEVER (USUALLY 102° OR MORE) **AND** (2) VOMITING **OR** DIARRHEA.

If you have these signs during your period, you should remove the tampon at once, discontinue use, and see your doctor right away.

There may be other signs such as a sudden drop in blood pressure, dizziness, or a rash that looks like a sunburn. If you have any of these signs also, you may need emergency medical care.

The Food and Drug Administration (FDA) offers this advice:

1. **You can almost entirely avoid the low risk of getting TSS by not using tampons.**
2. **If you choose to use tampons, you can reduce your risk by using them on and off during your period.** For example, you may want to use tampons during the day and napkins at night.
3. **About one in every three girls or women who have had Toxic Shock Syndrome have gotten it again. So, if a doctor has told you you have had Toxic Shock Syndrome, or, if you believe you have had the disease, do not use tampons until you check with your doctor.**

Toxic Shock Syndrome is believed to be a recently identified disease caused by a bacterium called Staphylococcus aureus. Further research is being done to find out more about this disease and why tampons have been associated with it.

Reporting of this disease has been increasing. It is estimated currently that as many as 15 of every 100,000 girls and women who are menstruating will get this disease each year.

If you have any questions, please call your doctor.

10/21/80

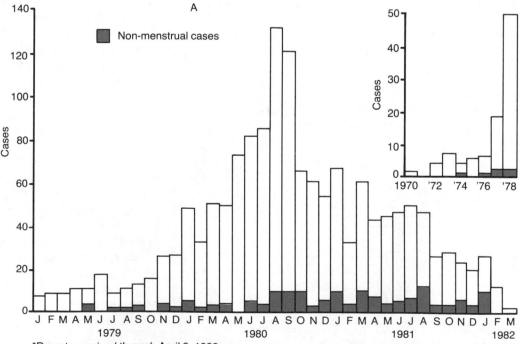

FIGURE 16.3
(A) This toxic shock syndrome warning is typical of the information made available to tampon users. (B) Confirmed cases of TSS, United States, January, 1970–March, 1982.* (Part A courtesy of the Food and Drug Administration. Part B courtesy of the Center for Disease Control, Atlanta.)

*Reports received through April 9, 1982.

*The current CDC case definition is the original case definition (1) with 2 modifications suggested by the Conference of State and Territorial Epidemiologists: (1) orthostatic dizziness is now considered sufficient evidence of hypotension, and (2) the presence of *Staphylococcus aureus* in blood cultures does not exclude a case from consideration. The change in case definition results in the reclassification of fewer than 5% of cases.

3 Motility: Nonmotile.

4 Source of Infection-Reservoir: Found in human mouth, throat, and respiratory tract, and in dust from sick rooms and hospitals.

5 Disease: Pneumonia, impetigo, scarlet fever, septic sore throat, cellulitis, erysipelas, puerperal fever.

6 Pathogenic Relatives: *S. pneumoniae, S. mutans, S. faecalis.*

7 Features: Produce a wide variety of toxins and necrotizing enzymes; responsible for hypersensitive immunogenic diseases, rheumatic fever and acute glomerulonephritis.

The streptococci are Gram-positive chains of cocci that grow in small, uncolored colonies on nutrient agar. This group of bacteria contains a great variety of pathogenic and non-pathogenic forms. The pathogenic streptococci usually have a thick capsule and are able to cause a greater variety of diseases than any other pyogenic bacterium. In an attempt to better understand these microbes, several classification systems have been developed. One older system is based on such characteristics as pathogenicity, cell products, and location of the microbe on the body. The four separate strep groups included in this system are pyogenic, viridans, enterococci, and lactic. **Pyogenic** strep can release a wide range of toxins and enzymes which cause host damage. Their most characteristic products are hemolysins (streptolysins). Microbes in this group include *S. pyogenes, S. pneumoniae,* and *S. scarlatinae.* The **viridans** strep are best represented by *S. viridans.* This microbe is an opportunistic pathogen that appears as normal flora on the surface of the skin. The **enterococci** (e.g., *S. faecalis*) are found in the gastrointestinal tract as normal resident flora. The **lactic strep** are nonpathogenic and very useful. This group includes *S. cremoris, S. lactis,* and *S. thermophilis,* all of which are used by the dairy industry for the fermentation of such products as yogurt and cultured buttermilk.

The more current and useful classification system is known as the Lancefield system, which is based on the antigenic nature of cell-wall carbohydrate known as the C substance. For example, Group A strep are primarily human pathogens; Group B cause mastitis (udder infection) in cattle; Group C may be pathogenic for lower animals; and Group D are opportunistic pathogens found as commensals in the gastrointestinal tract of humans. Virulence of Group A strep is directly related to the release of a variety of products (Table 16.2). Each has a specific action on host tissue, and in some instances is characteristic of a particular disease syndrome, for example, the effects of erythrogenic toxin during scarlet fever. The principal diseases caused by beta-hemolytic, Group A streptococci are septic sore throat, scarlet fever, impetigo, septicemia, erysipelas, cellulitis, pneumonia, and peritonitis.

Septic Sore Throat   This infection is also known as strep throat or streptococcal pharyngitis. It begins with the accumulation of large numbers of beta-hemolytic, Group A *S. pyogenes* and the release of toxic cell products. The incubation period is from one to three days. Their action in the pharynx stimulates an inflammatory response, and the lysis of leukocytes and red blood cells. An inflammatory exudate of cells and fluid is released from blood vessels and deposited in the surrounding tissues. The infection results in fever, sore throat, tonsillitis, and may progress to an inner ear infection. About 20 percent of all cases are asymptomatic, and it has been estimated that 40 percent of those experiencing the disease become carriers. As long as there is a nasal discharge from untreated people, the period of communicability may last for weeks or months. The incidence of the disease is highest in young school children. In the past, this strep infection would often progress to scarlet fever, which in turn stimulated the immunogenic disease rheumatic fever. Penicillin and other antibiotics have been very effective in the treatment of strep throat. The pathogenic strep have shown only a very slow change to drug-resistant forms. Immunity to the toxins released during the infection develops, but repeated infections result from the presence of different antigenic forms of the toxins.

Scarlet Fever   Scarlet fever is caused by the invasion of erythrogenic, toxin-producing *S. pyogenes* in the upper respiratory tract or pharynx of susceptible individuals. A toxin-produced rash develops as small "goose pimples" on the skin within 12 to 24 hours and spreads to include the entire body (excluding the face) in about 36 hours. During convalescence, the skin begins to slough off by a process known as **desquamation.** The lymph nodes become swollen and tender during the height of the infection and may remain sore for as long as three to six weeks. The tongue develops a "furry" white coating in the first few days that changes to the typical "strawberry tongue" after about four to five days. High fever, nausea, and vomiting also occur with scarlet fever. The harmful effects of the erythrogenic toxin seldom appear during subsequent streptococcal infections, since a high antibody titer is produced as a result of initial infection.

The **Dick test** may be performed to determine the presence of an immunity to scarlet fever. This involves the intradermal injection of a small amount of erythrogenic toxin. If no neutralizing antibody is available to react with this antigenic material, there will be a localized rash and skin damage at the site of injection. No reaction will occur if antibodies are present as a result of a previous sensitization. To determine whether a skin rash is caused by erythrogenic toxin, the **Schultz-Charlton test** is performed. A small amount of erythrogenic antitoxin is injected intradermally at the center of the rash. If the reddening is caused by the presence of the toxin, the antibody will neutralize the toxin and produce a blanching, or clearing, of the skin at the site

TABLE 16.2
Group A streptococcal products.

Streptococcal Product	*In Vivo* Effects	*In Vitro* Effects
Streptolysin S	Intravascular hemolysis; necrosis, induces arthritis and renal tubule necrosis.	Lyses erythrocytes, leukocytes, tumor cells, mesenchymal cells, and platelets by altering membrane permeability; kills leukocytes; releases enzymes from lysosomes; inhibits phagocytosis.
Streptolysin O	Lethal to mice, rabbits, myocardial necrosis; intravascular hemolysis; cardiotoxicity.	Lyses erythrocytes, leukocytes, tumor cells, mesenchymal cells, and platelets by altering membrane permeability; causes systolic contraction of perfused heart; constricts coronary arteries in perfused heart; releases lysosome enzymes.
Streptokinase	Enhances spread of infection.	Activates plasminogen to plasmin, which hydrolyzes various proteins, liberates chondroitin sulfate from collagen; activates complement components, and generates permeability and chemotactic factors.
Mitogen	May be responsible for arthritis.	Causes transformation of lymphocytes to blast cells; lymphocytes from rheumatic fever patients show diminished reactivity to mitogen.
Hyaluronidase	Acts as spreading factor and may enhance spread of infection.	Hydrolyzes hyaluronic acid.
Deoxyribonuclease	May enhance multiplication of virulent streptococci.	Cleaves DNA.
Erythrogenic toxin	Induces skin rash, probably by delayed hypersensitivity; pyrogenic; lethal to rabbits; causes myocardial necrosis; suppresses RES function; inhibits antibody formation; inhibits phagocytosis; increases host response to streptolysin O and endotoxin; causes carditis.	Cytotoxic in tissue culture.
Hyaluronic acid	Enhances invasiveness of streptococci.	Antiphagocytic.
M antigen	Enhances invasiveness of streptococci; protective antigen of streptococci; may cause glomerulonephritis in rats by immune-complex formation.	Toxic to leukocytes and platelets; antiphagocytic.
C polysaccharide	Causes granulomas when injected; causes necrotic lesions of myocardium; cross reacts with mammalian connective tissue antigens; in immune complex, causes toxic manifestations in animals.	Group antigen.

SOURCE: I. Ginsburg. *Journal of Infectious Disease*, 126:419–56, 1972. Reprinted by permission of the University of Chicago Press.

of injection. The incidence of scarlet fever has fallen in many countries because of the quick identification of strep throat infections and the use of antibiotics such as penicillin.

---

1 The Lancefield classification system is based on what cellular trait?

2 Approximately what percentage of the population are asymptomatic carriers of *S. pyogenes*?

3 What compound is responsible for the rash in scarlet fever?

4 Distinguish between the Dick test and the Schultz-Charlton test.

Impetigo Contagiosa This infection, also known as streptococcal pyoderma, typically occurs in young school-age children. It is caused by beta-hemolytic, Group A streptococci that are spread by direct contact from one person to another and from one place on the skin to another. The disease begins as small erythematous spots on the epidermis and develops into thin-roofed vesicles (vesicopustules) which rupture, discharge a serous exudate and dry to a thick, honey-colored crust (Figure 16.4). The initial stages of vesicle development may resemble chickenpox. The patient experiences burning and itching (pruritis) sensations at the site of infection. The disease is easily brought under control in most cases with the use of penicillin, erythromycin, or clindamycin. Topical antibiotics may be applied in addition to those given orally. Clothing or linens that have come in contact with the microbes should be washed separately. Spread of the disease to others may be reduced by keeping the child out of school until the infection has been brought under control. The most serious complication of the infection that may result is acute glomerulonephritis. This hypersensitivity, or immunogenic, disease may develop between two and four weeks after the impetigo and occurs in about 10–15 percent of the cases.

Pneumococcal Pneumonia About 90 percent of all upper respiratory diseases known as pneumonia are caused by *S. pneumoniae* and are referred to as pneumococcal pneumonia. The microbes occur as normal flora in 20–50 percent of the population and are typically found in the nasopharynx. They are easily spread by these carriers in droplet nuclei. The disease begins after the bacteria are acquired through the upper respiratory tract and become established in the alveoli of the lungs. First symptoms include severe chill, shaking, fever, and chest pain. As the bacteria reproduce in the pleural cavity, they cause an inflammation known as **pleurisy.** If the pleural cavity fills with fluid and becomes

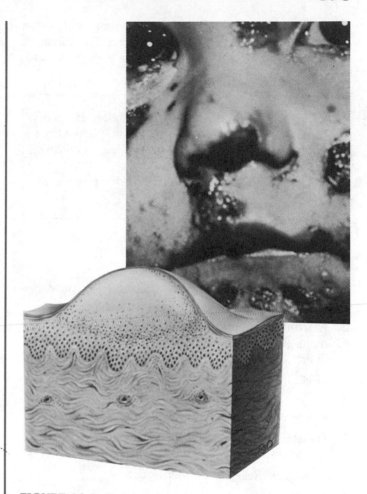

FIGURE 16.4

This child demonstrates the characteristic lesions associated with impetigo contagiosa. Note how the surface of the skin rises to form the vesicular "bubble." (From *Bacterial Infections of the Skin,* Abbott Laboratories.)

infected, a disease state known as **empyema** becomes established. As the body attempts to wall off the lesion by building fibrin containing serous fluid, lung tissue becomes consolidated, making expansion and contraction of the lungs difficult and painful. Because the bacteria readily move into the blood and establish a bacteremia, blood cultures from patients are often positive for *S. pneumoniae.* Penicillin and other related antibiotics may be used effectively in the treatment of pneumococcal pneumonia and have resulted in a greatly reduced mortality rate. The body also helps control the disease by responding immunologically to the presence of antigenic capsular polysaccharides. Even though the immunity that follows disease is not long-lived, it does prevent frequent reoccurrences. In 1977, a pneumococcal polysaccharide vaccine was released for use in preventing infection in the elderly and those with chronic respiratory and cardiac conditions such as asthma and emphysema. The vaccine is made from the fourteen most common pathogenic serotypes of *S. pneumoniae*.

---

1 What is the most serious complication that may result following a case of impetigo contagiosa?

2 Inflammation of the pleural cavity is known by what term?

3 From what cellular material has the pneumonia vaccine been manufactured?

NEISSERIA: N. GONORRHOEAE

1 Morphology and Staining Reaction: Gram-negative, diplococci.

2 Respiration-Fermentation: Facultative anaerobes; ferment glucose to acid. Cultured on Thayer-Martin medium.

3 Motility: Nonmotile.

4 Source of Infection-Reservoir: Other humans only, occurs in carrier and clinical cases.

5 Disease: Gonorrhea, opthalmia neonatorum, vulvovaginitis in children.

6 Pathogenic Relatives: *N. meningitis*.

7 Features: Humans are the only known reservoir; communicability period may last months or years; no immunity develops after infection; pili enable pathogen to attach to host cells; may be found as intracellular parasites.

These small, Gram-negative diplococci are responsible for the venereal disease known as **gonorrhea** that only occurs in humans. *Neisseria gonorrhoeae* is a very fastidious pathogen and requires very complex nutrients for growth. High moisture content and carbon dioxide are essential for culture and the optimum growth temperature is 37°C (98.6°F). The high degree of sensitivity is primarily responsible for determining the mode of transmission (i.e., intimate sexual contact), since *N. gonorrhoeae* can only survive outside the body for one to two hours. Gonococci enter the body primarily through the nonciliated columnar urethral or buccal epithelia. Initial infection is dependent on the presence of pili on the gonococcal surface that enable the cells to adhere to the host cells. These organelles appear to be essential for the infection process. The gonococci are then enveloped by the microvilli and become intracellular parasites. Phagocytes may also contain gonococci but are un-

able to destroy them. Since they are intracellular parasites, the immune system also has little effect. Following envelopment, the host tissue responds with infiltration of mast cells, PMNs, and plasma cells. The symptoms of the disease in males are considerably different than in females. Urethral infection in males is characterized by a burning sensation (dysuria) while urinating and the discharge of a yellowish pus. The infiltration may become so severe that the urethra closes (strictures develop), making urination very painful and difficult. This has led to the more common names for the disease, "clap" or "strain." The symptoms in females are much less obvious and may be limited to an increased vaginal discharge of yellowish, watery fluid. These minor symptoms may be unnoticed and result in infected females going untreated for longer periods than males. The absence of major symptoms enables females to serve as a major reservoir of infection.

Infection may progress through the urogenital system to include the prostate, epididymus, and testis in males, and the Fallopian tubes and ovaries in females. Strictures formed in the Fallopian tubes may result in sterility or an increased chance of embryo implantation in the tube, which is potentially fatal. In some cases, the gonococci enter the blood and produce fever, chills, and loss of appetite. These pyogenic bacteria may then become localized in the joints, causing arthritis; in the heart, resulting in endocarditis; or in the skin, where small red pustules are formed. In some females, the disease becomes chronic and results in an asymptomatic cervicitis. Other forms of gonorrheal infection are listed in tables 16.3 and 16.4.

Although gonorrhea is far less damaging or fatal than syphilis, the incidence of the disease is epidemic in the United States. Drug resistance has increased steadily since the introduction of penicillin in the 1940s, and new strains of penicillinase-producing *N. gonorrhoeae* (PPNG) have been identified. Because the antigenic nature of the microbe is weak and they are intracellular parasites, no natural active immunity has been shown to develop. Changes in sexual behavior in the United States, especially among teenagers, is thought to be primarily responsible for the maintenance of high morbidity rates (Figure 16.5). Another factor which also influences the rate has been the extensive use of birth control pills. Studies have shown that women on the "pill" are more susceptible to infection, since the drug alters the

---

1 What is the reservoir of infection for *N. gonorrhoeae*?

2 What is the reason for calling an infection of gonorrhea "clap"?

3 What does PPNG mean?

TABLE 16.3
Gonococcal diseases.

Disease	Mode of Transportation	Pathogenesis
Vulvovaginitis of children	Intimate, direct contact with infectious exudate on articles; thermometers, materials placed in the vagina; sexual contact.	Inflammation of urogenital system in prepubescent females; nonpurulent discharge, self-limiting disease with recovery in most cases within six months; carrier state may be formed; may involve the labia and thighs.
Gonococcal ophthalmia neonatorum	Contact with gonococci in birth canal during childbirth.	Acute swelling and inflammation of the conjunctiva of the eyes in newborns; mucopurulent discharge containing gonococci; may result in corneal ulcerations, perforations, or blindness.

pH of vaginal secretions, making them less acid and less antimicrobial. Contraceptives such as low-pH jellies and foams are better able to reduce infection, as is the use of condoms which prevent gonococcal transmission.

N. meningitidis   This close relative of *N. gonorrhoeae* is responsible for **bacterial meningitis,** which is inflamma- tion of the meninges of the central nervous system. The cocci are normally found on the mucosal epithelium of the nasal passageway and may enter a susceptible host through breaks in this tissue. The pathogens are passively carried through the circulatory system in the early stages of the dis- ease (within 24 hours) and may cause small hemorrhages on the skin. These **petechiae** usually form on the wrist and

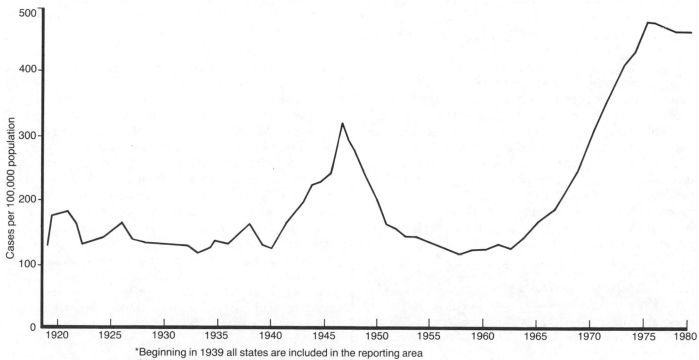

*Beginning in 1939 all states are included in the reporting area
(Military cases included 1919–40 excluded thereafter)
**1919–40 Fiscal years. Twelve month period ending June 30 of year specified — 1919–40
1941–80 calendar years

FIGURE 16.5
Reported cases of gonorrhea per 100,000 population by year, United States,* 1919–80.**
(Courtesy of the Center for Disease Control, Atlanta.)

TABLE 16.4
Locations with identified strains of beta-lactamase-producing *Neisseria gonorrhoeae.**

Africa	Americas	East Asia	Europe	South East Asia
Morocco	Canada	Philippines	France	Indonesia
Ghana	United States	Hong Kong	Belgium	Singapore
Mali	Mexico	Taiwan	Netherlands	Malaysia
Nigeria	Panama	Guam	United Kingdom	Thailand
Central African	Argentina	Japan	West Germany	India
Republic	Colombia	Republic of Korea	Denmark	Sri Lanka
Gabon		New Zealand	Poland	
Zaire		New Hebrides	Switzerland	
Madagascar		Australia	Sweden	
Zambia			Norway	
Senegal			Finland	

*Information obtained through WHO Epidemiological Surveillance System; adapted from PAHO Epidemiologic Bulletin.
SOURCE: *Morbidity and Mortality Weekly Report,* vol. 31, nos. 1 and 2, Jan., 1982.

ankles. During this time, chills, fever, headache, and drowsiness also occur. In some cases, the microbes invade the adrenal glands and cause the Waterhouse-Friderichsen syndrome, a septicemia which causes circulatory system collapse. This reaction is fatal but rare. The usual course of the disease results in the invasion of the meninges of the spinal cord and cortex of the brain. Bacteria and pus found in the spinal fluid can be used as a diagnostic characteristic. Symptoms of meninges infection include fever, stiffness of the neck, drowsiness, vomiting, and a slowed pulse. One of the characteristic signs of meningeal irritation is a positive Brudzinski test. This is the flexion of the ankle, knee, and hip when the patient's neck is bent forward.

The disease may be spread by direct contact or droplet nuclei from healthy carriers. Patient isolation is required for twenty-four hours after chemotherapy is started. In general, children over three months old and adolescents are most susceptible. The incidence of disease decreases with age, since normal resistance to infection increases. Both the sulfonamides and penicillins may be used for treatment. However, resistance to sulfa drugs has greatly increased, and penicillin has become the drug of choice. This antibiotic must be given in high doses during acute meningeal infection, since penicillin does not easily cross the meningeal membrane prior to this stage.

1 What are the symptoms of meningococcal meningitis?
2 How is this disease spread?

GRAM-POSITIVE, SPORE-FORMING BACILLI

BACILLI: *BACILLUS ANTHRACIS*

1 Morphology and Staining Reaction: Gram-positive bacilli in chains.

2 Respiration-Fermentation: Aerobes; fermentation of dextrose and sucrose results in acid production.

3 Motility: Nonmotile.

4 Source of Infection-Reservoir: Soil.

5 Disease: Anthrax in farm animals and humans.

6 Pathogenic Relatives: *B. cereus,* food poisoning.

7 Features: Highly refractile endospores formed centrally within the cell; endospore diameter does not exceed the width of the cell; colonies have a "Medusa head" appearance on nutrient agar.

Predominantly a disease of domesticated animals, **anthrax** is caused by a pathogenic, aerobic, spore-forming bacillus. Spores may remain viable in the soil for as long as thirty years. This disease has been studied by many researchers, including Robert Koch, who used it to demonstrate the relationship between microbes and infectious disease. Pasteur attempted to attenuate the pathogen for use as a vaccine; however, this effort met with only limited success. No useful vaccination against anthrax is presently available. The disease has been successfully brought under control in developed countries through isolation and eradication programs designed to destroy or isolate and cure diseased animals. Humans may acquire the microbe through inhalation, ingestion of spores, or more likely, through skin contact with spore-containing animals or their hides. Individuals most likely to acquire the disease are veterinarians, butchers, ranchers, and industrial workers who handle imported hides, wool, or bone products.

After entering through the skin, the pathogen produces a black eschar (scab) with surrounding edema and a "pearl wreath" of small, bacilli-containing vesicles. The infection may heal spontaneously or progress into the body as a septicemia. Virulence is due to the presence of a thick capsule and the release of toxins that have antiphagocytic activity. This disease is highly invasive. Symptoms of the septicemic form are fever, shock, tissue swelling, electrolyte imbalance, and kidney failure. Immunity to the toxins develops after the disease. Penicillin, erythromycin, and sulfa drugs are used to treat the few infections that occur each year. Only 15 cases were recorded during a recent six-year period.

CLOSTRIDIA: *CLOSTRIDIUM TETANI*

1 Morphology and Staining Reaction: Gram-positive bacilli in short chains with subterminal endospores showing a "drumstick" appearance.

2 Respiration-Fermentation: Anaerobes; acids and alcohols produced by fermentation of peptone.

3 Motility: Motile.

4 Source of Infection-Reservoir: Soil and gastrointestinal tract of humans and animals.

5 Disease: Tetanus (lockjaw).

6 Pathogenic Relatives: *C. perfringens* (gas gangrene), *C. botulinum* (food poisoning).

7 Features: "Drumstick" appearance; production of a specific exotoxin.

Tetanus, or "lockjaw," gets its name from the spasmic contractions (tetani) of muscles in the jaw, neck, and face, which prevent the mouth from being opened. This symptom is produced by a powerful nerve toxin, **tetanospasmin,** released by these obligate anaerobes as they grow in tissue that has been deprived of oxygen. A second toxin, **tetanolysin,** is a powerful hemolsyin that aids in tissue destruction. The spasms usually begin at the site of infection and progress throughout the body. The cause of death is asphyxia (suffocation) caused by the contraction of respiratory muscles. Tetanus has been feared by soldiers throughout history, since infection is most likely to occur in tissues that have been damaged by puncture wounds or in lacerations that have not been promptly cleansed and treated. During World War I, countless soldiers died from tetanus-infected wounds. The bacteria were picked up from the clostridia-laden battle fields of France that had been fertilized with manure. Puncture wounds caused by nails (not necessarily rusty), wood slivers, and broken glass are of particular concern. These materials may harbor *C. tetani* and inject the pathogens deep into tissue made anaerobic by the puncture. The incubation period of the disease ranges from two to fifty days, and symptoms first appear as a headache, stiff neck, and minor muscle spasms. *C. tetani* remains localized at the penetration site, producing toxins that spread through the body. The tetanospasmin chemically combines irreversibly with nerves, and its presence blocks the transmission of nerve impulses. Therefore, it is imperative that the toxin be neutralized as soon as possible. Antitoxin antibody may be used; however, immunization with toxoid is by far the most successful means of control. Antibody titer builds rapidly after a booster given either on a regular basis or immediately after tissue injury.

Clostridium perfringens   Another important species, *C. perfringens,* is the most frequent cause of **gaseous gangrene.** Like tetanus, gangrene has also been associated with war wounds but may occur in any injured tissue that has become anaerobic. The microbe produces subterminal endospores that may be found in the soil, intestinal tract, or on the skin. Their growth as anaerobes is restricted to tissues which have become damaged and anaerobic as a result of other diseases (for example, bowel obstructions), accidental injury, or medical procedures. Unlike most clostridia, *C. perfringens* is aerotolerant and can grow in conditions of low oxygen. Once established, the vegetative form releases gases, and a variety of toxins and enzymes such as hemolysin, collagenase, proteases, hyaluronidase, and deoxyribonuclease. These produce a severe, spreading, tissue necrosis that extends the anaerobic area, and encourages the growth and further spread of infection. Some toxins spread through the body causing heart and kidney destruction, and ultimately death. In addition to the use of antibiotics, superficial infections may be brought under control by elevating the oxygen level of the diseased tissue. A special chamber may be placed around the gangrenous area and infused with oxygen (Figure 16.6). Oxygen is forced into the tissue and causes the anaerobes to either die or revert to their harmless endospore form. Once brought un-

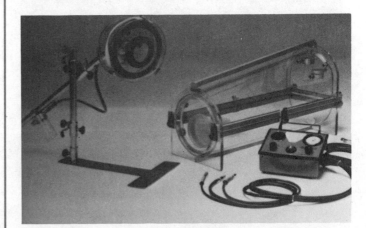

FIGURE 16.6
Treatment times for ulcerations and lesions originating from *Clostridium perfringens* are substantially reduced by using a pulsed oxygen chamber (hyperbaric chamber). The median healing time with such a unit is nineteen days, a significant improvement over conventional treatments. (Courtesy of the Topox Corporation.)

1 How might a human acquire anthrax?
2 What two toxins contribute to the virulence of *C. tetani?*
3 How does *C. tetani* produce its disease symptoms?

der control, tissue healing and revascularization occurs to reestablish the aerobic conditions.

C. perfringens also produces an enterotoxin responsible for **food poisoning.** This molecule is produced during the spore-formation stage of growth. Although not as strong as the toxin released by *Clostridium botulinum* (refer to Chapter 23), this enterotoxin acts on the lining of the intestine to cause diarrhea and nausea that last for one day or less. The poisoning occurs after ingesting foods contaminated by endospores from the soil or fecal material. Since the enterotoxin is easily destroyed by heating, all such foods should be thoroughly cooked and served hot. Foods typically identified in *C. perfringens* enterotoxin poisoning are meat pies, canned gravies, stews, and poultry stuffings (Figure 16.7).

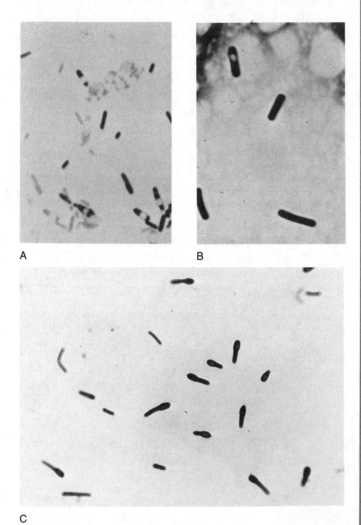

A

B

C

FIGURE 16.7
(A) *Clostridium botulinum;* (B) *C. perfringens;* and (C) *C. tetani.* (Photos A and B from S. S. Schneierson, *Atlas of Diagnostic Microbiology.* Abbott Laboratories. Photo C courtesy of Carolina Biological Supply Company.)

---

1 What does the term "aerotolerant" mean?
2 Which foods have the best growing conditions for *C. botulinum?*

ENTERIC GRAM-NEGATIVE BACILLI

The enteric Gram-negative bacilli are a large group typically found in the intestinal tract as normal flora (e.g., *E. coli*) or pathogens (e.g., *Salmonella*). As described in Chapter 1, many groups of microbes have "overlapping" genetic relationships which make them difficult to separate into distinct species. All members of this group have a close, complex, evolutionary relationship. Therefore, the enteric bacilli are divided on a functional basis into groups, subgroups, and serotypes. All are facultatively anaerobic, nonspore-forming rods and have colony characteristics that resemble one another very closely. Those with capsules form smooth colonies ("S" strains), while those which lack this outer covering appear rough on agar ("R" strain). Most members of the group contain endotoxic lipopolysaccharide (LPS) material in their cell walls that is responsible for their toxicity, can be used as serotyping antigens, and induces some nonspecific immunity to infection. The effects of the endotoxins are similar regardless of which bacterium produces them (Table 16.5). Some members of the group are also able to produce potent exotoxins responsible for gastroenteritis and related symptoms (Table 16.6). Based on genetic and symptomatic similarities, the enteric bacilli have been subdivided into several groups.

COLIFORMS

Members of this group, which includes *Escherichia, Enterobacter, Klebsiella,* and *Serratia,* make up a large part of the intestinal flora of humans and domesticated animals. Although they are normally kept under control, enteropathogenic forms may cause disease by the release of endotoxins, the invasion of the intestinal tract, or being displaced to other parts of the body. All have been found to be responsible for urinary and upper respiratory tract infections, gastroenteritis, and peritonitis. The incidence of nosocomial infections caused by these bacteria is increasing and of great concern to infection control personnel. Though normally associated with upper respiratory infections, *K. pneumoniae* is also responsible for a high incidence of urinary tract infections. *Serratia* has been found to be responsible for septicemia following surgery, and *Enterobacter* is involved in both types of infections.

TABLE 16.5
Endotoxins of the enteric Gram-negative bacilli.

Toxin-induced Symptoms	Description
Fever	Endotoxin acts on tissue cells to release pyogens that initiate fever response in brain; a 60–90 minute delay occurs after first injection of endotoxin; IgM is induced and immune tolerance can develop.
Leukopenia	A reduction in the number of leukocytes as a result of cell death; occurs along with fever and decrease in glucose blood level (hypoglycemia).
Hypotension	Vascular constriction (chill) occurs first, followed by dilation, increased vascular permeability, and shock; endotoxins stimulate the release of vasoactive amines.
Acidosis	A drop in pH in heart, kidney, liver, lungs, etc., due to poor circulation, decrease in oxygen level, and increase in acid production and accumulation.
Complement loss	Endotoxins can react with C3 of the complement series and stimulate a decrease in the serum level of this important substance; may also stimulate membrane damage as system is activated.
Disseminated intravascular coagulation (DIC)	The stimulation of clot formation and lysis throughout vascular system followed by necrosis of capillaries and skin hemorrhage; a special form is known as the Shwartzman phenomenon.

TABLE 16.6
Exotoxins of the enteric Gram-negative bacilli.

Source	Toxin	Action
E. coli	Enterotoxin	A plasmid-controlled, heat-labile toxin which acts at the surface of the intestinal villi to stimulate water and chloride ion release into the intestine; also inhibits sodium reabsorption; produces diarrhea.
Klebsiella pneumoniae	Enterotoxin	Stimulates excretion of fluids and electrolytes; causes diarrhea; may be similar to E. coli enterotoxin; may also be found in Citrobacter.
Pseudomonas aeruginosa	Exotoxin	Inhibits protein synthesis and causes necrosis; very similar to diphtheria exotoxin.
Shigella dysenteriae	Exotoxin	Produces diarrhea; inhibits sugar and amino acid absorption in small intestine; also acts as a neurotoxin, causing coma.
Vibrio cholerae	Enterotoxin	Causes excess excretion of water and electrolytes into intestine, causing diarrhea, dehydration, acidosis, and death.

ESCHERICHIA: E. COLI

1 Morphology and Staining Reaction: Gram-negative, short, individual bacilli.

2 Respiration-Fermentation: Facultative anaerobes; microaerophilic; acid and gas produced from glucose and lactose.

3 Motility: Motile and nonmotile strains found.

4 Source of Infection-Reservoir: Colon of humans, domesticated animals, and soil.

5 Disease: Enteritis, peritonitis, cystitis.

6 Pathogenic Relatives: Enterobacter, Serratia, Klebsiella.

7 Features: Normally a beneficial bacterium found as resident flora of the intestinal tract; able to survive well outside the body; fecal pollution indicator.

Probably one of the most discouraging discoveries has been the identification of enteropathogenic forms of E. coli responsible for infantile diarrhea and Gram-negative sepsis. **Infantile diarrhea** has been found to be caused by E. coli with somatic O antigens O-55, O-111, and O-127. The disease is caused by the release of enterotoxin and runs its course in about four days. The infection may affect children before or after weaning, but it is more prevalent in bottle-fed infants and after weaning. Malnutrition plays an important role in the infection. In severe cases, the great water loss, dehydration, and electrolyte imbalance cause death. For this reason, it is extremely important that nosocomial infantile diarrhea be quickly diagnosed and serotyped to locate the source of the enteropathogen. Traveler's diarrhea is also caused by enteropathogenic *E. coli*. However, the diarrhea, abdominal pain, and cramping are caused by the invasion of these new serotypes into the intestinal epithelium. These "foreign" pathogens enter the intestinal tract on foods and in water.

Gram-negative sepsis, or **endotoxic shock,** is an infection that may occur as a nosocomial infection after abdominal surgery. Invading *E. coli* produce a septicemia and release endotoxins, which produce chills, fever, nausea, vomiting, diarrhea, and ultimately prostration (exhaustion). A noticeable drop in blood pressure occurs when the endotoxins are released into circulation as large numbers of *E. coli* enter the stationary growth phase and die. Patients undergoing surgery involving the GI tract are in a high-risk category for this type of infection. Antibiotic therapy has been of great value in reducing the incidence of Gram-negative sepsis; however, there has been an increase in drug-resistant *E. coli* to such widely used broad-spectrum antibiotics as tetracycline, neomycin, chloramphenicol, and streptomycin. Multiple drug resistance is also increasing in *E. coli* populations.

Members of the *Klebsiella-Enterobacter-Serratia* group are very similar Gram-negative coliforms responsible for respiratory, urinary, and gastrointestinal tract infections. *Klebsiella pneumoniae* (also known as Friedländer's bacillus) is a nonmotile, heavily encapsulated bacteria typically associated with inflammatory conditions of the upper respiratory tract and is a cause of bacterial pneumonia. This form of pneumonia causes more tissue damage than strep or staph infections. The necrosis and formation of scar tissue may become extensive enough to require surgery to prevent the infection from becoming chronic. *Enterobacter aerogenes* (formerly known as *Aerobacter*) and *Serratia* are very similar to *K. pneumoniae*. In some clinical situations, it is almost impossible to distinguish among them. *Enterobacter aerogenes* is found in the intestinal tract and is an opportunistic pathogen. This bacterium is often motile, has a small capsule, and may become displaced from the intestinal tract, resulting in urinary tract infection. Septicemia from bacteria on the skin may occur after intravenous infusion of fluids. *Serratia marcescens* are free-living bacteria usually not considered to be pathogenic; however, this opportunistic pathogen is able to cause a nosocomial bacteremia, endocarditis, and pneumonia. Many times, hospital-acquired pneumonia cases have been traced to *Serratia*-contaminated respiratory therapy equipment. Infections caused by

all three of these bacteria respond to treatment with antibiotics, but multiple drug resistance is common (refer to Chapter 11).

---

1 How do the coliforms cause disease in a healthy individual?

2 Name two diseases that may be caused by *E. coli.*

The coliform bacteria known as *Arizona, Edwardsiella,* and *Citrobacter* are occasionally pathogenic to humans. They may cause gastroenteritis and be "super infections" in patients already hospitalized for other infectious diseases. The close physiological, antigenic, and genetic similarity to *Salmonella* has led some microbiologists to include members of this group in the genus *Salmonella;* for example, *Salmonella arizonae.*

PROTEUS: P. VULGARIS

1 Morphology and Staining Reaction: Gram-negative, slightly curved bacilli.

2 Respiration-Fermentation: Facultative anaerobes; acid and gas produced from sucrose and maltose.

3 Motility: Motile, forming swarming colonies on agar surface.

4 Source of Infection-Reservoir: Human intestinal tract and soil.

5 Disease: Eye, ear, peritoneal cavity (peritonitis), cystitis, and polynephritis.

6 Pathogenic Relatives: *Escherichia, Enterobacter, Klebsiella, Serratia.*

7 Features: Swarming colonies on agar surface, proteolytic activity high under aerobic conditions; classified according to both H (flagella) and O (somatic) antigens.

A close genetic relationship exists between coliforms and the genus *Proteus.* Some species once placed in the "Providence" or "paracolon" coliform subgroup have more recently been identified as members of the genus *Proteus. Proteus* species are motile, swarming cells known for their ability to produce proteolytic enzymes that cause putrefaction under aerobic conditions (Figure 16.8). They are found free-living in water, soil, sewage, and are part of the normal flora of the adult intestinal tract. Pathogenic species may cause ear, eye, and urinary tract infections as well as "summer diarrhea" in children, peritonitis, suppurative abscesses, cystitis (inflammation of the urinary bladder), and polynephritis. *Proteus* excretes an enzyme (urease) that necroses kidney tubules and stimulates the formation of kidney stones.

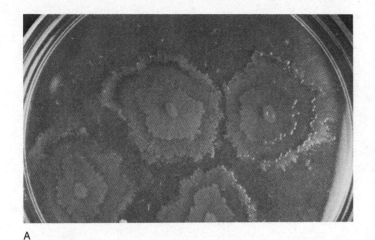

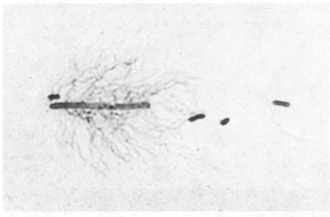

A   B

FIGURE 16.8
(A) *Proteus vulgaris* growing on Endo's agar medium. Note the characteristic "waves" of swarming cells. (B) *Proteus vulgaris* stained to reveal its peritrichous flagella. (Part A from S. S. Schneierson, *Atlas of Diagnostic Microbiology*. Abbott Laboratories. Part B photomicrograph courtesy of Carolina Biological Supply Company.)

The presence of this enzyme is a key diagnostic characteristic of the *Proteus* group. Five common species have been identified: *P. vulgaris, P. mirabilis, P. morganii, P. inconstans,* and *P. rettgeri.* Many of the species found in nosocomial infections demonstrate multiple drug resistance as do other enteric Gram-negative bacilli.

SALMONELLA: S. TYPHI

1 Morphology and Staining Reaction: Gram-negative bacilli.

2 Respiration-Fermentation: Facultative anaerobes; ferment lactose to acid.

3 Motility: Motile.

4 Source of Infection-Reservoir: Human clinical, subclinical, and carrier cases; transmitted through milk, water, and food.

5 Disease: Gastroenteritis, enteric fevers.

6 Pathogenic Relatives: *S. paratyphi* A, B, C.

7 Features: All members of the genus are pathogenic, able to survive phagocytosis, and remain alive within phagocytes causing intracellular infections.

All species of the genus *Salmonella* are pathogenic. Each member is distinguished on the basis of its antigenicity by either serotyping or phage typing (Table 16.7). Serological identification is important in determining the exact antigenic nature of a microbe, and the amount and type of antibody protection developed by an individual. Many diseases can occur repeatedly in the same individual, since each infection is caused by different antigenic forms of the same species. If the host has not been sensitized to the antigen of the second infecting microbe of the same species, there will be no serum antibodies to protect against infection. Serotyping is also important in developing maximally effective vaccines against infection. The most complete protection against infection can only be achieved by the injection of all antigenic forms of the microbe.

---

1 What enzyme does *Proteus* produce that causes necrosis in kidney tubules?

2 What special features of *Salmonella* members better enable them to survive natural body defense mechanisms?

These pathogens all contain endotoxic lipopolysaccharide molecules in their cell walls (refer to Table 16.5). Their primary action on the human host is the alteration of capillary permeability and fever production. An enterotoxin is also produced by some *Salmonella* and is similar to the enterotoxin of *E. coli* (refer to Table 16.6). *Salmonella* infections are acquired by ingesting microbes in contaminated food, water, or milk. The two general types of infection recognized are an acute infection of the gastrointestinal tract and a more gradual infection that affects other tissues.

***Salmonella* Gastroenteritis**   Often called "food poisoning," this type of infection begins abruptly between 8 and 48 hours after ingesting the pathogen, and may last from 2 to 5 days. The two species primarily responsible are *S. typhimurium* and *S. cholerae-suis.* Once inside the gastroin-

TABLE 16.7
Salmonella typing serums.

Group	O Antiserums		H Antiserums	
	Antiserum	Antigen	Antiserum	Antigen
A	1, 2, 12	*S. paratyphi* A	a	*S. paratyphi* A
B	4, 5, 12	*S. paratyphi* B	b	*S. paratyphi* B, phase 1
C₁	6, 7	*S. thompson*	c	*S. cholerae-suis*, phase 1
C₂	8	*S. virginia*	d	*S. typhimurium* phase 1
D	9, 12	*S. gallinarum*	i	*S. typhimurium* phase 1
E	3, 10, 15	*S. anatum* and	1, 2, 3, 5	*S. thompson*, phase 2 and
		S. newington		*S. newport*, phase 2

SOURCE: B. A. Freeman, *Burrows Textbook of Microbiology*, 21st ed. W. B. Saunders, 1979.

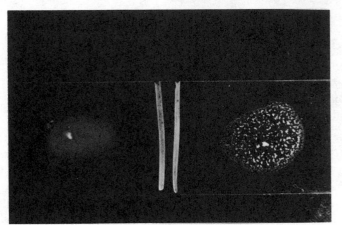

Salmonella species slide agglutination, negative (left) and positive (right). (From S. S. Schneierson, *Atlas of Diagnostic Microbiology*, Abbott Laboratories.)

testinal tract, the endotoxin stimulates fever and is followed by nausea, vomiting, and diarrhea. The severity of infection depends on the serotype of the *Salmonella*. Stool samples contain the pathogen soon after the symptoms occur but *Salmonella* rarely enter the blood during gastroenteritis. If septicemia develops, there is a rapid rise in body temperature, and blood cultures for the presence of *Salmonella* are positive. The bacteria become dispersed throughout the body and may form local foci of infection in the lungs (pneumonia), bone marrow (osteomyelitis), heart (endocarditis), and central nervous system (meningitis).

Microbes lodged in the gall bladder result in the formation of a carrier state, and bacilli are excreted in the feces. Carriers with poor personal hygiene serve as a source of further infection. However, most new infections result from the ingestion of food contaminated with *Salmonella* from infected animals, such as rodents, fowl, turtles, dogs, and cats. Foods made from contaminated poultry, raw eggs, unpasteurized milk, and raw sausage are common sources of gastroenteritis. Although antibodies are formed against the endotoxins, no significant active immunity develops after infection.

Salmonella Enteric Fevers   Typhoid fever is caused by *S. typhi*. About fifty different serotypes (based on the Vi-

surface antigens) have been identified. All are transferred to susceptible individuals by water or food, and from carriers or infected individuals. Unlike *Salmonella*-induced gastroenteritis, typhoid fever is an infection of the lymphatic system and other tissues. The disease begins with the invasion of the mucosal epithelium and rapid movement of the pathogens to lymphoid tissue associated with the gastrointestinal tract. The invading pathogens multiply in the lymphoid tissue, move to the blood, and spread through the body. Blood cultures remain positive for only a short period as the bacilli become localized in various tissue. The typical typhoid symptoms of headache, fever, malaise, spleen enlargement, and constipation result from the necrosis of lymphoid tissue and the liver. Necrosis of Peyer's patches in the intestine produces hemorrhages. The larynx, periosteum, gallbladder, and bone marrow may also be foci of infection. If the intestinal wall becomes perforated, the infecting pathogens enter the peritoneal cavity, causing peritonitis and death.

Although infection does stimulate the formation of antibodies, the degree of active immunity produced is not always sufficient to prevent a second or even third infection. Vaccines have been produced from killed typhoid bacilli but appear to be effective only when the number of infecting bacilli are few in number.

The incidence of typhoid fever has dropped steadily in the United States since the early 1940s. This is primarily because of the effective control of waterborne and milkborne *S. typhi*, the primary source of epidemic typhoid fever (Figure 16.9). Endemic typhoid is caused by the transmission of bacilli from those who are ill, or from convalescent or healthy carriers. This epidemiological form also declined as a result of these same sanitary control measures.

A more mild form of typhoid known as **paratyphoid** is caused by *S. paratyphi* A, B, and C. The symptoms begin with chills and run the same basic course as those caused

1 In what organ can *Salmonella* become localized and established in a carrier state?
2 What are the typical symptoms of typhoid fever?
3 What efforts are chiefly responsible for the sharp decrease in typhoid fever in the United States?

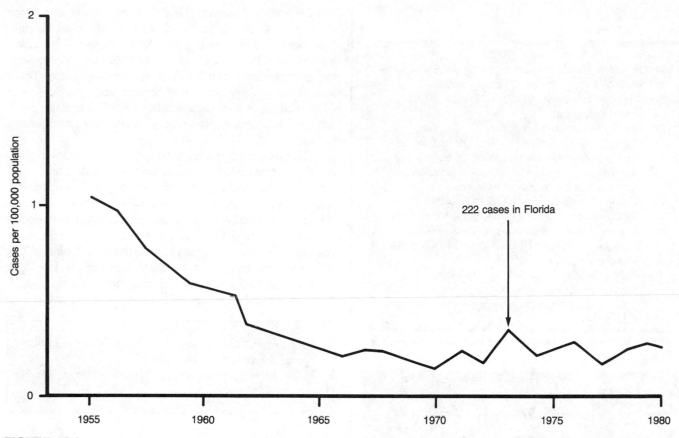

FIGURE 16.9

Reported cases of typhoid fever per 100,000 population by year, United States. (Courtesy of the Center for Disease Control, Atlanta.)

by *S. typhi.* The only way to accurately distinguish between the two diseases is to isolate and identify the bacteria. This disease is less severe, rarely fatal, and has declined in incidence since reaching a high point at the turn of the century.

SHIGELLA: S. DYSENTERIAE

1 Morphology and Staining Reaction: Gram-negative bacilli.

2 Respiration-Fermentation: Facultative anaerobes; ferment glucose to produce acid but not gas.

3 Motility: Nonmotile.

4 Source of Infection-Reservoir: Humans, clinical, subclinical cases.

5 Disease: Shigellosis, or bacillary dysentery.

6 Pathogenic Relatives: *S. flexneri, S. boydii, S. sonnei.*

7 Features: Release a powerful exotoxin through intestinal mucosa.

Members of the genus *Shigella* are divided into four Groups (A, B, C, and D) based on the serotype of their O antigens (Table 16.8). Groups A, B, and C contain thirty-one different serotypes, but Group D contains only one serotype. Virulence is correlated with serotype. All members of this genus are able to cause a disease syndrome called **bacillary dysentery,** or **shigellosis.** In recent years, bacillary dysentery shifted from a disease primarily associated with periods of war to a significant diarrheal disease of infants and children. This intestinal infection is characterized by diarrhea, abdominal cramping, pain, fever, vomiting, and in severe cases, bloody, pus-containing stool. Unlike typhoid, these Gram-negative bacilli rarely enter the circulatory system. *Shigella* have been found to possess endotoxin lipopolysaccharide cell walls, and in some species, (i.e., *S. dysenteriae*) an exotoxin and enterotoxin are produced.

The pathogens enter the gastrointestinal tract and have an incubation period of about 48 hours. Once established in the mucosal lining, they may become the only type of bacteria found in the feces during the disease. The infected mucous membranes become necrotic and ulcerated. This leads to bloody diarrhea and the formation of a "pseudo-membrane" over the tissue. *Shigella* may remain in the epithelium for weeks after the infection and serve as a source of infection to others. Fecal-contaminated fingers, food,

TABLE 16.8
Shigella O antigen groups.

> **1** The non-mannitol-fermenting bacilli:
> Group A. *Shigella dysenteriae.* Including the Shiga dysentery bacillus, the Schmitz bacillus, and the Large-Sachs group of parashiga bacilli as immunologically independent serotypes.
> **2** The mannitol-fermenting bacilli:
> (a) Group B. *Shigella flexneri.* The Flexner dysentery bacilli, having characteristic cultural reactions and antigens in common, i.e., group antigens, but differentiable into numbered serotypes.
> (b) Group C. *Shigella boydii.* Culturally similar to the Flexner bacilli, but serologically unrelated to them or to one another, differentiable as independent numbered serotypes and including the Newcastle-Manchester bacilli.
> (c) Group D. *Shigella sonnei.* The slow lactose-fermenting dysentery bacilli, or Sonne-Duval bacilli, and including the bacilli formerly known as *Shigella ceylonensis* A.

SOURCE: B. A. Freeman, *Burrows Textbook of Microbiology,* 21st ed. W. B. Saunders, 1979.

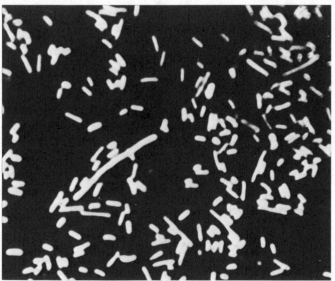

Shigella dysenteriae. Courtesy of the Armed Forces Institute of Pathology.

fomites, and flies serve to spread the bacilli through a population. A healthy carrier state has not been recognized.

The formation of IgA and IgM antibodies occurs after shigellosis. The production of IgA secretory antibody may be helpful in preventing subsequent infections only if a small number of bacilli enter the intestinal tract. However, since these pathogens remain localized in the tract, the rise in IgM antibody titer is of little value in preventing further infection and greatly limits the use of killed bacilli as a form of active immunization. However, there is available an attenuated vaccine for shigellosis that is better than the killed vaccine. Controlling the disease is primarily accomplished by isolating patients with acute cases and establishing effective sanitary procedures which will reduce disease transmission. These include the sanitary disposal of human feces, purification of water, control of flies, pasteurization of milk and dairy products, and the sanitary preservation, preparation, and serving of food.

The treatment of shigellosis requires antibiotics that are capable of penetrating the intestinal wall and making contact with the pathogen. Sulfonamides, chloramphenicol, ampicillin, and tetracycline have been effective agents of control. R plasmid-acquired drug resistance has increased in the past twenty years. This resistance may affect a single antibiotic or result in multiple-drug-resistant strains. Along with antibiotic therapy, it is essential to replace lost body fluids to maintain electrolyte balance and prevent dehydration and acidosis that may be fatal, especially to children.

> **1** The symptoms of shigellosis are produced by what cellular products and materials?
> **2** In addition to antibiotics, what other measures control shigellosis?

VIBRIO: V. CHOLERAE

1 Morphology and Staining Reaction: Gram-negative, short, slightly curved bacilli.

2 Respiration-Fermentation: Facultative anaerobes, oxidize dextrose, galactose, sucrose, and mannitol to acid but not gas.

3 Motility: Motile.

4 Source of Infection-Reservoir: Human clinical and subclinical cases; a carrier state is rare.

5 Disease: Cholera.

6 Pathogenic Relatives: *V. cholerae* biotype *eltor* and *V. cholerae* biotype *cholerae.*

7 Features: Survive well outside the body; easily grown in water containing peptone; easily killed by high temperatures (10 minutes @ 55°C) and chemical disinfectants; may survive in cool, moist places outside the body for days.

Cholera is another diarrheal disease caused by enteric Gram-negative bacilli. Originally localized in India and other parts of Asia, this pathogen has only recently spread throughout the world. Seven major pandemics have been recognized since cholera first moved through Europe in 1817. The bacillus was identified in 1883 by Robert Koch. During his investigations, two co-workers voluntarily swallowed broth cultures of what was then known as *"Spirillum cholerae asiaticae"* and became ill with cholera. This incident, along with laboratory research, confirmed that the comma-shaped bacilli were responsible for the disease.

The first U.S. cases were reported in New Orleans and New York City in 1832. Since that time, several waves have passed through all parts of the world, causing injury and death. The seventh pandemic spread from Indonesia in 1937, and approximately 90,000 cases continue to be identified each year from reporting nations. Transmission of the bacteria occurs through water, contaminated foods, and from chronic carriers. After the microbes are ingested, there is an incubation period of approximately two or three days before symptoms appear. *Vibrio cholerae* adhere to receptor sites on intestinal epithelial cells aided by an adhesive factor associated with the single, long flagellum. Once established, the bacilli release a powerful enterotoxin and mucinase. The enterotoxin is responsible for the excessive chlorine and water loss from surrounding tissues and the inhibition of sodium uptake. Mucinase released during the disease causes the sloughing off (desquamation) of surface epithelium. These changes result in severe water loss and an electrolyte imbalance that ultimately leads to diarrhea, dehydration, acidosis, shock, and death.

Treatment of cholera victims requires both antibiotic therapy and replacement of lost body fluids. In some severe cases, as much as 25 liters of fluid per day are required to adequately rehydrate tissues and reestablish the electrolyte balance. Antibiotic therapy using tetracylines is preferred in comparison to chloramphenicol, which produces more harmful side effects. After infection, both IgA and IgG serum titer levels rise; however, the protection provided by these immunoglobulins is only short lived. Within a few months, the protective level may drop, leaving the individual susceptible to infection. Immunization with killed vibrios has been attempted with some success, but the expense and difficulty of vaccinating large numbers of people does not make this control method practical. Routine sanitary control measures are far superior in preventing epidemics.

The establishment of a cholera infection is limited by the pathogen's susceptibility to stomach acids and the ease with which it is destroyed by normal intestinal flora. The communicable period of a convalescent carrier lasts only a few days. However, in some cases, cholera bacilli may continue to be found in feces for as long as five weeks. A healthy carrier state has been found in rare incidences of biliary tract infection. These cases have little influence on disease transmission and epidemics.

PSEUDOMONAS: P. AERUGINOSA

1 Morphology and Staining Reaction: Gram-negative bacilli.

2 Respiration-Fermentation: Facultative anaerobes; glucose is fermented to acid.

3 Motility: Motile.

4 Source of Infection-Reservoir: Soil and individuals with clinical, subclinical, and carrier stages of infection.

5 Disease: Urinary tract, systemic, and skin infections.

6 Pathogenic Relatives: *P. pseudomallei* (melioidosis), *P. mallei* (glanders disease).

7 Features: Produce a sweet "grape juice" odor in culture and a water-soluble, yellow pigment that fluoresces yellow-green under ultraviolet light. A blue pigment (pyocyanin) is also produced that may be seen in pathological lesions.

Pseudomonas aeruginosa is part of the normal flora of the intestinal tract and skin. This bacterium is normally considered to be harmless but functions as an opportunistic pathogen when body defenses are lowered. A loss of competitive intestinal flora due to antibiotic therapy or a decreased responsiveness of the immune system during cancer therapy may allow *P. aeruginosa* to establish a parasitic infection. Some of the more common infections involve the eye, urinary tract, peritoneal cavity, lungs, heart, central nervous system, blood, and skin. Surgical wounds and burns are typical sites of infection. The formation of a blue-green pus caused by the excretion of pyocyanin pigment is found in these infections. Although *Pseudomonas aeruginosa* does not produce a capsule, certain strains release an extracellular slime that is lethal to mice and protects the pathogen from phagocytosis. The slime is a virulence factor for *P. aeruginosa*. The lipopolysaccharide layer of the cell wall is endotoxic and able to cause host damage similar to other Gram-negative pathogens (refer to Table 16.5). More importantly, this pathogen produces an exotoxin capable of inhibiting protein synthesis and causing necrosis. As the most common cause of death in burn victims, *Pseudomonas* infections are very difficult to control, due to the presence of multiple-drug-resistant strains. The aminoglycosides, polymyxins, and chloramphenicol are used, but resistant strains have been identified. Antibiotic resistance has probably de-

1 The adhesive factor which enables infection of *V. cholerae* is associated with what cell structure?

2 Why is an immunization program to control cholera impractical?

veloped by the transfer of R plasmids from other drug-resistant, enteric Gram-negative bacteria.

An alternative to controlling *Pseudomonas* burn infections utilizes immunization with vaccines made from endotoxic lipopolysaccharides. These vaccines have had positive results in animals and provide some protection against sepsis in humans. This therapy provides protection for only about ten days in cases of burns, leukemia, and others in which immunosuppressive therapy is used. Different antigens located in the lipopolysaccharide layer and flagella have allowed microbiologists to classify *Pseudomonas aeruginosa* into nineteen antigenic types.

SUMMARY

Though only about five percent of all bacteria are responsible for human illness, the suffering and death they cause are significant. Each year, millions of people become infected with pathogenic bacteria belonging to one of the three major groups described in this chapter:

1 Pyogenic cocci
  (a) staphylococci
  (b) streptococci
  (c) *Neisseria*
2 Gram-positive, spore-forming bacilli
  (a) bacilli
  (b) clostridia
3 Enteric Gram-negative bacilli
  (a) coliforms
    (i) *Escherichia, Enterobacter, Klebsiella,* and *Serratia*
    (ii) *Arizona, Edwardsiella,* and *Citrobacter*
  (b) *Proteus*
  (c) *Salmonella*
  (d) *Shigella*
  (e) *Vibrio*
  (f) *Pseudomonas*

The members of each group have common genetic, antigenic, physiological, structural, and chemical characteristics which aid in their identification. Each group is also able to cause similar disease symptoms. Controlling an infectious disease requires a knowledge of the responsible microbe, the disease process, and epidemiology. Many other members of these groups are also pathogenic but have not been presented, because they produce more mild or uncommon diseases.

WITH A LITTLE THOUGHT

Many pathogenic microbes described in this chapter cause intestinal infections with very similar symptoms. Identify these microbes, describe the fundamental differences among them, and explain how each induces symptoms. Also describe the epidemiological differences associated with the transmission of these pathogens through a population.

STUDY QUESTIONS

1 Which pathogenic bacteria are responsible for the formation of pus during infection?

2 Name three diseases caused by streptococci.

3 What extracellular products may be produced by pathogenic *Staphylococcus aureus*?

4 What tissue is most likely the site of attachment for *Neisseria gonorrhoeae*? Is a strong, lasting immunity developed after gonorrhea?

5 What genera of bacteria are endospore producers? What diseases might they cause?

6 What effects might an endotoxin from an enteric Gram-negative bacillus infection have on host tissues? Name three bacteria that might induce such symptoms.

7 What is the difference between an exotoxin and an enterotoxin? List two enterotoxin-producing bacteria.

8 How can beneficial *E. coli* be distinguished from enteropathogenic *E. coli*?

9 What are the chief differences between *Salmonella*-induced gastroenteritis and typhoid fever?

10 Describe the difference between the diseases shigellosis and cholera.

SUGGESTED READINGS

BENENSON, A. S., ed. *Control of Communicable Diseases in Man,* 13th ed. The American Public Health Association, Washington, D.C., 1981.

DAVIS, B. D., R. DULBECCO, H. N. EISEN, H. S. GINSBERG, and W. B. WOOD, Jr. *Microbiology,* Harper & Row, New York, 1980.

DAVIS, J. P., et al. "Toxic Shock Syndrome." *New England Journal of Medicine,* 303(25), Dec., 1980.

FIELDSTEEL, A. H. "Cultivation of Virulent *Treponema pallidum* in Tissue Culture." *Infection and Immunity,* 32 (2). May, 1981.

FREEMAN, B. A. *Burrows Textbook of Microbiology,* 21st ed. W. B. Saunders Company, Philadelphia, 1979.

OLDS, R. J. *A Colour Atlas of Microbiology.* Year Book Medical Publishers, Inc. Chicago, 1975.

TOP, F. H., Sr. and P. F. WEHRLE, editors, *Communicable and Infectious Diseases,* 8th ed. C. V. Mosby, St. Louis, 1976.

KEY TERMS

purulent	desquamation	gangrene
pyogenic	pleurisy	putrefaction
exudate	empyema	

Bacterial Pathogens and Diseases (II)

SMALL GRAM-NEGATIVE BACILLI
Brucella
Yersinia
Haemophilus
Bordetella

CORYNEBACTERIA

ACID-FAST BACTERIA

SPIROCHETES

GRAM-NEGATIVE ANAEROBES
Bacteroides

MYCOPLASMAS

RICKETTSIAS

CHLAMYDIA
Legionellosis

Learning Objectives

☐ Know the eight major groups of pathogenic microbes described in this chapter.

☐ Be familiar with the seven basic characteristics of the representative bacteria in this chapter including morphology and staining, respiration-fermentation, motility, reservoir of infection, disease, pathogenic relatives, and features.

☐ Be able to describe the symptoms, mode of transmission, virulence characteristics, and chemotherapeutic agents used in the treatment of the disease-causing microbes presented.

☐ Learn the medical terms presented in the Key Terms.

This chapter is a continuation of Chapter 16 and describes an additional eight major groups of pathogenic bacteria. These groups are distinguishable from one another on the basis of chemical, genetic, antigenic, physiological, and symptomatic differences.

SMALL GRAM-NEGATIVE BACILLI

BRUCELLA: B. ABORTUS

1 Morphology and Staining Reaction: Gram-negative bacilli.

2 Respiration-Fermentation: Strict aerobes.

3 Motility: Nonmotile.

4 Source of Infection-Reservoir: Cattle, swine, goats, sheep.

5 Disease: Brucellosis, Malta fever, undulant fever, Bang's disease.

6 Pathogenic Relatives: *B. suis, B. melitensis.*

7 Features: Facultatively intracellular parasites; some require 5–10 percent CO_2 in environment for growth in isolation; no communicability among humans is suspected.

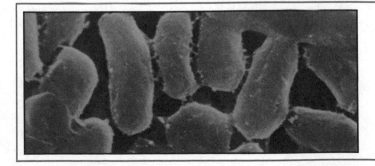

17

Brucellosis has been recognized as an infectious disease of cattle, sheep, goats, and swine since the late 1800s. The three species characteristically found in these animals are *B. abortus* in cattle, *B. suis* in swine, and *B. melitensis* in goats and sheep. Although only occurring in 30–50 percent of infected cattle, the most familiar symptom is abortion of the fetus. Abortion, or reproduction failure, occurs after the pathogen is established in fetal membranes, where it is encouraged to grow by the presence of a growth factor known as erythritol. The chemical is not present in human fetal membranes and abortion in humans is rare. Loss of the fetus releases pathogens from the infected host for dispersal to others and can also result in sterility. During the active stage of infection, the bacteria may be found in the blood and lymph glands, later becoming localized in the mammary gland udder. Chronically infected animals serve as long-term reservoirs of infection by harboring *Brucella* in their udders and transmitting the bacteria through the milk.

Milkborne infections are the most common form of brucellosis in humans, and they may demonstrate several symptoms. The common name, undulant fever, caused primarily by *B. melitensis,* is derived from the fact that the body temperature rises daily in steps until it reaches a maximum, gradually declines, and then regularly repeats the rise and fall. Brucellosis has symptoms of rheumatism, nighttime sweating, and weakness, with near normal temperatures in the morning becoming elevated to 103°F in the evening. During the malignant form, which may be fatal, the temperature rises, remains high, and the patient becomes emotionally disturbed and possibly neurotic.

The most frequent cause of human infection is *B. melitensis* and *B. suis,* and both can last from one to four months. Pathogens can be acquired by ingestion, inhalation, or through breaks in the skin. The organisms move through the lymphatic system and blood to all parts of the body. They become localized as intracellular parasites in lymphatic tissue, liver, spleen, bone marrow, and macrophages. Once inside, the disease stimulates an increase in mononuclear cells, the release of fibrin with coagulation and necrosis, and the formation of fibrous tissue. Although no known exotoxins are produced, these pathogenic changes most likely result from the endotoxic effects of the lipopolysaccharide cell-wall layer. Those most likely to become infected are employees in slaughterhouses, meat packing companies, veterinarians, butchers, and farm workers (Figure 17.1). Recent cases have been reported in the southwest United States where raw, unpasteurized goat's milk has been consumed as a "health food." Milkborne *Brucella* may be easily controlled by pasteurization, since these bacteria are susceptible to heat.

Treatment of brucellosis requires such antibiotics as streptomycin or tetracycline. Because the pathogens are intracellular, therapy must be prolonged and can require a minimum of three weeks. Antibodies are produced in humans after infection, and their presence is used in serological diagnosis; however, they may be of little value in preventing further infection because of the intracellular nature of the microbe.

---

1 What is the reservoir of infection of *Brucella abortus?*

2 How can brucellosis be transmitted?

3 Why is prolonged antibiotic therapy important in the treatment of brucellosis?

YERSINIA: Y. PESTIS

1 Morphology and Staining Reaction: Gram-negative bacilli, exhibit bipolar staining.

2 Respiration-Fermentation: Facultative anaerobes, microaerophilic, ferment sugar to form a small amount of acid but not gas.

3 Motility: Nonmotile.

4 Source of Infection-Reservoir: Wild rodents (black house rats, ship rats, and gray sewer rats); have also been found in squirrels, prairie dogs, dogs and cats, bobcats, and coyotes.

5 Disease: Plague, "black death," bubonic plague.

6 Pathogenic Relatives: *Pasteurella septica* (septicemia), *Francisella tularensis* (tularemia).

7 Features: Rods vary greatly in size and shape after prolonged incubation; symptoms include formation of buboes—infected, inflamed hemorrhagic lymph nodes; capsule helps prevent phagocytosis.

Plague is a disease that probably originated in Asia or central Africa and has spread throughout the world as humans have become increasingly mobile. One of the earliest recorded pandemics occurred in 542 B.C. during the reign of the Roman Emperor Justinian. Occasional epidemics occurred throughout Europe during the Middle Ages; however, during the "great plague" of 1348–49, an estimated twenty-five million Europeans (one-quarter of the population) died from the disease. A second wave swept through London in 1665 killing more than 70,000. Since that time, epidemics have been reported in Hong Kong and Bombay (1876); Santos, Brazil (1899); San Francisco (1900); Manchuria, China (1910); Vietnam (1965). In the United States, five deaths occurred among the 18 reported cases of plague in 1980. Thirteen of those cases were acquired in New Mexico. This rate appeared to be holding steady as by mid-1982, 6 cases of human plague were confirmed in New

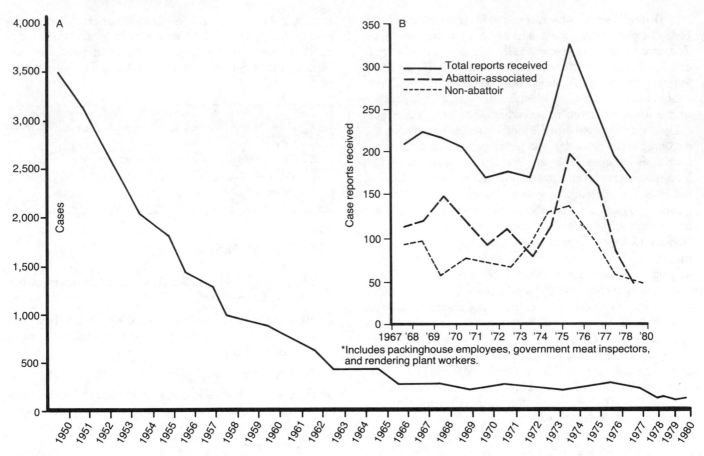

FIGURE 17.1

(A) Reported cases of brucellosis in the United States. (B) Total brucellosis cases and proportion association with abattoirs (slaughterhouses) in the United States. (Courtesy of the Center for Disease Control, Atlanta.)

Mexico (Figure 17.2). The reservoir of infection is wild rodents that are resistant to the disease. The disease is known as sylvatic (forest), or wild rodent, plague when it occurs in prairie dogs, ground squirrels, wood rats, and mice. The infection is spread from rodent to rodent and occasionally to humans by biting fleas. The bacilli are taken into the insect's intestine, where they multiply and reach populations large enough to clog the intestinal tract. When the starving fleas jump to another rodent or susceptible human, they are unable to ingest the blood meal which is regurgitated along with thousands of plague bacilli into the new host.

Bubonic plague has an incubation period of from two to five days during which time the pathogens enter the lymph system and spread through the body. Many bacilli lodge in lymph nodes of the groin and axilla, causing enlarged, hemorrhagic, inflamed nodes called "buboes." Necrosis of the nodes may occur, allowing the bacilli to spread to other parts of the body. Once inside the blood stream, they can invade any organ. If localized in the skin, the hemorrhaging capil-

laries will form dark patches that join together and cover large areas. This symptom was very predominant in the "great plague" of the 14th century and the disease became known as the "black death." If the bacilli lodge in the lungs and cause secondary plague pneumonia, microbes are expelled from the upper respiratory tract in droplet nuclei. This form of the disease is highly fatal. Droplet nuclei transmission is the primary means of human transfer in epidemics. Primary pneumonic plague results from direct droplet nuclei infection (not from flea bites) and is responsible for high morbidity rates during epidemics.

Two toxin types are responsible for disease symptoms. The first is an endotoxin of the lipopolysaccharide layer and its actions strongly resemble those of many enteric Gram-negative bacilli. The second type, murine toxin, has properties similar to both endotoxins and exotoxins. These toxins are able to produce swelling, necrosis, and electrolyte imbalance. Treatment with tetracyclines early in the disease is essential for control. If chemotherapy is delayed, the plague

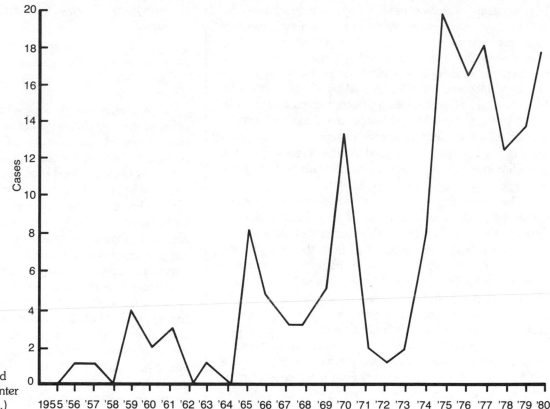

FIGURE 17.2
Reported cases of human plague by year in the United States. (Courtesy of the Center for Disease Control, Atlanta.)

bacilli can reach such high populations that antibiotic-induced bacilli death will lead to severe endotoxic shock in the patient. Measures to eliminate plague-infected rodents and fleas are also of great importance in controlling this disease. Rat control programs have been attempted in many cities throughout the world to reduce the probability that bacilli will be spread from wild rodents to city rats and finally to humans. Since rat fleas, human fleas, and human body lice can all spread this microbe, insect control and body delousing are also of value. Those who recover develop a strong, long-lasting immunity. The most important antibodies appear to be opsonins that stimulate phagocytosis of the encapsulated pathogens. Use of live, attenuated, or killed bacilli as vaccines has been attempted in Vietnam, the Soviet Union, South Africa, and elsewhere. However, these efforts to induce actively acquired immunity have not been very successful. In many cases, the immunity is short-lived

> 1 What is a buboe?
> 2 How are fleas involved in the transmission of plague?
> 3 Why must antibiotic therapy be administered as soon as possible in a case of plague?

and induces both unpleasant local and generalized side effects.

HAEMOPHILUS: H. INFLUENZAE

1 Morphology and Staining Reaction: Gram-negative bacilli, often pleomorphic.

2 Respiration-Fermentation: Facultative anaerobes.

3 Motility: Nonmotile.

4 Source of Infection-Reservoir: Human clinical, subclinical, and carrier cases.

5 Disease: Meningitis, especially in infants; pneumonia; urinary tract infections, chronic bronchitis.

6 Pathogenic Relatives: 20 species including *H. haemolyticus*, *H. suis*, *H. parainfluenzae*, *H. vaginalis*.

7 Features: Require X factor (a hemoglobin or hematin-related chemical) and V factor (possibly NAD or NADP) for growth; virulence appears to be enhanced if found as secondary invader to viral infections (e.g., measles, influenza virus); capsule inhibits phagocytosis.

The genus and species names of this pathogen provide some important, but often misleading, information. The term "haemophilus" indicates that this microbe requires heme for growth. This can be supplied in the form of whole blood or chocolate agar to successfully culture *H. influenzae*. However, the species name is misleading, since this pathogen is not responsible for epidemic influenza of the upper respiratory tract. Influenza ("flu") is caused by a virus and not by *H. influenzae*. *H. influenzae* and other species have been found in association with such cases, and it has been learned that establishment and virulence is enhanced by the presence of a prior viral or bacterial infection. *Haemophilus* is often found along with such diseases as tuberculosis, measles, whooping cough, and viral influenza.

The most common infection caused by *H. influenzae* is meningitis. This accounts for about 15 percent of all meningitis cases in children during the second six months of life, when antibodies passively acquired from the mother have fallen to unprotective levels. An increase in meningitis in children less than six months old has occurred in recent years and is probably due to a decrease in bactericidal antibody titer in the mother's serum. *Haemophilus* may be found as part of the normal flora on mucosal epithelium and enters the circulatory system through breaks in these membranes. Once in the meninges, they produce symptoms of meningitis that do not differ significantly from those caused by *N. meningitidis* or other bacteria. Occasionally, the pathogen may also cause genital and urinary tract infections, pneumonia, and sinusitis. Disease is the result of endotoxin release after cell death. No exotoxins are produced.

After infection, a carrier state may develop in about five percent of the cases. The influenza bacilli are carried in the nasopharynx and transmitted by droplet nuclei. Immunity to the microbe centers on the capsular antigens and is serotype specific. The antibodies formed are bactericidal and function in conjunction with the complement system. Since formation of these antibodies may also be induced by certain strains of *E. coli* normally found in the intestinal tract, a high antibody titer is usually maintained throughout life. Antibiotic therapy using penicillin and ampicillin has, for the most part, been discontinued, and replaced with tetracycline and chloramphenicol due to the development of drug-resistant strains.

BORDETELLA: B. PERTUSSIS

1 Morphology and Staining Reaction: Gram-negative bacilli, may form chains.

2 Respiration-Fermentation: Strict aerobes.

3 Motility: Nonmotile.

4 Source of Infection-Reservoir: Human cases, droplet nuclei, and fomites.

5 Disease: Whooping cough, pertussis.

6 Pathogenic Relatives: *B. bronchiseptica*, *B. parapertussis*.

7 Features: Strong antigenic nature provides for effective vaccine production; capsule is present; hemolysins are produced.

Whooping cough, or **pertussis,** is a very common upper respiratory disease found throughout the world. It has been estimated that as many as 95 percent of the world population has experienced either mild or severe symptoms. The disease occurs most frequently in young children, and females have higher morbidity and mortality rates than males. Unlike *H. influenzae* meningitis, no protection is afforded newborns from maternal antibodies. *Bordetella pertussis* enters the upper respiratory tract of children in droplet nuclei from carriers or diseased individuals. Once inside, the disease passes through three distinct stages after a one-week incubation period: (1) catarrhal (inflamed mucous membranes), (2) paroxysmal (spasms), and (3) convalescent. During the catarrhal stage, which lasts from one to two weeks, the patient develops symptoms typical of most upper respiratory infections. *Bordetella pertussis* can be found during this stage in large numbers on the epithelial surface adhering to the cilia and interfering with normal ciliary movements. The release of endotoxins similar to most Gram-negative bacilli irritates the epithelium to produce a mild cough, sneezing, edema, and lymphocytosis, which is an excess of lymphocytes. Patients in the catarrhal stage are highly infectious and expel great numbers of microbes. Toxins also release large amounts of leukocytic exudate as the catarrhal stage develops into the paroxysmal stage. During this stage, which lasts from four to six weeks, the patient experiences repeated and uncontrolled coughing. Series of explosive coughs occur in a single exhalation followed by long inhalations with the typical "whooping" sound. The frequency of coughs and difficulty in breathing interferes with swallowing. During the paroxysmal stage, large amounts of *B. pertussis*-laden saliva flow from the mouth and nose. Coughing becomes so frequent and spasmotic that exhaustion, vomiting, and convulsions may occur. As the second stage continues, pathogens disappear from the epithelium, but whooping and coughing continue. This is unique to *B. pertussis* upper respiratory infections and may be a residual effect of the endotoxins. The third, or convalescent, stage develops as the frequency of whooping and coughing decreases. Convalescence is usually slow and may last several weeks. *Bordetella pertussis* rarely enters the blood during the disease but is often associated with other bacterial and viral pathogens. Although the fatality rate for whooping cough is very low, secondary invaders cause complications such as meningitis, pneumonia, or influenza.

Pertussis bacilli are susceptible to many antibiotics, including erythromycin, ampicillin, and chloramphenicol. When these chemotherapeutic agents are used during the catarrhal stage, *B. pertussis* is eliminated from the ciliated epithelium. However, this type of therapy is of little help to those who have entered the paroxysmal stage since, unlike other bacterial infections, the symptoms continue in the absence of the pathogens.

The development of a strong, long-lasting immunity takes place after infection. Second infections may occur in children but are more mild. Adults who become infected and have experienced a decrease in protective antibody titer may suffer severe cases of whooping cough. The pertussis vaccine has been highly effective in reducing the number of whooping cough cases in the United States (Figure 17.3).

---

1 What diseases can be caused by *H. influenzae*?
2 What virulence factor is responsible for symptoms caused by *H. influenzae*?
3 Name the three stages of infection for whooping cough.
4 What measures have brought about the control of whooping cough?

The vaccine is given to infants in combination with tetanus and diphtheria toxoids three times during the first year. If a nonimmunized infant is exposed to *B. pertussis*, passive immunity can be provided with hyperimmune immunoglobulin.

CORYNEBACTERIA

CORYNEBACTERIUM: C. DIPHTHERIAE

1 Morphology and Staining Reaction: Gram-positive bacilli, very pleomorphic.

2 Respiration-Fermentation: Facultative anaerobes, ferment glucose to produce acid but not gas.

3 Motility: Nonmotile.

4 Source of Infection-Reservoir: Human clinical, subclinical, and carrier cases, cattle.

5 Disease: Diphtheria.

6 Pathogenic relatives: Three biotypes are recognized: *C. diphtheriae* var. *gravis*; *C. diphtheriae* var. *mitis*; and *C.*

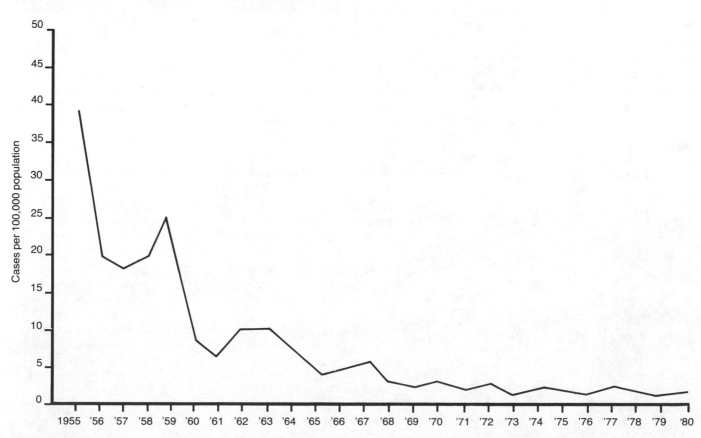

FIGURE 17.3
Reported cases of pertussis per 100,000 population by year in the United States. (Courtesy of the Center for Disease Control, Atlanta.)

diphtheriae var. *intermedius*. *C. acne* associated with acne or pimples.

7 Features: Cells are irregular in shape with swelling at ends; possess metachromatic granules giving a beaded appearance, and cells may resemble Chinese alphabet characters; toxicity controlled by bacteriophage.

Diphtheria, or **diphtheritic pharyngitis,** is caused by the Gram-positive bacterium *Corynebacterium diphtheriae*. The genus name is derived from the fact that the cells show swelling at the ends which gives the rods a club shape ("koryne" means *club*). These microbes were first discovered in 1883 by Edwin Klebs, but they were not suspected of being the cause of diphtheria, since they were found in both healthy and diseased individuals. The significance of this discovery became evident at a later date. *Corynebacterium diphtheria* and other species are normal flora on the skin and mucous membranes. Three biotypes have been identified and are distinguished by their growth on potassium tellurite medium (Table 17.1). All three types are pathogenic and show no significant difference in degree or severity.

Bacteria enter the upper respiratory tract on droplet nuclei and become lodged on the tonsils. Their virulence is attributed to a mild but possibly fatal degree of invasiveness, and the release of a powerful exotoxin capable of inhibiting protein synthesis and causing nerve damage. As the bacteria invade the mucosal epithelium, they destroy host cells and stimulate the formation of a thick, fibrinous exudate filled with dead cells, leukocytes, and fibrin. This tough, gray tissue forms a characteristic "pseudomembrane," which spreads and may fatally involve the larynx (Figure 17.4). The "membrane" is not separated or on top of the epithelium but is a part of the tissue. Any attempt to remove it results in bleeding, and this feature is used to diagnose diphtheritic pharyngitis in comparison to other throat infections. The pathogens remain localized in the membrane and rarely move from this area. As the membrane develops, it can interfere with breathing and swallowing, and may ultimately lead to suffocation. During this disease phase, pa-

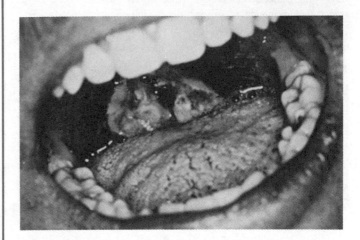

FIGURE 17.4
Localized invasion of the oropharynx mucous membranes stimulates formation of the diphtherial pseudomembrane. The *Corynebacterium diphtheriae* remain localized, while the exotoxin they release spreads through the body. Any attempt to remove the membrane will result in bleeding. (From *Bacterial Pharyngitis.* Abbott Laboratories.)

TABLE 17.1
Corynebacterium: characteristics of the diphtheria bacillus types.

Type			*mitis*	*intermedius*	*gravis*
Morphology	Microscopic		Usually long, with many metachromatic granules—80 percent typical	Usually barred, club forms common—80 percent typical	Short, evenly staining—50–60 percent typical
	Colonial	Tellurite	Small, round, smooth, convex, black with grayish periphery	Small, flat, dull, gray, raised center	Large, irregular, dull, gray, raised center, radial striations
		Chocolate	Smooth, semiopaque, glistening	Flat, dry, opaque, slight greenish zone	Flat, dry, matt, opaque
		Broth	Uniform turbidity, sometimes slightly granular, soft pellicle	Finely granular turbidity	Granular, flakes, pellicle—variable
Physiology	Fermentation of	Glycogen	−	−	+
		Starch	−	−	+
	Hemolysis		−	−	±

tients have a low-grade fever, cough, and swelling of the neck known as "bull neck." The membrane can remain in place for up to three weeks before it begins to slough off and healing begins.

The two types of toxins released during the infection are an exotoxin and a cord factor. The exotoxin spreads from the site of infection, causing necrosis of the heart, liver, kidneys, and adrenal glands. In the acute disease, damage may become severe and heart failure is the immediate cause of death. The exotoxin is released by pathogens that contain a lysogenic bacteriophage with the gene for toxin production. *Corynebacterium diphtheria* that lack this phage do not produce the exotoxin. The presence of the gene appears to provide no advantage for the bacteriophage. The cord factor is a toxic glycolipid that contributes to the disease process by disrupting mitochondria in host cells. A similar substance is also released by *Mycobacterium tuberculosis*. The effect of both these toxins may be felt for several months after the disease. Heart murmurs, abnormal rhythms, and permanent heart damage develop in about 20 percent of the patients with severe cases of diphtheria. Motor and sensory nerves can also become damaged, leading to loss of pupillary accommodation in the eyes, paralysis of the soft palate, which alters the voice, and difficulty in swallowing. These complications may last as long as a year.

Diphtheria is typically thought of as a childhood disease; however, susceptibility varies with the level of serum antibody protection. Infants are passively protected with maternal antibodies but become susceptible as the titer decreases with age. In recent years, the incidence of diphtheria has increased in older individuals, peaking between the ages of 30 and 50. The change is a result of the loss of acquired immunity from a decrease in subclinical (not apparent) infections or vaccination. As the age distribution of the population changes and more individuals fall into this highly susceptible age range, epidemiologists will need to be concerned with maintaining an adequate level of herd immunity. This may be a problem, since a high percentage of older individuals react adversely to vaccination with the commonly used alum-precipitated toxoid. This form of toxoid is used to vaccinate children in combination with tetanus and pertussis toxoid (DPT immunization). A highly purified toxoid (designated Td) is required if severe side effects are to be avoided in adults.

Immunity or susceptibility is determined by performing the **Schick test.** A single dose of diphtheria toxin is injected intradermally in one forearm and a control dose of heat-treated toxin is injected into the other forearm. The test is read at 24 and 48 hours, and again in 6 days. Four reactions are possible:

1 Positive reaction. Indicates absence of neutralizing antitoxin. Redness will occur on arm at site of injection; no reaction on control arm. Not protected.

2 Negative reaction. No reaction at either site of injection: antibody present. Protected.

3 Pseudoreaction. Hypersensitivity to material other than toxin. Pseudoreactions may show redness and swelling on both arms; this constitutes a negative reaction. Protected.

4 Combined reaction. Like pseudoreaction, redness and swelling occur at both injection sites; however, toxin continues to exert effects while control rapidly subsides. This reaction denotes both hypersensitivity and susceptibility to toxin. Not protected.

Treatment of diphtheria requires antibiotics to destroy the pathogens and antibodies to neutralize the toxin. Administration of diphtheria antitoxin produced by sheep, horses, goats, or rabbits effectively neutralizes circulating exotoxins. However, the incidence of severe side effects is high; therefore, it is essential that the patient is infected with an exotoxin-producing strain of *C. diphtheriae* and that no hypersensitivity exists. The toxigenicity of the pathogen can be tested in tissue culture by mixing the bacteria with susceptible monkey cells. Death of the monkey cells indicates toxin production. A simple skin test can be used to determine hypersensitivity to the antitoxin. The antibiotics penicillin and erythromycin inhibit the growth of *C. diphtheria* but do not affect the course of the disease as long as circulating exotoxin is present.

---

1 What is a pseudomembrane?
2 Virulence of *C. diphtheriae* is attributed to what factors?
3 What is the Schick test, and how is it performed?

ACID-FAST BACTERIA

MYCOBACTERIUM: M. TUBERCULOSIS

1 Morphology and Staining Reaction: Acid-fast, Gram-positive bacilli.

2 Respiration-Fermentation: Obligate aerobes, growth enhanced with CO_2.

3 Motility: Nonmotile.

4 Source of Infection-Reservoir: Human clinical and subclinical cases; infected cattle.

5 Disease: Tuberculosis (consumption): pulmonary, bones, joints, eyes, lymph nodes, kidneys, larynx, skin.

6 Pathogenic Relatives: *M. leprae* (leprosy); atypical mycobacteria: photochromogens, scotochromogens, nonphotochromogens, rapid growers.

7 Features: Encapsulated; strains identified: *M. tuberculosis* var. *bovis*, *M. tuberculosis* var. *hominis*; resistant to chemical agents but highly susceptible to photooxidation; very resistant to drying; very slow growing; have a thick, protective waxy coat.

Tuberculosis, or "consumption," has been and continues to be one of the most common communicable diseases in the world. At the turn of the century, it was the leading cause of death in the United States. In mid-1982 cases of tuberculosis were reported at an average of 25,000 for the year. The highest incidence of the disease occurs in areas with large black, American Indian, Asian, or Hispanic populations. The case rate varies with sex, race, age, and location of residence. Males are affected at about twice the rate of females; noncaucasians may experience a rate five times that of Caucasians; persons over fifty and under five years of age are most susceptible; and urban residents have a case rate that is approximately twice that of the national average. It has been estimated that 30,000,000 cases currently exist in the world and that 2,000,000 new clinical cases continue to develop each year (Figure 17.5). Among the more notable sufferers of tuberculosis have been John Keats, Chopin, Goethe, Edger Allen Poe, and Moliére.

The two strains of *M. tuberculosis* known to cause human infection are *M. tuberculosis* var. *hominis* and *M. tuberculosis* var. *bovis*. The primary mode of transmission of the bovine strain has been contaminated milk; however, the incidence of milkborne infection in the United States and other developed countries has been greatly reduced as a result of pasteurization. Today the most common pathogen is the *hominis* variety transmitted to susceptible individuals through the upper respiratory tract. Virulent strains of *M. tuberculosis* are very slow growing; and because of their thick, waxy coat, they are more resistant than other bacteria to chemical agents and drying. The microbe can survive for six to eight months outside the body in cool, dark dust. Tubercle bacilli do not release toxins, rather they cause disease by their growth and interaction with the host. Infection depends on the number and location of pathogens entering the body, and host susceptibility. Individuals suffering from poor nutrition, overcrowding, and inadequate immune responses are most likely to become infected. Although the most familiar type of tuberculosis occurs in the lungs, infection may occur through breaks in the skin, gastrointestinal tract, and conjunctival membranes.

The two recognized forms of tuberculosis are **exudative** and **productive.** Exudative infections that originate outside the body (exogenous) are known as primary infections. The exudative form usually occurs in children and adults who have not previously been exposed. Primary infections (exudative or productive) may occur anywhere in the lung. Upon entering the lungs, the tubercle bacilli stimulate migration of the PMNs, monocytes, and macrophages to the focus of infection. The initial response resembles most bacterial pneumonia types. If the body's defenses are not sufficient to handle the infection, the bacilli cause massive necrosis at the focus of infection. During exudative tuberculosis, patients discharge bacilli in sputum and in later stages, the tuberculin skin test becomes positive. Bacilli enter the lymph and become concentrated in the lymph nodes. The name "Ghon complex" refers to both the infected lymph nodes and the primary lung infection. In severe cases, microbes can either spread to other sites in the body or develop into a productive type of lesion. Spreading occurs through the lymphatic and blood systems. When the pathogens become established in several organs, for example, kidneys, liver, and meninges, the disease is known as **acute miliary,** or **chronic disseminated, tuberculosis.**

The productive type of lesion may form at the initial pulmonary infection site or at disseminated foci. Productive lesions consist of cells without the formation of exudate. The lesion is easily recognized as a nodule composed of three distinct zones or layers. The three-layered nodule is known as a **tubercle** (Figure 17.6). The central zone, or core, contains large "giant cells" derived from fused macrophages. The tubercle bacilli are located inside these cells as intracellular parasites. Their intracellular nature makes therapy with antibiotics more difficult. Surrounding this central area of infection is a layer of epithelioid cells that have migrated to the site. The outermost layer is composed of monocytes, fibroblasts, and lymphocytes. As the tubercle develops, the core begins to degenerate and becomes necrotic. The tissue takes on a "cheesy" appearance that has resulted in the term **caseation necrosis.** If caseation necrosis continues, the tubercle may break through the lung wall and release bacilli into the pleural cavity or bronchiole. This process is called **cavitation.** The most common clinical symptoms of pulmonary tuberculosis are fatigue, weight loss, fever, a chronic cough, and blood in the sputum.

Occasionally calcium salts are deposited in the tubercle, and this walls-off the bacilli from surrounding tissue. The tubercle becomes a hard, dry mass that enables the encased pathogens to survive for years. If calcification occurs, symptoms of infection will pass and the patient will recover. However, **active reinfection tuberculosis** caused by these encased, or endogenous, microbes can occur at a later time if the tubercle ruptures. Disease caused by endogenous bacilli that have survived from a primary lesion is the most common form of second infection.

Reinfection tuberculosis is characterized by the establishment of new tubercles, fibrosis, and caseation necrosis. Unlike primary cases, tubercle formation is at the apex of the lung. The tissue response during reinfection is usually more severe as a result of the development of delayed hypersensitivity. This has been demonstrated in mice and is known as **Koch's phenomena.** The animal is first injected subcutaneously with virulent *M. tuberculosis,* which stimu-

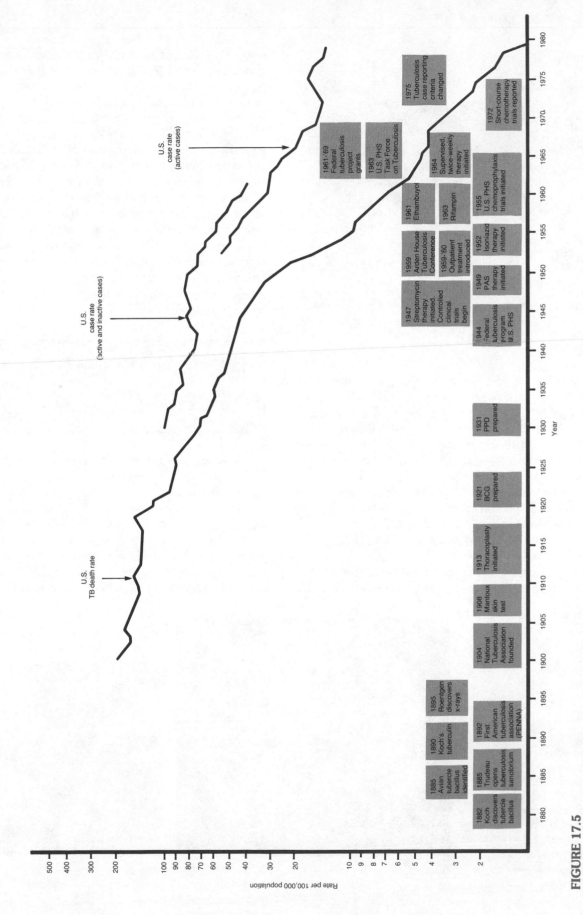

FIGURE 17.5

Tuberculosis case rates in the United States have declined drastically, and many believe immunization is no longer necessary. However, the case distribution map shows that there still is need for concern. (Courtesy of the Center for Disease Control, Atlanta.)

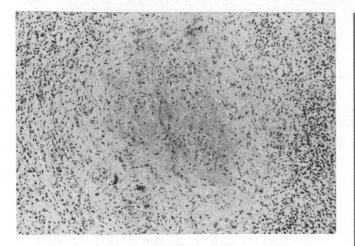

FIGURE 17.6
This tubercle is a lesion of pulmonary tuberculosis with central caseation surrounded by epithelial cells. (Courtesy of E. V. Perrin, Children's Hospital of Michigan.)

lates an exudative lesion. Several weeks later, a second injection stimulates Koch's phenomena, a massive inflammatory response at the site of injection that tends to be walled off and often leads to necrosis. This reaction does not require live *M. tuberculosis* but can be produced by either the injection of old tuberculin (OT) or purified protein derivative (PPD), which are cell extracts also used for the TB skin test. OT is a concentrated filtrate of six-week-old *M. tuberculosis* broth culture. The solution contains materials from the broth as well as tuberculin proteins. PPD is OT that has been chemically purified and is the preferred material used for the tuberculin skin test.

Although certain internal components of the cell are able to induce delayed hypersensitivity, the molecules chiefly responsible for this response are proteins, lipids, and carbohydrates found in the cell wall. Lipids injected into test animals stimulate caseation necrosis, an increase in surrounding epithelial and connective tissue, and delayed hypersensitivity. These complex lipids are also, in part, associated with the acid-fast staining characteristics of mycobacteria. A unique lipid found in *M. tuberculosis* is known as the **cord factor.** The presence of this molecule enables tubercle bacilli to form "serpentine cords" or chains (Figure 17.7). There is a correlation between cord formation and virulence, since the cord factor is leukotoxic, inhibits leukocyte migration, and causes the formation of granulomas, which are tumorlike masses surrounding the tubercle. Tuberculin proteins found in the wall stimulate delayed hypersensitivity, which is visible in the skin test, and antibody formation. Polysaccharides attached to the wall also appear to be involved in antibody formation.

---

1 What methods of diagnosis are used in cases of TB?
2 What is a tubercle?
3 What is Koch's phenomena?

The complex nature of the cell wall and its thick, waxy coat leads to two separate immune responses after first infection. The delayed hypersensitivity seen in the tuberculin skin reaction is stimulated by different cell-wall components than the production of a protective immune response. Humoral antibody production is of little value, and the major defense is cell-mediated immunity (refer to Chapter 13). Attempts at inducing immunity through vaccination with attenuated *M. tuberculosis* have been made with some success.

FIGURE 17.7
(A) *Mycobacterium tuberculosis* var. *hominus* in a smear from a sputum sample. (B) Virulent tubercle bacilli, strain H37R$_V$. (Part A courtesy of the Center for Disease Control, Atlanta. Part B courtesy of the National Tuberculosis Association, American Lung Association.)

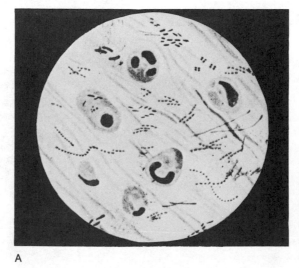

A

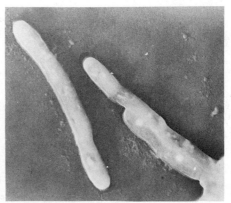

B

Attenuated, live BCG (bacille Calmette-Guérin) bovine bacilli were first used in the early 1920s as an oral vaccine. However, attenuation was not complete, and a number of vaccinated individuals developed fatal tuberculosis. Since that time, attenuation methods have improved and the vaccine is now given intracutaneously as a substitute for primary infection. The vaccination appears to provide protection against tuberculosis in childhood but does not last into adulthood. Many still fear that attenuation is not complete and disease will result from widespread use of the vaccine. However, thousands of people in Scandinavia and Europe have been vaccinated, with only a few instances of induced disease. In the United States, BCG is only used on a limited basis for those in high-risk situations.

Chemotherapy for patients with active tuberculosis requires the use of antibiotics that penetrate the tubercle and enter the giant cells. The difficulty of penetrating host barriers and the outer waxy bacterial coat demands that drugs be given on a regular basis for several months. Because of the development of drug-resistant strains of *M. tuberculosis*, antimicrobials are usually given in combination. Rifampicin (RIF), isoniazid (INH), ethambutol (EMB), para-aminosalicylic acid (PAS), and streptomycin (SM) are the most widely used drugs.

Atypical Mycobacteria These mycobacteria grow rapidly and are normally found in the soil as saprobes, but they can also function as opportunistic pathogens. They have been called atypical because the course of lung infections

caused by them resembles tuberculosis but is not identical. Atypical mycobacteria are relatively resistant to antibiotics used to treat *M. tuberculosis* infections, and require a culture and susceptibility test to determine proper treatment. Four groups of atypical mycobacteria have been recognized (Table 17.2).

Leprosy Another mycobacterium, *M. leprae,* is the cause of leprosy, or Hansen's disease. Ancient Egyptian, Chinese, Japanese, Asian, European, and African writings all have accounts of this skin disease. In the middle ages, lepers were prevented from attending public meetings, forbidden to talk with children, and required to ring a bell to let everyone know they were in the area. The term "leper" became synonymous with outcast.

Isolation of infected individuals in leper colonies to prevent others from becoming "unclean" was standard practice. There is no doubt that many who were forced into these colonies did not have Hansen's disease but were infected with other types of pathogens. Today there are approximately 11 million cases of leprosy in the world, mostly in the tropics. An estimated 4200 cases are known to exist in the United States, an increase of 500 percent since 1960. The main reason for the increase appears to be changes in the traditional pattern of immigration from countries such as Mexico, the Philippines, Vietnam and Cuba. An accurate morbidity rate is difficult to determine, because Hansen's disease symptoms can appear up to 20 years after infection. Acute and chronic forms of leprosy can develop, depending

TABLE 17.2

The atypical mycobacteria.

Group	Characteristics	Example	Disease
I Photochromogens	Produce bright yellow pigment in presence of light. Rough (R) colonies produced after 14 to 21 days.	*M. kansasii* *M. marinum (balnei)*	Pulmonary disease resembling tuberculosis, Swimming pool granuloma; lesions usually on skin of extremities.
II Scotochromogens	Produce reddish-orange pigment, independent of light. Smooth (S) colonies produced after 10 to 14 days.	*M. scrofulaceum* *M. szulgai*	Lymphadenitis in children. Pulmonary disease.
III Nonphotochromogens	Usually nonpigmented; if produced, pigment is weak and light-independent. Require 14 to 21 days for growth.	*M. avium* *M. xenopi*	Chronic pulmonary disease. Pulmonary disease.
IV Rapidly growing	Some pigmented. Colonies produced in 5 to 7 days.	*M. fortuitum*	Pulmonary disease; skin abscesses.

on the nature of the cell-mediated immune response. Tuberculoid leprosy is the more mild form, and lasts as long as twenty years. A low-level, cell-mediated immunity reduces the severity of this type of infection. Skin lesions, known as macula, appear but contain few pathogens. Nerve cells become infected and damage results in the loss of some feeling. Tuberculoid leprosy can progress to the more acute form, known as the lepromatous type, or it may begin as the primary infection. If the immune response is unable to control a lepromatous infection, the disease becomes generalized. The typical lesion, known as a leproma, occurs on different parts of the body. Several may fuse to cause severe distortions and disfigurement. Lepromas contain numerous intracellular pathogens, which can move easily through the circulatory system to other parts of the body. Kidneys, spleen, liver, and eyes may also become infected over a long period. The lesions destroy nerve endings and stimulate the reabsorption of bone causing a loss of feeling and shortening of extremities.

Susceptibility to leprosy is highest among infants, and infection requires prolonged, close contact with an infected individual. This usually occurs in family situations that include an infected adult. The incubation period can extend over several years, and symptoms may not be demonstrated until later in life. The exact mode of transmission has not been confirmed, and research into the disease's course has been made difficult by the fact that this pathogen has not been grown in pure culture. In recent years, armadillos have been used as laboratory research animals, and human cases have been studied at the National Leprosy Hospital in Carville, Louisiana. Diagnosis of the infection is made from skin scrapings stained by the acid-fast method. Rifampicin as well as dapsone (diaminodiphenylsulfone, DDS) have been

used to successfully treat leprosy. Disfigurement may be averted if treatment is started in the early stages (Figure 17.8).

In recent years, researchers have been attempting to develop a vaccine against leprosy. Bacteria from armadillos are killed and used in the production of the vaccine. It is speculated that the recent successes in animal studies will soon lead to human tests, but because of the long incubation period, several years may pass before results are known.

---

1 What is the BCG vaccine?
2 Name the two forms of leprosy.
3 How is leprosy transmitted?

SPIROCHETES

TREPONEMA: T. PALLIDUM

1 Morphology and Staining Reaction: Spirochetes do not respond to Gram stain; silver nitrate staining darkens cells for viewing in light microscope; usually visualized in dark-field microscopy.

2 Respiration-Fermentation: Cells have not been grown in pure culture but have been maintained for up to seven days under anaerobic conditions at 25°C.

3 Motility: Motile by axial fibrils.

4 Source of Infection-Reservoir: Humans.

5 Disease: Syphilis (leus, the French disease, French pox).

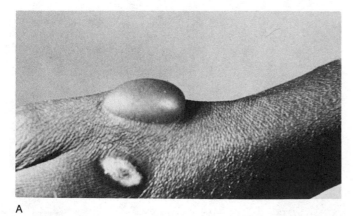

A

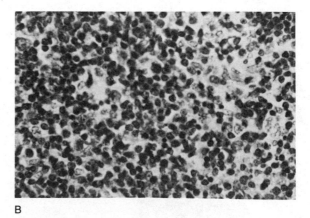

B

FIGURE 17.8
(A) These lesions result from an infection by *Mycobacterium leprae*. (B) *M. leprae* in a lymph node. (Courtesy of the Center for Disease Control, Atlanta.)

6 Pathogenic Relatives: *Borrelia recurrentis* (relapsing fever); *Treponema pertenue* (yaws); *T. carateum* (pinta); *T. microdentium* (ulcerative gingivitis); *Leptospira interrogans* (leptospirosis); *Spirillum minor* (rat-bite fever).

7 Features: Highly susceptible to soaps, drying, and temperature changes; easily controlled with penicillin; can cross the placenta; never grown in pure culture.

Syphilis is one of the most familiar venereal diseases, and over the years it has become known by a number of different names. When the English were at war with the French (late 1400s), the disease was called either the French pox or Englishmen's disease, depending on whose side you were on! The name syphilis comes from a poem written in 1530 by Fracastorius called *Syphilis sive Morbus Gallicus* (Syphilis, or the French Disease). In this story, a Spanish shepherd is punished for being disrespectful to the gods by being cursed with the disease. Several years later, Fracastorius (also a well-known scientist) published a serious paper in which he described the possible mode of transmission of the "seeds" or "germs" of syphilis through sexual intercourse. This was pure speculation on his part since he had never seen a microbe, and it would be 130 years before Anton van Leeuwenhoek (1676) would make his first discoveries. Since publication of the poem and the article, the disease has become known as syphilis. The term "venereal disease" is derived from the name Venus (Roman goddess of love) and is used to describe diseases that are transmitted by intimate contact. Syphilis is only one of several venereal diseases (VD) (Table 17.3) or sexually transmitted diseases (STD).

Syphilis progresses through three stages in the untreated adult: primary, secondary, and tertiary. *Treponema pallidum* enters the body through breaks in the skin or intact mucous membranes. Upon entering the body, the spirochetes quickly spread from the portal of entry and may be found in distant lymph nodes within an hour. The **primary stage** is characterized by the formation of a lesion known as a chancre (destructive sore) at the portal of entry (Figure 17.9). This will appear between 10 and 30 days following infection. A chancre is painless, has a soft center, hard floor, and numerous spirochetes. Contact with this lesion during sexual intercourse will result in disease transmission. A chancre may appear on the skin or mucous membranes but can go unnoticed if it is small and formed on the labia, vaginal wall, or cervix of the female. Depending on the nature of sexual contact, chancres may also develop in the mouth, penis, rectum, or on the lips or palms of the hands. Spirochetes can be isolated from the blood during primary syphilis. If untreated, a chancre will heal spontaneously but infection will continue to the secondary stage. **Secondary syphilis** develops between four and eight weeks

TABLE 17.3

The venereal diseases (VD) or sexually transmitted diseases (STD).

Disease	Causative Organism
Spirochetal	
Early syphilis	*Treponema pallidum*
Nonspecific infection	Multiple
Viral	
Herpes genitalis	HSV-2 (DNA) virus (10–15 percent HSV-1)
Condyloma acuminatum	Papova (DNA) virus
Molluscum contagiosum	Pox (DNA) virus
Bacterial (other than syphilis)	
Gonorrhea	*Neisseria gonorrhoeae*
Chancroid	*Haemophilus ducreyi*
Donovanosis	*Donovania granulomatis*
Lymphogranuloma venereum (LGV)	Group A *Chlamydia*
Protozoal	
Trichomoniasis	*Trichomonas vaginalis*
Animal	
Scabies	*Sarcoptes scabiei* (mite)
Pediculosis ("crabs")	*Phthirus pubis* (lice)

after the initial infection. Lesions are typically formed on the skin (maculopapular rash) or in the mouth, but can also occur in other tissues, giving rise to meningitis (meninges), periostitis (bone), hepatitis (liver), nephritis (kidney), and conjuntivitis (eye). The patchy lesions on the skin and mucous membranes of the vagina and oral cavity contain pathogens which are a source of further infection to others. During the secondary stage, spirochetes may also cross the placenta in women who are at least four months pregnant, causing **congenital syphilis** in the unborn child. Because of the extreme susceptibility to damage, a fetus infected during the fourth or fifth month of development is likely to cease development and be aborted or stillborn. If the infection occurs at a later time, the child may survive but demonstrate other symptoms of tissue damage including saddle nose, Hutchinson's teeth (Figure 17.10), periostitis, mental retardation, scars at the angle of the mouth, or the "old man" look. In cases where the child is born with no symptoms of infection, tertiary syphilis symptoms may appear months or years later.

Secondary syphilis in adults will disappear within four weeks to twelve months and may enter a latent period.

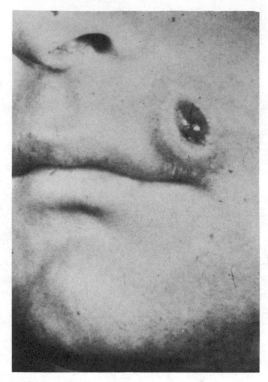

FIGURE 17.9
The chancre is a soft, craterlike lesion that is characteristic of primary syphilis.

Latency can be interrupted by reoccurrences of secondary symptoms within three to five years, causing further tissue damage and increasing incidence of disease transmission. About one-third of untreated patients with early syphilis re-

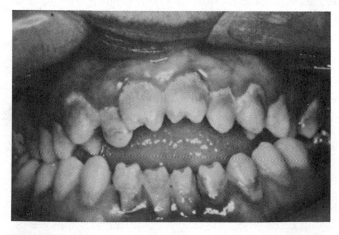

FIGURE 17.10
Hutchinson's teeth are a manifestation of congenital syphilis. Note the characteristic "points." (From R. Colby et al, *Color Atlas of Oral Pathology*, 3rd ed. Harper & Row, 1971.)

cover spontaneously (become serologically negative), another one-third become latent (remain serologically positive), and one-third progress to tertiary (third degree) syphilis.

Tertiary, or **late, syphilis** might not appear for years (five to twenty) after the initial infection. This stage is characterized by a drastic decrease in the number of spirochetes and the degeneration of tissues resulting from hypersensitivity reactions. These destructive, noninfectious lesions are known as gummas and may be formed in the skin, liver, central nervous system, and circulatory system. This stage has become known as the "great impostor," because many of the symptoms resemble other types of infections. Cardiovascular tertiary syphilis is a chronic inflammatory disease of the aorta causing a weakening of the arterial wall that may eventually lead to rupture (aneurysm) and death. Central nervous system involvement can result in tissue loss leading to mental retardation, a "shuffle" walk (tabes dorsalis), blindness, hearing loss, or insanity. Many of these symptoms have been documented in such notables as Kaiser Wilhelm, Beethoven, Nietzsche, Franz Schubert, Henry VIII, Hitler, and the composer Scott Joplin. With increased understanding of syphilis and more rapid identification and treatment, tertiary syphilis is an uncommon occurrence in the United States and other developed countries, but according to current reports, the number of cases is increasing.

---

1 Name three pathogenic relatives of *T. pallidum* and the diseases they cause.
2 What are the symptoms of primary syphilis, secondary syphilis, and tertiary syphilis?

Humans are universally susceptible to *T. pallidum* infection and respond with the formation of antitreponemal antibody and a complement-fixing, antibody-like substance known as reagin (not to be confused with IgE). Reagin is able to react with cardiolipin, a material derived from mitochondrial membranes, and is useful in the serological diagnosis of syphilis (Table 17.4). It is not presently known whether antibody formation is responsible for the noted decrease in the severity of the disease, but there is a definite increase in the ability of infected individuals to resist superinfection, which is a second infection during the course of a first infection. Immunity is not complete and second infections can occur once the first infection has spontaneously disappeared or been eliminated as a result of antibiotic therapy. The long generation time of *T. pallidum* (33 hours) makes the long-acting penicillins, such as procaine or benzathine, the drugs of choice. A single injection should be sufficient; however, patients should be monitored for several

TABLE 17.4
Diagnosis of syphilis.

Microscopic
    Dark-field microscopy, chancre specimen examined before antibiotic therapy is begun.

    Ultraviolet microscopy, using immunofluorescent antitreponemal antibody with tissue specimens.

Serological Tests for Syphilis (STS)
    VDRL (venereal disease research laboratory test)
        An antibody (patient reagin)-antigen (beef heart cardiolipin) flocculation reaction.

    RPR (rapid plasma reaction)
        Similar to VDRL.

    Wasserman
        A test showing test cell lysis in the presence of reagin containing serum in the presence of beef cardiolipin as an antigen

    FTA-ABS (fluorescent treponemal antibody test)
        An indirect immunofluorescence reaction involving killed *T. pallidum*, patient's serum, and immunofluorescent antihuman gamma globulin.

    TPI (*Treponema pallidum* immobilization test)
        Patient's serum antibodies stop movement of rabbit-cultured *T. pallidum*.

weeks in cases of late neurosyphilis, since microbes present in these stages may survive treatment.

Another form of syphilis known as **nonvenereal syphilis,** or bejel, occurs largely in children and families living in underdeveloped parts of the world. This form of the disease is similar to venereal syphilis; however, chancres are rarely formed. The most common symptoms are secondary, occurring on the skin and mucous membranes of the mouth and anogenital area. Transmission occurs primarily through close contact between mothers and their children. Spirochetes are frequently found on the nipples of nursing mothers. Nonvenereal syphilis rarely occurs in the United States.

Several spirochetes of the genus *Borrelia* are able to cause the disease known as **relapsing fever.** Infection results in a three- to four-day period of high fever, chills, and severe headache. This is followed by a period of relief that may last a few hours or days. The cycle may be repeated from three to ten times during the course of the infection. Pathogens are found in the spleen, liver, kidneys and urine, gastrointestinal tract, and brain. Blood cultures are positive during the febrile (fever) period but negative during the relapse. *Borrelia* species can be transmitted directly through contaminated urine or feces but most frequently by body

lice or ticks. The cause of epidemic relapsing fever in humans *(B. recurrentis)* is transmitted by the body louse *Pediculus humanus.*

GRAM-NEGATIVE ANAEROBES

BACTEROIDES: B. FRAGILIS

1 Morphology and Staining Reaction: Gram-negative bacilli, pleomorphic.

2 Respiration-Fermentation: Strict anaerobes; produce lactic, formic, acetic, and other acids.

3 Motility: Nonmotile.

4 Source of Infection-Reservoir: Human normal flora, intestine.

5 Disease: Appendicitis; abscesses in lung, pelvis, and liver; peritonitis.

6 Pathogenic relatives: *B. melaninogenicus* (intestinal or oral cavity); *Fusobacterium nucleatum* (trench mouth).

7 Features: Grow best with 10 percent CO_2; show bipolar staining; classified according to biochemical features and drug sensitivity; cell will autolyse (self-destruct) in culture.

Bacteroides fragilis is a nonspore-forming, nonmotile anaerobe found as the most frequent bacterium of normal intestinal flora. It has been estimated that 95 percent of normal flora found in feces belong to the genus *Bacteroides.* Other members of this group are found as normal flora of the pharynx, upper respiratory tract *(B. melaninogenicus),* and nasopharynx *(B. pneumosintes).* Growth as opportunistic pathogens occurs in tissues that become both infected by this rapid grower and anaerobic due to trauma caused by accident or surgery. Should *Bacteroides* enter the circulatory system, a generalized septicemia will develop and bacteria might be established in other tissues, forming abscesses. An **abscess** is a small, pus-containing cavity developed as a result of tissue disintegration. The pus produced in the abscesses is foul-smelling, and surgical draining is an important part of therapy in certain situations. Abscesses may be formed in the lungs, brain, heart, peritoneum, mucous membranes, and skin. Several strains of *B. fragilis* have been identified, but only a few are hemolytic. Identification of *Bacteroides* species must be done in anaerobic culture in order to make an accurate diagnosis. A relatively stable pattern of antibiotic resistance and susceptibility makes possible the use of this characteristic for the classification of *Bacteroides* species and varieties. Generally, *B. fragilis* is resistant to penicillin, kanamycin, and vancomycin. Most, however, are susceptible to tetracyclines, erythromycin, clindamycin, chloramphenicol, and metronidazole.

MYCOPLASMAS

MYCOPLASMA: M. PNEUMONIAE

1 Morphology and Staining Reaction: Gram-negative, highly pleomorphic, primarily coccoid; staining reactions poor.

2 Respiration-Fermentation: Aerobes-anaerobes.

3 Motility: Motile by possible release and reattachment of terminal cell organelle; no flagella present.

4 Source of Infection-Reservoir: Humans.

5 Disease: Pleuropneumonia, atypical pneumonia in humans and cattle.

6 Pathogenic Relatives: Several other species of *Mycoplasma*.

7 Features: Highly irregular in size and shape; smallest cellular life form; lack a cell wall; lipid in cell membrane provides great elasticity; require cholesterol for growth; produce H_2O_2.

Mycoplasmas were first discovered early in the 1900s as the cause of pleuropneumonia in cattle. Since that time, it has been demonstrated that humans can also be infected by members of this group. Because microbes isolated from some types of human pneumonia resembled the pathogens responsible for pleuropneumonia in cattle, the as yet unidentified cells were named the PPLOs, or pleuropneumonialike organisms. Since that time their characteristics have become more clearly understood, and the PPLOs have been separated from the bacteria and placed in the class Mollicutes. Members of the genus *Mycoplasma* are unique in that they require cholesterol and lack the typical cell wall found in bacteria. They have a highly pleomorphic appearance and have been described as coccoid, teardrop shaped, and ringlike. Their great elasticity and ability to withstand osmotic pressure changes without the aid of a cell wall results from the presence of unique combinations of cell membrane lipids. One end of *M. pneumoniae* contains a special terminal structure that apparently is responsible for locomotion. These smallest of cellular forms are suspected of moving by the repeated attachment, release, and reattachment of the terminal structure.

Mycoplasmas may be found in the mouth, pharynx, and genitourinary tract. The most frequent species of the upper respiratory tract, *M. pneumoniae*, is the agent responsible for approximately 90 percent of the disease known as **primary atypical pneumonia (PAP).** PAP received its

name from the fact that a number of upper respiratory infections resembling pneumonia were found to be caused by microbes other than *Streptococcus pneumoniae*, including adenoviruses and fungi. In humans, mycoplasmal PAP has an incubation period of from nine to twenty-one days, and the disease may last up to three weeks. Headache, cough, and malaise usually appear just before a fever develops, which can last for the duration of the untreated infection. The severe spasmodic cough is nonproductive (lacks exudate), but contains a whitish sputum with *M. pneumoniae* cells. Sputum cultures for identification require up to three weeks incubation. The fever seldom rises above 103°F (39.5°C) and may be lowered very quickly with antibiotic therapy. This disease does not follow the seasonal pattern found in pneumonia caused by the influenza virus. The fact that mycoplasmal pneumonia occurs regularly in the population throughout the year is used as a diagnostic feature. Transmission is most likely through droplet nuclei.

During the course of infection, *M. pneumoniae* attach to ciliated epithelial cells of the upper respiratory tract. These bacteria have the ability to release hydrogen peroxide that is thought to cause hemolysis and the release of cilia from the host cell surface. Hydrogen peroxide may also be involved in stimulating the formation of antibodies that lead to the agglutination of erythrocytes in cold temperatures. High concentrations of these **cold agglutinins** can be found in serum, and their presence is used in the diagnosis of the disease. This serological test may be done at bedside and is known as a cold agglutinin test. A small amount of blood is placed in a tube with anticoagulant and cooled in ice for one to two minutes. The presence of the cold antibodies will cause the erythrocytes to clump. If the tube is taken from the ice and warmed in the hands, the clumps will disappear, confirming the presence of the antibodies.

Therapy with penicillins, cephalosporins, and vancomycin is not effective, since these pathogens lack cell walls. Erythromycin and tetracyclines will quickly bring most infections under control. Infection results in the formation of complement-fixing and growth-inhibiting antibodies, but whether they are able to prevent further infection is doubtful. Immunization with *M. pneumoniae* is being investigated in hopes of stimulating IgA formation for the purpose of inhibiting replication of the pathogens early in the disease.

Certain mycoplasma types are also believed to be responsible for, or at least play a part in, other human diseases. *Ureaplasma urealyticum* (also known as T strain for their tiny colonies) have been found in women experiencing repeated or habitual reproduction failure and in women who consistently give birth to underweight children. T strains and *M. hominis* are frequently found in the genitourinary tract and may be responsible for some cases of **nongonococcal urethritis (NGU),** a disease primarily caused by members of the genus *Chlamydia*. Mycoplasma have also been found in cases of arthritis and otitis media (inflammation of the middle ear).

---

1 What diseases are caused by *B. fragilis*?
2 What is the cause of the primary atypical pneumonia?
3 What serological test is used to diagnose PAP?

RICKETTSIAS

The extremely small size and obligate intracellular nature of rickettsias are characteristics that have, in the past, led researchers to believe these parasites were closely related to the viruses. However, detailed examination with the electron microscope has revealed the presence of true procaryotic cell structures and reproduction by binary fission. Rickettsial diseases are primarily transmitted by arthropods and have as their major reservoir of infection humans, rats, mice, and small mammals. Mites, ticks, and body lice are the primary vectors of rickettsial infections. Only the Q-fever microbe, *Coxiella burnetii*, is able to be transmitted without the aid of a vector. *C. burnetii* is transmitted by a tick or through contaminated milk. Pasteurization at 71.5°C for 15 seconds will ensure the destruction of this pathogen. The extremely small size does not allow rickettsias to maintain all biochemical pathways essential to an independent life. They do have many of their own pathways but rely on their host for others that are necessary for their survival. All rickettsias are cultured in host tissues and only *Rochalimaea quintana*, the cause of trench fever, has been grown in pure culture.

Rickettsial infections are characterized by fever, rash, and other, more specific, species-related symptoms. They are divided into five main groups based on different clinical features (Table 17.5). Grouping is also done by the **Weil-Felix test** developed in 1915. This test is based on the fact that rickettsial diseases stimulate the production of antibodies capable of agglutinating certain strains of the bacterium *Proteus vulgaris*. *Proteus* have been designated by the initial "X" followed by another letter or number, for example, X-2 or X-K. Since the agglutinating antigen is actually a portion of the somatic, or "O," group, these strains have become identified by the letters OX. This serological test is not a specific reaction but does allow the pathogens to be grouped and aids in diagnosis. (Table 17.6)

Rickettsial diseases are caused by the multiplication of the pathogens within the endothelial cells of blood vessels in the skin and brain. The release of endotoxins and hemolysin furthers the disease. Vessels rupture and become inflamed and necrotic, if they are severely damaged. Two

TABLE 17.5
Major rickettsial groups.

Typhus Group:
  Epidemic typhus, louseborne: *R. prowazekii*.
  Endemic typhus, murine flea-borne: *R. typhi*.

Spotted Fever Group
  Rocky Mountain spotted fever (RMSP): *R. rickettsii*.
  Mediterranean fever (boutonneuse fever), South African tickbite fever, Kenya tick typhus, Indian tick typhus: *R. conorii*.
  North Asian tickborne rickettsiosis: *R. sibirica*.
  Queensland tick typhus: *R. australis*.
  Rickettsialpox, Russian vesicular rickettsiosis: *R. akari*.
  R. canada: Transmitted by ticks in North America; causes a disease resembling Rocky Mountain spotted fever.

Scrub Typhus (Tsutsugamushi Fever): *R. tsutsugamushi*.

Q Fever: *Coxiella burnetii*.

Trench Fever: *Rochalimaea quintana*.

of the most important diseases that have occurred throughout the world since ancient times are epidemic louse-borne typhus fever and murine typhus.

RICKETTSIA: R. PROWAZEKII

1 Morphology and Staining Reaction: Gram-negative coccobacilli, pleomorphic.

2 Respiration-Fermentation: Must be cultured as intracellular parasites; aerobic.

3 Motility: Nonmotile.

TABLE 17.6
Weil-Felix grouping of rickettsial diseases.

Immunological Group	Type	Disease
OX-19	OX-19 + + + + OX-2 + OX-K −	Classic European typhus; Brill's disease; endemic murine typhus.
OX-K	OX-19 − OX-2 − OX-K + + + +	Tsutsugamushi disease; scrub, or rural, typhus; Sumatran mite fever.
Indeterminate	OX-19 + OX-2 + OX-K +	Spotted fever, São Paulo typhus; Fièvre boutonneuse; South African tick fever; Kenya typhus; Indian tick typhus; North Asian tick typhus.

4 Source of Infection-Reservoir: Humans and arthropods.

5 Disease: Epidemic louseborne typhus fever.

6 Pathogenic Relatives: Spotted fever group: Rocky Mountain spotted fever *(R. rickettsii)*; Scrub typhus group: scrub fever *(R. tsutsugamushi)*; Q fever *(Coxiella burnetii)*; Trench fever: *Rochalimaea quintana.*

7 Features: Obligate intracellular parasites; transmitted by arthropods to humans; cultured in yolk sac of embryonated eggs.

Epidemic louseborne typhus fever and **murine typhus** are caused, respectively, by two closely related species, *R. prowazekii* and *R. typhi*. Murine, or endemic, typhus follows the same course of infection but is more mild, rarely fatal, and has the rat as a reservoir of infection. The fatality rate of epidemic typhus may range as high as 70 percent in individuals over 40 years of age. The disease is transmitted from the fecal material of the body louse *Pediculus humanus* (Figure 17.11). This arthropod defecates as it feeds from the skin of its human host. The fecal material contains large numbers of rickettsias that may be scratched into the skin before the arthropod dies from the infection. Once inside the host, the rickettsias establish themselves in the endothelial cells of the blood vessels. The incubation period is from 5 to 18 days, and the disease appears very suddenly. Fever and chills are quickly followed by a severe headache and the formation of a rash that will last until the fever subsides. The peak of the disease occurs near the twelfth day, and recovery may require an additional two weeks. During the fever, pathogens can be found in the blood, and they move to other host cells throughout the body. It is during this time that body lice acquire the pathogens while taking a blood meal and may transmit the pathogens to other susceptible individuals. Many complications can arise from infection, and the fatality rate is high in untreated cases. Gangrene, bronchial pneumonia, otitis media, and typhus encephalitis are among the most important secondary complications.

Control of the active disease is accomplished with antibiotic therapy. However, the sulfonamides should be avoided since this class of drugs actually stimulates the pathogen and makes the symptoms more severe. The drugs most frequently used are the tetracyclines and chloramphenicol. Following infection, a stable, long-lasting immunity is developed, but the intracellular nature of *R. prowazekii* may enable it to "hide" in the host for years. A recurrence of the disease due to the emergence of these hidden pathogens is known as latent typhus, or Brill's disease. During the latent period, the host may serve as a source of infection, if conditions were to become favorable for vector transmission by body lice. Crowding, unsanitary conditions, and malnourishment found in slums, labor camps, and during wartime have contributed to the development of epidemics of louseborne typhus fever. Bringing the epidemic under control requires the use of insecticides for delousing infected humans. Vaccines are available to stimulate an active immunity. These may either be attenuated pathogens or inactivated cell extracts that had been cultured in yolk sacs of eggs or cell cultures. Vaccination is an important control method in many epidemic situations since quick and complete control of body lice may not be possible (Figure 17.12).

---

1 Name the vectors for rickettsias.
2 Of what value is the Weil-Felix test?
3 What are the symptoms of murine typhus?

CHLAMYDIA

There are only two currently recognized species in the genus *Chlamydia*: *C. psittaci* and *C. trachomatis*. These two species are responsible for several diseases (Table 17.7). More recently, research has indicated that the cause of legionnaire's disease *(Legionella pneumophila)* may also be a member of the *Chlamydia*. All members are obligate intracellular parasites and similar to Gram-negative bacteria; however, they lack the biochemical pathways necessary for the production of ATP. These pathways are provided by the host after the pathogens have entered the cell. Infection of the host cell and growth of the pathogens follow a sequence known as the developmental cycle. This cycle is unique to the *Chlamydia* and begins with the microbe's adsorption to receptor sites on the host cell. They are then taken inside by phagocytosis and encapsulated in a vacuolar membrane. Many of these small vacuoles may join to-

FIGURE 17.11

Pediculus humanus var. *capitis*, the female head louse. (Courtesy of the Center for Disease Control, Atlanta.)

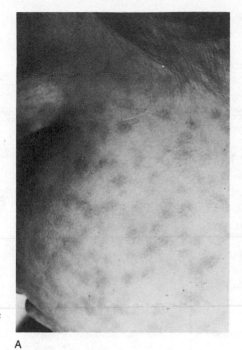

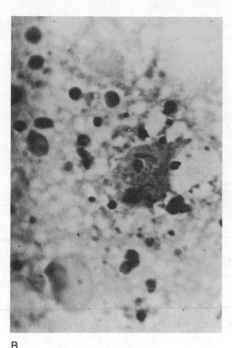

FIGURE 17.12

(A) Skin rash of Rocky Mountain spotted fever, nine days old. (B) Intracellular rickettsia of Rocky Mountain spotted fever. (Courtesy of the Center for Disease Control, Atlanta.)

A

B

gether to form larger "initial bodies." The pathogens reproduce inside by a form of binary fission and fill the vacuoles. They are then referred to as inclusion bodies. *Chlamydia* produce no toxins but cause host cell death as a result of severe host cell membrane damage at the time of phagocytosis. Later in the disease, they also cause the release of lysosomal enzymes, killing the eucaryotic host and releasing the pathogens.

Chlamydia infections can occur in psittacine birds (e.g., parrots), wild birds (e.g., pigeons, gulls, turkeys), some lower animals (e.g., mice), and humans. In humans, infection typically occurs in the eye, urogenital tract, and lungs. The most interesting clinical feature of these infections is the long-term, host-parasite balance established between *Chlamydia* and human host. This balance develops into illnesses that become subclinical and last for years.

TABLE 17.7

Chlamydia infections.

Chlamydia	Reservoir Hosts	Principal Diseases
C. trachomatis (TRIC agent; LGV agent, Chlamydia subgroup A)	Humans mice	Trachoma, inclusion conjunctivitis, lymphogranuloma venereum, urethritis.
C. psittaci (Psittacosis agent, Chlamydia subgroup B)	Birds, mammals (both domestic and wild), arthropods(?)	Psittacosis-ornithosis.

CHLAMYDIA: *C. TRACHOMATIS*

1 Morphology and Staining Reaction: Gram-negative; however, this stain is of little diagnostic value.

2 Respiration-Fermentation: Obligate intracellular parasites that rely on the respiratory pathways of the host cell.

3 Motility: Nonmotile.

4 Source of Infection-Reservoir: Humans and mice.

5 Disease: Lymphogranuloma venereum (LGV); trachoma; inclusion conjunctivitis; nongonococcal urethritis (NGU).

6 Pathogenic Relatives: *C. psittaci.*

7 Features: Obligate intracellular parasites; may remain infective for years when frozen at $-50°C$ to $-70°C$; can produce CO_2 from glucose; growth is inhibited by antibiotics.

Trachoma is an infection of the conjunctiva of the eye acquired by direct contact with infectious nasal or ocular discharge from others. The disease has a 5 to 12 day incubation period, appears suddenly, and can last in a subclinical form for months or a lifetime. Trachoma begins as an inflammation of the conjunctiva with the formation of small, tumorlike masses. As the infected cells die, the surrounding area becomes filled with a thick discharge. Scar tissue containing fibroblast cells develops and can extend into the eyelid. The eyelid can become deformed, and scarring may occur that is so extensive that blindness results. Trachoma is reported to be the leading cause of blindness. The microbe remains localized in the eye during the disease.

Inclusion conjunctivitis is a disease similar to trachoma that occurs in newborns and adults. Infection of newborns involves the lower lid which becomes inflamed for several weeks, then slowly disappears. In many cases, the disease takes the form of a typical trachoma infection and may lead to scarring and blindness. In adults, the infection may appear as a mild conjunctivitis, such as "swimming pool conjunctivitis," and disappear very quickly. In others, the disease is indistinguishable from trachoma. The similarity of these diseases has lead to the use of the term "TRIC agent" to designate trachoma and inclusion conjunctivitis caused by *C. trachomatis.* (Figure 17.13) Antibodies are formed in response to these infections, but they appear to provide little protection. The antibiotics most helpful in controlling infection are the tetracyclines and sulfonamides.

Chlamydia trachomatis is also responsible for a venereal disease known as LGV, or lymphogranuloma venereum. This disease is transmitted by intimate sexual contact and enters the system in the urogenital or anal region. The pathogens spread quickly through females and establish vesicular lesions within the urethra, labia, vaginal wall, or cervix. In males, the vesicular eruptions may occur on the penis. The vesicles rupture and form shallow ulcers that quickly heal without scarring. Following this stage, the pathogens invade the lymph nodes of the pelvic region and form buboes. These infected nodes become suppurative and may continue to drain for a long period. Generalized symptoms during this stage include fever, body ache, conjunctivitis, and central nervous system disruptions. When the disease enters the final stage, there is swelling of the genitals and surrounding tissues. The term "elephantiasis" (not to be confused with the roundworm-induced disease of the same name) has been used to describe the massive enlargement of the penis and scrotum of males, and the labia and clitoris of females. The infection is treated with sulfonamides and tetracyclines, as are other *Chlamydia* infections.

Another venereal disease of increasing importance is **nongonococcal urethritis (NGU).** This disease has become so extensive throughout the United States that it probably exceeds gonorrhea in frequency. Evidence points to the TRIC agent as the cause. The infection may begin as a first infection from *Chlamydia* normally found in the urogenital tract or as a secondary infection resulting from penicillin therapy used to treat gonorrhea. The symptoms are similar to those found in true cases of gonorrhea, but they are more mild and do not respond to treatment with penicillin. They are controlled with tetracycline. Symptoms include an opaque discharge, burning sensation upon urination, and a urethral itching.

---

1 Name two diseases caused by the *Chlamydia.*
2 What is LGV and how is it transmitted?
3 What is the cause of NGU?

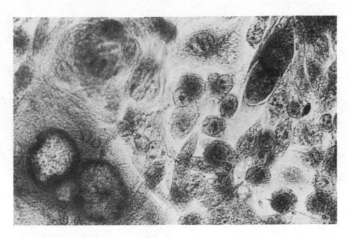

FIGURE 17.13
Chlamydia trachomatis. (Center for Disease Control, Atlanta.)

LEGIONELLOSIS

The exact cause of legionellosis, or legionnaire's disease, has not been clearly established as yet but is being included at this point because of its similarity to the *Chlamydia.* The bacterium responsible for this disease has more recently been named *Legionella pneumophila.* These bacteria are slow growing (generation time of about 30 minutes), Gram-negative cells that require iron (from hemoglobin) and the amino acid L-cysteine; they are aerobic with optimum growth occurring in 2.5 percent CO_2 at 35°C. The disease is a form of atypical pneumonia that may also involve the central nervous system, kidneys, and gastrointestinal tract. The symptoms of the disease include fever, malaise, headache, muscle pain, diarrhea, confusion, and in some cases, memory loss. Older patients who smoke are more susceptible to infection. The incubation period is two to ten days after initial exposure. Although the most notorious outbreak took place in 1976 at the American Legion Convention in Philadelphia, the disease occurs worldwide. Cases have been reported in Scotland, Spain, Canada, the Netherlands, and England. Four strains of *L. pneumophila* have been identified: Togus, Los Angeles, Bloomington, and Philadelphia-Knoxville. Diagnosis of the disease is done by direct and indirect fluorescent antibody testing with specimens taken from sputum, bronchial washings, and pleural fluid.

The reservoir of infection has been identified as lakes, rivers, cooling towers, and other aquatic environments. Investigators have found suitable conditions for transmission from soil contaminants in evaporative condensers of air conditioning systems. The water released from the system may contain droplet nuclei with *L. pneumophila.* These pathogens have also been isolated from water in cooling towers. Culture and susceptibility testing has demonstrated that a variety of antibiotics inhibit growth; however, complete testing with human patients has not indicated which drug is most effective. It has been demonstrated that all strains of

L. pneumophila produce penicillinase. The two antibiotics recommended are erythromycin and rifampicin.

The discovery of *L. pneumophila* has been an exciting experience for microbiologists and poses many unanswered questions for epidemiologists, immunologists, and clinicians. Researchers are continuing to explore the nature of the microbe as well as other diseases that may be caused by this pathogen or its unidentified, related forms.

SUMMARY

Eight groups of pathogenic bacteria have been described in this chapter. These representative organisms and the diseases that they cause occur throughout the world. As the sciences of microbiology, immunology, and epidemiology have expanded our understanding of disease processes, the incidence of each of these human illnesses has decreased. Once the leading cause of death, many bacterial infections have been successfully brought under control in developed countries. A continued effort to identify, control, and prevent disease is being supported by such agencies as the World Health Organization, national public health agencies, and local health departments.

The eleven major groups of pathogenic bacteria presented in Chapters 16 and 17 include:

1 Pyogenic cocci

2 Gram-positive bacilli

3 Gram-negative enteric bacilli

4 Small Gram-negative bacilli

5 Corynebacteria

6 Acid-fast bacteria

7 Spirochetes

8 Gram-negative anaerobes

9 Mycoplasmas

10 Rickettsias

11 Chlamydia

Morphological, physiological, chemical, and antigenic differences are used as the basis for this classification system. This system is not meant to be a substitute for the formal taxonomy presented in Chapter 2, but should be considered only as a functional alternative used in the more specialized field of medical microbiology.

STUDY QUESTIONS

1 What diseases discussed in this chapter are transmitted by water? Milk? Air?

2 Which disease-causing bacteria produce symptoms by releasing exotoxins? Endotoxins?

3 List three bacteria that are highly susceptible to drying during transmission from one host to another.

4 List two bacterial pathogens that are primarily found in the upper respiratory tract during illness.

5 Which pathogens are intracellular parasites?

6 Which pathogens cause venereal diseases?

7 What is the difference between "pneumonia" and "atypical pneumonia"? What organisms are responsible?

8 Explain the following: DPT, Schick test, BCG, PPD, OT, VDRL, PPLO, NGU, LGV, TRIC.

9 What are the symptomatic differences among primary, secondary, and tertiary syphilis? Between exudative and productive tuberculosis?

10 Which pathogenic "bacteria" were once thought to be other forms of microbial life?

KEY TERMS

exudate	caseation	gumma
tubercle	necrosis	

Fungal and Protozoan Disease

MEDICAL MYCOLOGY
Candida albicans
Dermatophytes

SUBCUTANEOUS MYCOSES
Sporothrix (Sporotrichum) schenckii
Mycotic Mycetoma and Chromomycosis

SYSTEMIC MYCOSES
Histoplasmosis
Coccidioidomycosis
Blastomycosis
Aspergillosis
Cryptococcosis

ACTINOMYCETES: FILAMENTOUS BACTERIA
Actinomycosis
Nocardiosis

PROTOZOA OF MEDICAL IMPORTANCE
Subphylum: Sporozoa
Subphylum: Ciliata
Subphylum: Mastigophora
Subphylum: Sarcodina

Learning Objectives

☐ Be familiar with basic fungal characteristics and the three clinical types of mycoses.

☐ Know the classification, morphology, diseases, tissue reactions, and reservoir of infection of *Candida albicans.*

☐ Be familiar with the medical terminology found in the Key Terms.

☐ Know the names, morphology, nature of disease, tissue reactions and reservoir of infection of dermatophytes.

☐ Recognize the nature and causes of subcutaneous mycoses.

☐ List the most common causes of systemic mycoses.

☐ Be familiar with the causes, symptoms and reservoir of infection of the actinomycetes.

☐ Be able to describe the classification system used for the protozoa.

☐ Know the causes, symptoms, treatment, and reservoir of infection for diseases caused by members of the Sporozoa, Ciliata, Mastigophora, and Sarcodina.

The parasitic relationships described in chapters 16 and 17 occur between bacteria and eucaryotic hosts. Two other groups of microbes have members that may also establish harmful relationships. Fungi and protozoa are eucaryotes that have for the most part evolved as free-living saprotrophs. However, a small number of these microbes are of medical importance because of their parasitic nature. The study of pathogenic fungi is known as **medical mycology** and dates back to 1839, prior to the bacteriological work of Robert Koch and Louis Pasteur. The study of parasitic protozoa has become a branch of **medical parasitology,** since these microbes are considered by many to be single-celled animals. Other pathogens studied in medical parasitology are the parasitic round and flat worms (liver flukes, tapeworms, pinworms, etc.) and arthropods (lice, ticks, and fleas). Worms and arthropods are true multicellular animals and will not be described. This chapter will focus on medical mycology and the parasitic protozoa.

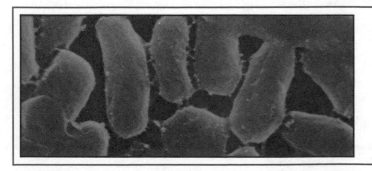

18

MEDICAL MYCOLOGY

The scientific study of pathogenic fungi began in 1839 when scrapings from a skin disease known as favus were grown in pure culture on potato slices. The fungal growth was then used to inoculate the scalp of a healthy child and resulted in a second case of the disease. This procedure demonstrated the fungus *Trichophyton schoenleinii* to be responsible for the disease. Over 100 years of research has uncovered more than 100,000 species of fungi. Most are nonpathogenic, saprotrophic, yeastlike or filamentous microbes found in the soil. A few have evolved to be true pathogens and are able to actively infect, cause harm, and be transmitted from one suitable, susceptible host to another. Most fungi only demonstrate clinically recognizable symptoms of infection when, by chance, large numbers of spores enter and overwhelm a host; or when normal, opportunistic fungal flora are not kept under control by defense mechanisms. The extensive use of antibiotics and immunosuppressive drugs has led to a sharp increase in infections caused by fungi normally found in or on the body.

Classification of fungi is based on cell morphology and the kinds of sexual spores produced (refer to Chapter 2). Cells may be single and yeastlike, or they may be filamentous. Each filament is known as a **hypha,** while an interwoven mat of hyphae is called a **mycelium.** Some fungi have the ability to change from one morphological type to the other depending on the growth environment. These species are known as **dimorphic.** There are four classes of fungi: (1) Phycomycetes are sexually reproductive, producing zygospores; (2) Ascomycetes are sexually reproductive, producing ascospores; (3) Basidiomycetes are sexually reproductive, producing basidiospores; and (4) Deuteromycetes do not yet have an identified sexual stage.

Infections caused by fungi are known as **mycoses** (*sing., mycosis*). Three clinical types of mycoses have been identified: cutaneous, subcutaneous, and systemic (Table 18.1). Fungi known as **dermatophytes** are able to cause **cutaneous** infections. Members of this group have the ability to use the protein keratin and become localized in the surface layers of the skin, hair, or nails. Seldom do they penetrate or extend into the deep tissues of the body. The dermatophytes are true pathogens capable of establishing an infection in a healthy host and may be transmitted from one host to another. The group of fungi responsible for **subcutaneous** infections is only acquired from the soil and enters the body through breaks in the skin caused by traumatic experiences. These grow very slowly once inside the skin and may show no visible signs of host damage for many years. Fungi responsible for **systemic,** or deep, infections may be acquired from the soil or are opportunistic, normal flora of the body. Minor infections caused by these fungi occur frequently but are kept in check by body defenses. Occasionally they are not controlled and cause the death of their host and themselves. These pathogens are not contagious under normal circumstances.

Fungi are poor antigens and stimulate little or no protective humoral immunity. However, there is a significant degree of natural immunity in healthy, normal individuals. Infections are primarily controlled by nonspecific host defense mechanisms and cell-mediated immunity. The type of host damage seen as a result of fungal infection depends on the nature of the infecting species and the tissue response of the host (Table 18.2). Delayed (cell-mediated) hypersensitivities occur frequently and may be the direct or indirect result of exposure or infection. Other tissue responses include the formation of pus (pyogenic response), granulomas (tumorlike masses), thick fibrous scars, and extensive in-

TABLE 18.1
Examples of fungal infections.

Cutaneous infections	Candidiasis of skin, mucous membranes, and nails	*Candida albicans* and related species
	Ringworm of scalp, glabrous skin, nails	Dermatophytes (*Microsporum* sp., *Trichophyton* sp., *Epidermophyton* sp.)
Subcutaneous infections	Chromomycosis	*Fonsecaea pedrosoi* and related forms
	Mycotic mycetoma	*Petriellidium boydii, Madeurella mycetomii,* etc.
Systemic infections	Sporotrichosis	*Sporothrix schenckii*
	Histoplasmosis	*Histoplasma capsulatum*
	Blastomycosis	*Blastomyces dermatitidis*
	Coccidioidomycosis	*Coccidioides immitis*
	Cryptococcosis	*Cryptococcus neoformans*
	Aspergillosis	*Aspergillus fumigatus*
	Candidiasis, systemic	*Candida albicans*

1 What are the two morphological types of fungi?
2 What are the three clinical types of mycoses and how do they differ?
3 The type of host damage caused by a fungal infection depends on what factors?

flammation. Since many of these symptoms are also caused by bacteria known as Actinomycetes, these microbes will also be described in this chapter.

Accurate diagnosis of a fungal infection requires extensive laboratory time. Four to six weeks may be necessary to culture specimens. Once isolated in pure culture, identification is based on morphology, pigmentation, size, spore structure, sexual stages, and other features. This work requires keen observation and a thorough knowledge of fungal characteristics. Since fungi are poor antigens, serological identification is of little value but may be used to determine the course of infection. A more simple and rapid clinical test to identify the presence of dermatophytic infections is done by mixing potassium hydroxide with skin scrapings for microscopic analysis. This base dissolves excess necrotic tissue away from the hyphae and makes them more visible. The test may also be performed on specimens taken from infected lung tissue. Some fungal infections can be diagnosed

by using a Wood's light. This lamp emits a wavelength of 3650 Å that will stimulate a green or pink fluorescence in dermatophytic infections caused by *Microsporum*.

Although countless infections occur in human populations, only in cases of massive spore contact or heightened susceptibility do individuals display disease symptoms. However, when fungal infections do become evident, they are difficult to control and require extensive, extended therapy. This may include the use of potentially lethal drugs or surgery. The eucaryotic nature of the pathogens makes treatment with typical antibacterial antibiotics almost useless. A few antifungal antibiotics less selective in their action are available but their use demands careful monitoring in order to prevent host harm. Maintaining a high degree of host health is by far the best way of preventing all clinical forms of mycoses.

CANDIDA ALBICANS

1 Classification: Deuteromycetes; may have a sexual stage.
2 Morphology: Yeastlike; form pseudohyphae (false filaments) under certain growth conditions.

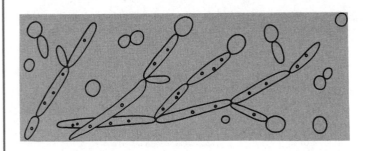

3 Diseases: Cutaneous:
　　Skin infections, candidiasis (moniliasis)
　　Diaper rash
　　Fingernail and toenail candidiasis
　Mucous membranes:
　　Thrush, stomatitis
　　Vulvovaginitis
　　Pulmonary candidiasis
　　Alimentary candidiasis
　Systemic:
　　Endocarditis
　　Meningitis
　　Septicemia
　　Urinary tract candidiasis
　Hypersensitivities:
　　Eczema
　　Asthma
　　Gastritis

TABLE 18.2
Tissue reactions in fungal diseases.

Disease Responses	Fungal Diseases
Chronic inflammation	*Rhinosporidium seeberi*
	Entomophthoromycosis
Pyogenic reaction	*Actinomyces israelii*
	Acute aspergillosis
	Acute candidiasis
Mixed pyogenic and granulomatous reaction	*Blastomyces dermatitidis*
	Coccidioides immitis: neutrophils
	Sporothrix schenckii
	Mycetoma
Pseudoepitheliomatous hyperplasia	*B. dermatitidis*
	Chromomycosis
Histiocytic granuloma	*Histoplasma capsulatum*
Granuloma	*H. capsulatum*
	Coccidioides immitis
	Cryptococcus neoformans
	Occasionally *H. capsulatum*
Fibrocaseous pulmonary granuloma; "tuberculoma"	*H. capsulatum*

SOURCE: J. W. Rippon, *Medical Mycology.* W. B. Saunders, 1974.

4 Tissue Reaction: Erythema, inflammation, "scalded skin reaction," vesicle formation.

5 Reservoir of Infection: Humans, soil.

Candida albicans is one of several types of pathogenic yeastlike fungi (e.g., *Torulopsis, Rhodotorula, Cryptococcus*). While found on the skin as normal flora, cells are also commonly found on the mucous membranes and in the gastrointestinal tract. This fungus is kept under control by the presence of lactic acid, nonspecific host defense mechanisms, and a balanced, normal bacterial flora. *Candida* is an opportunistic pathogen that is able to establish infections of the skin, mucous membranes, and intestinal tract if control mechanisms fail. Individuals who are in poor health, diabetic, undergoing immunosuppressive therapy or prolonged antibacterial therapy are more susceptible to candidiasis. Infection occurs on the moist, warm folds of the skin in the area of the groin, axilla, and breasts. Individuals who have their hands, feet, or whole bodies in prolonged contact with warm, moist environments (for example, cooks, dishwashers, car wash employees) are more likely to develop candidal skin infections. The site becomes inflamed and resembles the "scalded skin reaction" of *Staphylococcus aureus* (refer to Chapter 16). Vesicles form and erupt, leaving raw, exposed tissue. The infection may also occur on fingernails and toenails, spreading beneath the nail and causing painful swelling and loss of the nail and nail-producing tissue. In rare instances, the skin infection can lead to the formation of granulomas. This is more likely to occur on the face but has been found on the scalp, trunk, and nails.

---

1 Why are mycoses difficult to control?
2 Name three mycoses caused by *Candida albicans.*
3 What factors contribute to mycoses caused by *C. albicans?*

Thrush, or "black hairy tongue," is similar to other candidiasis infections of mucous membranes. Infections occur in infants during the establishment of normal flora (within the first three days after birth) or in older individuals after prolonged antibacterial antibiotic therapy. Fungal growth leads to the formation of white, curdy patches of pseudomembrane on the tongue and buccal mucosa (Figure 18.1). If the infection destroys papillae, the tongue takes on a blackened appearance. The disease also occurs in adults who experience mucosal irritation from poorly fitting dentures or in those who fail to maintain good oral hygiene. The infection may spread, causing chelitis (scattered foci of infection

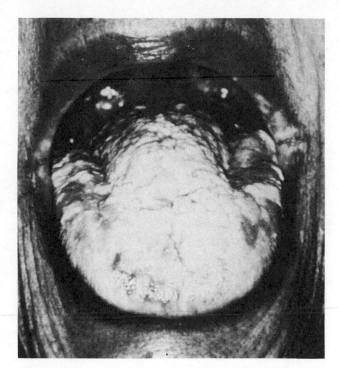

FIGURE 18.1
Thrush. (Courtesy of the American Society of Microbiology.)

on the lips) or angular stomatitis seen as pursed lips with grooves at the corners of the mouth. Vulvovaginitis is a candidal infection that runs a course similar to thrush but occurs on the mucosal membranes of the vagina. Symptoms include intense itching, irritation, and vaginal discharge. The condition is more likely to develop as the pH of vaginal secretions becomes more alkaline. Under normal circumstances, bacterial flora help maintain a low pH that limits the growth of *C. albicans* (refer to Chapter 12). *Candida* is part of the normal vaginal and internal flora; most individuals contain IgG antibodies against this microbe, and it is not considered contagious.

C. albicans causes a great variety of symptoms upon infection. These symptoms are as diverse as those seen in syphilis and may involve any organ of the body. Systemic infections can involve the lungs, kidneys, liver, heart, meninges, and other internal organs. The mortality rate from such infections is high; however, the morbidity rate is very low, occurring only after prolonged illness and antibacterial or steroid therapy. Amphotericin B (fungizone) is used to control systemic infections. This drug causes a great many side effects and must be monitored to prevent severe damage to kidneys, liver, and blood-forming tissues. Infections of the skin and mucous membranes can be controlled by many drugs, including nystatin, gentian violet, trichomycin, sodium propionate, and miconazole.

DERMATOPHYTES

1 Classification: Genus name based on asexual reproductive stage; class
Deuteromycetes

| Trichophyton | Microsporum | Epidermophyton |

2 Morphology: Hyphal growth form

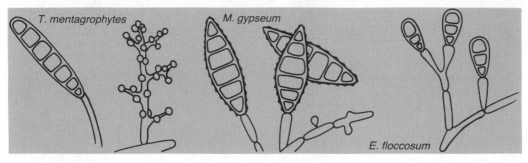

3 Diseases: Tinea pedis (athlete's foot); tinea cruris (jock itch); tinea barbae (barber's itch); tinea unguium (nail infections)

Tinea corporis (body ringworm); tinea capitis (scalp ringworm)

Tinea pedis; interriginous dermatitis; tinea cruris; tinea unguium

4 Tissue Reaction:

"Scalded skin reaction," erythema, vesicle formation, peeling, cracking, inflammation

5 Reservoir of Infection:

Humans, animals (dogs, cats, cattle)

Members of the dermatophytic fungi are found in the soil, on humans, and on animals. The three most common dermatophytes belong to the genera *Epidermophyton, Trichophyton,* and *Microsporum.* While some genera members are true pathogens, many others have never been found in association with disease. The pathogens are capable of utilizing skin protein, keratin, and become localized in the superficial, dead, keratinized layers of the skin. They seldom move inward to become systemic mycoses. During the disease, hyphae can be found in scrapings taken from the infection site. These scrapings rarely show only a single type of dermatophyte. In many cases, two dermatophytes are identified, and secondary bacterial infections are common.

The nature and names of superficial mycoses are determined by their location on the body. Although they resemble genus and species names, the following are clinical terms indicating the body part affected and do not identify an exact etiological agent: tinea pedis ("tinea" means *worm,* "pedis" means *foot*), t. corporis (body), t. cruris (perineal folds, groin), t. barbae (face and neck), t. capitis (head), and t. unguium (nails).

Dermatophytic infections are acquired through breaks in the skin which become infected with fungal spores. The site of infection becomes inflamed and itchy. Vesicles form and rupture, leading to peeling and cracking of the skin

(Figure 18.2). As the infection develops, there is a central clearing as hyphae break up and fungal spores are released. This is seen most clearly in tinea corporis, since the infection occurs on the flat, hairless areas of the body (Figure 18.3). Hyphae continue to grow and spread from the infection margins. They might become chronic if they move below the keratinized epithelium into the deeper layers of epidermis. Some dermatophyte infections, especially tinea capitis, lead to a hypersensitivity known as a **dermatophytid reaction.** This is an allergic reaction to fungal products that move from the infection site to other, noninfected parts of the body. The reaction results in the formation of itchy, inflamed vesicles on the hands or fingers, or as an eczema reaction.

Dermatophytes also occur in association with infections of hair and hair-producing tissue. *Microsporum* grows around the outside of the hair shaft, while *Trichophyton* may actually penetrate the shaft. This last growth pattern can result in breaking of the hair or temporary baldness. Characteristics of tinea capitis include erythema, edema, peeling, and severe inflammation that resembles a pyogenic bacterial infection.

Several antifungal agents can be used to control body, foot, and scalp infections. Athlete's foot is the most common dermatophytic infection, but it rarely occurs in societies where

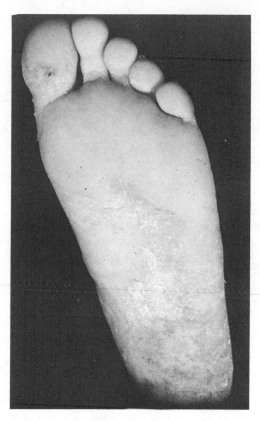

FIGURE 18.2
Tinea pedis, or ringworm of the foot, caused by *Trichophyton rubrum.* (Courtesy of the Center for Disease Control, Atlanta.)

the people wear closed shoes. Acute cases can be treated by soaking the feet in potassium permanganate to relieve acute symptoms. Chronic symptoms can be treated by applying antifungal agents containing such chemicals as undecylenic acid, salicylic and benzoic acid, or micronazole cream. Scalp infections can be controlled by removing and treating the infected hair with micronazole cream or selenium sulfide. The antifungal antibiotic griseofulvin is given orally in more extreme cases.

---

1 What drugs may be used to control fungal infections?
2 Define the terms "tinea pedis" and "tinea cruris."
3 What is a dermatophytid reaction?

SUBCUTANEOUS MYCOSES

SPOROTHRIX (SPOROTRICHUM) SCHENCKII

1 Classification: Deuteromycetes

2 Morphology: Dimorphic

3 Diseases: Subcutaneous skin infections—lymphocutaneous sporotrichosis; more rarely involves mucous membranes, lungs, and meninges.

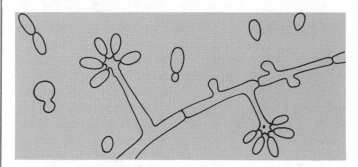

4 Tissue Reaction: Granuloma formation, chronic inflammation, necrosis-forming gummas (lymphocutaneous form).

5 Reservoir of Infection: Soil and plants.

Sporotrichosis occurs most frequently as an infection of cutaneous lymphatic tissue, and if untreated, may spread to other tissues. Many people acquire the fungus while work-

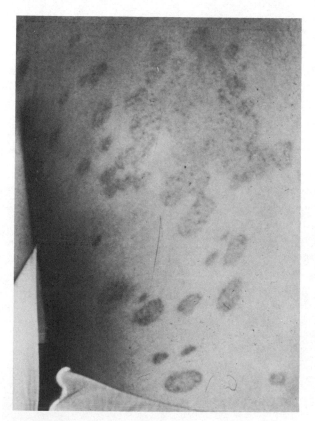

FIGURE 18.3
Ringworm lesions caused by *Trichophyton verrucosum.* (Courtesy of the Center for Disease Control, Atlanta.)

ing with soil or gardening. Thorns, thistles, and many weeds are contaminated with this common soil fungus. However, person-to-person transmission is not known to occur. Once inside the skin, a pus-containing nodule (pustule) develops deep within the skin. As the hyphae penetrate and spread, they cause a discoloration of the skin and necrosis similar to that in a syphilitic chancre. This type of lymphocutaneous sporotrichosis may disappear completely or reappear in adjacent areas. The infection may demonstrate itself for years. A chronic condition affecting localized lymphatic tissue may develop if the primary lesion is left untreated or treated only with topical antifungal drugs. The disease's chronic form begins as a series of new pustules along lymphatic ducts. These lesions are characteristically a cordlike series of firm pustules that later become necrotic, suppurative, and drain. This is the gumma-type lesion (Figure 18.4). In rare instances, the infection involves bones, periosteum, and synovial membranes.

Infections can be controlled with potassium iodide given orally or with amphotericin B. However, since this pathogen is a common soil microbe that only causes infection in susceptible, traumatized hosts, wearing gloves and shoes when working with plants and soil is the best way to prevent sporotrichosis.

MYCOTIC MYCETOMA AND CHROMOMYCOSIS

Mycetoma and chromomycosis are subcutaneous infections that grow slowly on the feet and hands. These diseases are found primarily in tropical and subtropical areas among barefoot workers who acquire pathogens through breaks in the skin. These diseases are not contagious. Mycetoma is caused by a number of filamentous actinomycete bacteria

commonly found in the soil, including *Actinomyces israelii, Nocardia asteroides,* and *Streptomyces somaliensis.* A few true fungi also stimulate this disease sometimes known as Madura foot. Several months' time and continued trauma may be required before the microbes establish a parasitic relationship. The disease begins as a small, localized tumor at the site of injury. The tumor softens, ulcerates, and extends to the skin's surface, eventually rupturing. The lesion then extends deeper into the localized tissue causing swelling, pain, and tissue distortion. Abscesses formed during the disease process become suppurative and drain through open "sinus tracts," or cracks in the infected tissue (Figure 18.5). Colored microcolonies of pathogens are typically found in the draining pus. The presence of these pink, red, or dark brown "grains" are of diagnostic value and help distinguish actinomycelial mycetoma from fungal mycetoma and sporotrichosis. Tissue distortion can become so severe that hands or feet lose function. If treated early, a person experiences only minor general health problems; however, therapy must be prolonged to ensure that the drug penetrates the infection site. The actinomycetes responsible for this subcutaneous infection can be controlled with sulfa drugs. Fungal mycetoma can be treated with amphotericin B but surgery may be required.

Chromomycosis is a true subcutaneous fungal infection possibly caused by several soil microbes, including members of the genera *Phialophora, Cladosporium,* and *Fonsecaea.* The disease occurs on the feet, hands, and arms of farm workers unprotected by shoes, gloves, or clothing. The fungal spores enter the body as a result of traumatic tissue damage and become localized in the lymphatic tissue. A wartlike growth develops in the subcutaneous tissue and forms a pustule that later ruptures on the skin. The surface becomes scaly, raised, and gray in appearance. The fungi slowly

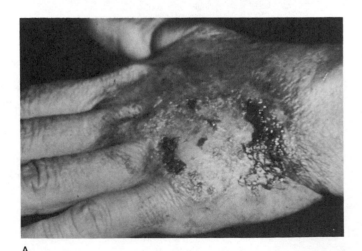

A

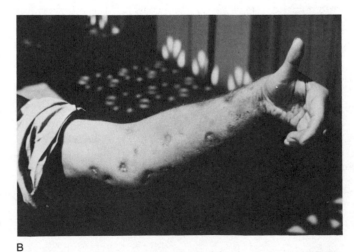

B

FIGURE 18.4

(A) Sporotrichosis lesion of the hand. (B) Lesion on the arm. (Courtesy of the Center for Disease Control, Atlanta.)

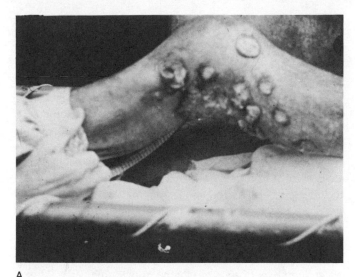

A

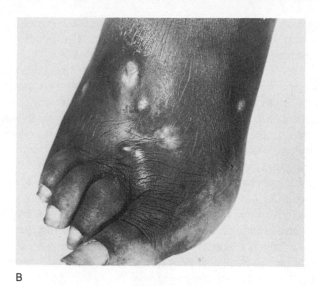

B

FIGURE 18.5
(A) Actinomycotic mycetoma of the leg. (B) Mycetoma localized lesion of the foot due to *Actinomyces israelii*. (Courtesy of the Center for Disease Control, Atlanta.)

spread from the initial infection site through the adjacent lymphatic ducts, establishing additional wartlike growths and skin disruptions. The feet and hands may become swollen as lymph ducts are made impassible by fungal hyphae and granulomatous tissue. This process requires months or years but seldom extends beyond a localized area. The fungus will rarely spread to become systemic and fatal. As the disease progresses, individual lesions may fuse to form cauliflowerlike granulomas with crusting on the skin's surface. These granulomas rise 1–3 centimeters above the surface. Treatment of the infection may require surgical removal of granulomatous tissue and the use of one of three drugs: amphotericin B, flucytosine, or thiabendazole.

---

1 What is the reservoir of infection of *Sporothrix*?
2 How do you treat a mycetoma?

SYSTEMIC MYCOSES

HISTOPLASMOSIS

1 Name: *Histoplasma capsulatum*

2 Morphology: Dimorphic; growth form change regulated by temperature—budding yeast cells at 37°C, hyphal growth at 25°C.

3 Disease: Three forms: (1) acute pulmonary, (2) chronic pulmonary, (3) progressive, disseminated histoplasmosis.

4 Tissue Reaction: Granuloma, necrosis, and calcification.

5 Reservoir of Infection: Soil, dried bird excrement, and small rodent fecal material.

Like all other causes of systemic mycoses, *Histoplasma capsulatum* is a dimorphic soil inhabitant. This fungus is able to change form and grow in host tissues as a yeastlike intracellular parasite. Spores and hyphae are acquired through inhalation. The microbes commonly occur in soils of the Appalachian Mountains, Mississippi and Ohio river valleys, and in Central America. Birds and small rodents are notorious sources of *H. capsulatum* spores. The spores are easily spread from dried fecal material found in nests, burrows, barns, chicken coops, and caves. Most infections are asymptomatic, and only a select few individuals have the physiological characteristics which enable *Histoplasma* to establish a parasitic infection. The traits have not as yet been identified. Histoplasmosis is an infection of the reticuloendothelial system usually localized in the lungs but may on rare occasion spread to other parts of the body. The three disease forms are termed acute pulmonary; chronic pulmonary; and progressive, disseminated histoplasmosis. *H.*

capsulatum may be found as an intracellular parasite of macrophages and is seldom extracellular. The infection is endemic in the Ohio and Mississippi river valleys, where it causes a "summer flu" in children. Symptoms of acute pulmonary disease strongly resemble tuberculosis and include fever, chills, cough, chest pains, and nightly sweating. Patients usually recover quickly and have an immunity to reinfection. In the lungs, granulomas are formed which undergo calcification similar to that occurring with tuberculosis. In a small percentage of cases, the disease progresses to the chronic pulmonary form. This form of the disease may also develop as the primary infection in older, susceptible individuals. Chronic pulmonary histoplasmosis follows a course that also parallels tuberculosis. Symptoms of acute pulmonary infection occur; but, in addition, there is caseation necrosis and cavitation of lung tissue. Both acute and chronic forms of the disease are rarely fatal.

The fatality rate for untreated cases of progressive, disseminated histoplasmosis is very high. Infants, older persons, and severely debilitated individuals are most susceptible to this infection. Once inside these individuals, *H. capsulatum* spreads quickly and establishes foci of infection in many organs of the body. Bone marrow, spleen, liver, heart, meninges, and mucous membranes of the mouth, nose, and larynx can all be involved. The tissue undergoes rapid necrosis, leading to death in untreated individuals. As with other mycoses, amphotericin B is used as the therapeutic drug.

COCCIDIOIDOMYCOSIS

1. Name: *Coccidioides immitis*
2. Morphology: Dimorphic; morphology changes from hyphal form to yeast form when grown in tissue; endospores are characteristically formed within a small sphere (spherule).

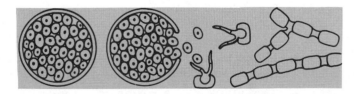

3. Disease: Coccidioidomycoses (San Joaquin Valley fever, valley fever, coccidioidal granuloma desert rheumatism). Primary pulmonary and secondary, or progressive, coccidioidomycoses.
4. Tissue Reaction: Granuloma formation, tissue necrosis; hypersensitivity (erythema nodosum, erythema multiforme).
5. Reservoir of Infection: Soil; may infect wild desert rodents; cattle, dogs, cats, horses, and other animals.

Coccidioidomycosis is the most feared and dangerous of all systemic mycoses. *Coccidioides immitis* occurs commonly in arid (desert) regions of the southwestern United States, and Central and South America (Figure 18.6). Spores are found in the soil surrounding the burrows of small desert rodents and are easily carried on the wind to susceptible hosts. They may be inhaled or enter the body through breaks in the skin. Most infections go unnoticed except for an increase in protective antibody titer. Others experience mild upper respiratory "flulike" symptoms known as **primary coccidioidomycosis.** The symptoms of this disease form include fever, pain, coughing, and malaise. Primary infections are usually self-limiting and brought under control by natural defense mechanisms in a few days or weeks. Some of those experiencing primary infection can become hypersensitive to the fungus and develop allergic reactions known as erythema nodosum, erythema multiforme, and "desert rheumatism." Erythema nodosum appears on the legs and feet as itching, painful, reddened nodules deep within the skin and lasts about a week. Erythema multiforme typically appears on the upper body as inflamed vesicles, pustules, or nodules that may hemorrhage. This allergic reaction usually lasts only four to seven days. The rheumatic reaction results in heat, pain, swelling, and stiffness of joints. Those

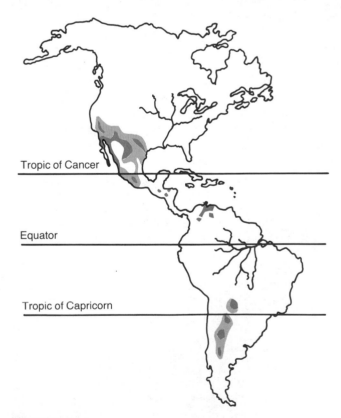

FIGURE 18.6
Distribution of *Coccidioides immitis* in the western hemisphere. Darker color indicates regions of greater frequency.

who demonstrate hypersensitivity rarely develop secondary coccidioidomycosis.

Secondary infections are similar to histoplasmosis in that the disease symptoms strongly resemble tuberculosis. Infection is not limited to the lungs and may spread to involve the skin, internal organs, bones, mucous membranes, and nervous system (Figure 18.7). This form of the disease is highly fatal. Tissue reactions result in the formation of granulomas that later undergo necrosis. Spores containing spherules can be isolated from the infection sites, and in some cases, their presence may be essential in distinguishing secondary coccidioidomycosis from miliary tuberculosis. If diagnosed early, the infection might be controlled by amphotericin B.

1 Name two important sources of *Histoplasma.*
2 What bacterial lung disease resembles pulmonary histoplasmosis?
3 How is miliary TB distinguished from secondary coccidioidomycosis?

BLASTOMYCOSIS

1 Name: *Blastomyces dermatitidis*
2 Morphology: Dimorphic, growth form change regulated by temperature: budding cells at 37°C, hyphal growth at 25°C.

3 Disease: Blastomycosis, or Gilchrist's disease; can involve the skin, lungs, and internal organs.
4 Tissue Reactions: Granuloma formation and pyogenic reaction.
5 Reservoir of Infection: Soil.

Blastomyces dermatitidis is found predominantly in soil in the north-central United States. The disease becomes es-

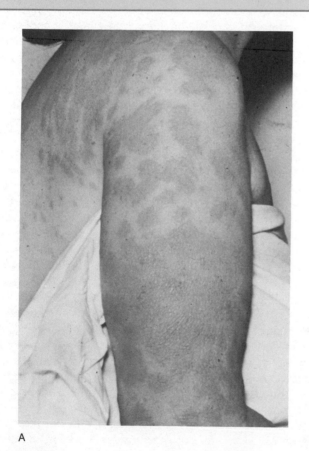

A

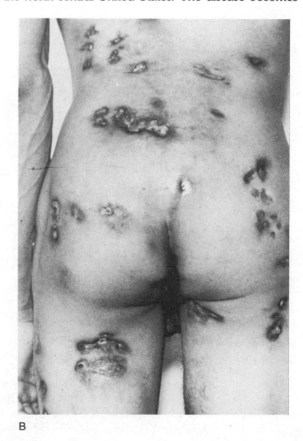

B

FIGURE 18.7
(A) Coccidioidomycosis lesion on the arm. (B) Coccidioidomycosis ulcerating lesions on many areas of the body. (Courtesy of the Center for Disease Control, Atlanta.)

tablished in the lungs by inhaling airborne spores. Primary lesions begin in the lungs, and in the early stages, they strongly resemble tuberculosis and histoplasmosis. As the lung infection progresses, dark, "crab-claw" shaped shadows may be seen on chest x-rays that are similar to those found in patients with lung cancer. Pulmonary lesions are rarely self-limiting and may spread through the circulatory system to other internal organs or the skin. Fungi that become lodged in the skin can produce either cutaneous or subcutaneous lesions. Secondary cutaneous skin lesions begin with the formation of small nodules that enlarge, join together, and undergo suppuration. Subcutaneous infections differ from the pus-producing, cutaneous granulomas. The subcutaneous form begins with marked inflammation and tissue necrosis. Large amounts of pus can be drained from this deeper lesion. Granuloma formation usually does not take place. The subcutaneous type of lesion may also develop in skeletal muscles, bones, heart, and the prostate gland in males. The course of infection is progressive, slow, and may last over a period of years.

ASPERGILLOSIS

1 Name: *Aspergillus fumigatus*
2 Morphology: Hyphal growth form.

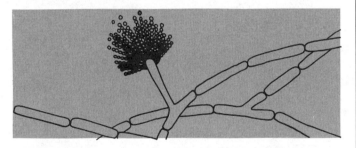

3 Disease: Two categories: allergic reactions (e.g., asthma and malt worker's disease) and tissue infections.

4 Tissue Reactions: Based on type of disease. Allergic reactions include Type I immediate hypersensitivities; Type III immune-complex hypersensitivities (refer to Chapter 14). Tissue infections include aspergilloma, granuloma, pyogenic abscess formation, and tissue necrosis.

5 Reservoir of Infection: Soil, damp hay, decaying vegetation, molding cereal grains.

Aspergillus fumigatus is one of over three hundred species of *Aspergillus* found throughout the world. The fungus grows in environments with a minimum of nutrients, and is responsible for the decay and decomposition of many foods and materials, both synthetic and natural (Figure 18.8). In the reproductive stage, large numbers of airborne spores are released, which can be found in household dust, on refrig-

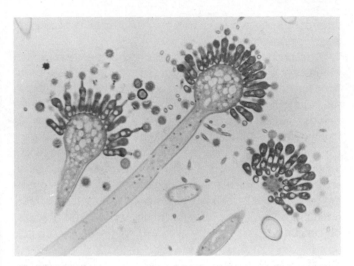

FIGURE 18.8
Aspergillus showing a portion of the fungal hypha and its spores. (Courtesy of Carolina Biological Supply Company.)

erated foods, and on the surfaces of almost all objects. *Aspergillus fumigatus* is an opportunistic fungus capable of causing allergies and tissue infections in humans. Inhalation of spores may stimulate an immediate Type I hypersensitivity characteristic of bronchial asthma, or a Type III immune-complex allergic response known as malt worker's disease or farmer's lung. These hypersensitivities can be controlled by avoiding environments that are heavily laden with *Aspergillus* spores, using drugs to counteract the harmful immune reactions, or undergoing desensitization with *Aspergillus* allergen injections.

Aspergilloses are tissue infections caused primarily by *A. fumigatus;* however, seven other species have also been identified as opportunistic pathogens. Disease results from inhaling spores and is most likely to occur in individuals who have a lowered resistance to infection caused by previous infections, immunodeficiency diseases, prolonged immunosuppressive therapy, or antibiotic therapy. An increase in aspergillosis in recent years is correlated with the more extensive use of immunosuppressive drugs to treat certain cancer types. Physicians have been alerted to this fact, and they pay special attention to the possibility that leukemia patients are especially susceptible to aspergillosis during therapy. After inhalation, spores can become lodged in lung cavities formed by previous infections such as tuberculosis and pneumonia. Fungal growth on the cavity's inner surface takes the form of an aspergilloma, or "fungus ball." These masses of hyphal filaments invading host tissue have been found as both primary and secondary lung infections, but they may also occur in the eyes or nasal sinuses. Infection stimulates severe inflammation, capillary breakdown, and release of blood and hyphal filaments from the cavity. Blood and fungi may be found in the sputum and fluid discharge of infected individuals. Necrosis can become so extensive in untreated cases that death results.

Aspergillomas remain localized on the cavity surface, but tissue infections can penetrate into the lung, skin, central nervous system, or elsewhere. Pulmonary infections display the pneumonialike symptoms of fever, chills, and coughing. Chest x-rays of the infected site strongly resemble cancerous lesions and may be mistaken for tumors. Tissue invasion results in severe necrosis caused by fungal endotoxins that stimulate formation of pyogenic abscesses. Invasion of the central nervous system causes a pyogenic fungal meningitis, while skin infections take the form of a granuloma composed of thick nodules that resemble leprosy lesions. Cutaneous infections can be controlled with the antifungal nystatin, while amphotericin B is used for internal infections. Surgical removal of aspergillomas may be necessary in addition to drug therapy.

---

1 A pulmonary blastomycosis resembles what other type of abnormality?

2 What types of hypersensitivities are associated with aspergillosis?

3 What correlation exists between the use of immunosuppressive drugs and aspergillosis?

CRYPTOCOCCOSIS

1 Name: *Cryptococcus neoformans*

2 Morphology: Yeast with a thick capsule.

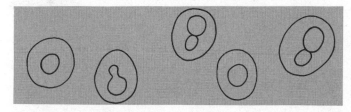

3 Disease: Cryptococcosis (torulosis, European blastomycosis) of the lungs, meninges, kidney, bone, and liver.

4 Tissue Reaction: Granuloma formation and mild inflammation.

5 Reservoir of Infection: Soil and fecal material of pigeons and chickens.

Cryptococcosis is a disease that occurs in apparently healthy individuals, but its presence usually signals a hidden deficiency of body defense mechanisms. The most common symptomatic form of the disease is cryptococcal meningitis; but pulmonary, cutaneous, and visceral forms are also found. Infections occur by inhaling spores from contaminated soil or fecal material of birds. Pulmonary cryptococcosis is asymptomatic but progresses to an easily identifiable fungal meningitis or infections of the skin, spleen, kidneys, or liver. Cutaneous infections are seen as ulcers formed from *C. neoformans* that has become localized in lymph nodes of the face and neck. In untreated cases, the fungus will continue to spread through the circulatory system, ultimately becoming fatal. Cryptococcal meningitis is a slow, progressive disease that may undergo remission and recurrence over a period of several years. This form begins as an asymptomatic pulmonary infection that moves to the central nervous system. Symptoms resemble brain tumors, abscesses, and meningococcal meningitis. Thick granulomas of the brain and meninges develop, and macrophages engulf yeast cells found in the spinal fluid. Cryptococcosis is usually treated with two drugs, fluorocytosine and amphotericin B, given orally over a period of months.

ACTINOMYCETES: FILAMENTOUS BACTERIA

ACTINOMYCOSIS

1 Name: *Actinomyces israelii*

2 Morphology: Filamentous bacteria; a Gram-positive, branching, anaerobic to microaerophilic, bacterium that resembles a true fungus.

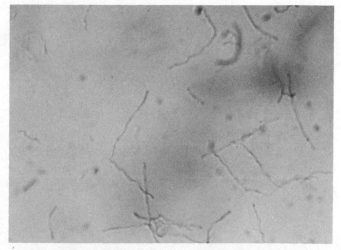

Courtesy of L. LeBeau, Department of Pathology, University of Illinois Medical Center.

3 Disease: Actinomycosis, or "lumpy jaw," in cattle: infection typically localized in jaw but may spread to abdomen and thorax; more rarely spreads to skin and internal organs.

4 Tissue Reaction: Granuloma, abscess formation, and pyogenic reaction.

5 Reservoir of Infection: Humans (normally found in the mouth, throat, and tonsils); found in soil and other external environments.

Actinomyces israelii has been identified as the cause of "lumpy jaw" in cattle and human actinomycosis. The microbe was suspected of being transmitted to humans from the soil or contaminated grain and hay. However, this opportunistic pathogen only occurs in the mouth, throat, and tonsillar areas. As part of the normal flora, it only causes illness following oral surgery, tooth extraction, accidental injury, or after chronic oral irritation. Infection begins as a hard, inflamed node that eventually ruptures from the mucous membrane surface. Lesions occur in the gingiva, around carious teeth, or in tonsillar sinuses. The necrotic abscesses become surrounded by fibrous tissue and drain through sinus tracts. The pus characteristically contains "sulfur granules," which actually are small colonies of the filamentous bacteria. The healing process is very slow. Constant draining from the oral cavity can result in aspiration or swallowing of pathogens, leading to thoracic or abdominal infections. Pathogens rarely spread through the circulatory system. These pathogens are now known to be transmitted from person to person. Actinomycoses are very rare today as a result of the widespread use of antibiotics such as penicillin and the sulfa drugs. *A. israelii* is susceptible to these antibiotics, and their prophylactic use following tooth extraction and oral surgery controls asymptomatic infections without difficulty. Those experiencing the disease do respond immunologically; however, the protective value of this response is not sufficient to prevent further infection.

NOCARDIOSIS

1 Name: *Nocardia asteroides*

2 Morphology: Filamentous bacteria; a Gram-positive, partially acid-fast, branching, aerobic bacterium that resembles a true fungus.

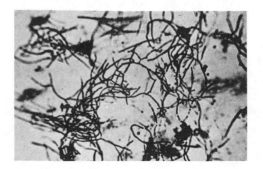

3 Disease: Pulmonary nocardiosis, may spread through the circulatory system to the skin, brain, and other organs; may also be a cause of mycetoma.

4 Tissue Reaction: Abscess formation and pyogenic reaction; rarely granuloma formation.

5 Reservoir of Infection: Soil.

Nocardiosis has become a common disease among individuals predisposed to infection as a result of the use of immunosuppressive drugs. Transplant patients and cancer patients show a significantly higher incidence of this disease than is found in the general population. *N. asteroides* is most likely acquired from external sources such as soil and dust, but it is often found as normal flora of the trachea and bronchioles. Primary infection begins in the lungs and resembles a form of pneumonia or tuberculosis. Pathogens cause severe necrosis to lung tissue and may spread to other organs. The brain is the most likely site of secondary infection, while the kidneys are the next most commonly infected organs. The lesions are typically suppurative and draining, but they can become granulomatous if localized in the skin. Since the disease is caused by an opportunistic pathogen, person-to-person transmission is unlikely. No specific immunity is developed after infection, and prolonged therapy with sulfonamide drugs is effective in controlling the infection.

---

1 Cryptococcus infections are primarily associated with what tissues?
2 What drugs can be used to treat infection caused by *Actinomyces israelii*?
3 Name two genera of filamentous bacteria.

PROTOZOA OF MEDICAL IMPORTANCE

The morphology, physiology, and ecology of protozoa are considered in both zoology and protozoology texts, since members of this phylum are single-celled animals. Of the thousands of species currently recognized by zoologists, only about thirty-five are considered medically important. These few microbes are responsible for diseases in humans and many animals. They may be transmitted directly or by vectors from one type of host to another. Taxonomists have recognized four subphyla of Protozoa based on the nature of their motility (Table 18.3). The life cycles of protozoa vary from simple cell division to complex alternations of asexual and sexual phases. Asexual reproduction is generally a form of mitosis referred to as binary fission, but be careful

TABLE 18.3
Phylum: Protozoa.

Subphylum	Example (Genus)	Reproduction	Cyst Formation	Motility
Sarcodina	*Entamoeba**	Binary fission	Yes	
Ciliata	*Balantidium***	Binary fission and conjugation	Yes	
Mastigophora	*Giardia* *Trypanosoma* *Trichomonas*† *Leishmania*	Binary fission	Yes	
Sporozoa	*Plasmodium* *Toxoplasma*‡	Sexual and asexual alternate	Yes	

Pseudopod*
Cilia**
Flagella†
Nonmotile‡

not to confuse this process with binary fission of bacteria, which is not a mitotic process. The sexual phase of reproduction is known as conjugation. The exact nature of this genetic recombination depends on the species of protozoan.

Many of the medically important protozoa are able to form resistant structures known as **cysts** during harsh environmental conditions. The microbe rounds up, becomes dehydrated, and produces a thick protective covering, i.e., a cell wall, over the entire cell. This process differs markedly from endospore formation in bacteria. Pathogenic protozoa are found throughout the world; however, the more troublesome occur predominantly in tropical and subtropical areas. Although many are thought of as foreign, exotic diseases, travel to remote parts of the world has resulted in transmission of these diseases to northern areas including the United States (Table 18.4). In addition, some of these diseases historically have been found in the United States but have been reduced through the effective use of epidemiological control methods. For example, malaria existed in epidemic proportions on the eastern seaboard of the United States during the colonial period. Draining swampy areas known to be breeding areas for the mosquito vector quickly eliminated this disease.

Protozoa are responsible for diseases of the blood, intestinal tract, central nervous system, urogenital tract, and other organs. Preliminary diagnosis is accomplished by microscopic examination of specimens and confirmed by serological methods such as complement fixation tests, immunofluorescence, and precipitin reactions. The representative pathogens described in this chapter are presented according to the subphylum of each.

SUBPHYLUM: SPOROZOA

Genus: *Plasmodium* Several members of this genus are responsible for the disease malaria. Many epidemiologists speculate that it is the most frequently occurring infectious human disease. The nonmotile sporozoan is transmitted from person to person through the infected saliva of

TABLE 18.4
Malaria cases by category.

| Category | Year | | | | | | | | | | | |
	1969	1970	1971	1972	1973	1974	1975	1976	1977	1978	1979	1980
Military	3914	4090	2975	454	41	21	17	5	11	31	11	26
Civilian	139	134	148	160	18	302	431	405	470	585	863	1837
US	90	90	79	106	103	158	199	178	233	270	229	303
Foreign	49	44	69	54	78	144	232	227	237	315	634	1534

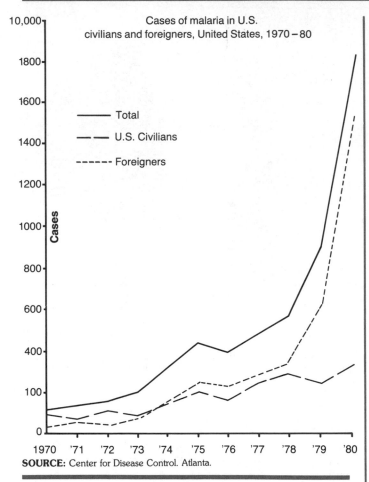

Cases of malaria in U.S. civilians and foreigners, United States, 1970–80

— Total
– – U.S. Civilians
----- Foreigners

SOURCE: Center for Disease Control. Atlanta.

and the use of larvacides, such as oil, and insecticides, such as DDT, lindane, and dieldrin, have been very successful vector control methods. A new, more ecologically sound method is being tried in areas where drainage or the use of chemicals is unwise. The small mosquito fish *Gambusie* has been introduced into mosquito-infested waters. The fish is a relative of the common tropical guppy and feeds on mosquito larvae.

> **1** How are pathogenic protozoa transmitted?
> **2** What methods are used in the preliminary diagnosis of protozoan infections?
> **3** What is the vector of *Plasmodium?*

Four species of *Plasmodium* are responsible for different clinical malaria forms: *P. vivax* (benign tertian malaria), *P. falciparum* (malignant tertian malaria), *P. malariae* (quartan malaria), and *P. ovale* (ovale malaria). *P. falciparum* causes the more severe and classical symptoms of fever, sweating, chills, headache, circulatory problems, shock, kidney failure, delirium, and coma. The length of the asexual portion of the disease-causing sporozoan's life cycle determines the frequency and severity of the disease. During the active clinical stage, *P. vivax* demonstrates recurrent symptoms every 48 hours, *P. malariae* every 72 hours, *P. ovale* every 48 hours, and *P. falciparum* is variable between 36 and 72 hours. Clinical malarial symptoms occur as clones of merozoites simultaneously burst from infected erythrocytes and enter the plasma to reinfect other red blood cells and repeat the cycle. Some hemoglobin-digesting forms do not change to the reinfectious merozoites. These forms, trophozoites, develop into cells called micro- and macrogametocytes that may be ingested by the *Anopheles* mosquito. Once inside the intestinal tract of the vector, they fuse to form a fertilized egg, or zygote, and begin the formation of the sporozoite. Sporozoites may pass through the mosquito's salivary gland and into a human host. Once inside, the parasite enters an exoerythrocyte stage known as a cryptozoite. This form is able to enter tissue cells such as in the kidney and liver, before changing into merozoites which can infect red cells and initiate clinical symptoms. The formation of cryptozoites from active merozoites is responsible for the remis-

certain species of the female *Anopheles* mosquito, while it takes a blood meal. Though most mosquitos in the United States, such as *Culex* and *Aedes,* do not harbor or transmit this parasite, a species of *Anopheles* is found in southern California. Thus, malaria is endemic only in subtropical and tropical areas of the world. All cases reported in malaria-free areas must have been imported. Within the past ten years, there has been a marked increase in civilian malaria among tourists, teachers, students, and businesspeople who have traveled to malarious areas of the world (refer to Table 18.4). Because the only reservoirs of infection are mosquitos and humans, epidemiological control and erradication of malaria is based on quick recognition and treatment of human cases and elimination of mosquitos. The draining of swampy areas

sion of symptoms. Remission and recurrence may last up to three years but will ultimately be eliminated as the parasites are destroyed.

A partial acquired immunity develops after infection. This depends on activating phagocytes which engulf the parasites. Unfortunately, the immunity is not long lasting. Malaria treatment may consist of drugs that reduce the numbers of parasites to a point that clinical symptoms are not produced. Other drugs actively destroy dormant cryptozoites in fixed tissue cells. The classical medication quinine is derived from the bark of the cinchona tree and reduces numbers of circulating parasites. Chemically modified and stronger forms of this drug include chloroquine, camoquin, and atabrine. The drug primaquine will destroy cryptozoites. As with bacteria, *Plasmodium* species have undergone natural selection for drug resistance. A number of drug-resistant *P. falciparum* strains were found in cases of malaria acquired by American soldiers returning from the Vietnam War.

Genus: *Toxoplasma* *T. gondii* is a sporozoan found throughout the world in soil and many animals. Serological tests show that a very high percentage of people have antibodies against this parasite, indicating previous contact or infection. These infections, however, are usually asymptomatic in adults or display only minor symptoms that resemble viral influenza or infectious mononucleosis. The most significant form of the disease occurs in pregnant women and leads to congenital abnormalities or stillbirths. Although the exact nature of the life cycle of *T. gondii* is not yet known, the human infective stage can apparently be ingested or inhaled as an oocyte released in fecal material of cats, horses, cattle, swine, sheep, or goats. Therefore, pregnant women should be especially careful around cats, and these pets should be checked for the presence of oocytes in fecal material. Infected cats should be treated and have their litter boxes moved outside.

Congenital toxoplasmosis is a rare but acute disease that occurs when parasites from an infected mother are transmitted *in utero* to the fetus. Approximately 1 percent of all pregnant women acquire the infection during pregnancy, and 25 percent of the infants born to these mothers show clinical symptoms of infection. In most cases, the mother is asymptomatic or may only display a slight rash or joint and muscle pain. Once inside the fetus, the pathogens migrate as intracellular macrophage parasites which become centered in the central nervous system (Figure 18.9). The most common symptoms are encephalitis and chorioretinitis. If the parasite crosses the placenta within the first three months of pregnancy, tissue damage may be so severe that death results. Infections occurring later in pregnancy result in less severe symptoms. The antiprotozoan drug pyrimethamine can be given in combination with sulfonamides to control the infection. Antibodies present in the serum ap-

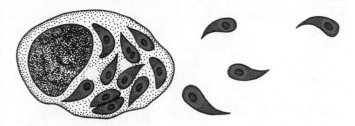

FIGURE 18.9
Diagram shows *Toxoplasma gondii* protozoan parasites outside and inside of macrophage.

pear to persist for life; however, little is known about their protective value against reinfection.

> **1** What drugs are used to treat malaria?
> **2** What animals may be sources of *Toxoplasma gondii?*
> **3** What measures could be followed to limit congenital toxoplasmosis?

SUBPHYLUM: CILIATA

Genus: *Balantidium* Balantidiasis is a rare, waterborne intestinal disease caused by the ciliated protozoan *B. coli* (Figure 18.10). This largest of intestinal protozoans causes abdominal cramping, nausea, vomiting, and abscess formation of the large and small intestines. Symptoms vary from constipation to diarrhea. The disease is primarily confined to tropical areas, and a carrier state can be formed that helps maintain the pathogen in endemic pro-

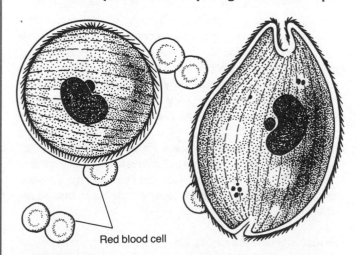

Red blood cell

FIGURE 18.10
The red blood cells next to the *Balantidium coli* cyst (left) indicate the relative size of the pathogen. The active protozoan is seen on the right. This is the only member of the Ciliata known to infect humans.

portions. Epidemics of balantidiasis have occurred in areas where sanitary disposal of contaminated fecal material has not been effective. Active cases of the disease are treated with antiprotozoal drugs such as metronidazole and paramomycin.

SUBPHYLUM: MASTIGOPHORA

Genus: *Trichomonas*   The flagellated protozoan *T. vaginalis* is the cause of the veneral disease known as trichomoniasis or "trich" (Figure 18.11A). The parasite is transmitted by sexual intercourse but may also be acquired from contaminated objects, such as towels, catheters, douche equipment, and examination tools. Infection results in mild inflammation of the vagina, cervix, and vulva. A foul-smelling, yellow or cream-white discharge is produced from the sloughing of infected surface tissue. In males, the infection may extend to the prostate, seminal vesicle, and urethra, but only a thin, white discharge occurs. Infection can also result if *T. vaginalis,* which is found as opportunistic normal flora of the vagina, becomes the predominant microbe due to uncontrolled growth. An increase in the pH of vaginal fluids is primarily responsible for this form of infection. These pH changes may be due to a loss of normal acid-producing bacterial flora (refer to Chapter 12), hormonal imbalance, or the effects of drugs such as oral contraceptives. Infection is controlled by topical or oral treatment with flagyl (metronidazole), other antiprotozoal drugs, and reestablishment of the normal vaginal pH between 3.8 and 4.4.

Genus: *Giardia*   Another important flagellated parasite is *G. lamblia,* the cause of giardiasis (Figure 18.11B). This protozoan occurs throughout the world and is the most frequently identified intestinal parasite in the United States. The primary reservoir of infection is human clinical and carrier cases. Infection is acquired from contaminated water or direct contact with feces-contaminated individuals or objects. Giardiasis is an infection of the small intestine resulting in cramping, bloating, fatigue, weight loss, and chronic, greasy diarrhea. The greasy consistency of the feces is characteristic of this disease and results from poor fat absorption. *G. lamblia* does not invade intestinal tissue but is present in high numbers in the lumen. Cysts are formed and released with the feces. Controlling the spread of this waterborne disease is made more difficult by the fact that cysts survive the standard concentration of chlorine used by most water and wastewater treatment plants.

Genus: *Trypanosoma*   Other important members of the subphylum Mastigophora are known as hemoflagellates because they are found in the blood during the course of infection. *Trypanosoma gambiense* and *T. rhodesiense* are responsible for slightly different clinical forms of African sleeping sickness, while *T. cruzi* of Central and South America is the cause of Chaga's disease (Figure 18.11C).

Trypanosoma gambiense and *T. rhodesiense* enter the host through a bite from *Glossina,* the tsetse fly, and form a primary lesion known as a chancre. Chaga's disease is transmitted by *Cimex lectularius,* commonly called bedbugs, and other closely related insect species. There are few symptoms during the primary stage; however, a complex series of reproductive phases occurs which may last a few weeks or several months. Following this period, flagellated protozoa spread from this lesion through the lymph and blood systems centering in the central nervous system and heart. Clinical symptoms of infection develop during the blood phase of the disease and may last for several years or only a few weeks. There is a progressive loss of tissue (wasting) that leads to weakness, loss of appetite, and apathy. Periods of unconsciousness are initially short but become progressively longer, until the patient lapses into a coma and dies.

There is no evidence of an acquired, long-lasting immunity, if the patient recovers. Treatment during the early stages of the disease requires use of the drugs suramin and pentamidine. During later infection, other more powerful drugs may be necessary. These drugs are available from the Center for Disease Control, Atlanta, Georgia, and through the World Health Organization. To date, several methods of preventing epidemic African sleeping sickness and Chaga's disease have been attempted and center on elimination of arthropod vectors. None has been successful.

---

1   What are the symptoms of trichomonas infections in males?

2   What is the relationship between pH change and vaginitis caused by trichomonas?

3   Giardiasis is an infection of what human tissue?

4   What is the vector of *Trypanosoma gambiense*?

SUBPHYLUM: SARCODINA

Entamoeba histolytica   The name of this pathogen provides some basic information about the nature of the disease it causes: "ent" means *inside,* "amoeba" means *varied shape,* "histo" means *tissue-associated,* and "lytica" means *bursting. E. histolytica* is the cause of waterborne and foodborne amoebic dysentery. The amoeboid form of the parasite is highly susceptible to environmental influences and easily destroyed when released from the intestinal tract in fecal material. Transmission of this stage to other persons is highly unlikely; however, *E. histolytica* does form a more resistant cyst that is easily transmitted in water, food, or by

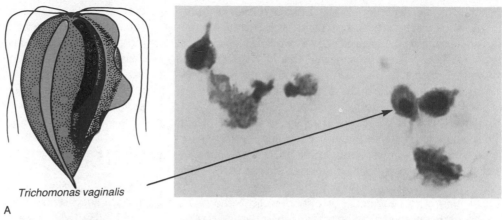

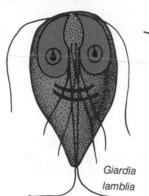

Trichomonas vaginalis

A

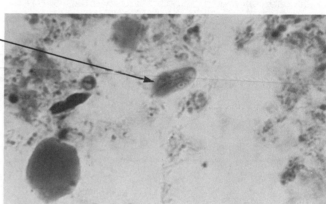

FIGURE 18.11
These three pathogenic members of the subphylum Mastigophora are all in the trophozoite stage of the life cycle. (A) *Trichomonas vaginalis;* (B) *Giardia lamblia;* and (C) *Trypanosoma gambiense.*

Giardia lamblia

B

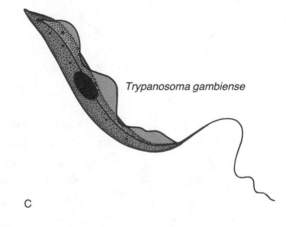

Trypanosoma gambiense

C

houseflies. Normal levels of chlorine in drinking water do not destroy the cyst. Ingested cysts release four amoeboid trophozoites capable of establishing an intestinal tract infection. The amoeba multiply in the large intestine by binary fission and feed on epithelial tissue and intestinal bacteria. The severity of the disease depends on how many lesions there are and the extent of tissue damage. Many infected individuals are asymptomatic and become carriers. Those who develop symptoms may demonstrate only a mild, prolonged case of diarrhea, or they may have moderate diarrhea containing mucus and blood. Patients become weakened; and in extreme cases, amoeba may enter tissues, causing abscesses in the liver, lungs, or other organs. The usual course of infection involves a cycle of tissue destruction and repair at various points on the intestinal mucosa. This prolongs the disease, which may last for months.

Active acquired immunity has not been demonstrated. The drug flagyl is one of the most common treatments to control infections, but it can produce side effects in patients who have developed liver abscesses. Controlling the disease and preventing epidemics is based on eliminating water contaminated with cysts and the development of good personal hygiene.

SUMMARY

Diseases caused by fungi and protozoans are studied in the fields of medical mycology and medical protozoology. Although only a relatively few species are capable of establishing active infections in healthy hosts, many function as opportunistic pathogens. Debilitated, older individuals and those receiving immunosuppressive therapy are more susceptible to infection by these commonly occurring microbes. In general, fungal infections are classified as cutaneous, subcutaneous, and systemic mycoses. Many of the disease symptoms occurring with fungal infections are the result of host tissue reactions. They demonstrate that disease is truly a process of interaction between host and parasite. Controlling mycoses is difficult because of the complexity of these interactions and the eucaryotic nature of pathogens. Preventing mycoses is based on maintaining optimum health and limiting contact with known pathogens. Person-to-person transmission of mycoses is highly unlikely and most infections are acquired from an external source such as the soil.

A closely related group of opportunistic pathogens is the genus *Actinomyces*. Although the members of this genus are bacteria, their filamentous nature and interaction with host tissues strongly resemble true mycoses. The pathogenic protozoa are also eucaryotic and responsible for countless cases of debilitating or fatal infections. Protozoa are classified according to their means of locomotion as Ciliata, Sarcodina, Mastigophora, and Sporozoa. Fortunately, only a few species are associated with human disease. Some of the most common infectious diseases are protozoan in nature. Malaria continues to be a major concern throughout the world, while in the United States, one of the most frequent causes of vaginitis is *Trichomonas vaginalis,* a flagellated protozoan.

WITH A LITTLE THOUGHT

No actual question will be presented for this chapter, instead a study suggestion may be of more value. In chapters 16 and 17 (Bacterial Diseases), a brief outline was presented in the chapter summaries. This chapter was more difficult to outline, since the microbes presented belong to such different groups. Prepare an outline similar to those of chapters 16 and 17, and expand on its content to help in your understanding of the material and relationships presented.

STUDY QUESTIONS

1 Why are the fields of medical mycology and medical parasitology separated from bacteriology?

2 Define the following terms: dimorphic, hyphae, mycelium, granuloma, pyogenic, and tinea capitis.

3 Give two examples of dermatophytic fungi. What diseases are they associated with?

4 Give two examples of fungi that are opportunistic pathogens. What is their major reservoir of infection? What circumstances might lead to infection?

5 What is the difference between a true mycoses and an actinomycoses? Give examples.

6 Give two examples of systemic mycoses. Why are they so difficult to treat? What medications might be used?

7 What is a cyst? How does it differ from a bacterial endospore?

8 List the protozoan infections that may be transmitted by vectors. Name the vectors.

9 What are some basic differences among the following stages in the *Plasmodium* life cycle: sporozoite, oocyte, and trophozoite?

10 Which protozoans might be transmitted in water? In foods? By contact?

SUGGESTED READINGS

BROWN, H. W. *Basic Clinical Parasitology,* 4th ed. Appleton-Century-Crofts, New York, 1975.

EMMONS, C. W. et al. *Medical Mycology,* 3rd ed. Lea & Febiger, Philadelphia, 1977.

GOLDSMITH, R. S. "Infectious Diseases: Protozoal." *Current Medical Diagnosis & Treatment 1978.* M. Krupp and M. J. Chatton, editors. Lange Medical Publications, Los Altos, Calif., 1978.

NETO, DOS SANTOS. "Toxoplasmosis—Historical Review, Direct Diagnosis Microscopy and Report of a Case." *American Journal of Clinical Pathology,* 63 (6), June, 1975.

RIPPON, J. W. *Medical Mycology—The Pathogenic Fungi and the Pathogenic Actinomycetes.* W. B. Saunders Company, Philadelphia, 1974.

SZABO, T. "The Present State of Toxoplasmosis." *Mt. Sinai Journal of Medicine,* 41(6), Nov.– Dec., 1974.

KEY TERMS

hyphae	granuloma	stomatitis
mycelium	dermatophyte	mycetoma (mi″se-to′mah)
mycosis	eczema	
erythema		cyst

Pronunciation Guide for Organisms

Trichophyton (tri-kof′i-ton)

Microsporum (mi″kros′po-rum)

Torulopsis (tor″u-lop′sis)

Rhodotorula (ro″do-tor′u-lah)

Cryptococcus (krip″to-kok′us)

Epidermophyton (ep″i-der-mo-fi′ton)

Sporothrix (spo′ro-thriks)

Nocardia (no-kar′de-ah)

Streptomyces (strep″to-mi′sez)

Histoplasma (his″to-plaz′mah)

Blastomyces (blas-to-mi′sez)

Actinomyces (ak″ti-no-mi′sez)

Plasmodium (plaz-mo′de-um)

Toxoplasma (toks″o-plaz′mah)

Culex (ku′leks)

Aedes (ah-e′dez)

Balantidium (bal″an-tid′e-um)

Giardia (je-ar′de-ah)

Viruses

GENERAL CHARACTERISTICS OF VIRUSES
CULTIVATION AND COUNTING
CLASSIFICATION OF VIRUSES
BACTERIOPHAGE LIFE CYCLES
ANIMAL VIRUS LIFE CYCLES
The Lytic Cycle
Latent Infections
Persistent Infections
TRANSFORMATION
TRANSMISSION OF TRANSFORMING VIRUSES
CANCER: VIRUS-HUMAN INTERACTION?
UNCONVENTIONAL VIRUSES
"Slow" Viruses
Viroids

Learning Objectives

☐ Be familiar with the general characteristics of viruses, including their classification, structure, and replication cycle.

☐ Know how viruses are cultivated and the methods by which they are counted.

☐ Diagram the lytic and lysogenic bacteriophage life cycles.

☐ Recognize the various life cycles of animal viruses: lytic, latent, and persistent.

☐ Define the term "transformation" and explain this process as it relates to oncogenic viruses.

☐ Know the difference between horizontal and vertical viral transmission.

☐ Be able to explain the Oncogene Theory.

☐ Recognize the theoretical relationship between cancer and viruses.

☐ Be able to name the unconventional viruses and describe their structures.

Virology had its origin in the study of slimy, poisonous fluids thought to cause plant, animal, and human diseases. In this century, these disease-producing fluids have been studied extensively. As more sophisticated research procedures were developed, it became clear that not all infectious diseases were caused by such cellular microbes as bacteria and fungi. Many resulted from agents lacking characteristics of true cells. The term **filtrable virus** ("virus" means *poisonous fluid*) was originally used to describe chemical-like agents that remained in infected tissue fluid after being passed through a porcelain filter capable of screening out all cellular material. These acellular pathogens remain infective after drying, resist many chemicals able to destroy cellular pathogens, and are now known to be submicroscopic, obligate intracellular parasites, commonly referred to as **viruses.** Viruses can be cultured in host cells using special methods. They are so small that only a few types can be seen with the compound light microscope. The electron microscope has been extremely valuable in providing further knowledge of viruses' existence and nature.

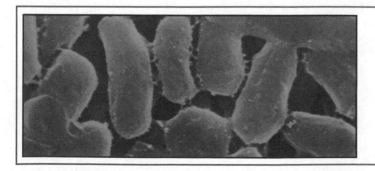

19

Viruses found to infect plant, animal, and cellular microbes are all protein-nucleic acid complexes that vary in their morphology and function. Their unknown evolutionary relationship to cellular life and other viruses has made virus classification difficult. Classification schemes are based on the chemical composition of viruses and not on hypothetical evolutionary lines. Because viruses are acellular, their physiological characteristics and life cycles are very different from those of cellular microbes. Some of these microscopic invaders replicate more of their own kind and quickly kill the host cells. Others enter but are latent until they harm their host, while still others transform their host from a specialized cell to one that does little else but reproduce both itself and the parasitic virus. The transformation of normal host cells to cancer cells as a result of viral infection has been demonstrated in a number of animals, and evidence of virus-induced cancer in humans is increasing. The difference in symptoms caused by a viral infection relates directly to the nature of the life cycle. Understanding viral life cycles has resulted in the control of many plant, animal, and human diseases. However, the complexity of virus-host interaction is great, and much remains to be uncovered and explained. In recent years, research has led to the discovery of two groups of microbes referred to as "unconventional" viruses. The "slow" viruses and viroids have been shown to cause disease in many organisms, but their exact nature and host interactions are not fully known.

GENERAL CHARACTERISTICS OF VIRUSES

Viruses are generally subdivided into three major groups based on the kind of host they infect: bacterial viruses (bacteriophage), animal viruses, and plant viruses. The term "bacteriophage" ("phage" means *eating*) refers to viruses that are able to infect bacterial cells. The ease with which bacteriophage (commonly referred to as phage) are handled in the laboratory has made them one of the most widely studied groups of microbes. Much of the information concerning their interactions with their microbial hosts is applied to both animal and plant viruses. Animal viruses have been identified as obligate intracellular parasites of insects, fish, reptiles, birds, and many mammals. They are responsible for such human diseases as chickenpox, mumps, measles, rabies, infectious mononucleosis, polio, and the common cold. Current research has also implicated animal viruses in many other diseases, such as multiple sclerosis and cancer. Plant viruses have been identified primarily as parasites of flowering plants; but ferns, fungi, algae, and cone-bearing plants may also be victims of infection. Some of the most important plant infections cause severe economic losses to growers of potatoes, beets, barley, tomatoes, cauliflower, carnations, chrysanthemums, beans, and tobacco. Most plant viruses are transmitted by the chewing or sucking insects, such as aphids, whiteflies, and beetles. Although they infect different hosts, members of all three virus groups have many common properties.

All virions (virus particles) are composed of protein and a nucleic acid core. Some characteristically have a membraneous **envelope** that surrounds the particle. Those that lack this envelope are referred to as "naked." Phospholipid envelopes are usually not produced by virions but come from host membranes such as the cell or nuclear membrane. The coat, or **capsid,** is composed of separate protein segments known as **capsomeres.** Capsomeres chemically bond with one another to surround the core, forming either a helical or polyhedral (many-sided) shape. Some bacteriophage have a combination of helical and polyhedral capsomeres with additional parts forming leglike and platelike structures (Figure 19 1). The virion's genetic material contains information necessary to synthesize the capsid proteins and enzymes required for completing its life cycle. The few complete enzymes associated with viruses are involved with penetrating host cells and synthesizing viral nucleic acid. The capsid of naked viruses and the envelope of others serve several important purposes. They aid in viral attachment to the host cell, determine the antigenic nature of the virus, and protect the enclosed nucleic acid.

The **core** of a virion contains genetic material that is either DNA or RNA. Several forms have been found, including double-stranded DNA, single-stranded DNA, double-stranded RNA, and single-stranded RNA. In some viruses, the nucleic acid is circular, while in others it is a straight molecule, or it may be fragmented. The smallest viruses (about the size of a ribosome) contain about 10 genes, while the largest (about the size of an extremely small bacterial cell) have about 500 genes. These genes are only expressed after the virion has penetrated a suitable host cell and enters the process known as **viral replication.** This reproductive process differs greatly from that carried out by cellular forms. Viruses do not "grow" inside the host as do other obligate intracellular parasites. Instead of simply "feeding" off of the host, viruses take command of their host's metabolic pathways and direct them in the production of essential enzymes and products required for the production of more virions. The five stages of viral replication that have been identified are: (1) adsorption, (2) penetration, (3) replication of nucleic acid, (4) maturation, and (5) release of new virions. The exact way in which each stage occurs differs among viruses.

Adsorption involves a specific chemical bonding between capsid or envelope and mucoprotein attachment sites found on the host cell's surface. If these recognition sites are masked or absent, the virus will not be able to attach, and infection will not take place. Once attached, bacteriophage typically penetrate their host by injecting their nucleic acid

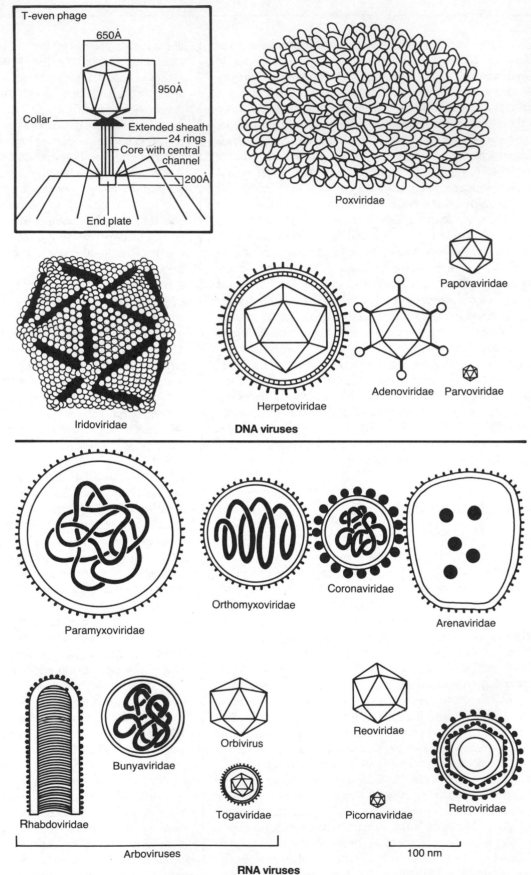

DNA viruses

T-even phage

650Å
950Å
Collar
Extended sheath
24 rings
Core with central channel
200Å
End plate

Poxviridae

Papovaviridae

Iridoviridae

Herpetoviridae

Adenoviridae   Parvoviridae

Paramyxoviridae

Orthomyxoviridae

Coronaviridae

Arenaviridae

Rhabdoviridae

Bunyaviridae

Orbivirus

Togaviridae

Reoviridae

Picornaviridae

Retroviridae

100 nm

Arboviruses

RNA viruses

FIGURE 19.1
These representative animal viruses show their variety of shapes and sizes. (From F. Fenner and D. O. White, *Medical Virology,* 2nd ed. Academic Press, 1976. Reprinted by permission.)

452

core through the host cell surface, leaving their protein coat outside. Many animal, and a few plant viruses enter the host by a form of phagocytosis or by passing through the cell surface in some as yet unknown manner. Most plant viruses are unable to enter without the aid of an insect vector. These animals break through the plant cell wall, allowing the virus to come in contact with the host cell. Once inside, the animal and plant viruses are uncoated (i.e., the protein is removed), and then nucleic acid replication begins. Gene replication varies from one kind of virus to another and depends on the number of strands and type of nucleic acid (Figure 19.2). While inside the host, the nucleic

acid serves as a template for the production of messenger RNA molecules necessary for viral protein synthesis (Figure 19.3). During the maturation process, the parasitic virus synthesizes viral enzymes using host ribosomes and amino acids. The host's metabolism is redirected for the production of capsomere proteins, which are attached to one another around the newly formed viral nucleic acid. Although not all proteins and nucleic acids are used in this self-assembly process, each phage replicates a relatively constant number of virions during each host infection. Once the maturation stage is complete, virions are released from the host cell. Some viruses destroy their host by enzymatically rupturing the cell membrane. These are known as **lytic viruses** and emerge "naked" from the host. Many of the animal and plant viruses release new virions without lysing the host cell. These viruses are pushed out, or extruded, from the host and may acquire an outer envelope from a portion of the host cell membrane (Figure 19.4). The cell membrane from which the virus will bud is first modified by the addition of viral-specified proteins, before the viral particle can "bud" through it. This "pinching off" process does not kill the host and may last indefinitely, or until the host cell is killed by severe damage to its metabolism. Once outside, new virions are capable of infecting other suitable host cells and repeating the life cycle.

1 Phospholipid envelopes found on some viruses come from what host cell structure?
2 What purpose does a capsid serve?
3 List the five stages of viral replication.

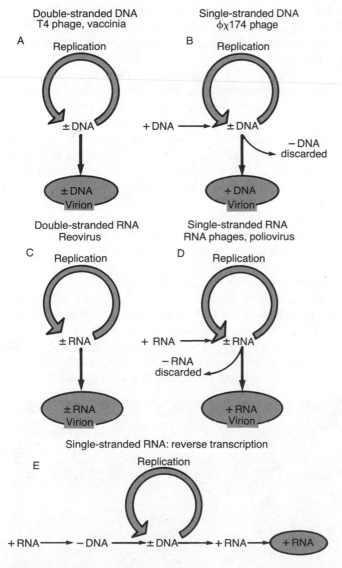

FIGURE 19.2

Gene replication in viruses. The pattern is dependent on the composition of the virus' nucleic acids. (From Brock, *Biology of Microorganisms*, 3rd ed. Copyright © 1979, p. 318. Reprinted by permission of Prentice-Hall, Inc., Englewood Cliffs, New Jersey.

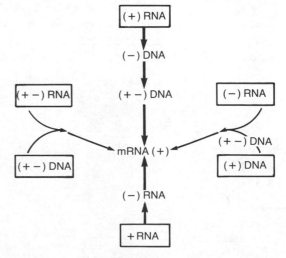

FIGURE 19.3

Viruses can be classified according to their method of mRNA transcription. (From Baltimore. *Bacteriological Review,* 35:236, 1971.)

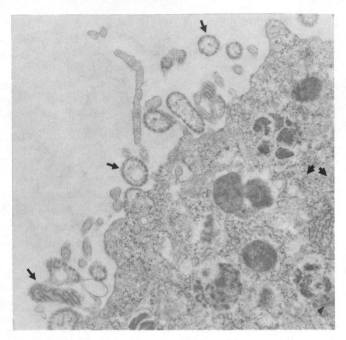

FIGURE 19.4
A virus can be released from a host cell without destroying the host by a process known as budding. (From G. Dethe, C. Becker, and J. Beard. *Journal of the National Cancer Institute,* 32:201, 1964.

CULTIVATION AND COUNTING

Knowledge of intracellular activity and host-virus interactions is obtained from laboratory cultivation of these microbes in suitable host cells. Bacteriophage are grown in pure cultures of host cells that have been plated on the surface of nutrient agar. As lytic phage complete their life cycles, they destroy their host cells and produce a clear spot, or **plaque,** on the plate surface (Figure 19.5). Many phage form such typical and unique plaques that the plaques' characteristics can be used as a means of viral identification. Phage which are not lytic do not form plaques but do produce other identifiable changes on the culture plate. Samples of nonlytic phage can be transferred to a broth culture of host cells for further culturing and experimentation. Since lytic phage destroy their hosts, broth tubes of phage-infected bacteria change from a cloudy, turbid culture to a clear solution resembling sterilized broth. Population estimates are made by doing serial dilutions from these tubes. Samples from the dilution series are mixed with host cells and plated on the semisolid nutrient agar surface. The number of virions per milliliter in the original sample is determined by counting the plaques and multiplying this number by the dilution factor of the countable plate. This method is very similar to the standard plate count used for bacterial cultures.

Virus concentrations can also be determined by the **hemagglutination test.** Many phage, animal, and plant viruses contain surface glycoproteins that are able to combine with erythrocytes and cause agglutination. These glycoproteins appear as "spikes" on the surface of enveloped viruses such as influenza and rabies. The test is run by making a serial dilution of the viral suspension and adding erythrocytes to the samples. Agglutination appears as a pinkish net of cells on the bottom of the tube containing the minimal number of virions necessary to cause the reaction. The hemagglutination titer (relative number of virions) is given as the reciprocal of the last dilution to show agglutination. Another method used to quantify viruses is known as **endpoint assay.** This method determines the infectivity of a virus sample by measuring host cell infection, damage, or death. Although this method does not yield the actual number of virions, it does provide the relative number of infective units per sample. The test is run by inoculating suitable host cells with samples taken from serial dilutions of the virus. The test is based on the fact that the most dilute sample

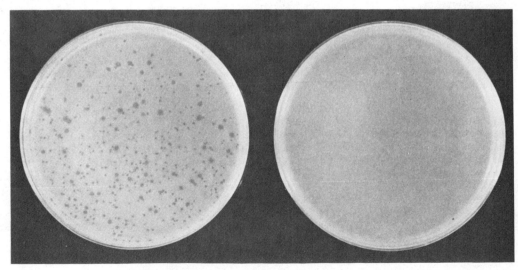

FIGURE 19.5
(R) This petri plate contains a bacterial culture of *E. coli* without bacteriophage. (L) This plate shows numerous points of clearing formed by the destruction of *E. coli* host cells infected by the bacteriophage known as λ (lambda). (From Demerec and Fano. *Genetics* 30:119, 1945.)

which infects 50 percent of the test cells (ID_{50} = infective dosage) or kills 50 percent of the test cells (LD_{50} = lethal dosage) must contain at least one virion. This dilution is the last concentration or endpoint of the series. Endpoint assays are expressed as the reciprocal of the minimal dilution that results in an ID_{50} or LD_{50}.

The plaque method is also used to count animal viruses. Suitable host cells are grown in a thin, single layer (monolayer) by using special methods known as tissue culturing. Many types of cells can be used for this purpose. African green monkey kidney cells and human fibroblasts are widely used as viral hosts. **Primary cultures** are made from cells taken directly from an infected animal or human. These types are usually short-lived and can only be maintained for a limited time. **Secondary cultures** maintain their regular diploid chromosome number but are derived from cells that have undergone biochemical changes which allow them to survive for much longer periods. Chromosomal content is not maintained in tissue cultures known as **continuous cell lines.** These cells have varied chromosome numbers and can be maintained in the lab for an extended period. Tissue cultures inoculated with animal viruses may show plaque formation, metabolic changes, or abnormal growth patterns which are all identifiable by various methods.

---

1 How is the number of virions per milliliter determined in the plaque counting method?
2 What is the purpose of endpoint assay?
3 What cell types can be used as hosts in the laboratory cultivation of viruses?

Viruses can also be cultured in laboratory animals or embryonated eggs. Since viruses are not able to penetrate the skin, animals are inoculated by injection or by infection of mucous membranes. A successful infection is one that demonstrates typical disease symptoms, which may include generalized body responses, tissue and cellular changes, and most importantly, immunological responses. If symptoms are not produced, tissue specimens suspected of being infected can be isolated and transferred to another host. Embryonated chicken eggs contain developing embryos with fetal membranes (chorioallantoic membranes) capable of serving as viral host tissue. These lesions are known as **pocks** (pox) and appear as small pustular patches on the membrane surfaces. Embryonated eggs are regularly used by pharmaceutical companies for culturing animal and human viruses (Figure 19.6). Replicated virions are isolated from embryo cultures and attenuated for use as vaccines to protect against diseases such as mumps, rabies, and influenza.

Plant viruses are also cultured by direct inoculation into host organisms. The specimen is either rubbed onto a leaf

FIGURE 19.6
Harvesting virus-laden tissue from embryonated eggs after inoculation with vaccinia virus. (Courtesy of the Center for Disease Control, Atlanta.)

or injected into the plant. Lesions formed may appear on leaves, stems, or roots and take the form of colored patches, streaks, rings, necrotic spots, tumors, or leaf curling (Figure 19.7).

CLASSIFICATION OF VIRUSES

The wide variety of cultural and research methods used by virologists enable them to determine many properties of viruses. This data has contributed to more complete and accurate knowledge of the chemical and physical nature of viruses. However, little is known about viral evolution. Although several hypotheses have been proposed, no single pattern has been accepted as the most probable evolutionary course. One hypothesis speculates that viruses arose independently before cellular life; another proposes that they are fragments of early cells, possibly ribosomes or free genes; while still another contends that viruses resulted from extreme specialization of pathogenic bacteria, i.e., they were bacteria that became the "ultimate parasite." Due to the uncertainty of the relationship between viruses and cellular life as well as the relationships among viruses themselves has made nomenclature (naming of viruses) and taxonomy (classification) problem areas in virology. The current classification scheme is based on known chemical and physical properties of viruses, not on the more classical "evolutionary tree" used by most biologists (refer to Chapter 2), (Table 19.1). Traditionally, plant and animal viruses have been named for the diseases they cause (e.g., smallpox, rabies, tobacco mosaic virus), while bacteriophage have been

FIGURE 19.7
Local necrotic lesions on plant leaves caused by viruses: (A) Apple mosaic virus symptoms on apple leaves show degrees of severity; (B) Symptoms of ringspot virus on tobacco leaves. (Courtesy of G. N. Agrios, University of Massachusetts.)

assigned number and letter codes (e.g., T_2, C_{16}, and $\phi\chi$ 174). To eliminate such confusion, the International Committee on Nomenclature of Viruses has proposed an alternate system, but the system has not yet been accepted. Although evolution, nomenclature, and classification remain problem areas in virology, research into the life cycles has revealed valuable information enabling virologists, epidemiologists, and members of the medical profession to better understand and control these obligate intracellular parasites.

BACTERIOPHAGE LIFE CYCLES

Bacteriophage are primarily polyhedral viruses with either DNA or RNA as their genetic material. These microbes have been identified as infectious agents of many bacterial species, but the most widely studied phage are those infecting the bacterium *E. coli*. Depending on the phage, the two life cycles that occur are called lytic and lysogenic. In the **lytic cycle,** viruses, such as the T phages, adsorb to the surface

TABLE 19.1
Classification of viruses.

Nucleic Acid	Capsid Symmetry	Presence of Envelope	Diameter of Helical Capsid or Number of Capsomeres	Typical Members
RNA	Helical	−	100–130 Å	Potato χ
		−	170–200 Å	TMV
		−	250 Å	Barley stripe mosaic
		+	100 Å	Influenza; fowl plague; measles
		+	170 Å	
	Polyhedral	−	32 units	RNA phages; turnip yellow mosaic
		−	60 units	Polio;
		−	92 units	Reovirus; wound tumor
DNA	Helical	+	90–100 Å?	Vaccinia
	Polyhedral	−	12 units	φχ174
		−	42 units	Polyoma, papilloma
		−	252 units	Adenovirus
		−	812 units	*Tipula* iridescent
		+	162 units	Herpes, pseudorabies
	Polyhedral and helical	−		T-even and only tailed phages

SOURCE: Modified from Lwoff et al, *Cold Spring Harbor Symp.* 27, 51 (1962).

of the host at special receptor sites and inject their nucleic acid core through the outer covering (Figure 19.8). Once inside, the parasitic nucleic acid takes command of the host's metabolism to synthesize more of its own kind. During this period known as the **eclipse,** or **latent, phase,** many biochemical changes occur within the host. Viral DNA is immediately transcribed by host enzymes to mRNA molecules, which code for other enzymes required for replicating viral DNA, mRNA, and capsomere proteins. Enzymes are also produced which selectively destroy host genes not required by the parasite for replication. After all of the viral components have been made, the virions begin to self-assemble, the host is ruptured by phage lysozyme, and the new virions are released. The lytic cycle takes place very quickly, and only a short time elapses between the time of initial penetration to bursting. It is possible that during the assembly process a case of "mistaken identity" occurs, and a small segment of the destroyed host DNA may be fashioned into the new virion. This segment may be carried to another cell, placed inside, and integrated into the endogenote in a process called generalized transduction (refer to Chapter 11).

Lytic life cycles are characterized by a one-step growth curve (Figure 19.9). Unlike the population growth curve of bacteria (described in Chapter 8), this curve shows no lag, log, stationary, or death phases. When virions are introduced into a host cell population, there is no initial change in their numbers, since replication is impossible until they penetrate. A period of time occurs (the eclipse phase) during which no virions are found free in the culture. During the eclipse phase, the virions have entered the host cells and are undergoing maturation. Following this brief period when no free virions are found, there is a sudden increase in their number as the newly replicated virions burst from their host cells. This period is known as the rise period, and the number of virions released from each host cell is called the **burst size.** Burst size is relatively constant and characteristic for each type of virus.

In the latent phase, or **lysogenic cycle,** there is a delay after the nucleic acid has penetrated the host. Once inside, the virus integrates into the host DNA loop and is called a **prophage** (before the virus), (Figure 19.10). In this condition, the host cell is called a **lysogenic bacterium,**

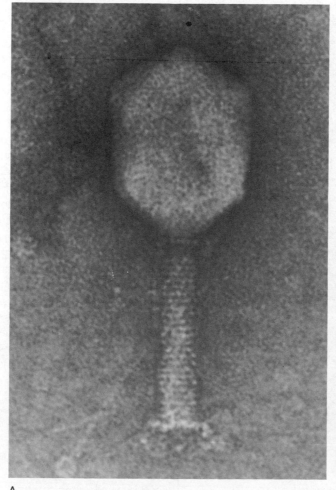

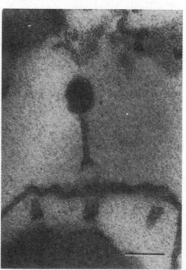

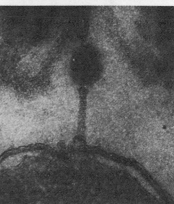

FIGURE 19.8

(A) This T-even virion is capable of infecting *E. coli.* (B) Attachment of T-even virions to membrane of *E. coli.* (Part A courtesy of the Center for Disease Control, Atlanta. Part B reprinted by permission of Academic Press. From J. Bayer. *Virology,* 2:346, 1968.)

A

B

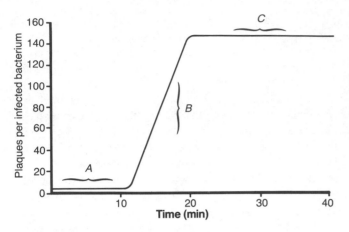

FIGURE 19.9

After a virus is introduced into a population of susceptible host cells, an eclipse period occurs (A) when no intact viruses are recoverable from the mixture. This is followed by a rise (B) in the number of recoverable viruses. Some rupture from the host cells. After all susceptible cells have been lysed, the maximum number of infectious phage particles is recovered (C).

---

1 What problems have made it difficult to classify viruses?

2 During what phase of the viral growth curve is viral nucleic acid transcribed?

3 During what phase of a viral growth curve is it possible to determine burst size?

---

because it may be destroyed at a later time when the virus prophage becomes active and reenters the lytic cycle (Figure 19.11). Bacteriophage capable of entering into a lysogenic life cycle are called **temperate phage.** The delay of host destruction that is characteristic of the lysogenic cycle results in the formation of many generations of lysogenic

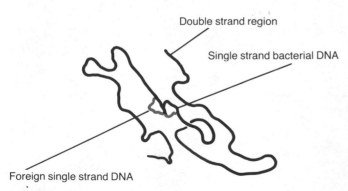

Double strand region

Single strand bacterial DNA

Foreign single strand DNA

FIGURE 19.10

This illustrates how viral DNA integrates into the host DNA.

bacterial cells. During the period between viral infection and lysis, these virus-infected bacteria can function as normal cells. Temperate phage are able to remain as quiet passengers within their host cells due to formation of virally produced **repressor protein** (refer to Chapter 7, Box 13). These molecules prevent mRNA synthesis used in the production of viral enzymes required for entering the lytic cycle. As long as the repressor is active, the virus will remain in the prophage state. Entry into the lytic life cycle is affected by chemical, physical, or biological changes in repressor action. By controlling these various environmental factors, it is possible to study temperate phage and experimentally regulate their reentry into the lytic phase.

In this research, a form of **phage immunity** has been discovered in lysogenic bacteria. This form of immunity prevents the host bacterium from being destroyed by another phage of the same type or of a closely related group. The presence of repressor molecules inside the lysogenic host prevents the second phage from replicating and lysing the cell. However, phage which are not inhibited by the repressor are able to lyse the cell. Research has also revealed that some bacteria are **phage resistant.** These bacteria do not prevent phage replication through the action of repressor proteins but are unable to be infected as a result of mutations that modify phage receptor sites. Phage attempting to infect such a cell are unable to adsorb to the cell surface.

Mutations of phage genes have also been identified. Phage with mutations that make complete viral replication impossible are called **defective** phage. If such a mutation occurs it does not affect the viruses' ability to adsorb and penetrate another host but does prevent it from being released from the prophage state and entering the lytic phase. Completion of the life cycle can occur, however, if a lysogenic host becomes "superinfected" by a second phage. The second invader may assist the defective prophage in completing its life cycle and escaping from the host. This can happen if missing enzymes and other proteins are made available to the defective phage by the secondary invader. Experimentation with superinfections has revealed a high percentage of defective phage in what were once thought to be uninfected host cells. Defective phage are known to play an important role in the specialized transduction of bacterial genes (refer to Chapter 11) and are responsible for the development of drug resistance in some bacteria, and the incorporation of the gene for toxin production into *C. diphtheriae.*

ANIMAL VIRUS LIFE CYCLES

Viruses have been found to infect insects, birds, reptiles, and mammals, including humans. Although their generalized life cycles parallel one another, bacteriophage and animal vi-

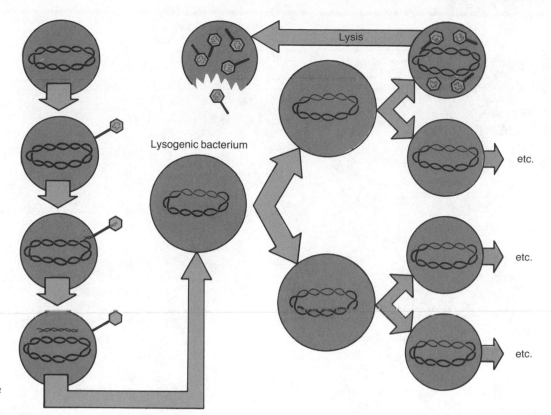

FIGURE 19.11
The lysogenic bacteriophage cycle.

ruses differ greatly in structure and function. Dissimilarities occur between size, chemical composition, and the rate at which they complete their life cycles. Animal viruses have either DNA or RNA as their genetic material (Table 19.2). After adsorption, penetration, and uncoating of the virion, most virions containing DNA replicate their genetic material after entering the host's nucleus. The poxviruses and RNA viruses carry out this stage of their life cycle in the cytoplasm. Capsids of animal viruses vary widely in size and morphology (Figure 19.12). Viruses such as poliovirus, echovirus, equine encephalitis, and the herpesviruses have polyhedral capsids. The poxvirus, mumps, measles, and

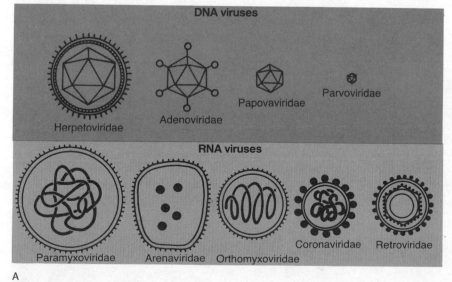

A

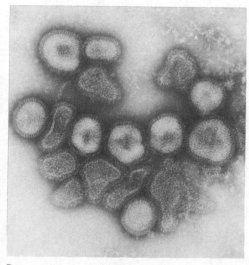

B

FIGURE 19.12
(A) Variety exhibited by animal viruses. (B) Electron micrograph of type A influenza virus. (Courtesy of the Center for Disease Control, Atlanta.)

TABLE 19.2
The animal viruses.

Group	Capsid	Number of Capsomeres	Virion Size (nm)	Type of Nucleic Acid	Number of Genes (Approximate)
Picornavirus	Icosahedral (20 equal sides)	32	20–30	Single-stranded RNA	12
Togavirus, or "arbovirus"	Icosahedral	32?	40–70	Single-stranded RNA	15
Paramyxovirus	Helical	—	150–300	Single-stranded, "negative-strand" RNA, plus virion polymerase	30
Orthomyxovirus	Helical	—	80–120	Negative-strand RNA; virion polymerase	15
Rhabdovirus	Helical	—	70–175	Helical, single-stranded, negative RNA and virion polymerases	20
Reovirus	Icosahedral	?	60–80	Double-stranded RNA	40
Retrovirus	Unknown or complex	?	100	Helical RNA, single-stranded, RNA segmented	50
Bunyavirus	Helical	—	90–100	Negative-stranded RNA in helical capsid	15
Coronavirus	Unknown or complex	—	80–130	Single-stranded RNA	30
Arenavirus	Unknown or complex	—	50–300	Single-stranded, RNA segmented	15
Parvoviruses	Icosahedral	32	18–26	Icosahedral, naked, single-stranded DNA	7
Papovavirus	Icosahedral	72	45–55	Circular, double-stranded DNA	10
Adenovirus	Icosahedral	252	70–90	Linear DNA	50
Herpesvirus	Icosahedral	162	100+	Double-stranded linear DNA	180
Poxvirus	Complex	—	230–300	Double-stranded DNA	400

Disease	Features
Human: poliovirus, Coxsackie Rhinovirus Foot and mouth disease; virus of cattle	Can cause paralysis; cause variety of symptoms; over 100 serotypes "common cold" viruses.
Infect animals, birds, humans, insects Type A: eastern and western encephalitis Type B: dengue, yellow fever, St. Louis encephalitis	About 20 viruses; some mosquito-borne, some tick-borne.
Mumps; Newcastle disease virus of chickens Measles; respiratory syncytial virus of humans, distemper of dogs and others	Contain hemagglutinin in single protein; infectious for many tissues.
Influenza A: humans, swine, birds, cattle Influenza B: humans Influenza C: humans	Hemagglutinin; a strain most important for human disease; undergoes constant antigenic variation.
Infect mammals, insects, plants Vertebrate: vesicular stomatitis virus Rabies	Infects insects and mammals (cattle); humans, bats, dogs, and other mammals subject to neurological destruction.
Reovirus of humans, many other mammals, birds, cause mild illness of respiratory and GI tract Cytoplasmic polyhedrosis virus of insects Plant viruses also in this group	
Leukosis group in birds (Rous sarcoma virus) Mouse viruses and other mammalian viruses Visna virus Foamy agents Cause leukemia, sarcoma, various others	Numerous antigenic subgroups, not all tumorigenic; no definite human representative but cause leukemias in cats, mice, and cattle; cause slow infection in sheep resulting in demyelination in central nervous system.
Bunyamwera viruses Arthropod-borne	Hundreds of members can cause encephalitis; only recently recognized as different from togaviruses.
Human strains	Common colds and possible GI disease.
Lassa virus Lymphocytic choriomeningitis viruses	Rare, serious, generalized infection in humans; spread from rodents; causes chronic infections in rodents.
Minute virus of mice (nondefective) Adeno-associated virus (AAV, defective)	All defectives and nondefectives depend on coinfection with adenovirus for replication.
Polyoma virus of mice Simian virus 40 Shope papilloma (rabbits; many other papilloma viruses known)	Wide distribution in nature, natural tumors unknown; occur frequently in nature in association with papillomas of skin.
Human adenovirus Mammalian adenovirus subgroups CELO virus	Cause upper respiratory disease; infect cattle, dogs, mice, monkeys; formerly "chicken orphan virus."
Human herpes simplex, types 1, 2 Epstein-Barr Zoster (varicella) Pseudorabies Cytomegalovirus Lucké's Marek's disease	Type 1 causes "fever blisters"; type 2 causes genital herpes (? carcinoma, cervix uterus); causes infectious mononucleosis, associated with Burkitt's lymphoma; causes "shingles" (chronic infection of neural ganglia) in adults and "chickenpox" in children; causative agent of "mad itch" in swine and cattle; infects humans and many other species, each host-specific; associated with fetal damage; causes frog adenocarcinoma; causes tumors in birds.
Variola (human) Vaccinia (human)	Causes smallpox; provides immunity to smallpox; some tumor-transforming.

influenza viruses are helical. Capsomere proteins used in manufacturing capsids are synthesized on host cell ribosomes in the cytoplasm. The synthesis of complete virions may result in the formation of **inclusion bodies,** a characteristic of certain viral infections (Figure 19.13). Negri bodies found in the cytoplasm of rabies-infected nerve cells are accumulations of virions. Guarnieri's bodies of vaccinia (cowpox) are sites of virion assembly in the cytoplasm and Cowdry's type A inclusions are replicated viral DNA strands of herpesvirus. Many capsids have surface projections or "spikes." The synthesis of these glycoproteins is controlled by viral genes. Spikes play a part in the adsorption of the virion to lipoprotein or mucoprotein receptor sites on host cell surfaces. In addition, many viruses are covered by an envelope. This is a characteristic directly related to host-virion interaction at the time of release. The phospholipid portion of the envelope is derived from the cell membrane of the host as the virion is "pinched off," or excluded from the cell. The envelope may also contain protein from the membrane and viral proteins that attach prior to release (refer to Figure 19.4). Since the way in which a virus is released from the host is genetically determined, the presence or absence of an envelope can be of use in classifying viruses.

---

1 What are the characteristics of a temperate phage?
2 List the changes that must occur for a phage to become defective.
3 Name two animal virus infections which typically show inclusion bodies.

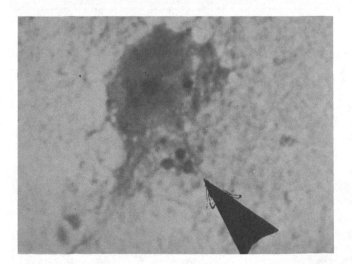

FIGURE 19.13
This brain cell contains Negri bodies of rabies (arrow). These viral structures were discovered in 1903 by the Italian physician Adelchi Negri. (Courtesy of L. LeBeau, University of Illinois Medical Center.)

Not only do animal viruses differ in the way they are released, but variations also occur in other stages of the life cycle. The four forms of viral infection of animal cells are: lytic, latent, persistent, and transforming.

THE LYTIC CYCLE

All lytic life cycles begin with the adsorption of the virion to the cell surface followed by penetration and a complex series of reactions which uncoat the viral nucleic acid. The freed genetic material, however, replicates using different biochemical pathways depending on whether the genetic material is DNA or RNA. Double-stranded DNA, such as that in poxvirus, or single-stranded DNA that quickly forms a second complementary strand, such as that in parvovirus, begins transcription and translation of mRNA and proteins needed for the manufacture of new viral DNA. The process follows the protein synthesis mechanism described in Chapter 3 and is known to occur in both procaryotic and eucaryotic cells. These first proteins are called early proteins, and their production continues until DNA replication is complete. The synthesis of **late proteins** is then initiated. These are capsomeres and proteins used for virion assembly and regulation. In many cases, proteins are formed which integrate into the host's plasma membrane, changing its antigenic nature. These changes are specific for each virus type and can be detected by immunological methods. During early and late periods, other viral genes suppress the synthesis of host proteins, enhancing the ability of the cell to replicate virions. The production of DNA virus is made easier because the enzyme necessary for the formation of new DNA (DNA polymerase) is already present in the host. However, RNA viruses do not have such an advantage, since the host cell lacks a mechanism for replication of RNA. In order to replicate, RNA-containing viruses use alternative pathways that include the rapid production of **RNA polymerase** necessary for duplicating of viral RNA.

Single-stranded RNA viruses, such as polioviruses, carry this essential enzyme as a part of their core. Following the uncoating process, RNA polymerase is released into the host cell and initiates viral gene replication (refer to Figure 19.2). More RNA polymerase is produced as "early protein" to speed the process. If viral RNA is able to function as both mRNA (a positive strand) and genetic material, a second complementary, or negative($-$), strand is formed. This double-stranded RNA molecule, known as the **replicative form (RF),** then begins the replication of more $+$RNA for incorporation into new virions and, at a later time, serves as genetic material for the production of "late proteins." Another group of $+$RNA viruses, known as the **retroviruses,** does not follow this same pattern but directs the production of a complementary strand of $(-)$DNA. The $(-)$DNA is then duplicated to form a $(+-)$DNA, which functions as a typical double-stranded DNA virus. However, when virions are produced from the retroviruses, mRNA(+) is used as core material. A third group of single-stranded RNA viruses is known as the negative-strand viruses, since their core nucleic acid is not able to be directly used as mRNA. A copy of this strand must be made in order for protein synthesis to occur. The last group of viruses, such as the reoviruses, contain double-stranded (+ −) RNA. This group produces its mRNA in much the same manner as double-stranded DNA before "late proteins" are synthesized. The construction of complete virions is well ordered and regulated. Each protein segment is systematically set into its proper position, encasing the appropriate nucleic acid. At the completion of the synthesis, enzymes (endolysins) rupture the cell from the inside, releasing virions and cell contents into the surroundings.

LATENT INFECTIONS

Latent viral infections occur in many common human diseases. Viruses such as those responsible for infectious mononucleosis, chickenpox, and shingles are able to establish a balanced host-parasite relationship for extended periods before they enter the lytic cycle. On first examination, this process appears to resemble the lysogenic life cycle of bacteriophage; however, it differs in that integration of viral nucleic acid into the host endogenote seldom occurs.

Probably the best example of a latent infection is the "cold sore" caused by **herpes simplex** (Figure 19.14). After host invasion, these DNA virions persist for long periods without demonstrating symptoms of infection. However, virions are replicated and sent to daughter cells at the time of cell division. If a change in the physiology of the host upsets this balance, the latent herpes shifts into the lytic cycle and demonstrates acute vesicular eruptions of a "fever blister" or "cold sore." Following the acute lytic stage, the virus infects surviving cells and reestablishes an asymptomatic latent infection. Because the virion is located inside the cell nucleus, the chances of getting rid of the cold sore virus is unlikely.

---

1 What are "early" and "late" proteins?
2 What function does the replicative form (RF) of RNA serve an RNA virus?
3 Name the key differences between a latent animal virus and a lysogenic bacteriophage infection.

Latent infection can also occur following infection with rubeola, the measles virus (Figure 19.15). Defective viruses lacking the ability to form a complete capsid have been found in inclusion bodies taken from brain cells of patients with subacute sclerosing panencephalitis (SSPE). The incubation period for this rare disease may last from 2 to 20 years after a measles infection and results in death from a progressive loss of brain function. The reason that some individuals are more susceptible to this disease may be that their immune systems are unable to recognize and destroy measle-infected cells that have had their membranes altered as a result of infection. (Refer to Chapter 14).

This same type of sclerosis, or hardening of brain and spinal cord tissue, occurs in the chronic disease multiple sclerosis (MS). Although no measles virus has been isolated from MS patients, they do demonstrate an elevated

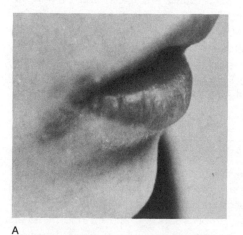

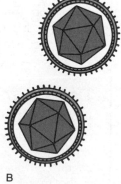

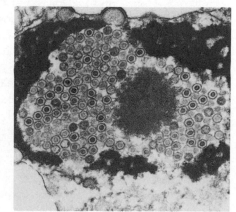

A   B   C

FIGURE 19.14
(A) A cluster of blisters around the lip is the first sign of a recurrent herpes infection. (B) Herpesvirus structure. (C) Cell infected by herpesvirus. (Electron micrograph by C. McLaren and F. Siegel, Courtesy of Burroughs Wellcome Company, Research Triangle Park, N.C.)

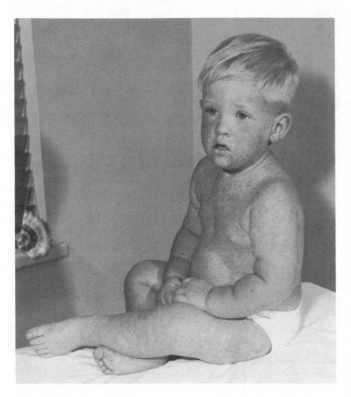

FIGURE 19.15
The measles vaccine introduced in 1963 has been used worldwide to protect millions of children from the potentially fatal after-effects of the disease. Unfortunately for this child, he was not vaccinated. (Courtesy of the Center for Disease Control, Atlanta.)

antimeasles antibody titer similar to that found in SSPE patients. Further research may eventually reveal MS to be a case of latent viral infection caused by a defective measles virus or by another virus. A hypothesis being investigated by Drs. S. Cooke and P. Dowling at the New Jersey College of Medicine and Dentistry proposes that the scarlike sclerotic patches formed on myelin sheaths are the result of a latent infection of dog distemper virus. This is based on epidemiological studies involving MS cases and the incidence of distemper. Studies have revealed that MS patients have had a higher incidence of contact with small house dogs showing neurological illnesses prior to the onset of MS. It has been speculated that fondling or prolonged close contact with the subclinically infected animals results in transfer of the distemper virus and development of a human infection which becomes acute only after an extended latent period of about five years. Whether the actual cause of MS is the result of measles or distemper virus remains to be discovered. The incidence of MS is at relatively low levels as a result of the efficient operation of both nonspecific and specific host defense mechanisms in a high percentage of the human population. Immunization against measles in humans and distemper in dogs also helps in maintaining the low morbidity of MS.

PERSISTENT INFECTIONS

The inevitable destruction of the host cell is not the only outcome of viral infection. Many viruses enter into a host-parasite relationship which allows them to be carried and released for long periods. In some cases, the host will multiply and transmit the contained virion to daughter cells. This life cycle results from a steady-state relationship between the host and cytoplasmic viral nucleic acid in which virions are extruded from the cell without lysis. Viruses which are able to establish this balance are known as **persistent viruses.** Once inside a host, persistent viruses continually release virions but need not reinfect other cells to maintain their existence. Infection is maintained by the transfer of virions to daughter cells at the time of cell division.

One of the best-known examples of a persistent viral infection is serum hepatitis caused by hepatitis B virus (HBV). This double-stranded DNA virus is primarily transmitted in blood, blood products, and on blood-contaminated instruments. HBV infects liver cells and enters the lytic stage after an incubation period that may range from 8 to 22 weeks. During the lytic phase, host cell destruction results in necrosis, inflammation, fever, protein and blood in the urine, gastrointestinal upset, jaundice, and abdominal pain. HBV enters the circulatory system during this period and may be transmitted to others by blood transfusions, during surgery, or on instruments that break the skin or mucous membranes (e.g., razors, toothbrushes, and hypodermic needles). Following the acute phase, about 10 to 12 percent of the patients shift into a persistent infection. Persistence develops in liver cells which have been infected but not lysed. Virions replicate on a slow continuous basis over a period of months or years. The extruded virions circulate through the body and can be spread in the same manner as those present during the acute phase. HBV can also be released in feces, saliva, and breast milk. Infant serum hepatitis acquired from breast milk of carrier mothers may occur during the first few months after birth. The persistent extrusion of virions from carriers is a major epidemiological problem in the world. It has been estimated that approximately one million people in the United States are carriers of HBV.

TRANSFORMATION

The fourth possible virus-host interaction is known as **viral transformation,** a process which converts normal cells to tumor cells.* Viruses which have this ability are called **oncogenic viruses** (tumor-producing) and include members of the papovaviruses, poxviruses, adenoviruses, herpesviruses, and retroviruses. The first oncogenic virus was discovered by F. Peyton Rous in 1911 and is now known as the

---

*Do not confuse viral transformation with the genetic recombination method found in bacteria (refer to Chapter 11).

Rous sarcoma* virus. The hosts of this oncogenic virus are connective tissue cells of chickens. Since that time, cancer research has expanded into a number of other fields, including immunology, bacteriology and microbial genetics. Sparked by an increasing knowledge of latent and persistent virus activity in animal and human hosts, virologists have intensified their investigation into the transformation process. Most research has concentrated on the Rous sarcoma virus, papilloma viruses (causing benign warts in humans), polyoma viruses (causing a variety of tumors), Simian virus, or SV_{40} (infecting cultured rhesus monkey cells), Epstein-Barr virus (suspected of causing human cancer), and hepatitis B virus (suspected of causing liver cancer in humans). Most of the animal viruses used in cancer research are able to function as latent or persistent viruses. Therefore, to demonstrate their oncogenic abilities, it is necessary to use genetically modified (defective) virus forms which are unable to replicate, or to use host cell cultures that are unable to support viral replication (nonpermissive cells). Both techniques have been used in the investigation of such oncogenic viruses as SV_{40} and polyoma viruses.

Transformation begins with adsorption, penetration, and the uncoating of an oncogenic virus in a susceptible cell. Once inside, the virus initiates the synthesis of "early proteins" and soon integrates into the host's genetic material. Integration does not occur at a specific point on the host DNA but all viruses except the Epstein-Barr virus (EBV) have been shown to attach (Figure 19.16). When integrated into the host nucleic acid, the transforming virus is known as a **provirus**—only bacteriophage are known as prophage. The process of integration is controlled by enzymes produced by viral genes as "early proteins." In order for RNA viruses (retroviruses) to become proviruses, they must be converted to their complementary form of DNA. This synthesis is controlled by a viral enzyme known as **reverse transcriptase.** The presence of this enzyme in cells can be used as evidence of RNA virus infection. Early proteins are also responsible for the alteration of normal cell structure and behavior typically seen during the transformation process. It is the presence and effects of "early proteins" which begin and maintain the transformed state of the host cell. In the polyoma virus, this may possibly be done by the direct or indirect action of only one or two "early proteins." "Late proteins" are not produced during transformation, since their synthesis controls the replication and release of completed virions. A transformed cell is identified by the changes it undergoes as a result of virus-host interactions. These include structural and behavioral changes, cell-surface modifications, and the production of new proteins. One of the most outstanding behavioral changes occurring in cancer cells is the loss of **contact inhibition.** Under normal circumstances, cells grow and reproduce until contact with surrounding cells inhibits mitosis. Transformed cells do not respond in this fashion but continue to reproduce in an unregulated manner, piling on one another to form a mass. Structural changes are also evident.

---

1   What evidence exists linking multiple sclerosis to viruses?
2   Why is HBV considered a persistent viral infection?
3   What is an oncogenic virus?
4   How does reverse transcriptase function?

---

*The suffix "-oma" added to a word describes a tissue that has formed a tumor; e.g., "lymphoma" means *tumor of lymph tissue,* "sarcoma" means *tumor of connective tissue.* The important exception to this rule is the term used to describe cancer of the white blood cells, leukemia.

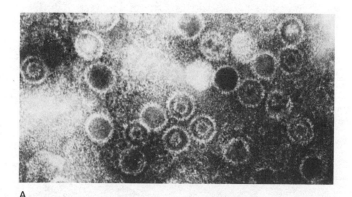

A

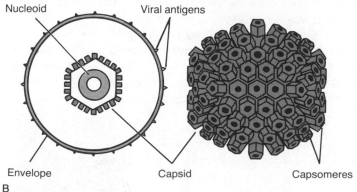

B

FIGURE 19.16

(A) This electron micrograph shows a thin section through a group of immature, intracellular Epstein-Barr virus particles. When appropriately orientated to the plane of the section, the particles show a hexagonal outline; they are either empty or contain central, ring-shaped or dense "nucleoids." (B) Diagram of the EB virus. (Electron micrograph courtesy of M. A. Epstein, Department of Pathology, University of Bristol, England.)

Internally, the cytoskeleton (composed of microtubules and microfilaments) disappears, resulting in a fluffy or ruffled external cell surface (Figure 19.17). Cancer cells also demonstrate changes in their biochemical nature. Differentiated cells which had specialized in the synthesis of particular products have these biochemical pathways modified or closed down, and genes more characteristic of their embryonic period become activated. In some way, the presence of viral DNA stimulates the transcription of repressed fetal genes and the production of carcinoembryonic antigen (CEA) and tumor-specific antigen (TSA). Both antigens play a key role in the diagnosis and control of cancer. These internal and cell-surface changes make transformed cells easily distinguished from normal cells. Culturing on semisolid medium is made easy by the fact that cancer cells do not have a finite, or limited, life expectancy as do primary or secondary tissue cultures. As long as the cells are "thinned" and transferred to fresh medium, they will grow indefinitely. Table 19.3 presents a more complete list of alterations found in virus-transformed cells.

TRANSMISSION OF TRANSFORMING VIRUSES

During transformation, many animal viruses leave the provirus state, and they are replicated and extruded from cancer cells. These virions are then able to infect other susceptible cells, enter the provirus state, and initiate transformation. The movement of transforming viruses from one cell to another by release and reinfection is known as **contagious, or horizontal, transmission.** Horizontal transmission has been found in many animal studies. An RNA-oncogenic virus that causes mammary cancer in mice has been discovered in the body cells and breast milk of cancerous female mice. When this oncogenic virus was investigated, it was found that virus-free strains of susceptible mice had a higher percentage of mammary tumor transformation if they had been nursed by cancerous females, in comparison to mice born to and nursed by virus-free females. There is as yet no conclusive evidence that human oncogenic vi-

TABLE 19.3
Properties altered in transformation by viruses.

Morphological and Behavioral Changes

Mutual orientation more random, lose contact inhibition of movement;
Grow on top of each other;
Grow to high or indefinite saturation density, kill themselves rather than stop growth;
Become invasive;
Become more rounded, have looser attachment to medium;
Grow in suspension, lose anchorage dependence;
Have decreased serum requirement.

Surface Alterations

Hyaluronic acid increased;
Sugar transport increased;
More easily agglutinated by plant lectins;
Surface proteins more mobile;
Lipid fluidity not changed;
Microfilaments (actin) cables disappear but diffuse actin remains;
Myosin-like filaments disappear;
Microtubules disaggregate;
Fetal antigens become evident;
Virus-specific transplantation-rejection antigens appear.

Nonsurface Biochemical Changes

Release of proteases;
Transcription of fetal genes.

SOURCE: Modified from S. E. Luria, J. E. Darnell, D. Baltimore, and A. Campbell, *General Virology*, 3rd ed, 1978. Reprinted by permission of John Wiley & Sons.

ruses may be transmitted horizontally. The question, however, remains to be studied in light of the fact that RNA particles similar to the oncogenic mouse virus for mammary tumors have been isolated from human breast milk and body cells. Even though this virus has not been shown to be oncogenic for humans, its presence leads to the speculation that such viruses may be contagious.

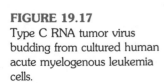

FIGURE 19.17
Type C RNA tumor virus budding from cultured human acute myelogenous leukemia cells.

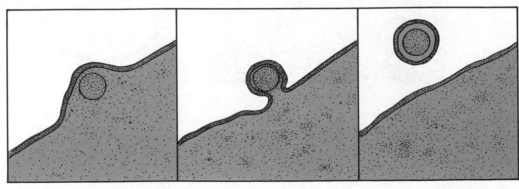

The alternative method of virus transmission is based on the fact that oncogenic viruses are able to integrate into the genetic material of their hosts. The passage of proviruses from one generation to another during reproduction is known as **congenital,** or **vertical, transmission.** Vertical transmission of bacterial prophage from one cell generation to another during the lysogenic cycle is well-established, and strong evidence is growing that oncogenic proviruses of vertebrates are regularly transmitted to daughter cells at the time of mitosis and meiosis. Research into possible cases of vertical transmission has involved mice, rats, hamsters, pigs, chickens, many primates, cats, and mosquitos. Data from humans is lacking because of ethical considerations and difficulties involved in gathering precise information from past generations. According to the **Oncogene Theory** of R. J. Huebner and G. J. Todaro, all vertebrates, including humans, contain "virogenes" acquired by mutation or by viral infection and provirus integration. It is proposed that a portion of these genes, the oncogenes, code for the production of "early proteins" responsible for tumor transformation. Other segments of the virogene code for tumor-specific antigens and enzymes used for replication. The reason that all cells containing virogenes do not transform is theoretically due to the presence of repressor proteins produced by the host. Only under certain circumstances are these genes "triggered," resulting in the transformation of host cells and the release of latent oncogenic virions.

Evidence in support of oncogenes and this triggering action comes from studies on supposedly "virus-free" cells from pigs, cats, primates, and numerous other vertebrates. When these cells were cultured and exposed to radiation and carcinogenic chemicals, several changes occurred. In some experiments, chicken and mouse (murine) cells were found to produce viral DNA; while in others, RNA virus was released. Murine mammary tumor virus (MMTV) was also produced from similar experiments with "virus-free" strains of mice. These experiments provide support for the oncogene theory in vertebrates. A case for vertical transmission has also resulted from epidemiological research into the arbovirus responsible for western equine encephalitis (WEE). The vector for this disease is the mosquito *Culex tarsalis*. After the female mosquito has taken a blood meal from an infected horse, the virus passes through her body, infects her cells and is eventually incorporated into eggs during meiosis. If these eggs become fertilized, vertical transmission takes place and the developing larva contains WEE in its body cells inherited from its parent. Once the mosquito matures and begins feeding, the WEE can be transmitted to other healthy horses. Whether similar types of vertical transmission occur in humans or other vertebrates remains to be discovered. Much additional evidence must be accumulated before the oncogene theory can be proven valid. However, it would help explain why cancer appears in some families

more than others and enable researchers to develop methods of controlling oncogenic proviruses.

1 Name the primary differences between horizontal and vertical virus transmission.
2 State the Oncogene Theory.
3 What viral infection supports the Oncogene Theory?

CANCER: VIRUS-HUMAN INTERACTION?

Thus far, no virus has been confirmed as the cause of human cancer. However, several are strongly suspected because of (1) their ability to transform human cells under laboratory conditions, (2) the restricted geographic concentration of certain cancers, suggesting horizontal transmission, and (3) a close immunological similarity between certain viruses and cancer. Burkitt's lymphoma is among the most studied cancers suspected of being caused by a virus, although no virions have been identified inside tumor cells. This malignant cancer of the lymphoid tissue results in the formation of massive tumors of the jaw and neck (Figure 19.18). Tumor growth is very rapid, doubling in size every two to four days, with fatal metastasis to other parts of the body. Burkitt first identified this lymphoma (1958) among children aged six to eight living in a restricted area of Africa and New Guinea. More detailed epidemiological studies have demonstrated the disease (although rarely) in North and South America, Europe, Asia, and Australia. Immunological

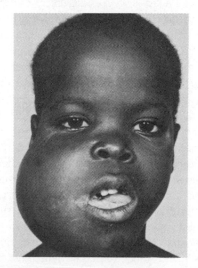

FIGURE 19.18
Burkitt's lymphoma. (From A. Altman and A. Schwartz, *Malignant Diseases of Infancy, Childhood and Adolescence,* 1978. Reprinted by permission.)

studies strongly suggest that Burkitt's lymphoma is caused by Epstein-Barr virus (EBV), the agent of infectious mononucleosis. Several lines of evidence support this link.

EBV commonly occurs throughout the world and the antibody titer against this virus is normally very high. Eighty-five percent of the adult population in the United States has EBV antibody, indicating previous exposure. Protection from infectious mononucleosis during infancy is provided by maternal antibody, which is later replaced as a result of inapparent, subclinical contact with the virus. Clinical cases of the disease probably occur in those who have not experienced inapparent infection as children. Infection of B-lymphocytes during the acute phase results in lysis or persistent infection. The lysed cells release virions which infect other B-cells, are responsible for transmission of the "kissing disease," and stimulate antibody production. Before lysis, there is also an increase in the number of white blood cells which show changes similar to those seen during leukemia. However, this increase is soon brought under control and symptoms subside. Control of mononucleosis is primarily the result of killer T-cells (cell-mediated immunity) which dramatically increase in number and activity. Their ability to identify infected B-cells is based on the presence of new cell-surface proteins produced as "early proteins." B-cells which experience a persistent infection may, on occasion, be destroyed, releasing Epstein-Barr-determined nuclear antibody (EBNA) and stimulating production of antibody characteristic of EBV infections. EBNA may be found in the sera for years after mononucleosis because of its persistent nature.

Investigations have revealed many virus-host interactions that strongly resemble preliminary stages in tumor transformation. This has led many to think of infectious mononucleosis as a self-limiting leukemia. A number of these interactions and host responses are common to both mononucleosis and Burkitt's lymphoma. The diseases are centered in the B-cells, show the same antigenic cell-surface changes and other tumor-transforming characteristics. Both diseases show elevated EBNA and EBV serum antibody levels. Killer T-cell activity is also increased in both. Burkitt's lymphoma is suspected of being caused by EBV because of the host's inability to control the viral infection. This may be the result of weak T-cell activity caused by inheritable genetic abnormalities, severe malnutrition, or the effects of other diseases such as malaria. Evidence for association of Epstein-Barr virus with Burkitt's lymphoma is summarized in Table 19.4. Another form of cancer strongly suspected of being caused by EBV or a closely related form is nasopharyngeal carcinoma (NPC). This disease also demonstrates elevated EBNA and transformation characteristics similar to Burkitt's lymphoma. Genetic factors play a part in this disease and are no doubt responsible for its greater frequency among the Chinese.

TABLE 19.4
Evidence for the association of EBV with Burkitt's lymphoma.

1 Epidemiologic association with viral infection.
2 Enhanced titers of EBV antibody.
3 Presence of virus nucleic acid.
4 Presence of virus-associated antigens.
5 Release of infectious virus from Burkitt cell lines.
6 Virus stimulation of cell DNA synthesis *in vitro*.
7 Virus transformation of normal lymphocytes *in vitro*.
8 Induction of malignant lymphoma in nonhuman primates.

SOURCE: R. Rapp and B. McCarthy. "Persistent Herpes Virus Infections and Cancer." *Viruses and Environment.* Proceedings of the Third International Conference on Comparative Virology. Quebec, Canada, 1977, Academic Press.

Evidence is mounting which implicates herpes simplex virus type 2 (HSV-2) as a cause of cervical cancer. This virus is transmitted by sexual contact and may invade cervical tissue cells, causing a persistent venereal disease for which there is no cure. The connection between this disease and cancer is being traced through serological, epidemiological, and biochemical research. Thus far, evidence indicates that cervical cancer is linked to early, first sexual contacts, poor socioeconomic conditions, and sexual promiscuity. Laboratory experiments have resulted in the isolation of HSV-2 DNA from a tumor biopsy, cell cultures demonstrating the presence of HSV-2 transforming ability, virus stimulation of host DNA synthesis, elevated HSV-2 antibody levels, and the induction of cervical cancer in mice. If this virus is found to be responsible for cervical cancer, there is a strong likelihood that HSV-2 is also involved with cancer of the prostate.

1 What virus has been associated with Burkitt's lymphoma?
2 Control of virus-infected cells is primarily mediated by which immune system (AMI or CMI)?
3 What is the link between HSV-2 and cervical cancer?

One other virus also being considered as a tumor transformer is hepatitis type B. The persistent nature of this virus and epidemiological evidence has led to the speculation that certain types of liver cancer (hepatomas) may result from previous serum hepatitis infections. The exact nature of the shift from persistence to tumor-transforming has not as yet been explained but may be affected by outside forces such as carcinogenic chemicals, alcohol, and radiation. In recent years, investigation into host-virus interactions and the disease process has led to many new discoveries. Among these has been the identification of particles which do not fit the picture of a "conventional virus," composed of a nucleic acid core and protein coat. For this reason, they have become known as the "unconventional" viruses.

UNCONVENTIONAL VIRUSES

"SLOW" VIRUSES

"Slow" virus infections require a long latent period before clinical symptoms are demonstrated. The disease which follows is usually fatal and results from degeneration of the central nervous system. "Conventional viruses" that follow this clinical pattern are congenital rubella, subacute sclerosing panencephalitis (SSPE), Crohn's disease, and infectious hepatitis. However, the term **slow virus** is becoming more closely associated with "unconventional" agents which do not have typical virion structure. The slow viruses do not form virions, and they demonstrate many different physical and chemical properties. Their unconventional resistance to ultraviolet light, nuclease enzymes, and other agents that normally destroy conventional viruses has led to the speculation that these particles lack a nucleic acid. Currently, research is being conducted to determine the nature of "slow" viruses with the idea that they might be self-replicating membranes. If this turns out to be the case, the concepts of self-replication and the nucleic acid gene would be open to revision.

In the past, slow virus infections have been classified as spongiform virus encephalopathies, and included animal (scrapie and mink encephalopathy) and human diseases (kuru and Creutzfeldt-Jakob disease). Kuru (the trembling disease) is the name given to a chronic, degenerative disease of the central nervous system found among cannibals known as the Fore people in the eastern highlands of New Guinea. The victims of this disease were primarily women who, according to Dr. Gajdusek, became infected during their ritual cannibalistic consumption of their dead relatives as a rite of respect and mourning. Brain tissue from the dead was removed, squeezed into pulp, and packed into bamboo tubes for steaming and consumption. Those who ate the tissue were not the only ones who became infected. Children became infected from mothers who had not washed their hands before washing their child or wiping their nose or eyes. Men in the tribe seldom became infected, since they lived in separate houses and did not celebrate this ritual. Once infected, the victim would experience the slow and progressive onset of trembling and uncontrollable shivering lasting approximately a year. During this time, they were known as kuru "dancers" and held favored status in the tribe. However, as the disease progressed and the trembling became worse, the kuru dancers became unable to walk, stand, or swal-

low—they entered their "dance of death." Kuru was identified as an infectious disease by chimpanzee inoculation by D. Carlton Gajdusek, who was later awarded a Nobel Prize for his discovery. The government of New Guinea has since banned the cannibalistic ritual; and within a short time, the disease has almost entirely disappeared.

Unlike kuru, the slow virus infection known as Creutzfeldt-Jakob disease (CJD) is not confined to one geographic area. It is also a rare disease of the central nervous system and has symptoms similar to kuru. In both cases, it is likely that the degenerative changes are the result of humoral and cell-mediated hypersensitivities. Transmission of CJD in nature has as yet not been identified; however, cases of medically related transmission have been documented. One individual developed the disease after a corneal transplant, while two others became infected from contaminated electrodes used to perform an electroencephalogram (EEG). The intracellular nature of the slow viruses has led to speculation that vertical (congenital) transmission may occur, but a valid case has not been identified.

VIROIDS

Fundamental research into the nature of these unconventional viruses began in 1967 by Diener and Raymer. **Viroids** (viruslike particles) are the cause of infectious diseases of some higher plants. Only seven viroids have been identified, and each has been named for the symptoms it causes: potato spindle viroid, citrus exocortis viroid, chrysanthemum stunt viroid, chrysanthemum chlorotic mottle viroid, coconut cadangcadang viroid, cucumber pale fruit viroid, and hop stunt viroid (Figure 19.19). Viroids are extremely small,

FIGURE 19.19
Chlorotic mottle disease on leaves of *Chrysanthemum morifolium*. (Courtesy of R. K. Horst, Cornell University.)

nonencapsulated nucleic acids present in the plants demonstrating disease symptoms but absent from healthy specimens. The nucleic acid is circular RNA and remains free in the cell. When these agents are inoculated into healthy cells, they replicate and induce symptoms of the disease. Like the slow viruses, they are resistent to ultraviolet light and ionizing radiation. These naked loops of nucleic acid are apparently capable of self-replication, but in some cases are aided by conventional viruses. Their independent nature makes viroids the smallest known agents of infectious disease. All viroids discovered so far are pathogens of higher plants and none has been found in animal or human cells. However, the unknown nature of the slow viruses leads many to speculate that these two unconventional viruses may be very closely related.

---

1 What is the difference between a conventional and unconventional virus?
2 How is kuru transmitted?
3 What type of cells serve as hosts for viroids?

SUMMARY

The viruses have been under investigation since the turn of the century, but not until the development of more sophisticated research techniques has their nature been made clear. These submicroscopic, acellular microbes function as obligate intracellular parasites of microbes, plants, and animal cells. All conventional viruses are composed of an RNA or DNA core and a protein coat known as a capsid made up of subunits called capsomeres. Some viruses are also covered by an envelope derived, for the most part, from their host cell membranes. Morphologically, viruses are helical, polyhedral, or combinations of these shapes. Reproduction is accomplished by a process called viral replication and occurs in five stages: adsorption, penetration, replication of nucleic acid, maturation, and release. If the release stage results in the destruction of the host, the life cycle is called lytic. However, many viruses do not destroy their host but release by an extrusion, or "pinching off," process. Cultivation of microbes can be done on cell cultures, in host organisms, or in embryonated eggs. The hemagglutination test and endpoint assay methods enable virologists to determine the number of virions in culture.

Classification of viruses is based on the nature of their nucleic acid, morphology, type of host, and life cycle. The unknown nature of their evolutionary relationship to cellular life forms makes the present taxonomic system a functional one. The bacteriophage infect bacteria and other procaryotic cells, and are either lytic or lysogenic. Plant and animal viruses may be lytic, latent, persistent, or tumor transforming. The one-step growth curve is typical of lytic life cycles. Disease resulting from viral infection is directly correlated with the nature of the life cycle. Many human illnesses may be the result of delayed host-virus interactions. Multiple sclerosis, subacute sclerosing panencephalitis, and others might be due to latent viral infection, while some human cancers are suspected of being caused by oncogenic viruses. Although no confirmed case of viral human cancer has as yet been identified, there is strong evidence that viruses may be the cause of Burkitt's lymphoma, cervical cancer, and liver cancer. The route of virus transmission, either vertical or horizontal, has an influence on the likelihood that tumor transformation by viruses can occur in humans and may be inherited.

Research has also revealed the existence of the unconventional viruses. These microbes are known as the "slow" viruses and viroids. They have been identified as the cause of many rare but fatal diseases in humans, animals, and higher plants. However, their exact nature and relationship to the conventional viruses and their host cells remain to be understood.

WITH A LITTLE THOUGHT

Viruses are acellular microbes that function as obligate intracellular parasites. Their life cycles are divided into four separate types: lytic, latent, persistent, and tumor transforming. However, this may turn out to be an artificial separation used merely for the convenience of virologists and clinicians. As more valid and detailed information becomes available, these "life cycles" may have to be linked together as a spectrum of possible host-virus interactions. What viral diseases and characteristics are "typical" of each of these four cycles and which show features that might lead virologists to believe that a "shift" from one life cycle to another may "normally" take place in nature?

STUDY QUESTIONS

1 What are the major structural and functional differences between viruses and cellular forms of life?
2 Describe the five stages of viral replication.
3 How are viruses cultured and counted in the laboratory?

4   What are the differences between: lytic and lysogenic cycles; latent and persistent cycles; persistent and transforming life cycles?

5   Compare and contrast the population growth curve for cellular microbes and the single-step growth curve for viruses.

6   What is a defective virus? Phage immunity? Phage resistance?

7   What are the functions of "early" proteins? "Late" proteins?

8   How do the initial life cycles of RNA and DNA viruses differ?

9   What is meant by horizontal transmission? Vertical transmission? Give examples of each.

10 What are the "unconventional" viruses? How do they differ from the "conventional" viruses?

SUGGESTED READINGS

BALDWIN, R. W., M. J. EMBLETON, J. S. P. JONES et al. *Int. J. Cancer,* 12:73–83, 1973.

BAXT, W. G., and S. SPIEGELMAN. "Nuclear DNA Sequences Present in Human Leukemic Cells and Absent in Normal Leukocytes." *Proc. Natl. Acad. Sci. U.S.,* 69, 3737, 1972.

CAIRNS, J. "The Cancer Problem." *Scientific American,* Nov., 1975.

DALES, S. "Early Events in Cell-Animal Virus Interactions." *Bact. Rev.* 37:103, 1973.

DE THÉ, G. "Epstein-Barr Virus and Human Cancers: A Multidisciplinary Epidemiological Approach." *Viruses and Environment;* Proceedings of the Third International Conference on Comparative Virology, held at Mont Gabriel, Quebec, Canada, May, 1977. Edited by Edouard Kurstak and Karl Maramorosch. Academic Press, New York, 1978.

DIENER, T. O. *Viroids and Viroid Diseases.* John Wiley & Sons, New York, 1979.

————. "Viroids." *Scientific American,* Jan., 1980.

GAJDUSEK, D. C. *"Unconventional Viruses and the Origin and Disappearance of Kuru."* Science 197:943–60, 1977.

————. "Kuru and Creutzfeldt-Jakob Disease." *Ann. Clin. Res.* 5:254–61, 1973.

HENLE, W., G. HENLE, and E. T. LENNETTE. "The Epstein-Barr Virus." *Scientific American,* July, 1979.

HOLLAND, J. J. "Slow, Inapparent and Recurrent Viruses." *Scientific American,* Feb., 1974.

KLEIN, G. "The Epstein-Barr Virus and Neoplasia." *New England Journal of Medicine,* 1975.

————. "Cancer, Viruses, and Environmental Factors." *Viruses and Environment;* Proceedings of the Third International Conference on Comparative Virology, held at Mont Gabriel, Quebec, Canada, May, 1977. Edited by Edouard Kurstak and Karl Maramorosch. Academic Press, New York, 1978.

KNIGHT, C. A. *Molecular Virology,* McGraw-Hill, New York, 1974.

LISAK, R. P. "The Enigma of Multiple Sclerosis." *Viruses and Environment;* Proceedings of the Third International Conference on Comparative Virology, held at Mont Gabriel, Quebec, Canada, May, 1977. Edited by Edouard Kurstak and Karl Maramorosch. Academic Press, New York, 1978.

LURIA, S. E., J. E. DARNELL, D. BALTIMORE et al. *General Virology,* 3rd ed. John Wiley & Sons, New York, 1978.

OLD, L. J. "Cancer Immunology." *Scientific American,* May, 1977.

PERLMANN, P., C. O'TOOLE, and B. UNSGAARD. "Cell-Mediated Immune Mechanisms of Tumor Cell Destruction." *Fed. Proc.,* 1973.

RAFFERTY, K. A., "Herpes Viruses and Cancer." *Scientific American,* Oct., 1973.

RAPP, F., and B. A. MC CARTHY. "Persistent Herpesvirus Infection and Cancer." *Viruses and Environment;* Proceedings of the Third International Conference on Comparative Virology, held at Mont Gabriel, Quebec, Canada, May, 1977. Edited by Edouard Kurstak and Karl Maramorosch. Academic Press, New York, 1978.

SAMBROOK, J. "Transformation by Polyoma Virus and SV40." *Advances in Cancer Research,* 1972.

TEMIN, H. M. "RNA-Directed DNA Synthesis." *Scientific American,* Jan., 1972.

KEY TERMS

capsid

capsomere (kap'so-mer)

plaque (plak)

hemagglutination (hem"ah-glu"ti-na'shun)

bacteriophage

lytic cycle

lysogenic

oncogenic (ong"ko-jen'ik)

viroid

Viruses and the Disease Process

CYTOPATHIC EFFECTS AND SYMPTOMS
INTERFERON
IMMUNITY AND SEROLOGY
IMMUNIZATION AND CHEMOTHERAPY

Learning Objectives

☐ Know the eight cytopathic effects of viruses.

☐ Be able to list the six stages of a generalized viral infection.

☐ Define the term "interferon" and explain its most likely mode of action.

☐ Be familiar with immunology and serology as associated with viral infections.

☐ Recognize the two types of vaccines used to induce active acquired immunity against viral infections.

☐ Name several problems associated with vaccination procedures.

☐ Be familiar with the difficulties in the use of antiviral drugs.

Virus-host interactions can take many forms; however, all result in some form of damage or death to the host cell. For centuries, researchers have attempted to learn the nature of these disease processes in order to cure those infected and prevent the disease's recurrence. Virologists are only now beginning to understand the intricate biochemical defense mechanisms found in many types of cells and organisms. With this knowledge, many new developments in the areas of immunology, chemotherapy, and epidemiology have led to significant advancements in the control of viral diseases. A detailed description of all the information available in the field of virology is not possible. Therefore, this chapter will provide background on some currently important areas of virology.

CYTOPATHIC EFFECTS AND SYMPTOMS

Cell damage or death of a host cell as a result of viral infection and replication is known as the **cytopathic effect (CPE).** The most familiar CPE is **cytolysis,** which results from a lytic life cycle. However, fatal damage to the host cell can also result from latent, persistent, and tumor-transforming infections. Other viral-induced CPE changes in-

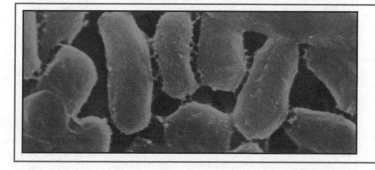

20

clude: (1) changes in membrane permeability, (2) loss of essential metabolic products, (3) loss of essential enzymes due to repression or destruction of host genes, (4) development of hypersensitivities to virus-specific antigens or viral products, (5) fusion of infected cells to form a giant, multinuclear cell known as a syncytium, (6) cell clumping, and (7) inclusion body formation. If these changes involve a significant number of cells, clinical symptoms of infection will result. Although observable changes in the whole organism are the "symptoms" of viral infection, it must be kept in mind that the actual host is a single cell. Overt symptoms can be either localized or generalized. Localized symptoms usually develop at the portal of entry and only involve a limited area; for example, warts and cold sores occur on the skin, and "common colds" and influenza involve the respiratory tract. Generalized viral infections demonstrate initial but mild symptoms at the portal of entry and spread through the body in the following pattern:

1   Invasion at the portal of entry;

2   Multiplication in local, adjacent lymph nodes;

3   Spread through the blood (primary viremia);

4   Multiplication in susceptible internal organs;

5   Further spread through the blood (secondary viremia), and

6   Localization of the virus in target cells, causing CPE and characteristic disease symptoms.

Symptoms of viral infection are unique to each type of virus and follow a course which reflects the nature of its particular life cycle (lytic, latent, persistent, or transforming). However, many infections occur without demonstrating clinical symptoms. These **inapparent infections** are important for several reasons. First, inapparent infections can stimulate the immune system, conferring life-long immunity to the virus. For instance, contact with the Epstein-Barr virus in early childhood stimulates this type of immunity to infectious mononucleosis. Second, although symptoms can be controlled by the immune system, subclinical CPE may continue, resulting in host harm. This occurs in cases of rubella (German measles) infections of pregnant women. The administration of gamma globulin controls clinical symptoms, but if not given in time, the fetus may develop a subclinical infection known as congenital rubella syndrome. Third, generalized symptoms may not be produced, if the virus is unable to reach the target organ. However, inapparent infection at other sites will result in the continual release of infectious virions. Immunization against poliomyelitis prevents the virus from reaching the central nervous system and causing paralysis, but replication occurs in cells of the gastrointestinal tract. The constant release of these infective virions serves to maintain the virus in the environment and makes poliovirus a continuing public health threat. Protec-

tion from inapparent and apparent viral infection involves both nonspecific and specific host defense mechanisms (refer to Chapter 13). The skin, organic acids, cilia, lysozyme, properdin and complement systems, inflammation, fever, and phagocytosis all play important nonspecific roles in defense. In addition, the production of molecules known as **interferon (IF)** specifically aid in the control of viral infections.

---

1   Name three types of cytopathic effects that can be induced by a virus.

2   After entering through the appropriate portal, what is the general pattern of a viral infection?

3   Why are inapparent viral infections of such great importance?

INTERFERON

The interferon system is a group of glycoproteins discovered in 1957 by A. Isaacs and J. Lindemann while investigating the process of viral interference. Interference occurs in cells infected with two viruses; the presence of one interferes with the replication of the other. Since that time, interferon has been found in many organisms and is probably produced by all human cells in response to certain stimuli. Contact with viruses, rickettsias, protozoa, bacterial endotoxins, double-stranded RNA, and certain antibiotics (e.g., cycloheximide and kanamycin) stimulates the release of small amounts of these inhibiting molecules. The most effective interferon-producing cells are leukocytes, fibroblasts, and T-cells; however, more than forty types have been identified. Table 20.1 lists four of the more well-known forms. Recognizable increases in the amount of IF can be detected within hours of stimulation, but they soon decrease. The lag period before IF production is relatively short when compared to the hours or days needed to increase serum antibody levels. Unlike antibodies, IF is host species-specific and not virus-specific. Interferons are not directly antiviral but are indirectly effective against a variety of viruses. For example, IF capable of inhibiting influenza type A virus is also effective against equine encephalitis and vaccinia. As a species-specific molecule, IF produced by one species is only effective in members of the same species. Therefore, IF produced in one species cannot be used to produce passive immunity in another species. This means, for example, IF produced in horses cannot be isolated and used in humans to protect against infection.

The IF system has the ability to inhibit viral replication, slow the growth of cancer cells, and enhance the action of phagocytes. The three mechanisms of action proposed for the effects are: (1) inhibition of viral mRNA translation, (2) blockage of viral nucleic acid transcription, and (3) alteration of host cell membranes to prevent the release of virions. At

TABLE 20.1
A few of the more than forty known types of interferon.

Type	Uses	Source
Leukocyte (alpha)	Most common form; used in herpes research. Found in small tumors of blood-forming tissues.	B-lymphocytes
Fibroblast (beta)	Effective against skin and muscle tumors, and epidermal viral infections.	Connective tissue
Lymphoblastoid	Similar to leukocyte (alpha).	Cultured Namalva cells (from Burkitt's lymphoma patients)
Immune (gamma)	Unknown as yet. Possibly stimulates killer cell activity.	T-lymphocytes

this time, the first hypothesis has the most support. Once stimulated, a cell releases small amounts of IF which act like a cellular hormone; that is, it is produced at a distance from its site of action (Figure 20.1). Molecules bind to receptor sites on the surface of uninfected cells, triggering the manufacture of enzymes that produce the antiviral effect. This is no doubt a case of induction and repression. Two such enzymes have been identified which selectively prevent the translation of viral mRNA into essential viral proteins.

Interferon used for experimentation is extracted from human cell cultures, from prepuce tissue taken from male infants after circumcision, or from genetically engineered bacterial cell cultures. The administration of this human-specific IF has proven effective in experimentally treating hepatitis B, herpes simplex, and common cold infection. However, the cost of providing the large quantities necessary to bring about reasonable control currently prevents the use of IF as a chemotherapeutic agent. IF has also demonstrated the ability to affect the growth of cancer cells and alter the activity of macrophages. Several forms of lymphoma,

Hodgkin's disease, and certain cases of leukemia have undergone inhibition as a result of administering interferon. It is likely that the molecule stimulates the production of enzymes, which in some as yet unknown manner interfere with spindle fiber formation during mitosis. A marked increase in macrophage motility and killing activity has also been seen, which aids in the destruction of abnormal cells. The enhanced cancer control effects of the BCG vaccine may be related to the action of IF. The presence of BCG may stimulate the release of IF which, in turn, intensifies the anticancer activity of macrophages.

The antiviral role of macrophages is also related to activities of the immune system.

---

1 Name the cells most active in IF production.
2 What is meant by the phrase "IF is species-specific"?
3 How might IF control a viral infection?

FIGURE 20.1
The production of interferon is stimulated by viral infection. This antiviral chemical is synthesized and transmitted to adjacent cells, stimulating them in turn to produce antiviral proteins, which inhibit viral replication.

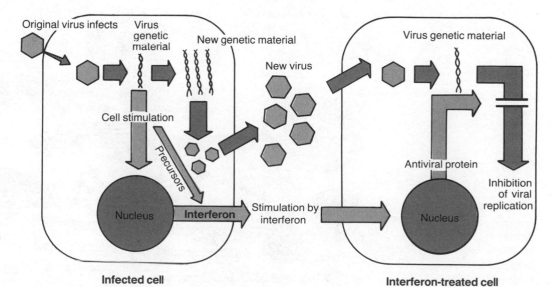

IMMUNITY AND SEROLOGY

Both cell-mediated and humoral immune systems are involved in the control of viral infections. Cell-mediated responses of virus-sensitized lymphocytes primarily aid in the control of latent, persistent, and transforming-virus infections. Sensitized lymphocytes may respond directly to the presence of viral antigens (e.g., coat material) or antigenic changes in the surfaces of infected cells. Stimulated by the presence of these antigens, sensitized T-lymphocytes release interferon, cytotoxins, macrophage inhibiting factor (MIF), and chemotactic factors. The result is the destruction of infected host cells, phagocytosis of free virions, and inhibition of viral replication (Figure 20.2). Immune control of extracellular viruses also involves antibodies released from sensitized B-cells. Of the five classes of immunoglobulins, IgG, IgM, and IgA demonstrate antiviral activity. IgG (gamma) globulins contain the greatest number and variety of antiviral antibodies produced after sensitization. The response time for IgG production varies from hours to days after the first effective exposure. The most important feature of IgG is its ability to cross the placenta and provide passive immunity for newborns. However, the presence of maternal IgG in infants restricts the use of attenuated vaccines for the production of active immunity. The maternal antibodies will neutralize the vaccine, preventing sensitization and the development of the infant's own antibodies.

Immunoglobulins of the IgM class contain a group of antibodies known as the **Forssman, or heterophile, antibodies.** These are produced in response to a variety of antigens typically found on a number of unrelated viruses. Heterophile antibodies are found in the serum several days after initial infection and are valuable in diagnosis. The presence of heterophile antibody in the serum of individuals suspected of Epstein-Barr virus infections (infectious mononucleosis) is determined by performing an agglutination test.

The two forms of IgA that have been identified are serum and secretory. The production of IgA shows the longest lag period and is not found for days or weeks after infection. Both forms demonstrate antiviral activity; however, IgA in saliva, tears, and other external secretions plays a more effective role in preventing initial infection. Attenuated poliovirus is better able to stimulate secretory IgA production, if administered orally. An increase in secretory IgA is the favored response, since this virus establishes an infection through the oral portal of entry. Because secretory IgA production is increased by bringing viral antigens in direct contact with the mucous membranes, virologists are attempting to develop intranasal vaccines against upper respiratory viruses such as the common cold virus and influenza. These vaccines would be administered through inhalers similar to those used by asthmatic individuals to atomize adrenaline directly into their lungs.

Since inapparent and apparent viral infections stimulate both cellular and humoral immune responses, the laboratory diagnosis of disease requires serological studies, in addition to the presence of typical symptoms and the occurrence of cytopathic effects. A number of serological tests used for this purpose have been described in Chapter 13 (Immunity): complement fixation (CF), complement immu-

---

1 Which immune system is most effective in controlling viral infections if the virions are intracellular?

2 What are Forssman, or heterophile, antibodies and why are they important?

3 Which class of immunoglobulins is best able to prevent viral infections by interfering with adhesion?

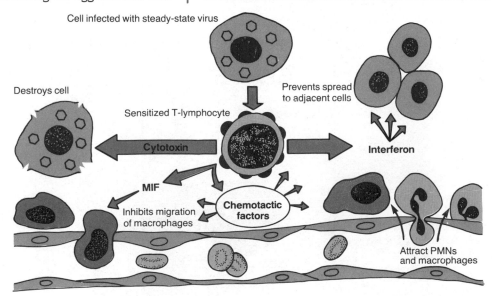

FIGURE 20.2
Not only does interferon aid in the elimination of viral infections, but a complex, cell-mediated immune response is involved in the process.

noelectrophoresis (CIE), cell culture (CC), flocculation tests (Ft), neutralization tests (Nt), fluorescent antibody tests (FA), zone electrophoresis (ZE), immunoelectrophoresis (IEP), radioimmunoassay (RIA), immunodiffusion (ID), and skin hypersensitivity tests. Table 20.2 lists common animal and human viral infections as well as the serological and cell-culturing methods used in their diagnoses.

IMMUNIZATION AND CHEMOTHERAPY

An understanding of immunological responses has made it possible in recent years to more efficiently control viral infections by immunization. The two types of vaccines used to induce active acquired immunity are live, attenuated vaccines and killed virus vaccines. Attenuated viruses have been used since Edward Jenner introduced the idea of vaccination in 1795. His concept of inoculating with a modified, noninfective form of cowpox virus to control smallpox has been expanded. Attenuated vaccines are now available for poliovirus (Sabin vaccine), measles, rubella, yellow fever, adenovirus (types 3, 4, and 7), and mumps. Live, atten-

uated viruses are advantageous in that they are easy to administer, induce higher antibody levels, and in some cases, stimulate secretory IgA production because they enter the body through the normal portal of entry. Killed virus vaccines are available for poliovirus (Salk vaccine), measles, rabies, adenoviruses, mumps, influenza, and respiratory syncytial virus. The high level of individual and herd immunity that has developed as a result of mass immunization has significantly reduced morbidity and mortality rates. However, vaccination procedures are not without drawbacks, and several problem areas have been identified:

1 Protection provided by some vaccines is short-lived, and immunoglobulins can fall to unprotective levels in only a few months.

2 Viruses may escape modification during heat killing or formalin treatment, enabling them to induce disease.

3 In some situations, for example, influenza type A vaccination, immunoglobulin levels rise sharply but provide little long-term protection because of regular changes in the antigenic nature of the viruses known as **antigenic drift.** For this reason new flu shots have to be given each year.

TABLE 20.2
Laboratory diagnosis of viral infections.

Disease	Human Specimens to Be Tested	Cell Culture or Animals to Be Inoculated	Diagnostic Serologic Tests
Encephalitides California St. Louis Western equine Eastern equine	Brain, blood	Mice	Nt in mice or CC CF
Yellow fever	Blood, viscera	Monkeys Mice	Nt
Dengue	Blood	Mice (difficult)	Nt, CF
Colorado tick fever	Blood	Hamsters Mice	Nt CF
Poliomyelitis	Spinal cord, feces, throat swabs	CC	Nt, CF
Rabies	Brain	Mice	FA
Lymphocytic choriomeningitis	Brain, blood, spinal fluid	Mice Guinea pig	Nt CF
B virus infection (herpes B)	Brain, spleen	Rabbit Rabbit kidney CC	CF, Nt CF, Nt
Measles	Brain	Direct exam	FA
Variola (smallpox)	Skin lesions, vesicle fluid, blood	Embryonated egg	CF ID
Vaccinia	Skin lesions, vesicle fluid	Embryonated egg Rabbit CC	CF Nt in eggs, rabbits, mice, CC;
Varicella (chickenpox)	Vesicle fluid	CC	CF, Nt

Zoster	Vesicle fluid	CC	CF
Measles	Nasopharyngeal secretions, blood	CC	CF, Nt
Rubella (German measles)	Nasopharyngeal secretions, blood, amniotic fluid	CC	Nt, CF, FA ID
Congenital rubella syndrome	Throat swab, urine, feces, spinal fluid, blood, bone marrow, conjunctival swab	CC	Nt, CF, FA
Exanthem subitum	Blood	Monkeys	
Herpes simplex	Skin lesions, brain	Mouse (newborn)	Nt in mice, eggs, or CC
		Embryonated egg	CF
		CC	Nt, CF
Influenza A	Throat swab, nasal washings, lung	Eggs	Hemagglutination
Influenza B		Ferrets, mice	CF
Influenza C		CC	Nt in eggs, mice, or CC
Parainfluenza		CC	Nt, CF
Respiratory syncytial (RS) infection		CC	Nt, CF
Common cold (rhinovirus group)	Nasopharyngeal washings, swabs	CC	Nt
Mumps	Saliva, spinal fluid, urine	Monkeys	CF
		Eggs	Hemagglutination
		CC	Nt in CC
Adenovirus group	Throat swab, pharyngeal washings, stool	CC	CF, Nt
Infectious hepatitis (type A)	Blood, feces	(No satisfactory method)	Microtiter RA
Serum hepatitis (type B)	Blood, feces, urine	"	CIE, CF, ID, RA
Coxsackie infection	Feces, throat swab, spinal fluid, vesicle fluid	Infant mice	Nt in infant mice or CC
		CC	Nt, CF
Echovirus infection	Feces, throat swab, spinal fluid	CC	Nt, CF
Reovirus infection	Feces, throat swab	CC	Nt, CF
Molluscum contagiosum	Skin lesions	Electron microscope	None
Verrucae (warts)	Skin lesions	"	"
Encephalomyocarditis (Col. SK-Mengo)	Blood	Mice	Nt in mice
Epidemic keratoconjunctivitis	Conjunctivas	CC	Nt test for adenovirus 8
Foot-and-mouth disease	Skin lesions	Guinea pigs Newborn mice	CF
Cytomegalic (inclusion) disease	Oral swabs, urine, Various organs	CC	Nt, CF

SOURCE: Modified from E. Jawetz, J. L. Malnick, and E. A. Adelberg, *Review of Medical Microbiology*, 13th ed, 1978. Lange Medical Publications.

4 Severe side effects caused by some vaccines, such as with the original attenuated measles vaccine, precludes general use in many cases.

5 Adenovirus type 7 induces immunoglobulin production but is only rarely used because of its oncogenic character demonstrated in laboratory animals.

6 The use of some vaccines is questionable in light of their ability to stimulate a protective elevation of antibody titer before the onset of disease symptoms.

The successful control of viral diseases by vaccination has been proven throughout the world. However, disease prevention by this method is not always possible due to the problems previously noted, and the fact that there are areas in the world where vaccination is not regularly practiced. These difficulties have stimulated research into the development of clinically effective, antiviral, chemotherapeutic

drugs. Because viruses are obligate intracellular parasites and dependent upon host-cell metabolic pathways, effective drugs must be highly specific in their action and only interfere with pathways essential to the virus. The normal metabolism of the host cell should not be fatally disturbed. The success of an antiviral drug is based on its ability to selectively interfere with one of the five stages of the viral life cycle: (1) adsorption, (2) penetration and uncoating, (3) replication of nucleic acid, (4) synthesis of viral proteins, (5) assembly and release. To date, several drugs have been found that are effective in each of these stages (Table 20.3), but most have not been widely accepted by the scientific or medical community for use by the general public. A great deal of controversy surrounds these drugs due to their questionable selective properties and possible side effects. For example, the National Institute of Allergy and Infectious Diseases (NIAID) only recommended the use of amantadine for prevention and therapy of influenza type A after twenty years of inves-

TABLE 20.3
Chemotherapeutic antiviral drugs.

Drug	Action	Viruses Affected
Amantadine (Symmetrel®)	Inhibits penetration and uncoating; inhibits adsorption.	Influenza type A rubella, some oncogenic RNA viruses, myxoviruses.
Rifamycins	Inhibit reverse transcriptase and maturation.	Oncogenic RNA viruses, poxviruses.
Iododeoxyuridine (IUdR or IDU); Bromodeoxyuridine (BUdR); Fluorodeoxyuridine (FUdR) (Stoxil®)	Inhibit viral DNA synthesis; toxic to humans.	Herpesvirus.
Adenosine arabinoside (ARA-C; ARA-A) (Vira-A® opthalmic ointment)	Inhibits viral DNA replication.	All DNA viruses except polyomaviruses and adenovirus type 3.
Hydroxybenzyl-benzimidazole (HBB); Guanidine	Inhibit RNA replication and protein synthesis.	Rhinoviruses, Coxsackie, echoviruses, polioviruses, foot-and-mouth.
Thiosemicarbazone (IBT and mathisazone)	Inhibits protein synthesis.	Poxviruses.
Acyclovir (zovirax; acycloguanosine)	Activated by a herpesvirus enzyme and converted into a compound that inhibits viral DNA synthesis.	Herpesvirus.
Inosiplex (isoprinosine)	Enhances the body's immune system against a number of viruses, including HSV, flu, and some nasal viruses.	
Ribavirin (virazole)	Effective against influenza type A and others.	

tigation. Dr. Albert Sabin (developer of the polio vaccine of that name) has argued that this particular drug is not only ineffective but unsafe. The side effects of amantadine include insomnia, light-headedness, nervousness, and mental dulling. This cautious position is taken by many scientists with regard to the introduction of new drugs as a part of their ethical and moral responsibilities to the general public. If this drug receives wide acceptance, it may not only prove effective against influenza type A (provided it is given within 48 hours of the onset of symptoms) but may also give relief for those suffering from Parkinson's disease.

1 Name two types of vaccines used to prevent viral infections.
2 Describe two problems associated with vaccination procedures used to control viral infections.
3 Why is it so difficult to develop an antiviral antibiotic?

SUMMARY

The various cytopathic effects (CPE) seen in viral infections can result from alterations in both cell structure and function. These affect the membranes, metabolic products, hypersensitivity, cell fusion and clumping, and the formation of inclusion bodies. Localized infections result in CPE at or near the site of invasion, while generalized infection results in viremia, involvement of various internal organs, and localization of the virus in a target tissue. The symptoms of both types of infections are characteristic of a particular virus and are an important diagnostic aid. The limiting of infection and CPE are the result of immune mechanisms and the production of interferon. Interferon is a system of glycoproteins with the ability to inhibit viral replication, slow the growth of cancer cells, and enhance the action of phagocytes. The immune system also plays an important role in limiting viral infections. Sensitized T-lymphocytes release several chemical factors which destroy virus-infected cells and stimulate phagocytes into action. B-cells release immunoglobulins IgG, IgM, and IgA, which are able to combine with viral antigens and prevent or limit infection. The localized production of secretory IgA at the infection site prevents adsorption of many virus types, such as those of the influenza group. The presence of virus-specific antigen and antibody is used to diagnose infections and determine appropriate therapy. Many serological tests have been developed which are highly specific in their action. Prevention of viral infection is also made possible by vaccination with live, attenuated viruses or killed viruses. Although vaccination is not able to protect against all viral diseases, many have been successfully brought under control using this method. Recently, the development and use of antiviral chemotherapeutic drugs have also been explored. Although a few types are available, their use is limited because of the obligate intracellular nature of viruses and the subtle differences between host and parasite metabolic pathways.

WITH A LITTLE THOUGHT

Most people have very little knowledge of exactly what happens to them from the time they come in contact with a virus, until the time they have recovered from the disease. From the information in previous chapters and this one, give a scenario of "what happens."

STUDY QUESTIONS

1 What cellular and organismic cytopathic effects result from viral infections?
2 What is interferon? How might its function limit viral diseases?
3 What role is played by T-cells in limiting viral disease? B-cells?
4 Give three reasons why vaccination against viral infections may not be effective or possible.
5 Describe four serological tests used in the diagnosis of viral diseases.
6 During what stage of a viral infection are virions likely to be found in the blood?
7 What factors are necessary for a virus to successfully infect an individual cell? Host?
8 Why are inapparent infections important?
9 What is the heterophile antibody? Why is it important?
10 List three laboratory tests used in the diagnosis of viral diseases.

KEY TERMS

| cytopathic effect | interferon | heterophile antibody |

Viral Diseases I: The DNA Viruses

PARVOVIRUSES

PAPOVAVIRUSES

POXVIRUSES

HERPESVIRUSES
Herpes Simplex Viruses
Cytomegalovirus
Varicella-Zoster Virus
Epstein-Barr Virus

ADENOVIRUSES

Learning Objectives

☐ Know the major classes of DNA and RNA viruses.

☐ Be able to list the groups of DNA viral diseases and give examples of each.

☐ Be familiar with symptoms, mode of transmission, complications, and immunizations of representative DNA viral infections.

Viral diseases are classified according to the nature of their nucleic acid core into two major groups, the DNA viral diseases and the RNA viral diseases. Further subgrouping is based on morphological, immunological, and epidemiological differences. The DNA viral diseases described in this chapter are parvoviruses, papovaviruses, poxviruses, herpesviruses, and adenoviruses.

The classification scheme for this information is functional and subject to constant change, as new information is made available through more detailed and advanced research efforts.

PARVOVIRUSES

These helical, single-stranded viruses have not been recognized as human pathogens. The minute virus of mice is a parvovirus that is able to replicate in its host; however, others in this group are defective viruses and require coinfection with an adenovirus for replication.

One of the most troublesome members of the parvovirus group to evidence itself in recent years is a pathogen of dogs known as canine parvovirus. This disease was first identified in Louisville, Kentucky, at a dog show in the spring of 1978. The disease spread rapidly throughout the country and has become epidemic in proportion. The symptoms of infection are most severe in puppies and include loss of appetite, followed by severe diarrhea or a loose stool that can vary in color from chalky gray to a dark excretion that contains variable amounts of blood. The animals become rapidly dehydrated in 24 to 48 hours, after which they may die due to electrolyte imbalance. The canine parvovirus is spread easily from one dog to another through feces. Humans can transmit the viruses to healthy animals from fecal contami-

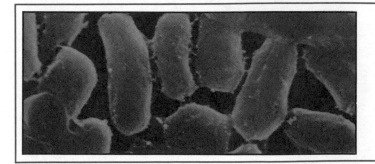

nation of shoes or clothing. Treatment of infected animals is by intravenous feeding and methods which will maintain electrolyte balance. By treating the symptoms effectively for three or four days, the dogs usually recover. A vaccine is available for use and effective in establishing a high antibody titer against the viral antigens.

PAPOVAVIRUSES

Some members of this group are regularly used in the study of transformation. Polyomavirus of mice and simian virus 40 have a wide distribution in nature but are not known to naturally produce tumor transformation, except under laboratory conditions. The one member of this group of human importance is the human wart virus. This virus causes a benign outgrowth of the skin typically occurring on the feet, hands, and more rarely on other parts of the body. Although painful and time-consuming to remove, no significant harm results from the growth of the infected Malpighian layer of the skin. However, infectious wart virus has been found to be the cause of genital warts, which may become malignant. The incubation period for this papilloma (branching, benign tumor of the skin) virus is several months and results from infection through breaks in the epithelium.

POXVIRUSES

Viruses of this group are responsible for smallpox, vaccinia, cowpox, monkeypox, mousepox, and infectious viral diseases of rabbits. The name "pox" is derived from the typical, eruptive pustular lesion formed during the course of infection. Many poxviruses produce recognizably different symptoms but are immunologically related; they are among the largest of the animal viruses. The most significant human poxvirus is smallpox (variola), which has been a scourge throughout the world since ancient times. The virus is acquired by close contact with infected individuals or contaminated articles such as linens, towels, and clothing. Susceptibility is universal, and once established in a population the disease is responsible for a high mortality rate. The two forms of the disease are variola major (malignant smallpox or blackpox) and variola minor (Cuban itch, cottonpox, or paravariola). These two forms follow the same course and differ only in severity. Having entered through mucous membranes of the upper respiratory tract, the virus moves passively through the lymph system and becomes a primary viremia once it enters the blood. Smallpox virus invades the viscera (e.g., spleen, liver), where it multiplies during its 7–17 day incubation period. The release of replicated virions into the blood produces a secondary viremia with the following typical symptoms: chills, fever, vomiting, and the development of pox on the skin, mucous membranes, and viscera. The pox begin and are most predominant on the face and extremities (a centrifigal distribution) but later form on the trunk. In cases of variola major, the pox changes from a macular (thickened, stained spot) lesion to a pustular lesion, causing hemorrhaging into the surrounding tissue. Pox can fuse with one another to form large hemorrhagic lesions, resulting in severe necrosis, becoming fatal 5 to 7 days after their appearance. The mortality rate of variola major is between 25 and 40 percent. By comparison, the less severe variola minor has a fatality rate of only 1 percent.

Immunity resulting from infection is strong and long-lasting; seldom have second cases of smallpox been reported. The severe scarring and high mortality rate of smallpox led to many disease control attempts (refer to chapters 13 and 15). Previously known as variolation (to prevent variola), current vaccination practices have been developed on the basis of a close immunological relationship among variola, vaccinia, and cowpox viruses. Smallpox, vaccinia, and cowpox are all pox diseases having cross-reacting antigens that stimulate the formation of protective antibodies. Cowpox virus was originally used to provide active acquired immunity; however, this virus was replaced by vaccinia, since it is more closely related to smallpox and results in less severe side effects. As a result of the worldwide immunization campaign, smallpox has been eliminated as a human infectious disease (Figure 21.1). The only remaining virus is contained in a limited number of research laboratories. Whether these cultures will be destroyed is a question of major importance. The cultures are maintained for experimentation and as a source of attenuated vaccines. However, their escape could result in an epidemic of major proportions, since immunization programs in many countries have been discontinued for several years and the level of herd immunity may have dropped to unprotective levels. One escape occurred from a London laboratory (1978) resulting in two deaths, and the protective quarantine and vaccination of an additional 300 people.

---

1 Give an example of a parvovirus disease.
2 What is the portal of entry of the poxviruses?
3 Why was the vaccinia poxvirus used instead of cowpox virus as vaccination against smallpox?

HERPESVIRUSES

Included in this group of DNA viruses are a number of pathogens able to infect many animals, including humans. The word "herpes" is derived from a greek word meaning *to creep*. It was originally applied to what we now call herpes, as well as to shingles. It is really more applicable to shingles,

CENTER FOR DISEASE CONTROL

MMWR

MORBIDITY AND MORTALITY WEEKLY REPORT

October 26, 1979 / Vol. 28 / No. 42

International Notes
497 Smallpox Certification — East Africa
500 Pneumococcal Meningitis — Australia
Epidemiologic Notes and Reports
498 Shigellosis in a Children's Hospital — Pennsylvania
Current Trends
505 Urban Rat Control — United States, April-June 1979

International Notes

Smallpox Certification — East Africa

Two years ago today the world's last known patient with endemic smallpox had onset of rash (Figure 1). Ali Maow Maalin, a cook at the district hospital in Merka, Somalia, developed smallpox on October 26, 1977. Since then, intensive surveillance has failed to identify any additional cases of naturally transmitted smallpox.*

Separate International Commissions have been assessing campaigns in the last 4 countries requiring certification of eradication: Somalia, Kenya, Ethiopia, and Djibouti. Today in Nairobi, the chairpersons of these commissions will make their reports to the Director-General of the World Health Organization. It is anticipated that the Horn of Africa will be certified to be smallpox free.

This meeting will complete the documentation required to certify global eradication of smallpox. This documentation will be reviewed by the Global Commission for Smallpox Eradication from December 6-9, 1979, in Geneva. If the criteria for global eradication have been met, documentation will be presented to the World Health Assembly in May 1980 for final global certification of smallpox eradication.

Reported by the Bureau of Smallpox Eradication, CDC.

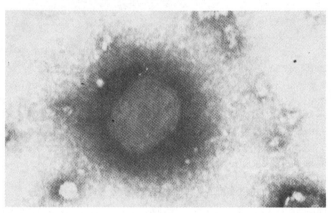

FIGURE 1. Variola minor virus isolate from world's last endemic case, Merka, Somalia, 1977. Magnification, 115,000 X.

*Two cases of laboratory-acquired smallpox occurred in Birmingham, United Kingdom, on August 13 and September 8, 1978, but the World Health Organization's Global Commission for Smallpox Eradication has determined that these cases should not alter the plans for certification of naturally transmitted smallpox.

U.S. DEPARTMENT OF HEALTH, EDUCATION, AND WELFARE / PUBLIC HEALTH SERVICE

FIGURE 21.1
A cause for celebration. (Courtesy of the Center for Disease Control, Atlanta.)

in which the blisterlike sores move along in successive waves. The four forms responsible for human disease are herpes simplex virus (e.g., cold sores), cytomegalovirus (e.g., salivary gland disease), varicella-zoster (e.g. chickenpox and shingles), and Epstein-Barr virus (e.g., infectious mononucleosis).

HERPES SIMPLEX VIRUSES

The herpes simplex group (also known as HSV, herpes labialis, herpes genitalis, and herpes fibrilis) is comprised of many polyhedral viruses subdivided into the two major groups HSV-I and HSV-II. The major differences between these immunologically distinct viruses are outlined in Table 21.1. These viruses are transmitted by oral-respiratory and urogenital (venereal) routes, and symptoms follow a general pattern of **primary** and **recurrent** infections. Primary infection is first noted as a burning sensation to a localized cluster of nodules. They progressively fuse to form a large, thin-walled vesicle filled with fluid, virions, and epithelial cells. The vesicle ruptures and heals without scarring. The infection then becomes latent but may recur when the "dormant" viruses are activated by changing environmental factors such as variations in ultraviolet light (sunbathing), hormones (menstrual cycle or birth control pills), or metabolic upset due to other viral or bacterial infections. Since there are no cross-reacting antigens among the herpesviruses, infection with one form does not provide immunity to others. Primary infection usually is an HSV-I cold sore on the lip but may involve adjacent mucous membranes, and on rare occasions, the tonsils. If neutralizing antibodies are not produced, the virus can spread through the body and result in other HSV-I diseases that may be fatal. However, this is a rare occurrence because of the presence of protective maternal antibodies and a high incidence of subclinical infections in adults. Primary HSV-II infections typically occur at the time of puberty, but they are also found in infants born to infected mothers.

HSV types I and II have both been implicated in skin diseases resulting from traumatic damage. Athletes, dentists, physicians, and nurses who suffer skin damage and breaks can experience traumatic herpes (herpes Whitlow) of the skin seen as tender, inflamed vesicles that fuse to form typical eruptive vesicles. They are often painful, inducing fever and malaise which can last as long as two or three weeks. As noted in Chapter 19, HSV-II is strongly suspected of being an oncogenic virus and might be responsible for certain forms of cervical cancer.

Treatment of HSV infections is recommended only in cases involving infection of slowly dividing host tissues, such as in keratoconjunctivitis. This is made necessary by the fact that idoxuridine (IUdR) and arabinoadenosine (ARA-A) can interfere with both viral DNA and host DNA, causing severe side effects. Severe systemic and neonatal HSV infections have been successfully treated with ARA-A, but extreme caution must be exercised in its use in order to prevent damage to blood-forming organs. Acyclovir is also approved for use against initial infections of herpes genitalis. The drug is used to control localized herpes simplex infections for both genital and labial herpes in patients whose natural defenses are impaired and unable to control the spread of the infection. This drug reduces the length of time that live virions are present in the vesicles. This viral shedding in the lesion allows the virus to be transferred from one person to another through sexual contact, and the disease is highly contagious when sores are present. Another control

TABLE 21.1
A comparison of herpes simplex viruses.

	Location	Heat Sensitivity	Culture Characteristics	Virulence in Neurons	IUdR Resistance	Diseases
HSV-I	Found primarily in nongenital areas.	Less temperature sensitive.	Forms small poxs on embryonated egg membranes.	Less virulent in neurons.	Less resistant to this antiviral drug in cell culture.	Acute herpetic gingivostomatis (Vincent's stomatitis) Eczema herpeticum Keratoconjunctivitis Meningoencephalitis Herpes labialis (cold sores)
HSV-II	Primarily genital or genital in origin.	More temperature sensitive.	Forms large poxs on embryonated egg membranes.	More virulent in neurons.	More resistant to this antiviral drug in cell culture.	Genital herpes (herpes progenitalis) Neonatal herpes

attempt centers on the use of homologous vaccines made from attenuated HSV along with heterologous vaccines prepared from attenuated forms of viruses with cross-reacting antigens. For example, heterologous vaccination has been attempted with smallpox and polio vaccines. These have been given to patients with severe recurrent HSV infections such as cold sores, but they have benefited only a few individuals by reducing the recurrence rate. The value of this procedure has been questioned from several points: (1) the population already demonstrates a high HSV antibody titer; (2) recurrent infections appear in the presence of high antibody titers; (3) those with high antibody titers demonstrate the greatest recurrence rate; and (4) HSV-II has suspected oncogenic properties that could possibly be demonstrated under certain circumstances.

CYTOMEGALOVIRUS

Cytomegalic inclusion disease (CID) occurs most frequently in infants and is a viral disease of the salivary glands and other tissues. It is caused by the intrauterine (congenital) or postnatal transmission of cytomegalovirus (CMV). Although rare, CID may also occur in adults receiving immunosuppressive therapy. Like other herpesviruses, CMV infections result in vesicular eruption of host tissues. The virus is not only found in the salivary glands but also the kidneys, liver, brain, lungs, and eyes. Tissue destruction can be fatal or result in severe brain damage, blindness, deafness, and heart defects. CMV is believed to be able to enter a latent period, as do other herpesviruses, resulting in a delay in the onset of symptoms of up to two years. At present, there is no treatment for this disease. A vaccine of attenuated CMV has been developed, but its value in preventing infection is questionable.

VARICELLA-ZOSTER VIRUS

This is a single virus responsible for two human diseases, varicella (chickenpox) and zoster (shingles). At one time, these skin diseases were thought to be caused by two different viruses, but isolation, electron microscopic analysis, and immunological studies have demonstrated that a zoster infection is a reactivated, latent varicella infection.

Varicella (chickenpox) is a highly contagious, common skin disease occurring as primary infection of children and, more rarely, adults. Approximately 75 percent of all children under the age of 15 have experienced this virus acquired by exposure to infected individuals. The most likely transmission route is by direct contact, droplet nuclei, or contact with contaminated articles. The disease is communicable five days before the eruption of pox, but not more than six days after the last vesicles appear. Scabs do not contain the virus and are not infectious. Once inside the body, there is an incubation period of two to three weeks before the development of a maculopapular rash, which changes to a vesicular form in three or four days (Figure 21.2). The vesicles become pustular, erupt, and scab leaving no scars. Lesions typically occur on the trunk but are also found in the mouth, ears, axilla, vagina, and conjunctiva. The pox usually develop in successive "crops," but infections may be either mild (only a few pox) or inapparent (no pox formed). Children with immunodeficiency diseases or those receiving immunosuppressive drugs are likely to experience more severe symptoms and complications such as varicella pneumonia or encephalitis. Another complication receiving greater attention is **Reye's syndrome.** This is an acute encephalopathy with severe fatty infiltration and degeneration of the liver, kidneys, and other organs. The symptoms appear two or three days after seeming recovery from the viral infection. The child experiences a low-grade fever, repeated vomiting,

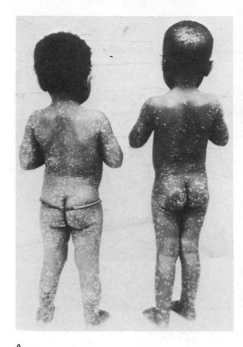
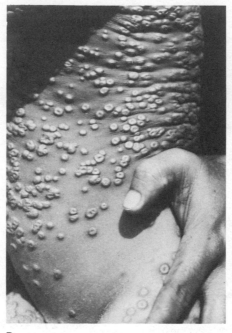
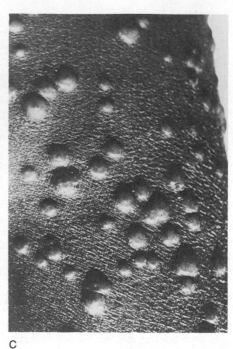

A B C

FIGURE 21.2

(A) Smallpox, twenty-first day of the rash. (B) Frontal view of the pox, taken in Bangladesh, 1973. (C) Smallpox pustules: round, deep, and all in the same state. (Photos A and B courtesy of the Center for Disease Control, Atlanta. Photo C courtesy of the World Health Organization, Diagnosis of Smallpox Slide Series, 1968.)

and quickly progresses from mental disorientation and delirium to a comatose state. Although the reactions are reversible, the fatality rate is as high as 50 percent. Reye's syndrome has not only been associated with varicella-zoster virus but also influenza type B infections, common colds, and gastrointestinal upset. No doubt the increase in cases is the result of better reporting and diagnosis, and not an actual increase in the incidence of the disease.

> 1 List four diseases caused by herpesviruses.
> 2 Name two environmental factors which can induce an active herpes infection.
> 3 What is the relationship between chickenpox and shingles?

Varicella infections are usually self-limiting and pass without complications. However, during the course of the disease, the virus establishes a latent infection that recurs in adulthood as zoster (shingles) (Figure 21.3). Shingles typically occurs in individuals who have experienced apparent or inapparent varicella disease as children, but more rarely it can develop as a primary infection in adults or children. Zoster may be initiated by a lowered immune state, heavy

metal poisoning, or injury. These factors are likely to occur and be more severe in the elderly. The disease begins with fever and malaise, followed by skin pain and itching of the

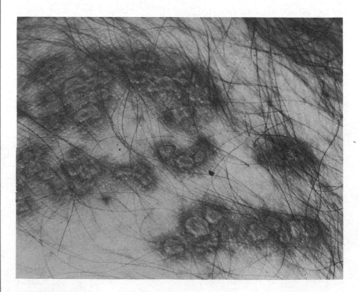

FIGURE 21.3

Shingles (herpes-zoster) lesions on the leg caused by varicella-zoster virus. (Courtesy of the National Institute of Health.)

affected region. In most cases, zoster appears on the torso and neck, but the throat, eyes, and mouth may also be involved. Within a few days, vesicular eruptions appear on the skin, which later heal with scabbing. Although both result in vesicle formation, chickenpox and shingles do not resemble one another symptomatically (Table 21.2). Since the majority of varicella-zoster infections are mild and self-limiting, there is no need for treatment other than aspirin and calamine lotion to relieve symptoms. In more severe cases involving patients with depressed immunological conditions, ARA-A has been of some value. Zoster immuno-globulin (ZIG) prepared from the plasma of convalescent patients is also available on a limited basis. ZIG is only used in the case of high-risk children with conditions such as leukemia and lymphoma. No chickenpox vaccine is yet available; however, an experimental, live, attenuated varicella vaccine has been developed and is being field tested.

EPSTEIN-BARR VIRUS

Early EBV infection of children results in a mild, almost unrecognizable, case of infectious mononucleosis commonly called "mono." However, individuals between the ages of 14 and 30 who have not previously been exposed to the virus experience more severe symptoms and a prolonged disabling illness. Symptoms include nausea, vomiting, fever, loss of appetite, sore throat, fatigue, chills, sweating, and most typically, tender, enlarged cervical lymph nodes. Swollen nodes may also be present in the axilla and groin. The liver and spleen become enlarged and susceptible to rupture. Infectious mononucleosis is considered by many to be a self-limiting form of leukemia, since during the course of the disease there is a marked increase in the white blood cell count (i.e., B-cells). The name of the disease stems from this noted increase. It is also known as the "kissing disease," since it is thought to be transmitted by the oralpharyngeal route. The virus may be present in oral excretions and the blood for up to a year following infection, but the virus does not appear to be highly infectious. Diagnosis of the disease is made from blood smears and serological tests. EBV has also been strongly implicated as the cause of Burkitt's lymphoma and nasopharyngeal cancer (refer to Chapter 19).

Infectious mononucleosis occurs on a worldwide basis, but at this time, no preventive measures are available. Treatment involves the use of aspirin as well as other medications that relieve symptoms and provide support to the patient.

ADENOVIRUSES

The adenoviruses consist of animal and human viruses. The animal forms have been found in mice, birds, cattle, swine, primates, and dogs. The group name stems from the fact that these viruses are commonly found as latent infections of the adenoid and tonsillar tissues. Thirty-three antigenic forms are responsible for the three human diseases **acute respiratory disease, epidemic keratoconjunctivitis,** and **pharyngoconjunctival fever.** Adenoviruses are easily isolated from tonsillar and adenoidal tissue of most children and adults, where they have become located following droplet nuclei infection. Upon entering the lytic cycle, antigenic types 1, 2, and 5 cause acute respiratory disease of the mucous membranes and display fever, pharyngitis, and cough lasting about five days. Infection results in a high specific serum antibody titer. Immunization with live, attenuated adenoviruses of these serotypes has been used in the military. The ingestible vaccine stimulates a mild intestinal infection but induces the formation of protective antibodies. Adenoviruses types 3 and 7 are responsible for pharyngoconjunctival fever characterized by fever, conjunctivitis, pharyngitis, fatigue, and swollen cervical nodes. No corneal damage results from this short-lived illness. Types 18 and 19 cause epidemic keratoconjunctivitis, an illness similar to pharyngoconjunctival fever but also resulting in enlargement of the auricular (behind the ear) lymph nodes and clouding of the cornea. The vision impairment can last for as long as two years. An epidemic of this infection occurred among ship-

---

1 What are the symptoms of Reye's syndrome?
2 In what age group do most cases of the "kissing disease" occur?
3 Name three diseases caused by adenoviruses.

TABLE 21.2
A comparison between varicella and zoster.

	Varicella	Zoster
Age	Childhood	Adult
Epidemiology	Late winter, early spring epidemic	No seasonal variation
Contagiousness	Highly contagious	Less contagious
Spread	Hematogenous	Via nerves
Host	Normal host	Compromised host
General characteristics	Primary infection, no burning	Past history of varicella, burning common

yard workers on the West Coast of the United States in 1941 and 1942. This outbreak led to the disease becoming known as "shipyard eye."

SUMMARY

The viruses are divided into two major groups, those containing DNA and those containing RNA. Within the DNA viruses, five subgroups have been described in this chapter: parvoviruses, papovaviruses, poxviruses, herpesviruses, and adenoviruses. These microbes and the diseases they cause occur throughout the world. As the science of virology ex-

pands, our understanding of disease processes increases. Through this research, great strides have been made in the control and even elimination of what were once unknown and unseen microbes.

WITH A LITTLE THOUGHT

Several of the viral diseases presented in this chapter demonstrate similar symptoms (for example, chickenpox, measles, and rubella). Describe the different features of these diseases according to nature of the virus, symptoms, treatment, and immunity developed.

STUDY QUESTIONS

1 What are the major differences between chickenpox and shingles? Between herpesviruses I and II?

2 What methods and materials were necessary to erradicate smallpox?

3 List two diseases caused by herpesvirus I. How are they transmitted?

4 Why are viruses sensitive to ultraviolet light?

5 What factors make the use of vaccines against herpes of limited value?

6 Describe a "typical" course of infection for chickenpox.

7 What is Reye's syndrome? How might it be induced?

8 Why is the common name "the kissing disease" a misnomer for mononucleosis?

9 What is the proposed relationship between Epstein-Barr virus and cancer?

10 List two diseases caused by the adenoviruses.

SUGGESTED READINGS

Refer to the Suggested Readings at the end of Chapter 22.

KEY TERMS

| papilloma (pap″i-lo′mah) | pharyngitis (far″in-ji′tis) | conjunctivitis |

Viral Diseases II: The RNA Viruses

PICORNAVIRUSES
Polioviruses
Coxsackie Viruses
Rhinoviruses
RHABDOVIRUSES
HEPATITIS VIRUSES
PARAMYXOVIRUSES
Rubella Virus
ARBOVIRUSES
Togaviruses
ORTHOMYXOVIRUSES

Learning Objectives

☐ Be able to list the groups of RNA viral diseases and give examples of each.

☐ Be familiar with symptoms, mode of transmission, complications, and immunizations of representative RNA viral infections.

This is a continuation of Chapter 21 and describes an additional six major groups of viruses. These groups are subdivisions of the RNA viruses and include picornaviruses, rhabdoviruses, hepatitis viruses, paramyxoviruses, arboviruses, and orthomyxoviruses.

PICORNAVIRUSES

These are small (pico), RNA-containing viruses (pico-RNA-viruses) with a polyhedral morphology. The group is subdivided into the enteroviruses and the rhinoviruses (Table 22.1). One of the oldest and best-known members of the enteroviruses is poliovirus.

POLIOVIRUSES

Poliomyelitis has been found throughout the world since ancient times and occurs only in humans. Most infections are asymptomatic, but clinical cases result in both nonparalytic and paralytic disease. Although commonly called "infantile paralysis," the symptoms are more severe in adults than in children. The more common nonparalytic, or abortive, poliomyelitis ("summer flu") is acquired through the oral-respiratory route and causes fever, constipation, headache, nausea, vomiting, and stiffness of the back and neck. Diagnosis of the abortive form is made by serological tests and by identification of the virus shed in the feces, after its replication in Peyer's patches and other lymphoid tissues. If the virus enters the circulatory system, an asymptomatic viremia will occur that can progress to paralytic poliomyelitis. Symptoms begin after the virus has infected and lysed motor neurons of the central nervous system. The body areas that develop flaccid paralysis depend on which regional nerves are infected. Paralysis reaches a maximum involvement in

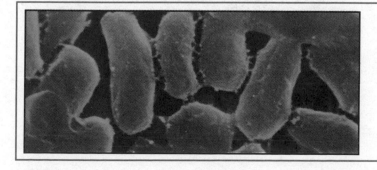

22

TABLE 22.1
Picornaviruses

Virus	Important Characteristics	Virus Subgroup	Disease
Enterovirus	Resistant to pH 5.5: relatively heat-stable	Polioviruses	Causes poliomyelitis in humans and 3 serotypes.
	Transient flora of the GI tract and nose	Coxsackie viruses	Variety of illnesses produced, including herpangina, febrile pharyngitis, epidemic pleurodynia, aseptic meningitis, and myocarditis. Two groups: A (23 serotypes) and B (6 serotypes).
		Echoviruses	Most serious disease produced is meningitis. Enterovirus 70 causes acute hemorrhagic conjunctivitis. 33 serotypes of echoviruses: 4 serotypes of unspecified enteroviruses.
		Encephalomyo-carditis virus	Produces meningoen-cephalitis in humans: endocarditis and myelitis in monkeys. Rodents are probably the natural host.
Rhinovirus	Sensitive to pH 5.5: relatively heat-stable	Rhinoviruses	Viruses responsible for most common colds. 90 or more serotypes.
	Transient flora of the nose and throat	Foot and mouth disease virus	Causes foot and mouth disease in cattle, swine, sheep, and goats: rarely infects humans. 7 serotypes.

SOURCE: Modified from B.A. Freeman, *Burrows Textbook of Microbiology,* 21st ed., W.B. Saunders Co., 1979.

about three days. If the infection progresses to the spinal cord and brain stem, paralysis may fatally involve motor neurons of the respiratory system (Figure 22.1).

The presence of poliovirus in fecal material of clinical and subclinical patients led to the belief that the disease could be transmitted through contaminated water. During the early 1950s, epidemiologists in the United States closed many public pools and swimming areas because they were believed to be sources of poliovirus. However, more current research indicates that the virus is transmitted by direct contact with pharyngeal secretions and feces of infected individuals. Epidemics of polio are correlated with socioeconomic conditions. In underdeveloped regions of the world, polio is endemic and only rarely epidemic, due to the ease and frequency of virus transfer to infants undergoing a transition from maternal antibody protection to a state of naturally acquired immunity. These infants develop only mild symptoms and a strong, long-lasting immunity. In more developed countries, epidemic polio appears in older children and adults, because they have not experienced the early and more mild, subclinical infections. Control of epidemic polio has been achieved through the use of the Salk and Sabin vaccines developed in the United States during the mid-1950s (Figure 22.2). The original vaccines were monovalent (containing only one of the three attenuated antigenic types), but these have been replaced by a trivalent vaccine to ensure the development of an artifically acquired immunity to all three viruses. Currently no treatment of the active disease is available, since antibodies are present during the clinical phase.

FIGURE 22.1

This equipment was commonly used to help polio victims regain neuro-muscular control. Today, few people would recognize an "iron lung." (Courtesy of the Center for Disease Control, Atlanta.)

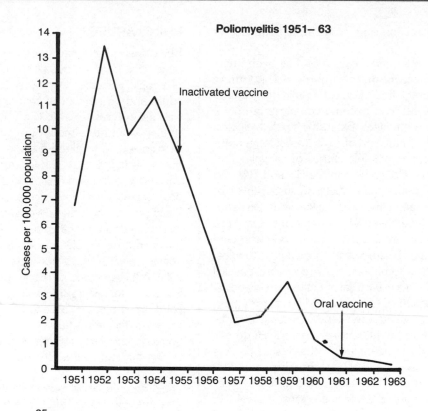

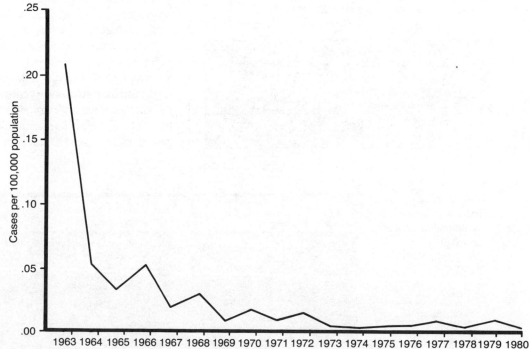

FIGURE 22.2
Reported case rates of poliomyelitis per 100,000 population in the United States. Note that the bottom graph uses a much smaller vertical scale to reflect the drastic reduction in the incidence of polio following the introduction of the oral vaccine. (Courtesy of the Center for Disease Control, Atlanta.)

COXSACKIE VIRUSES

This antigenically diverse group of picornaviruses is subdivided into serotypes A and B. Group A Coxsackie viruses (named for Coxsackie, New York, where they were first iso-

lated) consist of 23 serotypes able to affect striated muscle, causing weakness and a flaccid paralysis resembling poliomyelitis. There are only six serotypes in group B, which primarily affects nervous tissue but can also cause necrosis of the pancreas and liver. Diseases caused by both groups A and B are of short duration and occur most frequently in children.

Recently a Coxsackie virus called B4 has been implicated as a cause of **juvenile-onset diabetes.** This form of diabetes begins abruptly and requires insulin for control. Children with the disease can become comatose in only a few days and die if untreated. Extensive epidemiological evidence and laboratory infection of mice implicated a number of common viruses as possible causes of juvenile-onset diabetes. Mumps virus, Coxsackie viruses B3 and B4, and reovirus type 3 have all induced diabetes in susceptible laboratory mice. Confirmation that Coxsackie virus B4 is an etiological agent of juvenile-onset diabetes resulted from analysis of a B4 serotype from a ten-year-old boy who died of the disease. The viral specimen was inoculated into mice, monkey, and human cell cultures, then recovered and injected into susceptible mice who later developed diabetes. Isolation and identification of Coxsackie B4 virus from these mice fulfilled Koch's postulates and established this virus as an etiological agent. Factors influencing susceptibility to this type of infection and the exact nature of the disease process have not as yet been determined. It is clear, however, that not everyone is susceptible. Research is also being conducted in the development of a vaccine against this virus in hopes of reducing the incidence of infection.

---

1 Name the two clinical forms of a poliovirus infection.
2 What is the transmission route of poliovirus?
3 What is the relationship between Coxsackie virus B4 and juvenile-onset diabetes?

RHINOVIRUSES

Rhinoviruses have been isolated more frequently than any other virus from the nose ("rhino" means *nose*) and throats of people with the "common cold." About 70 percent of adults and 35 percent of children with "colds" are infected with rhinoviruses. Over 100 of these RNA viruses have been serotyped and are known to cause edema, excessive nasal secretions, and localized tissue sloughing (desquamation) of the submucosa and surface epithelium. The viruses are transmitted through droplet nuclei, pharyngeal secretions and nasal secretions. Habits such as touching the face and nose, biting pencils and other objects, and failing to wash contaminated hands are probably more important transmission routes than airborne droplet nuclei. It has also been found that "getting a chill" does not increase an individual's susceptibility to these viruses or result in disease. However, it is very important to restrict the activity of children with common colds, since they are able to transmit these viruses to others more easily than adults. On an average, households with younger school-age children experience twice as many colds during the early fall and winter compared to other families (Box 26).

The incubation period of the rhinoviruses is between two and four days, after which the illness can last up to seven days. The all too familiar symptoms include headache, bodyache, "watery eyes," chilling, sneezing, malaise, excessive nasal secretions, and cough. The nasopharynx becomes swollen, reddened, and tender; coughing may persist for two to three weeks following recovery. The seasonally high morbidity rate and great economic loss associated with this disease have encouraged research into possible prevention and control of the common cold. However, several apparent inconsistencies have been revealed: (1) A natural,

---

TABLE 22.2
Routine poliomyelitis immunization schedule summary.

Dose	OPV* age/interval	IPV** age/interval
Primary 1	Initial visit, preferably 6–12 weeks of age	Initial visit, preferably 6–12 weeks of age
Primary 2	Interval of 6–8 weeks	Interval of 4–8 weeks
Primary 3	Interval of ≥ 6 weeks, customarily 8–12 months	Interval of 4–8 weeks
Primary 4		Interval of 6–12 months
Supplementary	4–6 years of age† (school entry)	4–6 years of age† (school entry)
Additional supplementary		Interval of every 5 years††

*OPV: oral polio vaccine
**IPV: inactivated polio vaccine
†If the third primary dose of OPV is administered on or after the fourth birthday, a fourth (supplementary) dose is not required. If the fourth primary dose of IPV is administered on or after the fourth birthday, a fifth (supplementary) dose is not required at school entry.
††Supplementary doses are recommended every 5 years after the last dose until the 18th birthday or unless a complete primary series of OPV has been completed.
SOURCE: Center for Disease Control, Atlanta.

BOX 26

THE
COMMON
COLD—
A
NEW
PERSPECTIVE

WHILE research has not as yet found a cure for the common cold, a great deal of information has been learned that contradicts many old ideas and gives new insights into this problem. Epidemiologists have discovered that the common cold moves through the population three times each year. The first "cold season" occurs in September when children reenter school. The second epidemic typically appears in midwinter and the third in spring. Not only is there a "cold season" but research has also revealed a "cold day"! More people develop symptoms on Monday than on the other day. Fewer cases occur on Tuesday and Wednesday, and the lowest number for the week is on Thursday. After Thursday, cases begin to increase to their high point on the following Monday.

The phrase "common cold" is a good one because colds occur in all people at all ages. Infants have a cold (although in most cases they are minor) almost every two months. Children under the age of four have an average of eight colds each year while adults suffer from about two. Since the common cold is not a reportable disease, there is no way to accurately determine the total number of people who have colds each year, but it has been estimated that the number is probably in the billions!

Colds also show a sex preference related to age. Boys under three years of age get more colds than girls. However, there is a reversal after the age of three, when girls suffer more often than boys.

It has also been learned that people who travel "catch cold" more than those who are less mobile. This is no doubt related to the fact that they become exposed to a greater variety of cold viruses during their travels. Statistics show that smokers do not catch colds more than nonsmokers, but the fact that smoking already irritates the upper respiratory tract makes the "smoker's cough" more pronounced and longer in duration.

Another phrase commonly used is to "catch cold." While it was thought for years that the most likely transmission route for cold viruses was through droplet nuclei, more recent studies have demonstrated that the viruses are most effectively transmitted by hand contact. The entry points are the nose and eyes. The simple, unconscious actions of touching the nose or rubbing an eye will transfer viruses to the upper respiratory tract. Viruses placed in the eye move to the nose through the tear duct. The most effective way to reduce the likelihood of transmission is to wash hands frequently and keep them away from the face.

short-lived immunity exists, but persons with circulating serum antibodies are more likely to be infected by the sensitizing virus than those lacking these antibodies; (2) people in isolated areas "catch" more severe colds that last longer than those in areas where exposure occurs regularly; (3) older adults experience fewer colds than younger people; (4) a person rarely becomes reinfected with the same serotype; and (5) resistance is apparently not related to serum antibody levels. These observations can likely be explained by the activity and level of secretory IgA found in nasal secretions. These antibodies are produced locally after antigenic stimulation and do not persist for long periods. Their production is also independent of serum IgA levels. The possibility of developing a common cold vaccine based on secretory IgA protection is currently being explored. Although there is reason to believe an effective vaccine is possible, several problem areas have been identified. Production of an effective vaccine is made difficult by the fact that the rhinoviruses are difficult to grow in culture, secretory IgA only provides short-term immunity, and the variety of serotypes causing colds makes a multivalent vaccine costly to produce. However, there is a possibility that a vaccine could reduce the incidence of colds caused by certain serotypes, if they tend to be found more frequently in localized geographical areas. The naturally occurring antiviral agent interferon is also being explored as a common cold preventative. Research has indicated that the administration of large amounts of interferon in a nasal spray limits infection. Since this form of treatment is very costly, an alternative to the direct use of interferon is also being explored. When the interferon-inducing agent propanediamine was experimentally used in a nosedrop preparation with patients intentionally exposed to rhinoviruses, not a single person in this study caught a cold. There is a significant drawback to this method—the propanediamine must be used before exposure to the virus, which would be very difficult.

> 1   What are the most common symptoms of a rhinovirus infection?
> 2   Explain the relationship among serum IgA, secretory IgA, and the common cold.

RHABDOVIRUSES

Rabies is known worldwide as a viral disease of major human importance, but there are others in this group responsible for **vesicular stomatitis,** a gum disease, (Figure 22.3), and several other diseases of plants, animals, and birds. The fatal nature of rabies has been recorded throughout history. Writing by Aristotle (4 B.C.) and many others warned of the severe consequences of being bitten by animals. The term

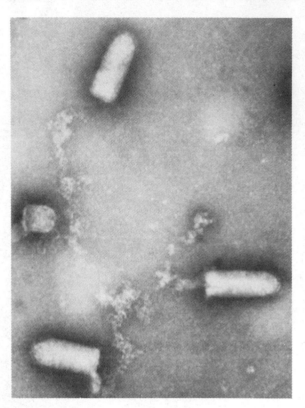

FIGURE 22.3
These bullet-shaped virions cause vesicular stomatitis virus, a mild disease in humans but more severe in some animals. (Courtesy of J. Holland, University of California, San Diego.)

"rabies" derives from the latin *rabere,* which means *to rage.* Cases of suicide have been documented involving people bitten by rabid animals who wished to avoid the frenzied trauma they knew would precede their death. Although it was well known in the 1800s that the disease was transmitted in the saliva of rabid animals, it was not until modern research techniques that the actual etiological agent was identified as a virus. Rabies is a bullet-shaped RNA virus with spikes and an envelope derived from its host as the virus is pinched off from the cell. Several closely related but antigenically different forms have been identified. The main reservoirs of infection are wild animals and domesticated dogs. In the United States, more than 50 percent of wild animal rabies is found in the striped skunk, while insectivorous bats account for an additional 15 percent of the cases. In South America, the vampire bat has a reputation for transmitting rabies while taking a blood meal from sleeping animals. In other parts of the world, the fox, wolf, wild dog, jackal, and mongoose serve as a constant source of rabies virus.

Virions enter a susceptible host through breaks in the skin or open wounds caused by animal bites, scratches, or abrasions. One rare, confirmed case of human rabies was acquired through the oral-respiratory route from airborne virions in a cave heavily populated with bats. A bite from a rabid animal does not necessarily mean that a clinical case of the disease is inevitable. In fact, less than 50 percent of humans and dogs bitten by rabid animals develop this highly fatal disease. Factors which influence transmission are (1) the number of virions in the wound, (2) the severity of the wound, (3) the amount of clothing covering the injured area, and (4) the cleaning of the wound. If effective transmission occurs and clinical symptoms appear, the disease is almost 100 percent fatal. To date, only two individuals are known to have recovered after intensive care was provided.

The two clinical forms of the disease generally seen in animals and humans are **furious rabies** and **dumb rabies.** The traumatic course of furious rabies was described by the Italian physician Girolamo Fracastor (1546):

> Its incubation (following a bite by a rabid animal) is so stealthy, slow and gradual that the infection is very rarely manifest before the 20th day, and in most cases after the 30th, and in many cases not until four or six months have elapsed. There are cases recorded in which it became manifest a year after the bite! Once the disease takes hold, the patient can neither stand nor lie down; like a madman he flings himself hither and thither, tears his flesh with his hands and feels intolerable thirst. This is the most distressing symptom, for he so shrinks from water and all liquids that he would rather die than drink or be brought near to water; it is then that they bite other persons, foam at the mouth, their eyes look twisted, and finally they are exhausted and painfully breathe their last.

Clinically, victims of furious rabies display fever, headache, a tingling sensation around the wound, malaise, and nausea. This is followed by alternating periods of depression and excitability. As the disease progresses, the periods of heightened sensitivity and frenzy increase, and the patient experiences painful spasmodic contractions of throat muscles. Difficulty in swallowing leads to drooling (foaming at the mouth), and slight noises or changes in light cause exaggerated reactions. Eventually, the patient becomes comatose, and death usually takes place in the throes of uncontrollable, frenzied convulsions. The difference between furious and dumb rabies lies in the length of the excitation phase. Dogs which experience dumb rabies become depressed and do not display the long periods of excitation during which they uncontrollably growl and snap at everything in the immediate area, even nonexistent objects. "Dumb" animals act cowardly and cough as if they have something caught in their throats. These overt symptoms are directly correlated with cellular events.

There is recent evidence that after the virus has entered the body, replication occurs in muscle cells near the wound. If not destroyed by natural defenses or vigorous cleaning (soap and water, iodine, alcohol, or quaternary ammonia compounds), virions enter nerve endings and move through axons toward ganglia located next to the spinal cord.

Degeneration and demyelination of neurons occur at this time. Transfer across synaptic junctions allows the virus to spread through the spinal cord toward the brain and to other parts of the body, including the salivary glands. Once inside the brain and glands, virus replication causes abnormal behavior, release of virions through saliva, and ultimately death. The incubation periods varies from two weeks to several months. Delay in the onset of clinical symptoms is a bewildering event. It is not known whether the virus enters a latent stage at this time, or if some other form of host-parasite balance is established. The communicable period begins three to five days before symptoms appear and lasts throughout the course of the disease.

The effective use of vaccines to control and prevent rabies is made possible by a long incubation period averaging 20–30 days. During this time, immunity can be artificially acquired by stimulation with the attenuated virus. Louis Pasteur was the first to successfully produce a vaccine by attenuating, or "fixing," canine virus, which he called "street" virus. "Fixing" was performed by passing the virus through a series of 20–25 rabbit incubations. The result was Pasteur's rabbit vaccine of a virulent "street" virus. Until recently this same technique was performed in many countries, but the viruses were passed through more than 2000 inoculations and inactivation with phenol to better ensure loss of virulence. The original Pasteur vaccine was given as a series of 21–28 inoculations following contact with a rabid animal. Impurities in the vaccine and the possibility of persistent virulent viruses led to a high percentage of serum sickness, anaphylactic shock, and questionable cases of vaccine-induced rabies. These drawbacks were reduced by the development of the duck embryo vaccine (DEV). In this technique, the rabies virus is passed through over one hundred duck egg incubations to eliminate virulence. This high egg passage (HEP) vaccine is unable to cause infection of brain tissue in dogs and rabbits, and its high level of purity results in only minor side effects. Fourteen to twenty-one inoculations of DEV are given abdominally to ensure better and more rapid absorption into the body. Modified HEP vaccines are used to immunize dogs, cats, and other domesticated animals. More recently, Wiktor, Plotkin, Kaplan, and Koprowski have developed a rabies vaccine from virus grown in human tissue culture and then chemically inactivated.

The human diploid cell strain of rabies vaccine (HDCV) is highly purified and the most potent available. Only four to six inoculations are required to induce immunity. The

---

1 List two factors which influence the transmission of rabies virus.
2 Why is rabies also known as hydrophobia?
3 What is "street" virus?

HDVC vaccine is currently used throughout the world in preference to the DEV vaccine. In the United States, DEV is no longer manufactured and HDVC is used in all cases of rabies requiring treatment.

Several types of antisera are also employed in rabies control (Figure 22.4). The presence of preformed antibodies in the wound and throughout the body better ensures elimination of the virus. Antiserum is also used to infiltrate a cleansed wound to neutralize the virus before it replicates or enters nerve endings. Originally, horse and mule antiserum was administered along with rabies vaccine; however, serum sickness and shock occurred regularly. Human rabies immunoglobulin G produces fewer and less severe side effects and is made from the blood of volunteers vaccinated against rabies. A third form of antiserum has been developed from tissue cultures by a process known as the monoclonal-antibody technique. This is a highly specific procedure that detects different antigenic strains of rabies virus. The antibody released from the cell culture is highly effective against antigens found on specific strains of rabies, can be produced in large amounts, and is relatively inexpensive.

HEPATITIS VIRUSES

The **hepatitis virus** group is comprised of several complex viruses or viruslike particles able to cause acute infections of the liver. There are three different subgroups known as hepatitis type A virus (HAV), hepatitis type B virus (HBV), and non-A non-B hepatitis agents (NANB). Clinicians and researchers have only recently become aware of the fundamental differences among these groups through serological, epidemiological, and electron microscopic analyses. A detailed comparison of the two best-known viruses, HAV and HBV, is presented in Table 22.3. It is clear that these two viruses are not related but both demonstrate a high level of resistance to disinfectants; poor growth (or none) in cultures; and similar clinical symptoms. By comparison, little is as yet known about NANB, but it appears that there is more than one agent in this group responsible for hepatitis similar to that caused by HBV.

Hepatitis type A is commonly known both as epidemic or infectious hepatitis because of its mode of transmission (Figure 22.5). The virus is easily spread through a population; outbreaks are frequently associated with summer camps, dormitories, military barracks, households, and institutions. The usual mode of transmission is by the fecal-oral route, but large common-source epidemics have also resulted from HAV transmission in water, food, and milk. One of the largest waterborne outbreaks occurred in New Delhi, India, in 1955–56, and encompassed an estimated 4000 people. Foodborne outbreaks have involved salads, raw and steamed clams and oysters, and bakery products. Food handlers discharging HAV in feces serve as common sources of epidemics, if they fail to follow the simple personal hygiene practice of washing their hands before food

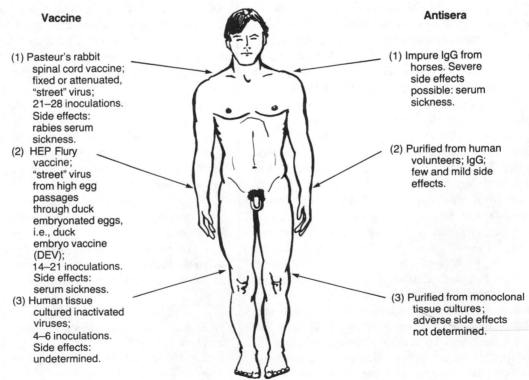

Vaccine

(1) Pasteur's rabbit spinal cord vaccine; fixed or attenuated, "street" virus; 21–28 inoculations. Side effects: rabies serum sickness.

(2) HEP Flury vaccine; "street" virus from high egg passages through duck embryonated eggs, i.e., duck embryo vaccine (DEV); 14–21 inoculations. Side effects: serum sickness.

(3) Human tissue cultured inactivated viruses; 4–6 inoculations. Side effects: undetermined.

Antisera

(1) Impure IgG from horses. Severe side effects possible: serum sickness.

(2) Purified from human volunteers; IgG; few and mild side effects.

(3) Purified from monoclonal tissue cultures; adverse side effects not determined.

FIGURE 22.4
Rabies prophylaxis.

TABLE 22.3
A comparison of HAV and HBV.

Features	HAV	HBV
Common name	Short-incubation infectious, or epidemic, hepatitis.	Long-incubation, or serum, hepatitis.
Susceptible population	Highest morbidity among children and young adults.	Adults.
Mode of transmission	Infection by ingestion (fecal-oral) or with blood (parenteral); also milk, water, food. Found in feces and also in blood, not in urine or nasal secretions. An institutional disease, i.e., schools, military, dormitories, etc.	Parenteral transmission; possibly by sexual contact (semen, menstrual blood, saliva), and gastrointestinal routes. Occurs among those handling blood, blood-related products (plasma, serum) or cutting-piercing tools.
Nature of virus	27 nm, spherical virus, no envelopes. May have RNA?	Dane particles? Or HBV: spherical, 42 nm, enveloped, DNA identified.
Culture	Not as yet grown in culture.	Same.
Incubation period	1–6 weeks.	8–22 weeks.
Onset	Abrupt or insidious onset.	Same.
Symptoms	Fever, GI upset, headache, anorexia, malaise, cervical nodes swollen and tender. Before jaundice appears: lasts up to 3 weeks, tenderness of liver and spleen—jaundice, liver damage but with complete regeneration. With jaundice: may persist for weeks, cases of relapse: 1–18 percent but mild. Rare complications: pneumonia, meningitis.	Same; may occur as acute serum hepatitis, persistent or chronic hepatits, or asymptomatic hepatitis.
Distribution	Approximately 60 percent of all hepatitis cases.	25 percent (15 percent NANB).
Mortality	0.5 percent.	4–5 percent.
Communicable period	During later portion of incubation period and early clinical stage.	Weeks before onset of symptoms, through acute stage and for years in chronic carriers.
Immunity	Long-lasting active immunity occurs after disease.	Several antibodies present in serum.
Prevention	Passive immunity provided from pooled IgG: Immune serum globulin (ISG).	Same but with more limited success.
Control of epidemics	Public and private sanitary procedures and habits.	Screening of blood-bank blood; sterilization of blood-contaminated equipment and instrument materials.
Vaccines	No vaccines acceptable.	Vaccine developed from HBV capsid proteins harvested from chimpanzees.
Oncogenic properties	Not observed.	Suspected in association with liver cancer.

preparation and serving. Unlike HBV, short-incubation hepatitis is only rarely transmitted parenterally, i.e., not through the intestinal tract but by infection. Hepatitis caused by HAV ranges from an asymptomatic infection to a disabling disease that requires hospitalization and long-term care. However, most patients recover without difficulty and develop a strong, active immunity. Children seldom develop jaundice, and accurate diagnosis depends on liver function and serum enzyme tests. Although no vaccine is available, passive immunity with immune serum globulin (ISG) provides short-term protection. ISG is made by combining or pooling the plasma of normal adults and is available for those in high-risk situations. This includes travelers to countries with a high incidence of HAV, doctors and nurses, and others in a known area of endemic hepatitis.

Hepatitis type B, formerly known as serum hepatitis, is most frequently transmitted in blood, blood products, or on blood-contaminated instruments (Figure 22.6). Whole blood, menstrual blood, plasma, hypodermic needles, scalpels, dental instruments, toothbrushes, razors, blood-contaminated towels, and tattooing needles have all been identified in the transmission of HBV. Only rarely is the virus

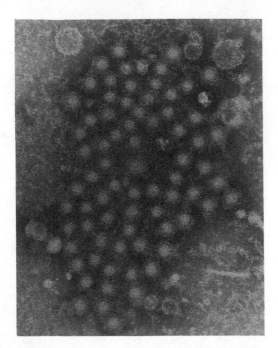

FIGURE 22.5
Hepatitis A virus is responsible for epidemic, or infectious, hepatitis. (Courtesy of the Center for Disease Control, Atlanta.)

transmitted by other routes. The symptoms of HBV closely resemble those found in hepatitis caused by HAV; however, chronic, persistent cases of HBV can develop. These individuals can serve as sources of HBV for years and must be counseled to prevent them from infecting others. Recent evidence also points to the fact that patients who have experienced an HBV infection also run a higher risk of liver cancer. Although a great deal of information is known about HBV, the exact events associated with virus-host interactions are still under investigation.

The surface of the HBV virus contains an antigen called HB_sAg that is 22 nm in diameter and found free in the serum of infected patients. HB_sAg particles can combine to form larger units recognizable in the electron microscope. Two other antigens have also been identified: core antigen, HB_cAg, and *e* antigen, HB_eAg. The actual hepatitis B virus, also known as the Dane particle, is believed to be composed of DNA, HB_sAg, and HB_cAg. The relationship of HB_eAg to the virus has not as yet been made clear, but *e* is found during both the acute and persistent stages of the infection. The presence of various HBV antigens and their serum antibodies are used in the diagnosis of the disease. A vaccine against HBV has been produced and protective levels of anti-HB_sAg have been induced by vaccination with purified human HB_sAg and chimpanzee HB_sAg. The success of this procedure indicates that the HB_sAg antigen plays an important role in the course of the disease. The vaccine is able to provide protection for about five years. Its use will

be of great value in preventing hepatitis B infections for those working in high risk areas such as hospitals and blood banks, and in the long run will significantly reduce the number of carriers. Currently the vaccine is only recomended for those who are most likely to come in contact with the virus.

Non-A non-B hepatitis, NANB, is a form of the disease recently discovered among individuals who had received transfusions with blood that was carefully screened for the presence of HBV and HB_sAg. An estimated eighty to ninety percent of the posttransfusion hepatitis cases in the United States are the result of NANB infections. This liver disease closely resembles HBV both clinically and epidemiologically. Little is currently known about this new form of

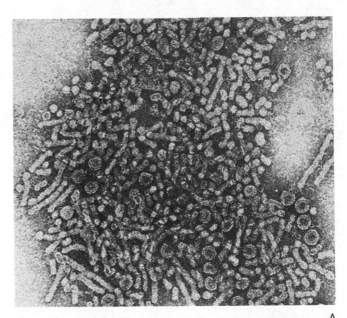

A

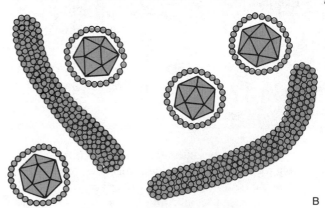

B

FIGURE 22.6
(A) Hepatitis B virus is responsible for serum, or long-term incubation, hepatitis. (B) The structure of HBV.
(Photomicrograph courtesy of G. Cabral, Medical College of Virginia, Virginia Commonwealth University. Part B from J. Melnick, G. Dreesman, and F. Hollinger. "Viral Hepatitis." *Scientific American,* July, 1977.)

hepatitis; however, research using chimpanzees has indicated that there may be more than a single agent involved in NANB infections. Chimpanzees apparently can develop acute and persistent infections. The viruslike particles are in the range of 25–30 nm in diameter and have been found to remain in the blood as infectious agents for as long as six months following the acute disease (Figure 22.7). There is no protection from the presence of HAV and HBV serum antibodies.

---

1 How is human rabies immunoglobulin produced?
2 How do the transmission routes between HAV and HBV differ?
3 What are the symptoms of viral hepatitis?

PARAMYXOVIRUSES

This group contains some of the largest, enveloped RNA viruses known, and several are responsible for human disease. **Parainfluenza virus** is a common virus infecting children and some adults, and demonstrates symptoms similar to the common cold. **Respiratory syncytial virus (RSV)** causes an upper respiratory infection of children that resembles pneumonia and bronchitis. The other, more familiar paramyxoviruses are the mumps and measles viruses.

The **mumps virus** is a highly contagious paramyxovirus found throughout the world. It causes epidemic parotitis (infection of the parotid salivary glands) primarily in children; but on rare occasions, adults may develop the disease. Two antigens have been found in association with the virus. Soluble, S, antigen is free in the serum of infected patients; and viral, V, antigen is a component of the capsid. The virus is spread by droplet nuclei or direct contact with saliva from infected individuals. The incubation period ranges from twelve to twenty-six days followed by the appearance of fever, swelling, and tenderness of one or more salivary glands. The parotid glands are primarily involved but maxillary and submaxillary glands may also become infected. Swelling reaches a peak in 48 hours (Figure 22.8) and may last as long as two weeks. Patients are communicable about two days before the onset of swelling and can remain infective for nine days. Inflammation of the testis (orchitis) occurs in 15–25 percent of the male cases but rarely leads to sterility. In females, inflammation of the ovaries (oophoritis) occurs in only 5 percent of the mumps cases and does not cause sterility, since the ovaries are not bound by a membrane and can swell without rupturing. Other rare complications of mumps infection are aseptic meningitis and pancreatitis.

Recovery from mumps induces a strong, long-lasting immunity. Antibodies to S and V antigens are found after infection, but V antibody persists for a longer period. Immunization against the mumps virus may be given as a

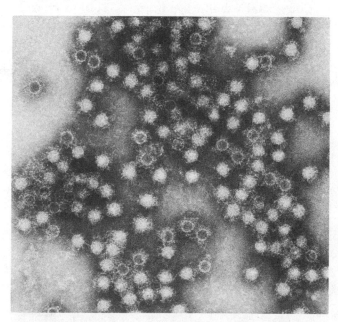

FIGURE 22.7
While not fully classified and analyzed, non-A non-B hepatitis (NANB) virions resemble those of hepatitis B. (Courtesy of the Hepatitis and Viral Enteritis Division, Center for Disease Control, Phoenix.)

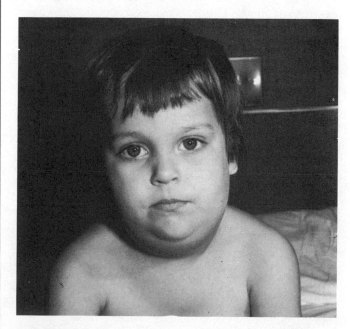

FIGURE 22.8
This child exhibits typical mumps symptoms. Although this viral infection usually is mild, such severe side effects as sterility can result. (Courtesy of the Center for Disease Control, Atlanta.)

single, live, attenuated vaccine or in combination with measles and rubella (MMR vaccine). The vaccine should be given any time after one year of age with no fear of an adverse reaction. This will ensure protection from the virus and eliminates the possibility of orchitis in males. Passive immunity provided by human immunoglobulin is also available to those who have not been vaccinated. These antibodies usually do not prevent mumps, but do prevent orchitis.

Hard measles, rubeola, or **"measles,"** was a common childhood disease prior to the introduction of the measles vaccine (Figure 22.9A). Like mumps, this paramyxovirus is spread by droplet nuclei or direct contact with nasal and other secretions from an infected individual. The incubation period lasts about ten days before the onset of fever, muscle pain, and photophobia which is an abnormal sensitivity to light. The classical measles rash usually begins on the fourteenth day. The communicable period begins three or four days before and lasts about four days after the skin eruptions appear. A characteristic and diagnostic feature of hard measles is the presence of Koplik spots on the buccal mucosa (Figure 22.9B). These spots are local necrotic lesions containing serum, endothelial cells, giant cells, and virus antigen. The skin rash begins on the face and spreads as the virus moves through the blood. Virions are eliminated through nasal, conjunctival, and urinary excretions but not from the skin eruptions. The disease is more severe in adults, but the virus does not cross the placenta to cause congenital disease. Complications from the disease are rare; however, a strong possibility exists that a defective measles virus may be responsible for subacute sclerosing

panencephalitis, SSPE. A major breakthrough in the control of measles came in 1954 when the first live, attenuated vaccine was produced from virus grown in tissue culture. Since that date, the more effective Swartz vaccine has become licensed (1965) and a worldwide measles eradication program initiated. Today immunization against measles is performed at the age of 15 months in a combined vaccine with attenuated mumps and rubella viruses and has been shown to induce immunity in over 95 percent of the children.

RUBELLA VIRUS

This RNA virus has not yet been classified but is responsible for a mild illness commonly known by such names as three-day measles, soft measles, German measles, and rubella. Although symptoms in children and adults are more mild than rubeola, the virus is able to cross the placenta and cause severe congenital malformations known as the rubella syndrome. Symptoms of rubella infection in adults and children (postnatal rubella) appear after an incubation period of from 14 to 21 days. The patient experiences malaise, slight fever, and a mild rash that begins on the face and trunk but shifts to the arms and legs. No Koplik spots develop during rubella, but lymph nodes become swollen and tender. The virus can be transmitted through nasal secretions about four days before and after the appearance of the rash (Figure 22.10). Antibodies (IgM) are produced within three weeks and provide a lifelong immunity.

Congenital rubella syndrome is likely to occur in 25 percent of the children born to mothers exposed to the

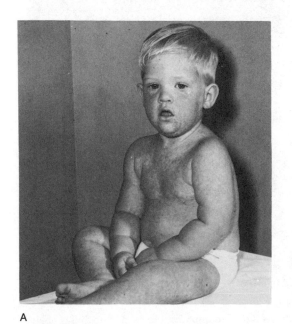

A

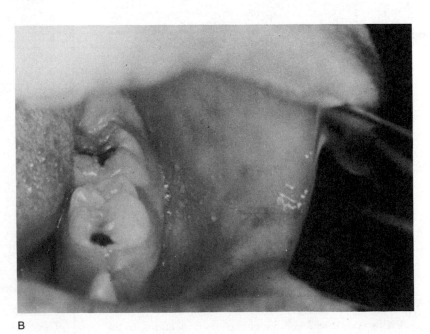

B

FIGURE 22.9
(A) This child displays the rash typical of a measles infection. (B) Inside the oral cavity, Koplik spots appear on the buccal surface. (Courtesy of the Center for Disease Control, Atlanta.)

virus during the first four months of pregnancy. If the virus crosses the placenta within the first month, 80 percent will show malformations that typically involve tissues of the eyes (cataracts, blindness), ears (deafness), heart (septal defects), and brain (retardation).

The mortality rate is 20 percent for those demonstrating abnormalities at birth. The degree of distortion caused

1 Name two diseases caused by paramyxoviruses.
2 List two important complications associated with mumps.
3 How are the measles and mumps viruses transmitted?

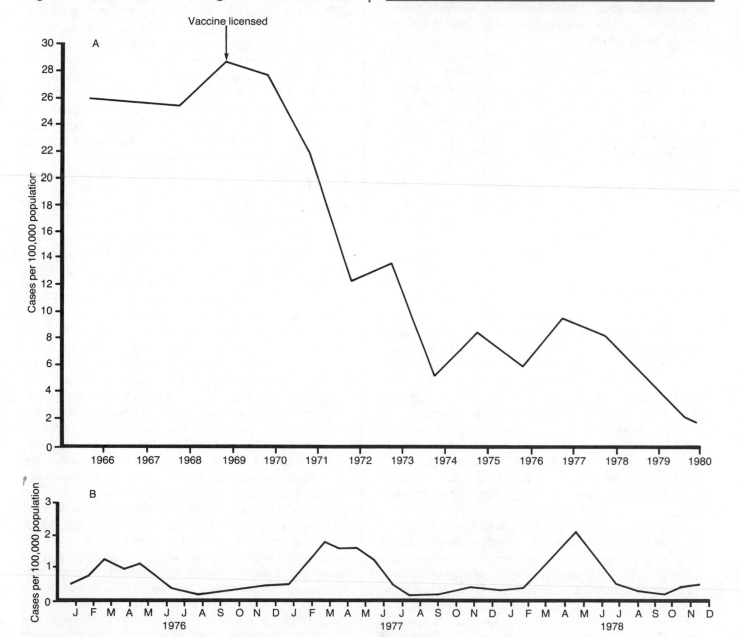

FIGURE 22.10

(A) Reported case rates of rubella (German measles) per 100,000 population in the United States. The incidence of rubella has declined drastically since 1969, the year of rubella vaccine licensure. To further illustrate this dramatic decrease, the number of reported cases for a one-week, midyear period in 1960 was 1234. For the same period in 1970, the number was 157, and for 1982 the number had fallen to 4. (B) Case rate by month for a three-year period. Note that the highest incidence rates occur in late winter and spring. (Courtesy of the Center for Disease Control, Atlanta.)

by the cytopathic effects of rubella is directly related to the level of tissue differentiation at the time the virus enters the fetus. Malformations result from a delay in cell division, differentiation, and specialization, rather than from cell destruction (cytolysis). A fetus experiencing infection during the final portion of the last trimester of pregnancy has less than a 15 percent chance of demonstrating abnormalities but may be born with the virus. A persistent infection occurs in such children, and virions may be released for one to one and one-half years after birth.

Since the introduction of the rubella vaccine in 1969, there has been a marked decrease in the incidence of post-natal and congenital rubella cases. The vaccine induces immunity in over 95 percent of those vaccinated, but its use is not advised for previously unvaccinated pregnant women or those who anticipate becoming pregnant in the near future. Evidence indicates that the fetus can become infected with the vaccine strain of virus, but it has not been demonstrated that malformation occurs from this infection. Therefore, control of congenital rubella syndrome by vaccination is directed toward immunization of children and developing a high degree of herd immunity (refer to Chapter 15).

ARBOVIRUSES

Several groups of viruses are regularly discussed together because they have several common morphological and epidemiological characteristics. Their most outstanding feature is their transmission and multiplication in arthropods, which serve as both hosts and vectors. This characteristic has resulted in the popular group name arbovirus (arthropod-borne viruses). These viruses multiply within an arthropod host without fatal results and are considered to be accidental pathogens of humans. Arboviruses are transmitted vertically in many ticks and some mosquitos through eggs to offspring, and a vertebrate host is unnecessary to maintain the life cycle: insect → vertebrate → insect. Person-to-person

↳ insect ↵

transmission does not occur among the arboviruses.

The three major subgroups of arboviruses are togaviruses, bunyaviruses, and orbiviruses. Members of these groups are responsible for three clinical types of human illnesses: (1) **encephalitis,** inflammation of the brain leading to degenerative changes that may be fatal; (2) **hemorrhagic fever,** involving extensive external and internal disease with acute fever and high fatality rate; and (3) **febrile illnesses,** demonstrating short benign fever that lasts only a short time. Although over 300 of these RNA viruses have been identified, only one member of each group will be described, since many occur only rarely and in remote tropical areas of the world.

TOGAVIRUSES

This large group includes several important mosquito-borne (Culex and Aedes) diseases that have many common features and all cause encephalitis. The most familiar diseases are eastern equine encephalitis (EEE), western equine encephalitis (WEE), Venezuelan equine encephalitis (VEE), St. Louis encephalitis (SLE), and Japanese equine encephalitis (JEE). Their names give some indication of their distribution and the nature of an alternate vertebrate host, the horse. However, about twenty species of birds and several rodents, bats, and reptiles can also harbor these viruses. Infected mosquitos transmit the virus to susceptible humans and other animals as they feed. The incubation period varies from one day to several weeks. Once localized in the brain, spinal cord, and meninges, degenerative changes can result in a mild, almost asymptomatic infection or an acute illness that could be fatal within twenty-four hours. Symptoms of a severe infection appear suddenly and include headache, high fever, disorientation, and drowsiness. They progressively change to tremor, amnesia, convulsions, and coma. In nonfatal cases, the acute stage can last up to ten days before complete recovery and the establishment of a strong, long-lasting immunity. Individuals in areas of endemic equine encepalitis have a high protective antibody titer as a result of mild and inapparent infections.

Yellow fever is a hemorrhagic disease transmitted by the infected mosquito Aedes aegypti. It occurs in endemic and epidemic proportions in Central America, South America, Africa, and in the past, North America. The incubation period of yellow fever is from three to six days after a susceptible individual has been bitten by a virus-laden mosquito. Like encephalitis, the symptoms may be mild or severe and occur suddenly. Fever, headache, nausea, and vomiting occur in mild cases and during the early stages of a severe illness. During the acute phase, which occurs three or four days later, the pulse slows, the fever increases, and veins in the kidneys, spleen, liver, and heart weaken and rupture. Recovery from the infection results in a solid, long-lasting immunity, and individuals in endemic areas of yellow fever experience inapparent infections that stimulate the formation of protective antibodies within a week. Active immunization can be performed with a viable strain of yellow fever virus. However, use of the vaccine has been limited because of the difficulty in reaching significant numbers of susceptible people in remote areas and the frequent occurrence of fatal side effects. The most effective control measure is the elimination of adult and larval mosquitos (refer to Chapter 15). This method has successfully reduced the morbidity rate of yellow fever in many parts of the world.

Colorado tick fever occurs in two phases. The first begins after a four to five day incubation period and demonstrates symptoms similar to Rocky Mountain spotted fever.

The patient experiences fever, headache, bodyache, nausea, loss of appetite, and occasionally, a rash. The symptoms disappear for two to three days but recur, lasting another two to three days. The disease is acquired from the bite of the tick *Dermacentor andersoni,* while taking a blood meal from humans, ground squirrels, mice, and chipmunks. The virus is not transmitted among people. Antibodies against the virus appear within two weeks of the disease and establish a long-lasting immunity. Second attacks are seldom seen.

---

1   What is the transmission route for the arboviruses?
2   Name two diseases caused by the togaviruses.
3   List the symptoms of yellow fever.

ORTHOMYXOVIRUSES

Influenza viruses are members of this important group. They are small, enveloped virions, each containing several strands of RNA. Virions have a ribonucleoprotein core surrounded by several layers of viral and host origin (Figure 22.11). The outermost cover contains two important antigens, hemagglutinin antigen (HA) "spikes" and neuraminidase antigen (NA) "spikes." Although influenza viruses resemble one another morphologically, there are significant differences in their HA and NA spikes which enable virologists to classify them into three major antigenic types. **Type A** influenza viruses are more frequently involved in "flu" epidemics and undergo frequent antigenic changes known as antigenic drift or shift. These changes occur as a result of frequent genetic recombination with coinfecting viruses and may occur while type A virus is in nonhuman host cells

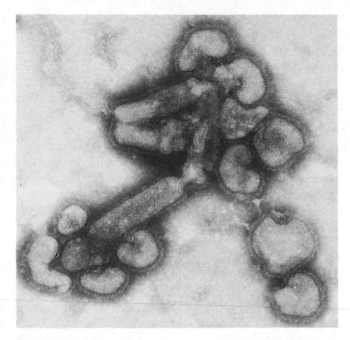

FIGURE 22.11
Influenza type A virus. (Courtesy of the Center for Disease Control, Atlanta.)

between epidemics. Type A influenza has been responsible for several major epidemics worldwide (Table 22.4). **Type B** is antigenically more stable than type A and has only rarely been implicated in epidemics. Influenza **type C** is considered to be a single and stable antigenic form that occurs sporadically. The naming of flu viruses is based on (1) their antigenic types: A, B, or C; (2) the most likely point of origin; for example, Hong Kong, Victoria, Asia; (3) their

TABLE 22.4
A history of "flu."

Year	Status	Origin	Type
1890	Epidemic	Asia	A
1918–19	Pandemic	Swine-type	A
1929	Epidemic		A
1947	Pandemic		A
1957	Pandemic	Asia	A
1959	Epidemic	Maryland	B
1962	Epidemic		B
1963	Epidemic		B
1964	Epidemic	Singapore	B
1965–66	Epidemic	Singapore	B
1968	Pandemic	Hong Kong	A
1972	Epidemic	London	A
1973	Epidemic	Port Chalmers	A
1975	Pandemic	Victoria	A
1976	Epidemic	Victoria and Swine-type	A
1977	Epidemic	Russia and Texas	A

discovery date, for example, 8/76; and (4) the nature of their HA and NA spikes, for example, H3N2. Thus, the Hong Kong flu epidemic of 1968 was technically caused by influenza virus B/Hong Kong/68 (H3N2) (Figure 22.12).

Influenza viruses are easily transmitted by droplet nuclei from individuals with clinical and inapparent infections. The viruses enter and are dispersed from the upper respiratory tract. The incubation period lasts from one to two days in susceptible people. Some form of natural immunity may exist in light of the fact that many people exposed to the viruses do not develop a clinical case of the disease. The virions remain localized on the lining of the upper respiratory tract and no viremia develops. The neuraminidase on the virions' surface decreases the viscosity of the mucus lining the tract, enabling the particles to reach the surface of goblet cells and ciliated epithelial cells, where they cause necrosis. Symptoms of infection include chills, muscle ache, malaise, heavy mucous discharge from the nose and throat, fever (lasting about three days), and prostration. Complications resulting from infection are relatively rare, but those that occur may be fatal. Secondary bacterial pneumonia caused by streptococci and staphylococci can occur in severely debilitated or elderly persons; this was the principle cause of death during the flu epidemic in 1918–19. Cardiovascular and renal diseases can also develop as complications. More recently, Guillian-Barré syndrome and Reye's

syndrome have been noted with increasing frequency. Reye's syndrome can occur after various viral infections but occurs more frequently after type B influenza infections. The mortality rate of this secondary complication has been reported to be as high as 50 percent in those experiencing the syndrome.

Immunity to the flu develops shortly after the disease but is short-lived, lasting only about six months. Secretory IgA plays an important role in preventing establishment of the virus in host cells. Individuals with high secretory IgA titers rarely become infected, even after deliberate exposure to viruses. The short life of IgA and the rapid antigenic drift of influenza has made it difficult to develop a flu vaccine that will control all types of viruses. Immunization to prevent type B infections has been more successful because of their more stable antigenic character. Influenza vaccines are prepared from type A and B viruses grown in embryonated eggs. They can be whole, killed viruses or antigenic subunits of complete virions (subunit vaccines).

---

1 Of what value are the "spikes" on the surface of the influenza viruses?
2 Which type of influenza virus is primarily responsible for "flu" epidemics?
3 Why has the immunization program to prevent type B "flu" infections been more successful than the program against type A infections?

The specific strains used in the vaccines are selected on the basis of epidemiological forecasting. For instance, epidemics of influenza in Australia will make their way to the United States over a six-month period and arrive during the following winter. The effectiveness of a vaccine is greatest when administered between September and November, just prior to the peak "flu season" of January through March. Immunization is especially important for children, the elderly, and individuals with chronic diseases, since they are more likely to experience harmful complications. However, the degree of protection depends on the additional amount of IgA produced, and the length of time between vaccination and exposure. It has been estimated that only 40 percent to 70 percent of individuals receiving "flu shots" are actually protected because of these variables. A more effective method of stimulating secretory IgA production is by intranasal inoculation with live, attenuated influenza viruses. It is hoped that the elevated secretory IgA levels will interfere with neuraminidase activity and prevent adsorption of the viruses to host cells. This form of vaccine is currently under investigation and has not been made available to the general public.

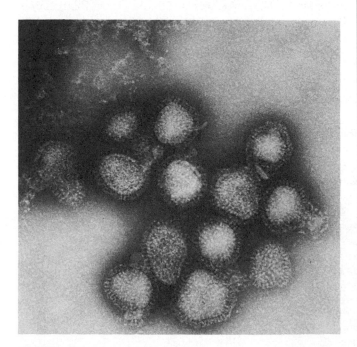

FIGURE 22.12
Hong Kong influenza virus. This infamous "flu" virus spread throughout the world in 1968. (Courtesy of the Center for Disease Control, Atlanta.)

SUMMARY

Viruses can be divided into two major groups based on whether they contain DNA or RNA. Within the RNA group are six subgroups characterized by their sizes, shapes, antigenic properties, and host cytopathic effects. While many of these disease organisms have been brought under control, they continue to cause death and harm to those who become infected. The groups described in this chapter are the picornaviruses, rhabdoviruses, hepatitis viruses, paramyxoviruses, arboviruses, and orthomyxoviruses. Members of these groups are responsible for such diseases as polio, juvenile-onset diabetes, the common cold, rabies, hepatitis, and mumps.

WITH A LITTLE THOUGHT

Box 26 in this chapter lists several statements that appear to be fallacies about the "common cold." Review these statements in light of your background in virology and propose reasons why these "fallacies" might still be considered valid.

STUDY QUESTIONS

1 Name the two forms of polio infection and describe how one is responsible for maintaining the virus population.

2 What causes the "common cold"? Describe the course of a typical infection.

3 How are antibodies and antigens used to prevent rabies?

4 By what routes are the three forms of hepatitis transmitted?

5 What is congenital rubella syndrome?

6 List three viral diseases transmitted by insects.

7 What is the proposed connection between viruses and juvenile-onset diabetes? Between measles and subacute sclerosing panencephalitis?

8 Why would a nasal spray of attenuated common-cold virus or influenza virus be more effective than "flu shots"?

9 What are some important complications of the following viral diseases: measles, rubella, influenza, polio?

10 What is the role of NA spikes on the surface of influenza viruses?

SUGGESTED READINGS

BURKE, D. C. "The Status of Interferon." *Scientific American,* vol. 236, no. 4, April, 1977.

Center for Disease Control. "Perspectives on the Control of Viral Hepatitis, Type B." *Morbidity and Mortality Weekly Report,* vol. 25, no. 17, May, 1976.

HENDERSON, D. A. "The Eradication of Smallpox." *Scientific American,* vol. 235, no. 4, Oct., 1976.

HENLE, W., G. HENLE, and E. T. LENNETTE. "The Epstein-Barr Virus." *Scientific American,* July, 1979.

HOLLAND, J. J. "Slow, Inapparent and Recurrent Viruses." *Scientific American,* vol. 230, no. 2, Feb., 1974.

"Interferon (I): On the Threshold of Clinical Application." *Science,* vol. 204, June, 1979.

"Interferon (II): Learning How It Works." *Science,* vol. 204, June, 1979.

KAPLAN, M. M., and H. KOPROWSKI. "Rabies." *Scientific American,* Jan., 1980.

LANGER, W. L. "Immunization Against Smallpox Before Jenner." *Scientific American,* Jan., 1976.

MELLSTEDT, H., M. BJORKHOLM, B. JOHANSSON et al. "Interferon Therapy in Myelomatosis." *The Lanet,* Radiumhemmet, Karolinska Hospital and Department of Medicine, Seraphimer Hospital, Stockholm, Sweden.

"Panel Urges Wide Use of Antiviral Drug." *Science,* vol. 206, Nov., 1979.

ROSS, M. E., T. ONODERA, K. S. BROWN et al. "Virus Induced Diabetes Mellitus—Genetic and Environmental Factors Influencing the Development of Diabetes after Infection with M. Variant of Encephalomyocarditis Virus." *Diabetes,* vol. 25, no. 3, March, 1976.

SPECTOR, D. H., and D. BALTIMORE. "The Molecular Biology of Poliovirus." *Scientific American,* vol. 232, 1975.

YOON, J. W., M. AUSTIN, T. ONODERA et al, "Virus Induced Diabetes Mellitus—Isolation of a Child with Diabetic Ketoacidosis," *New England Journal of Medicine,* vol. 300, no. 21, May, 1979.

KEY TERMS

desquamation	jaundice	orchitis (or-ki′tis)
stomatitis	syncytial (sin-sish′al)	

PART 5
Environmental Microbiology

MICROBES are found throughout the environment, and their activities vary from the decomposition of organic and inorganic compounds to food preservation. Because microbes shape and influence the lives of all other organisms, microbiologists have focused their attention on several areas of environmental microbiology. The fields of food and dairy microbiology concentrate on understanding and preventing microbial food spoilage, developing better methods to preserve foods, using microbes to improve available food, and reducing the incidence of foodborne infections and poisoning. Soil and agricultural microbiologists deal with the role of microbes in soil formation, the recycling of elements, decomposition of organic material, soil fertility, and the relationships that exist among microbes, plants, and animals. Knowledge of these areas has enabled researchers, technicians, and farmers to maintain microbes and benefit from microbial activities.

Those working in the area of water and wastewater microbiology are attempting to ensure a consistent source of high-quality water. Research involves the identification, removal, and control of harmful chemicals and biological contaminants found in the ground and surface water before it is made available for use. In addition to medical microbiology, this field is one of the most applied of all the biological sciences. Microbiologists, engineers, and businesspeople have come together in this field in order to use microbes for the large-scale production of microbial cells or their products. By controlling the activities of certain microbes, they have been able to produce more economically such products as enzymes, amino acids, vitamins, antibiotics, and alcohol. Expensive equipment and well-trained personnel are essential to the smooth operation of this multi-million dollar microbial industry.

◀ *The fungus* Aspergillus sp.; *a scanning electron micrograph. Magnification* × *6000. (Courtesy of R. M. Pfister and J. M. Hansen, Ohio State University.)*

Food and Dairy Microbiology

MICROBES AND THEIR GROWTH IN FOOD

MICROBIAL PROBLEMS: SPOILAGE, POISONING, AND INFECTION
Staphylococcal Food Poisoning
Botulism Food Poisoning
Mycotoxins
Fish and Shellfish Food Poisoning

NATURAL PRESERVATION

DRYING

REFRIGERATION AND FREEZING

IRRADIATION

CANNING

CHEMICAL PRESERVATIVES

FOOD MICROBIOLOGY: FERMENTED FOODS

THE STAFF OF LIFE

Learning Objectives

☐ Be familiar with the types of microbes that grow in foods and those factors which influence growth.

☐ Know definitions of the terms "spoilage," "poisoning," and "infection"; and give examples of microbes associated with each problem.

☐ Be able to name the organisms associated with fish and shellfish poisoning.

☐ Recognize the natural preservative properties of various foods.

☐ Be able to describe how various drying methods are used to preserve foods.

☐ Be able to explain how to refrigerate and freeze foods.

☐ Recognize the advantages and disadvantages of preserving by irradiation.

☐ Define the terms "cold-pack" and "hot-pack" canning and explain how these methods preserve foods.

☐ List several types of chemical preservatives and how they function.

☐ Be able to explain the fermentation process as it relates to the preservation of milk in the form of cheese, yogurt, and other dairy foods.

☐ Define the term "leavening" and explain how it is done.

Microbes are widely distributed throughout the environment in water, air, soil, and ice. They can establish different types of symbiotic relationships with humans, animals, other microbes, and plants. The key role of microbes in nature is the decomposition of complex organic materials into simpler compounds and atoms, which are then made available for reuse by all other life forms. This is a process known as geochemical cycling and will be considered in chapters 24 and 25. The survival and continuation of any living organism depends on the availability of a suitable source of energy, building materials, and environment. Since many of these requirements are available in the same form to microbes and humans, there is competition for these nutrients. The battle is continuous, far-reaching, and requires human attention on many fronts. It has been estimated that as much as one-third of the world's annual food supply is lost due to

23

microbial **spoilage,** insect damage, and poor distribution methods. In order to maintain a competitive edge, the fields of food and dairy microbiology have evolved with the goals of (1) preventing food spoilage and loss due to the activities of contaminating microbes or their harmful toxins, (2) developing better methods to preserve foods against microbial decomposition during distribution and storage, (3) using beneficial microbes to improve the nutritional value, texture, aroma, and flavor of foods, and (4) reducing microbial infections and poisoning from the ingestion of harmful foods.

MICROBES AND THEIR GROWTH IN FOOD

Microbes belonging to the major categories described throughout this text have been found as food contaminants. However, the three most important are bacteria, yeasts, and molds. Table 23.1 lists microbes from these groups and their known activities in foods. Their ability to cause harmful or beneficial changes depends on several chemical and physical properties of food. Foods with properties that readily support destructive microbial growth are called **perishable;** foods that can be kept longer are called either **nonperishable** or **semiperishable.** One of the most important factors affecting this quality is the amount of water in the food. The amount of moisture in a food available to microbes for growth is called the **water activity level, (A_w).** Each microbe has an optimum A_w for growth, and reducing the available water below minimal A_w requirements will lengthen the lag phase of microbial growth. In general, bacteria require more water than yeasts and molds (bacteria 18 percent, yeasts 13 percent, molds 15 percent). Exceptions to this include the halobacteria and some xerophilic ("xero" means *dry*) fungi. The A_w is highest in most fresh foods but can be reduced by salting, drying, gelling, or acidifying. For example, the following dried foods have different but low amounts of water: dried milk (8 percent), flour (10–14 percent), dried beans (15 percent), and dehydrated vegetables (14–20 percent).

Another property that varies among different foods is the pH range. Those with a near neutral pH, such as meats, fish, and soups, support the growth of the greatest variety of microbes. Foods with a more acid nature, such as cheese, sausage, soft drinks, fruits, and sauerkraut, restrict the growth of most microbes but can be damaged by acidophilic yeasts and molds. The more alkaline foods, such as corn, beans, egg whites, and peas, are also restrictive, but during preparation or preservation, there may be a change to the neutral pH range. This change supports growth of toxin-producing clostridia and other proteolytic (spoilage) bacteria.

The type of nutrients and level of oxygen in foods determine which microbes will grow and what products they

TABLE 23.1
Microbes in food.

Microbe	Effect
Bacteria (genus)	
Pseudomonas	Spoilage, slime on surface of meats.
Halobacterium	Spoilage in salty foods.
Acetobacter	Ferments apple juice to vinegar.
Alcaligenes	Spoilage of dairy products.
Staphylococcus	Spoilage and enterotoxin production in high carbohydrate foods.
Clostridium	Food poisoning.
Pediococcus	Meat preservation by fermentation.
Lactobacillus	Fermentation of yogurt, kefir, and buttermilk.
Propionibacterium	Fermentation in Swiss cheese manufacture.
Bacillus	Fermentation in cheddar cheese manufacture; some species produce toxins.
Streptococcus	Fermentation in cottage cheese and buttermilk manufacture.
Molds (genus)	
Aspergillus	Fermentation of rice for rice wine; production of fungal toxins.
Penicillium	Fermentation of blue, Roquefort, and Camembert cheese; antibiotic production.
Rhizopus	Spoilage of bread and vegetables.
Yeasts (genus)	
Saccharomyces	Fermentation for wine, beer, and liquor; leavening of bread; single-celled protein as ironized yeast.
Candida	Pathogenic yeast; spoilage of butter and margarine.
Zygosaccharomyces	Osmophilic yeast, cause spoilage of honey, syrups, and molasses; also used in fermentation of soy sauce.
Torulopsis	Spoilage of milk, fruit juices, and acid foods.

will produce during metabolism. Foods containing little or no oxygen may support the growth of strict anaerobes (e.g., *Clostridium botulinum*) or aero-tolerant anaerobes, such as members of the genus *Streptococcus*. Their growth results in food decomposition and the production of toxins. Proteins are often broken down under anaerobic conditions to foul-smelling compounds. As the oxygen level increases, aerobes find a suitable growth environment. Foods with only the minimal amount of available oxygen allow many bacteria to break down carbohydrates and release organic acids that give food an acid, fruity smell. With higher oxygen levels, the food can be completely broken down into carbon dioxide and water. The balance of nutrients available for metabolism also influences microbial growth. High protein foods, such as meat and fish, are more likely to support the growth of proteolytic bacteria, such as *Pseudomonas;* while foods with a higher relative concentration of carbohydrates favor the growth of less proteolytic aerobic bacteria, yeasts, and molds. Because microbes vary widely in their vitamin requirements, many are only able to grow in foods that contain specific compounds. For example, foods with a high concentration of B vitamins are readily attacked by many types of spoilage bacteria. The presence of a variety of growth factors in meat and milk supports the growth of pathogens, since these microbes characteristically require certain vitamins or other growth factors.

The physical structure of food also plays an important part in its nonspoiling qualities. Many fresh foods naturally contain more microbes per gram or square centimeter because of their growth patterns. For example, the smooth skin of an unwashed tomato contains about 2000 microbes per square centimeter, while cabbage leaves may contain 1–2 million microbes in the same area. Foods which naturally harbor more microbes are more likely to become spoiled in a shorter time. By grinding or chopping foods, the likelihood of contamination increases as the surface area available for microbial growth increases. These food types must be carefully handled, since their keeping quality is reduced in comparison to unprocessed forms. For example, hamburger contains many times more microbes per gram in comparison to an equal weight of steak. In order to limit the destruction of foods that have the most favorable chemical and physical growth characteristics, several preservation methods are used. Each is selected for its ability to inhibit or destroy contaminating microbes and to ensure maintenance of an acceptable, high-quality product.

---

1 What do the halobacteria and xerophilic fungi have in common?
2 Name three properties of foods that contribute to microbial growth.
3 Name three properties of foods that inhibit microbial growth.

MICROBIAL PROBLEMS: SPOILAGE, POISONING, AND INFECTION

Microbial growth can render a product unacceptable in three ways. First, the food may become **spoiled** as a result of enzymatic decomposition of the product, an increase in microbial numbers, or the release of compounds which result in changes in flavors, colors, and aromas. Spoiled foods are usually not eaten because of their unappetizing appearance and odor. Second, microbes can release toxins into the food during their growth. Ingestion of these toxins results in a brief but violent illness only a few hours after the chemical agents contact susceptible tissues. This type of foodborne illness is known as **food poisoning,** or food intoxication. Third, foods may be unsuitable for consumption because of the presence of viable pathogens that can establish an infection after they have been ingested. In this case, the food serves to carry the pathogens into the host where they reproduce and cause disease. Outbreaks of this type of foodborne disease are known as **food infections.**

The exact number of people who experience food poisoning each year will probably never be known, since most victims fail to report the disease to their physicians. Although the illness is reportable (refer to Chapter 15), the brief period of illness experienced by the majority of people makes it difficult to identify the etiological agent and almost impossible to determine the morbidity rate. One attempt was made by the California State Health Department between 1959 and 1964. During this period, 705 outbreaks of food poisoning were investigated involving a total of 14,831 cases. Of the total, no specific etiological agent could be identified in 48 percent of the cases, 20.8 percent were caused by *Staphylococcus* enterotoxin, 8.1 percent by *Salmonella* strains, 4.6 percent by *Clostridium perfringens,* and 1.4 percent resulted from a variety of bacteria, including *Bacillus cereus, Clostridium botulinum, Enterobacter aerogenes,* and streptococci. The remaining cases were those resulting from chemical and plant poisonings. Because these illnesses are so common and in some cases fatal, it is important that proper food processing, preservation, and preparation procedures be followed to avoid foodborne poisoning.

STAPHYLOCOCCAL FOOD POISONING

Certain strains of *Staphylococcus aureus* are able to release an exotoxin known as **enterotoxin** while growing in high carbohydrate foods. Even though the bacteria are easily destroyed by heating, the toxin is heat stable and capable of withstanding boiling for up to one hour. The enterotoxin is produced at room temperatures and has been isolated for *S. aureus* growing in such foods as cream pies, custards, mayonnaise, potato salad, salad dressings, ham, turkey, sliced meats, gravy, dressings (stuffing), milk, and milk products. Contamination usually occurs by droplet nuclei or direct inoculation from food handlers during preparation and serving.

The time between ingestion of contaminated food and onset of symptoms typically ranges from two to four hours, but may be as long as six hours. Ingestion of the toxin results in an abrupt onset illness characterized by severe nausea, cramps, vomiting, diarrhea, and prostration. These symptoms may be mild, lasting only an hour, or they may be severe and continue for several hours. Recovery usually occurs in a day, and deaths are rare. The most important method of controlling or preventing staphylococcal food poisoning is by practicing good personal hygiene. Individuals with known infections should not prepare or serve food, especially to large groups of people. The proper handling of high-risk foods, such as poultry and stuffing, should receive special attention. A higher incidence of food poisoning typically occurs during holidays when such foods are served to large groups. Frozen turkeys should be thawed under constantly running cold water or in a refrigerator to prevent the growth of poison-producing bacteria on the surface before the bird is completely thawed in the center (Figure 23.1). Stuffing should not be placed inside the cavity of the bird and refrigerated before roasting. When this cold bird is placed in the oven for roasting, contaminating *S. aureus* in the stuffing may grow, producing dangerously high levels of enterotoxin. An inside temperature of 165°F is necessary to ensure toxin destruction. After the meal is finished, leftovers should be refrigerated immediately to prevent further enterotoxin production by bacteria that may have entered the food during the meal.

---

1 What is the difference between food poisoning and food infection?
2 Name the toxin responsible for staphylococcal food poisoning.
3 Is this toxin heat stable or heat labile?

BOTULISM FOOD POISONING

Clostridium botulinum is the anaerobic bacterium responsible for food poisoning known as botulism (Figure 23.2). This exotoxin is one of the most potent neurotoxic substances produced by microbes. Four serotypes (A, B, E, and F) have been found to produce the toxin when grown under anaerobic conditions. Contamination of foods by *C. botulinum* spores occurs frequently because of the wide distribution of the microbes. However, the release of toxin occurs only when microbes actively grow in the vegetative stage, which is fortunately a rare occurrence. Toxin production is least likely when the environment is aerobic, cold (below 5°C), acid (pH of 5 or less), and contains nitrite ions (as in smoked meats and fish). Poisoning may be prevented by establishing these conditions or by boiling suspect foods for 15 to 30 minutes, since the toxin is heat labile. Foods most commonly responsible for botulism outbreaks are those with a near-neutral pH, such as corn, beans, peas, chili peppers, string beans, and spinach. The more alkaline pH (7–8) of these fresh vegetables is decreased during the canning process, making these foods favorable growth media for surviving spores. Because of the development of new, less acid strains of tomatoes and other vegetables, special precautions should be taken to ensure the maintenance of a low

FIGURE 23.1
Thawing instructions are provided for frozen turkeys and many other fowl. Thawing may be a lengthy process, but the chances of food spoilage will be greatly reduced.

FIGURE 23.2
The contents of this can supported the growth of contaminating anaerobic microbes that have released gases, causing the can to become distended.

pH during home canning. This is easily done by adding lemon juice or commercially available organic acids.

The most frequent outbreaks of botulism involve serotypes A and B; serotype F is more rare; and type E cases appear to be confined to the ingestion of contaminated fish and seafoods. When the toxin is ingested, it enters through the intestinal tract and interferes with the release of acetylcholine at myoneural junctions. Symptoms of poisoning usually appear after a period of 12 to 36 hours or as long as several days. The victim progressively experiences a swollen tongue, difficulty in swallowing and speaking, double vision, dizziness, acute nausea, vomiting, diarrhea, fatigue, and respiratory failure. If untreated, death occurs in three to seven days. These symptoms have also been noted in a number of infant deaths and have led investigators to suspect that some cases of sudden infant death syndrome (SID) may be caused by the actual growth of *C. botulinum* in the intestinal tract of infants. This is unusual, since this anaerobe has not been found to actively grow in adults. One of the most common baby foods containing *C. botulinum* spores is honey. Recently, it has been recommended that infants be breastfed and that honey not be given because of the possible connection between botulism and SID (refer to Chapter 12).

Botulism toxin may cause irreversible damage if not neutralized early in the illness. Neutralization is easily and effectively accomplished by the administration of botulism antitoxin. A trivalent antitoxin containing A, B, and E is most commonly used, but a monovalent type E form is also available.

Another form of clostridial food poisoning is caused by *C. perfringens*. This disease is more mild than botulism with symptoms of abdominal pain, diarrhea, and possibly nausea and vomiting which appear about 12 hours after ingestion. These symptoms last only about 24 hours and are the result of a mild necrosis of the intestinal lining. On occasion this illness has been fatal to elderly persons and infants. Foods associated with outbreaks are boiled, stewed, or lightly roasted meats and poultry. Stews, sauces, pies, salads, and casseroles have also been implicated.

The aerobic bacterium *Bacillus cereus* is also capable of producing an enterotoxin with symptoms similar to those seen in *C. perfringens* poisonings. While increasing in importance in the United States, incidences of *B. cereus* poisoning have also been reported in Canada, England, and Europe from contaminated cereals, rice, vegetables, and meats.

---

1 Under what conditions is *C. botulinum* least likely to produce toxin?
2 What are the symptoms of botulism poisoning?
3 How are the symptoms of botulism counteracted?

MYCOTOXINS

Poisoning can also result from the ingestion of fungi or their toxic products known as **mycotoxins.** Several species of mushrooms are known to be highly toxic. The most notorious fungus is the mushroom *Amanita phalloides* var. *verna,* known as the "Angel of Death." However, many "edible" species of mushrooms may become poisonous to humans, if they grow in environments suitable for toxin production (Figure 23.3). Molds also have the ability to produce mycotoxins. One of the most important mycotoxins, known as **aflatoxin,** was first recognized in the 1960s in Great Britain and the United States. Aflatoxin is produced by members of the genus *Aspergillus* and at least one species of *Penicillium, P. puberulum.* The toxin is extremely heat stable and can withstand autoclaving. *Aspergillus* species commonly grow on grains such as peanuts, cottonseeds, cornmeal, oats, buckwheat, rice, and soybeans. Originally identified as the source of fatal poisoning of turkeys, fish, and sheep that had been fed moldy grain, suspicion is growing that aflatoxins are also responsible for cases of human intoxication. Aflatoxins have been found in milk from cows that were fed aflatoxin-containing grain and in human breast milk. Research has shown that low levels of aflatoxin are carcinogenic, producing liver tumors in rats and trout. In light of these facts, many private food corporations producing cereal products and other grain foods such as peanut

FIGURE 23.3
Members of the genus *Amanita* produce a deadly mycotoxin. Edible mushrooms are very difficult to identify growing in the wild. (Courtesy of J. R. Waaland, University of Washington, Biological Photo Service.)

butter or oatmeal have initiated research into the control and elimination of aflatoxin-producing mold.

FISH AND SHELLFISH FOOD POISONING

Two other important foodborne diseases have been found to significantly affect humans. **Shellfish poisoning** caused by toxins of the dinoflagellate algae *Gonyaulax catenella* have been recorded for hundreds of years. This foodborne illness occurs after eating mussels *(Mytilus californianus* and *M. edulis)* or Alaskan butter clams *(Saxidomus nattallii* and *S. giganteus)* that have concentrated the dinoflagellates by filtering them as food from the surrounding sea water. If the concentration exceeds 200 cells per ml in the surrounding water during a "bloom" (refer to Chapter 2), the clams and mussels will contain sufficient toxin to cause illness to those eating the shellfish. Alaskan butter clams can remain a potential source of illness for more than a year after they have concentrated poisonous algae.

The second foodborne illness results from ingesting contaminated fish belonging to the scombroid group—tuna, mackerel, and skipjack. **Scombroid food poisoning** is caused by a histaminelike substance released into the fish as a result of the possible activity of *Proteus* bacterial species. The illness has symptoms that resemble botulism: headache, dizziness, nausea, facial swelling, burning of the throat, difficulty in swallowing, and throbbing of the carotid and temporal blood vessels. Fish contaminated with this toxin has an unusual bitter or peppery taste. Symptoms may appear within five minutes to two hours after ingestion and usually last less than 6 hours, but the fatigue and weakness may linger for 24 hours. The food most frequently associated with these outbreaks has been prepared from dolphins. Confirmation of the disease is difficult because of the as yet unclear relationship between the toxin and the *Proteus* species. To avoid the possibility of scombroid food poisoning, fish of this group should be eaten soon after they are caught or refrigerated as quickly as possible. Control of this illness requires the use of antihistamines.

---

1 Name two fungi that produce mycotoxins.
2 What type of microbe is responsible for shellfish poisoning?
3 What bacterium is currently thought to be the cause of scombroid food poisoning?

Many foodborne illnesses do not result from microbial toxins but are caused by the growth and activities of microbes ingested from contaminated foods. These illnesses are known as **food infections.** Many human pathogens are transmitted in this manner. Table 23.2 lists several of the most widely occurring food infections, their causes, and a brief description of how to control them. Since these diseases have for the most part been described in chapters 16 through 22, this information will not be repeated. Refer to the appropriate chapter for information regarding their epidemiology, symptoms, and prevention. Controlling the transmission of these illnesses is accomplished by many of the same techniques and procedures that prevent food poisoning. In addition to being a source of human disease, many of these microbes are responsible for spoilage. To prevent spoilage, poisoning, and infection, and maintain an ade-

TABLE 23.2
Food infections.

Disease	Etiologic Agent	Foods Involved	Control Measures
Bacterial Infections			
Shigellosis (bacillary dysentery)	*Shigella sonnei,* *S. flexneri,* *S. dysenteriae, S. boydii*	Moist, mixed foods; milk, beans, potatoes, tuna, shrimp, turkey, macaroni salad, and apple cider.	Practice personal hygiene, chill foods rapidly in small quantities, prepare food in sanitary manner, cook foods thoroughly, protect and treat water, dispose of sewage in sanitary manner, control flies.
Enteropathogenic *Escherichia*	*Escherichia coli*	Coffee substitute, salmon (?), cheese.	Chill foods rapidly in small quantities, cook foods thoroughly, practice personal hygiene, prepare foods in sanitary manner,

			protect and treat water, dispose of sewage in sanitary manner.
Beta hemolytic streptococcal infections (scarlet fever, septic sore throat)	*Streptococcus pyogenes*	Milk, ice cream, eggs, steamed lobster, potato salad, egg salad, custard, and pudding.	Chill foods rapidly in small quantities, practice personal hygiene, cook foods thoroughly, pasteurize milk, exclude workers from handling food if suffering from respiratory illness or skin lesions.
Yersiniosis (*Yersinia enterocolitica* infection; pseudotuberculosis)	*Yersinia pseudotuberculosis, Y. enterocolitica*	Pork and other meats, or any contaminated foods.	Cook foods thoroughly, protect foods from contamination, control rodents.
Vibrio parahaemolyticus infection	*Vibrio parahaemolyticus*	Raw foods of marine origin. Saltwater fish, shellfish, crustacea, and fish products. Cucumbers and other salty foods have been implicated.	Cook foods thoroughly, chill foods rapidly in small quantities, prevent cross-contamination from saltwater fish, sanitize equipment.
Cholera	*Vibrio cholerae* and *V. cholerae* biotype *eltor*. Enterotoxin (exotoxin) elaborated in small intestine	Raw vegetables, mixed and moist foods. Raw mussels, shrimp, raw fish, cucumbers. Foods made up of, washed, or sprinkled with, contaminated water; foods prepared in utensils that were rinsed in contaminated water.	Dispose of sewage in sanitary manner, protect and treat water, practice personal hygiene, cook foods thoroughly, isolate cases. Immunization provides incomplete protection.
Brucellosis	*Brucella melitensis, B. abortus,* or *B. suis*	Raw milk, raw goat cheese.	Eradicate brucellosis from livestock (immunize young animals, restrict movement, test, segregate, or slaughter). Cook foods thoroughly, pasteurize milk and dairy products.
Tuberculosis	*Mycobacterium tuberculosis* and *M. bovis*	Raw milk.	Eradicate tuberculosis in animals (test and slaughter reactors), pasteurize milk. Isolate and treat cases. Immunize with BCG in high prevalence areas.
Diphtheria	*Corynebacterium diphtheriae*	Raw milk.	Immunize, pasteurize milk, prevent contamination by humans after heat treatment of milk, practice personal hygiene, isolate cases.
Tularemia	*Francisella tularensis*	Rabbit meat.	Cook meat of wild rabbits thoroughly, use rubber gloves when dressing rabbits, wear protective clothes and use tick repellent in endemic areas.
Campylobacter (Vibrio) fetus infection (Vibriosis)	*Campylobacter (Vibrio) fetus*	Raw beef liver and meat, raw milk.	Cook meats thoroughly, pasteurize milk.

Viral and Rickettsial Diseases

Hepatitis A (infectious hepatitis)	Hepatitis virus A (virus of infectious hepatitis)	Shellfish, milk, orange juice, potato salad, cold cuts, frozen strawberries, glazed doughnuts, whipped cream cakes, sandwiches.	Cook foods thoroughly, prevent pollution of shellfish growing areas, dispose of sewage in sanitary manner, treat water by coagulation-settling-filtration-chlorination. Practice personal hygiene, isolate cases for 7 to 10 days after jaundice, clean and sterilize needles, syringes, and other instruments used for parenteral injections. Give gamma globulin to contacts.
Poliomyelitis	Poliovirus	Milk, cream-filled pastry, lemonade(?).	Immunize against all three types of poliovirus, cook foods thoroughly, practice personal hygiene.
Q fever (Query fever)	Coxiella (Rickettsia) burnetii	Milk (rarely transmitted by this source.)	Pasteurize milk at 145°F for 30 minutes or 161°F for 15 seconds, practice personal hygiene (animal workers), vaccinate animals.
Coxsackie infections Herpangina (summer grippe)	Coxsackie group A viruses (including types 2, 4, 5, 6, 8, 10, 22)	Unknown, could be any contaminated food.	Practice personal hygiene dispose of sewage in sanitary manner, cook foods thoroughly.
Echovirus infections	Echovirus (enteric cytopathogenic human orphan virus)	Unknown, could be any contaminated food.	Practice personal hygiene, dispose of sewage in sanitary manner, cook foods thoroughly.

Protozoan Infections

Amebiasis (amoebic dysentery)	Entamoeba histolytica	Raw vegetables and fruits.	Practice personal hygiene (food handlers), cook foods thoroughly, dispose of sewage in sanitary manner, protect and treat water, control flies, avoid using human excreta for fertilizer (night soil).
Balantidiasis (balantidial dysentery)	Balantidium coli	Pork, raw foods.	Practice personal hygiene, cook foods thoroughly, treat cases, control flies.
Giardiasis	Giardia lamblia	Raw foods.	Practice personal hygiene, cook foods thoroughly, dispose of sewage in sanitary manner.

Mycotic Infections

Phycomycosis (mucormycosis, zygomycosis)	Absidia, Rhizopus Mortierella, Basidiobolus, Mucor, and Cunninghamella spp.	Possibly any moldy food.	Avoid eating moldy foods. Optimum control of diabetes mellitus.

SOURCE: Center for Disease Control, Atlanta.

quate supply of nutritious, acceptable food for the world population, several methods of food preservation have been developed. Some of these methods originated in ancient times, while others are more recent.

NATURAL PRESERVATION

Foods are whole or parts of plants and animals that were at one time alive. As living things, each type had its own defense mechanisms against harmful microbes. After plants or animals are harvested for consumption, their defenses are no longer maintained. The tissue begins to die, decompose, and become increasingly susceptible to the destructive actions of microbes. Many foods, however, have chemical or physical qualities that allow them to resist microbial decomposition for long periods. These foods can be preserved by maintaining their natural defense mechanisms. Coverings such as shells, membranes, husks, and skins are natural packaging materials that are barriers to microbes (Figure 23.4). Organic acids such as citric acid (fruits), benzoic acid (cranberries), and lactic acid (cheeses) act as chemical preservatives. Acidic foods (pH 3.7–4.5) such as tomatoes, pears, and pineapples have a longer shelf life than medium-acid foods (pH 4.5–5.3) such as beets, spinach, and pumpkins, or low-acid foods (pH above 5.3) such as peas, corn, meats, fish, poultry, and milk. Other natural antimicrobial chemicals serve to preserve low-acid foods. Lysozyme is found naturally in egg white and inhibits bacteria that may enter through small cracks in the shell. Milk has lysozyme and many other antimicrobial agents that keep it from spoiling: (1) immunoglobulins are available to combine with and neutralize antigens of contaminating microbes; (2) a fatty acid has been demonstrated to have antistaphylococcal properties; (3) poliovirus, influenza virus, enteropathogenic bacteria, and several protozoans are known to be inhibited by as yet unidentified chemical agents; and (4) an antibacterial agent similar to the peroxidase system found in saliva is also present in milk.

DRYING

One of the oldest food preservation methods is drying. It is inexpensive and results in products that are not only long-lasting but flavorful, light-weight, and easy to store. Apricots, raisins, coconuts, beans, peas, peppers, pork, and fish are regularly preserved by drying. The decrease in the water activity level (A_w) that takes place during this process inhibits the growth of most microbes. However, some osmophilic fungi and halobacteria may find the high osmotic pressure of this environment suitable for growth. Sun drying is the oldest, most basic method and can be done in the home (Figure 23.5). On a commercial scale, ovens are used to quickly extract water and concentrate salt. An already high

FIGURE 23.4
Many foods last for long periods after they are harvested due to their natural coverings, which prevent microbes from reaching and decomposing their interiors.

FIGURE 23.5
For centuries, many foods have been preserved by drying. (From C. Elliott, *Care of Game Meats and Trophies.* Outdoor Life Books, 1975.)

osmotic environment can be further increased by dipping the food into a hot brine solution before drying. Brine is a solution containing table salt (15 percent NaCl), sugar, sodium nitrate, and sodium nitrite. One of the most familiar foods preserved by this method is beef jerky (Table 23.3). Salt pork differs from jerky in that jerky is "cured" in the brine. **Curing** is preservation of food by salting, and no refrigeration is required. Although this product is not "dry," it has an equally effective, high osmotic pressure because of a decreased A_w.

> 1 Name three naturally occurring barriers to food spoilage.
> 2 What types of microbes have the ability to spoil dried foods?
> 3 What is brine?

Many products available in food stores have been produced by a rapid drying process in which a liquid food is forcefully sprayed into a large, round container. The cyclone force created in the drum quickly dries the product, which is then scraped off and packaged (Figure 23.6). Powdered milk is made by this process. Another method utilizes a heated cylinder that is rotated through liquid food products such as coffee, milk, fruit drinks, soups, and juices. Like the cyclone method, the food is scraped off for packaging. Freeze drying, or **lyophilization,** is also used by food processors. Lyophilization is quick freezing under a high vacuum that enables water to be extracted from the food before ice crystals form to cause cell damage. This method is used to preserve food

TABLE 23.3
Recipe for beef jerky.

Ingredients:
  2 cups water
  2 tsp. salt (table salt, onion salt, or garlic salt, or any
    pleasing combination)
  3 tsp. vinegar
  ⅙ cup soy sauce
  ½ lb. lean beef
Optional:
  Worcestershire sauce
  Red pepper
  Cut the meat with the grain into thin baconlike strips. Place the meat strips in the solution of water, vinegar, salts, and soy sauce. Boil for ten minutes. For "hot" jerky, add some Worcestershire sauce and red pepper to taste. Remove the strips and separate them on a wire rack. Heat the strips in a 200 °F oven for one hour or until the strips are dry. For best results, leave the jerky in the cold oven for several hours or overnight before storing in a paper bag. Do not store in a plastic bag; the jerky will become moldy in only a few days.

FIGURE 23.6
A spray-drying unit is designed to dehydrate such foods as milk, eggs, and coffee. The foods are sprayed into the cylindrical unit on top. The cold temperature of the unit causes the food to adhere to the inner walls, dry out, and flake off into the catch basin below.

for long-term storage. When the product is rehydrated by adding water, its flavor, aroma, and nutritive value are very similar to the original food. It should be kept in mind that after water is added, contaminating microbes originally found in the food will again be able to grow, causing spoilage and food infections or poisoning. Freeze-dried foods should be treated as cautiously as if they were fresh.

REFRIGERATION AND FREEZING

Low temperatures slow chemical reactions and microbial growth. Refrigeration of food reduces self-decomposition by enzymatic action and the likelihood of undesirable changes caused by most contaminating microbes. The normal operating temperature of a refrigerator should be approximately 5°C (40°F) but can vary between 0°C and 10°C. Home freezers should maintain a temperature of −18°C (0°F). In general, refrigeration slows microbial growth and freezing prevents growth. However, cold can also act as a selecting agent and affect the type of microbes found on foods during storage. As the temperature decreases, some bacteria slow or stop growing before others. *Pseudomonas* spp. can be the predominant bacteria at 1°C, while *E. coli* will have become inactive at 15°C. Many common fungi, such as species of *Penicillium, Candida, Cladosporium,* and *Sporotrichum,* are psychrophilic and grow at temperatures of −4°C. One yeast has been found to grow at temperatures as low as −34°C.

In order to maintain food quality, the optimum temperature, relative humidity, and ventilation should be determined. Different foods have different optimum food storage temperatures. Although most are preserved just above freezing, some are damaged at this temperature. For example, bananas keep best at 15°C (59°F), while sweet potatoes should be stored at 10°C (50°F). Foods also have an optimum relative humidity for storage. Too low of a relative humidity can cause moisture and weight loss. If the water forms ice crystals that evaporate from the surface, the food will appear dry, brown, and granular—a condition commonly known as "freezer burn." Too high of a relative humidity will foster microbial growth on the "sweating" food. Crisper compartments are designed for foods which are best kept at a higher relative humidity than found in other parts of the refrigerator. A high relative humidity is maintained in "crispers" by preventing rapid and frequent air circulation.

Blanching, steaming, and scalding are methods used in preparing fresh produce for home and commercial freezing. Foods such as broccoli, pears, string beans, corn, and spinach are immersed in boiling water for a short time then drained, packaged, and frozen. The blanching-freezing technique better preserves the food for several reasons: (1) many heat-sensitive microbes are destroyed during blanching; (2) plant enzymes are destroyed before they can "tenderize" the food, giving it a mushy character and producing changes in flavors, colors, and aromas; and (3) vitamin C (ascorbic acid) is retained in the food for a longer time.

Freezing results in the formation of ice crystals and does not guarantee the death of all contaminating microbes. When frozen foods are thawed, microbes can resume spoilage activities, and lytic enzymes of the food will begin to cause decomposition of ice-crystal damaged tissue. These actions increase as the temperature increases and result in foods that drip, leak, and become mushy. Because this damage is irreversible, frozen foods that have been thawed should not be refrozen.

IRRADIATION

Ultraviolet, beta, x-, and gamma radiation are lethal forms of energy used to destroy microbes. Microwaves emitted by certain types of broadcasting equipment and ovens are also able to kill microbes, but only because they produce high temperatures when focused on a material. The limited ability of ultraviolet light to penetrate deep into food restricts its use as a means of food preservation. UV radiation used in

cold food lockers helps limit spoilage on meat surfaces before it is cut and packaged. This control of surface decomposition does not affect the slow, lytic enzyme activity occurring deeper in the meat. In meats that are "hanging" for about ten days to two weeks, these enzymes begin to hydrolyze tough connective tissue fibers and tenderize the meat. Beta radiation is a controlled beam of moving electrons, and like UV light, it is only effective on surfaces. Beta rays directed onto the surface of potatoes destroy their "eyes," which prevents spoilage by budding. Many attempts have been made to use gamma and x-radiation as a means of food preservation because they easily penetrate deep into metal and glass containers, and theoretically they are able to destroy *Clostridium botulinum* spores. Canned foods, bacon, and potatoes are just a few foods that have been irradiated as a means of preservation, but this method is no longer permitted in the United States. Foods that have been exposed to gamma and x-radiation undergo significant, undesirable physical and chemical changes. They become mushy and lose their characteristic color, flavor, and aroma without the assurance of *C. botulinum* spore destruction.

---

1 What temperatures should be maintained for home refrigeration and freezing?

2 How does the process of blanching-freezing preserve foods?

3 How do microwaves kill microbes?

CANNING

One of the most important goals of food preservation is the destruction of *Clostridium botulinum.* This toxin-producing, anaerobic bacterium can be a natural or introduced contaminant of many foods. Canning techniques are designed to eliminate the problem and produce foods that are bacterially inactive, commercially sterile, or practically sterile. "Practical sterility" is a term used to describe foods that have been sealed in a canister (glass or metal) and heated to ensure the destruction of *C. botulinum.* Canning is practical, though not necessarily complete, because some thermoduric, aerobic bacteria may survive the process but are unable to grow in the absence of oxygen.

The success of canning is well known. First made practical in the late 1700s by the Frenchman Nicolas Appert, food preservation by canning has progressed from his method of heating cork-stoppered glass jars in boiling water to an industrial process of enormous proportions (Figure 23.7). Although the commercial canning business accounts for the largest source of preserved foods, it has been estimated that 40 percent of all families in the United States practice this

FIGURE 23.7
Modern commercial canning operation. (Courtesy of the Campbell Soup Company.)

method of food preservation in their own homes. To ensure that foods achieve practical sterility, it is necessary to select a proper processing time and temperature. These factors vary among high-acid and low-acid foods. In general, as the pH decreases (acidity increases), the time and temperature for preserving food also decreases. Table 23.4 lists several examples of high-acid and low-acid foods as well as their average sterilizing times.

Foods to be canned may be packed cold and raw into containers, or they may be preheated (though not cooked)

and packed hot. **Raw** or **cold-packed** foods are placed into the container and covered with boiling hot syrup, juice, or water before they are sealed and heat-processed. **Hot-packed** foods are first heated in water, steam, or syrups before being placed in a preheated container and sealed. To exhaust the air and seal the container before heat processing, the cans or jars are immersed in a 76°C (170°F) water bath. To destroy *C. botulinum* in high-acid foods, the cans are next placed in boiling water for approximately 20 minutes. A longer period may be necessary at elevations above sea level. Low-acid foods, however, require temperatures above boiling to achieve practical sterility. These foods are usually processed at 115°C (240°F) in pressure canners. This method is especially important for canning low-acid foods, since they are likely to have an optimum pH for growth of *C. botulinum*. The seal is checked and containers cooled by air or running water. Pressurized canned foods, such as dessert toppings and foam cheese spreads, are also preserved using this same basic method. However, the pressurized gas (carbon dioxide, nitrogen, or nitrous oxide) injected into the can might contain spoilage microbes, such as *Bacillus coagulans* or *Streptococcus faecalis*. In most instances, however, the gas is presterilized and chosen for its antimicrobial characteristics. As with other commercially canned foods, it is rare to find a spoiled package.

In recent years, plant geneticists have introduced many new tomato hybrids that many claim are less acid, and this feature prevents them from being safely canned by standard methods. Therefore, if tomatoes or other foods are suspected of being too low in acid content, the pH may be further decreased to a safe level by adding lemon juice, vinegar, or citric acid to the food before it is packaged.

TABLE 23.4
High-acid and low-acid foods, and their average practical sterilizing times.

High-Acid	Boiling Water Bath (minutes/quart)	Low-Acid	Pressure Canner (minutes/quart)
Tomatoes	45	Meat	90
Rhubarb	10	Poultry	90
Apples	25	Fish	100
Cherries	20	Low-acid tomatoes	15
Grapes	20	Asparagus	30
Peaches	25	Beans (string,	50
Plums	25	lima)	
Strawberries	15	Broccoli	40
Cranberries	10	Brussel sprouts	55
Figs	30	Carrots	30
Pears	30	Corn	85
		Greens	90
		Hominy	70
		Mushrooms	35
		Peas	40
		Peppers, green	35
		Squash	80
		Turnips	25

CHEMICAL PRESERVATIVES

Chemical preservatives added to foods can be classified into the four general groups of "salts," pyroligneous acids, spices, and acids. **Salting** foods with table salt as a means of preservation without refrigeration is as old as drying. However, sodium chloride is not the only chemical able to increase the osmotic pressure to concentrations that inhibit vegetative and spore forms of microbes.

The preservative action of sucrose is probably best known in foods such as "preserves," jellies, and jams. Fruits are prevented from spoiling because the sugar makes water unavailable to contaminating microbes. However, some osmophilic yeast can grow in this environment and spoil these products, as well as honey and natural syrups.

---

1 What happens to the pH of foods during the canning process?
2 What is the difference between "cold" and "hot" packing?
3 Name the four general groups of chemical preservatives.

Sugar, sugar-salt combinations, and salts other than NaCl are also used for this purpose. Foods to be preserved are allowed to "cure" after being immersed, injected, or coated with a brine solution. Chemicals permitted in the brine are sugar, sodium chloride (NaCl), sodium nitrite, (NaNO$_2$), sodium nitrate (NaNO$_3$), and vinegar. Table salt and sugar lower the A$_w$ and preserve the flavor of the product. Sugars (sucrose and glucose) not only enhance the flavor but act as a source of energy for contaminating bacteria (nitrate reducers) that convert nitrate to nitrite. The sodium nitrate is an added source of sodium nitrite which fixes the natural meat color by changing myoglobin to a stable red color. It also acts to inhibit the germination of spores. The bacteriostatic action of sodium nitrite is well known; but in recent years, its use has been brought into question since being identified as a carcinogenic agent. For this reason it has been recommended that sodium nitrate and nitrite not be used in the curing of foods such as bacon and hams.

Another old-fashioned, dry method of preserving foods is "smoking" with wood smoke. This method not only inhibits microbial growth but gives food an additional flavor enjoyed by many people. Smoked meat acquires a pleasing gloss and more tender surface. Smoking preserves by depositing **pyroligneous acids** on the surface of fresh meats and fish. This group of antimicrobial chemicals includes such well-known compounds as formaldehyde, phenol, and cresols. In many cases, smoking is performed in conjunction with fermentation. Meats are hung in the smokehouse after they have been inoculated with fermenting bacteria such as *Pediococcus cerevisiae* (Figure 23.8). These microbes release preservative organic acids inside the salami, sausage, or bologna as its surface is being covered with inhibiting pyroligneous acids. A product known as "liquid smoke" is available but has no preservative qualities. It is a combination of chemicals that only imparts a "smokehouse flavor" to meat or fish.

Spices are another group of chemical preservatives derived from plant materials. Originally the idea of "spicing foods" or eating "spicy foods" had little to do with flavor. Spices were added to preserve the food, since refrigeration was not available, and the incidence of food infections and

FIGURE 23.8
When meat is hung in the smokehouse, space must be left between each piece for the smoke to come in contact with all surfaces. (Courtesy of the PEET Packing Company, Chesaning, Michigan.)

poisonings was very high. Although today many think Christopher Columbus went on his voyage in search of table salt and pepper, these spices were readily available in Spain, Italy, and other Mediterranean countries. He was really seeking a short route to the Far East where he could buy preservative spices. Spices with antimicrobial properties include garlic, onions, mustard flour, cloves, and cinnamon (Figure 23.9). Extracts of cabbage, thyme, bay leaves, and turnips also have antibacterial properties.

Many foods have a natural **acidity** capable of inhibiting microbial decomposition and spoilage (refer to Table 23.4). This preservative action can be artificially added or developed in many products. One of the most familiar organic acids used to preserve foods is vinegar. The 4 – 6 percent acidity of white vinegar inhibits the growth of most microbes, when used as a preservative in such foods as fresh-packed dill pickles, pickled pigs feet, and other foods (Table 23.5). Other foods are acid-preserved with salts of organic acids such as sodium benzoate (benzoic acid), calcium sorbate (sorbic acid), sodium propionate (propionic acid), sodium citrate (citric acid), and sodium lactate (lactic acid). Table 23.6 lists some of the foods and their acidic additives. Developed acidity results from the action of fermenting bacteria inoculated into the food. One of the most important groups of bacteria responsible for the production of preserved foods such as cheeses, yogurt, and cultured buttermilk are the lactic acid bacteria of the genus *Lactobacillus*. While growing in the food, these bacteria release lactic acid into the surroundings at a rate that inhibits the growth of spoilage microbes. However, these are not the only kinds of microbes used in the food industry. Controlled fermen-

TABLE 23.5
Fresh-pack dill pickles (yields 7 quarts).

Cucumbers, 3 to 5 inches long, packed 7 to 10 per quart jar	17 to 18 pounds
5 percent brine (¾ cup pure granulated salt per gallon of water)	about 2 gallons
Vinegar	1½ quarts
Salt, pure granulated	¾ cup
Sugar	¼ cup
Water	2¼ quarts
Whole mixed pickling spice	2 tbs.
Whole mustard seed	2 tsp. per quart jar
Garlic, if desired	1 or 2 cloves per quart jar
Dill plant, fresh or dried (or)	3 heads per quart jar
Dill seed	1 tablespoon per quart jar

Wash cucumbers thoroughly; scrub with vegetable brush; drain. Cover with the 5 percent brine. Let sit overnight, drain.

Combine vinegar, salt, sugar, water, and mixed pickling spices that are tied in a clean, thin, white cloth; heat to boiling. Pack cucumbers into clean, hot quart jars. Add mustard seed, dill plant or seed, and garlic to each jar; cover with boiling liquid to within ½ inch from top of jar. Adjust jar lids.

Process in boiling water for 20 minutes* (start to count the processing time as soon as the hot jars are placed in the actively boiling water).

Remove jars and complete seals if necessary. Set jars upright on a wire rack or folded towel to cool. Place them several inches apart.

*Processing time is given for altitudes less than 1000 feet above sea level.
SOURCE: *Home and Garden Bulletin*, No. 92. USDA Prepared by Agricultural Research Service.

tations are carried out using different lactic acid-producing microbes or combinations of microbes in order to improve the shelf-life, flavor, texture, and appearance of the product (Table 23.7).

---

1 How does "smoking" preserve food?
2 Name four spices that have antimicrobial activity.
3 How does the fermentation process preserve foods?

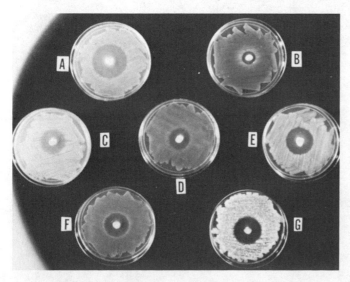

FIGURE 23.9
The natural antibacterial properties of garlic. The garlic is placed in the center of each dish. (From D. A. Anderson and R. J. Sobieski, *Introduction to Microbiology*, 2nd ed. C.V. Mosby Co., St. Louis, 1980.)

FOOD MICROBIOLOGY: FERMENTED FOODS

There is no way of knowing exactly how long ago fermentations were discovered. Sumerian tablets from ancient Mesopotamia refer to cheese as early as 4000 B.C., but there are strong suspicions that prehistoric cave dwellers may have made cheese as a means of preserving milk more than 8000

TABLE 23.6

Foods preserved by the addition of acids.

Mayonnaise	Acetic acid
Pickles	Acetic acid
Pickled pigs feet	Acetic acid
Jams, jellies, and canned fruits	Sodium citrate, sodium benzoate, calcium sorbate
Soft drinks	Sodium benzoate, sodium citrate
Baked goods	Sodium sorbate, calcium propionate
Cheeses	Sodium sorbate, propionic acid
Olives	Lactic acid
Syrups	Sodium citrate
Margarine	Sodium benzoate

Courtesy of the PEET Packing Company, Chesaning, Michigan.

TABLE 23.7

Fermented foods and their microbes.

Food	Starting Material	Microbes
Leavened bread	Flour	Saccharomyces cerevisiae
Cottage cheese	Cow's milk	Streptococcus lactis
Roquefort cheese	Sheep's milk	Penicillium roquefortii
Blue cheese	Cow's milk	Penicillium roquefortii
Brick cheese	Cow's milk	Bacillus linens
Swiss cheese	Cow's milk	Propionibacterium shermanii and Streptococcus thermophilus
Parmesan and romano cheeses	Cow's milk	Lactobacillus bulgaricus and Strep. thermophilus
Yogurt	Cow's or goat's milk	Strep. thermophilus and Lactobacillus bulgaricus
Bulgarian buttermilk	Cow's or goat's milk	Lactobacillus bulgaricus
Filia	Cow's milk	Strep. lactis var. hollandicus
Kefir	Cow's milk	Strep. lactis, Strep. cremoris, and several yeasts
Kumiss	Horse's milk	Lactobacillus spp.
Cultured sour cream	Cow's milk	Strep. diacetilactis, Strep. cremoris, or Strep. lactis
Cultured buttermilk	Cow's milk	Strep. cremoris, Strep. lactis, Leuconostic cremoris, Strep. diacetilactis, Leuconostoc citrovorum
Sauerkraut	Cabbage	Lactobacillus spp. and Leuconostoc mesenteroides
Soy sauce	Soybeans	Aspergillus spp.
Fermented tea	Tea leaves	Acetobacter spp. and Saccharomyces spp.
Fermented olives	Olives	Lactobacillus spp.
Cacao	Cocoa beans	Candida krusei
Vanilla	Vanilla beans	Unknown
Citron	Citron fruits	Saccharomyces spp. and Bacillus citri
Tempeh	Soybeans	Rhizopus spp.
Poi	Taro plant	Lactobacillus pastorianus Strep. lactis and others
Wine	Grape	Saccharomyces cerevisiae

524

years ago. The Egyptians are credited as being the first to ferment grains in the production of beer. The wine industry no doubt dates back to at least 3000 B.C. in what is now northern Iran where grapes grow wild. Many legends describe the accidental discovery of fermented products such as wine, yogurt, and cheese (Box 27). The Roquefort cheese story dates back more than 2000 years to a lovesick shepherd who left his lunch of sheep's milk and bread in one of the caves of Mount Cambalou to meet with his beloved. But, all did not go well. He lost his lover, and his sheep ran away. However, on returning to the cave, he found his milk ripened to a blue-veined cheese. The unique circulation and atmosphere of the cave had imparted an incomparable flavor, texture, and aroma to the cheese. It was named Roquefort after the village of its origin and became known worldwide. Pliny, the Elder of Ancient Rome and later Charlemagne of France, praised its uniqueness and qualities. The cheese still has a well-founded reputation; however, the small business has grown. Today over fifty million pounds of Roquefort cheese are aged in the caves each year.

According to J. G. Davis:

> Cheese is the curd or substance formed by the coagulation of the milk of certain mammals by rennet or similar enzymes in the presence of lactic acid produced by added or adventitious microorganisms from which part of the moisture has been removed by cutting, warming, and/or pressing, which has been shaped in moulds and then ripened by holding for some time at suitable temperatures and humidities*.

The process is divided into two portions: souring and ripening (aging). During the **souring** process, the enzyme rennet (taken from the stomach of nursing calves or derived from certain bacteria) is added to facilitate the coagulation of the milk protein (casein), butterfat, and other components into a formed material known as **curd.** (The French word for cheese, fromage, means *formed milk.*) Bacteria occurring naturally in the milk and others are added as **starter cultures.** During this process they ferment the milk sugar, releasing lactic acid into their surroundings. The steady, slow increase in lactic acid along with the coagulative action of rennet results in the separation of curd from the water and dissolved materials called **whey.** In most cheese-making operations, the curd is banked on the sides of the fermenting tube, cut, and the whey is drained away (Figure 23.10). However, in some Scandinavian countries, the whey is cooked to concentrate the remaining lactose and milk solids. These are fermented a second time into Getmesost, a butterscotch-colored, sweet dessert cheese. If the curd is salted and packaged without aging, it is known as unripened cheese.

*From J. G. Davis, *Cheese, Basic Technology,* vol. 1. Elsevier, 1975.

Cottage cheese, farmer cheese, cream cheese, and Neufchâtel are unripened cheeses. Curd which undergoes a second microbial fermentation during the aging or ripening process is called ripened cheese. A variety of cheeses can be produced using different combinations of milk microbes and environmental conditions. **Ripening** begins with the inoculation of the curd with the desired bacteria or fungi. These microbes can be placed inside the curd or spread on the surface. The curd is molded in cheese cloth and placed in a suitable environment (ripening room or house) for incubation. As the microbes ferment the curd, they produce changes in the flavor, color, aroma, and texture of the cheese. Depending on their texture and water content, ripened cheeses are divided into the four classes of soft, semisoft, hard, and grating cheeses. To ensure consistency and a high-quality product, all stages in the cheese-making process are carefully controlled to prevent defects. The most likely problems are the growth of unwanted molds and the development of fermentation by-products such as gases and chemicals that give cheese undesirable flavors and aromas.

> **1** What materials are necessary for souring milk in the cheese-making process?
> **2** What changes occur during the ripening process?

Yogurt is the product of milk sugar (lactose) fermentation by the bacteria *Lactobacillus bulgaricus* and *Streptococcus thermophilus*. Both of these bacteria convert lactose to lactic acid (Figure 23.11). Since acids have a sour taste, the result of their activity is a sour-tasting dairy product. The production of the acid is responsible for the changes in the consistency of the milk. The greatly lowered pH causes the milk protein to coagulate and thicken. *Streptococcus thermophilus* also produces other compounds that are important in determining the final flavor.

In order to provide a consistently palatable product, commercial yogurt producers must have a thorough understanding of the biology of the bacteria they use. They must be aware that changes in the environment will alter the activities of the bacteria. Fresh milk normally contains a mixture of bacterial species that will eventually ferment the milk (turn it sour). Since these bacteria produce unwanted flavors and compete with the desired bacteria, it is desirable to reduce the natural bacterial population as much as possible. This is usually done by heating the milk for a specified time.

Substances that inhibit the growth of bacteria are also problems for the yogurt producer. One type of inhibitor is produced naturally by the cow. The second kind of inhibitor is a much greater problem that results from cows treated

A

B

C

D

E

F

with antibiotics. If cows are being treated with penicillin or some other antibiotic that slows down the growth of disease-causing bacteria, this antibiotic can get into the milk and prevent the bacteria from making yogurt. Heating also rids the milk of antibiotics.

Researchers in Israel have developed what is known as "instant yogurt." A powder of lyophilized bacteria, fermenting agents, dehydrated milk, and fruit flavor is mixed with

milk or water and allowed to incubate at room temperature. In one hour, the suspension is converted to yogurt and is said to taste the same as regular yogurt.

Cheeses, yogurt, and many other dairy products are fermented with starter cultures of lactic acid bacteria added to milk. However, many foods are fermented by lactic acid bacteria found naturally on their surfaces. Among this group, one of the most unique fermented foods is sauerkraut. It is

G

H

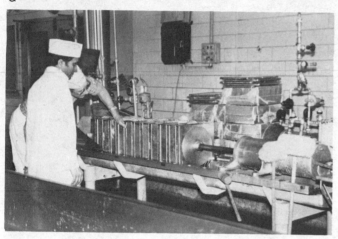

I

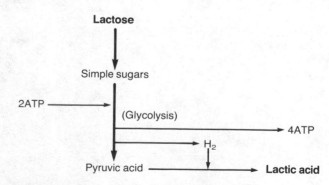

J

FIGURE 23.10

(A) Milk used in the cheese-making process is first pasteurized in this bulk pasteurizer. (B) The milk is then pumped into the vat where paddles suspended from above travel back and forth to constantly stir it. After the vat is filled, renin from the fungus *Mucor* is added along with lactic acid to produce bacterial starter and begin the curding process. (C) After the milk has "setup," wire forms are used to cut the curd, which is then firmed to separate the whey. (D) A bank, or dam, is moved to one end of the vat to separate the curds and whey. The whey is drained from the vat through a spigot at the base. (E) The curds are "ditched" and banked" to drain more of the whey. (F) After draining, the curds are cut into large chunks and "cheddared" by gently heating and rotating them in the vat. (G) The cheddared blocks are "milled" into smaller blocks that can be salted, which improves the flavor and acts as a preservative. (H) The curds are removed from the vat and "hooped." The frames are lined with cheese cloth or a plastic disposable wrapper that allows more whey to drain. The cheese curds are enclosed and the frame is covered. (I) Twenty-pound block frames of cheese are placed in a hydraulic press, which squeezes the desired amount of whey from the curd and helps knead the curd into a "wheel," or block. (J) The blocks of cheese are removed from the frames and placed into a ripening room, where the bacteria "age" the cheese. Finally, the cheese is cut, wrapped, and marketed. (Courtesy of K. Nakrani and D. Henning.)

unusual in that it is eaten as a main dish and not as a relish. Sauerkraut has been an important food in the diet of millions for hundreds of years. Besides being a flavorful food, it is an excellent source of vitamin C, containing amounts equal to that found in many citrus fruits. "Kraut" was regularly eaten on long sea voyages to prevent scurvy. In addition, the high lactic acid content of the food aids in stabilizing normal intestinal flora (refer to Chapter 12) and preventing gastroenteritis.

FIGURE 23.11

As lactose, or milk sugar, is metabolized to lactic acid, the milk sours and milk protein coagulates to form curd.

BOX 27

CURDS
AND
WHEY

AS the human population has grown, advances in food technology have introduced many low-cost, high-quality products. Because these items are readily available and need not be homemade any more, many families no longer hand down food-making traditions to the younger generation. An example is the tradition and vocabulary of cheese making.

Miss Muffet's terrifying experience with the spider is one of Mother Goose's most famous nursery rhymes, but few children and not many more adults know what she was eating while sitting on her tuffet. Most think she was eating oatmeal or porridge (coarse oatmeal), but she was really enjoying a bowl of fresh cheese curd still in its whey.

As parents and teachers read fairy tales and rhymes, children often ask if the stories are true. In this case, there is a good chance that Miss Muffet was an actual child. She was probably Patience Muffet, the daughter of Sir Thomas Muffet (1553–1604), who was a well-known English physician, entomologist (specializing in spiders), and poet. Among his most famous books of poetry were *Silkworms and Their Fleas* and *The Theater of Insects or Lesser Living Things*. In addition, he wrote a textbook on human nutrition, *Health's Improvement or Rules Comprising and Discovering the Mature Method and Manner of Preparing all Sorts of Food Used in This Nation*. In the text he advised eating fresh cheese and avoiding old cheese since it would cause constipation. Curds and whey are the freshest form.

Making sauerkraut at home is a long-standing tradition in many countries and probably originated in Germany or Poland. The outer leaves of fresh cabbage are removed, and the head is quartered and shredded. Using a large crock or glass container (metal will inhibit the bacteria), three tablespoons of pickling or canning salt is thoroughly mixed with five pounds of the shredded cabbage. No water is added to produce a liquid brine, since the cabbage contains a great deal of fluid that will be extracted from the cells to form its own brine. This wilts the cabbage and makes it easier to pack into the crock without breaking. It is essential that a 2–3 percent salt solution be maintained to discourage the growth of unwanted microbes and encourage fermentation by three types of lactic acid bacteria: *Leuconostoc mesenteroides*, *Lactobacillus plantarum*, and *Lactobacillus brevis*. The crock is filled almost to the top with several five-pound layers of salted and shredded cabbage, covered with a snug-fitting plate, and weighted down with a heavy weight or water-filled plastic bag (Figure 23.12). This maintains an anaerobic environment by preventing air from entering the brine. The cabbage is allowed to ferment at room temperature (68–72° F) for 9 to 14 days. If a scum or slime forms on the surface, it is most likely due to the growth of an encapsulated form of *L. plantarum* and can simply be scraped off. The fermentation is over when gas bubbles no longer rise to the surface. The sauerkraut can then be served or canned. Other thin-sliced vegetables, such as carrots, cucumbers, and turnips, can also be pickled using this method. A more mild form of sauerkraut can be made from head lettuce.

Pickling is a method of preserving vegetables by lowering the pH with organic acid. This can be done by lactic acid fermentations lasting several weeks or by the fresh-pack method. Fresh-packed dill pickles (cucumbers), cauliflower, carrots, and other vegetables are brined for only a few hours, drained, and covered with boiling vinegar. The four percent acetic acid serves the same purpose as the lactic acid of brined pickled products. Vinegar is also a product of microbial fermentation. Several members of the bacterial genus *Acetobacter* oxidize ethyl alcohol under aerobic conditions to acetic acid. This fermentation is called acetification. Many alcoholic solutions are acetified for the production of vinegar. Wine, hard apple cider, and alcoholic malt mash are typically used in the commercial preparation of vinegar. Although Louis Pasteur was hired by the French wine industry to inhibit the action of these microbes, one of the most popular specialty products today are wine vinegars used in salad dressings.

---

1 What kinds of substances can inhibit bacteria in a yogurt culture?
2 Name the bacteria responsible for sauerkraut production.
3 What two methods can be used to pickle vegetables?

Commercially, white distilled vinegar can be produced by spraying alcohol over the surfaces of an *Acetobacter*-inoculated material that provides a large surface area for acetification. Beechwood chips, corncobs, or charcoal can be used in the fermenter, called a vinegar generator. As the solution trickles over the surfaces, the *Acetobacter* converts the alcohol to acetic acid, and vinegar is drained from the tank. Using this controlled acetification method, a high-quality, consistent product is produced. Today many industrial producers of vinegar use another type of fermenter called an **acetator** (Figure 23.13). This unit is small, fully automated, and needs no packing material such as beechwood shavings. Unlike the vinegar generator, an acetator is able to handle a variety of raw materials such as wine or fruit. These fermenters can be run on either a batch or continuous culture basis and have automated temperature and aeration control mechanisms.

THE STAFF OF LIFE

The title of this section is only one of many phrases that reflects the nutritional and symbolic importance of bread in the lives of people throughout the world. Bread is considered in many cultures as a basic human requirement.

The fermentation of moist, ground grain, such as wheat and rye, does not preserve the food but produces other distinctive and desirable qualities. Dough, or pre-bread, is a mixture of flour made from ground grain, salt, and water.

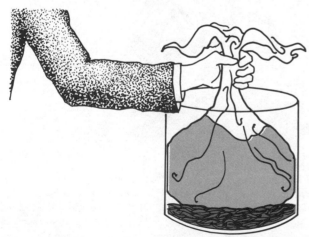

FIGURE 23.12
To ferment sauerkraut at home, place a water-filled plastic bag on top of the shredded cabbage. The water bag will fit snugly against the sides of the container to prevent exposure to air. (From "Making Pickles and Relishes at Home." *Home Garden Bulletin #92*, U.S. Department of Agriculture.)

FIGURE 23.13
A vinegar acetator, or aerator, is used throughout the world in the manufacture of vinegar. (Courtesy of Heinrich Frings GmbH & Co., Bonn, West Germany.)

Microbes can also be added to the dough before it is baked to ferment the sugars and produce carbon dioxide gas and alcohol. As gas accumulates in small pockets, the dough rises, or is leavened. Bread baked without a leavening agent does not have a porous texture, and it is more compact and heavy.

Leavening can be accomplished by microbes or chemicals. Chemical leavening may be done by incorporating CO_2 directly into the dough or by adding chemicals that release CO_2 when heated. Baking powder is often added to ensure adequate CO_2 production, and self-rising flour contains chemical components which react to form baking powder when moistened. The most common microbe used to leaven bread is *Saccharomyces cerevisiae*, or baker's yeast. Yeast is added in the form of cake yeast, or active dry yeast. Cake yeast is used less often in home baking because of its short shelf life and bulk. Active dry yeast is lyophilized *S. cerevisiae*, which has a longer storage life and does not have to be refrigerated. During raising, the yeast cells metabolize sugar to produce CO_2 and alcohol. The alcohol is lost during baking, but the CO_2 is trapped among the gluten (flour protein), which has become "conditioned" by proteolytic enzymes and acids in the flour. Conditioning gives the dough its elastic or "spongy" texture that allows it to retain more of the gas. Dough is sometimes called the "sponge" be-

cause of its texture. Dough conditioners, or "yeast food," are sometimes used to develop this quality at a faster rate. Many commercial bakers do not add these conditioners, so that they can label their products "natural." Amylase enzyme produced from bacterial cultures is added by commercial bakers to hydrolyze the starch amylose. This releases more sugar for yeast metabolism, releases a larger amount of CO_2, makes kneading easier, and gives bread a soft, moist texture. Home-baked bread is more coarse, because this enzyme is not usually added to the dough.

Due to the difficulty in maintaining active yeast cultures, many people throughout history have utilized other microbes to leaven their bread and baked goods. One of the most familiar and popular breads leavened by bacteria is "sourdough." Lactobacilli and other lactic acid bacteria found in flour and milk ferment lactose sugar, releasing CO_2 and organic acids. These combine to leaven the bread and provide a sour or acid flavor. Sourdough bread pancakes and muffins are made from a "starter" of fermenting flour and milk that is used in much the same way as cake yeast. Many interesting stories have been told that describe the origin and use of sourdough (Figure 23.14). Sourdough starter is prepared in a number of ways, but all require the souring of milk. Souring by lactic acid bacteria can take place naturally in an open pan, or it is encouraged by the addition of yogurt or a small amount of commercially available organic

FIGURE 23.14
The University of Alaska's publication *Sourdough* mentions several legends about sourdough: " 'Sourdough Pete,' when a young man, went to Alaska at the turn of the century to seek his fortune. His grandmother, who had pioneered in the Michigan woods, knew a thing or two about hardships in a new land. Her parting gift, a crock of starter for sourdough hotcakes and bread, made Pete famous over the land. With the help of a sack of flour, 'Sourdough Pete' always had hotcakes to eat whether he struck it rich or not. He shared them with friends who, the story goes, walked miles to renew or get a starter of the original product. Pete became known for his generosity and his name 'Sourdough Pete' originated."

acids. Lactic, pyruvic, and propionic acids are usually sold in powdered form. The sour milk is combined with the flour and incubated in a covered container. The sponge can be kept without refrigeration, but it lasts longer if kept cool. Before it is used for baking, the starter sponge is built-up with additional flour and milk, and incubated for several hours (Table 23.8). This provides a sufficient amount of active culture for leavening and enough to save for a later time. Many people maintain sourdough sponges for years by following some basic microbiological principles: (1) if it is refrigerated between uses, it will last longer; (2) the sponge is a live culture of lactic-acid fermenting bacteria and must be "fed" milk sugar to be maintained; (3) the culture should not be kept in a metal container because it may be antibacterial; and (4) the container should not be filled with milk and flour before incubating. The gas produced by an active fermentation will cause the sponge to double in size.

---

1 How can dough be leavened?
2 What is the advantage of using bacteria as a leavening agent?

TABLE 23.8
Sourdough wheat bread recipe.

2 cups sourdough starter (set the previous night)
1 cup whole wheat or graham flour
1½ teaspoons salt
2 tablespoons sugar
1 cup white flour

Combine ingredients and mix well with a fork—this sponge will be sticky. Set in a warm cupboard for 2 hours or more. Turn out on a warm, well-floured board. Knead 1 or more cups white flour into the dough for 5 to 10 minutes. Shape into a round loaf and place in a well-greased pie pan. Grease sides and top of loaf, cover with a towel and let rise 1 hour or until doubled. Bake in a preheated oven at 450°F. for 10 minutes, then reduce heat to 375°F. and bake 30 to 40 minutes longer. Makes one large loaf.

SOURCE: *Sourdough Cookin'*, Michigan State University Cooperative Extension Service, East Lansing, Michigan, 1976.

SUMMARY

The fields of food and dairy microbiology have four important goals: (1) preventing food spoilage and loss due to the activities of contaminating microbes or their harmful toxins; (2) developing better methods to preserve foods against microbial decomposition during distribution and storage; (3) using beneficial microbes to improve the nutritional value, texture, aroma, and flavor of foods; and (4) reducing micro-

bial infections and poisoning from the ingestion of harmful foods. To achieve these goals, microbiologists have studied the chemical and physical properties of foods and the microbes that contaminate them. The pH, water activity level, available oxygen supply, and nutrient balance all influence the likelihood of spoilage, food poisoning, or infection. Spoilage can result from enzymatic decomposition, microbial growth in excessive numbers, or the release of microbial products that produce undesirable flavors, colors, or aromas. Foods containing microbes that release toxic substances can be sources of illness known as food poisoning. *Staphylococcus, Clostridium,* and *Bacillus* are the genera most involved in bacterial food poisonings. Some species of *Aspergillus* and *Penicillium* release mycotoxins (fungal poisons), such as aflatoxin, that can cause foodborne intoxication. Shellfish and fish are also sources of illness, since they can concentrate toxic algae or carry poisons generated by contaminating bacteria. Foodborne infections are caused by the growth and activities of microbes ingested from contaminated foods. Among the most notable are shigellosis, septic sore throat, brucellosis, cholera, typhoid, hepatitis, and amoebic dysentery.

The prevention of foodborne illnesses is made possible by a number of food-preserving methods. Natural preservation can be the result of protective husks, skins, shells, or membranes. The high acidity of many foods prevents microbial growth, and antimicrobial agents such as lysozymes and fatty acids are also effective. Other methods include drying, refrigeration, freezing, irradiation, canning, and the use of chemical preservatives such as salt, spices, pyroligneous acids, and organic acids. The generation of organic acids in foods by fermentation not only increases the shelf-life of a product but results in enhanced flavors, aromas, and textures not found in the starting material. Among the many foods fermented by fungi and lactic-acid producing bacteria are cheeses, buttermilk, sauerkraut, vinegar, and leavened bread.

WITH A LITTLE THOUGHT

Sourdough Pete had a staphylococcal infection when he was hired as a baker. The infection made him nauseated and lethargic while preparing his famous sourdough bread. Several days later, Pete was hospitalized, the starter sponge turned green, and the bread failed to raise properly. One of his coworkers is worried that Pete's microbes may cause illness from eating the bread. He feels that the sponge will have to be destroyed, and the restaurant will be closed by public health officials.

As the investigating microbiologist, what connection might exist among these various problems, and what recommendation might you make in order to best resolve the problem?

STUDY QUESTIONS

1 What are the major goals of food and dairy microbiology?

2 What is the water activity level, A_w? How might this level be changed in a food?

3 Describe two chemical-physical properties of food that influence the growth of microbes and make them more perishable.

4 What is the difference between food poisoning and food infection? Give examples and disease symptoms of each.

5 Why should canned food be boiled for 15 to 20 minutes before it is served?

6 What is aflatoxin? Under what circumstances is it produced and how might it be controlled?

7 List several natural food preservatives and explain how they protect the food from microbial damage.

8 Why should frozen foods not be refrozen after they have been thawed?

9 Why are canning procedures different for high-acid and low-acid foods? Give examples of these foods.

10 Name the microbes associated with the following food products: Swiss cheese, cultured buttermilk, yogurt, sauerkraut, leavened bread, sourdough bread, wine, honey, jams and jellies, and potato salad.

SUGGESTED READINGS

American Public Health Association. *Recommended Methods for the Microbiological Examination of Foods,* 2nd ed. New York, 1966.

AYRES, J. C., J. O. MUNDT, and W. E. SANDINE. *Microbiology of Foods.* W. H. Freeman and Company, San Francisco. 1980.

BEUCHAT, L. R., ed. *Food and Beverage Mycology.* AVI Publishing, Westport, Conn., 1978.

BRYAN, F. L. *Diseases Transmitted by Foods.* U.S. Department of Health, Education, and Welfare; publication no. (CDC) 75–8237, Feb., 1975.

CAMPBELL, T. C., and L. STOLLOFF. "Implication of Mycotoxins for Human Health." *Journal of Agricultural Food Chemistry* 22(6):1006–15, 1974.

FRAZIER, W. C., and D. C. WESTHOFF. *Food Microbiology,* 2nd ed. McGraw-Hill, New York, 1978.

GOLDBLATT, L. A., ed. *Aflatoxin: Scientific Background, Control, and Implications.* Academic Press, New York, 1969.

JAY, J. M. *Modern Food Microbiology,* 2nd ed. Van Nostrand Reinhold, New York, 1978.

WEISER, H. H., G. J. MOUNTNEY, and W. A. GOULD, *Practical Food Microbiology and Technology.* AVI Publishing, Westport, Conn., 1976.

KEY TERMS

water activity level, A_w	aflatoxin	pyroligneous acids (pi″ro-lig-ne-us)
mycotoxin		

Pronunciation Guide for Organisms

Cladosporium (klad″o-spo′re-um)

Sporotrichum (spo-rot′ri-kum)

Soil and Agricultural Microbiology

SOILS, SOIL FORMATION, AND MICROBES

DECOMPOSERS AND BIODEGRADABLE MATERIALS

BIOGEOCHEMICAL RECYCLING

NITROGEN CYCLE

CROPS AND MICROBES

ANIMALS AND MICROBES

Learning Objectives

☐ Become familiar with the roles microbes play in soil development.

☐ Know the terms "humus," "microbial gum," "water-holding capacity" and "soil-filtration rate."

☐ Be able to explain the concepts of biodegradable and nonbiodegradable as they relate to microbial action.

☐ Give examples of biological amplification, biological control, and biogeochemical recycling.

☐ Be able to trace the sequence of events in the nitrogen cycle and list the names of microbes that play key roles in this cycle.

☐ Be able to describe the processes of eutrophication and composting.

☐ Explain the significance of mycorrhizal associations.

☐ Be able to list various methods of plant disease control.

☐ Be familiar with the process of silage production.

☐ Know the microbial activities that occur in animals known as ruminants.

Today many people use the words "ecology" and "environment." Students, politicians, and microbiologists speak of "ecological" or "environmental" issues. **Ecology** is the study of organisms in relationship to their environment. This is a simple statement for a very complex study. The total of all living or nonliving factors that influence an organism is its **environment.** An **ecosystem** is the interaction between living and the nonliving environment in which they exist. The populations of an ecosystem that interact with one another constitute a **community** of living things. Many types of ecosystems exist, but all of these together make up a single worldwide **biosphere.**

Microbes play many important roles in all ecosystems found throughout the biosphere. Their presence and activities, along with chemical and physical agents, result in the formation of soil from solid rock. There are many types of soils, and each has different chemical, physical, and microbiological properties. These differences directly affect the kinds of plants and animals that grow in and on the soil. The

24

activities of soil microbes prevent complex molecules and compounds from becoming locked into the soil and unavailable for reuse by other life forms. By decomposing compounds or altering their form, microbes release many beneficial compounds and elements into the soil as fertilizers. Their activities also result in the breakdown of toxic substances that, if concentrated, would cause harm to many plants, animals, and humans.

Microbes also exist in a number of symbiotic relationships with other organisms. Plants, animals, and humans can be affected by parasitic microbes. These relationships have resulted in significant crop and animal losses, as well as innumerable cases of human illness and death. However, research has revealed the existence of many plant-microbe and animal-microbe relationships that are mutualistic, providing benefits to both kinds of organisms. In many situations, this mutualism is so complete that the loss of the microbes will result in death of the host. In many cases, mutualism goes beyond the immediate microbe-plant interaction and provides benefits to other plants, animals, and humans that cannot be equaled by human effort. Knowledge of these beneficial relationships has enabled researchers, technicians, and farmers to benefit from these microbial activities.

SOILS, SOIL FORMATION, AND MICROBES

Soil is the loose mineral material on the earth's surface that is the growth medium for many microbes, plants, and animals. The formation of soil is the result of chemical, physi-

cal, and biological **weathering** that occurs over long periods. Repeated freezing and fracturing of large sections of rock provide an increased surface on which the first, or **pioneer plants,** take hold. These are usually algae (green and blue-green), lichens (fungal-algal symbionts), and mosses. Working together over a long period, they excrete various organic acids capable of dissolving the parent rock. As they die and are replaced by other life forms, a soil base is built on which more complex organisms grow. The first soil contains large rock fragments and a minimum of organic material derived from dead, decaying plants and microscopic animals. The acids of decay percolate deeper into the freeze-fractured rock and extend the weathering process. Eventually, these processes result in the formation of distinct soil layers called **horizons.** Geologists and soil scientists recognize differences among horizons and use these characteristics to classify soils into types and ages (Table 24.1).

The size, shape, and chemical composition of the inorganic parent material influences the texture and biological activities of different soil types. Clay soils have very small particles which pack tightly together. This limits the amount of pore space between the particles. When water is poured on clay, it adheres to the particles, fills the pore spaces, and leaves little room for air. This gives clay a high **water-holding capacity** and a low **soil-filtration rate** (Figure 24.1). The lack of air in soil provides little opportunity for the growth of aerobes but a favorable environment for anaerobes. When a soil becomes waterlogged, it quickly becomes anaerobic and begins to develop a putrid odor caused by the release of foul-smelling by-products from anaerobic bacteria. Drainage ditches, shallow ponds, and stagnant pools of water are well known for this characteristic odor. It is also difficult to

TABLE 24.1
New soil orders and approximate equivalents in great soil groups of 1949 system.

	Order	Meaning	Approximate Equivalents
1	Entisol	Recent soil	Azonal soils and some Low Humic Gley soils
2	Vertisol	Inverted soil	Grumusols.
3	Inceptisol	Inception, or young soil	Ando, Sol Brun Acide, some Brown Forest, Low Humic Gley, and Humic Gley soils.
4	Aridisol	Arid soil	Desert, Reddish Desert Sierozem, Solonchak, some Brown and Reddish Brown Soils, and associated Solonetz.
5	Mollisol	Soft soil	Chestnut, Chernozem, Brunizem (Prairie), Rendzinas, some Brown, Brown Forest, and associated Solonetz and Humic Gley soils.
6	Spodosol	Ashy (Podzol) soil	Podzols Brown Podzolic soils, and Ground-Water Podzols.
7	Alfisol	Pedalfer (Al-Fe) soil	Gray-Brown Podzolic, Gray Wooded, Noncalcic Brown, Degraded Chernozem, and associated Planosols and Half-Bog soils.
8	Ultisol	Ultimate (of leaching)	Red-Yellow Podzolic, Reddish-Brown Lateritic (of United States), and associated Planosols and Half Bog soils.
9	Oxisol	Oxide soils	Laterite soils Latosols.
10	Histosol	Tissue (organic) soils	Bog soils.

SOURCE: Modified from H. Foth, *Fundamentals of Soil Science,* 6th ed. John Wiley & Sons, 1978.

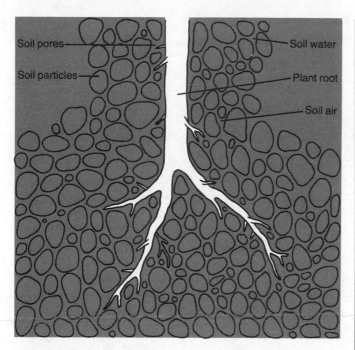

FIGURE 24.1
Whether a soil serves as food medium for plant and animal growth depends on soil particle size, water-holding capacity, and air-holding capacity. Proper pore space enables roots to receive adequate water and air for growth.

grow plants in clay because root cells are aerobic. Sandy soils have very large, granular particles which do not pack well and have large pore spaces. Their water-holding capacity is very low and soil-filtration rate high. Water poured on sand percolates through and is quickly lost. The low moisture content of sand makes it difficult for microbes and plants to survive. Soil types that are optimum for microbial growth are those that contain humus. **Humus** is the dark-colored organic material of soil derived from natural and microbial decomposition of microbes, plants, and animals. When humus is mixed with clay or sand, it optimizes the water-holding capacity and soil-filtration rate, enabling many life forms to grow. The natural development of humus occurs as soils mature from their basic parent material to a mature soil known as **mollisol** (Figure 24.2). Mollisol has a maximum amount of humus in the surface layer and only a moderate amount of clay in the lower levels. This rich, black soil is found in many parts of the world, and in the United States it is the best growth medium for corn. Constant microbial activity in the upper layer helps maintain the high quality of this soil type.

Soil contains many kinds of organisms that contribute to its fertility. The root systems of higher plants and animals, such as worms, nematodes, and insects, penetrate to break up clumps, weather, loosen, and aerate the soil. Microbes constantly interact with these higher organisms and contribute to the total weight and quality of the soil. Microbes are

essential to the formation of **microbial gum,** which is composed of fungal mycelia and microbial products such as waxes, fats, and electrically charged products. The gum binds particles in **aggregates** of size and shape characteristic of each soil type. The amount and nature of the aggregate found in different soils determines the important properties of consistency, weight, pore space, and density (Table 24.2).

> **1** Distinguish between water-holding capacity and soil-filtration rate.
> **2** How does humus optimize growth conditions in soil?
> **3** What is microbial gum?

The most numerous microbes in soil are the bacteria, followed by the actinomycetes, fungi, algae, and protozoa. Members of the genus *Arthrobacter* are the most abundant bacteria (5–35 percent), while the rest of the microbial community is comprised of members of the genera *Pseudomonas, Clostridium, Bacillus, Micrococcus, Flavobacterium, Chromobacterium,* and *Mycobacterium.* The actinomyces are represented by several species of *Streptomyces* and *Nocardia.* The smell of fresh "dirt" is caused by volatile compounds from the activities of *Streptomyces griseus,* a microbe well known for its ability to produce antibiotics. Commercially available potting soil that has been sterilized lacks this characteristic smell, until it becomes inoculated with *S. griseus* from the roots of plants. Among the representative fungi are *Aspergillus, Penicillium, Trichoderma, Mucor,* and others. Although many other types are no doubt present, they have been difficult to accurately identify and quantify because of complex culturing problems. Typically associated with aquatic environments, algae and protozoa are also present in moist soils, but their number and kind vary

TABLE 24.2
The weight of soil organisms per acre-foot of fertile agricultural land.

Organism	Weight (lb/acre-ft)	
	Low Fertility	High Fertility
Bacteria	500	1000
Fungi	1500	2000
Actinomycetes	800	1500
Protozoa	200	400
Algae	200	300
Nematodes	25	50
Other worms and insects	800	1000
Total weights	4025	6250

SOURCE: Data from O. H. Allen, *Experiments in Soil Bacteriology.* Burgess Publishing, 1957.

Soil development in relation to time

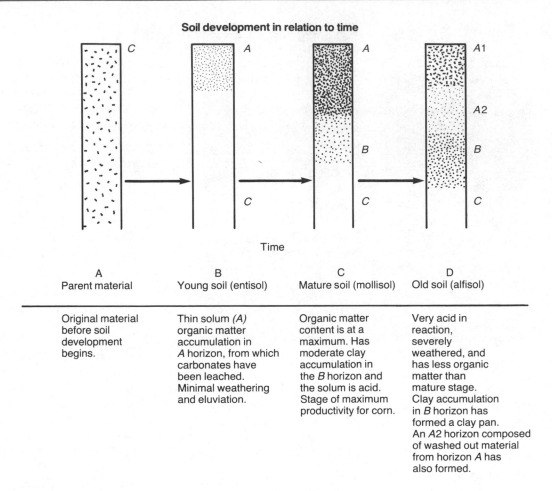

A	B	C	D
Parent material	Young soil (entisol)	Mature soil (mollisol)	Old soil (alfisol)
Original material before soil development begins.	Thin solum (A) organic matter accumulation in A horizon, from which carbonates have been leached. Minimal weathering and eluviation.	Organic matter content is at a maximum. Has moderate clay accumulation in the B horizon and the solum is acid. Stage of maximum productivity for corn.	Very acid in reaction, severely weathered, and has less organic matter than mature stage. Clay accumulation in B horizon has formed a clay pan. An A2 horizon composed of washed out material from horizon A has also formed.

FIGURE 24.2
As time passes, the microbial action in soil results in change. (Modified from H. Foth, *Fundamentals of Soil Science,* 6th ed. John Wiley & Sons, 1978.)

greatly. For example, the green algae population increases significantly after a lengthy springtime with heavy rains. The activity of all microbes influences the fertility of soil in several ways: (1) they produce humus by decomposing dead plant and animal material; (2) release minerals (e.g., calcium, magnesium, iron) from organic compounds for recycling; (3) interconvert ions and molecules to other beneficial forms; and (4) decompose toxic substances to harmless forms.

DECOMPOSERS AND BIODEGRADABLE MATERIALS

Most soil microbes are heterotrophs and derive their nutrients from organic molecules present in the soil. When a plant or animal dies, the energy and matter within its body are finally released to the environment by soil microbes that decompose the body into its most basic molecules, such as carbon dioxide and water and heat energy. For this reason soil bacteria and fungi are known as **decomposers.** This process is the oldest and most effective form of recycling.

As long as the sun provides the necessary energy for photosynthesis, elements are recycled over and over. This process is almost as old as life on the earth and is essential if life is to continue. Growth of soil microbes is limited to those that can use the available nutrients, and by the fact that there is a limited amount of useful organic material. Increasing the amount of nutrients by the addition of organic fertilizers, such as manure or crop cuttings, markedly increases microbial populations and enriches the soil. Nutrients are also more available in the form of plant-root products. The zone immediately surrounding the root and containing these organic materials is known as the **rhizosphere.** Because there is a higher concentration of nutrients in this zone, the rhizosphere is a place of great microbial diversity and activity. Even though soil microbes metabolize a diversity of organic molecules, there are many synthetic compounds that are unable to be decomposed or require extensive periods of microbial activity before they are degraded. Materials that are unable to be broken down before they cause harm to living organisms are called **nonbiodegradable,** or recalcitrant. Among the thousands of manmade chemicals released into the environment each year, some of the most detrimental are pesticides and herbicides.

DDT is an abbreviation for the chemical name dichlorodiphenyl trichloroethane. This structure is diagrammed in Figure 24.3. DDT is one of the group of organic compounds called chlorinated hydrocarbons. DDT was first produced in 1873 but was not used as an insecticide until 1944. It was discovered to have insect-killing properties by Dr. Paul Mueller, and for his discovery he was awarded the Nobel Prize in Physiology and Medicine in 1948.

DDT was a very valuable tool used by the U.S. Armed Forces during World War II. It was sprayed on clothing and dusted on the bodies of soldiers, refugees, and prisoners to kill body lice and other insects. Lice, besides being a nuisance, carry the bacteria that cause typhus fever, *Rickettsia prowazekii*. Body lice can be passed from one person to another by contact or through infested clothing. This use of DDT was so effective in the control of lice and typhus that peacetime uses were planned. Some envisioned the end of pesky mosquitos and flies, as well as the elimination of many disease-carrying insects.

When DDT is applied to an area to get rid of pests, it is usually dissolved in an oil or a fatty compound. It is then sprayed over an area and falls on the plants that the insects use for food or sprayed directly on the insect. Eventually, the insect takes the DDT into its body, where it interferes with the normal metabolism of the organism. If small quantities are taken in, the insect will begin to digest and break down the DDT as if it were any other organic chemical compound. Since the DDT is soluble in fat or oil, the insect stores the DDT or its degraded products in body tissues. Sometimes the insect manages to break down and store all of the DDT and, therefore, survives. If an area has been lightly sprayed with the biocide, some insects will die, some will tolerate the DDT, and some will survive because they will be able to store the DDT in their fat cells. As much as

one part of DDT per one billion parts of insect tissue can be stored. If an area was sprayed with a small concentration of the chemical to kill insects, the algae and protozoa of the area might accumulate as much as 250 times the concentration of DDT sprayed, since they also concentrate the biocide. Imagine the following sequence: algae and protozoa are eaten by insects that are, in turn, eaten by frogs. If the concentration of DDT in frogs is measured, it may be 2000 times the concentration originally sprayed. Birds that feed on the frogs and fish in the area could accumulate as much as 80,000 times the original amount. What was originally a very small concentration of biocide used to kill insects has now become so concentrated in certain carnivores and other organisms that they are also threatened. For example, birds generally cannot tolerate these high levels of DDT, and in several well-documented cases, many birds have died from the biocide. DDT is known to interfere with eggshell production by limiting the amount of calcium deposited in the shell. These thin-shelled eggs are more easily broken. This problem is more common in carnivorous birds, because they are at the top of the food chain. One of the dangers of using a chemical as a biocide is that it tends to become concentrated in an ecosystem. This situation is referred to as **biological amplification** (Table 24.3). As a result of the buildup of pesticide in the food materials that humans eat, many humans have up to 35 parts per billion DDT in their fat cells. A great deal has been written about the dan-

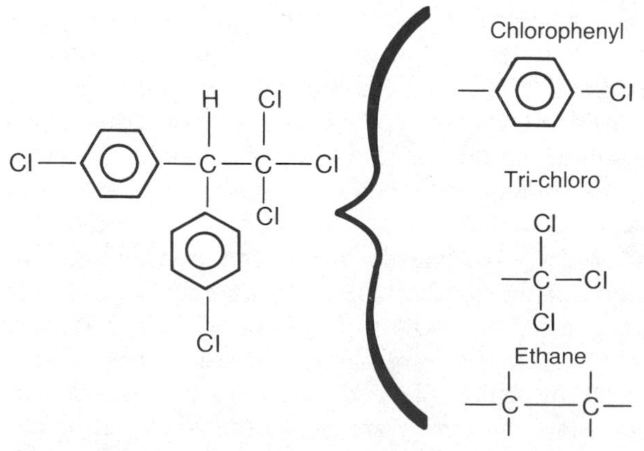

---

1 How do microbes influence soil fertility?
2 What type of microbes play an ecological role as decomposers?
3 What is the difference between a biodegradable and a nonbiodegradable chemical? Give an example of each.
4 How are microbes important in biological amplification?

FIGURE 24.3

The arrangement of atoms in a molecule of DDT. (From Eldon D. Enger et al, *Concepts in Biology,* 3rd ed. © 1982, 1979, 1976 by Wm. C. Brown Company Publishers, Dubuque, Iowa, Reprinted by permission.)

TABLE 24.3

Food chain concentration of DDT.

	Parts per Million DDT Residues
Water	0.00005
Plankton	0.04
Silverside minnow	0.23
Heron (feeds on small animals)	3.57
Herring gull (scavenger)	6.00
Fish hawk (osprey) egg	13.8
Merganser (fish-eating duck)	22.8

SOURCE: E. P. Odum, *Fundamentals of Ecology,* 3rd ed. Copyright © 1971, 1959, 1953 by W. B. Saunders Company. Reprinted by permission of Holt, Rinehart and Winston, CBS College Publishing.

gers of this quantity of DDT in human fat cells. The fact remains that little is known about the potential danger of DDT. It has been noted that nursing mothers need to be particularly conscious of the problem, since breast milk contains a great deal of fat and dissolved DDT. If the milk sold commercially had as much DDT in it as some mother's milk, it would be illegal to sell.

What was originally used to kill insects has proven to have many undesirable effects. Instead of harming only the insect pests, DDT accumulates in other forms of life and causes reduced reproductive ability in birds and many health concerns in humans. Therefore, in the early 1970s, the sale of DDT in the United States was prohibited.

DDT is not the only nonbiodegradable biocide. Many others have also been shown to persist in the soil and undergo biological amplification (Table 24.4). In addition, many have demonstrated a direct, harmful effect on bacteria important to crop production. Members of the genus *Rhizobium* are symbionts of leguminous plants, such as clover, alfalfa, soybeans, and cowpeas. They grow in root nodules and provide the host plant with usable nitrogen from the atmosphere. Legumes without these mutualistic bacteria show decreased vigor and lower productivity. When pesticides such as thiram and phygon are sprayed on fields of these leguminous crops, root nodules fail to develop, usable nitrogen is unavailable to the plants, and crop yields decrease. In order to overcome this problem, new strains of *Rhizobium* have been isolated and are being used to "seed" fields regularly treated with pesticides.

There are other methods of producing sufficient food without dangerous accumulations of chemical poisons. Much work is being done with **biological control,** which is the introduction of predators of the pests into the community to naturally control the pest population. Two species of the aerobic spore-forming bacterium *Bacillus* are used as biological control agents. *Bacillus thuringiensis* has been dem-

TABLE 24.4
Selected data on persistence of pesticides.

Chemical	Detectability*	Approximate Half-life
Chlordane	21 yr	2–4 yr
DDT	24 yr	3–40 yr
Dieldrin	21 yr	1–7 yr
Heptachlor	16 yr	7–12 yr
Loxaphene	16 yr	10 yr
Dalapon	10 wk	—
DDVP	—	17 days
Methyl demeton S	—	26 days
Thimet	—	2 days

*Period of time after which measurable amounts of the pesticide were detected. The actual periods of persistence of the first four chemicals are probably much longer.

onstrated to be over 90 percent effective in the control of Japanese beetles, flour moths, and alfalfa caterpillars. These bacteria are sold under the trade names of Biotrol, Dipel, and Thuricide. *Bacillus popilliae* can infect the larvae of Japanese beetles, causing the fatal disease known as milky spore disease. Bacterial spores can be sprayed on infested fields and will be effective for at least two seasons.

BIOGEOCHEMICAL RECYCLING

The earth is a closed biosphere, since no significant amount of new matter enters. Only light energy comes to the earth in a continuous stream and, ultimately, this energy is released back into space. Light energy is used on earth to drive all biological processes. Living systems have evolved ways for using this energy to continue life through growth and reproduction. Since no new atoms are being added to the earth, living systems must use the available atoms over and over. In this **biogeochemical recycling,** inorganic molecules are combined to form the organic compounds of living organisms. If there were no way to recycle this organic matter back into its inorganic form, the organic material would build up in the form of dead organisms. The decomposers are vital to this process when they are given the chance. If they are kept from doing their job, organic material builds up. This happened on earth in the past, and we identify this absence of recycling as deposits of oil, natural gas, and coal.

Living systems contain many different kinds of atoms, but some are more common and important than others. Carbon, nitrogen, oxygen, hydrogen, and phosphorus are found in all living organisms and must be recycled when an organism dies. One example of a biogeochemical cycle is the **carbon-oxygen cycle.** Carbon and oxygen are combined to form the molecule carbon dioxide, which is found as a gas in the atmosphere. During photosynthesis, the carbon dioxide is combined with water to form complex organic molecules. At the same time, oxygen molecules are released into the atmosphere. This additional organic matter present in the bodies of plants may be used by herbivores as food. When an herbivore eats a plant, it breaks down the complex organic molecules into simpler molecules that are used for building the herbivore's body. These carbon and oxygen molecules may be transferred to a carnivore for its use, if the herbivore is preyed upon. Finally, decomposers may completely break down waste products and the organic molecules of dead organisms into carbon dioxide and water. The carbon atoms that started as carbon dioxide in the atmosphere passed through a series of organisms as organic carbon and were finally returned to the atmosphere as carbon dioxide (Figure 24.4).

Soil microbes involved in the carbon-oxygen cycle are bacteria and fungi. Cellulose from plants is decomposed by

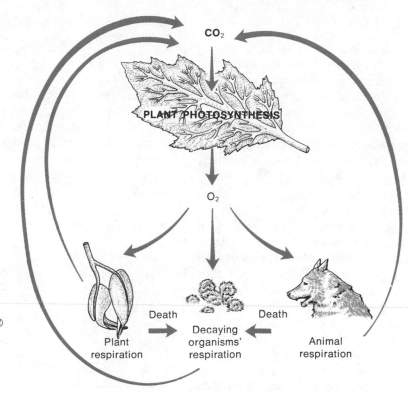

FIGURE 24.4

The carbon and oxygen cycle. (From Eldon D. Enger et al, *Concepts in Biology,* 3rd ed. © 1982, 1979, 1976 by Wm. C. Brown Company Publishers, Dubuque, Iowa. Reprinted by permission.)

the enzyme cellulase that is released by fungi. Few heterotrophic bacteria release this enzyme but are chiefly responsible for the decomposition of animal tissue. Under optimum aerobic conditions, carbon dioxide and water are produced during decomposition; but if the oxygen level becomes restrictive, the methanogenic bacteria, *Methanococcus* and *Methanosarcina* (Figure 24.5), become predominant and anaerobically generate methane gas:

$$C_2H_5OH + CO_2 \longrightarrow CH_3COOH + CH_4$$

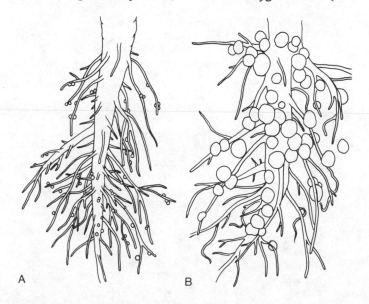

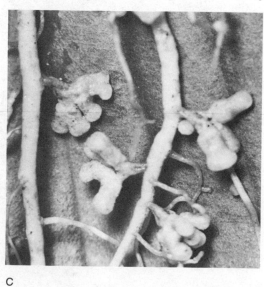

FIGURE 24.5

(A) A red clover root system with nodules of the nitrogen-fixing bacteria *Rhizobium.* (B) The nodulated roots of a soybean plant. (C) Root nodules on a clover plant. (Photo C courtesy of Carolina Biological Supply Company.)

NITROGEN CYCLE

Another very important atom that undergoes biogeochemical recycling is nitrogen. This atom is essential in the formation of proteins and nucleic acids. Nitrogen is the most abundant gas in the atmosphere (79 percent). However, only a few kinds of bacteria are able to change the gas into a form that other organisms can use. The amount of usable nitrogen available to plants is an important factor that limits their growth. Therefore, nitrogen is known as a **limiting factor.** The usable forms of nitrogen are NH_4^+, NO_3^-, NO_2^-, and nitrogen-containing organic compounds such as amino acids. Small changes in the amount of available, usable nitrogen cause significant changes in plant and animal growth. There are two ways in which plants and animals can get usable nitrogen compounds. First, **symbiotic nitrogen-fixing bacteria** in the root nodules of leguminous plants can convert molecular nitrogen (N_2) into a form that the host plant can use to make amino acids. There are many different leguminous plants capable of establishing this mutualistic relationship, however, the most important from a nutritional viewpoint are edible plants and forage, or cover, crops (Table 24.5). Once inside root cells and protected from the destructive effects of atmospheric oxygen, *Rhizobium*

reduces molecular nitrogen to ammonia with the enzyme nitrogenase:

$$N_2 \xrightarrow{H} NH_3$$

The ammonia may then be interconverted by host-cell enzymes to other usable forms, such as nitrite and nitrate ions (NO_2^-, NO_3^-), amino acids, and nitrogenous bases. Leguminous plants with *Rhizobium*-containing root nodules are able to fix ten times more nitrogen than free-living heterotrophs. This improves crop yield and enriches the surrounding soil. In addition, there are several nonleguminous plants that fix nitrogen in root nodules (Table 24.6). These are advantageous since they grow in nitrogen-poor soils, but

1 What types of pests are able to be biologically controlled with the bacterium *Bacillus thuringiensis?*
2 Cellulose from plant material is decomposed by what types of microbes? What enzyme is needed for this reaction?
3 Why are limiting factors of such great importance?

TABLE 24.5
Leguminous plants.

Edible plants
  Arachis (peanut)
  Cajanus (pigeon pea)
  Cicer (chickpea)
  Lens (lentil)
  Pachyrhizus (yam)
  Phasolus (string bean, mung bean, lima bean, etc.)
  Pisum (pea)
  Vicia (vetch)
  Vigna (cowpea)
  Ceratonia (carob)

Forage or cover crops
  Glycine (soybean)
  Lotus (crefoil)
  Medicago (includes alfalfa)
  Melilotus (sweet clover)
  Onobrychis (holy clover)
  Trifolium (clover)

Garden flowers and plants
  Anthyllis (Jupiter's Beard)
  Astragalus (milk vetch)
  Baptisia (false indigo)
  Coronilla (crown vetch)
  Desmodium (tick clover)
  Galega (goat's-rue)
  Hedysarum (honeysuckle)
  Lathyrus (sweet pea)
  Lespedeza (bush clover)
  Lupinus (lupine)
  Ononis (restharrow)
  Petalostemon (prairie clover)
  Psoralea (Indian breadroot)
  Thermopsis (Aaron's-rod)

Shrubs, trees, woody vines
  Cladrastis (yellowwood)
  Colutea (common bladder-senna)
  Cytisus (broom)
  Genista (woodwaxen)
  Laburnum (golden chain tree)
  Maackia
  Wistaria (wisteria)
  Prosopis (mesquite)
  Robinia (locust)
  Sophora (Japanese pagoda tree or scholar tree)
  Ulex (furze)
  Cercis (redbud)
  Gleditsia (honey locust)
  Gymnocladus (Kentucky coffee tree)

TABLE 24.6
Nonleguminous plants that fix nitrogen in root nodules.

Alnus (alder)	33 of 94 species
Casuarina (Australian pine)	18 of 40 species
Ceanothus (snowbrush, New Jersey tea)	31 of 56 species
Cercocarpus (mountain mahogany)	3 of 15 species
Colletia	1 of 6 species
Comptonia (sweet fern)	1 of 100 species
Coriaria (shrub)	12 of 80 species
Discaria (shrub)	1 of 10 species
Dryas (evergreen)	3 of 75 species
Elaeagnus (autumn olive)	14 of 31 species
Hippophae (sea buckthorn)	1 of 33 species
Myrica (wax myrtle)	20 of 57 species
Purshia (bitterbrush, deerbrush)	2 of 100 species
Shepherdia (buffalo berry)	2 of 67 species

because they yield comparatively small amounts of usable nitrogen, they are insignificant contributors to soil fertility.

The second way in which plants and animals can get usable nitrogen compounds involves nitrogen fixation by free-living microbes. Common soil microbes of the genera *Clostridium, Azotobacter, Enterobacter, Chlorobium,* and *Rhodospirillum* reduce molecular nitrogen to usable forms. The anaerobic clostridia and the photosynthetic *Chlorobium* and *Rhodospirillum* are responsible for nitrogen fixation in waterlogged or flooded soils. Rice paddies need not be fertilized due to the nitrogen fixation carried out by these microbes.

The nitrogen content of the soil is one of the major determinants in the growth rate of plants. Farmers can make usable nitrogen available to their crops in a number of ways. One year, they may plant crops that have symbiotic nitrogen-fixing bacteria; the next year in the same field, they may plant a nitrogen-demanding crop. Rotating the crops in this way helps maintain nitrogen in the soil. The farmer can also add nitrogen directly to the soil. Manure can be spread on the field, and the soil bacteria convert the nitro-

gen in the manure to a usable form. Industrially produced fertilizers can also be added. These are usually nitrate compounds or ammonia. Commercial fertilizers are made by the Haber process, in which molecular hydrogen (H_2) and nitrogen (N_2) are heated under pressure to form anhydrous ammonia (NH_3). Because this process requires the expenditure of large amounts of energy, the price of fertilizer has risen steadily, along with the price of oil and natural gas. If energy continues in short supply, commercial fertilizers may soon be financially out of reach for farmers in underdeveloped countries, something which could have a disasterous impact on world food production. To avoid this, researchers are investigating several ways of expanding the nitrogen-fixation process. One method is the development of hybrid strains of nonleguminous plants that will be able to form symbiotic relationships with *Rhizobium* or other nitrogen-fixers. Some preliminary success has been found with corn and the bacterium *Spirillum lipoferum*. This bacterium can fix nitrogen in association with cereal grains or when growing freely under microaerophilic soil conditions. Field studies are currently being conducted to explore the possibility of inoculating seeds or entire fields with this microbe in order to reduce the need for large amounts of commercial nitrogen fertilizer. But even if industrially produced fertilizers are added to the field, the soil bacteria are still needed to make nitrogen in the fertilizer available to the plants.

Once molecular nitrogen has been fixed into a usable form, it is transmitted through the **nitrogen cycle** (Figure 24.6). This biogeochemical cycle has five major stages: (1) proteolysis, (2) ammonification, (3) nitrification, (4) nitrate reduction, and (5) denitrification. Proteolysis (protein splitting) is an enzymatic process carried out by many types of soil bacteria, including members of the genera *Clostridium, Proteus,* and *Pseudomonas;* fungi; and actinomycetes. In this portion of the cycle, complex proteins are broken into smaller polypeptides and finally into individual amino acids. These free amino acids serve as nutrients for worms, insects, and other animals living in the soil. If not utilized by animals,

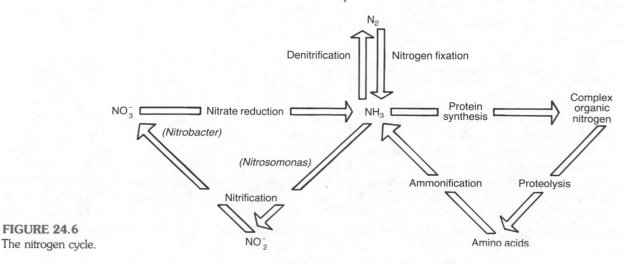

FIGURE 24.6
The nitrogen cycle.

the amino acids undergo ammonification by heterotrophic soil microbes. This reaction results in the oxidation of amino acids to ammonia, organic acids, carbon dioxide, and water. In alkaline soils, the ammonia is released into the atmosphere and lost as a source of usable nitrogen. Under more normal circumstances, however, the ammonia is utilized during nitrification by chemoautotrophic bacteria as a source of energy. Members of the genus *Nitrosomonas* oxidize ammonia in its ionic form, ammonium ion (NH_4^+). The reaction results in the release of nitrite ion (NO_2^-) and usable energy:

$$NH_4^+ + O_2 \longrightarrow NO_2^- + H_2O + H^+ + energy$$

Low levels of nitrite may be used by some microbes and other organisms in their metabolic pathways, but this ion is usually oxidized by the soil bacteria *Nitrobacter*. This is important since the accumulation of nitrite ion may have toxic effects. For example, the ingestion of water or foods contaminated with nitrite can result in a form of poisoning. This occurs in infants and some farm animals that drink well water contaminated with nitrite runoff from surrounding fertilized fields. Because infants and animals have a different form of hemoglobin than adults (alpha- and gamma- rather than alpha- and beta-hemoglobin), the nitrite ion chemically and irreversibly combines with it to form methemoglobin. This produces symptoms similar to carbon monoxide poisoning and requires much the same treatment. *Nitrobacter* in the soil is able to complete the oxidation and extraction of energy from nitrite, reducing the level of this harmful ion and producing nitrate ion (NO_3^-) as a product:

$$NO_2^- + O \longrightarrow NO_3^- + energy$$

Nitrates not utilized for metabolism by other organisms may be reduced to ammonia:

$$NO_3^- \longrightarrow NH_4^+$$

This is accomplished by many different microbes and helps maintain a constant supply of usable nitrogen in the soil. When anhydrous ammonia is injected into the soil to fertilize crops, these microbes convert NH_4^+ to all the other forms and establish an optimum balance of these most usable forms of nitrogen.

Nitrogen completes its cycling when it is returned to the atmosphere as molecular nitrogen or oxides of nitrogen during denitrification. This portion of the cycle involves the oxidation of organic nitrogen and inorganic nitrogen by *Pseudomonas, Thiobacillus, Serratia,* and other soil bacteria. The key to maintaining soil fertility is developing a balance in the whole cycle. If stagnation (pollution) occurs at any point, the accumulated nitrogenous material will interfere with the normal, healthy operation of the ecosystem, causing harm to its

members. This is seen in aquatic ecosystems that have been "fertilized" with nitrogen runoff from adjacent farm lands or residential areas overcrowded with septic systems. High nitrogen content in water causes premature aging, or **eutrophication,** of lakes and ponds. The "bloom" of growth stimulated by fertilization clogs the water with vegetation that cannot be decomposed as fast as it is being produced. The aerobic decomposers demand more oxygen than can be supplied, and they die, leaving only the anaerobes to continue the process. Anaerobes decompose the vegetation at a slower rate and release foul-smelling products, such as hydrogen sulfide gas. Because of the indiscriminate use of large amounts of nitrogen fertilizers and overdevelopment around small lakes, eutrophication has turned many picturesque recreational areas into foul-smelling, unattractive cesspools.

---

1 Give three examples of leguminous plants which can form a symbiotic relationship with *Rhizobium.*
2 What is meant by the phrase "usable nitrogen"?
3 List the five stages in the nitrogen cycle and a microbe involved in each stage.

Fertilizers usually contain more than just nitrogen compounds. The numbers on the fertilizer bag indicate the percentage of nitrogen, potassium, and phosphorus in the fertilizer. These other elements also cycle through ecosystems, and as crops are removed from the land, so are these elements (Figure 24.7 and Figure 24.8). Therefore, the farmer must replace them by adding more fertilizer. In a more natural ecosystem, the bacteria would decompose the dead plants and animals in the field, and recycle the elements.

The decomposition of organic material through biogeochemical recycling can be controlled in a process known as **composting.** The original compost heap was probably started accidentally in garbage piles containing such things as leafy vegetable material, sawdust, and manure. Today a compost is made from much the same material; however, added incentive for microbial growth and decomposition is provided by starting the pile with equal parts of manure and peat moss, along with the "garbage" just listed. The materials are mixed together and moistened. The rich, organic nature of the compost provides the optimum growth medium for most microbes. As they grow, reproduce, and decompose the organic matter, they generate large amounts of heat. It is not unusual for a compost pile to reach internal temperatures of 80–100°C, especially in the summer. When this happens and the pile becomes dry, it may spontaneously ignite. To control the decomposition and prevent fires, the compost is turned with a rake or shovel, and moistened during the hot months to release the generated

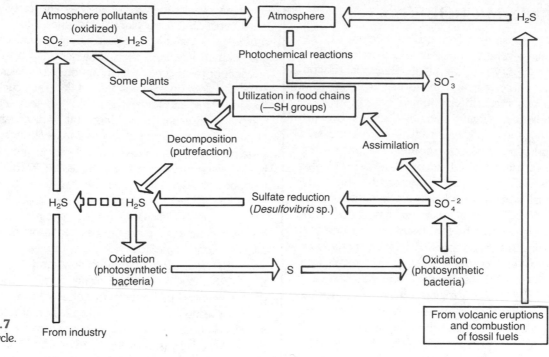

FIGURE 24.7
The sulfur cycle.

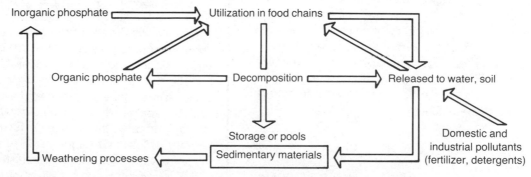

FIGURE 24.8
The phosphorus cycle.

heat. The self-heating process serves as a natural selecting agent for thermophilic microbes which become the predominant form within a few days after the pile has been started. Mesophilic bacteria are killed during heating but return once the temperature of the pile falls to a more moderate level (Figure 24.9). Decomposition of fats and carbohydrates occurs at a steady rate. Although starch and cellulose are broken down with little difficulty, the more resistant plant lignins remain in the compost at relatively high levels and give the product a coarse texture. The humuslike product is only about one-fourth to one-half the original volume; it makes an excellent soil conditioner and fertilizer. Compost piles can be maintained for years by continuously adding new plant material and removing some of the decomposed humus.

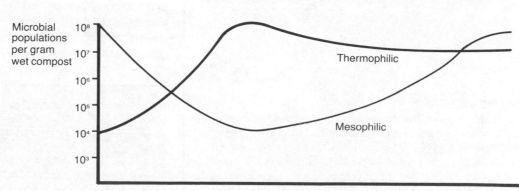

FIGURE 24.9

As time passes, there is an increase in the thermophilic population of the compost pile. The curve levels off, since the population is able to withstand the higher temperatures generated in the pile.

CROPS AND MICROBES

Producing and maintaining a high-quality medium for crop production requires careful attention to soil microbes and their interrelationships with higher plants. Beneficial relationships should be enhanced, and those that result in plant harm and crop loss should be prevented. One of the most beneficial associations already described occurs between members of the genus *Rhizobium* and leguminous plants. An equally important symbiotic association has been established between certain kinds of fungi and the roots of higher plants. This association is called a **mycorrhiza** (root fungus). If the fungal member of the duo grows on the outside of the roots, it is called an ectomycorrhizal association. If the fungus grows entirely inside the root tissue between the cells, it is known as an endomycorrhizal association. Ectomycorrhizal associations enhance the absorption of minerals from soils with low levels of these essential nutrients. They also regulate the transfer of these minerals into the crop roots. Ectomycorrhiza are formed on most forest tree roots, especially conifers, oaks, and beeches. If these plants are grown in an environment that prevents this symbiotic relationship, they will be shorter, slower growing, and have less vigor (Figure 24.10). In many endomycorrhizal associations, the fungus benefits the plant by providing nutrients to seeds which contain only a minimal amount of stored food material (endosperm). Orchid seeds are extremely small, deficient in endosperm, and do not develop in nature unless fungal hyphae infect the seed and coil through the tissue to supply the necessary micronutrients.

FIGURE 24.10

Nine-month-old seedlings of white pine, *Pinus strobus,* raised for two months in a sterile nutrient solution and then transplanted to Carrington (Plano) prairie silt loam. (A) These seedlings were transplanted directly from hydroponics to prairie soil. (B) These seedlings were inoculated with mycorrhizal fungi prior to transplantation. (Courtesy of J. G. Iyer, University of Wisconsin, for the late S. A. Wilde.)

Relationships between microbes and crops are not always beneficial. Many plants experience infectious diseases caused by a variety of bacteria, fungi, viruses, and viroids. Like human infection, these diseases may result from the invasion of pathogens or from opportunistic microbes found as normal flora on different parts of the plant, such as leaves, flowers, stems, and roots. Table 24.7 lists several of the most common and economically important plant diseases. As with human diseases, plant pathogens can establish superficial or systemic infections. Superficial infections are more easily treated since antimicrobials, such as the Bordeaux mixture (copper sulfate and lime) can be dusted or sprayed on the stems and leaves. Those that are systemic or located on the roots are more difficult to control and require the use of special techniques or chemotherapeutic agents. The most effective control methods for plant disease are: (1) exclusion, (2) eradication, (3) protection, and (4) "immunization." By excluding infected plants from disease-free areas of the country, pathogenic microbes cannot become endemic or epidemic. Exclusion is regularly practiced at a statewide level and by the federal government in the United States at designated ports of entry. This method has been relatively successful in preventing the spread of house plant and citrus fruit diseases. The eradication of pathogens is carried out on already diseased plants by using chemical sprays or dusts. Crop rotation is also an effective eradication method, since it reduces the number of available hosts on which the microbe can become established. Plants that have not become diseased may be protected by using chemical sprays or dusts prophylactically, or by destroying insect vectors before they transmit the pathogens. Plants are also immune, resistant, or tolerant to some pathogens. Here the term "immune" does not refer to the stimulated production of antibodies but to the genetic ability of plants to produce agents which inhibit the invasion, multiplication, or spread of disease-causing microbes. Through selective plant breeding, geneticists have developed many hybrid strains capable of resisting infection. In a corn population, genes may be present for resistance to corn blight (a fungal disease) and resistance to attack by insects. Corn plants that possess both of these genes are going to be more successful than corn plants that have only one or none of these genes. They will probably produce more offspring (corn seeds) than the others and will tend to pass on this same combination of resistant genes to their offspring.

1 What prevents the material in a compost pile from being completely decomposed?
2 Of what value is a mycorrhizal association to the host plant?
3 What methods can be used to control plant pathogens?

TABLE 24.7
Common microbial diseases of plants.

Disease	Pathogen	Susceptible Plants	Control
Bacteria:			
Halo blight of bean	*Pseudomonas phaseolicola*	Bean	Disease-free materials, resistant
Bacterial blight of cotton	*Xanthomonas malvacearum*	Cotton	plants, and crop rotation is
Crown gall	*Agrobacterium tumefaciens*	Apple, cherry, prune, grape, alfalfa, cotton, etc.	the only effective means of controlling bacterial infections.
Ascomycetes:			
Chestnut blight	*Endothia parasitica*	Chestnut	None.
Ergot of grains and grasses	*Claviceps purpurea*	Rye	Crop rotation, use of disease-free seed, sanitation.
Powdery mildew of cereals	*Erysiphe graminis*	Wheat	Resistant varieties.
Basidiomycetes:			
Stem rust of grains and grasses	*Puccinia graminis*	Wheat	Resistant varieties; sulfur dusting; eliminate barberry.
Cedar apple rust	*Gymnosporangium juniperi-virginianae*	Juniper	Eliminate one host; sulfur dusting
Common smut of corn	*Ustilago maydis*	Corn	Crop rotation, sanitation.
Phycomycetes:			
Late blight of potato	*Phytophthora infestans*	Potato	Bordeaux mixture as a spray.
Damping off of seedlings	*Pythium debaryanum*	All	Seed treatment with Ceresan or Semesan soil disinfestation.
Viruses:			
Mosaic diseases	Tobacco mosaic virus (TMV) and others	Tobacco	Development of resistant plants.
Tomato bushy stunt	Bushy stunt virus	Tomato	Development of resistant plants.
Sugar beet curley top	Curley top virus	Sugar beet	Development of resistant plants.
Viroids:			
Cucumber pale fruit disease	CPFD viroid	Cucumbers	Development of resistant plants.
Hop stunt disease	HSD viroid	Hops	Development of resistant plants.
Chrysanthemum chlorotic mottle disease	CCMD viroid	Chrysanthemum	Development of resistant plants.

The success of these disease prevention techniques is vividly demonstrated by increased crop production throughout the world. However, microbes are not limited to the destruction of living plants. Many common soil microbes are saprotrophs responsible for the decomposition and spoilage of grains used for human consumption and animal feed. One of the oldest methods of preserving animal food (fodder) is by the production of ensilage or silage. **Silage** is a combination of green hay, cereals, grass, and other suitable vegetation that has been preserved in a silo by anaerobic acid fermentation (Figure 24.11). Naturally occurring lactic acid bacteria, such as lactic streptococci and lactobacilli, which are on the fodder, ferment carbohydrates, releasing lactic acid that inhibits the growth and activities of spoilage microbes. Other aromatic by-products are also released during the fermentation, giving silage its characteristic smell. The

Ensilage (e·nsiledj), *sb.* [a. F. *ensilage*, f. *ensiler*: see ENSILE *v.*]
1. The process of preserving green fodder in a silo or pit, without having previously dried it.
1881 *Salem* (Mass.) *Gaz.* 10 June 1/2 On ensilage of Green Forage Crops in Silos. **1882** *Macm. Mag.* No. 278. 114 Ensilage is the packing of green forage in air- and water-tight structures. **1882** *Times* 30 Nov. 11 The object of ensilage is to maintain the sap as nearly as possible in its original state. **1884** *Boston* (Mass.) *Jrnl.* 20 Nov. 2/4 Norfolk is the county where the ensilage of fodder is most practised.
2. The material resulting from the process.
1883 *Echo* 11 June 1/6 Ensilage . . . is produced by cutting green fodder of different kinds when well matured . . . and pressing it down in water-tight pits, subsequently also made air-tight. **1882** *Times* 30 Nov. 11 About 3 in. of the ensilage was found to be mouldy.
3. *attrib.*
1883 *Edin. Rev.* Jan. 150 Five separate manufacturers advertised ensilage cutters. **1888** *Times* 24 July 13/1 Those who were prepared to make ensilage stacks.
Ensilage (e·nsiledj), *v.* [f. prec. sb.] *trans.* To subject to the ensilage process; to convert into ensilage. Hence **E·nsilaged** *ppl. a.*
1883 *West Chester Pa. Republican* VI. No. 37.4 An ensilage crop. **1883** *Chamb. Jrnl.* 274 Pease, oats, maize, and vetches might be ensilaged together. **1883** *Edin. Rev.* Jan. 149 Preserving green fodder by ensilaging it.
The Oxford English Dictionary, Being a Corrected Re-issue with an Introduction, Supplement, and Bibliography of
A New English Dictionary on Historical Principles. Vol III D–E, Oxford, Clarendon Press, 1933

FIGURE 24.11
The origin of silage.

production of high-quality silage requires following the same basic procedures used for culturing microbes in the laboratory. A sufficient amount of inoculum must be used in order to initiate the fermentation, and environmental conditions must be selective for the growth of lactic acid bacteria. These conditions are achieved by using vegetation that provides the optimum nutrients for lactic acid bacteria and reducing available oxygen by carefully packing the silo and regulating the moisture content by using only fresh, undried vegetation. Anaerobic bacteria exhaust the oxygen supply within the silo to create the necessary anaerobic environment for acid production. When these conditions are present, the lactic acid bacteria are in a selective growth medium that allows them to metabolize and reproduce more rapidly than spoilage microbes. The silage is used as animal feed after it has reached a pH of approximately four.

ANIMALS AND MICROBES

Domesticated animals have been a part of the human environment for thousands of years, serving as a source of food, clothing, labor, and companionship. The loss of animals such as cattle, sheep, swine, and horses as a result of infectious disease would severely affect the life styles of people all over the world. As with crops, it is essential that beneficial animal-microbe relationships be enhanced and disease be brought under control as quickly as possible. Many of the symbiotic relationships known to exist between humans and microbes are also found among the domesticated animals and their normal flora. *E. coli* in the intestinal tract of

cattle and swine out-compete pathogens, release vitamins and antibiotics (colicins), and produce organic acids that limit gastroenteritis. Animals belonging to a group known as the **ruminants** (for example cattle, sheep, goats, deer, and antelopes) have a unique relationship with microbes that is not found in other animals.

Ruminants are even-toed, grazing animals that have a stomach with four compartments: the rumen, reticulum, omasum, and abomasum (Figure 24.12). As grazers they feed on foliage and grasses that contain a great deal of cellulose. However, they cannot utilize this carbohydrate as a source of energy since they lack the genetic ability to produce the enzyme cellulase. They derive benefit from cellulose because the enzymatic hydrolysis of this molecule is accomplished by symbiotic microbes residing in the rumen. Ruminants eat a great deal of bulky vegetation and swallow it without much chewing. The food enters the rumen and is mixed with the anaerobic, fermenting culture by a churning action of the stomach. While inside the rumen, cellulose is broken down into more basic disaccharide and monosaccharide subunits by enzymes released from bacteria and protozoa. The sugars are used by the microbes as an energy source in fermentation reactions that release organic acids and gases as final products. The gases are expelled. The organic acids enter the circulatory system of the animal and are later used in the production of long-chain fatty acids, or they are respired aerobically as a source of energy. These activities take place during the several hours that the food is in the rumen. During this time, some of the material is moved into the reticulum, and small amounts are regurgitated into the mouth for a second chewing of the "cud."

Cud contains partially digested plant material and a very high concentration of microbes due to the rich medium in which they have grown rapidly. Because of their rapid reproduction rate, the large numbers of microbes serve as a continuous source of high-quality protein for the animal. When the cud is swallowed, it passes through the omasum and abomasum, where it is further digested. There, amino acids from the microbes are absorbed into the blood stream. Because ruminants have a continuous source of high-quality microbial protein from their rumen, they can survive well on a diet composed almost entirely of cellulose.

Ruminants and other domesticated animals experience a variety of infectious diseases. It is important to prevent and control these diseases, since they result in great economic loss, and many are zoonotic infections able to be transmitted to humans. Table 24.8 lists several of the most common and economically important animal diseases. The control and prevention of animal diseases are carried out using the same techniques and principles applied to humans. However, one additional option is available to veterinarians: eradication by slaughter. Although this epidemic control measure was used more frequently in the past, it remains a viable alternative in many undeveloped countries or in situations that are uncontrollable with antibiotics or passive immunization. The need for this alternative has been reduced in advanced countries because of the prophylactic use of antibiotics. Many companies producing animal feed supplement their products with antibiotics such as tetracycline, aureomycin, and terramycin (Figure 24.13). However, the development of antibiotic-resistant strains of animal pathogens able to be transmitted to humans has increased

at an alarming rate. This situation resulted in an order in 1977 by the U.S. Food and Drug Administration to cut back the routine use of penicillin, chlortetracycline, and oxytetracycline in animal feed. The FDA does allow tetracycline to be used in animal feed for five specific diseases because there are no effective substitutes at this time.

> **1** What is the difference between compost and silage?
> **2** What roles do microbes play as symbionts of ruminants?

SUMMARY

The field of agricultural microbiology involves the study of microbes and their association with soil formation, biogeochemical recycling, and symbiotic plant and animal relationships. Through the combined action of microbial, chemical, and physical occurrences, parent rock material is weathered to a fundamental soil able to support basic forms of plant and animal life. Prolonged microbial activities in this soil improves its chemical, physical, and biological properties by decomposing organic material to humus, releasing minerals for recycling, interconverting ions and molecules to other beneficial forms, and decomposing toxic substances to harmless molecules. In certain situations, pesticides and herbicides go unaffected by microbial action. These nonbiodegradable materials may become concentrated through the

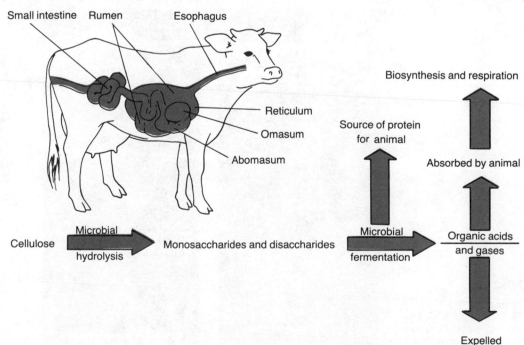

FIGURE 24.12

Anatomy of a ruminant.

TABLE 24.8
Common diseases of domesticated animals.

Disease	Pathogen	Host
Bacterial:		
Anthrax	*Bacillus anthracis*	Cattle, horse, sheep, goat, swine, fowl
Brucellosis	*Brucella* spp.	Cattle, horse, sheep, goat, swine, fowl
Glanders	*Pseudomonas mallei*	Horse
Fowl cholera	*Pasteurella multocida*	Chicken, turkey, etc.
Tuberculosis	*Mycobacterium tuberculosis*	
	var. *bovis*	Cattle, horse
	var. *avium*	Fowl
Mastitis	*Steptococcus agalactiae*	Cow
Leptospirosis	*Leptospira interrogans*	Cattle, horse, swine
Psittacosis	*Chlamydia psittaci*	Fowl
Hog cholera	*Salmonella* spp.	Swine
Fungal:		
Histoplasmosis	*Histoplasma capsulatum*	Fowl
Ringworm	Several species	Cattle, horse, mules
Viral:		
Equine encephalitis	Togavirus	Horse, fowl
Rabies	Rhabdovirus	Cattle, horse
Foot-and-mouth	Picornavirus	Cattle, sheep, goat
Newcastle disease	Paramyxovirus	Fowl
Cowpox	Poxvirus	Cattle
Influenza	Orthomyxovirus	Cattle, horse, sheep, goat
Protozoan:		
Tick fever	*Piroplasma bigemina*	Horses, sheep, cattle
Anaplasmosis	*Anaplasma morginale*	Cattle
Coccidiosis	*Eimeria* spp.	Fowl, cattle

TOP DRESS WITH TERRAMYCIN A/D FORTIFIED CRUMBLES IN THE
STRESS PROGRAMS

Grab a bag at the first suspicion of trouble. It could put you days ahead in fighting scours, shipping fever, and other disease conditions that respond to Terramycin, including many that are triggered by stress. Feed for 3 to 5 days at the **high-level**, disease-fighting rate, and in most cases, that's it.

But it's a flexible program:

—If the problem is especially tough, or if symptoms persist, feed continuously until it's under control.

—And because it's Terramycin A/D Fortified Crumbles, you can feed continuously at the **low-level** rates as an aid in:

1. Increasing rate of gain.

2. Improving feed efficiency.

3. Reducing incidence and severity of bloat.

4. Reducing the incidence of liver abcesses in cattle over 400 pounds.

How much do you use?

High-level, disease-fighting rates vary according to the severity of the problem and age of the animal—anywhere from ½ to 6 cups per head daily for bacterial diarrhea or scours . . . 1 to 4 cups per head daily for early stages of shipping fever.

Low-level rates for gains, feed efficiency, and liver abcess or bloat protection are 2 tablespoonfuls per head daily.

A 50-lb. bag holds 3200 tablespoonfuls or 200 cups of Crumbles. That means one bag holds 33 to 400 individual daily treatments at the **high-level**, disease-fighting rate. Or 1600 daily treatments for the **low-level**, continuous, "nutritional" rate.

FIGURE 24.13
Approximately one-half of the antibiotics used in the United States to treat animal infections require no prescription. (Courtesy of the Pfizer Chemical Company.)

food chain to levels that are harmful to living members of the ecosystem. A suitable alternative to the use of these toxic chemicals is often found in biological control agents. These are parasitic organisms used to keep a population of unwanted organisms in check.

The activities of soil microbes as decomposers ensures the reprocessing of materials so that they can be reused many times. The cycling of carbon, oxygen, nitrogen, sulfur, and phosphorus are all controlled by microbial activity in the soil. Since nitrogen is a limiting factor, any change in the activity of the microbes controlling this biogeochemical cycle will be easily seen as altered plant and animal growth. Soil fertility can be maintained or improved by stabilizing these cycles, and by enrichment with manure, composted materials, or commercially available fertilizers.

Among the symbiotic relationships found between plants and microbes are those involving the nitrogen-fixing bacteria, the mycorrhiza, and numerous plant pathogens. One of the main goals of agricultural microbiology is to maintain the beneficial symbiotic relationships and prevent those that are harmful. Animal-microbe relationships can also be beneficial or harmful. Microbes found in the intestinal tract of many domesticated animals, especially the ruminants, provide significant advantage to the host and symbiont. Diseases in both plants and animals are controlled or prevented by exclusion, eradication, protection, or immunization. A variety of biological, chemical, and chemotherapeutic agents are available for these purposes.

WITH A LITTLE THOUGHT

A unique form of food poisoning has just been discovered, and a team of researchers is attempting to "put the pieces together" in an attempt to control the problem. The data that follows has been collected from researchers in the areas of agricultural, food, and water microbiology. As leader of the research team, it is your responsibility to explain all the data and show its relevancy to the total problem and its solution.

Data:

1 The poison (toxin) is a nitrogenous, organic compound.

2 Human cases have one thing in common: all persons have consumed one of the following locally prepared foods: Italian salami, pepperoni, or hot bar sausages. Note: Examination of these foods revealed no toxin.

3 Cases of poisoning have only occurred in areas adjacent to an agricultural valley which has been heavily fertilized.

4 City workers have been on strike for the past month because of poor working conditions and the need for more accurate chlorine metering equipment.

5 Thus far only humans have demonstrated symptoms of the illness, but domesticated animals fed the prepared foods will show induced symptoms.

STUDY QUESTIONS

1 Describe how microbes and higher plants contribute to soil formation.

2 Explain the following: humus, microbial gum, water-holding capacity, and filtration rate.

3 Why do certain nonbiodegradable materials become harmful even though they are initially used in extremely low concentrations?

4 Give an example of a microbe used as a biological control agent.

5 What is a limiting factor? Give an example.

6 What would be the consequences of eliminating all nitrogen-fixing bacteria? What evidence is there that these microbes are being damaged by human activities?

7 What are the major stages in the nitrogen cycle? Explain the events of each.

8 What is the difference between composting and ensilage?

9 What is the difference between mycorrhiza and rhizosphere?

10 By what methods are microbial plant diseases controlled?

SUGGESTED READINGS

ALEXANDER, M. *Introduction to Soil Microbiology,* 2nd ed. John Wiley & Sons, New York, 1977.

The Biosphere. A Scientific American Book, W. H. Freeman and Company, San Francisco, 1970.

BORMANN, F. H., and G. E. LIKENS. "The Nutrient Cycles of an Ecosystem." *Scientific American,* Oct., 1970.

BRILL, W. "Biological Nitrogen Fixation." *Scientific American,* March, 1977.

FOTH, H. D. *Fundamentals of Soil Science,* 6th ed. John Wiley & Sons, New York, 1978.

HUNGATE, R. E. *The Rumen and Its Microbes.* Academic Press, New York, 1966.

JUKES, T. H. "Antibiotics in Animals' Feeds and Animal Production." *BioScience,* 22, 1972.

New York Academy of Sciences. "Microbial Insecticides." *Annals,* vol. 217, A symposium. June 22, 1973.

POINCELOT, R. P. *The Biochemistry and Methodology of Composting.* Connecticut Agricultural Experiment Station, bulletin no. 727, New Haven, Conn., 1972.

QUISPEL, A., ed. *The Biology of Nitrogen Fixation.* Elsevier, New York, 1974.

TSUTOMU, H. *Microbial Life in the Soil: An Introduction.* Marcel Dekker, New York, 1973.

KEY TERMS

humus	biological amplification	eutrophication (u"tro-fi-ka'shun)
microbial gum	biogeochemical recycling	mycorrhiza (mi"ko-ri'zah)
decomposer	limiting factor	silage
rhizosphere		

Pronunciation Guide for Organisms

Methanococcus (meth"ah-no-kok'kus)

Rhizobium (ri-zo'be-um)

Nitrosomonas (ni-tro"so-mo'nas)

Nitrobacter (ni"tro-bak'ter)

Thiobacillus (thi"o-bah-sil'us)

Water and Wastewater Microbiology

TYPES OF WATER
CLASSIFICATION OF WATER
WATER QUALITY STANDARDS
IDENTIFYING AND COUNTING CONTAMINANTS
WATER PURIFICATION
HOME SEWAGE DISPOSAL
PUBLIC WASTEWATER TREATMENT
Primary Sewage Treatment
Secondary Sewage Treatment
Tertiary Treatment

Learning Objectives

☐ Know the differences between ground water and surface water.

☐ Be able to describe how water is classified and define the terms "safe," "potable," "polluted," and "contaminated."

☐ Recognize the medical significance of heavy metal, nitrite contamination, and chlorine contamination.

☐ Be able to explain the categories of water quality standards.

☐ List the steps involved in the bacteriological examination of water in the detection of coliforms.

☐ Name the various methods used to purify a domestic water supply.

☐ Describe the fundamental operation of a septic tank.

☐ Recognize the advantages and disadvantages of a separate as opposed to a combined sewage system.

☐ Be able to define and explain the significance of the terms "dissolved oxygen" and "biological oxygen demand."

☐ Be able to explain the stages in primary sewage treatment, including sedimentation and anaerobic digestion.

☐ Be able to explain the stages of secondary sewage treatment, including trickling filter operation and oxidation processes.

☐ List various methods of tertiary sewage treatment.

All organisms require a regular source of useful water. If this essential factor is not available, some organisms die, while others form resistant stages. Often, an organism spends a great deal of time and energy to conserve its present water supply and search for a new source. Humans are not exempt from this requirement. Civilizations that have developed methods of providing a constant supply of fresh water have experienced a lower incidence of waterborne disease and have maintained a higher standard of living than those cultures unable to solve this problem. Cities could not exist without a central, reliable source of water. Advanced

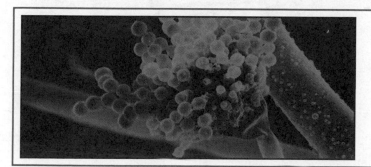

25

ancient civilizations developed water handling and purifying systems that lasted for thousands of years and are still being discovered or used today. The Roman baths and aquaducts (100 B.C.) are probably the most familiar artifacts of city water systems; but as early as 1250–1450 B.C., the Minoan Palace on the island of Crete had indoor plumbing with hot and cold running water, flush toilets, and drainage tiles. Following its destruction by a volcanic eruption, survivors of this great civilization migrated to Greece and took with them their engineering expertise. As the "dark ages" approached, less attention was given to water supplies and waste treatment; therefore, the incidence of waterborne disease increased and the level of civilization fell. It was not until the late 1800s that many countries throughout the world exerted a concentrated effort to deal with this problem. This is illustrated by the fact that outhouses and chamber pots were not replaced by flush toilets until less than 80 years ago. The explosive population increase and corresponding demand for household, commercial, and industrial water supplies in the past century has reduced the turnover time between the use and reuse of water. Today, it is essential to put forth an even greater effort to purify water and waste water.

In order to ensure a consistent source of high-quality water, it has become necessary to identify, remove, and control harmful contaminants which are found in both ground and surface water. This requires very accurate chemical and biological methods of water purification. After water has been used for drinking, bathing, washing, chemical processes, and cooling in manufacturing processes, it again becomes contaminated with agents that are pathogenic, biocidal, or biostatic. Because the release of these agents directly into the environment can have serious and far-ranging consequences, wastewater treatment facilities have been developed to separate these harmful materials from the water before it is discharged for recycling. Once removed, these potentially dangerous materials are also transformed by natural or artificial processes into simpler compounds found naturally in the environment. As is the case with medical microbiology, water and wastewater microbiology is one of the most applied of all the biological sciences.

TYPES OF WATER

There are two principle sources of water, **surface water** and **ground water** (Figure 25.1). Surface water comes from streams, lakes, rivers, shallow wells, and reservoirs created by damming. Because surface water is likely to come in contact with humans, animals, and soil microbes, it is more likely to contain harmful chemicals and microbes than ground water (Box 28). However, many of the microbes found in surface water are natural inhabitants. The number, location, and type of microbes in surface water vary depending on the dissolved or suspended salts, minerals, and organic nutrients. Algae and protozoa make up the majority of microbes found in fresh water, while the bacteria, fungi, and viruses are a smaller part of the aquatic microbial biomass. Aerobic microbes tend to predominate in the upper layers of a lake where atmospheric oxygen easily dissolves. Both facultative and strict anaerobes are more likely to be found in the lower layers or bottom sediment. Table 25.1 lists some of the bacteria found as normal flora in a freshwater lake. Pathogens found in fresh water are contaminants from

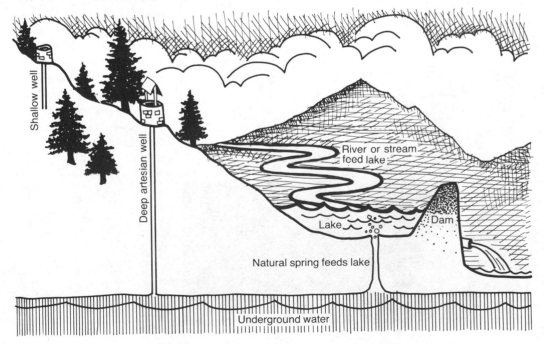

FIGURE 25.1
After rainfall, water either flows over the surface of the land to run off into rivers and streams, or it may slowly make its way deep into the ground.

IN at least one way, the world is going to the dogs.

How? Dog litter...Lots of dog litter.

Take New York City, for example. Dr. Alex Beck, head of the bureau of Animal Affairs, estimates there are up to one million dogs in the city. And that—he claims—means about 150 to 250 tons of litter dumped on city streets everyday. Everyday.

"Of course, it's distributed over a fairly large surface area," he added.

But that's little consolation for those who walk to work.

So the city is holding a hearing on the mounting problem.

Barbara Weiss and Cheryl Johnson, leaders of the 170-member Carl Schurz Park Association, argued for tighter controls and stricter enforcement of existing health and environmental laws.

But their plea has aroused more than a little anger from dog lovers.

Mrs. Weiss said she had been verbally abused and physically threatened by dog owners who refused to curb their animals.

"For many people the dog becomes a family member," she explained. "They equate it with 'Please put your child on a leash.' Also, it's the image. You know, you never see Lassie on a leash."

"You never see Lassie littering, either," objected one councilman.

But if New Yorkers think they're getting it all dumped in their lap, they can rest assured the problem is universal.

Paris is about to build the world's first public flush toilets for pets.

Sanitation officials will soon begin installing the roofless, concrete structures, complete with flowers and perfume, at key locations throughout the French capital.

All a dog will have to do is sniff his way inside one of the "Vespachiennes," as they will be called, and let nature take its course.

Pet owners will then simply press a button and flush away what until now has been one of the hottest political issues in the capital.

"The dog problem is the most irritating one in Paris from our point of view," said Pierre Roger, director of the Municipal Action Center for Cleanliness. "It's the problem we get the most complaints about."

The first Vespachienne will be placed at the entrance to a tiny park at Clignancourt Square.

The toilet will be equipped with a walkway, a 2½-foot post to replace the traditional fire hydrant and a flush button.

To encourage dogs to make the most of the opportunity, the facility will exude a special fragrance to the accompaniment of running water. Two flower boxes—which officials assure firmly are for decorative purposes only—will border the structure.

"We've got to do this with a sense of humor and not be too serious about it, or else we're liable to look ridiculous," Roger said.

But for New Yorker Fran Lee, its no laughing matter.

"You're fooling around with manure," she said at the New York hearing. "I'm talking about a clean and safe environment. Children are dying from disease. The streets are a disgrace. Perhaps. But not for long. New York is not about to be left holding the bag."

Carl Shurz Park will soon be the site of a two-month pilot project to get dogs to defecate in specially constructed sewer drains.*

---

*Reprinted by permission of United Press International, Aug. 21, 1975.

BOX 28
LASSIE
UNLEASHED
NOT
LIKED

TABLE 25.1
Bacteria found naturally in fresh water.

Type	Form or Genera
Cocci	Pigmented (*Micrococcus*) and nonpigmented forms.
Bacilli	Pigmented (*Serratia*); nonpigmented forms: *Pseudomonas*, sulfur bacteria, thermophiles, *Bacillus*, *Clostridium*, and pathogenic *Salmonella*; *Caulobacter* and *Cytophaga* able to concentrate minimal nutrients.
Spirilla	*Leptospira*, *Vibrio*, and others.
Stalked and sheathed bacteria	*Gallionella* and *Leptothrix* able to convert inorganic iron compounds to insoluble ferric compounds.
Nitrogen-fixing bacteria	*Clostridium*, *Azotobacter*, and *Aerobacter*.
Nitrifying bacteria	*Nitrosomonas* and *Nitrobacter*.
Photosynthetic bacteria	*Chromobacterium* and *Chlorobium* oxidize sulfur from hydrogen sulfide during photosynthesis.
Methane bacteria	*Methanobacterium* and *Methanococcus* produce methane gas from organic matter.

infected individuals or are derived from untreated sewage discharged into a lake or river. Their survival and transmission is made possible by the warmer temperatures of surface water and the presence of organic material that modifies the environment and protects them from destruction by oxidation.

Ground water, or **underground water,** varies only slightly in temperature and has many of its contaminants removed or oxidized as it percolates downward through many different soil horizons. This water is often obtained from springs or deep artesian wells. Although not completely free of microbes, ground water usually contains very few bacteria (one or two per milliliter), because of the effective natural filtering action of the soil and several other controlling factors.

Water that is extremely cold or hot contains very few of the microbes typically found in most surface waters of moderate temperature, such as lakes, shallow wells, and rivers. The cold waters of the oceans (most of which average less than 5°C), polar regions, and deep wells are favorable environments for psychrophiles. Thermophiles are adapted to environments that have high temperatures. The water of hot mineral springs and boiling geyser basins (70–90°C) may contain microbes such as *Thermus aquaticus*, *Thermoplasma acidophilum*, and *Sulfolobus acidocaldarius*. Some of these thermophiles have even been found growing in hot water heaters. The high mineral or salt concentrations of some water inhibit the growth of many microbes. Sea water has an average salt concentration of 3.5 percent compared

to fresh water, which has a salinity near zero. However, some osmophilic or halophilic species grow well in water with salt concentrations as high as 10 to 15 percent. The hydrogen ion concentration (pH) of fresh water is usually between 6.5 and 8, a range favorable for the growth of many microbes. However, some bodies of water have an acid pH that acts as a natural selecting agent favoring the growth of such unique microbes as *Sulfolobus acidocaldarius*. This bacterium is able to grow under extremely high sulfuric acid concentrations found in the hot acid springs of Yellowstone National Park. The temperature ranges from 60 to 92°C and at the same time has a pH as low as 0.8. All these factors influence the type of microbes that occur naturally in surface and ground water. As a result of human activities, some of these influences have been altered and new factors have been introduced that affect water quality.

---

1 What is the origin of surface water?
2 Which type of water is most likely to contain pathogens?
3 What type of microbes are most likely to be found in water from deep wells?

CLASSIFICATION OF WATER

In order to develop water quality standards and formulate plans for the effective control of chemical and biological contaminants in domestic supplies of water, several broad water-use categories have been developed in many states. Table 25.2 describes the categories used by states surrounding and having access to the Great Lakes. The placement of a water source in one of these categories depends on such factors as coliform bacteria count, dissolved oxygen, toxic and deleterious substances, dissolved nutrients, temperature, pH, and radioactive material present in the water at the time of sampling.

Within the category of domestic water, four classifications are used to describe its quality. **Safe water** has no evidence of fecal contamination but may have undesirable tastes or odors due to the presence of inorganic chemicals or organic compounds. **Potable water** is clear, sparkling, free from offensive odors and tastes, and is safe to drink. **Polluted water** is unsafe to drink or use for domestic purposes, since it may contain intestinal pathogens. This water is usually found in lakes or streams that have been used as dumps for sewage or industrial wastes and is clouded with sediment or organic material. **Contaminated water** also contains pathogens but can be clear, sparkling, and free of obnoxious odors and flavors. The presence of microbes makes it unsafe for drinking and domestic use. Among the more toxic chemicals found as water contaminants are (1)

TABLE 25.2
Use designation categories of water.

Water Supply	Description	Uses
Domestic	Raw water source intended for use as a potable supply. It can be made suitable for human consumption by conventional treatment methods.	Drinking water, food processing such as cooking, and a liquid ingredient in such items as carbonated beverages and beer.
Industrial	Intended for use in manufacturing processes other than food processing. It is not intended that this water will be used as a raw water source for a potable supply. Since most industries will accept municipal water as a source for industrial water, the standards are similar to domestic raw water sources, except from the public health standpoint.	Cooling water, a liquid ingredient other than in food products, and equipment washing.
Recreation Total body contact	Intended for uses where the human body may come in direct contact with water to the point of complete submergence. The water may be accidentally ingested and also certain body organs, such as the eyes, ears, etc. will be exposed to the water. Not intended that this source be used as a potable supply unless treatment is applied.	Swimming, water skiing, and skin diving.
Partial body contact	Intended for uses where the human body may come in direct contact with the water but not normally to the point of complete submergence. Not likely to be ingested nor will critical organs (eyes, ears, nose) normally be exposed to the water.	Fishing, boating, hunting, trapping, and equipment cleaning.
Fish, wildlife and other aquatic life	Intended for use by fish, wildlife, aquatic life and semiaquatic life as their natural habitat in which to not only exist, but propagate and grow.	Intolerant fish (cold-water species), intolerant fish (warm-water species), tolerant fish (warm-water species), and wildfowl and fur-bearing animals.
Agricultural use	Intended for general agricultural usage. It is directly used for livestock, the growing of crops, and not intended for direct human consumption.	Livestock watering, irrigation, and spraying.
Commercial and other uses	Intended for uses such as navigation and other uses not included elsewhere in the standards. It is distinguished from industrial use in that the water is not used directly in a process.	Hydroelectric power generation, commercial shipping, and electric power generation from steam.

SOURCE: Modified from Water Resource Commission, Department of Natural Resources, State of Michigan.

heavy metals, (2) nitrites, (3) chloramines, and (4) organic compounds.

Heavy metals frequently found in domestic water include cadmium, chromium, copper, lead, mercury, nickel, and zinc. All of these metals are capable of binding with enzymes and interfering with normal cell metabolism. Many industries regularly release these metals as wastes from manufacturing processes. Heavy metals may be discharged directly into the water or channeled through a municipal sewage treatment plant. In some cases, these metals concentrate through the food chain to levels that result in **heavy metal poisoning.** In recent years, the Environmental Protection Agency has required certain industries to install purification equipment to extract cadmium and other heavy metals from waste water, before they are released into the municipal sewage system. This measure is intended to keep the metal from concentrating in sewage used as fertilizer or landfill. Unacceptable levels of cadmium have been found in raw (untreated) water in many parts of the world, including New York, Wisconsin, and Japan. Cadmium poisoning is called itai-itai ("ouch-ouch") in Japan because it is a painful disease that can be fatal.

Metallic mercury is a relatively nontoxic metal; however, some bacteria are able to convert this heavy metal into more toxic methylmercury (CH_3Hg^+) and alkyl-mercury compounds. An estimated 23 million pounds of mercury metal are released into the environment each year from industrial, medical, and individual sources. A sizable portion of this mercury is converted by anaerobic bacteria in water to the toxic forms. Methylmercury has been found to be concentrated through the food chain and reaches humans through contaminated fish, such as tuna, swordfish, and ocean perch. The ingestion of these fish causes methylmercury poisoning. Symptoms include blindness, insanity, coordination loss, memory loss, deafness, and ultimately death. The most well-known outbreak of mercury poisoning occurred in Minamata City, Japan, in 1953, when more than 100 people died or suffered nervous system damage from eating fish taken from Minamata Bay. In the United States, high levels of methylmercury have been found in fish taken from the Great Lakes (pickerel) and the Atlantic Ocean (swordfish).

Nitrite poisoning, **methemoglobinemia,** described in Chapter 24, occurs from the ingestion of water or food containing high levels of this contaminant which forms a stable methemoglobin compound in infants (less than about three months of age) and farm animals. Bacteria normally found in the water are able to convert nitrate ions from fertilizers and organic wastes to nitrite. The concentration of nitrates and nitrites are reduced naturally by the action of the denitrifying bacteria in water and soil. Several attempts are currently being made to use these microbes in water and wastewater treatment plants to reduce harmful levels of toxic substances.

---

1 The placement of a water source in a particular category depends on what factors?
2 Differentiate among safe water, polluted water, potable water and contaminated water.
3 What kinds of metals are associated with heavy metal poisoning?
4 What ion is responsible for the formation of methemoglobin?

In recent years, concern has been voiced about the extensive use of chlorine in water and wastewater treatment. It was once thought to be a harmless method of oxidizing residual microbes and organic material before water and sewage were released to the environment. However, there is mounting evidence that residual amounts of chlorine are able to react with many compounds to form a group of toxins known as the **chloramines.** There is a higher level of chloramines found in waste water treated with chlorine than in drinking water not having a high organic compound load. Chloramines are known to be toxic to humans and animals, and act as spermicidal agents to several species of invertebrate animals. They are known mutagens that may also prove to be carcinogenic.

Pesticides, herbicides, cleaning agents, food additives, industrial materials, adhesives, and many other synthetic materials contain new chemicals that are constantly being introduced into the environment at levels that have never been previously experienced. Each year, approximately 70,000 kinds of organic chemicals are placed on the market that ultimately make their way through the environment to the water. The presence of these chemicals at increasing levels is of great concern because of their known toxicity, mutagenicity, and carcinogenicity. Although at the present time many occur at such low levels that it is unlikely that they are harmful, any significant increase may prove to make them health hazards. To avoid this, maximum permissible levels for certain types of organic materials such as pesticides have been established, and this concentration is being monitored on a regular basis. Organic contaminants are also tested for their mutagenic and carcinogenic properties before permissible levels are determined (refer to Ames test, Chapter 11).

Although water contains a variety of microbes naturally, those of greatest concern are the pathogenic bacteria, viruses, protozoa, and fungi. Transmission of these pathogens is most likely in surface water but ground water can also be a source of waterborne disease. This may occur in areas with a heavy concentration of septic tank fields improperly installed on very porous soil. This type of soil is ineffective in filtering and oxidizing pathogens before they percolate to the water table. The presence of large, underground water channels called aquifers serves to pipe contaminants directly to deep wells (Figure 25.2). Waterborne diseases result from improper water or sewage treatment and contamination through cracks and damaged pipes in the distribution system (see Table 25.5). Proof that pathogens are transmitted by water was first confirmed in 1854 by Dr. John Snow who investigated an outbreak of cholera that resulted in 700 deaths. His classic epidemiological investigation in the St. James Parish of London led to the discovery of a break in a sewer pipe that allowed sewage to seep into a shallow well used by all the local residents as a source of fresh water. With more improved methods, the incidence of waterborne disease has been more accurately identified as seen by the rise in the average annual number of waterborne outbreaks since the period 1951–55 (Tables 25.3 and 25.4). The microbes chiefly responsible for waterborne disease in the United States are listed in Table 25.7. The category designated "acute gastrointestinal illness" is caused by a variety of as yet unidentified microbes. This is

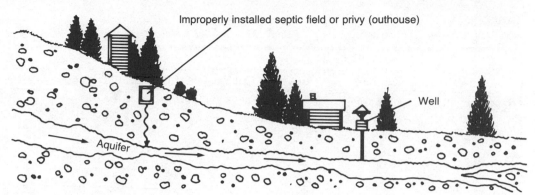

FIGURE 25.2
The placement of a well should not be downstream from a source of contamination. A pump might draw contaminants into the water supply and ruin it.

Improperly installed septic field or privy (outhouse)

Well

Aquifer

TABLE 25.3
Average annual number of waterborne disease outbreaks.

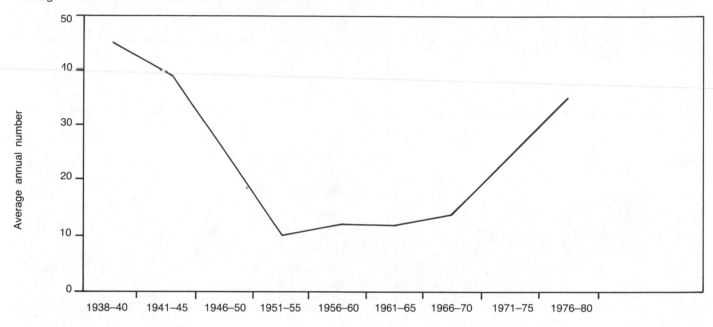

TABLE 25.4
Waterborne disease outbreaks.

	1972	1973	1974	1975	1976	1977	1978	1979	1980
Outbreaks	29	26	25	24	35	34	32	41	50
Cases	1650	1784	8363	10,879	5068	3860	11,435	9720	20,008

characterized by upper and/or lower gastrointestinal upset such as cramping, nausea, vomiting, and diarrhea. A detailed explanation of the many diseases that have been confirmed as waterborne may be found in chapters 16 through 22. Table 25.7 lists the major waterborne diseases.

WATER QUALITY STANDARDS

Because water can be a carrier of pathogenic microbes capable of causing disease in humans, domestic water supplies used for drinking and cooking should be checked for possible contamination on a regular basis. It is necessary to establish maximum limits or standards for microbial contamination to ensure a pure, wholesome water supply and sufficient protection from waterborne disease. Water quality standards have been established by every state. They include evaluation procedures and a maximum permissible limit for contaminating microbes. The primary intent of these standards is to limit the transmission of waterborne disease; but in most cases, the identification of pathogens in water and waste water on a routine basis is not practical. Most

TABLE 25.5
Waterborne disease outbreaks by type of system and cause of system deficiency for a recent year.

	Municipal		Semipublic		Individual		Total	
	Outbreaks	Cases	Outbreaks	Cases	Outbreaks	Cases	Outbreaks	Cases
Untreated surface water	1	200	1	55	1	12	3	267
Untreated ground water	0	0	9	547	2	22	11	569
Treatment deficiencies	4	1362	8	891	0	0	12	2253
Deficiencies in distribution system	6	718	1	47	0	0	7	765
Miscellaneous	1	6	0	0	0	0	1	6
Total	12	2286	19	1540	3	34	34	3860

SOURCE: Center for Disease Control, Atlanta.

TABLE 25.6
Waterborne disease outbreaks by etiology and type of water system for a recent year.

	Municipal		Semipublic		Individual		Total	
	Outbreaks	Cases	Outbreaks	Cases	Outbreaks	Cases	Outbreaks	Cases
Acute gastrointestinal illness	5	518	13	1396	2	24	20	1938
Chemical poisoning	4	612	1	11	1	10	6	633
Giardiasis	2	950	2	62	0	0	4	1012
Salmonellosis	1	206	1	7	0	0	2	213
Hepatitis	0	0	1	47	0	0	1	47
Shigellosis	0	0	1	17	0	0	1	17
Total	12	2286	19	1540	3	34	34	3860

SOURCE: Center for Disease Control, Atlanta.

waterborne pathogens find their way into the water from feces of infected individuals. Unless there is a massive epidemic, these pathogens will only be found in very low, difficult to isolate concentrations, since they are easily killed once outside the host. Therefore, microbiologists have adopted the practice of identifying **indicator microbes** that are normally found along with pathogens in the intestinal tract. Their presence in water indicates possible fecal contamination and, therefore, an increased probability of waterborne disease. The coliform bacteria such as E. coli, Klebsiella, and Enterobacter are most frequently used as indicators of fecal contamination for several reasons. First, they are normal flora of the human intestinal tract and the presence of high numbers in water would in all likelihood indicate human contamination. Second, the coliforms are very hardy bacteria able to survive for long periods outside the host. This makes it possible to isolate and identify them long after they have been released. Third, they are relatively easy to culture in the laboratory, requiring only the most basic microbiological materials and expertise. Fourth, they are found in high enough numbers in contaminated water to give a statistically valid count. Microbes that are scarce would not be as valuable, since their presence could be from any number of unrelated or nonhuman sources.

Another group of indicator bacteria used for the identification of fecal contamination are the **fecal streptococci** (for example, S. faecalis, S. faecium, S. bovis, S. equinus) (Figure 25.3). These bacteria are normally found in the intestinal tract of warm-blooded animals, including humans. Their identification in a water sample is a good indicator of human fecal contamination since (unlike the coliforms) members of the fecal strep are not normally found in nature. For waters that have a naturally high coliform level above state prescribed standards, the fecal strep count is a more valuable test. The water quality standard for domestic water in one Great Lake state (Michigan) as indicated by the presence of coliforms is given in numbers of bacteria per 100 milliliter or MPN (most probable number) sample:

A_1-1 Coliform Group (organisms per 100 ml or MPN) (domestic use): The monthly geometric average shall not exceed 5000, nor shall 20 percent of the samples examined exceed 5000 nor exceed 20,000 in more than 5 percent of the samples.

For industrial water (used for cooling and manufacturing processes) the water quality standard is:

A_2-1 Coliform Group (organisms per 100 ml or MPN) (industrial use):
The geometric average of any series of 10 consecutive samples shall not exceed 5000, nor shall 20 percent of the samples examined exceed 10,000. The fecal coliform geometric average for the same 10 consecutive samples shall not exceed 1000.

---

1 What microbes are chiefly responsible for water-borne diseases?
2 Why are indicator microbes identified in a bacteriological analysis of water?
3 Name two groups of indicator microbes.

IDENTIFYING AND COUNTING CONTAMINANTS

Confirming the presence and number of indicator organisms is based on standard methods for the bacteriological examination of water. These methods are listed in the American Public Health Association publication entitled

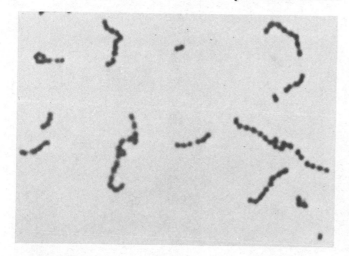

FIGURE 25.3
The presence of *Streptococcus faecalis* in a water sample indicates fecal contamination. (Courtesy of Abbott Laboratories.)

TABLE 25.7
Examples of waterborne diseases.

Agent	Disease
Bacteria:	
Vibrio cholorae	Cholera
Escherichia coli	Diarrhea
Leptospira sp.	Leptospirosis (Weil's disease, hemorrhagic jaundice)
Pseudomonas pseudomallei	Meliodosis
Salmonella paratyphi	Paratyphoid fever
Salmonella typhi	Typhoid fever
Protozoa:	
Entamoeba histolytica	Amoebic dysentery
Naegleria fowleri and *Acanthamoeba* sp.	Amoebic meningoencephalitis
Giardia lamblia	Giardiasis (lambliasis)
Viruses:	
Hepatitis virus, type A	Epidemic hepatitis
Respiratory syncytial virus	Respiratory syncytial disease
Fungi:	
Microsporum sp., *Trichophyton* sp. *Epidermophyton* sp.	Ringworm and other cutaneous fungal infections.

Standard Methods for the Examination of Water and Wastewater. The Environmental Protection Agency has specified that the methods and procedures noted in this publication will serve as the standard for drinking water. Water samples are taken for analysis in sterile sampling bottles using proper aseptic technique and sent to the laboratory for analysis. As with specimens for medical diagnosis, samples must be analyzed or refrigerated as quickly as possible (within 6 to 12 hours). Once in the laboratory, two procedures are followed: (1) counting and (2) identification of indicator microbes. Counts are done using one of several methods as described in Chapter 8. The standard plate count (SPC) provides the most accurate determination of viable cells per milliliter, but it is time-consuming and expensive. Membrane filtering (MF) can be done in a shorter time and used for coliforms. One of the most widely used counting methods is the most probable number (MPN) method. The MPN method is run in conjunction with the standard method of identifying the indicator organism. In the case of coliforms, the bacteriological examination of water is divided into three stages, the presumptive test, the confirmed test, and the completed test (Figure 25.4). To perform the **presumptive test,** three sets of three (or five) tubes containing lauryl tryptose broth are inoculated with 10, 1, and 0.1 ml sam-

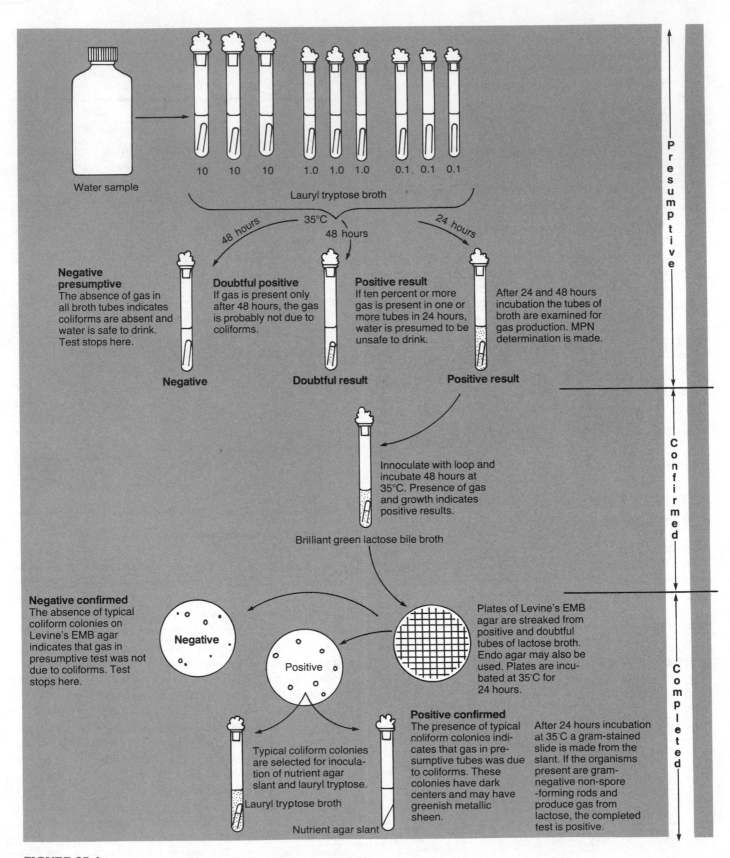

FIGURE 25.4
Three testing stages are used to verify the presence of fecal coliforms in a water sample: the presumptive test, the confirmed test, and the completed test.

ples, respectively, of the water to be tested. This sets up a dilution series on which a most probable number of bacteria per 100 milliliters can be determined. Lauryl tryptose broth Durham tubes are used as both a differential and selective growth medium for coliforms. Growth in this medium and gas collected in the inverted tubes after 24 and 48 hours incubation is presumed to be from the fermentation of lactose by coliforms. This test does not completely identify the water contaminant as a coliform, since many other bacteria or combinations of waterborne bacteria can give positive results. If the presumptive test shows no growth or gas production, it is safe to conclude that the water contained no coliforms, was presumably not contaminated with fecal material, and is bacteriologically safe to drink. Positive tubes (with gas and growth) verify the presence of coliforms by performing the **confirmed test.** Several selective and differential media types are used in this identification test. One of the most widely used media recommended in Standard Practices is brilliant green lactose bile broth. Brilliant green dye inhibits the growth of Gram-positive, nonconfirming organisms, and lactose encourages the growth of fermenting coliforms. The bile acts as a selecting agent that destroys bile-soluble bacteria but has no effect on bile-tolerant coliforms. If the confirmed test is negative (i.e., no gas produced in the broth culture), it is presumed that the organisms producing the growth and gas in the presumptive test were not coliforms. If the confirmed test is positive (i.e., gas is produced), samples are inoculated onto either Endo- or eosine-containing agar media. Eosine methylene blue agar (EMB) is a selective medium for Gram-negative bacteria, and their growth results in the formation of a green metallic sheen on the surface of coliform colonies. Endo agar may also be used, and the presence of coliforms is verified by a color change from light violet to a deep purple. If a positive test is found on either of these plates, samples are inoculated onto a nutrient agar slant for growth and later microscopic examination, and into another tube of lauryl tryptose broth to, again, demonstrate the organisms' ability to utilize lactose sugar and produce gas. The presence of Gram-negative, nonsporeforming, short bacilli that grow and produce gas in the lauryl tryptose tube completes the positive identification (**completed test**) of coliform contaminants. If the water tested is from a source of domestic water, it must be purified of potentially dangerous microbes before being used.

---

1   What medium is used in the presumptive test?
2   Does a positive presumptive water test prove that a water source is unsafe for drinking?
3   EMB is a selective growth medium for what type of microbes?

WATER PURIFICATION

Methods used to purify a domestic water supply are designed to remove large contaminants as well as destroy harmful bacteria and other microbes. **Sedimentation** removes the larger, more coarse particulate contaminants and microbes that may be attached to their surfaces. The process is made more effective by adding chemicals known as **flocculants** to the water. Water that is alkaline (or made alkaline by the addition of calcium hydroxide) will quickly form jellylike masses, or floccules, when aluminum sulfate or ferrous sulfate is added (Figure 25.5). The flocculant is insoluble and will settle to the bottom of the sedimentation basin, carrying with it suspended particles and microbes. The sediment formed at the bottom of the basin is later removed, treated, and used for filling land near the water treatment plant. After flocculation, the water is pumped through several tanks two to four feet deep that are filled with layers of gravel and sand of varying particle size. Residual floc is trapped in the sand and forms an effective filter able to remove most enteric bacteria and viruses. Water leaving the sand filter is clear and sparkling but may contain biological contaminants above acceptable limits. These are inactivated by the addition of chlorine to the water before it is pumped from the plant. Chlorine also eliminates foul odors that may be present in the water after treatment. The chlorine usually arrives at the plant as a liquid but is converted and injected into the water as a gas at a concentration that

FIGURE 25.5
A sedimentation basin. (Courtesy of the U.S. Environmental Protection Agency.)

will provide a chlorine residual level of about 0.2 to 0.6 part per million (ppm). At this time, **fluoride** can also be added to the water. At 1 ppm, fluoride helps prevent dental caries by combining with developing enamel in the teeth of children. This internal hardening does not occur in adults since their teeth are not growing, but fluoride in the water can combine with surface enamel for added protection against acid dissolution caused by *Streptococcus mutans* and other cariogenic bacteria.

Once pumped for public use, purified water is used for many purposes. Although most people are familiar with its use for drinking, it is also used for food processing, cooling, dilution of toxic and pathogenic materials, and as a carrier of wastes to a home or industrial sewage treatment facility. **Sewage** is a watery material carrying refuse and wastes in a drainage system. Sewage from a domestic source differs from that released by industry. Most domestic sewage has (1) a high content of organic fecal material, (2) a high microbial load, (3) a high percentage of organic vegetable material from garbage disposals and cellulose from toilet paper, and (4) chemicals such as soaps, detergents, and other household products. Industrial wastes usually contain a smaller portion of fecal material but higher percentages of oils, greases, lubricating detergents, and inorganic chemicals. In order to make the water safe for reuse, sewage must be treated to destroy or remove the pathogenic microbes and toxic chemicals, decompose the organic material into more oxidized and useful forms, and remove the nonbiodegradable materials for special treatment and handling. Several methods of sewage treatment are available. Home sewage disposal using a privy (outhouse) or septic tank and drain field system is effective, if used under the proper conditions. However, public sewage treatment facilities have become essential as the population has increased and reached densities beyond the ability of these private systems to adequately cleanse waste water.

HOME SEWAGE DISPOSAL

A **privy** is one of the oldest methods used to dispose of human waste. It consists of a pit or hole dug in the ground, over which is placed a small enclosure with a toilet stool and seat. Decomposition of excrement occurs by chemical and microbial oxidation but is very slow and often results in foul odors. Flies become a serious health hazard, since they may quickly pick up and transmit fecalborne diseases to those using the facility. Pathogens can also be transmitted through water that leaches from the site into nearby lakes, rivers, or wells. In order to prevent these problems, privies should only be constructed when there is no other suitable alternative. If they are built, they should be placed at least 50 feet from living units and 75 feet from wells. Earth should be banked

along the sides to make the pit dark and fly-tight. High nitrogen fertilizer may be added to the pit to help control odors, and the unit should be painted and well maintained. The privy should be filled with soil and reconstructed at a new site on a regular basis. A modern alternative to outhouses are chemical toilets used at construction sites and on motorhomes, boats, and remote campgrounds. These are merely holding tanks for waste, not treatment units. The chemicals added to the wastes mask odors and help dissolve solids for easier pumping to a treatment facility.

---

1 How do the sedimentation and flocculation processes for water purification differ?
2 What are the typical components of domestic sewage?
3 Why should a privy, or outhouse, be moved on a regular basis?

A more acceptable home sewage disposal system is a **septic tank** and drainage field. The tank is usually concrete, completely enclosed, and has an overflow pipe that connects to the field (Figure 25.6). Internal baffle plates separate the water from solid wastes. This system is basically the same as a public sewage treatment facility but operates less efficiently. Waste water from a home flows into the tank where bacteria and other microbes anaerobically digest the organic waste into simpler, more useful compounds. Heat generated by the anaerobic digestion helps destroy pathogens. Often the location of a septic tank can be seen in winter, because this heat melts the snow from the ground over the tank. The tank fills with partially digested solids that are periodically pumped out and removed to a landfill site. Water flowing through the tank leaves through an overflow pipe and enters a drain field constructed of separated clay tiles or perforated plastic pipe laid on a bed of porous gravel. As the water flows through the drain field, it percolates through the gravel and underlying soil where organic wastes and pathogens are further decomposed by the natural activities of soil organisms (refer to Chapter 22). In a sparsely populated area with soil of the proper water-holding capacity and filtration rate, a septic tank system is a reasonable means of treating domestic sewage from a private home. However, the water released from the system is unsuitable for drinking; and, therefore, the drainage field should be positioned away from wells or other sources of drinking water. To maintain the most efficient microbial digestion, care should be taken not to add high concentrations of microbicidal chemicals to the tank. The proper dilution can be maintained by selecting a tank with a holding capacity large enough to meet the needs of a family (Table 25.8). When the bacteria are killed by high concentrations of soaps, detergents, or other chemicals flushed down sinks and toilets

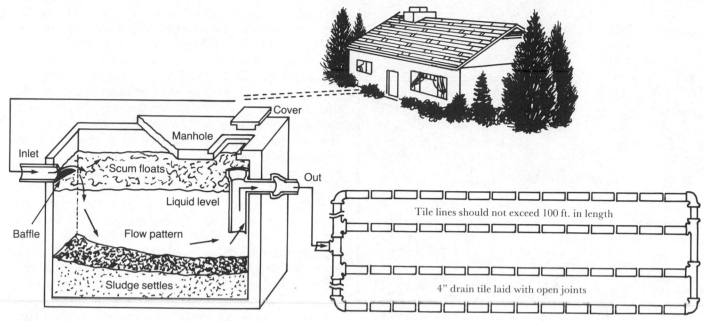

FIGURE 25.6

Sewage leaves the house and enters the septic tank where microbial activities begin. Sedimentation of heavier material begins here. The overflow drains into a field of tiles where the effluent percolates into the soil. There soil microbes complete the destruction of harmful materials.

without adequate dilution, the solid wastes accumulate too rapidly and escape into the field, causing foul odors. To restore the activity of the tank, cultures containing detergent-resistant, sporeforming and nonsporeforming bacteria may be added to the tank. These are commercially available and do not contain yeast.

PUBLIC WASTEWATER TREATMENT

As the population density increases in urban and suburban areas, home disposal systems become less efficient. The effluent (discharged waste) cannot be decomposed to harmless chemicals and microbes before it causes harm or unpleasant odors. To ensure the destruction of harmful mate-

TABLE 25.8

Recommended minimum septic tank sizes.

House Size	Tank Capacity Without Garbage Grinder	Tank Capacity With Garbage Grinder
2 bedrooms or less	750 gallons	1000 gallons
3 bedrooms or less	900 gallons	2–750 gallons
4 bedrooms or less	1000 gallons	1–1000 gallons plus 1–500 gallons

Each additional bedroom—400 gallons of capacity.

rials before discharge into the environment, public sewage treatment facilities have been developed throughout the country. A complex network of sewer pipes connects homes, businesses, and industries to treatment plants. There are two kinds of sewer systems, separate and combined (Figure 25.7). In the **separate system,** sanitary sewer pipes conduct only sewage from homes, businesses, and industries to the treatment plant; while a second, parallel system of storm-sewer pipes channels large volumes of rain water, melted snow, and land runoff to the plant. In a **combined system,** both types of water are carried through a common pipeline to the treatment plant. However, after a heavy storm or sudden snowmelt, the amount of raw waste entering the plant may be so large that the excess must be diverted through a bypass drain into a lake or stream. If this excess is not diverted, the waste treatment plant becomes overloaded, malfunctions, and must be shut down. If wastes are diverted too frequently, the receiving water will become depleted of its oxygen supply and become "stale," septic, or polluted.

Whether a body of water stays "alive" and "healthy" depends on the availability of **dissolved oxygen (DO).** The oxygen is used by fish, plants, and microbes in aerobic metabolism to maintain a stable ecosystem. Whenever the DO drops below the minimal level required by these organisms, eutrophication occurs, and the organisms die and are incompletely decomposed by anaerobes that release putrid compounds into the water and air. Along the shores of eutrophic lakes, streams, and oceans, the hydrogen sulfide level

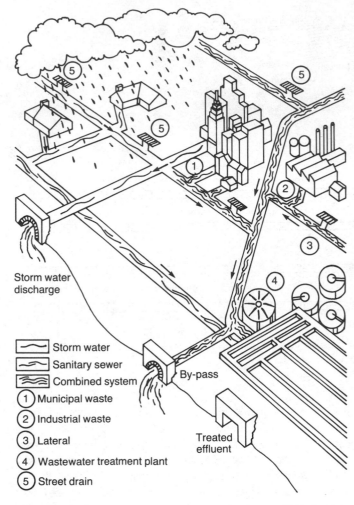

Storm water
discharge

⌇	Storm water	
⌇	Sanitary sewer	
⌇	Combined system	
①	Municipal waste	
②	Industrial waste	
③	Lateral	
④	Wastewater treatment plant	
⑤	Street drain	

By-pass

Treated
effluent

FIGURE 25.7

A public sewer system. (Courtesy of the U.S. Environmental
Protection Agency.)

can reach such high concentrations in the air that the paint on
nearby houses becomes blackened and beaches become
clogged with rotting organic material. In order to prevent local waters from becoming stale and polluted, the separate
sanitary and storm sewer system is preferred. It enables highly
fertile organic wastes to be continuously channeled into the
plant for treatment without being combined with excessive
amounts of storm water. In some cities, the discharge of storm
water is further controlled by the installation of large underground tanks that hold the excess water until after a storm,
when it can be discharged at a regulated flow.

Waste treatment plants operate on the same principles
that are found to decompose organic wastes and purify water in natural lakes and streams. However, oxidation of organic wastes is controlled and takes place at the most efficient rate possible. As in a natural environment, the amount
of dissolved oxygen available for decomposition must be
maintained at optimum levels. The measure of the amount

of dissolved oxygen used by microbes for the biological oxidation of wastes is called the **biological oxygen demand
(BOD).** The BOD is monitored regularly to ensure that wastes
will be decomposed into their most basic forms as quickly
as possible. In addition, BOD can be used as a measure of
the amount of organic material or pollution in water. If the
amount of wastes or toxic substances arriving at a treatment
plant increases sharply, aerobic microbes will demand more
oxygen to decompose the material. Since the amount of
oxygen used by the microbes is approximately proportional
to the amount of organic wastes in the sewage, there is an
increase in the BOD whenever this excessive amount of
biodegradable material enters the plant. If the demand for
oxygen is excessive, the microbes are not able to respire,
the system becomes anaerobic, and the treatment process
is ineffective. Effluent discharged from such a plant is high
in organic nitrogen, phosphorous, and other nutrients and
encourages eutrophication in the receiving lake or stream.
To prevent this, sewage treatment is set up in stages and
monitored closely to maintain an adequate dissolved oxygen supply for the microbial decomposition of biodegradable wastes. The three stages of sewage treatment that have
been developed are called primary, secondary, and tertiary.

---

1 List several types of material that should not be
   placed in a septic tank.
2 What are some advantages of a separate sewer
   system?
3 Why is important to measure the dissolved oxygen level at a sewage treatment plant?
4 What is the difference between DO and BOD?

PRIMARY SEWAGE TREATMENT

In **primary treatment,** solid wastes and grit are removed
from sewage by screening and settling (Figure 25.8). As the
raw sewage enters the plant, it passes through a series of
coarse and fine steel screening bars that trap solid wastes.
Paper, food debris, sticks, and heavy sediment collected on
the bars are raked off and used for landfill. The waste water
then flows into a grit chamber where sand, small stones,
and grit quickly drop to the bottom of the tank. This portion
of primary sewage treatment is especially important in cities
with a combined sewage system. When the grit chamber
fills to capacity, its contents are also removed to a landfill
site. The water flowing from the grit chamber contains a
great deal of suspended organic material, solids, and microbes that may be removed in the **sedimentation tank.**
Waste water moves slower as it enters this tank, allowing
smaller particles to settle to the bottom. To speed this process, many treatment plants add flocculants such as ferrous

A

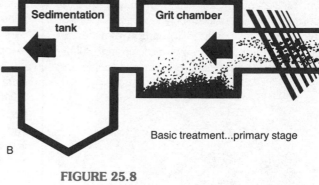

Basic treatment...primary stage

B

FIGURE 25.8

(A) A sedimentation tank. (B) Primary sewage treatment. (Part A courtesy of the U.S. Environmental Protection Agency.)

sulfate or aluminum sulfate to form an insoluble jellylike net that traps particles as it settles to the bottom. As much as one-third of the bacteria and organic wastes can be removed by sedimentation. The undigested sediment that forms on the bottom is called **raw,** or **settled, sludge** and is removed by hand after the tank is filled to capacity or on a continuous basis by a system of mechanically operated scrapers installed in the tank. The raw sludge (90 percent water) is disposed of by pumping it to an anaerobic digestor for further decomposition, or to a unit where it is dewatered, heat-treated, and bagged for use as a soil conditioner or fertilizer. The commercial fertilizer known as Milorganite is produced from municipal sludge of the city of Milwaukee, Wisconsin. In many cases, converting sludge to fertilizer is not possible because of the presence of toxic materials and pathogens. One way to avoid this problem is to treat the sludge with sulfuric acid followed by ammonia and other fertilizer chemicals that heat and dry the material. This process changes the sludge to a nonpolluting, nonhazardous granular fertilizer. Sludge may also be disposed of by filling land or incineration. Some cities are exploring the idea of burning sludge in "dedicated boilers" (those that will only burn sludge) for the production of electricity from steam. However, the efficient operation of such a plant requires large amounts of sludge not produced by most smaller cities.

Another sludge disposal method that generates energy is the **anaerobic digestion** process. Settled sludge from primary treatment is pumped into an anaerobic digestor unit and stirred for two weeks to a month (Figure 25.9). During this time, the high concentration of cellulose found in the sludge is digested into sugars, amino acids, and other soluble products. The freed carbohydrates are then fermented by the anaerobes to organic acids and alcohols, resulting in a drop in pH. This environment favors the growth of bacteria that utilize the acids and alcohols and produce hydrogen (H_2) and carbon dioxide (CO_2) gases. In the final stages of anaerobic digestion, the methanogenic bacteria convert the H_2 and CO_2 to methane gas (CH_4) and hydrogen sulfide (H_2S). These gases are piped from the digestor to a filter where the foul-smelling H_2S is dissolved in a water trap. The CH_4 is not removed but bubbles out and is piped to a compressor, where it is bottled. The methane gas may then be used to power the stirring paddles of the digestor, sold for use with bottled gas-operated appliances, or added to natural gas supplying local homes and industries. When the anaerobic digestion process is complete, up to 90 percent of the organic material originally found in the sludge is converted to water, CO_2, CH_4, H_2S, and other simple compounds. The small portion of undigested material is cleaned from the digestor and used as landfill, and the residual water is chlorinated and discharged, or it is piped into a secondary sewage treatment facility.

SECONDARY SEWAGE TREATMENT

Waste water discharged from the primary sedimentation is pumped into a trickling filter and/or an activated sludge processing unit. The microbial and chemical oxidation occurring in these **secondary treatment** processes removes as much as 90 percent of the suspended organic matter in sewage. A **trickling filter** is a large basin about six feet deep filled with stones, crushed rock, or coke (coal residue from dry distillation). The waste water is sprayed over the surface of the stone from pipes laid on the surface or from a rotating horizontal pipe that moves continuously over the surface (Figure 25.10). The system works like a small home aquarium filter to remove and oxidize contaminating organic material suspended in the water. The stones provide an extremely large surface area on which bacteria, fungi, protozoa, and invertebrate animals can grow in a thin slime.

As the effluent trickles over these microbe-laiden surfaces, the organic compounds are utilized by the microbes as nutrients. Organic carbon is converted to carbon dioxide; organic nitrogen is changed to ammonia and nitrates; sulfur-containing proteins are decomposed to hydrogen sulfide and sulfates; and organic phosphorus is converted to phosphates. All of these processes convert the organic contaminants in the effluent to inorganic matter, a process known as **mineralization.**

As with home charcoal aquarium filters, the porous surfaces of the stones become slimy and heavily coated with microbes after a period of time. When this happens, they become inefficient filters, and the stones must be removed and recharged. As the water reaches the bottom of the trickling filter, it is pumped into another sedimentation tank. In this stage of treatment, bacteria are removed with flocculants and the water is chlorinated before being discharged or pumped into an activated sludge unit.

Waste water from the trickling filter or primary sedimentation tank is pumped into an aeration tank (Figure 25.11). There it is mixed for several hours with air or pure oxygen and bacteria-laiden sludge. Oxygen from the air being injected from the base of the tank along with the bacteria decompose the organic contaminants. The actively growing microbes form flocks, or clumps, in the aeration tank that

A

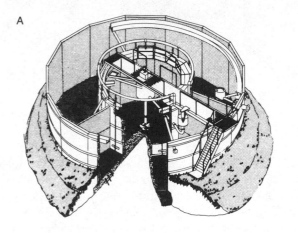

B

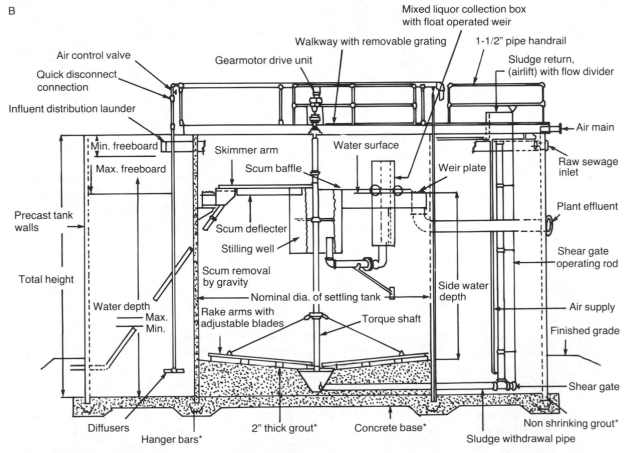

*Material supplied by contractor

FIGURE 25.9
(A) Sludge is pumped into this unit for anaerobic digestion by microbes. (B) Internal structure. (Courtesy of the TLB Corporation.)

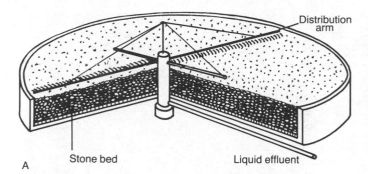

FIGURE 25.10

In secondary sewage treatment, microbes metabolize remaining compounds as the water trickles over the rocks. The water is then drained into a river or lake. (Part B courtesy of the Brinwell, Dow Chemical Company, Michigan Division of Publications.)

B

can be used again by recycling them to the aeration tank. The sludge formed in this process is known as **activated sludge,** since it contains large amounts of actively growing microbes. Among the microbes found in activated sludge is the Gram-negative bacillus *Zoogloea ramigera*. These bacteria form large, slimy colonies that aid in floc formation. Other microbes found in the floc are protozoa, fungi, filamentous bacteria, and autotrophic bacteria. After aeration, the effluent is pumped into another sedimentation tank where the activated sludge settles to the bottom. A portion of this sludge is pumped back to the aeration tank for reuse, and the rest is removed to be disposed of in a landfill area, or it may be treated with heat or chlorine and used as a soil conditioner or fertilizer.

One U.S. Department of Agriculture (USDA) study demonstrated that sludge contains approximately 172,000 tons of nitrogen; 86,000 tons of phosphorus; and 17,300 tons of potassium valued at over $100 million as fertilizer. In order to better utilize this natural resource, reduce the cost of commercial fertilizers, and dispose of sludge in a safe manner, the USDA is researching a composting operation that turns sewage sludge into useful fertilizer (Figure 25.12).

1 Describe what events occur during primary sewage treatment.
2 What valuable products are produced from anaerobic digestion?
3 Why is it unnecessary to "change" the crushed rocks in a trickling filter?

In this composting operation, sludge from municipal treatment plants is dewatered into "sludge cakes," and piled on top of a bed of wood chips and a network of large air ducts. As with home compost piles, the natural action of the bacteria in the pile decomposes the organic material and

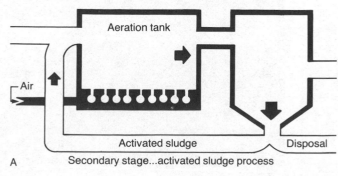

A   Secondary stage...activated sludge process

FIGURE 25.11

A secondary aeration tank. (Part B courtesy of the U.S. Environmental Protection Agency.)

B

FIGURE 25.12
(A) Sludge destined for composting rolls off the vacuum filters at the Blue Plains wastewater treatment plant near Washington, D.C. Samples are taken daily and analyzed for acidity, chemical content, and bacteria. (B) Sewage sludge mixed with wood chips as a bulking material is composted for use as a soil conditioner, fertilizer, or mulch on a fifteen-acre test site at Beltsville, Maryland. (Courtesy of R. C. Bjork, U.S. Department of Agriculture.)

A

B

produces heat that destroys harmful pathogens. Fans at the end of the ducts draw air into and through the piles. Because of the bacteria and natural filtering action of the pile, there is no odor released from the composting sludge. It has been estimated that only two acres of land are needed to compost all the sludge produced from a city with a population of 100,000 at a cost of only about $50 per dry ton of compost (as opposed to $175 per ton for incineration). Composting is economical and an ecologically sound alternative to dumping sludge in the ocean, rivers, and lakes, or using it as landfill.

In most cities, the water discharged from the secondary treatment plant is chlorinated to kill up to 99 percent of the harmful bacteria that may be left in the effluent. In some cities, this effluent is treated a third time by one of a number of tertiary treatment processes.

TERTIARY TREATMENT

Tertiary treatment processes are designed to remove or decompose the remaining organic and inorganic contaminates in water discharged from the secondary sewage treatment plant. The two least expensive and natural tertiary treatment processes are lagoonization and land applications. Artificial treatment is more costly and includes coagulation-sedimentation, adsorption, and electrodialysis.

Lagoons Lagoons are shallow stabilization or oxidation ponds (Figure 25.13). Lagoons are odor-free and should not be confused with cesspools. A cesspool is an open tank or well constructed of blocks or stone with open joints through which liquids can drain into the surrounding soil. Cesspools were used in the past as an alternative to a septic tank system for treating sewage from homes and small rural busi-

nesses. Their operation is very similar to a septic tank but less efficient and susceptible to greater variations in operating efficiency. A quick change in the weather often resulted in foul odors and incomplete decomposition of the sludge. Lagoons are ponds with depths of three to five feet that contain algae, bacteria, protozoa, and many other microscopic and macroscopic organisms. Their interactions with each other as well as their nonliving environment removes or oxidizes the contaminating organic and inorganic compounds entering the pond from the secondary sewage treatment plant. The water is allowed to remain in the lagoon for approximately 30 days before it spills over through a

FIGURE 25.13
A sewage lagoon. (Courtesy of the U.S. Environmental Protection Agency.)

standpipe or dam and flows into a lake, river, or ocean. In a well-maintained and ecologically balanced lagoon, as much as 95 percent of the suspended organic materials can be removed or altered during this month-long period.

Land Applications   Waste water from secondary treatment plants can be further purified and used by applying it to crop lands. The water is applied to the land by spraying, flooding, or furrow irrigation (Figure 25.14). As the water comes in contact with the air, sunlight, and soil, microbes and organic contaminants are decomposed or metabolized to useful forms. If properly regulated and only used on soil with adequate soil filtration rates and water-holding capacity, most of the water, nitrates, and other minerals will be absorbed by the crops, while the remaining water will evaporate or percolate through the soil to become ground water. If not controlled or applied to clay soils, the water does not percolate through at a sufficient rate to prevent the development of anaerobic conditions and stagnation of the soil. In such cases, the land becomes waterlogged and emits odors typical of a stagnant pond or cess-

pool. For this reason, land application of waste water has been most successful in arid parts of the world with sandy soils.

> 1 What role does the bacterium *Zoogloea ramigera* play during secondary sewage treatment?
> 2 Of what value is activated sludge?
> 3 What soil types are best for land applications of sewage?

Coagulation-Sedimentation   This advanced form of wastewater treatment is basically the same as the flocculation procedures used in primary and secondary treatment. Aluminum sulfate or ferrous sulfate is added to the water to form floc that traps contaminants as it settles to the bottom of a sedimentation tank. It is especially effective in removing phosphates before the water is discharged from the plant. In many cases, coagulation-sedimentation is a necessary

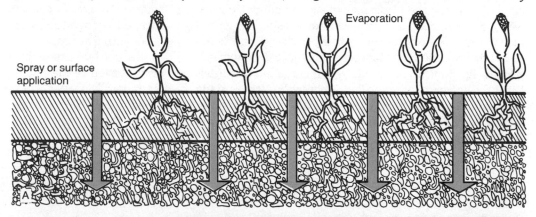

FIGURE 25.14

Spray and surface application of sewage. (Part B courtesy of the County of Muskegon, Michigan, Wastewater Management System.)

pretreatment before other forms of tertiary treatment are performed.

Adsorption   The adsorption process involves passing waste water over carbon granules whose surfaces serve as attachment sites for nonbiodegradable organic materials. Water passing through this system is able to be restored to a quality that is 98 percent free of difficult to extract organic molecules and compares in quality to tap water. This system is successful in removing chemicals that give water a bad taste and odor. It also helps remove compounds, which when concentrated in fish and sea food produce unacceptable flavors. This basic technique has been adapted to home water purifying units installed at sinks or on faucets. The charcoal granules in commercial units can be reused after they are heated to oxidize their surface contaminants, while in home units the internal charcoal cartridges are disposed of and replaced.

Electrodialysis   Electrodialysis is used to extract salts from water. Salts dissociated in water to ions are forced across an electrically charged plastic membrane and out of the waste water. Since municipal waste water has a salt concentration of between 300 and 400 milligrams per liter, electrodialysis allows the effluent to be desalinated to a level below that of purified tap water. Wastewater treatment plants using this tertiary treatment discharge effluent into freshwater lakes and streams without increasing their salt concentrations and disrupting ecosystems. Water pumped from such lakes and rivers for drinking will not have a salty taste.

SUMMARY

A continuous source of fresh water is essential for all life. Two main sources for use in commercial and industrial processes are surface water and ground water. Because surface water is likely to come in contact with pathogens and toxic materials, it is potentially more harmful than ground water. Factors such as temperature, salinity, and pH affect the number and kinds of microbes able to survive in surface water and ground water. Because of the variety of contaminants that can be in different sources of water, most states classify water into different categories such as domestic, industrial, recreational, and commercial. Some of the most important toxic chemicals found in contaminated water are heavy metals, nitrites, chloramines, and organic compounds. Microbial contaminants of greatest concern are pathogenic bacteria, viruses, protozoa, and fungi.

Since water can carry pathogens, it should be checked for possible contamination by using standard methods. The identification of microbes and their numbers can be deter-

mined indirectly by checking for indicator microbes such as the coliforms or fecal streptococci. The bacteriological examination for coliforms in water is divided into three tests: (1) presumptive, (2) confirmed, and (3) completed. If a domestic source of water contains harmful chemicals or pathogens, it must be purified before it is distributed for human consumption. This may be done by sedimentation with flocculants followed by chlorination. Once pumped for public use, it becomes contaminated and must be cleaned before discharge back into lakes or rivers in order to prevent pollution of the water source.

Several methods of sewage treatment are available. Home disposal units such as privies or septic tanks work well if installed and operated in the proper manner. However, to ensure the destruction of harmful materials and pathogens in areas with high population densities, public sewage treatment plants are essential. Sewage treatment may take place in three stages: (1) primary, (2) secondary, and (3) tertiary. Primary treatment consists of screening and sedimentation. The sludge produced may then be further digested anaerobically to produce methane gas in an anaerobic digestor. Secondary treatment may utilize trickling filters and/or aeration tanks. The tanks utilize chemical oxidation from atmospheric oxygen and biological oxidation carried out by bacteria in activated sludge. As long as the concentration of the organic contaminants is maintained at the proper level, the BOD (biological oxygen demand) will be adequate for the aerobic digestion of the sewage. Tertiary treatment processes are designed to remove or decompose the remaining organic and inorganic contaminants in water. Several natural and artifical processes have been developed, including lagoons, land applications, coagulation-sedimentation, adsorption, and electrodialysis.

WITH A LITTLE THOUGHT

For Sale:

> Beautiful three-bedroom ranch home located on the Belle River deep in del Magnificante Canyon only two miles down river from schools, shopping, churches, and the all new primary sewage treatment plant. This home has natural well water, a 500-gallon septic tank, washer, garbage grinder, and many other fine features. Slight crack in basement wall can be fixed easily to prevent minor basement flooding problem. Only minutes from Mt. Gastroenteritis Hospital that you might need in emergencies. Only $225,000!

A true handyman's special for those with some knowledge of water and wastewater microbiology! What is the problem here? Would you buy this home? What problems would have to be investigated and how would they be resolved?

STUDY QUESTIONS

1 Why is surface water more likely to be a source of waterborne disease than ground water?

2 What are the major differences among "safe water," potable water, polluted water, and contaminated water?

3 What types of toxic materials might be found in contaminated water? Describe what potential harm might result from ingesting each type.

4 Why have microbiologists adopted the practice of identifying indicator microbes instead of known pathogens?

5 Describe the three stages in the bacteriological examination of water for the presence of coliforms.

6 What is flocculation? How is it used to purify water and waste water?

7 How does a septic tank work?

8 What are the main events in primary sewage treatment?

9 What is meant by BOD, and why is it important in sewage treatment?

10 What are the major differences between secondary and tertiary sewage treatment?

SUGGESTED READINGS

ALEXANDER, M. *Introduction to Soil Microbiology,* 2nd ed. John Wiley & Sons, New York. 1977.

American Public Health Association. *Standard Methods for the Examination of Water and Wastewater,* 14th ed. Washington, D.C., 1976.

HUGHES, J. M. et al. "Outbreaks of Waterborne Diseases in the United States." *Journal of Infectious Disease,* 1975.

MAUGH, II, T. H. "New Study Links Chlorination and Cancer." *Science,* vol. 211, Feb., 1981.

PAUL, J. S. "Methane—Alternate Energy Source." *National Parks and Conservation Magazine,* Jan. 1979.

SMIL, B. "Energy Solution in China." *Environment,* Oct., 1977.

TABER, W. A. "Wastewater Microbiology." *Annual Review of Microbiology,* 1976.

"U.S. Drinking Water Regulations." *Federal Register,* vol. 40, no. 248, Dec. 24, 1975.

KEY TERMS

surface water	chloramines	dissolved oxygen
ground water	sewage	BOD
methemoglobinemia (met-he-mo-glo-bin-e'me-ah)		

Pronunciation Guide for Organisms

Thermus (ther'mus)

Thermoplasma (ther"mo-plaz' mah)

Sulfolobus (sul'fo-lo-bus)

Zoogloea (zo"o-glo-e'ah)

Industrial Microbiology

CULTIVATION OF MICROBES
CELL CULTURES
ALCOHOLIC FERMENTATION
ALCOHOLIC BEVERAGES
Wine
Beer
Distilled Alcohol as Beverage and Fuel
MICROBIAL ENZYMES IN INDUSTRY
VITAMINS AND AMINO ACIDS
ANTIBIOTICS AND STEROIDS

Learning Objectives

☐ Be able to explain the three fundamental industrial processes.

☐ Describe the differences between batch and continuous culturing.

☐ Be familiar with the method used to culture *Saccharomyces cerevisiae* and the goals of such cultures.

☐ Be able to explain alcohol fermentation as it is carried out in the production of wine, beer, and liquor.

☐ Name several microbially produced enzyme products and describe their production.

☐ List the microbes utilized in the production of various vitamins and amino acids.

☐ Explain the microbial production of antibiotics and steroids.

Since prehistoric times, humans have taken advantage of the beneficial activities of microbes. However, it has only been within the past forty years that these activities have been harnessed for the large-scale production of microbial cells or their products. Microbiologists, engineers, and businesses have come together to develop the field of industrial microbiology. By controlling the activities of certain microbes, they have made possible the simpler, more economical production of large quantities of useful products such as enzymes, amino acids, vitamins, antibiotics, organic acids and alcohol.

Yeast, bacteria, and molds are grown in containers with capacities as large as 50,000 gallons. For example, yeast cells are grown for use in ironized yeast tablets for human use, and in even larger quantities as a supplement to farm-animal feed. It is not an easy job to handle large quantities of growing microbes. Expensive equipment and well-trained personnel are essential to the smooth operation of an industrial process such as antibiotic or beer production. Industrial microbiologists must closely monitor the chemical activities of the microbes and be prepared to change or stop the organisms' activities to prevent product loss. Careful attention must also be paid to ensure that only one kind of microbe is being grown. Successful control of these factors has resulted in the development of multimillion-dollar microbial industries.

26

CULTIVATION OF MICROBES

Their small size, short generation time and rapid metabolic rate make microbes highly productive. Controlling microbial metabolic activities has enabled industrial microbiologists to develop three fundamental industrial processes. The first is the **cultivation** and **harvesting** of cells that may be used as food, food supplements, or inocula for other industrial processes. An example of this type of industrial process is the production of activated dry yeast and ironized yeast tablets. The second type of process utilizes select microbes as "factories" that enzymatically convert a particular substrate into a desired product, a process known as **bioconversion.** Steroids known as estrogens and progesterones used in birth control pills are produced by the action of microbes on specific substrates added to culture medium. The third industrial process utilizes microbes for their ability to generate large quantities of a specific **metabolic byproduct.** Alcohol, enzymes, and many types of organic acids are commonly produced by the microbial fermentation of inexpensive and readily available substrate material.

The industrial cultivation of microbes is carried out in large tanks known as fermenters (Figure 26.1). These units range in capacity from 5000 gallons (about 20,000 liters) to 50,000 gallons (about 200,000 liters). Some of the more basic and inexpensive substrates used as culture media are molasses (the thick syrup made from boiling sweet vegetables, sugar cane, or fruits), soybean meal, and malted grain. For specific bioconversion processes, more expensive synthetic media may have to be prepared from large quantities of purified ingredients. The fermenter is inoculated with a

FIGURE 26.1
This small industrial fermenter can be used in the production of numerous fermentation products. (Courtesy of the New Brunswick Scientific Company.)

volume of culture equal to 5 to 10 percent of the tank's volume. This means that a 15,000-gallon tank requires about 1500 gallons of pure culture to start the fermentation. This inoculum is built-up from a pure agar slant culture of microbes maintained in the laboratory, or it may be bought from a microbiological supply house. The initial culture is inoculated into ever increasing amounts of media to reach the final volume necessary for the processes. This repeated subculturing requires a great deal of space and close control to assure that contamination does not occur. In order to reduce these problems, lyophilized cultures can be purchased from suppliers who have concentrated the required volume into a more manageable container. Once the substrate has been inoculated, all the conditions necessary for growth and reproduction must be accurately maintained. The optimum temperature, pH, purity, salinity, and cell density must be checked regularly. Since many industrial processes are aerobic, an adequate air supply must be made available to these fermenting microbes. This is done by injecting sterilized air into the tank through inlet valves at the bottom of the fermenter and continuously stirring the culture with mechanically operated paddles installed inside. The rapid bubbling and stirring causes a layer of foam to form on the surface that is controlled by the addition of antifoaming chemicals.

The two different methods that have been developed for industrial fermentations are **batch** and **continuous** culturing (Figure 26.2). Batch cultures are made by adding the microbes to a limited quantity of medium and allowing the fermentation to run to completion without the addition of fresh nutrients or removing byproducts. Cheese is made by batch culturing. Once inoculated into the fermenter, the microbes reproduce and increase in number according to the population growth curve. The maintenance of cultures in the exponential (log) phase of growth is known as steady-state, or balanced, growth and allows microbes to be cultured on a continuous basis. Continuous cultures are maintained in an apparatus called a **chemostat.** Many antibiotics, organic acids, and steroids are produced by microbes in continuous culture. The log phase of growth is maintained by continuously adding fresh, sterile medium to the **fermenter** at a rate equal to the removal of culture and products. The exponential growth rate is regulated by maintaining the precise level of a limiting factor in the medium. This

1 Name the three fundamental types of industrial microbiological processes.
2 What are the main differences between batch and continuous culturing?
3 Microbes in a chemostat should be maintained in what growth phase?

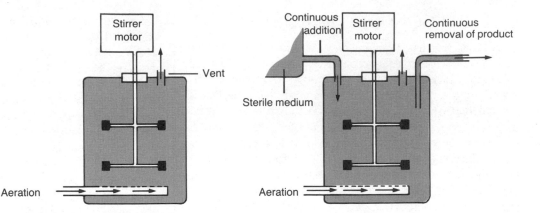

FIGURE 26.2
A typical batch fermenter (left) and a continuous culture unit (right).

practice keeps the cells from shifting into a lag, or stationary, phase. Continuous cultures have an advantage over batch cultures, since the product can be constantly manufactured in large amounts. However, minor chemical or physical changes in the medium or spontaneous mutations within the microbial population prevent this method from being carried out longer than a few days.

Once the fermentation is complete, the product must be separated from the culture medium and purified. Passing the culture through a filter separates the medium from the microbes. If cells are the desired product, they are washed and concentrated by centrifugation (Figure 26.3). If the product is dissolved in the medium, it is separated by mixing with a solvent which selectively extracts the product from unwanted compounds. Depending on the product type, it is then separated from the solvent and crystalized in pure form. In some cases, the chemical may not be suitable in a crystalline form and must be redissolved for chemical modification. Members of the penicillin family of antibiotics result from chemical modifications of the basic penicillin molecule crystallized from such a process.

With an increased understanding of organic chemistry, it has become possible to selectively modify many microbially produced compounds to the exact chemical structure needed for a particular purpose. However, successes in the chemical industry have not always benefited industrial microbiologists. Many of the organic compounds thought to be produced only by microbes are now being manufactured in the chemistry lab. These products are manufactured in larger quantities from low-cost starting materials in less sophisticated equipment, enabling them to be marketed at a lower price. To overcome this problem, industrial microbiologists constantly search for new microbes or techniques to produce complex, commercially important compounds. One method used to increase production and reduce costs has been the development of productive mutations. **Productive mutants** are microbes that have been produced by natural selection following exposure to mutagenic levels of radiation. This technique was successfully used to increase

penicillin production from the fungus *Penicillium chrysogenum* during World War II. Spores of this mold were isolated and irradiated with high dosages of ultraviolet light, which caused mutations in the genes. When some of these spores were germinated, the new mycelia were found to produce larger amounts of the antibiotic than the original

FIGURE 26.3
In the industrial production of antibiotics, quality control is vital. A constant, continuous monitoring is required to ensure product quality. (Courtesy of the Upjohn Pharmaceutical Company.)

nonmutated species. Another option being explored by industrial microbiologists is the development of new strains of microbes by **genetic recombination.** As a result of artificially introducing nonbacterial genes into a recipient bacterial cell, it is possible to microbially manufacture otherwise unobtainable proteins, for example, insulin and interferon, or to increase the amount of product being produced. For example, attempts are being made to extract DNA from conventional antibiotic-producing microbes, which grow slowly, and genetically recombine them into a recipient that is able to grow rapidly. The success of such a "genetic transplant" would mean that large amounts of antibiotic could be produced at a much lower cost in a shorter time. These types of technical advances have expanded the scope of industrial microbiology and hold great promise for the production of many medically important compounds not currently available at a reasonable cost. Today, industrial microbiology is a multimillion dollar business employing thousands of people in the research, development, and manufacture of microbial cells, alcohol products, enzymes, organic acids, vitamins, amino acids, antibiotics, and steroids.

CELL CULTURES

Of all the microbes used by industrial microbiologists, common baker's or brewer's yeast, *Saccharomyces cerevisiae,* are the most widely used. The more familiar uses of yeast are in the beverage industry for the anaerobic production of beer, wine, and liquors and in the baking industry for leavening bread. However, yeast are also cultured for their cell components and products released during metabolism. Yeast are an excellent source of B vitamins, vitamin D, protein, certain amino acids, ATP, NAD, RNA, and other complex organic compounds. Originally isolated from wild strains used for brewing beer, today's hybrid strains are maintained under carefully controlled conditions to ensure genetic stability. Many are bred from selective parental strains and are capable of producing large quantities of desirable byproducts.

The original wild strains of *S. cerevisiae* used by the brewing industry were only capable of releasing 5 to 6 percent alcohol, but some of today's hybrids can result in a product with 19 to 20 percent alcohol! The brewing industry uses yeast for the production of alcoholic beverages and in the past served as a source of yeast cells for the baking industry. However, bakers have developed their own unique strains of baker's yeast used for the leavening of bread. Not only do these yeast release large quantities of carbon dioxide that form gas pockets in dough, causing it to rise, but the presence of the yeast gives bread its characteristic flavor and aroma. Baked goods leavened with baking powder do not have the same flavor and are not technically called bread.

Yeast are also cultured as a source of food and animal feed. Because yeast are high in B complex vitamins, they are used as a supplement by many people who suffer from a vitamin deficiency disease or do not have sufficient amounts of these vitamins in their diet (Table 26.1). However, yeast do not contain all the essential amino acids required by humans and some farm animals; therefore they are an incomplete source of protein. For this reason, yeast are primarily used as a protein supplement in foods with the essential sulfur-containing amino acids, methionine and cystine.

The industrial process used to produce high concentrations of yeast cells is not the same used by the brewing industry. To reproduce cells efficiently, yeast must be cultured under aerobic conditions. The brewing industry cultures the cells anaerobically to ensure the development of carbon dioxide and alcohol as endproducts of yeast fermentation. The cultivation of yeast cells begins from a stock agar slant culture that is built-up under aseptic procedures to equal 10 percent of the volume used in the fermenter (Figure 26.4). The fermenter is well aerated and stirred to ensure optimum aerobic growth conditions. Yeast are inoculated into a dilute solution of molasses and cultured into the log phase of growth. As the sugar and other nutrients are utilized by the yeast for growth and reproduction, molasses is added to maintain the exponential growth rate. The addition of an excessive amount of molasses will cause the culture to become anaerobic, slow

TABLE 26.1

Source and uses of vitamins in the human body and vitamin deficiencies in humans.

Vitamin	Food Source	Use in Body	Symptoms of Deficiency
Thiamine (B_1)	Peas, beans, eggs, liver	Coenzyme use in Krebs cycle	Beriberi—breakdown of nerve cells, muscle and heart failure.
Riboflavin (B_2)	Milk, whole grain cereals, green vegetables, liver, eggs	Part of coenzyme used in electron transfer system (FAD)	Cracking of skin around the eyes and mouth; skin infections.
Pyridoxine (B_6)	Most foods	Coenzyme used in synthesis of amino acids	Vitamin is so readily available that no deficiency disease has been noted.
Cyanocobalamin (B_{12})	Meats, dairy products	Used in red blood cell formation	Pernicious anemia—defective formation of red blood cells.

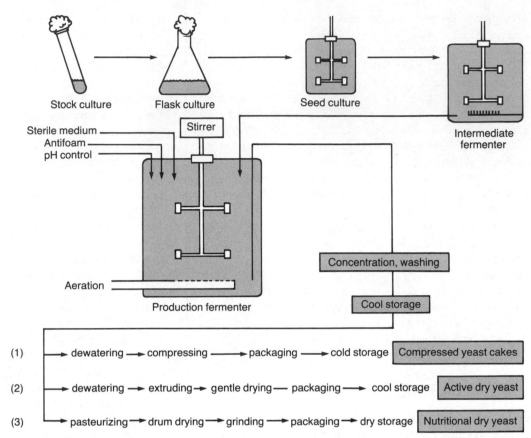

FIGURE 26.4
The industrial production of yeast, *Saccharomyces cerevisiae.*

the population growth rate, and result in an unwanted alcoholic fermentation. After the culture has reached its maximum density, the cells are filtered and washed for further processing into compressed yeast cakes, active dry yeast, or food supplements.

> **1** How are productive mutants created and what is their value?
> **2** Name three products of *Saccharomyces cerevisiae.*
> **3** What is the major difference between culturing yeast for alcohol production and culturing for high concentrations of cells?

ALCOHOLIC FERMENTATION

Historical accounts from cultures throughout the world describe the process of fermenting grains or fruits for the production of alcoholic beverages. As noted in the Bible, Genesis 9:20–21:"and Noah began to be a husbandman, and he planted a vineyard; and he drank of the wine and was drunken." While such accounts date wine-making to at least

3000 B.C., the production of beer (from the Anglo-Saxon word for barley, *baere*) dates even earlier (Figure 26.5).

Greek and Egyptian writings of the second century A.D. tell of the distillation of alcohol (from mild spirits such as wines and beer) in the making of liquor. Originally known as **aqua vitae** (water of life), whiskey was a distillation product of beer drunk as a medicine. It was an all-purpose cure containing such aromatic ingredients as cloves, anise, licorice, nutmeg, ginger, and caraway. Alcohol is not only produced for consumption but also as a solvent for many organic compounds, as a reactant in the production of other chemicals, and is found in such products as mouthwash, cough medicine, shaving lotion, antiseptics, perfume, cleaning agents, hair tonic, and vanilla flavoring. More recently, alcohol is also being investigated as an alternative fuel to supplement (gasohol) or replace gasoline. Alcohol is produced either from the breakdown of petroleum or by fermentation of sugar. The decomposition of petroleum by the action of heat is known as **pyrolysis,** or **cracking.** In this process, certain fractions of petroleum are separated from one another to isolate the portion that can be converted to alcohol. This fraction is simply passed through a catalyst-containing chamber heated from 400 to 600°F. At this high temperature, the organic petroleum molecules are split into smaller molecules which can be isolated, purified, and con-

FIGURE 26.5

Two Dionysian satyrs on a pottery vase from the sixth century B.C. demonstrate the ancient art of wine-making. (Courtesy of Musees Royaux d'Art et d'Histoire, Brussels.)

verted to alcohol by a simple chemical reaction involving sulfuric acid. However, this process requires the expenditure of a great deal of energy and is therefore very costly. The less expensive and probably oldest synthetic chemical process used is the fermentation of sugars by yeast to produce ethyl alcohol.

Large quantities of commercial grade alcohol (not used for consumption) are manufactured from the fermentation of molasses. Molasses is added to fermenters having capacities of 50,000 to 100,000 gallons along with ammonium sulfate and phosphates to serve as nitrogen, sulfur, and phosporous sources and buffered to a pH of 4.5 for optimum growth. After being inoculated with *Saccharomyces cerevisiae,* the culture is stirred and incubated for 48 hours at 25°C. The result is a solution known as "high wine" that has an alcohol concentration of between 10 and 14 percent. In addition to alcohol, there is a smaller amount of a substance known as **fusel oil.** This portion contains such compounds as isopentyl, propyl, isobutyl, and amyl alcohol. Fusel oil gives the product a "chemical" smell and taste that is often found in "home brew," and is the result of using an impure yeast culture. The ethyl alcohol is distilled off from the high wine and marketed for use in many commercially available products. Chemical companies use it as a solvent or reactant in the synthesis of other organic compounds. The carbon dioxide released from the fermentation is piped from the fermenter and compressed into a gas which may later be purified and sold, or converted into dry ice.

ALCOHOLIC BEVERAGES

WINE

Alcoholic beverages can be made by fermenting almost any grains, fruits, or vegetables that contain sugar. It was not uncommon for the early colonists in the United States to ferment pumpkins, potatoes, Indian corn, or carrots to make beer. They fermented and distilled honey to make a rum substitute known as "old methaglin." Apple cider (unfiltered and nonpasteurized apple juice) was fermented to "hard cider." Hard cider was sometimes placed in wooden barrels and frozen in a lake during the winter to separate the water and concentrate the alcohol. This "cold distilled" beverage was known as applejack and was very potent. Some prepared a mixture of applejack and gin in a drink known as "strip-and-go-naked" because of the strange behavior it caused in those who drank only a few mugfuls. Other fermented fruits included elderberries, plums, peaches, currents, raspberries, and grapes. Today grapes are the primary fruit used in the production of wine. **Wine** is the fermented juice of grapes and other fruits with an alcohol concentration of between 9 and 21 percent. Table or dinner wines contain no more than 14 percent alcohol, while dessert and appetizer wines may contain up to 21 percent alcohol (Table 26.2). **Beer** is the fermented extract of grains such as barley or rice with an alcohol concentration of between 3 and 6 percent. **Packaged liquors** are distilled fermentation products that have alcohol concentration as high as 85 percent.

Grape wine may be made from any of 5000 varieties of the plant *Vitis vinifera* or 2000 varieties of other *Vitis* species. Each has a characteristic chemical content that imparts a unique flavor, aroma (bouquet) and color to the wine made from each variety. Because the chemical content of the grape is greatly affected by the soil and climate in which it is grown, grapes used for making good wines are only grown within certain wine-growing regions around the world (Figure 26.6).

Grapes contain a variety of sugars but the two most important are glucose and fructose. Glucose is fermented by yeast to carbon dioxide and alcohol, and the fructose gives a sweet taste to the wine. Since fructose tastes twice as sweet as glucose, grapes with a high percentage of this sugar are excellent for the production of sweet white wines. Also found are organic acids, free amino acids, proteins, pigments, and various minerals. The particular balance of these nutrients and the wine-making process affects the characteristics and quality of a wine. Grapes are picked from the vine at a time when they are most likely to have the desired balance of sugar and other compounds. They are quickly taken to the winery where they are mechanically crushed (Figure 26.7).

TABLE 26.2
Taxonomy of wines.

Wines With No Added Flavors

Sparkling Wines (carbonated with an excess of CO_2)
  Champagne, Spumante, Sekt
  Pink champagne
  Sparkling burgundy and Cold Duck
  Sparkling muscat
Still Wines (having no excess CO_2)
  Table wine (less than 14 percent alcohol)
    White; Named by district: France: Chablis, Graves, Macon, Meursault, Montrachet, Pouilly-Fuisse, Sancerre

                                 Germany: Franken, Moselle (Piesport), Rheingan (Hattenheim), Rheinhessen
    Named by variety: Chardonnay, Riesling, Chenin Blanc
    Pink or Rosé   Grenache, Cabernet Rosé
    Red; Named by district: France: Beaujolais, Bordeaux, Burgundy, Chateauneuf-du-Pape, Hermitage, Medoc, Pomerol, Rhone, Saint-Emilion

                                 Italy: Barolo

                                 Spain: Rioja
    Named by variety: Cabernet Sauvignon, Pinot Noir, Zinfandel, Nebbiolo, Meriot
  Dessert Wines (having more than 14 percent alcohol)
    Sherry, dry to sweet
    Madeira
    Marsala
    Muscatel
    Port

Wines With Added Flavors

Herb-based
  Vermouth, dry
  Vermouth, sweet
  Dubonnet
Fruit-based and fruit flavor-based
  Blackberry, peach, gooseberry, cherry, apple wines

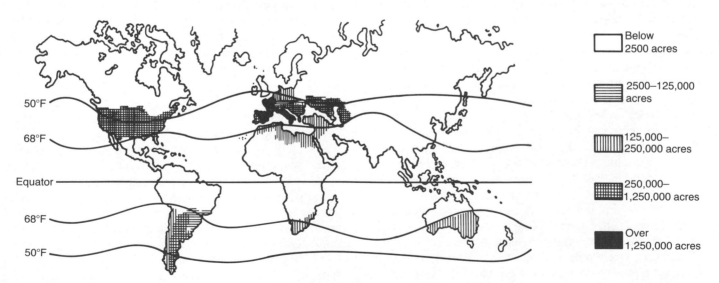

FIGURE 26.6
In the wine-growing regions of the world, if the combination of weather, soil, and plants is not perfect, the wine is not as good. (Courtesy of the Wine Institute, San Francisco.)

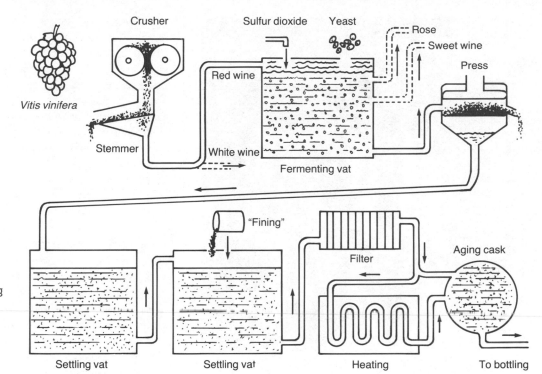

FIGURE 26.7
This is only one method among many that might be used commercially to prepare wine. (From M. Amerine. "Wine." *Scientific American*, Aug., 1964.)

The stems are separated from this mixture of pulp, skins, and seeds called "must." The skins are also removed if white wine is to be made. Red and rosé wines get their color and flavor from pigments and tannic acid in the skins. "Must" is pumped to the fermenting vat made of hardwood or glass-lined concrete. Once inside the vat, sulfur dioxide or sulfurous acid is added to inhibit the growth of wild, undesirable strains of yeast that occur naturally on the grapes. It also interferes with natural enzymes that "brown" the wine, destroys grape skin cells aiding in the release of red pigments, acidifies the must, and aids in clarifying the wine after fermentation. Pure cultures of *Saccharomyces cerevisiae* var. *ellipsoideus* are unaffected by the sulfur dioxide and are added to the vat to ferment the must for a few days to a few weeks (Figure 26.8). Heat generated by the yeast is carried away by cooling coils in order to maintain a fermentation temperature below 85°F for red wines and below 60°F for white wines. An increasingly popular wine in the United States is rosé. The production of rosé wines begins as with any other red wine, but the skins are removed early in the initial fermentation to prevent the development of a dark red color and heavier tannic acid concentration.

1 What kind of material can be used as culture media for commercial grade alcohol production?
2 What are the differences among wine, beer, and liquor?
3 Why is sulfur dioxide added to the wine "must"?

When this portion of the wine-making process is complete, the wine is pumped to a wine press to remove larger pieces of skins, seeds, and other solids. The wine is then sent to a settling vat where the fermentation is allowed to

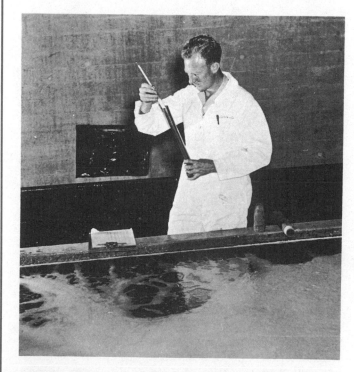

FIGURE 26.8
Sugar consumption during rapid fermentation of a white dessert wine "must" is followed by testing. (Courtesy of the Wine Institute, San Francisco.)

FIGURE 26.9
In this cool aging cellar at a historic California winery, wines remain for months and years before they are bottled for additional aging in glass. (Courtesy of the Wine Institute, San Francisco.)

Sparkling wines and champagnes are made in a unique way to capture and retain the natural effervescence of the carbon dioxide released during fermentation. The sparkling wine **Champagne** is only made in the Champagne region of France, while the uncapitalized name champagne is used as a generic name in the United States to describe any wine that contains an excess carbon dioxide level. When Champagne is made, the must is fermented and fined as with all wines. However, a second fermentation is initiated after the wine is bottled by the addition of more yeast and sugar. These specially designed bottles are wired shut to prevent the gas from escaping and allowed to ferment neck-down on a special rack for a few weeks to several months—a process called tirage (Figure 26.10).

When the fermentation is complete, dead yeast, tartar, and other compounds settle on the inside of the bottle. This sediment must be removed without disturbing the carbonation or quality of the wine. This is done by "riddling," a special bottle-turning done by hand that causes the sediment to drop into the neck of the bottle. A talented riddler can riddle, or turn, about 30,000 bottles of Champagne a day. Each bottle must be turned once a day for about a month to sediment all the yeast into the neck. The bottles are then cooled almost to freezing, so that they can be turned upright and the corks removed along with the frozen sediment. It is at this time that sugar syrup is added to give the Champagne a more appealing taste. Four different types of

continue, while smaller solids and yeast cells settle out. During this time, *Lactobacillus* species convert strong malic acid to weak acids, thus reducing the sharp flavor of the wine. When the wine is clear and has matured, it is "racked," or drawn off, to another settling tank. There the filtering process further clarifies the wine and reduces its oxygen content. Wines that lose their carbon dioxide during this process are called **still wines.** In many large wineries, clarification is aided by the addition of "fining" agents such as clay, gelatin, or egg white which precipitates extremely small particles in the wine. After clarification and fining, it is pumped into oak barrels or casks for aging (Figure 26.9).

During the aging process, chemical reactions occur which develop the wine's flavor, aroma, and color. The yeasty flavor is lost and red wines develop their typical rich red color, while white wines become amber. The aging process usually requires about two years for red wines and varies from several months to about two years for white wines. The decision to bottle is made by the wine master and is very crucial in the production of a fine wine. However, the aging process does not stop after bottling. Red wines continue to develop and mature for years after bottling. Most wine masters and connoisseurs feel that red wines are their best five to ten years after bottling and white wines reach their peak after about two years.

FIGURE 26.10
These racks hold hundreds of bottles of bottle-fermented California champagne, neck downwards. In this manner, sediment caused by fermentation moves to the neck of the bottle. From here it is disgorged by freezing the wine and sediment in the neck by briefly placing it in frozen brine. The clamp is released, and the pressure forces the cork out, along with its frozen sediment deposit. (Courtesy of the Wine Institute, San Francisco.)

Champagne are produced and vary according to the percentage of sugar added at this time. Brut is the driest (0.5 –1.5 percent sugar), sec has 2.5 – 4.5 percent sugar, demi-sec is more sweet (5 percent), and doux the "sweetest" with 10 percent added sugar. If riddling and tirage have been done well, the product is a clear, sparkling wine of high quality.

BEER

Another popular alcoholic beverage, beer, is believed to have been first produced by the Egyptians as the direct product of the fermentation of barley. As with wine, the production of a fine beer requires the skillful blending of science and art. Several different types of beer are made throughout the world and vary in their color, aroma, flavor, and alcohol content (Table 26.3).

The brewing of beer begins with the selection of high quality barley and a process known as malting. **Malting** involves steeping (soaking) the grain in water at 65°C to initiate germination, or sprouting. During seed germination, natural enzymes convert the stored food material in the seed (endosperm) from starch to glucose. This increases the percentage of glucose available for yeast fermentation and reduces the amount of starch to a level that prevents the development of fusel oil during fermentation. The malted grain is then dried in kilns or rotating drums at from 75° to 100°C. The actual temperature used for drying depends on the kind of malt used in the preparation of the brew. Lower temperatures are used for light color malt and higher temperatures are used for dark malt. Care must also be taken to preserve

the barley enzymes (amylase) needed to convert additional starch to sugar during the **mashing** process (Figure 26.11). In this process, the malt is cleaned, ground, weighed, and mixed with other cereal grains such as corn meal, soybean meal, corn grits, or corn flakes. These **adjuvants** help balance the protein and carbohydrate content to ensure a more flavorful product. The mixture is sent to a lauter tub, where it is purged with hot water to extract the nutrients from the mixed grains. The leftover of "spent" grain is dried and sold for use as animal feed. The protein content of this grain is very high, thus it is an excellent supplement to cattle and hog feed. The fluid extract, known as **wort** (pronounced "wert"), has the color of beer and is pumped to the brew kettle where it is cooked for 1½ to 2½ hours (Figure 26.12). While in the brew kettle, flowers of the female hop plant are added to give beer its characteristic aroma and inhibit the growth of contaminating bacteria that could later interfere with fermentation or cause defects. The most important contaminating bacteria are *Lactobacillus pastorianus, Pediococcus cerevisiae, Flavobacterium proteus,* and *E. coli.*

---

1 What changes occur in still wine during the aging process?
2 Why is it important to riddle champagne?
3 The malting process results in what chemical change in grain?
4 Why are adjuvants added to wort?
5 Why are hops added to wort?

TABLE 26.3
Types of beers and their characteristics.

Type	Fermentation	Yeast	Percent Alcohol by Weight	Characteristics
Lager	Bottom	*Saccharomyces carlsbergensis*	3.8	Most popular beers in U.S. lightly hopped, fermented at low temperature.
Bock	Bottom	*S. carlsbergensis*	5.5	Heavy sugar, malt, and hops; few adjuvants; dark color; a lager beer, fermented at low temperatures.
Pilsener	Bottom	*S. carlsbergensis*	3.8	A lager beer, light color, low sugar, medium hopped.
Malt liquor	Bottom	*S. carlsbergensis*	4.2	Lightly hopped, heavily malted.
Light beers	Bottom	*S. carlsbergensis*	3.8	A lager beer, double fermented to reduce sugar and diluted to reduce alcohol content.
Ale	Top	*S. cerevisiae*	4.7	Heavily hopped, high sugar, fermented at high temperatures.
Porter	Top	*S. cerevisiae*	4.7	Dark ale, sweet, lightly hopped, heavily malted.
Stout	Top	*S. cerevisiae*	4.7	Strong porter, heavily malted and hopped.

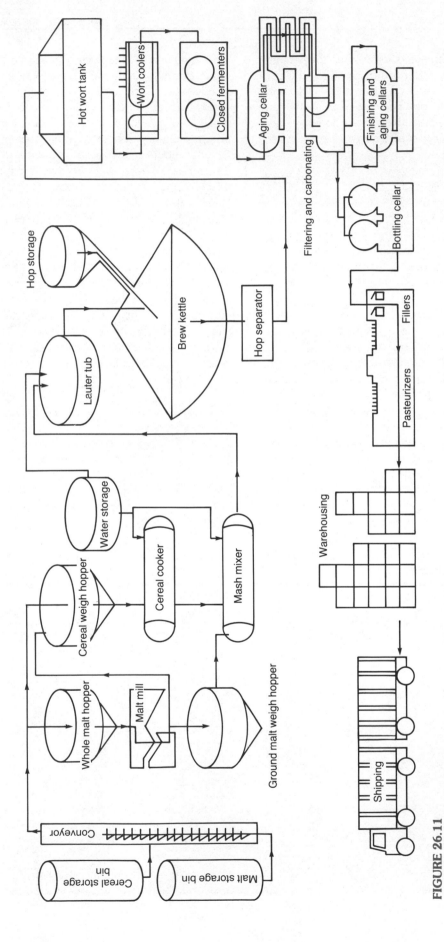

FIGURE 26.11
The brewing process. (Courtesy of the G. Heileman Brewing Company.)

582

A

B

FIGURE 26.12

(A) The copper brewing kettle is one of the most familiar pieces of equipment in the brewing industry. Wort is cooked with hops and adjuvants such as corn grits, before cooling and fermenting. (B) Breweries also have microbiology laboratories in which contaminants are monitored and yeast counts are determined on a daily basis. (Part A courtesy of the G. Heileman Brewing Company. Part B courtesy of D. P. Adler, quality control manager.)

After cooking in the brew kettle, the wort is passed through a hop separator that not only removes the hops but also extracts undesirable resins and residues. The clarified wort is then cooled and pumped into a closed fermenter where an active culture of *Saccharomyces* is "pitched," or inoculated, into the vat. Various species and strains of *Saccharomyces* are available for brewing. They are chosen for their unique genetic ability to ferment the wort at a desirable rate and determine the character of a beer. Brewery yeast are classified as either bottom-fermenting yeast (e.g., *S. carlsbergensis*) or top-fermenting yeast (e.g., *S. cerevisiae*). Bottom fermenters are used for the production of lager beer (the more common American beers) while top fermenters are used in the manufacture of ale. The fermentation usually lasts between seven to ten days (depending on whether it is top or bottom fermented) and is maintained at a temperature of about 40°C. As the fermentation procedes, alcohol, acetic acid, glycerol, and carbon dioxide are formed. The carbon dioxide is piped from the fermenter, compressed, and stored. It is later returned to the beer during the carbonating process. When the fermentation is complete, the yeast population has increased six times and settled to the bottom of the tank. A portion of this yeast is washed and reused for other fermentations, while the excess is dried and added to poultry and hog feed. These yeast may also be used in the fermentation of corn or sugar beets to produce alcohol for gasohol. The beer is pumped from the fermenter for aging (lagering), filtering, carbonating, and packaging. Beer in bottles and cans is pasteurized to lengthen its shelf life, but beer in kegs is not pasteurized and must be refrigerated.

In order to maintain production of a high quality beer, quality control measures are utilized throughout the brewing process. Equipment is cleaned and disinfected on a regular basis. Laboratory personnel maintain pure, active cultures of yeast, and a special effort is made to select for reuse only those yeast cells that will develop the highest quality beer. The final product is regularly checked to identify the presence of contaminating bacteria that may be responsible for beer defects.

DISTILLED ALCOHOL AS BEVERAGE AND FUEL

Alcohol produced during fermentation does not accumulate in the culture above a level of about 18 to 21 percent, since it is toxic to the yeast producing it. For this reason, a simple fermentation does not result in a beverage with an alcohol concentration greater than 18 to 21 percent. (The exact percentage depends on the strain of *Saccharomyces* used.) In order to produce an alcoholic beverage with a higher concentration, i.e., hard **liquor,** the beverage must be distilled from the original fermentation. **Distillation** is a heating process that drives alcohol vapor and some of its dissolved components from water and condenses it to a liquid that can then be purified, filtered, and clarified for consumption (Figure 26.13). Among the first alcoholic beverages distilled into liquors were wines (mentioned by Aristotle) and beer (described in early European writings). Today a variety of liquors are made by distilling alcoholic beverages fermented from many types of starch-based grains, and sugary fruits and berries (Table 26.4). Each beverage is made from

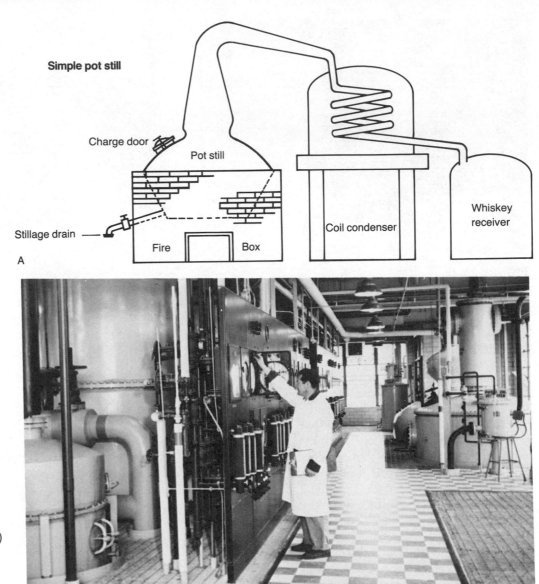

Simple pot still

Charge door

Pot still

Stillage drain

Fire   Box

A

Coil condenser

Whiskey receiver

B

FIGURE 26.13

(A) A simple pot "still" is suitable for "moonshiners." (B) For the industrial distillation of liquor, much complex machinery is needed. (Part B courtesy of Hiram Walker & Sons, Inc.)

special starting material, using fermentation techniques unique to each type of liquor. For example, Scotch whiskey is made from barley malt that has been kiln dried at a high temperature over a peat moss fire. This gives the whiskey its unique flavor. Bourbon is made from barley malt that has between 51 and 80 percent added corn, and some rye is always added. Canadian whiskeys are fermented mainly from rye. The final concentration of alcohol in a liquor (given in "proof" number) is controlled by the distillation process and dilution with water. Packaged liquors such as vodka and gin are usually sold with a final proof of 80 or 90, but may be as high as 170 proof. The proof number is twice the alcohol concentration, i.e., 170 proof vodka has an alcohol concentration of 85 percent. Grain alcohol has a higher final proof, and its sale is regulated by state liquor departments. In most

cases, this alcohol can only be purchased from a druggist and only with a valid letter of authorization describing its intended use.

In recent years the energy shortage has rekindled interest in ethyl alcohol as a liquid fuel to be used in automobiles, home heating systems, and elsewhere. The idea is not really old because alcohol lamps, stoves, and many other appliances have been fueled with clean-burning alcohol for hundreds of years. Ethyl alcohol may be used as a liquid fuel in combination with gasoline as gasohol (Box 29) or as the more purified grain alcohol (190 proof). As an engine fuel, grain alcohol has several advantageous characteristics. It can be burned in an engine with only minor modification and with the efficiency of a high octane gasoline that prevents engine "knock." The combustion is so complete that

G

ASOHOL, one of the hottest energy topics today, may be a partial answer to scarce petroleum supplies. The alcohol fermentation process has existed since 2300 B.C. when the Chinese made beer, kui, from rice. But not until the 20th century was alcohol used for much other than food and drink. During the 1920s and 1930s, Dr. Hale, an organic chemist, concentrated his studies on "farmer's alcohol," a fuel used at that time to power tractors and automobiles. Dr. Hale said, "when the apologists for the petroleum industry try to tell us that alcohol gasoline blends are not satisfactory, they seem not to recognize what the organic chemist knows."

By 1938 the Atchison Agrol Company was commercially producing farmer's alcohol. With feedstock such as barley, rye, corn, grain sorghum, and Jerusalem artichokes, the alcohol plant eliminated crop pileups in the surrounding area. The distillate was anhydrous (200 proof), denatured alcohol blended with gasoline. The remaining fiber was converted into concentrated livestock feed (40 percent protein, 90 percent digestible) and returned to the farmers.

During the 1930s, Dr. Hale said farmer's alcohol was "the only permanently renewable resource we have to replace the steadily disappearing and diminishing natural resource, gasoline." This statement is still true today.

MAKING GASOHOL

As petroleum supplies dwindle and gasoline lines grow longer, various alternative liquid fuels attract attention. One method used to produce a liquid fuel, ethyl alcohol, is shown in the diagram (p. 586). The mixture, nine parts unleaded gasoline to one part ethanol, is a trademark term, Gasohol.

One process to produce ethyl alcohol (ethanol) from corn starts with the grain being ground and mixed with water to form a slurry. Then the slurry is heated with steam in a continuous pressure cooker. This serves to gelatinize the starches by softening and disintegrating the grain. Next, the gelatinized slurry is pumped into a vat and cooled to 60°C (140°F). An enzyme, amylase, is then added to convert the starches to simple sugars, mostly glucose. Addition of the amylase increases the conversion of starch to sugar, producing up to 95 percent of the theoretical yield possible. This process is known as saccharification.

Once the sugar solution cools to about 26° to 32°C (80° to 90°F), it is transferred to the fermenting vats and yeast is added. The yeast converts the sugar mixture to an alcohol solution of about 8–12 percent. This solution is channeled through distillation columns which separate the alcohol from the residual grain. The illustration depicts a process using two distillation columns. The first column produces a solution of 50 percent alcohol (100 proof). The second column concentrates the solution to 95 percent alcohol (190 proof). The liquid is further dehydrated through extraction distillation to yield anhydrous (200 proof) ethyl alcohol. The residual grains can be dried to yield distiller's dried grains, a high protein (22 to 30 percent) feed for livestock.

Distilled 200 proof alcohol is drinkable (potable). Therefore, it must be denatured, i.e., made unfit to drink. Gasoline, a denaturing agent, is sometimes used and it has the advantage that no additional emission problems occur when the fuel is burned in an internal combustion engine. For further information about denaturing agents and permits to make ethanol,

BOX 29

MAKING GASOHOL

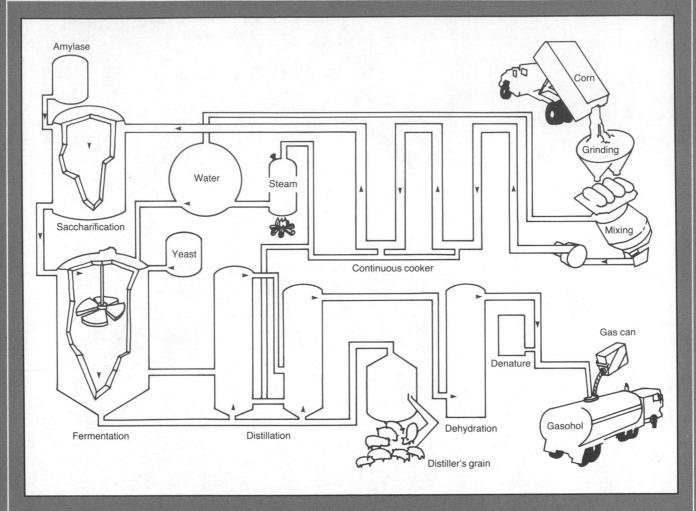

contact the Bureau of Alcohol, Tobacco, and Firearms of the Department of the Treasury in Cincinnati, Ohio.

FINAL REMARKS

Present technology poses no problems for alcohol production. Gasohol's fuel properties are good and no engine modifications are needed regardless of vehicle age or make. Anhydrous ethanol when mixed with 9 parts unleaded gasoline does not harm the fuel system, or present phase separation or starting problems in cold climates. Although the fuel properties are desirable, research is needed to improve the manufacturing process, the net energy gain, to increase the potential use of cellulose feedstock, and to increase the profitability of the operation.

Reprinted by permission of B. Ofoli, Agriculture Engineering Department, Michigan State University; and B. Stout, Texas A & M University.

TABLE 26.4
Examples of distilled alcoholic beverages.

Name	Starch Base
Scotch	Barley
Bourbon	Corn, rye, barley
Rye	Rye and barley
Vodka	Potato
Irish whiskey	Rye, wheat, barley
Gin	Grains or potatoes with juniper berries
Arrak	Rice
Cognac	Grape
Brandy	Grape
Sloe Gin	Plum
Tequila	Agave (century plant)
Rum	Sugar cane
Applejack	Apples
Toddy	Coconut milk
Framboise	Raspberries
Liqueur and cordial	Mixtures of distilled alcoholic beverages, herbs, sugar, and flavorings.

noxious exhaust emissions are much lower and less toxic than those released from gasoline. Engines operating on alcohol run from 20° to 40°F cooler, extending their life and reducing the chances of overheating. Some research with alcohol/water fuel mixtures has demonstrated as much as a 16 percent increase in mileage. Currently, industrial fermentation plants designed for the production of fuel alcohol are being constructed in many states. It is hoped that this will reduce the need for petroleum-derived gasoline by augmenting liquid fuel with alcohol, which is a renewable resource. An additional advantage to this type of fuel production is the generation of large amounts of high protein "spent" grain that can be used to supplement animal feed. Currently, research is underway to determine the average improved weight gain and health of farm animals fed this by-product of fermentation. There is strong indication that the increase will be as great as 10–20 percent.

MICROBIAL ENZYMES IN INDUSTRY

The enzyme industry is large and growing as more information relating to microbial metabolism becomes available through research. The industrial microbiologist applies this information in an attempt to increase the yield of a profitable product. Because these applications may be patented, much of the detailed information surrounding the actual growth of microbes and harvesting of enzymes is kept secret.

Enzymes are protein catalysts that speed the rate of specific chemical reactions and are produced through a complex sequence of protein synthesis reactions inside a living cell. The exact nature of an enzyme and its action is determined by the sequence of nucleotides found on the DNA of the cell in which it is manufactured. This fact makes it possible to select a particular genetic type of microbe for the production of an enzyme. The highly specific nature of enzyme action and the speed with which enzymes increase reactions make them excellent chemicals in many industrial, commercial, and medical processes. Microbes are excellent sources of enzymes, since they are easily grown in large aerated and submerged cultures, and produce enzymes on a regular basis. While in culture, bacteria, fungi, and filamentous bacteria synthesize and excrete **exoenzymes** that accumulate in the surrounding medium. Other enzymes are called **endoenzymes** since they are synthesized and utilized by the microbe inside the cell. To harvest these enzymes, the cultured cells must be separated, washed, and fragmented before the enzyme is isolated and purified.

---

1 How does distillation differ from fermentation?
2 What is "spent" grain and what advantages are there in using it as animal feed?
3 What is an exoenzyme? an endoenzyme?

Enzymes are highly specific in their actions and must chemically "fit" their substrate in order to function properly. Therefore, it is essential that they be produced and utilized in an environment that does not distort their shape or interfere with their action. One of the most widely used proteolytic enzymes that is frequently misused is found in household enzyme presoaks or as they are sometimes called "enzyme detergents." These enzymes are produced from cultures of the bacterium *Bacillus subtilis* (as seen on the cover of this text) and are able to hydrolyze many different proteins found in stains caused by chocolate, blood, and many foods. However, many people do not gain the maximum advantage from these products because they do not follow the product's directions, which are based on fundamental enzyme chemistry (Figure 26.14). These products should not be used in extremely hot water, because the high temperatures will denature the enzymes and prevent them from combining with substrate stains. Cold water slows their action and requires longer contact between the enzyme and substrate for effective stain removal. By rubbing the enzyme directly into the stain, more effective and rapid contact is made and hydrolysis of the stain is more efficient. Since enzymes also lose their effectiveness in environments which contain inhibitors, they should not be used in certain types of "hard" water (water containing inhibiting minerals) or along with detergents such as the dodecylbenesulfonate group, which includes most granular laundry detergents. Caution should also be taken with enzyme presoak products, since many people find them to be allergens. Wearing clothes

FIGURE 26.14
Enzymes produced by bacteria are found commercially in such products as this presoak laundry detergent. (Courtesy of the Colgate-Palmolive Company. Photo by Daria Smith.)

cleaned in these products may cause allergic skin reactions. Table 26.5 lists several other microbially produced enzymes used in many industrial, commercial, and medical processes and products. Their use must be carefully controlled. While both bacteria and fungi are used in the production of enzymes, it is the fungi that are primarily responsible for the production of large amounts of industrially important organic acids (Table 26.6). They are used to produce resins which are, in turn, used for the manufacture of such items as plastics, varnishes, electrical nonconductors, and medicines. Other organic acids are used for the production of such products as perfumes, oil additives, and textile dyes.

VITAMINS AND AMINO ACIDS

Any molecule that is required in minute amounts for the efficient operation of an organism and must be supplied from the environment since it is unable to be synthesized by that organism is known as a growth factor. Two of the most important classes of growth factors are the amino acids and vitamins. Although humans lack the ability to produce some of these compounds, many microbes are able to synthesize them. Therefore, what is a growth factor to a human may be only a normal metabolic byproduct to these microbes. Despite the fact that many microbes have the ability to synthesize these complex compounds, few produce excess amounts since they are essential to the normal metabolism of the microbe and are, in most cases, quickly utilized. Mi-

TABLE 26.5
Some industrially produced microbial enzymes and their uses.

Enzyme	Microbe Cultured (Genus)	Uses
Pectinase	Aspergillus	Separate fibers of the flax plant that may be used to make linen; same process used to make hemp rope. Used in green coffee processing. Used to clarify fruit juices.
Protease	Aspergillus Bacillus Streptomyces	Meat tenderizer; digestive aid; removes gelatin from photographic film for silver recovery; used in enzyme presoaks and detergents.
Amylase	Bacillus	Clarifies syrups; glucose production; softens dough in commercial bread making; removes starch and sizing from textiles; ingested as digestive aid, wallpaper remover.
Collagenase	Bacillus Clostridium	Digests necrotic tissue from burns.
Lipase	Rhizopus	Digestive aid.
Rennin	Mucor	Curd formation in cheese-making.
Cellulase	Trichoderma Streptomyces	Digestive aid. "Sweating" of animal hides to remove hair before tanning to leather.
Invertase	Saccharinyies	Soft-centered candies.
Urease	Fungal	Urea determination.
Lactase	Lactobacillus	Whole milk concentrates; lactose removal from milk for lactose-intolerant persons.
Streptokinase and streptodorase	Streptococcus	Digests necrotic tissue from burns.

TABLE 26.6
Microbially produced organic acids.

Acid	Microbe (Genus)	Uses
Lactic acid	*Lactobacillus*	Food preservative; in baking powder; as calcium lactate for food supplement for calcium deficiencies; bone development in pregnancies; used to make plastics.
Gluconic acid	*Aspergillus* *Penicillium*	Washing and softening agent; leavening agent; textile printing; As calcium gluconate for calcium deficiencies and in chicken feed to harden egg shells.
Citric acid	*Aspergillus*	In foods and beverages for taste and preservation.
Itaconic acid	*Aspergillus*	In manufacture of acrylic resins; plastic products; oil additives; dental resins.
Gibberellic acid	*Gibberella* *Fusarium*	Plant growth and flowering stimulant.
Kojic acid	*Aspergillus*	Analytical reagent; insecticide.
Fumaric acid	*Rhizopus*	Used as wetting agent and in manufacture of resins.
Gallic acid	*Aspergillus*	Ink and dye manufacture; skin disease cream.
Ustilagic acid	*Ustilago*	Used in "musks" for the perfume industry.

crobes selected for the industrial production of human vitamins and amino acids are defective mutants that have been isolated from wild cultures using special methods. These mutants lack the ability to use all the amino acids and vitamins they produce. The compounds accumulate inside the cells and are then released into the surrounding culture medium where they can be isolated and purified.

One of the most widely used amino acids produced industrially is glutamic acid. This amino acid is converted to the familiar compound monosodium glutamate (MSG) and added to many foods as a flavor enhancer. It is sold under the trade names Accent®, Glutavene® and Zest®. MSG is added to many commercially prepared foods including baby foods, canned soups, meats, seafoods, and some soy sauces. MSG has been identified as a cause of brain damage; acute degenerative lesions have occurred in the retina of infant mice because of their age-dependent ability to absorb this compound. Although as yet unconfirmed, many researchers suspect that MSG may have a comparable ability to harm human infants. Since this food additive confers no significant benefit and, as yet, no confirmed harm to human infants, the Food and Drug Administration cautiously permits its unrestricted use, except for foods specifically designed for infants. However, many researchers question the safety of this chemical and continue to discourage its widespread use, especially in light of the fact that MSG has been confirmed to cause harm to adults. In 1968, it was discovered that this additive was responsible for Chinese restaurant syndrome, a disease characterized by headache, burning sensations, facial pressure, and chest pain. These symptoms occur in adults after eating Chinese food prepared with high concentrations of MSG to intensify their flavors. In addition to glutamic acid, industrial microbiologists have produced other amino acids. Table 26.7 lists several of these along with the most frequently produced vitamins, B_2 and B_{12}.

The greatest variety of human vitamins are manufactured by the yeasts *Saccharomyces cerevisiae* and *Torula utilic*. These are sold in the form of compressed ironized yeast tablets and contain a mixture of many vitamins. The purified forms of vitamins are produced from cultures of bacteria such as *Streptomyces*, *Bacillus*, and *Propionibacterium*. One of the most important is vitamin B_{12} (cobalamin) which is produced in excess by *Streptomyces* growing in glucose or corn-steep broth with traces of cobalt. Vitamin B_2 (riboflavin) is produced from cultures of the fungus *Ashbya gossypii*. Although this fungus is a cotton plant pathogen, its vitamin product is one of the most widely used vitamin supplements in the commercial food industry.

ANTIBIOTICS AND STEROIDS

Antibiotics are unique molecules produced by microorganisms and they are able to destroy or inhibit the growth of other microbes. The industrial growth of large, submerged and aerated cultures of fungi, filamentous bacteria, and true bacteria marked a major breakthrough in the ability of the medical profession to treat and prevent infectious diseases. By culturing antibiotic-producing microbes in these large fermentation vessels, each cell is able to come in contact with nutrients essential to the production of the drug and maximize its output. The search for and detection of antibiotic-producing microbes relies on samples taken from water, soil, and other materials from many parts of the world. In one method, microbes suspected of being antibiotic producers are isolated from mixed samples and plated on nutrient media for growth in pure culture. After lawns of growth are formed on the surfaces of nutrient agar plates, small plugs are removed and placed on test plates inoculated with specific strains of *Staphylococcus aureus* or other pathogens (Figure 26.15). If any of the experimental microbes produce an antibiotic that is capable of destroying or inhibiting the pathogen, zones of clearing will form around the plug. These zones resemble those produced in a culture and susceptibility test

The search for antibiotics

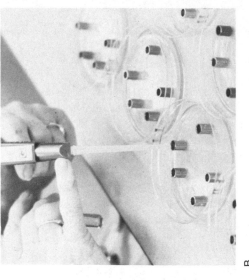

A

Samples of water, soil, and other materials from many parts of the world are sent by medical missionaries, professors, oceanographers, and others to the Lilly Research Laboratories, in Indianapolis, Indiana.

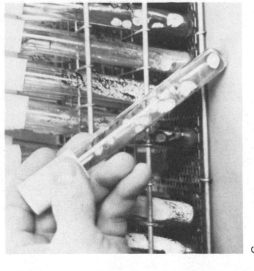

B

In the laboratory, each sample is treated so that the microorganisms it contains grow, reproduce, and form colonies.

C

The hidden microbial world in a grain of soil or a drop of water is now visible. Specially trained scientists examine the colonies and decide which should be tested for antibiotic production.

Streak test

D

A small portion of the selected colony is "seeded" in a streak on a test plate. Several kinds of infection-causing bacteria are also seeded on the plate in rows at right angles to the streak being tested
If growth is severely restricted, an antibiotic is being formed.

Antibiotic Manufacturing

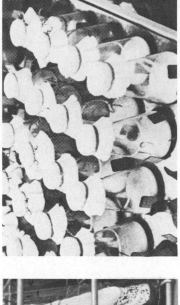

E

A selected strain is placed on an agar slant. After several days incubation, the organism from this slant is transferred to a shake flask containing prepared nutrient media. After continuous shaking, the growth will reach the optimum for transfer to larger vessels.

F

FIGURE 26.15

The production of antibiotics. (Courtesy of Eli Lilly and Company.)

Fermentation

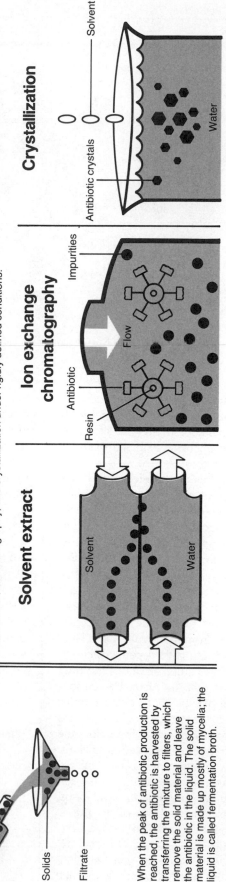

Water
Nutrients
Mixers

Inoculated with antibiotic–producing culture.

Aeration

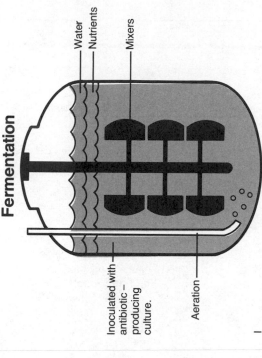

G

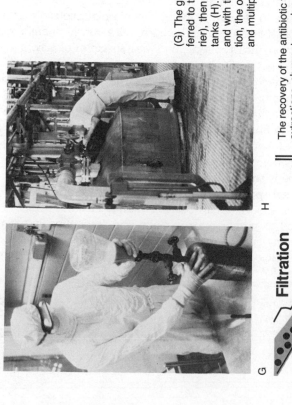

H

(G) The growth from flasks is transferred to the "bazooka" (transfer carrier), then from the bazooka to larger tanks (H). (I) In sterile nutrient media and with the aid of agitation and aeration, the organisms continue to grow and multiply and produce antibiotics.

I

The recovery of the antibiotic from the broth in a state of purity suitable for medicines involves many complex operations, including extractions, chromatography, and crystallization under rigidly defined conditions.

Crystallization

Solvent

Antibiotic crystals

Water

Ion exchange chromatography

Impurities

Antibiotic

Resin

Flow

Solvent extract

Solvent

Water

Filtration

Solids

Filtrate

When the peak of antibiotic production is reached, the antibiotic is harvested by transferring the mixture to filters, which remove the solid material and leave the antibiotic in the liquid. The solid material is made up mostly of mycelia; the liquid is called fermentation broth.

Chemical modifications

After crystallization, the antibiotic is exactly the same substance produced by the organism during fermentation except that it is very pure. It may be effective as a medicine for human or veterinary use, as is the case with many antibiotics. On the other hand, it may be modified chemically to give it properties it lacks or to make it more powerful in killing or halting the growth of bacteria. Some of the most useful antibiotics are chemical modifications and represent mankind's improvement on nature. The bulk crystalline antibiotic is then formulated into many lifesaving products. Capsules, tablets, liquids, and injectables are manufactured, filled, labeled, and packaged for shipments to all parts of the world.

Procaine penicillin

Penicillin V

Penicillin G

Penicillin nucleus

Side chain

591

TABLE 26.7
Vitamins and amino acids synthesized by microbes.

Product	Microbes	Uses
Amino acids		
Glutamic acid	*Micrococcus* spp. *Pseudomonas* spp. and *E. coli*	Flavor enhancer in foods as MSG (monosodium glutamate).
Lysine	*E. coli* and *Enterobacter* spp.	Food supplement, aids in preventing protein deficiency disease, kwashiorkor.
Threonine		
Methionine		
Tryptophan		
Vitamins		
B_2 (riboflavin)	*Ashbya gossypii*	Food supplement, prevents cracking of skin, aids in preventing infections.
B_{12} (cobalamin)	*Streptomyces* spp. *Pseudomonas* spp. *Bacillus* spp.	Food supplement, prevents pernicious anemia—defective formation of red blood cells.
Mixed vitamins and amino acids	*Saccharomyces cerevisiae* and *Torula utilis*	Food supplement.

performed in a medical laboratory. The microbes producing antibiotic are then grown in pure culture for further research to explore the nature of their metabolism and the likelihood that they will produce enough antibiotic to make them industrially profitable.

From agar slants the microbes are transferred to larger flasks to increase the number of cells. These are shaken continuously to ensure optimum aeration and growth. The shaker flask cultures serve as inocula for "bazookas" (transfer culture carriers) that are larger culture vessels used to increase the amount of inoculum to 10 percent of the fermenter volume. Once inside the fermenter, the microbes grow, reproduce, and release significant amounts of the antibiotic. When the maximum amount of antibiotic has been produced, the solid components of the culture (i.e., cells and other debris) are separated from the medium in a large rotary filter. The fermentation broth is then passed through a solvent extraction process that further separates the drug from unwanted chemicals. It is then transferred to an ion-exchange chromatography apparatus that contains resins which adsorb the antibiotic to their surfaces. This process separates the drug from the remaining impurities before the antibiotic is crystallized under rigidly controlled conditions. The drug is then packaged in the form of capsules, tablets, liquids, or injectables. However, an alternative course can be taken to chemically modify the pure drug to other forms. This is done primarily to avoid the problem of drug resistance. As a result, entire families of antibiotics have been produced from the fundamental compounds. The brief description of this process in no way reflects the years of intensive work needed to reach the point at which a microbe

may be cultured for its valuable, government-approved products. In some cases, 10,000 or more species or strains of microbes have to be investigated before one suitable for culturing is identified, and years of research and field testing are necessary before the drug is made available to the public. It is also important to realize that of the over 2000 antibiotics described in research literature, only a few are of economic and medical value. The rest are far too toxic or ineffective to be used in the treatment of infectious diseases. Table 26.8 lists many of the microbes involved in antibiotic production, their products, and their spectrum of action.

The last medically important group of compounds, the steroids, are produced by an industrial fermentation process known as bioconversion. **Bioconversion** is the use of microbial fermentation to enzymatically change one compound to another biologically active form. In this complex series of fermentations, biologically inactive steroid compounds from plants are used as starting materials. A specific microbe is selected for its genetic ability to produce an enzyme that specifically alters the compound to another form. A series of several fermentations using different microbes is required to enzymatically change one compound into another. The final product is a tailor-made molecule with the desired biological activity. Some of the medically important

1 What is MSG? How is it produced?
2 Name three microbes used to produce vitamins.
3 What kind of microbes are typically used to produce antibiotics?

compounds produced by bioconversion include bile salts, progesterone, testosterone, vitamin D, desoxycholate, cholesterol, estradiol, and cortisone. Among the more important genera of microbes used for bioconversions are *Rhizopus*,

Streptomyces, Aspergillus, Corynebacterium, and *Curvularia*. Figure 26.16 illustrates the complex fermentations required to convert one of the more basic compounds, progesterone, into a variety of other important molecules.

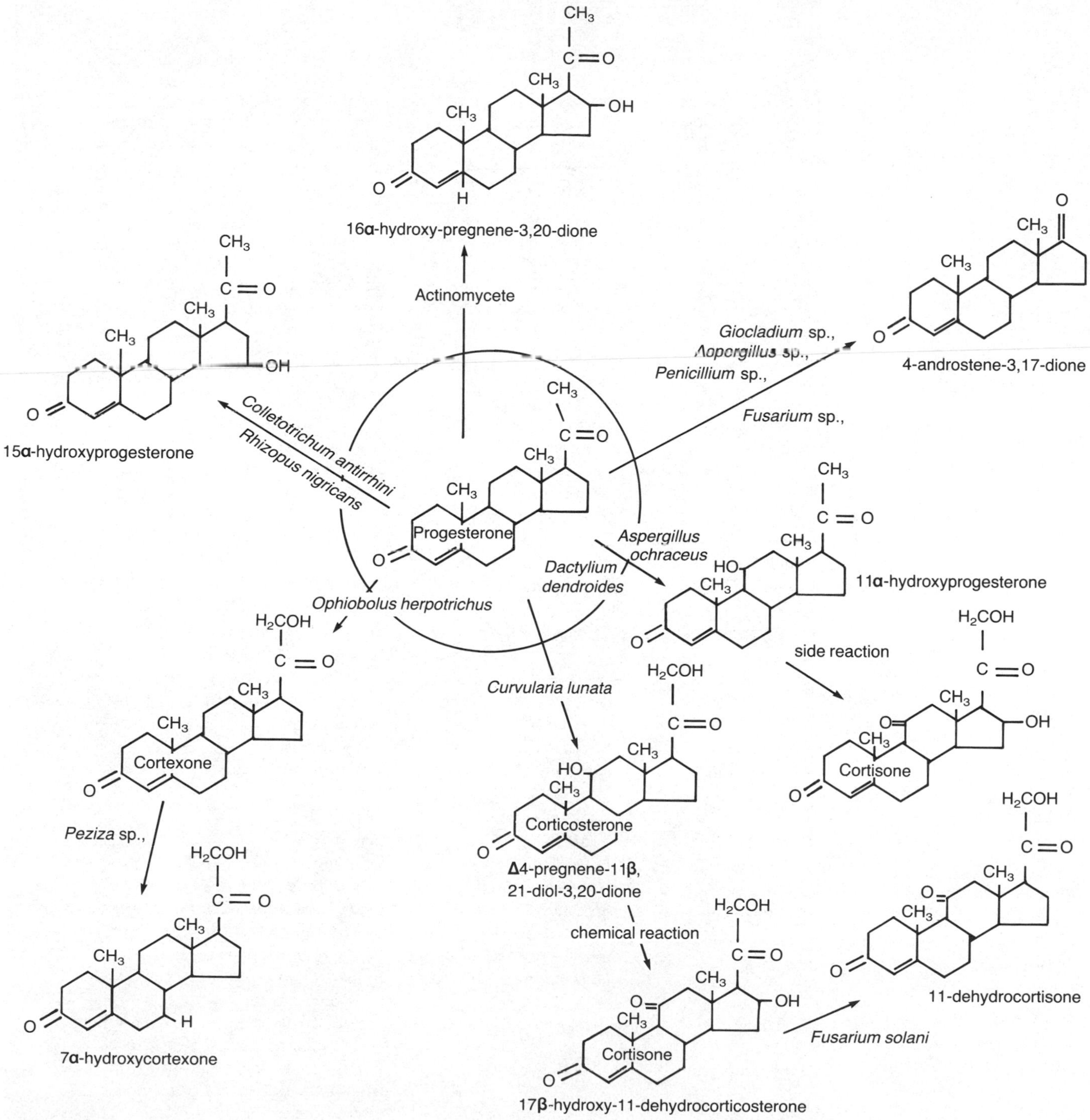

FIGURE 26.16

Bioconversion products. (From Frobisher et al, *Fundamentals of Microbiology,* 9th ed. W. B. Saunders, 1974.)

TABLE 26.8
Antibiotics of clinical importance.

Generic Name	Source	Antimicrobial Spectrum and Some Important Properties
Amphotericin B	*Streptomyces nodosus*	Fungi
Bacitracin	*Bacillus subtilis*	Gram-positive bacteria
Cephalosporins	*Cephalosporium* spp. and *Emericellopsis* spp.	Generally penicillinase-resistant but inactivated by cephalosporinase
Cephalexin	Semisynthetic	Gram-positive and some Gram-negative bacteria; orally absorbed
Cephaloglycin	Semisynthetic	Like cephalexin; orally absorbed
Cephaloridine	Semisynthetic	Like cephalexin; not orally absorbed
Cephalothin	Semisynthetic	Like cephalexin; not orally absorbed
Chloramphenicol	*Streptomyces venezuelae;* synthesis	Gram-positive and Gram-negative bacteria; rickettsias; *Entamoeba*
Colistin	*Aerobacillus colistinus*	Primarily Gram-negative bacteria
Erythromycin	*Streptomyces erythreus*	Gram-positive and a few Gram-negative bacteria; some protozoa
Gentamycin	*Micromonospora* spp.	Gram-positive and Gram-negative bacteria
Griseofulvin	*Penicillium* spp.; synthesis	Fungi
Kanamycin	*Streptomyces kanamyceticus*	Gram-positive and Gram-negative bacteria; some protozoa
Lincomycin	*Streptomyces lincolnensis*	Gram-positive bacteria
Neomycin	*Streptomyces fradiae* and other *Streptomyces* spp.	Gram-positive and Gram-negative bacteria; mycobacteria
Novobiocin	*Streptomyces* spp.	Primarily Gram-positive bacteria
Penicillins		
Ampicillin	Semisynthetic	Gram-positive and Gram-negative bacteria; penicillinase-sensitive; acid stable
Carbenicillin	Semisynthetic	Gram-positive and Gram-negative bacteria including *Pseudomonas* strains; penicillinase-sensitive; not orally absorbed
Cloxacillin	Semisynthetic	Gram-positive bacteria; penicillinase-resistant; acid stable
Dicloxacillin	Semisynthetic	Like cloxacillin
Methicillin	Semisynthetic	Gram-positive bacteria; penicillinase-resistant; acid labile
Nafcillin	Semisynthetic	Like cloxacillin
Oxacillin	Semisynthetic	Like cloxacillin
Penicillin G	*Penicillium* spp.; *Aspergillus* spp.	Gram-positive bacteria; penicillinase-sensitive; acid labile
Penicillin V	Biosynthetic	Gram-positive bacteria; penicillinase-sensitive; acid stable
Polymyxin	*Bacillus polymomyxa*	Primarily Gram-negative bacteria
Rifamycin	*Nocardia mediterranea*	Primarily Gram-positive bacteria and mycobacteria
Rifampin	Semisynthetic	Gram-positive and Gram-negative bacteria; mycobacteria; orally absorbed
Spectinomycin	*Streptomyces* spp.	Gram-positive and Gram-negative bacteria, specifically *Neisseria gonorrhoeae*
Streptomycin	*Streptomyces* spp.	Gram-positive and Gram-negative bacteria; *Mycobacterium tuberculosis*
Tetracyclines		Gram-positive and Gram-negative bacteria: rickettsias; coccidia; amoebae and balantidia; mycoplasmas

Chlortetracycline	*Streptomyces aureofaciens*	
Demeclocycline	*S. aureofaciens* mutant	
Doxycycline	Semisynthetic	
Methacycline	Semisynthetic	
Minocycline	Semisynthetic	Also inhibits staphylococci resistant to other tetracyclines
Oxytetracycline	*Streptomyces rimosus*	
Tetracycline	Catalytic hydrogenation of chlortetracycline; *S. aureofaciens* mutant	
Tyrothricin	*Bacillus brevis*	Gram-positive bacteria
Vancomycin	*Streptomyces orientalis*	Gram-positive bacteria

SOURCE: Modified from B. Miller and W. Litsky, ed. *Industrial Microbiology.* Copyright © 1976. Reprinted by permission of McGraw-Hill Book Co.

SUMMARY

The field of industrial microbiology uses the inherent ability of microbes to produce cells, cell products, or bioconversion products for personal or commercial use. The cultures are maintained in large vats or tanks known as fermenters operated by batch or continuous culture methods. The microbes selected for use may in many cases be genetically altered from the wild form by radiation or genetic recombination to strains that have greater productive abilities. Industrial processes may be used to produce cells as a product. The most notable is the production of yeast cells as a food supplement. However, these same microbes can be cultured anaerobically to produce alcohol. Alcohol for consumption is produced from the fermentation of grains or sugary fruits and is known as beer and wine. The distillation of these liquids concentrates the alcohol, its color, and flavors in the form of packaged liquor. A less pure form of commercial-grade alcohol has many other uses and can serve as an alternative to gasoline and oil as a liquid energy source.

A variety of bacteria, fungi, and filamentous bacteria also have the capability of producing useful enzymes, organic acids, vitamins, amino acids, and pharmaceutically valuable antibiotics and steroids.

WITH A LITTLE THOUGHT

Now that you are a well-educated microbiologist, you have decided to go into business for yourself with the goal of becoming wealthy as quickly as possible. Since your microbiological training is in the area of industrial fermentation, you have decided to develop and operate a large-scale operation that will generate a product that is in high demand by the public. Using your background, decide what product might be produced, what materials will be necessary to set the operation into motion, and describe the problems that you will have to resolve in order to make the operation profitable.

STUDY QUESTIONS

1 What three fundamental processes are carried out in the field of industrial microbiology?

2 What is the difference between a batch culture and a continuous culture?

3 What advantages are found in the use of productive mutants? How do they differ from genetic recombinants and wild microbes?

4 Define the following terms: commercial alcohol, malt, must, mashing, fusel oil.

5 Describe the differences between wine and beer fermentation.

6 What is distillation?

7 What are the differences among wine, beer, and packaged liquor?

8 What is gasohol and how is it produced?

9 What are enzyme presoaks and how are they produced?

10 Give five examples of vitamins and organic acids produced by industrial microbiological processes.

SUGGESTED READINGS

AMERINE, M. A. "The Search for Good Wine." *Science*, 154:1621, 1966.

AMERINE, M. A., and ROESSLER, E. B. *Wines: Their Sensory Evaluation.* W. H. Freeman, San Francisco, 1976.

CHEREMISINOFF, P. N. *Gasohol for Energy Production.* Ann Arbor Science Publishers, Inc., Ann Arbor, Mich., 1979.

CONVERSE, J. C. et al. *Ethanol Production from Biomass with Emphasis on Corn.* University of Wisconsin, College of Agriculture and Life Sciences, Madison, Wis., 1979.

GROSSMAN, H. J. *Grossman's Guide to Wines, Beers, & Spirits,* 6th ed. Revised by Harriet Lembeck. Charles Scribner's Sons, New York, 1977.

HANG, Y. D., D. F. SPITTSTOESSER, and E. E. WOODAMS. "Utilization of Brewery Spent Grain Liquor by *Aspergillus niger.*" *Applied Microbiology,* 30:879–80, 1979.

MILLER, B. M., and W. LITSKY. *Industrial Microbiology.* McGraw-Hill, New York, 1976.

Office of Technology Assessment. *Gasohol—A Technical Memorandum.* Washington, D.C., 1979.

UNDERKOFLER, L. A., ed. *Developments in Industrial Microbology,* vol. 16, Lubrecht & Cramer, 1976.

University of Nebraska. *Ethanol Production and Utilization for Fuel.* University of Nebraska, Cooperative Extension Service, Lincoln, Nebraska, 1979.

KEY TERMS

bioconversion	chemostat	gasohol

Pronunciation Guide for Organisms

Vitis (vi'tis)

Flavobacterium (fla"vo-bak-te're-um)

a-, ab-, abs-   away from; abductor
a-, an-   not, without; amitosis
achromo-   colorless; *Achromobacter*
ad-   to, towards; adductor
adeno-   gland; adenovirus
aer-   air; aerobe, anaerobic
albus-   white; *Staphylococcus albus*
allelon-   of one another; allele
amb-   both, on both sides; ambivalent
amoebe-   change; *Amoeba*, amoeboid
amyl-   starch; amylase
ana-   back, anew; anabolism, anaphase
andr-   man; androgen
anti-   opposite, against; antibody
aorta-   artery; aorta
apex-   tip, point; apical
aqua-   water; aquarium, aquatic
arthro-   joined; arthrospore
-ase   denotes an enzyme; oxidase
-ate   denotes the salt of an acid ending in -ic; succinate
atrium-   entrance; atrial
aur-   ear; auricle
auto-   self; autonomic, autotomy
azo-   pertaining to nitrogen; *Azotobacter*

bacterio-   bacteria; bacteriocin
basis-   bottom; basilar
bi-, bin-   double or two; binomial
bio-   living; bioluminescence
bronchus-   air tubes; bronchial

capill-   hair; capillary
cardio-   heart; cardiovascular
caryo-   nucleus; caryogamy
cata-   down; catabolism, catalyst
centrum-   center; centrifuge
cephale-   head; cephalic, encephalon
cerebrum-   brain; cerebral
cervix-   neck; cervical
cardia-   heart; cardial
chemo-   chemical; chemostat
chlor-   green; chloroplast
chole-   bile; cholesterol
chondrus-   cartilage; chondrocranium

chrom-   color; chromosome
-cide   kill; germicide
cilium-   eyelid; ciliata, ciliary
coel-   hollow; coelom
con-   with; connection
contra-   against; contralateral
corona-   crown; coronary
crypto-   hidden; *Cryptococcus*
cyst-   closed sac; cytocarp
cyt-   cell; leukocyte, cytoplasm
cyte-   state of the blood; leukocyte

de-   from; deoxyribose
dermato-   skin; dermatomycosis
di-   two; dicaryon
dia-   through; diaphragm, diarrhea
diplo-   double; diploid
dis-   apart, away; disinfect

ec-   out of; ectopic
ecto-   out; ectoplasm
-emia   state of the blood; toxemia
en-   in; encysted
encephalo-   brain; encephalitis
endo-   within; endospore
entero-   intestine; enteritis, enterotoxin
epi-   upon, above; epidermis, epidemic
erythro-   red; erythrocyte
eu-   good; eucaryote
ex-   out of, from, through; excretion
extra-   outside; extracellular

flavo-   yellow; flavoprotein
-form   shape; vermiform

gam-   marriage; gamete
gaster-   belly; gastroenteritis
gen, geny, geno, genesis   bring to life, create; gene, genetics, parthenogenesis
genic-   producing; pyogenic
gest-   carry; gestation, ingest, digest
glyco-   sweet, sugar; glycolysis
gram-   writing; diagram
gyn-, gynec-   woman; gynandrism, gynecology

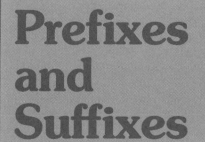

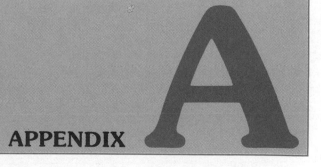

Prefixes and Suffixes

APPENDIX A

halo- salt; halophile
haplo- single; haploid
hapt- to grasp; hapten
hemi- half; hemicardius
hepat- liver; hepatic
hetero- other; heterozygote
histo- web; histology
homo- alike; homozygote
humor- fluid; aqueous humor
hydr- water; hydrophobia
hyper- above; hypertonic, hypertrophy
hypo- under; hypotonic, hypodermic

inferior- beneath
inter- between, among; interface
intra- within; intracellular
iso- equal; isotonic
-itis inflammation of; myelitis

lacto- milk; lactose, *Lactobacillus*
leuco- white; leucocyte
-logy study; biology, bacteriology
lys- loosening, decomposition; lysosome
-lysis breaking down, loosening; hemolysis, glycolysis

macro- large; macroscopic, macronucleus
mal- bad; malfunction
mega- great; megaspore, acromegaly
melan- black; melanin, melancholy
meso- middle; mesophile
meta- after; metabolism, metaphase
meter- measure; thermometer
micro- small; micronucleus, microscope
morph- form; morphology, amorphous
multi- many; multinucleate
mut- change; mutant, mutation
myco- fungus; mycology
myxo- mucus, slime; myxobacterium, myxovirus

necro- dead; necrosis
neo- new; neoplasm
nom- law; taxonomy, autonomic
nomen- name; nomenclature, bionomial
non- not; nondisjunction

oec- house; dioecious, ecology, economy
-oid like; neuroid, amoeboid
oligo- few; oligosaccharide
-oma tumor; polyoma
ortho- straight; Orthoptera, orthodontist
-ose denotes a sugar or carbohydrate; glucose
-osis action of or condition of; tuberculosis
osteo- bone; osteitis
ovum- egg; oviduct, ovarian

para- beside, near; parenteral
peri- around; peritrichous
-phag devour; phagocyte
-phage eating; macrophage
phago- to eat; phagocyte
-phase appearance; metaphase
-phil loving; hemophilia, thermophilic
-phile lover; thermophile
phob- hating; hydrophobia
-phobia fear; hydrophobia
photo- light; photosynthesis
phyco- seaweed, alga; phycochrome
phyl- tribe; phylum, phylogeny
phyto- plant; phytopathology
pleo- more; pleomorphism
pleur- side; pleural, pleuron
pneuma- breath; pneumatic, pneumon
pod- foot; pseudopod
poly- many; polymorphic
pre- before; prenatal, precaval
pro- before; procaryote
proto- first; protoplasm, protozoa
pseud- false; pseudopod
psychro- cold; psychrophile
pulmo- lung; pulmonary
pyo- pus; pyogenic

quadr- four-fold; quadruped

re- back; regenerate, react
-rhag burst; hemorrhage
rhodo- rose; *Rhodospirillum*

schizo- split; schizont
scler- hard; sclera, sclerosis
sect- cut; section, insect, dissect
semi- half; semilunar
sepsis- putrefaction, infection; aseptic
-soma body; chromosome
-some body; chromosome
spiro- coil, spiral; spirochete
spor- seed; spore
-stat stationary, set; bacteriostat
strepto- twined, chain-like; streptomycete
sub- under; subconscious
sulf- sulfur; *Desulfovibrio*
supra- above; suprarenal
sym-, syn- together; symbiosis, syncytium

-taxis movement in response to stimulus; phototaxis
telo- end; telophase
tetra- four; tetrad
therm- heat; thermometer, homothermal
thio- sulfur; *Thiobacillus*
trans- across, beyond, through; transformation

tricho-   hair; peritrichous
tropho-   nourishment; trophozoite
-trophy   nourishment; phototrophic

uni-   one; unicellular

xeno-   strange, foreign; axenic

zoo-   living being, animal; zoospore
zyg-   yoke; zygote, homozygous

ETYMOLOGY OF COMMON BACTERIA*

Aerobacter (G. gas-rod) *aerogenes* (L. gas-producing)
Bacillus (L. rodlet) *cereus* (L. waxen, wax colored)
Bacillus (L. rodlet) *megaterium* (L. big beast)
Escherichia (Escherich, who first isolated it) *coli* (L. of the colon)
Proteus (G. an ocean god who took many shapes) *mirabilis* (L. wonderful, surprising)
Proteus (G. an ocean god who took many shapes) *vulgaris* (L. common)
Pseudomonas (G. false unit) *aeruginosa* (L. full of copper rust, hence green)
Salmonella (Salmon, an American bacteriologist) *typhimurium* (L. stupor in mice)
Sarcina (L. package) *lutea* (L. yellow)
Serratia (Italian physicist) *marcescens* (L. decaying)
Staphylococcus (G. bunch of grapes) *aureus* (L. golden)
Staphylococcus (G. bunch of grapes) *epidermidis* (G. outer skin)

*Bergey's Manual of Determinative Bacteriology, 8th ed.

The metric system will be used throughout this text. These tables list the common metric units, their English equivalents, and the relationship among metric units.

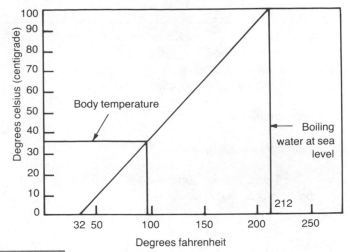

	Basic Measure	Units	Value	English Equivalent
Length	meter (m)			39.4 inches
		kilometer (km)	1000 m	0.621 miles
		centimeter (cm)	.01 m	0.394 inches
		millimeter (mm)	.001 m	0.0394 inches
		micrometer (μm)	.000001 m	
Mass	gram (g)			0.035 ounces
		kilogram (kg)	1000 g	2.2 pounds
		milligram (mg)	.001 g	
		microgram (μgm)	.000001 g	
Volume	liter (ℓ)			1.06 quarts
		milliliter (ml)	.001 ℓ	

Symbol	Meter m	Centimeter cm	Millimeter mm	Micrometer μm	Millimicron mμ	Angstrom Å
Meter (m)	1.0	100.0 10^2	1000.0 10^3	1,000,000.0 10^6	1,000,000,000.0 10^9	10,000,000,000.0 10^{10}
Centimeter (cm)	0.01 10^{-2}	1.0	10.0	10,000.0 10^4	10,000,000.0 10^7	100,000,000.0 10^8
Millimeter (mm)	0.001 10^{-3}	0.1 10^{-1}	1.0	1000.0 10^3	1,000,000.0 10^6	10,000,000.0 10^7
Micrometer (μm)	0.000001 10^{-6}	0.0001 10^{-4}	0.001 10^{-3}	1.0	1000.0 10^3	10,000.0 10^4
Nanometer (nm)	0.000000001 10^{-9}	0.0000001 10^{-7}	0.000001 10^{-6}	0.001 10^{-3}	1.0	10.0
Angstrom (Å)	0.0000000001 10^{-10}	0.00000001 10^{-8}	0.0000001 10^{-7}	0.0001 10^{-4}	0.1 10^{-1}	1.0

Metric Conversions

B

APPENDIX

To °C	←°F or °C→	To °F	To °C	←°F or °C→	To °F	To °C	←°F or °C→	To °F
−31.67	−25	−13	−3.89	25	77	23.89	75	167
−31.11	−24	−11.2	−3.33	26	78.8	24.44	76	168.8
−30.56	−23	−9.4	−2.78	27	80.6	25	77	170.6
−30	−22	−7.6	−2.22	28	82.4	25.56	78	172.4
−29.44	−21	−5.8	−1.67	29	84.2	26.11	79	174.2
−28.89	−20	−4	−1.11	30	86	26.67	80	176
−28.33	−19	−2.2	−0.56	31	87.8	27.22	81	177.8
−27.78	−18	−0.4	0	32	89.6	27.78	82	179.6
−27.22	−17	1.4	.56	33	91.4	28.33	83	181.4
−26.67	−16	3.2	1.11	34	93.2	28.89	84	183.2
−26.11	−15	5	1.67	35	95	29.44	85	185
−25.56	−14	6.8	2.22	36	96.8	30	86	186.8
−25	−13	8.6	2.78	37	98.6	30.56	87	188.6
−24.44	−12	10.4	3.33	38	100.4	31.11	88	190.4
−23.89	−11	12.2	3.89	39	102.2	31.67	89	192.2
−23.33	−10	14	4.44	40	104	32.33	90	194
−22.78	−9	15.8	5	41	105.8	32.78	91	195.8
−22.22	−8	17.6	5.56	42	107.6	33.33	92	197.6
−21.67	−7	19.4	6.11	43	109.4	33.89	93	199.4
−21.11	−6	21.2	6.67	44	111.2	34.44	94	201.2
−20.56	−5	23	7.22	45	113	35	95	203
−20	−4	24.8	7.78	46	114.8	35.56	96	204.8
−19.44	−3	26.6	8.33	47	116.6	36.11	97	206.6
−18.89	−2	28.4	8.89	48	118.4	36.67	98	208.4
−18.33	−1	30.2	9.44	49	120.2	37.22	99	210.2
−17.78	0	32	10	50	122	37.78	100	212
−17.22	1	33.8	10.56	51	123.8	38.33	101	213.8
−16.67	2	35.6	11.11	52	125.6	38.89	102	215.6
−16.11	3	37.4	11.67	53	127.4	39.44	103	217.4
−15.56	4	39.2	12.22	54	129.2	40	104	219.2
−15	5	41	12.78	55	131	40.56	105	221
−14.44	6	42.8	13.33	56	132.8	41.11	106	222.8
−13.89	7	44.6	13.89	57	134.6	41.67	107	224.6
−13.33	8	46.4	14.44	58	136.4	42.22	108	226.4
−12.78	9	48.2	15	59	138.2	42.78	109	228.2
−12.22	10	50	15.56	60	140	43.33	110	230
−11.67	11	51.8	16.11	61	141.8	43.89	111	231.8
−11.11	12	53.6	16.67	62	143.6	44.44	112	233.6
−10.56	13	55.4	17.22	63	145.4	45	113	235.4
−10	14	57.2	17.78	64	147.2	45.56	114	237.2
−9.44	15	59	18.33	65	149	46.11	115	239
−8.89	16	60.8	18.89	66	150.8	46.67	116	240.8
−8.33	17	62.6	19.44	67	152.6	47.22	117	242.6
−7.78	18	64.4	20	68	154.4	47.78	118	244.4
−7.22	19	66.2	20.56	69	156.2	48.33	119	246.2
−6.67	20	68	21.11	70	158	48.89	120	248
−6.11	21	69.8	21.67	71	159.8	49.44	121	249.8
−5.56	22	71.6	22.22	72	161.6	50	122	251.6
−5	23	73.4	22.78	73	163.4	50.56	123	253.4
−4.44	24	75.2	23.33	74	165.2	51.11	124	255.2

N	0	1	2	3	4	5	6	7	8	9
10	0000	0043	0086	0128	0170	0212	0253	0294	0334	0374
11	0414	0453	0492	0531	0569	0607	0645	0682	0719	0755
12	0792	0828	0864	0899	0934	0969	1004	1038	1072	1106
13	1139	1173	1206	1239	1271	1303	1335	1367	1399	1430
14	1461	1492	1523	1553	1584	1614	1644	1673	1703	1732
15	1761	1790	1818	1847	1875	1903	1931	1959	1987	2014
16	2041	2068	2095	2122	2148	2175	2201	2227	2253	2279
17	2304	2330	2355	2380	2405	2430	2455	2480	2504	2529
18	2553	2577	2601	2625	2648	2672	2695	2718	2742	2765
19	2788	2810	2833	2856	2878	2900	2923	2945	2967	2989
20	3010	3032	3054	3075	3096	3118	3139	3160	3181	3201
21	3222	3243	3263	3284	3304	3324	3345	3365	3385	3404
22	3424	3444	3464	3483	3502	3522	3541	3560	3579	3598
23	3617	3636	3655	3674	3692	3711	3729	3747	3766	3784
24	3802	3820	3838	3856	3874	3892	3909	3927	3945	3962
25	3979	3997	4014	4031	4048	4065	4082	4099	4116	4133
26	4150	4166	4183	4200	4216	4232	4249	4265	4281	4298
27	4314	4330	4346	4362	4378	4393	4409	4425	4440	4456
28	4472	4487	4502	4518	4533	4548	4564	4579	4594	4609
29	4624	4639	4654	4669	4683	4698	4713	4728	4742	4757
30	4771	4786	4800	4814	4829	4843	4857	4871	4886	4900
31	4914	4928	4942	4955	4969	4983	4997	5011	5024	5038
32	5051	5065	5079	5092	5105	5119	5132	5145	5159	5172
33	5185	5198	5211	5224	5237	5250	5263	5276	5289	5302
34	5315	5328	5340	5353	5366	5378	5391	5403	5416	5428
35	5441	5453	5465	5478	5490	5502	5514	5527	5539	5551
36	5563	5575	5587	5599	5611	5623	5635	5647	5658	5670
37	5682	5694	5705	5717	5729	5740	5752	5763	5775	5786
38	5798	5809	5821	5832	5843	5855	5866	5877	5888	5899
39	5911	5922	5933	5944	5955	5966	5977	5988	5999	6010
40	6021	6031	6042	6053	6064	6075	6085	6096	6107	6117
41	6128	6138	6149	6160	6170	6180	6191	6201	6212	6222
42	6232	6243	6253	6263	6274	6284	6294	6304	6314	6325
43	6335	6345	6355	6365	6375	6385	6395	6405	6415	6425
44	6435	6444	6454	6464	6474	6484	6493	6503	6513	6522
45	6532	6542	6551	6561	6571	6580	6590	6599	6609	6618
46	6628	6637	6646	6656	6665	6675	6684	6693	6702	6712
47	6721	6730	6739	6749	6758	6767	6776	6785	6794	6803
48	6812	6821	6830	6839	6848	6857	6866	6875	6884	6893
49	6902	6911	6920	6928	6937	6946	6955	6964	6972	6981
50	6990	6998	7007	7016	7024	7033	7042	7050	7059	7067

Four-Place Logarithms

APPENDIX

N	0	1	2	3	4	5	6	7	8	9
51	7076	7084	7093	7101	7110	7118	7126	7135	7143	7152
52	7160	7168	7177	7185	7193	7202	7210	7218	7226	7235
53	7243	7251	7259	7267	7275	7284	7292	7300	7308	7316
54	7324	7332	7340	7348	7356	7364	7372	7380	7388	7396
55	7404	7412	7419	7427	7435	7443	7451	7459	7466	7474
56	7482	7490	7497	7505	7513	7520	7528	7536	7543	7551
57	7559	7566	7574	7582	7589	7597	7604	7612	7619	7627
58	7634	7642	7649	7657	7664	7672	7679	7686	7694	7701
59	7709	7716	7723	7731	7738	7745	7752	7760	7767	7774
60	7782	7789	7796	7803	7810	7818	7825	7832	7839	7846
61	7853	7860	7868	7875	7882	7889	7896	7903	7910	7917
62	7924	7931	7938	7945	7952	7959	7966	7973	7980	7987
63	7993	8000	8007	8014	8021	8028	8035	8041	8048	8055
64	8062	8069	8075	8082	8089	8096	8102	8109	8116	8122
65	8129	8136	8142	8149	8156	8162	8169	8176	8182	8189
66	8195	8202	8209	8215	8222	8228	8235	8241	8248	8254
67	8261	8267	8274	8280	8287	8293	8299	8306	8312	8319
68	8325	8331	8338	8344	8351	8357	8363	8370	8376	8382
69	8388	8395	8401	8407	8414	8420	8426	8432	8439	8445
70	8451	8457	8463	8470	8476	8482	8488	8494	8500	8506
71	8513	8519	8525	8531	8537	8543	8549	8555	8561	8567
72	8573	8579	8585	8591	8597	8603	8609	8615	8621	8627
73	8633	8639	8645	8651	8657	8663	8669	8675	8681	8686
74	8692	8698	8704	8710	8716	8722	8727	8733	8739	8745
75	8751	8756	8762	8768	8774	8779	8785	8791	8797	8802
76	8808	8814	8820	8825	8831	8837	8842	8848	8854	8859
77	8865	8871	8876	8882	8887	8893	8899	8904	8910	8915
78	8921	8927	8932	8938	8943	8949	8954	8960	8965	8971
79	8976	8982	8987	8993	8998	9004	9009	9015	9020	9025
80	9031	9036	9042	9047	9053	9058	9063	9069	9074	9079
81	9085	9090	9096	9101	9106	9112	9117	9122	9128	9133
82	9138	9143	9149	9154	9159	9165	9170	9175	9180	9186
83	9191	9196	9201	9206	9212	9217	9222	9227	9232	9238
84	9243	9248	9253	9258	9263	9269	9274	9279	9284	9289
85	9294	9299	9304	9309	9315	9320	9325	9330	9335	9340
86	9345	9350	9355	9360	9365	9370	9375	9380	9385	9390
87	9395	9400	9405	9410	9415	9420	9425	9430	9435	9440
88	9445	9450	9455	9460	9465	9469	9474	9479	9484	9489
89	9494	9499	9504	9509	9513	9518	9523	9528	9533	9538
90	9542	9547	9552	9557	9562	9566	9571	9576	9581	9586
91	9590	9595	9600	9605	9609	9614	9619	9624	9628	9633
92	9638	9643	9647	9652	9657	9661	9666	9671	9675	9680
93	9685	9689	9694	9699	9703	9708	9713	9717	9722	9727
94	9731	9736	9741	9745	9750	9754	9759	9763	9768	9773
95	9777	9782	9786	9791	9795	9800	9805	9809	9814	9818
96	9823	9827	9832	9836	9841	9845	9850	9854	9859	9863
97	9868	9872	9877	9881	9886	9890	9894	9899	9903	9908
98	9912	9917	9921	9926	9930	9934	9939	9943	9948	9952
99	9956	9961	9965	9969	9974	9978	9983	9987	9991	9996
100	0000	0004	0009	0013	0017	0022	0026	0030	0035	0039
N	0	1	2	3	4	5	6	7	8	9

Abbé   10
  lens   146
Acanthamoeba   201
acetator   529, 530
Acetobacter   511, 529
acid   A substance that gives up hydrogen ions
  (H$^+$) when dissolved in water. Acids also
  act as hydroxyl ion (OH$^-$) acceptors.   64
acid-fast staining   A differential staining
  technique based on the ability of bacteria
  to resist decolorization with acid-alcohol,
  after they have been dyed with hot
  carbolfuchsin; typical acid-fast bacteria
  belong to the genus
  Mycobacterium.   141
acidogenic   171, 287–89
acidophile   An organism that grows best in
  acid environments.   160
acidophilus milk   287
Acinetobacter   379, 382
acne   279
acquired immunity   A specific host defense
  mechanism that is stimulated into action
  by contact with an antigen (active
  immunity) or by the transfer of previously
  formed antibody (passive
  immunity).   302
Actinomyces   436, 441, 442, 449, 535
  A. israelii   436, 441, 442
actinomycosis   441, 442
activated sludge   The sediment formed in an
  aeration tank during secondary sewage
  treatment; contains a high concentration
  of microbes that actively grow and
  decompose wastewater
  contaminants.   567
activation energy   67, 165
active site   31
active transport   A method whereby a cell
  transfers materials through its membranes,
  and in the process expends energy.   102,
  103
acute disease   204
acute period   205
acute respiratory disease   487
adaptation   The genetic response of an
  individual to a stimulus over a long period
  of time.   28, 183, 184
adenosine triphosphate (ATP)   An organic
  molecule that is able to store the energy
  needed for immediate use by a cell.   25,
  26, 68, 97, 102, 103, 112, 161–73, 426,
  575
adenoviruses   487–88
Aedes   444, 449
aerobic cellular respiration   A specific series of
  chemical reactions involving the use of

molecular oxygen in which chemical bond
  energy is released to the cell in a usable
  form.   25, 161–64
aerosol   363
aflatoxin   A mycotoxin (fungal poison)
  produced by the fungi *Aspergillus* and
  some species of *Penicillium,* while growing
  on stored grains such as peanuts and
  soybeans.   291, 514
agar-agar   A chemical from algae that is used
  to solidify liquid broth. Commonly
  referred to as agar. Microbes are grown
  on the surface of the solidified food.   10,
  50, 131, 210
agar diffusion test   309, 310
agar slant culture   135
agglutination   308, 454
aggregates (soil)   535
agranulocyte   297
Alcaligenes   511
alcohol   8, 235, 576–87
aldehyde   235, 236
algae   A form of Protist; has a cell wall and
  specific combinations of photosynthetic
  pigments in particular cell
  structures.   47–50
algin   49
alkalophile   An organism that grows best in
  more alkaline, or basic,
  environments.   160
alleles   Alternative forms of the genes for a
  particular characteristic.   250, 259
allergen   329, 588
allergy   *See* hypersensitivity.
allograft   347
allosteric site   173
Amanita verna   52, 60, 514
amensalism   A form of symbiosis in which
  one species is harmed but the other is
  neither benefited nor harmed.   275
American Type Culture Collection
  (ATCC)   132, 135
Ames test   253
amino acid   100, 101, 158, 159, 168, 171,
  173, 250, 276, 575, 588, 589
  L form   72, 73, 120, 121
  D form   72, 120, 121
Amoeba   39, 54, 103
amphitrichous   115, 116
anabolism   158, 173
anaerobic cellular respiration   A specific series
  of chemical reactions not involving
  molecular oxygen but alternative
  molecules in which chemical bond energy
  is released to the cell in a usable
  form.   25, 161, 162, 172, 220, 511
anaerobic digestion   The anaerobic

decomposition of sludge in which
  methane and hydrogen sulfide gases are
  released.   566
anamnestic (secondary) response phase   320,
  321
anaphylaxis   A shock reaction that occurs
  after second exposure to an allergen:
  symptoms include skin reddening and
  itching, capillary breakdown, headache,
  drop in blood pressure and smooth
  muscle contraction; the two forms are
  generalized and localized.   329, 331, 336
animal diseases   547, 548
Animalia   One of the four major kingdoms;
  examples include the higher animals such
  as worms, birds, and humans; their cells
  are eucaryotic.   37
ankylosing spondylitis   349
Anopheles   359, 374, 382, 444
antagonism   An interaction between two
  different microbes in which one releases a
  substance that may kill or inhibit the
  activities of the other but is completely
  ineffective against the species producing
  the material.   287
anthrax   A disease of animals and humans
  caused by the bacterium *Bacillus
  anthracis;* first investigated by Pasteur and
  Koch.   9, 301, 302, 396
antibiotic   A chemical produced by a microbe
  that is able to kill or inhibit the growth or
  activity of another microbe. They are
  selectively toxic to the microbe and not
  the host.   87, 104, 122, 182, 206, 221,
  228–46, 255, 259, 275, 277, 280, 288,
  301, 338, 389, 431, 526, 574, 575,
  589–94
antibody   A protein molecule manufactured
  by the body in response to the presence
  of a foreign molecule known as an
  antigen.   206, 244, 302–23, 328–54
antibody-mediated (humoral)
  immunity   Immunity as a result of
  antibody action in the humor (fluids) of
  the body; B-cells and plasma cells secrete
  these immunoactive protein
  molecules.   306, 307, 320, 321
anticodon   78
antigen   A large organic molecule, usually a
  protein, that is able to stimulate the
  production of a specific antibody with
  which it can chemically combine.   128,
  206, 244, 293, 302–23
antigenic determinant   The portion of an
  antigen that determines the specificity of
  the antibody-antigen reaction and
  combines with the antibody.   306

Glossary/
Index

antigenic drift   371, 503, 504

antihistamine   332, 336

antimetabolite   228

antimicrobial agent   Anything that kills or interferes with the multiplication, growth, or activity of microbes. These include disinfectants or antibiotics.   230

antiseptic   An agent that opposes sepsis. These are usually chemicals that are able to inhibit the growth and activity of microbes and may be used on skin or other tissues. They may be dilute disinfectants.   139, 230

antiserum   302

antistreptolysin O test (ASTO)   314

antitoxin   314, 415

API system   216

Appert, N.   520

arboviruses   502

Aristotle   387, 494

Arizona   400

Arthrobacter   535

Arthus, M.   338

Arthus response   338

asepsis   Any method that ensures freedom from infection by the prevention of contact with a microbe. Two forms are recognized: surgical and medical.   230

aseptic technique   Techniques and procedures performed in the laboratory which prevent contamination of the work area, worker, and culture.   132, 134, 377

Ashbya gossypii   589

aspergillosis   440

Aspergillus   51, 54, 60, 291, 511, 514, 535, 593

  A. fumigatus   440, 441

assimilation   A process characteristic of life in which molecules entering the cell become involved in biochemical pathways that maintain the life of the cell.   25, 182

asthma, bronchial   332

atom   The smallest particle of an element that still retains all the properties of the element.   61

atomic nucleus   The center of mass in an atom. The location of the protons and neutrons.   61

atomic number   61

attenuation   The loss of virulence in a population of pathogens as a result of laboratory culturing outside a living host.   292, 293, 302, 477, 490, 495

autoclave   An instrument used to sterilize glass, chemicals, or paper by exposing the materials to moist heat under pressure; the usual conditions are 121°C for 15 to 20 minutes at 15–20 psi (pounds per square inch) or in a high speed autoclave at 132°C for 3 minutes.   9, 136–38, 224

autoimmunity   306, 309

autolysin   122

autotroph   An organism that can manufacture its own complex organic compounds from simple inorganic molecules.   36, 161, 275

auxochrome   141

Azotobacter   541

B lymphocyte (B-cell)   A form of lymphocyte that has had its differentiation controlled by Bursa-like tissue; responsible for antibody-mediated immunity; they secrete antibodies.   306, 308, 329, 353, 468, 476

Bacille Calmette-Guérin (BCG) vaccine   293, 341, 352, 419, 475

bacillus   A rod-shaped microbe.   43, 114

Bacillus   116, 141, 160, 185, 219, 232, 254, 511, 521, 535, 589

  B. anthracis   143, 396

  B. cereus   128, 512, 514

  B. popilliae   538

  B. stearothermophilus   224

  B. subtilis (pictured on cover)   43, 184, 224, 587

  B. thuringiensis   88, 538

bacteremia   The presence of bacteria in the blood.   276

bacteria   A form of Procaryotae characterized by their shape and genetic ability to function in various environments.   36–40

bacteriology   The branch of biology that deals with the study of bacteria.   2

bacteriophage (phage)   A type of virus that infects bacterial cells.   11, 250, 260, 415, 451, 454, 456

Bacteroides   219, 245, 280, 284

  B. fragilis   423, 424

Balantidium coli   55, 60, 443, 445, 449

basal body   114

base   A substance that gives up hydroxyl ions, OH$^-$, when dissolved in water. Bases also act as hydrogen ion, H$^+$ acceptors.   64

Basidiomycetes   51

basophile   298

batch culture   A single, large culture of microbes made by adding an inoculum to a limited quantity of medium and allowing it to ferment to completion.   186, 573

Beadle, G.   11

beer   577, 581, 582

Beggiatoa   87

Bergy's Manual   40–42

beta propiolactone (BPL)   235

Bifidobacterium bifidus   287

bile   297

binary fission   A cell division process found in procaryotic cells, resulting in the formation of two daughter cells.   40, 118, 119, 254

binomial system of nomenclature   A system of naming organisms by assigning genus and species names.   39

biocide   537

bioconversion   The use of microbial fermentation to enzymatically change one compound to another biologically active form.   573, 592, 593

biogenesis   A theory that all living organisms can arise only from living parents.   6

biogeochemical recycling   The reprocessing of a material so that it can be used many times.   538, 543

biological amplification   Concentration of a chemical as it passes through the food chain of an ecosystem.   537

biological assay (bio-assay)   196, 197

biological control   The use of predators, parasites, or other biological agents to keep a population of unwanted organisms in check.   538

biological oxygen demand (BOD)   The amount of dissolved oxygen used by microbes for the biological oxidation of wastes.   564

biology   The science that deals with the orderly study or collection of knowledge of life.   1, 2

biomass   The total weight of a particular kind of organism in a culture; measured and correlated with the population growth curve as a population estimating technique.   192–96

biosphere   533

biosynthesis   172

biotechnology   265

Blastomyces dermatitidis   439, 440, 449

blastomycosis   439, 440

blocking antibody   337

blood typing   301, 308, 343–46

boil (carbuncle, furuncle)   281, 300

bonding energy   165

Bordeaux mixture   544

Bordetella pertussis   143, 226, 290, 412, 413

Borrelia recurrentis   423

botulism   513, 514

Brewer's plate or jar   220, 221

bright-field (compound) microscope   145, 146

brining   5, 519, 522

broad spectrum antibiotic   An antibiotic that is able to control a wide range of microbes, including Gram-negative and Gram-positive bacteria.   244

broth culture   135

Brown, R.   108

Brucella abortus   143, 300, 408, 409

brucellosis   409, 410

buboe   410

buffer   219

Burkitt's lymphoma   467, 468

burst size   457

C polysaccharide   392

Campylocabacter   43

cancer   337, 348–53, 451, 466, 467

Candida albicans   236, 248, 275, 289, 432, 433, 511, 520

canning   5, 520, 521

capsid   451

capsomere   451, 462

capsule   The outermost cover on certain microbes; a sticky layer composed of carbohydrates, protein, or organic acids; protects the cell from phagocytosis.   122, 123, 144

carbohydrate   94–97

carbon-oxygen cycle   538

carbuncle   99, 281

carcinoembryonic antigen (CEA)   Cell-surface antigens formed by the derepression of embryonic genes during neoplasia. Increases in CEA are characteristic of certain forms of cancer.   348, 350, 466

carcinogen   253, 522, 556

caries (dental)   282–84

cariogenic   283

carotene   111

carrageenan   50

carrier   An individual who has a particular pathogenic microbe but fails to show clinical symptoms or subclinical signs of infection; several types exist: chronic, casual, and incubatory (convalescent).   358, 402, 405

catabolism   158, 173

catheter 208, 209
cationic dyes 141
cell cultures 455
cell-mediated immunity (CMI) Controlled by T-cells which secrete lymphokines that kill pathogens or stimulate phagocytic macrophages into action; responsible for pathogen destruction, transplant tissue destruction, and cancer cell destruction. 306, 307, 318–20, 418, 431, 476
cellulase 120, 539
cellulose 50, 120, 538, 543
cell wall An outer covering on some microbes; may be composed of cellulose, chitin, or peptidoglycan. 50, 98, 120–22, 265
Center for Disease Control (CDC) 361, 374
Champagne 580
chancre 421, 422
Chang, A. 263
Chase, M. 11, 16
cheese 53, 524–28
chemical bond The attraction that one atom, ion, or molecule has for another atom, ion, or molecule. 63, 158
chemical reaction The rearrangement of chemical bonds between two or more chemicals, whereby new bonds are formed and energy is subsequently exchanged. 65
chemical sterilization 136, 138, 139
chemoautotroph 161, 542
chemostat An apparatus used to maintain a culture of microbes in the log phase of growth, i.e., a continuous culture. 186, 573
chemotaxis 115
chemotherapeutic agent A chemical that is used in the treatment of a disease. These agents should harm the infectious agent but not the host. 231, 238–46, 477, 479
chickenpox 322, 485–87
chimeras 263
Chinese restaurant syndrome 589
chitin 50, 120
Chlamydia 45
   C. psittaci 46, 426
   C. trachomatis 46, 60, 426–28
chloramines 556
Chlorobium 46, 47, 60, 541
chlorophyll 111, 112
chloroplast A membranous organelle of the plastid group; the site of photosynthesis; found in eucaryotic cells. 111–13
cholera 5, 404, 405
chromatophore 141
Chromobacterium 535
chromomycosis 436
chromosome Nucleoprotein that has become super-coiled during mitosis or meiosis and stains heavily. Formed in eucaryotic cells. 118, 119
chronic disease 204
cilia Hairlike structures that project through the cell membrane. They function like oars to move the cell through the environment or the environment past the cell. 55, 114, 115

Ciliata 55, 56, 443–46
Citrobacter 208, 226, 400
Cladosporium 436, 520, 532
classification 37, 41, 42, 161
   laboratory 202–3
   virus 455, 456, 460, 461
   water 554, 555
clean 136
clean-catch (midstream) method 208
clinical case 358
clinical stages of infection A series of four stages of an acute infection, including incubation, prodromal, acute, and convalescent. 204–6
clone Any population of organisms that has been produced by asexual reproduction from a single parent cell. 258, 263, 265, 318, 321
Clostridium 116, 141, 160, 185, 219, 232, 284, 290, 535, 541
   C. botulinum 26, 34, 143, 186, 235, 288, 290, 398, 511, 512–14, 520, 521
   C. perfringens 26, 34, 99, 143, 288, 292, 357, 397, 398, 512, 514
   C. tetani 143, 290, 321, 357, 397, 398
coagulase 280, 292, 388, 389
coagulation 528
coagulation-sedimentation 569, 570
Coccidioides immitis 204, 226, 438, 439
coccidioidomycosis 438, 439
coccus A spherical shape; morphology of certain microbes. 43
codon 78, 250
coenzyme 168
Cohen, S. 263
cold sore 463, 484
colic 288
coliform A group of bacteria that belong to the family Enterobacteriaceae; includes E. coli, Enterobacter, Hafnia, Serratia, and Klebsiella. 213, 219, 398, 558–60
collagenase 292
colony A population of microbes growing on the surface of solid medium that is the result of the reproduction of a single, isolated cell. 129, 131
comedo 279
commensalism A form of symbiosis in which one individual benefits and the other is not harmed. 275
common cold 492–94
communicable 205, 206, 373, 495
community 533
competent cells 254
complement system A complex series of reactions involving 11 serum proteins (not antibodies) which react with one another to produce different biologically active molecules; many of these enhance pathogen destruction. 310–12, 337
   fixation 312, 313
complementary base 75, 76
completed water test 561
complex medium A culture medium that is prepared from ingredients that have not been purified and precisely defined; also called nonsynthetic medium. 210
composting The decomposition of vegetable material by microbial action. 542, 567, 568

compound A combination of atoms of two or more kinds of elements that are chemically bonded. 63
confirmed water test 561
conjugation In procaryotic cells, the transfer of genes from one cell to another by direct contact. 115, 117, 253–58, 262, 264, 292
constitutive enzymes 174
contact dermatitis 341, 342
contact inhibition 465
continuous culture A culture of microbes that is maintained in the exponential (log) growth phase. 186, 573, 574
control agents 228–46
convalescence 206
Coombs, R.R.A. 329, 330
Coombs test 344, 345
coordination A characteristic of life that allows biochemical reactions to be established and linked to one another by the activities of enzymes. 30
cord factor 418
corticosteroid 301
Corynebacterium 593
   C. acne 279, 280
   C. diphtheriae 87, 106, 143, 290, 413–15
   C. parvum 352
Cosmarium 3
counterstain 142
covalent bonds 63, 64
cowpox 11
Coxiella burnetii 45, 60, 425
Coxsackie virus 490–92
Creutzfeldt-Jakob disease (CJD) 469
Crick, F. 11, 17, 250
crista 112
cryptococcosis 441
Cryptococcus 201, 499
   C. neoformans 441
Culex 444, 449, 467
culture characteristics 128, 133
culture and susceptibility The growth of a known pathogen in pure culture and its exposure to a variety of antibiotics of different concentrations to determine which drug will kill or inhibit the microbe's growth. 221–24
curd 525–28
curing 519, 522
Curvularia 593
cyanobacteria 46, 47, 112, 114
cyst 443, 445
cytochrome 111, 112, 168
cytolysis 473, 474
cytomegalovirus 485
cytopathic effect Damage or death of a host cell resulting from viral infection and replication. 473, 474
cytoplasm One of the two types of protoplasm; that portion of the cell excluding the nucleus. 84, 85
cytotoxic reaction 336–38

Dane particles 498
Danielli-Davson membrane model 89
dark-field microscopy 146
dark repair 252
Darwin, C. 29
death phase 184, 229

decomposers Saprotrophic bacteria and fungi that convert organic material into inorganic material. 536, 542
dehydration synthesis 68
delayed hypersensitivity 340–42
Dermacentor andersoni 359, 382, 503
dermatophyte 431, 434
desensitization 336
desquamation 391
detergent 232,233
Deuteromycetes 51
diaminopimelic acid (DAP) 120, 121
diapedesis 297–99
Dick test 391
differential medium A culture medium that is designed to show the differences among microbes. 213
differential stain 141, 142
differential test medium 213, 214
differentially permeable membrane A membrane that selectively allows some particles to pass through, while others are excluded. 98
dimer 252
dimorphic 431
diphtheria 301, 358, 372, 414, 415
dipicolinic acid (DPA) 116, 117
dirty 136
disease A process or event which results in damage to a living organism. 4, 276
    airborne 362–64
    classification 361, 362
    cycles 372
    foodborne 365–68
    waterborne 364, 365, 368
    zoonoses 369, 370
disinfectant Any agent, usually a chemical, that is capable of killing microbes when applied directly to an inanimate object. Two forms of disinfection are recognized: concurrent and terminal. 229, 230
dissolved oxygen 563, 564
distemper 464
distillation 583, 584
DNA (deoxyribonucleic acid) 11, 26, 42, 57, 80, 84, 86, 108, 111, 118, 119, 159, 182, 234, 250–70, 303, 351, 451, 458, 482, 575, 587
Döderlein's bacillus 289
DPT vaccination 290, 322, 338, 373, 415
drop therapy 336
droplet nuclei Small clusters of microbes surrounded by a thin layer of mucus that has been expelled from the respiratory tract. 362
drug resistance The ability of a population of microbes to withstand and survive the effects of an antibiotic. 244–46, 253, 255, 260, 265, 266, 389, 394, 432, 547
    chromosomal 259
    extrachromosomal 259
dry-heat sterilization 136, 138
drying 518, 519
dry-slide technique 186
dry weight 194–96
drug 104
duck embryo vaccine (DEV) 495
dysentery
    amoebic 54, 55
    bacterial 403, 404

early proteins 465
Echoviruses 490
eclipse phase 457, 458
ecology 533
ecosystem 533, 542
eczema 341, 342
edema 300
Edwardsiella 400
electrodialysis 570
electromagnetic spectrum 144, 145
electron 61, 62
electronic cell counting 187
electron microscopy 145–49
electron transporting system (ETS) The final series of biochemical reactions in aerobic cellular respiration; begins with the movement of hydrogen electrons through a series of cytochromes and results in the formation of ATP and water. 164–71
electrophoresis 315
element 61, 159
empyema 393
encephalitis, equine 467, 502, 503
encephalomyocarditis 490
endemic disease A disease which is constantly in a given geographical area at a low morbidity rate acceptable to public health authorities. 357, 358
endergonic 65
endocarditis 291
endocytosis The encasing of materials by membrane for transport into the cell; e.g., phagocytosis, pinocytosis, or engulfment. 103, 104
endoplasmic reticulum A membranous organelle found in eucaryotic cells; serves as a channel system for the movement of molecules throughout the cell and as a surface catalyst; two forms are found: rough and smooth. 108, 110
endospore A nongrowing, heat-resistant structure that may be formed within the vegetative cell of certain bacteria. 9, 116, 117, 141, 184, 229, 232, 235, 396–98, 522
endosymbiont 277
endotoxic shock (Gram-negative sepsis) 400
endotoxin A toxic component of many cells; lipopolysaccharide-protein in nature, heat stable, and weakly antigenic; released after cell destruction and responsible for fever. 120, 398, 399, 402, 405, 409, 410, 412, 441
endpoint assay 454, 455
enrichment medium A medium that contains chemicals to enhance the growth of specific microbes or group of microbes in a mixed culture. 210
Entamoeba histolytica 54, 60, 443, 446, 447
Enterobacter 208, 211, 226, 238, 260, 398, 512, 541, 558
enterococcus group 391
enterotoxin 365, 389, 401, 512, 513
Enterotube system 217
enteroviruses 489–91
envelope 451
enzyme A special kind of protein produced by living cells which controls and regulates the rate at which chemical reactions occur. 25, 29, 30–32, 67, 101,

120, 250, 252, 254, 259, 280, 283, 297, 300, 457, 463, 474, 575, 587–89
eosinophile 297
eosinophile chemotactic factor of anaphylaxis (ECF-A) 332
epidemic disease A disease which has a morbidity rate that is above a normal, expected level and results in an increase in cases, causing high mortality or great public harm and encompasses a larger than normal geographic area. 45, 357, 490
epidemiology The study of environmental and various other factors that affect host-parasite relationships and determine the frequency and spread of disease through a host population. 357–79
Epidermophyton 434, 449
Epstein-Barr virus (EBV) 308, 465, 468, 474, 476, 487
epithelium, keratinized and ciliated 297
erythroblastosis fetalis 343–46
erythrogenic toxin 392
Escherichia coli (E. coli) 3, 28, 34, 39, 86, 99, 102, 121, 128, 143, 147, 182, 201, 208, 211, 213, 223, 244, 245, 250, 251, 255, 259, 263, 264, 265, 276, 278, 284, 287, 288, 290, 291, 318, 358, 360, 378, 398–400, 454, 457, 515, 546, 558, 581
ethylene oxide (ETO) 235
etiological agent 362, 512
eucaryotic cell A type of cell having a true nucleus separated from the rest of the cell contents by a nuclear membrane. 36, 47, 84, 85, 87, 108–24, 229, 253, 263, 265, 431
eutrophication Aging of a lake or pond as a result of increased plant growth due to natural or manmade fertilizers. 542
evolution The response of a population that has undergone gene change and natural selection. The slow change in living organisms, enabling them to cope with changes in their environment over many generations. 28, 29, 38
exergonic 65
exfoliation 389
exogenote (extrachromosomal) 254, 292, 293
exotoxin A chemical toxin released from a microbe during its life; usually protein in nature, heat labile, and strongly antigenic. 291, 388, 399, 405, 409, 410, 415, 512
exponential notation 183
exudate 146, 300, 393, 414

F_c 329, 331, 337
F factor 255–58
facilitated diffusion The movement of molecules from an area of high concentration to an area of lower concentration through a differentially permeable membrane assisted by membrane proteins; no metabolic energy is required. 102
facultative anaerobe An organism that has the facility to either use or not use atmospheric oxygen in the oxidation process to manufacture ATP. 161, 162

FAD (flavin adenine dinucleotide) 158–60
fat metabolism 168
febrile illness 502
febrile period 204
feedback inhibition A metabolic control mechanism that takes place on the surface of enzymes; occurs when one of the end products of the biochemical pathway alters the three-dimensional structure of an essential enzyme in the pathway and interferes with its operation long enough to slow its action. 173
fermentation The anaerobic oxidation of an organic compound without the use of gaseous oxygen; another organic compound is used as the hydrogen acceptor. 171, 172, 212, 228, 523–31, 545, 576, 579, 583
fermenter A large vat or tank used in industrial microbiology for culturing microbes. 573, 574
fever 291
FIAX® system 316
filter sterilization 136, 138, 139
First Law of Thermodynamics Matter and energy are neither created nor destroyed but are converted from one form to another. 158
flagella Hairlike structures that are longer than cilia and project through the cell wall; they function to move the cell through the environment. Two types have been identified: eucaryotic and procaryotic. 10, 114, 115
flagellin 114
Flavobacterium 535, 581, 596
Fleming, A. 52
flocculation
    antibody 314
    water purification 561
flora 277
    ear 281, 283
    eye 281, 282
    gastrointestinal tract 284–89
    genital tract 288, 289
    oral cavity 281, 284, 285
    skin 278–80
    upper respiratory tract 281, 284, 285
fluorescent antibody test (FA) 315
fluorescent treponema antibody absorption test (FTA-ABS) 315
fluoride 283, 562
fluorochrome 147
Fonsecaea 436
food infection Illness acquired from ingesting food containing living microbes that continue to reproduce inside the intestinal tract and cause disease. 365, 512
food poisoning Illness acquired from ingesting food containing toxins; the toxins may be a natural compound of the food or produced by contaminating microbes. 365, 369, 401, 512–18
foot and mouth disease 490
formalin 235, 236
Fracastorius 6, 7, 495
free energy The amount of energy in a chemical system that is available to do work. Chemical reactions can only take place in a direction of lower free energy. 165
free radicals 234

freeze-etching 148–50
freeze-fracture 148–50
freezing 131, 519, 520
fumigant A gaseous agent used in the killing of microbes or other life forms. 231, 235
fungi A major group of Protists; lack chlorophyll, many have a cell wall of chitin; commonly called the yeast, molds, and mushrooms. 50–54, 430–40
fusel oil 577

Gajdusek, D.C. 469
gammaglobulin 303, 321
gas gangrene 99, 397, 398
gas sterilization 138, 140
gasohol 585, 586
GasPak jar 220
gassed-out tube 220
Gell, P.G.H. 329, 330
gene A portion of DNA molecule composed of a specific series of nitrogenous bases that chemically codes for the production of a specific protein or RNA molecule; or serves as an operator in controlling the transcription of RNA within an operon unit 250
generation time The time required for a cell to complete its life cycle; also called doubling time. 182, 250
genetic engineering 263
genetic recombination Any process that brings new combinations of genes together in a single individual, e.g., conjugation, transformation, transduction. 253, 257, 503, 575
genetics 249–70
genotype The catalogue of genes that an organism has, whether or not these genes are expressed. 250, 254
genus A classification name given to groups of species that are very similar; together with the species name, it forms the scientific name. 37
Germ Theory of Disease The theory that certain diseases are caused by the actions of a particular microbe which is able to be transmitted from one individual to another. 4
Giardia 3, 443, 449
    G. lamblia 446
gingiva 276
glomerulonephritis 338
glucose 72
glutaraldehyde (Cidex) 139, 236
glycogen 87
glycolipid 92, 103
glycolytic pathway (Embden-Meyerhof) The first series of biochemical reactions in the oxidation of glucose; begins with glucose and ends with the formation of pyruvic acid and the net formation of 2 ATP molecules. 164–73
glycoprotein 92, 103, 343, 347
gnotobiotic 284, 286
Golgi apparatus A membranous organelle in eucaryotic cells; interchangeable with the endoplasmic reticulum; its primary functions include concentration of synthesized materials and the activation of potentially dangerous molecules. Produces cell vacuoles or vesicles. 110, 111, 147

gonorrhea 394–96
Gonyaulax 49, 60, 515
Gram stain A differential staining procedure that divides bacteria into two major groups; Gram-positive and Gram-negative. 120–22, 131, 142, 143, 155, 156, 159, 160, 164, 211, 228, 244, 275, 279, 290, 362, 379, 561
grana 112, 113
granulocyte 297
granuloma 431, 432
Griffith, F. 254
ground water Water obtained from deep wells or springs; found deep within the ground and contains relatively few microbes and little toxic material. 552, 554
growth A characteristic of life in which new cell components increase in size through the process of assimilation. 26, 182, 183
growth factor Any molecule that is in a minute amount, required by a cell for efficient operation, and must be supplied from the environment, since it is unable to be synthesized by the cell; for example, vitamins and certain amino acids. 158, 228
growth factor analogue 239, 240
growth rate 182
Guillain-Barré syndrome 340, 504

Haemophilus 226, 254
    H. influenzae 143, 245, 315, 323, 411, 412
Hafnia 213, 226
Halobacterium 98, 511
halophile 99, 161
hanging-drop slide 140
hapten A molecule that is able to react and combine with an antibody but cannot stimulate the production of antibody; for example, antibiotic. 306
health 275
heavy metals 235, 555
hemagglutination test A measure of the concentration of viruses based on their ability to agglutinate erythrocytes. 454
heme 159
hemolysin A group of exotoxins that destroy cell membranes; alpha-hemolytic toxins result in incomplete red blood cell destruction, while beta-hemolytic toxins result in complete red blood cell destruction. 291, 388, 389
hemolysis The bursting of red blood cells; caused by chemical substances released by certain pathogenic bacteria; may be seen as a zone of clearing around a growing colony. 99, 313
hemolytic disease 343–46
hemolytic streptococci 391
hemorrhagic fever 502
heparin 298, 332
hepatitis 232, 236, 496–99,
    type A (HAV) 322
    type B (HBV) 323, 464, 468
herd immunity The development of active immunity in a population to a level that protects non-immune members. 370–74, 477, 502
herpesvirus 290, 462, 463, 483–87

Hershey, A. 11, 16

heterocyst 47

heterofermentation 171

heterophile antibody (Forssman) 308, 476

heterotroph An organism that cannot manufacture complex organic compounds but requires these molecules from living organisms or other sources. 36, 161, 275, 539

hexachlorophene 236, 238

Hfr 257

high egg passage (HEP) vaccine 495

Hippocrates 369, 387

histamine 298, 300, 332

histiocyte 299

histocompatibility antigen Complex glycoproteins that are components of cell membranes; their structure is genetically determined and unique to an individual; groups comprise the major histocompatibility complex (MHC) of an individual. 343, 347–52

histone 118

Histoplasma capsulatum 437, 449

histoplasmosis 437, 439

Hodgkin's disease 341

homofermentation 171

Hooke, R. 85

horizon 534

horizontal transmission The movement of transforming viruses from one cell to another by release and reinfection; also known as contagious transmission. 466

host An organism that provides the necessities for a parasite—generally, food and a place to live. 276, 277

Huebner, R. J. 467

human diploid cell vaccine 495

human leukocyte antigen (HLA) 347, 348, 349

humus Dark-colored organic material of soil derived from natural and microbial decomposition of microbes, animals, and plants. 535

hyaluronidase 292, 389, 392

hybridization theory 371

hybridoma cell 353

hydrogen bond 63, 65

hydrogen peroxide 67, 237, 284, 298

hydrolysis 68, 69

hydrophilic 89, 92, 93

hydrophobic 89, 92, 93

hypersensitive pneumonitis 338, 339

hypersensitivity (allergy) 52, 297, 298, 393, 416, 422, 431, 438, 440

hypha 50, 431

hypertonic 219

hypotonic 219

immersion oil 146

immune-complex reaction 338–40

immune serum globulin (ISG) 497

immune transfusion reaction 344–46

immunity (immune response) A specific host defense mechanism; two forms have been identified: antibody mediated and cell mediated. Both involve the sensitization of lymphocytes and specific responses to the antigens. 244, 275, 301–23, 329–53, 372, 476

immunization The controlled stimulation of immunity to a particular

antigen—vaccination. May be passive or active in nature; for example, DPT vaccination. 321–25, 373, 477, 493

immunocompetent cell 306

immunodeficiency 301, 351

immunoelectrophoresis (IEP) 315

immunoglobulin (Ig) A term used to refer to immunologically active serum proteins (antibodies). 284, 303–39, 476

immunologic hypersensitivity (allergy) An abnormal, harmful antibody-antigen response by a sensitized individual; some hypersensitivities are known as allergic reactions; four types are known: anaphylactic, cytotoxic, immune-complex, and delayed. 301, 329–53

immunologic tolerance 306

immunosurveillance The continual "looking over" of all cells and materials in the body by the immune system to detect foreign invaders. Important in the elimination of damaged cells, transplanted tissues, and tumors. 336, 337

impetigo 389, 393

IMViC test A differential test composed of four parts used to identify certain coliforms: I = test for indole production; M = test for acid production; Vi = test for acetylmethylcarbinol production; C = test for citrate utilization. 213

incineration 129, 234

inclusion Cytoplasm components, usually storage materials, that are not directly involved with the life processes of the cell; for example, starch and volutin. 87, 88

inclusion body 462, 474

incubation period 204, 205

indicator microbes 558

induction A regulation of metabolism that occurs on the gene; in induction, the action of a series of genes is turned "on" by the presence of a substrate, while its absence allows the genes to be turned "off." 174

induration 318, 340

infantile diarrhea 39, 399–401

infectious disease A disease of humans, animals, or plants caused by a microbe that enters, grows, and causes harm. 4

infectious mononucleosis 308, 314, 468, 487

inflammatory response A nonspecific host defense mechanism characterized by redness, swelling, heat and pain; involves macrophages, neutrophiles, histamines, and prostaglandins. 291, 297–301, 337, 338, 431

influenza 3, 232, 371, 459, 486, 503, 504

ingestion The process of taking nutrients into a living cell. One of the characteristics of living organisms. 25, 99

inhibitor 32

innate or inborn immune system 302

inoculating loop 129

inoculating needle 129

inoculum 129

inorganic molecules 69

interferon A system of glycoproteins produced in response to viral infection that are able to inhibit viral replication, slow cancer cell growth, and enhance the action of phagocytes; these molecules are host specific. 340, 474–76

intermediary metabolism A complex series of biochemical reactions centered around the glycolytic and Krebs pathways; different types of organic molecules necessary for cell functioning may be interconverted through this metabolic system. 173

intubation 207

invasiveness The property of a pathogen that enables it to enter, spread, and multiply in a susceptible host. Invasive factors include such agents as coagulase, collagenase, and streptokinase. 290, 292

iodine 236

iodophores 236

ionic bonds 63, 64

iris diaphragm 146

irradiation 520

irritability The response of a living individual to a stimulus in an immediate fashion that does not involve the action of genes. 28

Isaacs, A. 474

isoagglutinins 343

isoantigens 343

isolation 374, 375, 376

isotonic 219

isotopes 62, 63

Jenner, E. 11, 12, 477

juvenile-onset diabetes 492

keratoconjunctivitis 487

kingdom The largest division used in the binomial classification system of organisms. There are four major kingdoms: Procaryotae, Protista, Plantae, and Animalia. 37, 38

kinins 300, 301, 332

Kirby-Bauer disk susceptibility test 222, 223

Klebsiella 208, 226, 379, 398, 558

  K. pneumoniae 122, 143, 245

Koch, R. 9, 10, 11, 14

  phenomena 416

  postulates 11, 492

Koplick spots 500

Krebs cycle The second series of biochemical reactions in the oxidation of acetyl-CoA derived from pyruvic acid; begins with pyruvic acid and ends with the formation of carbon dioxide, 2 ATP molecules, and hydrogen released to the ETS. 164–72

kuru 469

L-form A bacterium that has lost its cell wall; these microbes are genetically capable of producing a cell wall. 42, 245

lactic acid 8

lactic streptococci 391

Lactinex 287, 288

Lactobacillus 171, 180, 244, 511, 523, 530, 580

  L. acidophilus 287, 290

  L. brevis 529

  L. bulgaricus 287, 525

  L. plantarum 529

lactoperoxidase 283, 284, 297

lag phase 183

lagoon 568

Lancefield, R. 318, 319, 391

Landsteiner, K. 301

late protein 463, 465

latent viruses 463

lattice network 308, 338

Lavosier 8, 12
leavening 530
van Leeuwenhoek, A. 7, 8, 12, 54, 85, 282
Legionella pneumophila 426–29
legionellosis 428
legumes 539, 541, 544
Leishmania 443
leprosy (Hansen's disease) 419, 420
Leptothrix 282, 295
Leuconostoc citrovorum 228
   L. mesenteroides 122, 126, 529
leukocidin 280, 292, 389
leukocytes 103, 112, 297, 298, 299, 306, 331
leukophoresis 345
leukostatin 292
lichen 54
life cycle 26, 182
limiting factor 184, 186, 540
Lindemann, J. 474
lipids 90, 91, 108, 300
lipopolysaccharide 120–22, 398
liquor 577, 583, 584, 587
Lister, J. 237
Listeria monocytogenes 315
logarithm 195
logarithmic growth (log phase) Also called exponential growth; an increase in number by a constant doubling of cells in a population. 183, 575
lophotrichous 115, 116
luminescence 27, 65
lymphogranuloma venereum (LGV) 428
lymphokines 306, 320, 340
lyophilization (freeze-drying) 131–34, 519
lysis 111, 184, 219, 228, 309, 311, 337
lysogenic cycle A bacteriophage life cycle with a period during which there is no lysis; phage are carried passively and transmitted at the time of host division. Lysis may occur in future cell generations. 260, 262, 457, 459
lysosome A single-membraned organelle produced by the Golgi; a vesicle containing powerful proteolytic and lipolytic enzymes. 111
lysozyme An enzyme produced by some eucaryotic cells that is capable of digesting the cell walls of certain bacteria. 111, 228, 279, 281, 297, 301, 457
lytic cycle A viral life cycle in which the host cell is ruptured at the time of virion release; characterized by a one-step growth curve. 260–62, 453, 456, 457

Mackarness, R. 336
macrophage 103, 104, 298, 299, 300, 351, 352
magnification 144, 145
major histocompatibility complex (MHC) 347
malaise 204
malaria 55, 56, 374, 444
malting 581
manitol salt agar 388
Mantoux test 340, 341
mass number 62
mast cell 298, 331, 336
Mastigophora 54, 114, 443, 446
matrix 112
measles 302, 322, 357, 373, 374, 463, 499–502
mechanical stage 146

Mechnikov, I. 287
media 209–19
membrane filter 191
membranous organelle An organelle of the cell that is composed of unit membrane. 88, 89, 108
memory cell 321
Mendel, G. J. 249
meningitis 395
mercury compounds 235, 555, 556
merozygote 254
mesophile A group of organisms that grow best at a midrange temperature from about 20°C to about 45°C. 160, 543
mesosome A membranous organelle of some procaryotic bacteria; it serves as a surface catalyst as well as the site of electron transport and other activities. 112, 113, 120
metachromatic granules (volutin) 87
Methanococcus 539, 550, 565
Methanosarcina 539, 565
methemoglobinemia 556
microaerophile A microbe that requires oxygen for respiration, but prefers a lower concentration (2–10 percent) than is found in the atmosphere (20 percent). 161
microbial gum Fungal mycelia and microbial products such as waxes, fats, and electrically charged products which bind soil particles into aggregates. 535
microbicide Also known as germicide; an agent that is designed to kill microbes; this is a general term used to describe different agents such as bactericides, virucides, and sporicides. 229, 230
microbiology The branch of biology that deals mainly with the study of microscopic organisms called microbes that are composed of only one cell. 2
   food and dairy 4, 510–31
   industrial 5, 572–95
   medical 4
   soil and agricultural 5, 533–50
   water and wastewater 5, 551–71
microbistat Any agent that is able to inhibit or halt the action of a microbe but does not necessarily kill it. This is a general term used to describe other more specific agents such as bacteriostats, virustats, and fungistats. 229, 230
Micrococcus (Sarcina) 535
   M. luteus 286, 295
micronutrient 158
microscope An instrument that is able to increase the size of an object so that it is easier to see; must be able to magnify and resolve the specimen. 128, 139 144–51
Microsporum 432, 434, 449
   M. audouinii 280, 282, 295
microtubule 114, 466
mildew 3
mineralization 566
minimal effective exposure 320
minimal inhibitory concentration (MIC) 218
mitochondrion A membranous organelle of the plastid group; the site of aerobic cellular respiration (Krebs cycle and ETS); found in eucaryotic cells. 111–13
mitogen 392

mixed culture 129
molecular biology A branch of science that deals with the kinds of molecules found in living cells, their behavior, and how they work together to make a cell alive. 2, 6, 12
monoclonal antibody 353, 496
monocytes 103, 298
monosodium glutamate (MSG) 589
MONOSPOT® test 308
monotrichous 115, 116
morbidity (incidence rate) The number of individuals that have been infected with a particular microbe in a given unit of time. 357
mordant 142
morphology 43, 128, 129
mortality rate The number of deaths occurring in a population during a stated period. 357
most probable number A statistical method of estimating the number of organisms in a sample of culture; done primarily with water samples to determine the level of contamination. 192, 194, 559
Mucor 535
mucus 297
multiple sclerosis (MS) 463, 464
mumps 302, 492, 499
must 579
mutation Any permanent change in the nitrogenous base sequence of DNA; two general types are recognized: point and chromosomal. Examples include deletion, inversion, insertion, and frame-shift mutation. 78, 80, 128, 250, 252, 259
   agents 78, 252
   productive 574
mutualism A form of symbiosis in which both species benefit. 275, 534
myasthenia gravis 349
mycelium 50, 431
Mycobacterium 201, 221, 234, 362, 535
   atypical 419, 420
   M. bovis 352
   M. leprae 141, 143, 152, 419, 420
   M. tuberculosis 11, 20, 141, 143, 182, 187, 191, 209, 228, 235, 236, 290, 340, 341, 415–19
mycology The branch of microbiology that deals with the study of fungi. 2, 430–42
Mycoplasma (PPLO) 42
   M. pneumoniae 154, 219, 424, 425
mycorrhiza A symbiotic relationship between certain fungi and the roots of plants; two forms occur: ecto- and endomycorrhiza. 544
mycosis 385
   cutaneous 431
   subcutaneous 431, 435
   systemic 431, 437
mycotic mycetoma 436, 437
mycotoxin Fungal poisons; found in some mushrooms and molds; for example, aflatoxins. 52, 291, 514

NAD (nicotinamide adenine dinucleotide) 158–60, 166–72, 575
NANB hepatitis 496–99
natural selection Genetically different organisms of a population reproduce at

different rates. This tends to transmit to the next generation the more favorable genes in conjunction with a changing environment. 253, 264

necrosis 300, 338, 416

Needham, J. 8, 12

needle aspiration 206–7

negative smear 140

Negri body 462

Neisseria 211, 220, 232

   N. gonorrhoeae 29, 34, 44, 45, 115, 143, 206, 245, 266, 290, 311, 357, 360, 394

   N. meningitidis 143, 201, 226, 266, 315, 323, 358, 395, 396

neoplasm The result of abnormal growth of a cell; the formation of a tumor which may develop into a malignancy. 336, 348–53

Neurospora 250, 270

neutralization 314

neutrophile A group of organisms that grow best in more neutral pH environments. 160

neutrophile (polymorphonuclear leukocytes) 103, 111, 297

nitrite ions 513, **522**, 542, 556

Nitrobacter 542, 550

nitrogen cycle 540–43

nitrogenous base A category of organic molecules found as components of the nucleic acids; there are five types: adenine, thymine, guanine, cytosine, and uracil. 74–76

Nitrosomonas 542, 550

Nocardia 449, 535

   N. asteroides 436, 442

nocardiosis 442

nonbiodegradable Materials that are unable to be broken down by microbial activities before they cause harm to living organisms. 536, 537

nongonococcal urethritis (NGU) 424, 425, 428

nonmembranous organelle Components of the cytoplasm that are directly involved in cell life processes; these are not composed of unit membrane. 89, 108

nonspecific host defense mechanisms A number of physical and chemical features of the body that inhibit or kill invading microbes and nonliving materials 297–301, 431

nonvenereal syphilis 423

normal flora 129, 256

nosocomial infection (hospital-acquired infection) 377, 378, 389, 398, 400

Nostoc 46

nuclear membrane A membranous organelle found in eucaryotic cells that identifies the structure known as the nucleus; this membrane is double-layered and interchangeable with the ER. 108, 109

nucleic acids 74–76, 306

nucleoid The location of the DNA in a procaryotic cell; also called the nuclear body. 108, 109, 262

nucleolus An accumulation of ribonucleic acid found within the nucleus of eucaryotic cells. 108, 109

nucleotide 74, 75, 250

nucleus 39, 84, 85, 87, 88, 108, 109, 111, 118

obligate aerobe An organism that must have oxygen to produce ATP through cellular respiration. 161

obligate anaerobe An organism that is unable to respire in the presence of oxygen. 161

obligate intracellular parasite Any organism that must live inside the cell of another species in order to carry out its life functions. 43, 56, 451, 479

ocular lens 146

old tuberculin (OT) 340, 418

Oncogene Theory 467

oncogenic virus Viruses that have the ability to cause tumor transformation. 464–67

operon theory 175–77, 250

opportunist 276, 278, 433

opsonin 310

opsonization The adsorption of certain antibodies or complement onto the surface of foreign material that results in the enhancement of phagocytosis. 310

optical density (OD) A method of determining the biomass of a culture in order to estimate the population of microbes. 193, 194

organelle Organized region in the protoplasm that has a specific structure and a specific function. 77, 87, 107–24, 148

organic molecules 69, 588, 589

orthomyxoviruses 503

Oscillatoria 47

osmosis Diffusion of water through a differentially permeable membrane. 98, 120, 122, 219, 228, 518

outbreak 364, 365

oxidation-reduction reaction An electron transfer reaction; molecules losing electrons become oxidized and those gaining electrons become reduced. 68, 69, 158, 159, 162, 164, 168, 170, 171, 228

oxidative phosphorylation 168

pandemic An epidemic that has grown to involve an exceptionally large geographic area or a situation in which several major epidemics have developed at the same time in many parts of the world. 358, 372, 404, 405, 409

para-aminobenzoic acid (PABA) 239, 240

parainfluenza 499

paramylum 48

paramyxoviruses 499

parasitism A form of symbiosis in which one species benefits and the other is harmed. 276, 534

   obligate 43, 56, 451

   facultative 43

paratyphoid 402

parfocal 146

parvoviruses 482, 483

passive diffusion The net movement of molecules from an area of high concentration to an area of lower concentration. 98

Pasteur, L. 2, 8, 9, 13, 430, 495, 529

pasteurization A heating process invented by Pasteur designed to reduce the number of harmful microbes in materials such as milk, wine, or cheese. 2, 4, 8, 233, 425

pathogen A microbe that is capable of causing disease. 4

pathogenicity The ability of a species to cause disease. 277

Paul-Bunnell test 308

Pediculus humanus 426

Pediococcus 228, 248, 511, 522, 581

penicillinase 29, 245, 266, 394

Penicillium 51, 511, 514, 520, 535

   P. chrysogenum 53, 60, 574

   P. claviforme 149

   P. italicum 53

   P. notatum 52, 53, 60

   P. roquefortii 53, 60

peptidoglycan 120, 121

Peptococcus 220

Peptostreptococcus 220

periodic table 66

periodontal disease 323

periplasmic space 120, 122

perishable 511

peristalsis 297

peritonitis 277

peritrichous 115, 116

permease 102

persistent virus A virus (usually animal) which replicates in the host cell and releases virions by extrusion over a long period. These viruses may be transmitted to future generations at the time of host cell division. 464

personal hygiene Personal measures taken to promote health and limit the spread of disease-causing microbes. 231

petechia 359, 395

petri plate 129

pH 64, 65, 160, 219, 232, 235, 283, 284, 287, 289, 302, 433, 511, 513, 521, 523, 525, 534, 554, 573

phage 458

phage typing An identification test based on the ability of known bacterial viruses to lyse different strains of bacteria; phage typing is performed to separate *Staphylococcus aureus* into groups. 388

phagocytosis A form of endocytosis leading to the formation of a vacuole or vesicle containing large materials or nutrients. 103, 111, 112, 123, 297

pharyngoconjunctival fever 487

phase-contrast microscope 146, 147

phenol (carbolic acid) 138, 237, 238, 495

phenol coefficient A comparative index showing the relative effectiveness of different agents in comparison to phenol. 238

phenotype 250, 252

Phialophora 436

phospholipase 99

phospholipid 89, 91, 92, 150

phosphorus cycle 543

photoautotroph 161

photocolorimeter 194, 196

photooxidation 234

photoreactivation 252

photosynthesis 46, 47, 65, 112

phycology The branch of microbiology that deals with the study of algae. 2

Phycomycetes 50

Phytophthora infestans 52, 60

picornaviruses 489–99

pili Tubular projections on some bacterial

cells that are either involved in conjugation or adhesion. 115, 116, 117

pinocytosis A form of endocytosis leading to the formation of a vesicle containing water or small ions. 103

pioneer plants 534

plague 276, 409–11

Plantae One of the four major kingdoms; includes mosses and trees. All are composed of eucaryotic cells. 37

plant diseases 545

plaque (dental) 283

plaque (viral) A clear spot on the surface of a cell culture resulting from the lysis of host cells by lytic viruses. 454, 455

plasma cell 306

plasma (cell) membrane The outermost limiting membrane of cells; composed of phospholipid, protein, and carbohydrates in a fluid mosaic pattern. 89, 93, 98, 99, 102, 103, 108, 111, 112, 124

plasmid A small, circular piece of exogenote DNA that is self-replicating and contains a limited number of genes; may be called F-factor, if they control conjugation or R-factor, if they contain genes for drug resistance. 255–69, 293, 389

Plasmodium 55, 60, 374, 443, 449
  P. falciparum 444
  P. malariae 444
  P. ovale 444
  P. vivax 359, 444

plasmolysis 98, 219

plastid A category of membranous organelles, including the mitochondria and chloroplasts; characterized by having pigment molecules such as cytochromes and chlorophyll attached to an inner membrane surrounded by an outer membrane. Found in eucaryotic cells. 111, 112

pleomorphic A cell that has many or varied shapes. 43

pleurisy 393

pneumonia 393

pocks (pox) 455

poison ivy 341, 342

polar molecule 64

polio (poliomyelitis) 302, 489–91

poliovirus 489–91

polyhydroxybutyric acid (PHB) 87

polymorphonuclear leukocyte (PMN) (neutrophiles) 297, 300, 338, 340

population growth curve A graph which displays the four phases of population increase. The phases are lag, log, stationary, and death. 183, 229

Porphyra 50, 60

portal of entry 369, 371

portal of exit 369, 371

potable water 554

pour plate A technique used in the lab to isolate a single type of microbe from a mixed culture; performed by making a serial dilution. 130

poxviruses 483

precipitin reaction 308, 338

preliminary incubation test (P.I.) 190

presumptive water test 559, 560

prevalence rate The total number of persons infected at any one time regardless of when the disease began; for example, 56 people infected with *Staphylococcus aureus* in a group of 100 during the week of August 14–21. 357

Priestly, J. 8, 12

primary atypical pneumonia (PAP) 424

primary response phase 320

primary sewage treatment The first stage in the decomposition or digestion of sewage; usually involves screening and sedimentation. 564

Procaryotae One of four major kingdoms; consists of the bacteria and the blue-green algae (cyanobacteria); these cells lack a true nucleus. 37, 39, 46, 183

procaryotic cell A cell having nuclear materials that are not separated from the rest of the cell contents by a nuclear membrane; they do not have a nuclear membrane. 36, 39, 43, 84, 85, 87, 108–24

prodromal 204

productive mutant 574

properdin system 310

prophage Bacteriophage viral nucleic acid which has integrated into the genetic material of the host cell. 260, 262, 263, 457, 465

Propionibacterium 511, 589

prostaglandin 297, 300, 332

protein 99, 100, 101
  synthesis 75–82, 250, 252, 303

protein metabolism 168

Proteus 208, 219, 226, 260, 288
  P. vulgaris 143, 400, 401, 425

Protista One of the four major kingdoms; includes protozoa, green algae, fungi 37, 47, 54

protoplasm A general term for the living contents of cells. 85

protozoa A form of Protista, lacking a cell wall and chlorophyll; single-celled animals. 54, 147

protozoology The branch of biology that deals with the study of protozoa. 2, 430, 442–47

provirus 465–67

pseudomembrane 403, 414

Pseudomonas 219, 254, 260, 288, 291, 379, 511, 512, 519, 535, 542
  P. aeruginosa 98, 104, 106, 143, 201, 245, 259, 378, 405, 406

pseudopod 103

Psilocybe mexicana 52, 60

psychrophile A group of organisms that grow best at low environmental temperatures ranging below 20°C. 160, 184, 554

ptomaine food poisoning 365

puerperal fever 290

pure culture (axenic) A population of microbes composed of only a single species. 128, 129
  pour plate 129, 132
  streak plate 129, 130, 131

purification, water 561–63

purified protein derivative (PPD) 340, 418

purines 74, 75, 159

pus 99, 279, 280, 298, 300, 387, 423

putrefaction 171

pyogenic 387–96

pyrimidines 74, 75, 159

pyrogen 292

pyroligneous acids 522

pyrolysis 576

pyruvic acid 164

quarantine 374

quaternary ammonium compounds (quats) 233

Quebec colony counter 188

R factor 255, 256, 259

rabies 9, 301, 323, 338, 360, 494–96

radioactive 62, 63

radio allergo sorbent test (RAST) 332

radioimmunoassay (RIA) 315

raw or settled sludge The undigested sediment that forms during primary sewage treatment on the bottom of a sedimentation tank. 565

reaginic antibody 314

recognition binding site 104

red tide 49

Redi, F. 7, 8, 12

refrigeration 131, 184, 519, 520

regulation A characteristic of life involving the control of the rate at which chemical reactions occur. 31

rejection response 342, 343, 347

replicative form (RF) 463

repression A regulation of metabolism that occurs on genes; in repression, the action of a series of genes is turned "off" by the presence of a substance, while its absence allows the genes to be turned "on." 174, 458

repressor protein 458

reproduction The characteristic of life which results in the generation of new living beings from preexisting parents. 26, 182

resazurin 162

reservoir of infection 357

resolving power (resolution) 144

respiratory syncytial virus 499

response Reaction to a stimulus. A charactistic of life; three responses are recognized: irritability, adaptation, and evolution. 28

reticulo-endothelial system A functional system of fixed phagocytes (macrophages) found in association with the endothelium of blood vessels, liver sinuses, GI tract, and lymphatic system. These cells are important in the inflammatory response and immune response. 299–301, 437

retroviruses 463

reverse transcriptase 465

Reye's syndrome 485, 486, 504

Rh blood 343–46

rhabdoviruses 494–96

rheumatoid arthritis 306, 308

rheumatoid factor (RF) 308

rhinoviruses 490–94

Rhizobium 538, 540, 544, 550

rhizosphere The zone immediately surrounding the roots and containing a high concentration of microbes and organic material. 536

Rhizopus 51, 60, 511, 593

Rhodospirillum 46, 47, 60, 541

RhoGAM 344

ribosome Small, nonmembranous organelles made of protein and RNA located in eucaryotic and procaryotic cells; the site of protein synthesis. 108, 117, 118, 245

Rickettsia prowazekii 45, 60, 425, 537

R. rickettsii 359

ringworm 280, 282, 435

ripening 525

RNA 108, 117, 118, 451, 466, 489, 575

RNA polymerase 463

Rochalimaea quintana 425

RODAC plate 224

Rose-Waaler test 308

rubella 303, 316, 322, 323, 499–502

rubeola 499–502

rumen 5, 546, 547

Sabin, A. 480

Saccharomyces 511, 593

S. cerevisiae 50, 51, 60, 235, 530, 575, 576, 577, 579, 583, 589

safety 134–35

saliva 208

Salmonella 219, 238, 244, 245, 253, 291, 365, 368, 400, 512

S. typhi 143, 182, 199, 229, 264, 300, 309, 315, 358, 401, 402, 403

salting 522

sanitizer An agent that is able to reduce the number of contaminating microbes to a safe level determined by public health agencies. 231

saprotroph (saprobe) Any organism that ingests decomposed nutrients by absorption through the cell boundaries. 184

Sarcodina 54, 443, 446, 447

scalded-skin syndrome 389

scarlet fever 391

Schick test 415

Schultz-Charlton test 391

science 1

scombroid food poisoning 515

Second Law of Thermodynamics Whenever energy is converted from one form to another, some energy is lost. 27, 158, 165

secondary sewage treatment The second stage in sewage treatment; usually comprised of trickling filter and/or digestion in an activated sludge unit. 565

sedimentation 561

selective medium A culture medium that is prepared with ingredients that inhibit the growth of unwanted microbes in a mixed culture and promotes the growth of others. 211

sensitization 320

sepsis 132

septic (strep) sore throat 391

septic tank 562, 563

septicemia A systemic disease with actively reproducing microbes and their products found in the circulatory system. 276, 280

septum 119

serial dilution 131, 132, 186, 189

serology A branch of microbiology that studies antibody-antigen reactions in a test tube (*in vitro*). 206, 318, 476, 489

serotonin 298, 332

serotype 318, 398, 494, 513

Serratia 213, 226, 260, 264, 398, 542

S. marcescens 379

serum 206

serum sickness 322, 338, 339

sewage The watery material carrying refuse and wastes in a drainage system. 562–70

shellfish poisoning 515

Shigella 210, 226, 238, 291, 315, 515

S. dysenteriae 143, 245, 403, 404

shigellosis (bacillary dysentery) 403, 515

shingles 322, 483–87

significant numbers 190

silage A combination of green hay, cereals, grasses, and other suitable vegetative material that has been preserved by anaerobic acid fermentation and used as animal food. 545, 546

silver compounds 235

simple stain 141

Singer-Nicolson fluid mosaic membrane model 89, 92

singlet oxygen 103, 298

skin sensitivity test 332, 333

sleeping sickness (African, Chagas' disease) 446

slides (microscopic) 139–44

slow-reacting substance of anaphylaxis (SRS-A) 332

"slow" viruses Unconventional viruses; probably of self-replicating membrane; responsible for spongiform virus encephalopathies such as kuru. 451, 469

sludge 565–67

smallpox 5, 11, 301, 302, 357, 361, 483–86

smear preparation 10, 140

smoking 522

soap 232, 233

soil 54, 534

soil filtration rate 534, 569

sourdough 530

souring 525, 530

Spallanzani, L. 8, 12

species A population of organisms other than bacteria that can interbreed naturally to produce fertile offspring; members of this group are very closely related; together with the genus name, it forms the scientific name. 37

specimen 129, 184, 201, 204, 206–8

spices 522

spirilla A group of rigid, spiral-shaped bacteria. 43

spirillum A curve-shaped microbe; three distinct forms are common: vibrios, spirilla, and spirochetes. 43

Spirillum lipoferum 541

spirochetes A morphological form of a spirillum; many curves make a spirochete resemble a corkscrew that is flexible and able to wiggle through the action of an axial filament. 43, 420–23

spoilage Changes in a food caused by natural self-decomposition that make it unacceptable; an increase in microbial numbers or the release of microbial compounds which result in changed flavors, colors, or aromas. 52, 98, 171, 184, 511, 512

spontaneous generation The theory of the origin of life stating that living organisms arise from nonliving material. 6

spores, fungal 51

Sporothrix schenckii 435, 436, 449, 520, 532

sporotrichosis 435, 436

sporozoa 55, 443, 444

spotted fever 359, 360, 427

sputum 191, 208, 209

staining technique 139–44

Staphylococcus aureus 11, 20, 43, 52, 53, 99, 121, 143, 160, 201, 219, 232, 234, 238, 244, 245, 260, 263, 275, 277, 280, 281, 290, 292, 300, 357, 365, 368, 369, 378, 388–90, 511–13, 589

staphylokinase 389

standard plate count (SPC) 188, 190, 559

starch 87

stationary growth phase 184

steady-state growth (balanced) 185

stem cells 298

sterility testing 223

sterilization Any process that destroys all living organisms, including the resistant endospores of bacteria. 134, 136–39, 231–38

steroid 589–94

stock culture 132, 214

stomatitis 494

streak plate A technique used in the lab to isolate a single type of microbe from a mixed culture; carried out with an inoculating loop or needle on solid medium. 130, 211

Streptococcus 99, 141, 148, 160, 211, 232, 245, 290, 378, 511, 558

S. cremoris 228

S. faecalis 288, 391, 521, 559

S. mutans 282–84, 290, 323, 562

S. pyogenes 32, 34, 37, 39, 41–43, 143, 201, 210, 390–94

S. salivaricus 282

S. sanguis 290

S. thermophilus 391, 525

S. viridans 275, 276, 391

streptokinase 392

streptolysin (O, S) 291, 392

Streptomyces 23, 128, 449, 535, 589

S. somaliensis 436

stroma 112, 113

structural formula 70, 71

structural gene 75, 175
subacute sclerosing panencephalitis (SSPE) 463, 464, 469, 500
subclinical case 358, 415, 468
substage condenser 146
substrate The material that the enzyme combines with in the reaction; the materials are changed into the end products. 30
substrate level phosphorylation 167
sudden infant death syndrome (SID) 288, 514
sulfa drugs 228, 239, 240
Sulfolobus acidocaldarius 554, 571
sulfur cycle 543
surface-active agent Soaps and detergents; a substance that is able to reduce surface tension by concentrating at the boundaries or interfaces between two materials. 232, 233
surface catalyst 108
surface tension A tight, strong layer at a surface capable of withstanding considerable force. 232
surface water Water from streams, lakes, rivers, and shallow wells used for human consumption and industrial processes; the water source most likely to be contaminated with pathogens or toxic substances. 552
surgical scrub 278
swab 206
Swartz vaccine 500
symbiosis A close physical relationship between two different species which may or may not be beneficial to the organism involved; mutualism, commensalism, amensalism, parasitism. 275–93, 510, 546
symbiotic, nitrogen-fixing bacteria 540
symptoms 201, 204, 205
synthetic medium A culture medium that is prepared from chemical compounds that are highly purified and accurately identified; also called defined medium. 210
syphilis 420–23
systemic lupus erythematosis (SLE) 308, 314, 337, 340

T lymphocyte (T-cell); A form of lymphocyte that has had its differentiation controlled by the thymus gland; responsible for cell-mediated immunity; secretes lymphokines and affects antibody-mediated immunity through contact with B-cells. 99, 306, 308, 318, 337, 340, 348, 468, 474, 476
Tatum, E. 11
teichoic acid 122
temperate phage 458
tertiary sewage treatment The third stage in sewage treatment designed to remove or decompose the remaining organic and inorganic contaminants in water discharged from the secondary stage; for example, lagoons, land application, coagulation-sedimentation. 568
tetanus 301, 397
thermal death time (TDT) 233

thermophile An organism that grows best at hot temperatures ranging above 45°C. 160, 543, 554
Thermoplasma acidophilum 554, 571
Thermus aquaticus 554, 571
Thiobacillus 542, 550
thioglycolate 162
Thiothrix 87
thrush 433
thylacoid 47, 114
tinea (capitis, pedis, corporis, barbae, unguium) 280, 434, 435
tine test 340
titer, antibody 206
Todaro, G. J. 467
Tomasi, T. 284
Torula utilis 589
Torulopsis 511
total cell count A number of different cell counting methods used to determine the number of cells in a culture; includes dry-slide and Petroff-Hausser counting chamber. 186
toxic shock syndrome (TSS) 389, 390
toxin 290, 300
toxoid A toxin that has been chemically or physically altered so that it is no longer harmful but is able to stimulate the immune system to produce antibody. 290, 397, 413, 415
Toxoplasma 443, 449
  T. gondii 55, 60, 445
toxoplasmosis 445
transcription The first of the two stages of protein synthesis; involves the formation of mRNA from a template of a structural gene in the DNA. 77, 78, 463
transduction The transfer of genes by viruses that carry them from one host cell to another; two forms are found: generalized and specialized. 250, 260, 263, 292, 457
transformation The transfer of a free, naked DNA to a living cell (a form of genetic recombination). 250, 254, 255, 292
  viral 464–66
translation The second of the two stages of protein synthesis; involves the pairing of mRNA with tRNA and results in alignment of the proper amino acids as determined by a structural gene. 77, 78, 463
transplantation immunology 342–52
Treponema 232
  T. pallidum 43, 146, 147, 160, 201, 234, 238, 266, 280, 315, 420–23
TRIC (trachoma inclusion conjunctivitis) 46, 427, 428
Trichoderma 535
Trichomonas 443
  T. vaginalis 55, 60, 289, 446
Trichophyton 434, 449
  T. schoenleinii 431
  T. verrucosum 435
trickling filter A large basin about two meters deep filled with stone, over which waste water is sprayed; microbes growing on the porous stone surfaces decompose organic contaminants; a secondary treatment unit. 565, 566

Trypanosoma 443
  T. gambiense 446
  T. rhodesiense 446
tubercle 416, 418
tuberculin 340, 341
tuberculosis 309, 318, 340, 341, 362, 415–19
tumor-specific antigen (TSA) Cell-surface antigens found on cancer cells; each tumor type has its own unique TSA; these antigens stimulate the immunosurveillance system. 350, 352, 353, 466
turnover number 31
tyndallization (spore-shocking) 138, 233
typhoid fever 5, 402
typhus, louseborne fever 426

ultrasound 234, 235
ultraviolet microscope 147
unconventional viruses 469, 470
United States Public Health Services 361
Ureaplasma urealyticum 424
urticaria (hives) 329
UV radiation 234, 252

vaccine A solution of disease-causing organisms or their products that has been specially treated so that they will not cause harm; it can be injected into animals or humans in order to stimulate the body to protect itself against that particular disease. 11, 393, 455, 476, 490
vacuole A membranous container inside a cell; specialized vacuoles may be termed food vacuoles, water vacuoles, etc. 111, 112
vaginitis 55, 289, 433, 446
valency (antibody) 303
varicella-zoster virus 485
Varro 6, 12
vasoactive amines A group of chemicals released from mast cells and basophiles during hypersensitivity responses; include: histamine, SRS-A, serotonin. 329, 332
vector 358, 359, 374, 410
vegetative cell The actively metabolizing cell; certain bacteria may synthesize endospores within their cytoplasm during this stage. 115–17
venereal disease research laboratory test (VDRL) 314
vertical transmission The passage of proviruses from one host generation to another during cellular reproduction; also known as congenital transmission. 467
vesicle 110
viable cell counting A number of different cell counting methods used to determine the number of live cells in culture; includes: standard plate count, membrane filter, most-probable number. 188
vibrio A morphological form of a spirillum; only a slight twist or curve is seen in these microbes. 43
Vibrio cholerae 4, 5, 11, 20, 43, 290, 295, 404, 405

V. fischeri 27

viral replication   The reproductive process carried out by viruses; five main stages: adsorption, penetration, replication of nucleic acid and protein, maturation, and release.   51

viral transformation   A process that converts normal cells to tumor cells.   464–68

viridans group streptococci   391

virion   The unit of a virus particle composed of a nucleic acid core and a protein coat.   56, 260, 264, 451

viroids   Unconventional viruses; probably composed of RNA; responsible for diseases of some higher plants such as chrysanthemums.   451, 469

virology   A branch of microbiology that deals with the study of viruses   2, 450–505

virulence   The degree to which a pathogen can cause host harm as a result of substances produced by or associated with the microbe during the course of the infection; may be due to toxigenicity or invasiveness.   123, 182, 201, 277, 292, 378

viruses   Acellular, obligate intracellular parasites; three main groups are bacteriophage, plant, and animal.   55, 56, 57, 450–505

vitamin   158, 160, 284, 285, 575, 588, 592

Vitis vinifera   577

volutin (metachromatic granules)   87

warts   483

Wassermann test   314

waste production   A characteristic of life which involves the elimination of useless or harmful materials and energy.   26, 99

water activity (A$_w$)   511, 518, 519, 522

water-holding capacity   534, 569

Watson, J.   11, 17, 250

weathering   534

Weil-Felix text   425

wet-mount slide   139

whey   525

whooping cough (pertussis)   4, 12, 413

wine   8, 577–80

World Health Organization (WHO)   45, 361, 376

wort (beer)   583

xerophile   511

x-radiation   234

yellow fever   502

Yersinia pestis   210, 226, 276, 409, 410, 411

Zoogloea ramigera   567, 571

zoonoses   Diseases of animals that on occasion are transferred to humans.   369, 370

zoster immunoglobulin (ZIG)   487

Zygosaccharomyces   511